Stem Cell Biology and Regenerative Medicine
Second Edition

RIVER PUBLISHERS SERIES IN BIOTECHNOLOGY AND MEDICAL TECHNOLOGY FORUM

Series Editors

PAOLO DI NARDO
University of Rome Tor Vergata, Italy

PRANELA RAMESHWAR
Rutgers University, USA

Biotechnology & Medical Technology (BTMT) Forum is an international initiative aimed at disseminating the results of studies interfacing basic and translational medicine, biomaterials and applied engineering. BTMT Forum is committed to rapidly communicate the results of the scientific studies to scientists and the wider public worldwide.

Peer-reviewed original comprehensive articles are published monthly on the basis of their novelty, relevance, interdisciplinary impact, and potential scientific and translational significance. Early outcomes from clinical trials are also published. The Forum publishes thorough authoritative reviews and commentaries on the most important issues and perspectives of cutting-edge research after a careful evaluation by the Editorial Board.

Monographic books are also printed on specific topics of relevant concern under invitation by the Editorial Board or after independent proposals by potential book editors.

Finally, BTMT Forum publishes a book series specifically intended for professionals eager to revitalize their acquaintance and students of different educational ages.

BTMT Forum activities also involve the promotion of Congresses, Seminars, etc.

Topics:

- basic and translational medicine
- biomaterials
- applied engineering

For a list of other books in this series, visit www.riverpublishers.com

Stem Cell Biology and Regenerative Medicine
Second Edition

Editors

Charles Durand

Sorbonne Université, France

Pierre Charbord

Inserm, France

River Publishers

Published, sold and distributed by:
River Publishers
Alsbjergvej 10
9260 Gistrup
Denmark

www.riverpublishers.com

ISBN: 978-87-7022-403-1 (Hardback)
        978-87-7022-402-4 (Ebook)

©2021 River Publishers

Contents

Preface   **xix**

List of Figures   **xxi**

List of Tables   **xxv**

List of Contributors   **xxvii**

List of Abbreviations   **xxxiii**

1   Stem Cell Concepts   **1**
Charles Durand and Pierre Charbord
1.1   Introduction 1
1.2   Embryonic and Adult Stem Cells 2
1.3   The Regulation of Stem Cells and the Stem Cell Niche   . . . 9
1.4   Stem Cell Models 13
1.5   Cell Therapy Using Stem Cells 17
1.6   Concluding Remarks 20

2   Transcription Regulation by Distal Enhancers: Dynamics of the 3D Genome   **33**
Eric Soler and Frank Grosveld
2.1   Introduction 33
2.2   Spatio-temporal Control of Gene Activity: The Pivotal Role of Enhancers 33
2.3   The Genome-wide Enhancer Landscape 37
2.4   Enhancer Dynamics 39
2.5   Enhancers and Promoters Are Distant Structures 40
2.6   Mechanisms of Enhancer Function: The 3D Genome 41
2.7   Dynamic versus Stable Chromatin Looping 44
2.8   Transcription Factories or Foci 47

2.9   Specificity of Enhancer–Promoter Contacts   48
2.10   TADs: Topologically Associating Domains   48
2.11   Functional Relevance of TADs and Their Borders   51
2.12   Alterations of Enhancers and 3D Genome Organization in
       Disease   53
2.13   Concluding Remarks   55

3 Repair of DNA Double-Strand Breaks in Adult Stem Cells   71
Leyla Vahidi Ferdousi, Haser Hasan Sutcu and Miria Ricchetti

3.1   Introduction   71
3.2   DNA Damage and Repair Mechanisms   72
       3.2.1   Homologous Recombination and Related Repair
               Mechanisms   73
       3.2.2   Classical and Alternative Non-Homologous End-
               Joining   74
       3.2.3   Context of Repair Processes   77
3.3   Cell Response to DSBs   78
3.4   Stem Cells Resistance to Genotoxic stress   81
3.5   Efficiency and Mechanisms of DSB Repair in Adult Stem
       Cells   81
       3.5.1   Cells in the Epithelial Tissue: Epidermal, Mammary,
               and Intestinal Stem Cells   81
       3.5.2   Cells in the Connective Tissue: Hematopoietic and
               Mesenchymal Stem Cells   84
       3.5.3   Skeletal Muscle Stem Cells   86
       3.5.4   Neural Stem Cells   89
       3.5.5   Germinal Stem Cells   89
       3.5.6   Induced Pluripotent Stem Cells   90
3.6   Other Responses to DNA Damage   91
       3.6.1   Apoptosis   91
       3.6.2   Senescence   92
       3.6.3   Differentiation   93
3.7   Concluding Remarks   94

4 Haematopoietic Stem Cell Niches in the Bone Marrow   107
Claire Fielding and Simón Méndez-Ferrer

4.1   Haematopoietic Stem Cell Niches   107
4.2   The Contribution of Structural Components   108
       4.2.1   Vascular Niche Components   108

	4.2.2	Mesenchymal Stromal Cells	109
	4.2.3	Mature Haematopoietic Cells	111
	4.2.4	Adipocytes	111
	4.2.5	Bone-associated Cells	112
	4.2.6	Innervation	112
4.3		Concluding Remarks	115

5 Computational Models of Spatio-temporal Stem Cell Organization 123

Torsten Thalheim and Joerg Galle

5.1	Introduction	123
5.2	Concepts of Stem Cell Organization	125
5.3	Intestinal Stem Cell Organization as a Paradigm	127
5.4	Models of Epigenetic Regulation	129
5.5	Stem Cells During Development and Aging	129
5.6	Stem Cells in Artificial Environments	132
5.7	Summary	133
5.8	Concluding Remarks	134

6 Transcriptomics Investigations 139

Christophe Desterke and Pierre Charbord

6.1	From Omics to Data	139
6.2	Data Analysis and Standardization in Omics	140
6.3	Study of Transcriptomes Using Microarrays	142
6.4	Study of Transcriptomes Using Next Generation Sequencing (NGS)	145
	6.4.1 Steps in Transcriptome Studies Using RNA-sequencing	146
6.5	Study of Single Cell Transcriptomes	147
6.6	Network Biology	151
6.7	Machine Learning Classifiers	154
6.8	Conclusion	155

7 The Regulatory Network of X Chromosome Inactivation 163

Jeffrey Boeren, Sabina Jašarević and Joost Gribnau

7.1	Introduction	163
7.2	The X:Autosome Ratio Dictates Initiation of XCI	165
7.3	Initiating X-Inactivation, the Regulatory Environment of the Xic	166

7.4   The Pluripotency Network and Other Trans-acting Factors of
XCI   169
7.5   The Balancing Act of Activators and Inhibitors in Initiation
of XCI   171
7.6   Concluding Remarks   173

**8   Directed Differentiation of Human-induced Pluripotent Stem
Cells into Hepatic Cells: A Transposable Example of Disease
Modeling and Regenerative Medicine Applications   181**
*Antonietta Messina, Eléanor Luce and Anne
Dubart-Kupperschmitt*
8.1   Introduction   181
8.2   Developmental Origins of the Liver   183
8.3   hiPSCs and Directed Differentiation In Vitro   187
8.3.1   Hepatoblastic Differentiation   187
8.3.2   Hepatocyte Differentiation   187
8.3.3   Cholangiocyte Differentiation   189
8.4   hiPSC-derived Liver Cell Applications   190
8.4.1   Disease Modeling   191
8.4.2   Drug Screening and Toxicology Studies   193
8.4.3   Bio Artificial Liver Devices   194
8.4.4   Cell Transplantation   195
8.4.5   Bioprinting of 3D Liver Tissues and Whole Organ
Assembly   195
8.5   Concluding Remarks   197

9   *Hydra* and the Evolution of Stem Cells   207
Alexander Klimovich and Thomas C.G. Bosch
9.1   Introduction   207
9.2   The "Mystery" of Hydra's Life Cycle   208
9.3   Revisiting Stem Cells in Hydra with Emerging   212
9.4   Stem Cells in Hydra Are Controlled by Conserved   215
9.5   Epigenetic Control of the Stem Cells in Hydra   217
9.6   Taxonomically Restricted Genes (TRGs) Have Their   218
9.7   Stem Cells Interact with the Environment   220
9.8   Microbiome Contributes to Tumor Formation   224
9.9   Open Questions and Future Perspectives   226
9.10   Concluding Remarks   228

10 Regeneration in Anamniotic Vertebrates   **243**
Prayag Murawala and Dunja Knapp   *243*
10.1 Introduction . . . .
10.2 Appendage Regeneration in Amphibians 245
    10.2.1 Launching Regeneration via Wounding 248
    10.2.2 Role of the Nerves in Inducing Blastema 248
    10.2.3 Maintenance of Blastema Outgrowth by the Shh–Fgf
        Activation Loop 249
    10.2.4 Cellular Sources and Differentiation Potential of
        Blastema Cells 250
        10.2.4.1 Becoming blastema cell—dedifferentiation
                versus stem cell activation 252
                10.2.4.1.1 Muscle 252
                10.2.4.1.2 Connective Tissue (CT) 253
    10.2.5 Specification and Re-specification of Positional
        Information 254
    10.2.6 Frogs—Regenerative Capacity Depends on Develop-
        mental Stage 255
10.3 Other Examples of Regeneration 256
    10.3.1 Brain Regeneration 257
    10.3.2 Heart Regeneration 258
    10.3.3 Lens Regeneration in Amphibians via Trans Differ-
        entiation 259
10.4 Regeneration and Metabolism 261
10.5 Concluding Remarks 262

11 Stem Cells and Regeneration in Plants   **273**
Marco Da Costa and Philippe Rech
11.1 Introduction 273
11.2 SAM Organization and Regulation 274
    11.2.1 The Core WUS/CLV3 Loop 276
    11.2.2 Role of Hormones in SAM Regulation 277
        11.2.2.1 Cytokinin 277
        11.2.2.2 Auxin 278
    11.2.3 Other SAM Regulators 279
11.3 RAM Organization and Regulation 280
    11.3.1 RAM Master Regulators 281
    11.3.2 Role of Hormones in RAM Regulation 282
    11.3.3 LR Formation 283

11.4 SAM Regeneration 284
    11.4.1 Tip Regeneration 284
    11.4.2 *De novo* Regeneration 286
    11.4.3 Key Players During DNSO (*de novo* Shoot Organogenesis) 287
11.5 Concluding Remarks 289
11.6 Take on Messages 289

12 Hematopoietic Development in Vertebrates   **297**
Hanane Khoury, Thierry Jaffredo and Laurent Yvernogeau
12.1 History of the Concept of HSCs 297
12.2 On the Origin of Blood 298
12.3 An Intra-embryonic Source of HSCs 300
12.4 Construction of the Aorta and Establishment of the Dorsoventral Polarity 304
12.5 Role of the Sub-aortic Mesenchyme 306
12.6 IAHC Formation is a Conserved Mechanism in Vertebrates . 309
12.7 BM ECs Can Generate a Transient Wave of HSPCs 311
12.8 Systems Used to Study Hematopoietic Cell Commitment . . 311
12.9 Concluding Remarks 313

13 Developmental Biology of Hematopoietic Stem Cells: Non-cell
Autonomous Mechanisms   **323**
Leslie Nitsche and Katrin Ottersbach
13.1 Introduction 323
13.2 Developmental Niches 324
13.3 Cell-intrinsic Factors of Hematopoietic Stem Cell Generation 327
    13.3.1 The Endothelial-to-Hematopoietic Transition 327
    13.3.2 Transcription Factor Dynamics During the EHT . . . 328
13.4 Supporting Nascent Hematopoietic Stem Cells 331
    13.4.1 Endothelial Cells 333
    13.4.2 Signals From Ventral Versus Dorsal Microenvironments 334
    13.4.3 Subaortic Mesenchyme 335
    13.4.4 The Sympathetic Nervous System 337
    13.4.5 Immune-derived Signals 337
    13.4.6 Macrophages 340
    13.4.7 Systemic Factors 341
        13.4.7.1 Blood Flow 341

13.4.7.2   Hormone Signaling 342

13.4.7.3   Metabolism 344

13.4.8   Hypoxia 344

13.5   Future Directions 345

13.6   Conclusions 347

14   Biology of Hematopoietic Stem Cells in the Adult   359

Rima Haddad, Francoise Pflumio and Marie-Laure Arcangeli

14.1   Definition, Concepts, History 359

14.2   Characterization of HSC 360

   14.2.1   Using Phenotype Analysis 360

     14.2.1.1   Mouse HSC (see also[4,5]) 360

     14.2.1.2   Human HSC 361

   14.2.2   Using Functional Assays (see also [31]) 363

     14.2.2.1   Colony-forming cells (CFCs) and long-term culture-initiating cells (LTC-ICs) . . 363

     14.2.2.2   Using liquid cultures 364

     14.2.2.3   Using in vivo transplantation models . . . 365

   14.2.3   Physiology of HSC 366

     14.2.3.1   Self-renewal and quiescence properties . . 366

     14.2.3.2   HSC potential versus HSC fate 368

14.3   Regulation of HSC Functions 369

   14.3.1   Extrinsic Regulators 369

     14.3.1.1   CXCL12/CXCR4 369

     14.3.1.2   Stem cell factor (SCF) and its receptor KIT   369

     14.3.1.3   Integrins and Adhesion molecules 370

   14.3.2   Intrinsic Regulators (see also [105]) 370

14.4   Ex Vivo Expansion of HSPC 371

   14.4.1   Extrinsic Factors for Ex Vivo HSPC Expansion . . . 372

     14.4.1.1   Usage of Cytokines and growth factors' combination 372

     14.4.1.2   Expansion in presence of stromal cells . . 372

   14.4.2   Developmental and Intrinsic Factors for Ex Vivo Human HSPC Expansion and or/Maintenance 373

     14.4.2.1   Approaches Using Delivery of Intrinsic Factors 373

     14.4.2.2   A. HOXB4-mediated expansion 373

     14.4.2.3   B. Other potential molecular targets (Notch and Wnt pathways) 373

14.4.2.4 Approaches using modification of human HSPC microenvironment: Focus on the Hypoxia/HIF pathway 374

14.4.3 Chemical Compounds in the Era of Advanced Technologies for Ex Vivo Human HSPC Expansion . . . 376

14.5 Conclusive Remarks and Perspectives 377

15 Epithelial Stem Cells in the Skin **393**

Romain Fontaine, Bénédicte Oulès, Mathieu Castela and Sélim Aractingi

15.1 Introduction 393

15.2 The Interfollicular Epidermis 395

15.2.1 Development of the IFE 395

15.2.2 IFE during Postnatal Growth 396

15.2.3 IFE in Adulthood 397

15.3 The Hair Follicle 399

15.3.1 HF Development 400

15.3.2 HF Cycling 401

15.4 Contribution of Bulge Stem Cells to the Epidermis 402

15.5 Contribution of Bulge Stem Cells to the Sebaceous Gland Lineage 404

15.6 The Sweat Gland 406

15.7 Concluding Remarks 407

16 Mammary Stem Cells **415**

Silvia Fre and Ulysse Cherqui

16.1 Introduction 415

16.2 The Mammary Epithelium and Its Stem Cells 416

16.3 Lineage Tracing Analysis to Study Stem Cells In Vivo and In Situ 418

16.4 Mammary Cell Plasticity 420

16.5 Molecular Signals Governing MaSCs and Gland Morphogenesis 422

16.5.1 The Notch Pathway 422

16.5.2 The Wnt Pathway 425

16.5.3 The FGF pathway 426

16.5.4 The TGF β Pathway 426

16.5.5 The Hedgehog Pathway 427

16.6 Concluding Remarks 428

17 The Intestinal Stem Cells in Homeostasis and Repair   **437**
Aline Stedman
17.1 Introduction 437
17.2 Different Overlapping Populations of Stem Cells Ensure
     Homeostasis of the Adult Intestine 438
     17.2.1 Stem Cells Lie Within Crypts 438
     17.2.2 Columnar Base Cells are Lgr5$^+$ Intestinal Stem Cells 440
     17.2.3 +4 Reserve Stem Cells 441
17.3 The Multiple Strategies to Protect and Repair the Intestinal
     Barrier 442
     17.3.1 Intestinal Stem Cells Under Stress 442
     17.3.2 A Highly Plastic Lineage 443
     17.3.3 The Extracellular Matrix Pulls the Trigger 444
     17.3.4 Mechano-sensitive Facultative Stem Cells 444
17.4 The Fine-tuning of Intestinal Stem Cells Fate 445
     17.4.1 Extrinsic Modes of Intestinal Stem Cell Regulation . 445
         17.4.1.1 Intestinal stem cells neighborhood 445
         17.4.1.2 The Wnt pathway 445
         17.4.1.3 The Notch pathway 446
     17.4.2 ISC Energy Metabolism: How to Boost Stemness . . 447
17.5 Bugs in the Niche 447
     17.5.1 Bacterial Metabolites: A Complex Language Still to
         be Deciphered 448
     17.5.2 Innate Immunity and ISCs-bacteria Crosstalk 448
17.6 Perspectives on ISCs-based Regenerative Medicine 449
17.7 Concluding Remarks 451

18 Neural Stem Cells   **461**
Nathalie Kubis and Martin Catala
18.1 Introduction 461
18.2 Evidencing Postnatal Neurogenesis in Mammals 462
     18.2.1 The S Phase 462
     18.2.2 Markers of Neurogenesis 463
     18.2.3 The Bomb and the Brain 465
     18.2.4 *In Vitro* Culture for Neural Stem Cells 466
18.3 Neural Stem Cells Lie in Specific CNS Regions 467
     18.3.1 Stem Cells in the Subventricular Zone 467
         18.3.1.1 Mouse brain 468
         18.3.1.2 Human brain 472

18.3.2 Stem Cells in the Subcallosal Zone 473
18.3.3 Stem Cells in the Dentate Gyrus of the Hippocampus 474
18.3.4 Neural Stem Cells in the Spinal Cord 475
18.4 Ecology of Neural Stem Cells 477
18.4.1 Neural Stem Cells Live in a Niche 477
18.4.2 Interaction With the Cerebrospinal Fluid (CSF) . . . 478
18.4.3 Interactions Within the Niche 480
18.4.4 When Endothelial Cells Meet Neural Stem Cells . . 481
18.5 Concluding Remarks 483

19 Non-hematopoietic Stem Cells of Bone and Bone Marrow   495
Pierre Charbord
19.1 Historical Background 495
19.2 Bona fide Murine Stem Cells 500
19.3 Stem Cells in Human Adult Bone and BM 506
19.4 A Stem Cell for Mesenchymal Stem Cells? 509
19.5 Models of Stem Cell Differentiation: Hierarchical or Plastic? 510
19.6 Concluding Remarks 513

20 Dental Stem Cells   523
Anne-Margaux Collignon, Caroline Gorin,
Catherine Chaussain and Anne Poliard
20.1 Introduction 523
20.2 State of the Art 524
20.3 Major Unsolved Problems and Ongoing Controversies in the
Field 528
20.3.1 Heterogeneity in Pulpal Cell Populations 529
20.4 True Nature of DPSCs 531
20.5 Fate of Implanted Dental Stem Cells in Tissue Repair 532
20.6 Contribution of the Team to the Field: DP Stem Cells and
Craniofacial Bone Repair 533
20.6.1 Optimization of the Scaffold to Support the DPSC-
promoted Tissue Repair 533
20.6.2 Fate of DPSCs/SHEDs Implanted in a Critical Defect
of the Calvaria 535
20.6.3 Pro-angiogenic Potential of DPSCs/SHEDs 537
20.7 Perspectives 538
20.8 Concluding Remarks 539

21 Stem Cells and Retina: From Regeneration   **553**
Olivier Goureau, Giuliana Gagliardi and Gael Orieux
21.1 Introduction 553
21.2 Eye Morphogenesis and Retinogenesis 554
21.3 Adult Retinal Stem Cells and Neurogenic Potential 556
21.4 The CMZ: a Retinal Stem Cell Niche 556
    21.4.1 The CMZ in Cold-blooded Vertebrates 556
    21.4.2 The CMZ in Birds and Mammals 557
21.5 The Neurogenic Potential of the RPE 558
    21.5.1 RPE Cell Transdifferentiation in Amphibian and
        Chick Embryos 558
    21.5.2 RPE Transdifferentiation in Mammals 558
21.6 The Müller Glial Cells: A Retinal Stem 559
    21.6.1 Fish 559
    21.6.2 Amphibians and Birds 560
    21.6.3 Mammals 561
21.7 Cell Replacement for Retinal Repair 561
    21.7.1 RPE Cells Derived From Human ES and iPS Cells . 563
    21.7.2 Photoreceptors Derived From Human ES and iPS
        Cells 563
21.8 Concluding Remarks 567

22 Glioblastoma Stem Cells   **579**
Nathalie Magne, Sandra E Joppé, Franck Bielle and
Emmanuelle Huillard
22.1 Introduction 579
22.2 GBM initiation 580
    22.2.1 Tumor Initiation and Evolution 580
    22.2.2 Cells of Origin for GBMs 581
22.3 Identification of GSCs 582
22.4 Regulation of GSC Activity 584
    22.4.1 Cell-intrinsic Regulation 584
    22.4.2 Interactions with the Tumor-associated Microenvi-
        ronment 585
22.5 GSC Heterogeneity and Plasticity 588
22.6 GSC Resistance to Therapies 589
    22.6.1 GSCs Are Resistant to Radiotherapy and
        Chemotherapy 589
    22.6.2 Strategies to Target GSCs 591

22.7 Concluding Remarks 592

**23 Cardiac Tissue Engineering for Repair and Regeneration of the
Heart   603**
Pierre Joanne and Onnik Agbulut
23.1 Introduction 603
23.2 Heart Diseases and Stem Cell-based Therapies 604
23.3 The Heart: A Complex Organ 606
23.4 Cell Types for Heart Repair: An Endless Story? 608
23.5 The Rise of Pluripotent Stem Cells-derived 611
23.6 Cardiomyocytes Subtypes of hPSC-derived Cardiomyocytes 613
23.7 Maturation of hPSC-derived Cardiomyo cytes 613
23.8 Biomaterials for Stem Cell 614
23.9 New Perspectives in Cardiac Tissue 616
23.10Concluding Remarks 617

24 Stem Cells for Red Blood Cell Production   627
Laurence Guyonneau-Harmand and Hélène Lapillonne
24.1 Introduction 627
24.2 Erythropoiesis 628
24.3 From RBCs to Transfusion and Back to Blood Substitutes . . 631
      24.3.1 Ex Vivo Generated RBCs From HSCs 632
      24.3.2 Ex Vivo Generated RBCs From hESCs 633
      24.3.3 Ex Vivo Generated RBCs From hiPSCs 633
24.4 The Cell Engineering Era 634
24.5 The Gene Editing Era 636
24.6 Conclusion 636

**25 Prospectives for Therapy With Stem Cells in Skeletal Muscular
Diseases   645**
Negroni Elisa, Butler-Browne Gillian and Mouly Vincent
25.1 Introduction 645
25.2 What Is Expected From a Good Cell Candidate? 645
25.3 Candidate Cells 647
      25.3.1 Intramuscular Delivery 647
              25.3.1.1 Myoblasts 647
      25.3.2 New Candidates for a Systemic Delivery 649
              25.3.2.1 BMSC and SP cells 649
              25.3.2.2 CD133+ cells 649

        25.3.2.3 Mesoangioblasts/Pericytes 650
25.4 Therapeutic Application and Future Optimization 650
25.5 Concluding Remarks 654

**26 Legal Framework for Research on Human Embryonic Stem
    Cells in France and in Europe   665**
    Arnaud de Guerra, Samuel Arrabal and
    Emmanuelle Prada-Bordenave
    26.1 Introduction 665
    26.2 How to Regulate hESC Research? The French Example . . . 667
        26.2.1 General Provisions Regarding hESC Research . . . 667
        26.2.2 Agence de la Biomédecine as the Central Piece of the
                 National Regulation 668
        26.2.3 Authorization Delivery 669
        26.2.4 Donation of Human Embryos for Research 670
        26.2.5 Translating hESC Research Into Clinics 672
    26.3 Examples of Regulations in Europe: France's Neighbors . . 673
        26.3.1 Spain 674
                26.3.1.1 Regulation Frame 674
                26.3.1.2 Criteria for Authorization 674
                26.3.1.3 Institutions 675
                26.3.1.4 Process for Approval 676
                26.3.1.5 Embryos for Research 676
        26.3.2 United Kingdom 677
                26.3.2.1 Regulation Frame 677
                26.3.2.2 Institutions 678
                26.3.2.3 Process for Approval 679
                26.3.2.4 Embryos for Research 680
        26.3.3 Germany 681
                26.3.3.1 Regulation Frame 681
                26.3.3.2 Institutions 681
                26.3.3.3 Process for Approval 682
                26.3.3.4 Embryos for Research 682
    26.4 Concluding Remarks 683

27 Stem Cell Conceptual Clarifications   685
    Lucie Laplane
    27.1 Introduction 685
    27.2 Stem Cell Definition 686

27.3 Self-Renewal and Differentiation Definition 687
27.4 Proving Stemness: The Uncertainty Principle 689
27.5 Natural Kind or Artificial Grouping 689
27.6 Does the Concept of Stem Cell Refer to an "Entity" or to a
      Cell "State"? 690
27.7 Stemness Ontology: Not Two but Four 692
27.8 Philosophy Matters 695
27.9 Concluding Remarks 697

28 Future Outlook   **705**
Olivera Miladinovic, Pierre Charbord and Charles Durand
28.1 Introduction 705
28.2 Understanding and Capturing Pluripotency 706
28.3 Integrating Signals of Intrinsic, Extrinsic and Systemic
      Origins 707
28.4 Stem Cell Heterogeneity and Plasticity 709
28.5 Interface Between Stem Cell and Computational Biology . . 710
28.6 Therapeutic Applications of Stem Cells 711

Index   **717**

About the Editors   **725**

Preface

The book 'Stem Cell Biology and Regenerative Medicine' takes its origin in the master programme (2nd year level) 'The Biology of Stem Cells' we have founded at the University Pierre and Marie Curie (now called Sorbonne Université) in Paris in 2008. It is also based on a unit of teaching 'Introduction to Stem Cell Biology' for students at the master 1 degree (Master Molecular and Cellular Biology). Students enrolled in the master 2 course represent a range of academic backgrounds (in biology, medicine and pharmacy) and are coming from French, European and other universities. They have dedicated one year of their education to the study of stem cells and are trained in this course with stem cell concepts and methods at the theorical and practical levels. In parallel to a 6-month internship in a host laboratory, master students follow a cycle of conferences coupled with a unique 3-week practical workshop allowing them to observe and manipulate mouse embryonic stem cells and adult-type stem/progenitor cells from the hematopoietic, neural, intestinal and muscle tissues. The conferences are given by worldwide recognized scientists in stem cell biology and they cover many facets in the field: pluripotent stem cells, cellular and nuclear reprogramming, regeneration, gene expression regulation, adult stem cell biology, stem cell niches, computational biology, cancer stem cells and therapies using stem cells. Students also participate in lectures on the regulatory and ethical aspects related to stem cell research. We thought that this high-level educational program should be of particular interest for young and confirmed scientists interested in stem cell biology and regenerative medicine.

This book is aimed not only at master students, but also at scientists in the field of stem cell biology, researchers in molecular and cellular biology, and clinicians. Lawyers and laymen interested in regulatory and philosophical issues implying stem cell therapy and concepts may also find useful information in this book.

In line with the first edition, this book contains five sections many of which have benefitted from novel contributions. Section 1 is related to concepts of critical importance for the study of stem cells: transcriptional regulation, DNA repair, epigenetics, stem cell niches, computational biology and transcriptomics investigations. Section 2 is dedicated to pluripotent stem cells and regeneration. Section 3 gives a large overview on adult stem cells in different tissues. Section 4 presents the concept of cancer stem cells using glioblastomas as models as well as the applications of stem cells or their derivatives for red blood cell production, and the treatment of cardiac and skeletal muscle diseases. Finally, section 5 discusses regulatory and philosophical aspects of stem cell biology.

We are very grateful to River Publisher and more particularly to Rajeev Prasad for accepting to consider this book proposal and to publish it, and to Junko Nakajima for the excellent editing process. We thank Sophie Gournet (CNRS UMR7622) for her excellent drawing assistance to design the cover of this book. We thank all the contributors of this book for their participation at the master course and for accepting to write a chapter on their current research topic. We thank also all the members of the master stem cell committee for all the time, support and advices they provide to the students enrolled in this master course. Finally, we thank all the students (past and present) enrolled in this programme for their strong enthusiasm, curiosity and investment.

Paris, May 18 2021
Charles Durand and Pierre Charbord

List of Figures

Figure 1.1   Embryonic and adult stem cells.   3

Figure 1.2   Factors involved in stem cell regulation.   10

Figure 2.1   Enhancers driving complex gene expression patterns.   35

Figure 2.2   Chromatin looping mechanisms.   42

Figure 2.3   TADs and their formation by the loop extrusion model.   50

Figure 3.1   Schematic representation of classical and alternative NHEJ.   76

Figure 3.2   Schematic representation of DNA damage response in general and in adult stem cells.   80

Figure 4.1   Schematic of anatomical BM HSC niches.   114

Figure 5.1   Concepts of stem cell organization.   125

Figure 5.2   Model of the intestinal crypt.   128

Figure 5.3   Epigenetic regulation of stem cell fate.   130

Figure 5.4   Simulated aging of stem cells.   131

Figure 5.5   Tissue shape feeds back on stem cell organization.   133

Figure 6.1   (A)From gene expression set to oriented gene network. Schematic flow chart: full details are in (Desterke et al., 2020).   156

Figure 7.1   The X inactivation centre (Xic).   165

Figure 7.2   Upon differentiation or development the concentration of the XCI-activator increases (*green*); and the concentration of the XCI-inhibitors drops (*blue*). .   172

Figure 8.1   Summary of liver development.   184

Figure 8.2   Principal steps of human embryogenesis and liver organogenesis (**left panel**) mimicked and recapitulated in protocols for human pluripotent stem cell differentiation into hepatocytes and cholangiocytes (**right panel**).   188

Figure 8.3   Applications for hiPSC-derived hepatic cells. . . .   192

Figure 9.1   Life cycle and anatomy of *Hydra*.   210

xxi

Figure 9.2   Three stem-cell lineages give rise to all the diversity of cell types in *Hydra*. 211

Figure 9.3   The conserved transcription factor FoxO is a molecular hub that mediates gene-environment interactions by integrating environmental signals with stem cell control, developmental pathways, and immune homeostasis. 221

Figure 9.4   Microbiome drives tumor formation in *Hydra*. . . . 225

Figure 10.1   Regenerative capacity as a function of developmental stage among vertebrate models of regeneration. 244

Figure 10.2   Limb regeneration in the urodeles. 246

Figure 10.3   Transdifferentiation during lens regeneration in newts. 260

Figure 11.1   Organization and regulation of the shoot and root meristems. 275

Figure 11.2   Schematic representation of regeneration systems in plants. 285

Figure 12.1   The journey of HSCsin different species: human, mouse, chicken, and zebrafish. 301

Figure 12.2   Dynamics of aorta formation and role of the dorsal and ventral structures. 308

Figure 12.3   Spatial localization of IAHC cells in different species: chicken, human, mouse, and zebrafish embryos. 310

Figure 13.1   The three hematopoietic waves and their developmental niches. 326

Figure 13.2   The endothelial-to-hematopoietic transition (EHT). 329

Figure 13.3   The aorta-gonad-mesonephros (AGM) niche. . . . 333

Figure 14.1   A functional way to get into the hematopoietic hierarchy. 363

Figure 14.2   Exposure to low doses of ionizing radiation (LDIR) alters HSC self-renewal potential (adapted from Henry et al 2020). 367

Figure 14.3   Hypoxia and HIF proteins are involved in the regulation of human early lymphoid progenitor functions (adapted from Chabi et al 2019). 375

Figure 15.1   Structure of the epidermis and stem cell compartments. 394

Figure 16.1   The mammary gland epithelium is composed of an outer layer of basal cells (in green) and an inner layer of luminal cells (in yellow). Images are whole mount staining of a mammary duct with the basal marker α-SMA (in green) or the luminal marker Keratin 8 (in yellow).   417

Figure 16.2   Multicolor lineage tracing of a lactating mammary gland (modified from Davis et al., 2016).   420

Figure 16.3   The mammary stem cell hierarchy.   421

Figure 16.4   Notch signaling promotes a luminal fate.   424

Figure 17.1   Functional organization of the intestinal epithelium.   439

Figure 17.2   Intestinal Organoids.   441

Figure 18.1   Anatomical localization of the cerebral zones containing neural stem cells.   469

Figure 18.2   Histological features of the different neurogenic zones of the mouse brain.   470

Figure 19.1   **A**: Schematic representation of bone during childhood.   505

Figure 20.1   The different tissues of an adult molar tooth shown in: (A) a micro- scanner view and (B) a schematic representation of a human molar tooth (C) a Masson trichrome staining of a mice molar section.   524

Figure 20.2   Stem-primed T-mDPSCs promote repair through an endochondral bone formation process.   536

Figure 20.3   Diagrams showing the action mechanisms of (A) hypoxic- (1%) or (B) FGF-2 priming on SHEDs SHEDs synthesize HGF, a pro-angiogenic factor in normal culture conditions.   538

Figure 21.1   Representation of key steps for the generation of retinal organoids from human ES or iPS cells. . . .   564

Figure 22.1   Model for the initiation and evolution of primary (IDH1-wildtype) GBMs.   583

Figure 22.2   GSCs' interactions within the different tumor niches.   586

Figure 23.1   Lineage specification of cardiomyocytes: from pluripotent stem cells to the formation of the heart.   610

Figure 24.1   Chronology of cultured red blood cells generation.   633

Figure 25.1   Grafting human myoblasts.   652

Figure 26.1   Diversity of legislations regarding hESC research in Europe (from European Science Fundation, Science Policy Briefing, May 2010). 674

Figure 27.1   Diagrammatic representation of the success or failure of therapeutic strategies according to stemness ontology. 696

List of Tables

Table 13.1 Cell populations analyzed in the outlined sequencing studies 332

Table 19.1 Non-hematopoietic Bone and BM Murine Stem Cells 502

Table 21.1 Major transcription factors involved in the specification/differentiation of each retinal cell type and subtype 555

Table 23.1 Principal heart diseases targeted in experimental studies that use cardiac tissue engineering as treatment . . 605

Table 27.1 What kind of property is stemness? 699

List of Contributors

Agbulut, Onnik, *Sorbonne Université, Institut de Biologie Paris-Seine (IBPS), CNRS UMR 8256, Inserm ERL U1164, Biological Adaptation and Ageing, 75005, Paris-France; E-mail: onnik.agbulut@sorbonne-universite.fr*

Aractingi, Sélim, *Institut Cochin, Inserm U1016-CNRS UMR8104, Paris, France; Université de Paris, France; Service de Dermatologie, Hôpital Cochin-Tarnier, Paris, France*

Arcangeli, Marie-Laure, *Team Niche and Cancer in Hematopoiesis, U1274, INSERM, 18 route du Panorama, 92260 Fontenay-aux-Roses, France; Laboratory of Hematopoietic Stem Cells and Leukemia/Service Stem Cells and Radiation /iRCM/JACOB/DRF, CEA, 18 route du Panorama, 92260 Fontenay-aux-Roses, France; Université de Paris and Université Paris-Saclay, Inserm, iRCM/IBFJ CEA, UMR Stabilité Génétique Cellules Souches et Radiations, F-92265, Fontenay-aux-Roses, France; E-mail: marie-Laure.arcangeli@inserm.fr*

Arrabal, Samuel, *Agence de la biomédecine, 1 avenue du Stade de France, SAINT-DENIS LA PLAINE Cedex 93212, France; E-mail: samuel.arrabal@biomedecine.fr*

Bielle, Franck, *Sorbonne Université, Institut du Cerveau - Paris Brain Institute - ICM, Inserm, CNRS, APHP, Paris, France; AP-HP, Hôpitaux Universitaires La Pitié Salpêtrière - Charles Foix, Département de Neuropathologie Escourolle, F-75013, Paris, France*

Boeren, Jeffrey, *Department of Developmental Biology, Oncode Institute, Erasmus MC, The Netherlands*

Bosch, Thomas C.G., *Zoological Institute, University Kiel, Germany; E-mail: tbosch@zoologie.uni.kiel.de*

Butler-Browne, Gillian, *Sorbonne Université, Inserm, Institut de Myologie, Centre de Recherche en Myologie, F-75013 Paris, France*

Castela, Mathieu, *Institut Cochin, Inserm U1016-CNRS UMR8104, Paris, France*

Catala, Martin, *Sorbonne Université, CNRS, Inserm U1156, Institut de Biologie Paris Seine, Laboratoire de Biologie du Développement/UMR7622, 9 Quai St-Bernard, 75005 Paris, France; E-mail: Mmartin.catala@sorbonne-universite.fr*

Charbord, Pierre, *Sorbonne Université, CNRS, Inserm U1156, Institut de Biologie Paris Seine, Laboratoire de Biologie du Développement/UMR7622, 9 Quai St-Bernard, 75005 Paris, France; E-mail: pierre.charbord@sorbonne-universite.fr*

Chaussain, Catherine, *Laboratoire Pathologies, Imagerie et Biothérapies Orofaciales, UR 2496, UFR d'Odontologie Montrouge, Université de Paris, 1 rue Maurice Arnoux, 92120 Montrouge, France*

Cherqui, Ulysse, *Department of Molecular Cell Biology, Weizmann Institute of Science, 234 Herzl Street, 7610001 Rehovot, Israel*

Collignon, Anne-Margaux, *Laboratoire Pathologies, Imagerie et Biothérapies Orofaciales, UR 2496, UFR d'Odontologie Montrouge, Université de Paris, 1 rue Maurice Arnoux, 92120 Montrouge, France; AP-HP, Dental Department, Charles Foix, Louis Mourier and Bretonneau Hospitals, France*

Da Costa, Marco, *Institut Jean-Pierre Bourgin, INRAE, AgroParisTech, Université Paris-Saclay, 78000, Versailles, France; Sorbonne Université, Centre National de la Recherche Scientifique (CNRS), Institut de Biologie Paris-Seine (IBPS), Laboratoire de Biologie du Développement (LBD), UMR7622, Paris, France; E-mail: marco.da_costa@sorbonne-universite.fr*

de Guerra, Arnaud, *Agence de la biomédecine, 1 avenue du Stade de France, SAINT-DENIS LA PLAINE Cedex 93212, France*

Desterke, Christophe, *Université Paris Sud, Hôpital Paul Brousse, Villejuif, France; E-mail: christophe.desterke@inserm.fr*

Dubart-Kupperschmitt, Anne, *INSERM unité mixte de recherche (UMR_S) 1193, Villejuif, F-94800, France; UMR_S 1193, Université Paris-Saclay, Villejuif, F-94800, France; Fédération Hospitalo-Universitaire (FHU) Hépatinov, Villejuif, F-94800, France; Email: anne.dubart@inserm.fr*

Durand, Charles, *Sorbonne Université, CNRS, Inserm U1156, Institut de Biologie Paris Seine, Laboratoire de Biologie du Développement/UMR7622,*

9 Quai St-Bernard, 75005 Paris, France; E-mail: charles.durand@sorbonne-universite.fr

Fontaine, Romain, *Institut Cochin, Inserm U1016-CNRS UMR8104, Paris, France; Université de Paris, France; E-mail: romain.fontaine@inserm.fr*

Fre, Silvia, *Developmental Biology and Genetics Unit, CNRS UMR3215/ Inserm U934, PSL Research University, Institut Curie Centre de Recherche, 26 rue d'Ulm, 75248 Paris, Cedex 05, France; E-mail: silvia.fre@curie.fr*

Gagliardi, Giuliana, *Institut de la Vision, Sorbonne Université, INSERM U968, CNRS UMR_7210, Paris, France*

Galle, Joerg, *Interdisciplinary Center for Bioinformatics, University Leipzig, Haertelstr. 16–18, 04107 Leipzig, Germany; E-mail: galle@izbi.uni-leipzig.de*

Gorin, Caroline, *Laboratoire Pathologies, Imagerie et Biothérapies Orofaciales, UR 2496, UFR d'Odontologie Montrouge, Université de Paris, 1 rue Maurice Arnoux, 92120 Montrouge, France*

Goureau, Olivier, *Institut de la Vision, Sorbonne Université, INSERM U968, CNRS UMR_7210, Paris, France; Email: olivier.goureau@inserm.fr*

Gribnau, Joost, *Department of Developmental Biology, Oncode Institute, Erasmus MC, The Netherlands; E-mail: j.gribnau@erasmusmc.nl*

Grosveld, Frank, *Department of Cell Biology, Erasmus Medical Centre, Rotterdam, The Netherlands; E-mail: f.grosveld@erasmusmc.nl*

Guyonneau-Harmand, Laurence, *Sorbonne Université, Inserm, Centre de recherche Saint-Antoine, CRSA, F-75012, Paris, France; Établissements Français du Sang Île-de-France, Unité d'ingénierie et de thérapie cellulaire, Créteil, F-75012, Paris, France; E-mail: laurence.harmand@sorbonne-universite.fr*

Haddad, Rima, *Team Niche and Cancer in Hematopoiesis, U1274, INSERM, 18 route du Panorama, 92260 Fontenay-aux-Roses, France; Laboratory of Hematopoietic Stem Cells and Leukemia/Service Stem Cells and Radiation /iRCM/JACOB/DRF, CEA, 18 route du Panorama, 92260 Fontenay-aux-Roses, France; Université de Paris and Université Paris-Saclay, Inserm, iRCM/IBFJ CEA, UMR Stabilité Génétique Cellules Souches et Radiations, F-92265, Fontenay-aux-Roses, France; E-mail: rima.haddad@universite-paris-saclay.fr*

Huillard, Emmanuelle, *Sorbonne Université, Institut du Cerveau - Paris Brain Institute - ICM, Inserm, CNRS, APHP, Paris, France; E-mail: emmanuelle.huillard@icm-institute.org*

Jašarević, Sabina, *Department of Developmental Biology, Oncode Institute, Erasmus MC, The Netherlands*

Jaffredo, Thierry, *Sorbonne Université, IBPS, CNRS UMR7622, Inserm U 1156, Laboratoire de Biologie du Développement; 75005 Paris; E-mail: thierry.jaffredo@sorbonne-universite.fr*

Joanne, Pierre, *Sorbonne Université, Institut de Biologie Paris-Seine (IBPS), CNRS UMR 8256, Inserm ERL U1164, Biological Adaptation and Ageing, 75005, Paris-France*

Joppé, Sandra E, *Sorbonne Université, Institut du Cerveau - Paris Brain Institute - ICM, Inserm, CNRS, APHP, Paris, France*

Khoury, Hanane, *Sorbonne Université, IBPS, CNRS UMR7622, Inserm U 1156, Laboratoire de Biologie du Développement; 75005 Paris; Department of Hematology, St. Jude Children's Research Hospital; 262 Danny Thomas, Memphis 38105 TN; USA*

Klimovich, Alexander, *Zoological Institute, University Kiel, Germany*

Knapp, Dunja, *DFG-Center for Regenerative Therapies Dresden/TU Dresden, Fetscherstraße 105, 01307 Dresden, Germany; E-mail: dunja.knapp@tu-dresden.de*

Kubis, Nathalie, *Université de Paris, INSERM U1148, Laboratory for Vascular Translational Science, F- 75018 Paris, France and Service de Physiologie Clinique – Explorations Fonctionnelles, DMU DREAM, APHP, Hôpital Lariboisière, F-75010 Paris, France*

Lapillonne, Hélène, *Sorbonne Université, Inserm, Centre de recherche Saint-Antoine, CRSA, F-75012, Paris, France; APHP, Hôpital Trousseau, département d'hématologie, F-75012, Paris, France*

Laplane, Lucie, *CNRS, University Paris I Panthéon-Sorbonne (IHPST – UMR 8590), Paris, France; Gustave Roussy Cancer Center (UMR 1287), Villejuif, France; E-mail: lucie.laplane@gustaveroussy.fr*

Luce, Eléanor, *INSERM unité mixte de recherche (UMR_S) 1193, Villejuif, F-94800, France; UMR_S 1193, Université Paris-Saclay, Villejuif, F-94800, France; Fédération Hospitalo-Universitaire (FHU) Hépatinov, Villejuif, F-94800, France; Email: eleanor.luce@inserm.fr*

Magne, Nathalie, *Sorbonne Université, Institut du Cerveau - Paris Brain Institute - ICM, Inserm, CNRS, APHP, Paris, France*

Messina, Antonietta, *INSERM unité mixte de recherche (UMR_S) 1193, Villejuif, F-94800, France; UMR_S 1193, Université Paris-Saclay, Villejuif, F-94800, France; Fédération Hospitalo-Universitaire (FHU) Hépatinov, Villejuif, F-94800, France; Email: antonietta.messina@inserm.fr*

Miladinovic, Olivera, *Sorbonne Université, CNRS, Inserm U1156, Institut de Biologie Paris Seine, Laboratoire de Biologie du Développement/UMR7622, 9 Quai St-Bernard, 75005 Paris, France*

Mouly, Vincent, *Sorbonne Université, Inserm, Institut de Myologie, Centre de Recherche en Myologie, F-75013 Paris, France*

Murawala, Prayag, *Mount Desert Island Biological Laboratory, PO Box 35, Salisbury Cove, ME-04672, USA; Clinic for Kidney and Hypertension Diseases, Hannover Medical School, Hannover – 30625, Germany; E-mail: pmurawala@mdibl.org*

Negroni, Elisa, *Sorbonne Université, Inserm, Institut de Myologie, Centre de Recherche en Myologie, F-75013 Paris, France; E-mail: elisa.negroni@sorbonne-universite.fr*

Nitsche, Leslie, *Centre for Regenerative Medicine, Institute for Regeneration and Repair, University of Edinburgh, Edinburgh, UK*

Orieux, Gael, *Institut de la Vision, Sorbonne Université, INSERM U968, CNRS UMR_7210, Paris, France*

Ottersbach, Katrin, *Centre for Regenerative Medicine, Institute for Regeneration and Repair, University of Edinburgh, Edinburgh, UK; E-mail: katrin.ottersbach@ed.ac.uk*

Oulès, Bénédicte, *Institut Cochin, Inserm U1016-CNRS UMR8104, Paris, France; Université de Paris, France; Service de Dermatologie, Hôpital Cochin-Tarnier, Paris, France*

Pflumio, Francoise, *Team Niche and Cancer in Hematopoiesis, U1274, INSERM, 18 route du Panorama, 92260 Fontenay-aux-Roses, France; Laboratory of Hematopoietic Stem Cells and Leukemia/Service Stem Cells and Radiation /iRCM/JACOB/DRF, CEA, 18 route du Panorama, 92260 Fontenay-aux-Roses, France; Université de Paris and Université Paris-Saclay, Inserm, iRCM/IBFJ CEA, UMR Stabilité Génétique Cellules Souches et Radiations, F-92265, Fontenay-aux-Roses, France*

Poliard, Anne, *Chaussain, Catherine, Laboratoire Pathologies, Imagerie et Biothérapies Orofaciales, UR 2496, UFR d'Odontologie Montrouge, Université de Paris, 1 rue Maurice Arnoux, 92120 Montrouge, France; E-mail: anne.poliard@parisdescartes.fr*

Prada-Bordenave, Emmanuelle, *Agence de la biomédecine, 1 avenue du Stade de France, SAINT-DENIS LA PLAINE Cedex 93212, France*

Rech, Philippe, *Institut Jean-Pierre Bourgin, INRAE, AgroParisTech, Université Paris-Saclay, 78000, Versailles, France; Institut de Systématique, Evolution, Biodiversité (ISYEB – UMR 7205 – CNRS, MNHN, SU, EPHE), Muséum national d'Histoire naturelle, 57 rue Cuvier, 75005 Paris, France; E-mail: philippe.rech@sorbonne-universite.fr*

Ricchetti, Miria, *Institut Pasteur, Team Stability of Nuclear and Mitochondrial DNA, Stem Cells and Development, Dept. of Developmental and Stem Cell Biology UMR3738 CNRS, 25, rue du Dr. Roux, 75724 Paris Cedex 15, France; E-mail: miria.ricchetti@pasteur.fr*

Soler, Eric, *IGMM, Univ Montpellier, CNRS, Montpellier, France; Laboratory of Excellence GR-Ex, Université de Paris, France; E-mail: eric.soler@igmm.cnrs.fr*

Stedman, Aline, *Sorbonne Université, UPMC Université Paris 06, IBPS, CNRS UMR7622, Inserm U 1156, Laboratoire de Biologie du Développement; Paris 75005, France; E-mail: aline.stedman@sorbonne-universite.fr*

Sutcu, Haser Hasan, *Institut Pasteur, Team Stability of Nuclear and Mitochondrial DNA, Stem Cells and Development, Dept. of Developmental and Stem Cell Biology UMR3738 CNRS, 25, rue du Dr. Roux, 75724 Paris Cedex 15, France*

Thalheim, Torsten, *Interdisciplinary Center for Bioinformatics, University Leipzig, Haertelstr. 16–18, 04107 Leipzig, Germany*

Vahidi Ferdousi, Leyla, *Institut Pasteur, Team Stability of Nuclear and Mitochondrial DNA, Stem Cells and Development, Dept. of Developmental and Stem Cell Biology UMR3738 CNRS, 25, rue du Dr. Roux, 75724 Paris Cedex 15, France*

Yvernogeau, Laurent, *Sorbonne Université, IBPS, CNRS UMR7622, Inserm U 1156, Laboratoire de Biologie du Développement; 75005 Paris; E-mail: laurent.yvernogeau@sorbonne-universite.fr*

List of Abbreviations

α-SMA	α-Smooth Muscle Actin
γH2AX	Phosphorylated histone H2AX
3D	three-dimensional
16S rDNA	16S ribosomal RNA gene
53BP1	Tumor suppressor p53 binding protein 1
A	adipocyte
A1AT	alpha-1-anti-trypsin
ABC	ATP-binding cassette
ACR4	ARABIDOPSIS CRINKLY4
ADMET	absorption, distribution, metabolism, excretion, and toxicity mechanisms
AEC	apical epidermal cap
AER	apical ectodermal ridge
AGM	aorta-gonad-mesonephros
AHK	ARABIDOPSIS HISTIDINE KINASE
Aibp2	Apolipoprotein A-I binding protein 2
AL	artificial liver
ALM	accessory limb model
Alt-NHEJ	Alternative Non-homologous end-joining
AMD	age-related macular degeneration
AMP	antimicrobial peptide
Ao	(dorsal) Aorta
AoD	Dorsal side of Aorta
AoV	Ventral floor of Aorta
ARF	Auxin Response Transcription Factor
ARR	Arabidopsis Response Regulators
ASC	Adult stem cells
ASCL1	achaete-scute family bHLH transcription factor 1
ASCs	adult stem cells
ATM	Ataxia telangiectasia mutated

atoh7	atonal BHLH transcription factor 7
ATR	Ataxia telangiectasia and Rad3-related
Aux/IAA	AUXIN/INDOLE-3-ACETIC ACID
BAL	bioartificial liver
BBB	Blood brain barrier
BC	Basal Cell
BEL-A	Bristol Erythroid Line Adult
BER	Base excision repair
bFGF	basic fibroblast growth factor
BFU-E	Erythroid Burst Forming Units
bHLH	basic helix-loop-helix
BIR	Break-induced replication
BL	basal/germinal layer
BM	bone marrow
BM	Basement Membrane
BMI-1	BMI1 proto-oncogene, polycomb ring finger
BMP	bone morphogenetic protein
Bmp4	Bone morphogenetic protein 4
BMSC	Bone Marrow Stem Cell
BRCA	Breast cancer susceptibility protein
BRD4	bromodomain containing 4
BSC	Bulge Stem cells
BSCP	Bone-Cartilage-Stroma Progenitor
C	Chondrocyte
CAM	chorioallantoic membrane
CAR-T	Chimeric Antigenic Receptor - T
CBC	Crypt base columnar cells
CDF4	CYCLING DOF FACTOR 4
CDK4	cyclin dependent kinase 4
CDKN2A/B	cyclin dependent kinase inhibitor 2A/B
CEN	Core Erythroid Network
CFDA-SE	carboxyfluorescein diacetate n-succinimidyl ester
CFU-E	erythroid colony forming units
CFU-f	colony-forming unit-fibroblasts
CFU-S	colony forming unit in the spleen
CGZ	circumferential germinal zone
CHK2	checkpoint kinase 2
CK	cytokinin

CLE	CLAVATA3/ESR-RELATED
CMP	Common myeloid progenitors
CMZ	ciliary marginal zone
C-NHEJ	Classical non-homologous end-joining (*alias* NHEJ)
CNS	central nervous system
CP	committed progenitors
CPE	ciliary body epithelium
Crx	cone-rod homeobox
CSC	columella stem cell
CSC	cancer stem cell
CT	connective tissue
CtBP	C-terminal binding protein
CtIP	CtBP interacting protein
CTLA-4	cytotoxic T-lymphocyte associated protein 4
CZ	central zone
DDR	DNA damage response
DEG	differentially expressed gene
DFSC	Dental Follicle Stem Cell
Dhh	Desert hedgehog
Dlk1	Delta-like homolog 1
Dll1	Delta-like1
D-loop	Displacement loop
DLP	dosal lateral plate
DMD	Duchenne Muscular Dystrophy
DNAPK	DNA dependent protein kinase
DNA-PKcs	DNA dependent protein kinase catalytic subunit
DNMT	DNA methyltransferase
DNSO	de novo Shoot Organogenesis
DP	Dental Pulp
DP	dermal papilla
DPSC	Dental Pulp Stem Cell
DSB	Double-strand break
E	embryonic day
EC	Endothelial Cell
EC	enterocyte
ECM	Extracellular matrix
ECM	Extra-Cellular Matrix
ECs	endothelial cells

EFTFs	eye-field transcription factors
EGF	epidermal growth factor
EGFP	enhanced green fluorescent protein
EGFR	Epidermal growth factor receptor
EHT	endothelial to hematopoietic transition
EPC	endothelial progenitor cells
EPO	erythropoietin
EPUs	epidermal proliferation units
ERα	Estrogen Receptor alpha
ERF115	ETHYLENE RESPONSE FACTOR 115
ERK	extracellular signal-regulated kinase
ES	Embryonic Stem Cell
ES cell	Embryonic Stem Cell
ES	embryonic stem
ESC	embryonic stem cell
EV	extracellular vesicle
EZH2	enhancer of zeste 2 polycomb repressive complex 2 subunit
FACS	Fluorescence-Activated Cell Sorter
FGF	Fibroblast Growth Factor
FGF2	Fibroblast growth factor 2
Flt3l	FMS-like tyrosine kinase 3 ligand
FOG-1	Friend of GATA1
FOXA	Forkhead Box Proteins A
GBM	glioblastoma
GC	Goblet cell
GC	Glucocorticoid
GCL	ganglion cell layer
GFP	green fluorescent protein
GGF	glial growth factor
GGT	γ-glutamyl transferase
GH	growth hormone
GMP	Good Manufacturing Practice
GMP	Granulocyte/macrophage progenitors
Gr	granular layer
GRN	Gene Regulatory Network
GSC	glioma stem cells
GSC	germ stem cell

GSEA	Gene Set Enrichment Analysis
GvHD	Graft-versus-Host Disease
Gy	Gray
HAM	HAIRY MERISTEM
HbF	fetal hemoglobin
HBOCs	Hemoglobin-Based Oxygen Carriers
HC	hematopoietic cell
HDA19	HISTONE DEACETYLASE 19
HDAC	Histone deacetylase
HE	Hemogenic Endothelium
HEC	Hemogenic Endothelial Cell
HEC1	HECATE1
hES	human embryonic stem cells
hESCs	human embryonic stem cells
HF	hair follicle
Hh	Hedgehog
HIF	Hypoxia Inducible Factor
HIF1	Hypoxia-inducible factor-1
Hif1a	Hypoxia-inducible factor 1α
hiPSCs	human induced pluripotent stem cells
HMG	high-mobility group
HMT	histone methyltransferase
HNF	hepatocyte nuclear factor
HPA/I	Hypothalamic–Pituitary–Adrenal/Interrenal
HPC	homeostatic property cluster
Hpf	hour post-fertilization
hPSC	human Pluripotent Stem Cell
HR	Homologous recombination
HS	hair shaft
HSPC	Hematopoietic stem/progenitor cells
HUVECs	human umbilical vein endothelial cells
IAHC	Intra-Aortic Hematopoietic Cluster
IBM	individual cell-based model
IDH1	isocitrate dehydrogenase 1
iDPSC	immature Dental Pulp Stem Cell
IFE	interfollicular epidermis
Ifna	Interferon-α

IGF	insulin-like growth factor
IGF1	insulin-like growth factor-1
iHeps	hiPSC-derived hepatocyte
Ihh	Indian hedgehog
IL-1β	interleukin-1β
IL-3	interleukin 3
IL-6	interleukin 6
ILs	Interleukins
INL	inner nuclear layer
IPL	inner plexiform layer
iPS	induced pluripotent stem
iPS cell	Induced Pluripotent Stem Cell
IR	Ionising radiation
IRS	inner root sheats
ISC	Intestinal stem cells
Jag1	Jagged-1
JAK	Janus kinase
K5	Cytokeratin 5
K8	Cytokeratin 8
K14	Cytokeratin 14
KGF	keratinocyte growth factor
klf4	kruppel like factor 4
Krt19	cytokeratin 19
LC	Luminal Cell
LDI	Low doses of irradiation
LIF	Leukemia Inhibitory Factor
LPM	lateral plate mesoderm
LR	lateral roots
LRM	lateral root meristem
LRP2	low-density lipoprotein receptor-related protein 2
MAPK	mitogen-activated protein kinase
MaSC	Mammary Stem Cell
MD	Muscular Dystrophy
MEF	Mouse Embryonic fibroblast
MEP	bipotent erythro-megakaryocytic progenitor
MeSCs	Melanocyte stem cells
MET	MET proto-oncogene, receptor tyrosine kinase
MGMT	O-6-methylguanine-DNA methyltransferase

mitf	melanocyte inducing transcription factor
MLL1	mixed lineage leukemia protein-1
MMEJ	Micro-homology mediated end-joining
MMP	matrix metalloproteinase
MMR	Mismatch repair
MMTV	Mouse Mammary Tumor Virus
MP	Monopteros
MRE11	Meiotic recombination 11
MRN	Mre11-Rad50-NBS1 complex
MRP-2	Multidrug resistance-associated protein 2
MSC	Mesenchymal Stem/Stromal Cell
MSC	Mesenchymal Stem Cell
MSC	Mesenchymal Stromal Cell
MyD88	Myeloid Differentiation primary response 88
MyoD	Myoblast determination protein
NAFLD	non-alcoholic fatty liver disease
NBS1	Nibrin, Nijmegen breakage syndrome
NC	Neural Crest
NER	Nucleotide excision repair
NF-κB	Nuclear Factor Kappa-light-chain-enhancer of activated B cells
NF1	neurofibromin 1
NG2	Neural/glial antigen 2
NGS	Next Generation Sequencing
NHEJ	Non-homologous end-joining (*alias* C-NHEJ)
NK	natural killer
NO	Nitric Oxide
NPC	neural progenitor cell
Nrl	neural retina-specific leucine zipper
NSC	neural stem cell
NSPC	Neural stem/progenitor cells
O	osteoblast
OC	organizing center
OLIG2	oligodendrocyte transcription factor 2
ONL	outer nuclear layer
OPC	oligodendrocyte precursor cell
OPL	outer plexiform layer
OPMD	Oculo-Pharyngeal Muscular Dystrophy

ORS	outer root sheat
OSM	oncostatin M
Otx2	orthodenticle Homeobox 2
PARP	Poly-(ADP-ribose) polymerase
PARP	Poly ADP-ribose polymerase
PAS	Periodic acid Schiff
Pax6	paired box 6
PBMCs	Peripheral blood mononuclear cells
PC	Paneth cell
PCA	principal component analysis
PCP	PreChondrogenic Progenitor
PD-1	programmed cell death 1
PDGFRA	Platelet-derived growth factor receptor alpha
PD-L1	Programmed death-ligand 1
PDLSC	Periodontal Ligament Stem Cell
PEC	pigmented epithelial cells
PFBOCs	Perfluorocarbon-Based Oxygen Carriers
PGE2	Prostaglandin E2
PHHs	primary human hepatocytes
PI3K	Phosphatidylinositide 3-kinase
PIK3CA	phosphatidylinositol-4,5-bisphosphate 3-kinase catalytic subunit alpha
PIK3R1	phosphoinositide-3-kinase regulatory subunit 1
PLT	PLETHORA
POLR1A$^+$	RNA polymerase I subunit A positive
PR	Progesterone Receptor
PSC	Pluripotent Stem Cell
PSM	presomitic mesoderm
P-Sp	para-aortic splanchnopleura
Ptch1	Patched 1
Ptch2	Patched 2
PTEN	phosphatase and tensin homolog
PZ	peripheral zone
QC	quiescent center
QC	quality control
RA	retinoic acid
RAD50	DNA repair protein (ATP-binding cassette-ATPase)
RAM	root apical meristem

RB	RB transcriptional corepressor 1
RBCs	Red blood cells
RGCs	retinal ganglion cells
RMG	retinal Müller glial
ROS	Reactive oxygen species
RPA	Replication protein A
RPCs	retinal progenitor cells
RPE	retinal pigmented epithelium
RTK	receptor tyrosine kinase
rx/rax	retina and anterior neural fold homeobox
RZ	rib zone
SAM	shoot apical meristem
SbG	sebaceaous glands
SC	stem cell
SCAP	Stem Cell from Apical Papilla
scATAC-seq	single cell ATAC sequencing
SCF	stem cell factor
Scf	Stem cell factor
SCID	Severe combined immunodeficiency
SCN	stem cell niche
SCR	SCARECROW
scRNA-seq	single cell RNA sequencing
SCs	Satellite cells
SG	Sebaceous gland
SHED	Stem cell from Human Exfoliated Deciduous tooth
Shh	Sonic hedgehog
SHR	SHORTROOT
SHY2	SHORTHYPOCOTYL2
Six3	six homeobox 3
Six6	six homeobox 6
SM	smooth muscle
SMC	Structural maintenance of chromosomes protein
SOX2	SRY (sex determining region Y)-box 2
Sox2	SRY-Box transcription factor 2
Sp	spinous layer
SP	side population
SSA	Single strand annealing
SSB	Single-strand break

SSC	Skeletal Stem Cell
SSCs	Spermatogonial stem cells
ssDNA	Single stranded Deoxyribonucleic acid
STAT	signal transducer and activator of transcription
STAT5	signal transducer and activator of transcription 5
StC	stratum corneum
STM	SHOOT MERISTEMLESS
SVZ	subventricular zone
SwG	sweat gland
TA	transit amplifying cells
TAM	tumor associated macrophage/microglia
TEB	Terminal End Bud
TERT	telomerase reverse transcriptase
TF	Tissue Factor
TF	transcription factor
TGF-α	transforming growth factor α
TGFβ	transforming growth factor β
TGFß	Transforming growth factor-ß
TGF-ß1	transforming growth factor ßeta 1
Thpo	Thrombopoietin
TIR1	TRANSPORT INHIBITOR RESISTANT1
TMZ	temozolomide
TNF-α	tumor necrosis factor-α
TOR	TARGET OF RAPAMYCIN
TP53	tumor protein p53
TPO	Thrombopoietin
TRG	taxonomically restricted gene
UV	Ultra violet
V(D)J	Variable diversity joining
VBI	ventral blood island
VDR	Vitamin D Receptor
Vegf	vascular endothelial growth factor
VEGF-A	Vascular Endothelial Growth Factor A
VSM	vascular smooth muscle
vsx2	visual system homeobox 2
WGCNA	Weighted Gene Correlation Analysis
WOX5	WUSCHEL-RELATED HOMEOBOX5
WRN	Werner syndrome protein

WUS	WUSCHEL
XLF	XRCC4-like factor
XPP	xylem pole-pericycle
XRCC	X-ray repair cross complementing protein
YAP	Yes-Activated Protein
YS	yolk sac
ZO-1	tight junction protein 1
ZPA	zone of polarizing activity

1

Stem Cell Concepts

Charles Durand and Pierre Charbord

Sorbonne Université, CNRS, Inserm U1156, Institut de Biologie Paris Seine, Laboratoire de Biologie du Développement/UMR7622, 9 Quai St-Bernard, 75005 Paris, France
E-mail: charles.durand@sorbonne-universite.fr; pierre.charbord@sorbonne-universite.fr

1.1 Introduction

Stem cell biology is an exciting and rapidly growing field in life science. What stem cells are, where they are, and how they are regulated, are major questions in fundamental biology with promising applications for human therapy. Not only is stem cell biology a scientific field under intensive investigations but also many issues that arise through the study and potential clinical use of stem cells have profound societal, philosophical, legal, and ethical impacts.

Stem cell researchers come from different scientific horizons: developmental biology, cell biology, genetics and epigenetics, systems biology, bioinformatics, biophysics, physiology, oncology, and other medical sciences. Thus, stem cell biology can be considered as a diverse and integrative branch of biology. Stem cell scientists in common utilize general and specific concepts and methods for understanding the complexity and diversity of stem cells. These concepts include "stemness" (the property of stem cells to both self-renew and differentiate), cell differentiation, cell proliferation, cell transformation, autocrine, paracrine, and systemic cell regulation, stem cell niches, tissue development, homeostasis, regeneration, and repair. The methods currently used by stem cell biologists involve procedures for cell isolation, *ex vivo* culture systems, clonogenic assays, transplantations, lineage tracing, high-throughput technologies, and computational analysis.

1

Historically, the experiments performed by James Till and Ernest McCulloch in the field of hematology were instrumental in the identification of stem cell concepts and methods. These authors, by injecting bone marrow cells into lethally irradiated mice, demonstrated the presence of macroscopic nodules in the spleen of the recipient mice. These nodules were called CFU-S for Colony-Forming-Units in the Spleen (Till and McCulloch, 1961). The authors showed that these modules had a clonal origin, contained several hematopoietic cell types and harbored self-renewal potential (Becker et al., 1963; Siminovitch et al., 1963). Together, these findings revealed the existence of blood-forming (hematopoietic) stem cells (HSCs) in the bone marrow.

1.2 Embryonic and Adult Stem Cells

Two types of stem cells have to be distinguished: embryonic and adult stem cells. Both types are able to self-renew and differentiate, but they considerably differ according to their origin and potentialities and thus represent distinct cellular entities (Figure 1.1). Self-renewal is a cardinal property of stem cells that defines their ability to divide without differentiation. In contrast, differentiation reveals the capability of an immature cell to acquire morphological, molecular, and functional characteristics of a specific cell lineage. Self-renewal, the capacity of mother cells to give rise to daughter cells with identical attributes, is not per se a sufficient condition to define a stem cell population (e.g., lymphocytes self-renew after antigen introduction to develop a clone that specifically recognizes the antigen). A bona fide stem cell population must simultaneously self-renew and commit to (a) specific lineage(s). This can be achieved by two mechanisms: either the mother stem cell divides asymmetrically giving rise to one identical daughter stem cell and to another committed daughter cell (lineage mechanism), or a stem cell divides symmetrically giving rise to two identical daughter stem cells while another stem cell gives rise to two committed cells (population mechanism) (Morrison and Spradling, 2008; Simons and Clevers, 2011; Spradling et al., 2001). Thus, asymmetric division is a property of stem cells, but is not the sole type of division in a stem cell population. Since expansion of the stem cell population can be achieved only by symmetrical division, stem cell populations use either lineage or population mechanisms to divide depending on the tissue requirements (steady-state versus stress conditions). It has been hypothesized that asymmetric division of stem cells might be accounted for by asymmetric segregation of chromosomes with an "immortal" DNA strand retained in the daughter cell retaining the stem cell properties (Cairns, 1975).

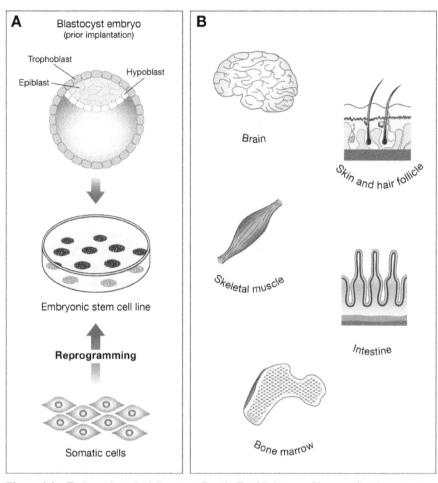

Figure 1.1 Embryonic and adult stem cells. (**A**) Establishment of immortal embryonic stem cell lines from pluripotent epiblast cells. Adult somatic cells can be reprogrammed into embryonic stem cell-like cells. (**B**) Adult tissues (such as the brain, skin and hair follicles, skeletal muscle, intestine, and bone marrow) harbor stem cell activity essential for their homeostasis and regeneration.

This non-random chromosome segregation might hold true for certain stem cells such as epithelial (Potten et al., 1978) or neural (Karpowicz et al., 2005), but is not in agreement with results using HSCs (Kiel et al., 2007; Wilson et al., 2008).

Totipotent cells are cells that give rise to all cells of the embryo including the extra-embryonic annexes. Such cells are the fertilized eggs and the cells generated by the subsequent 2–3 divisions. Shortly before implantation, a group of cells within the inner cell mass of the mammalian blastocyst embryos constitute the epiblast. These cells have the unique capability to give rise to all cell types of the organism including the germline, and are defined as pluripotent (Nichols and Smith, 2009). Importantly, naive epiblast cells exist during a short developmental window *in vivo* and switch to primed cells after implantation; they do not give rise to extraembryonic tissues that derive from the trophoblast and hypoblast. Thus, naive or primed epiblast cells do not form an entire organism by themselves: they are defined as pluripotent but not totipotent cells. Pluripotent embryonic stem (ES) cell lines were derived from the epiblast of pre-implantation blastocyst embryos, firstly in mice then in humans (Evans and Kaufman, 1981; Martin, 1981; Thomson et al., 1998). These lines can be propagated undifferentiated indefinitely when cultured in specific cell culture conditions. These include inhibitors of the mitogen-activated kinase (MAPK/ERK) pathway and glycogen synthase kinase-3 pathways, and/or the use of serum and mouse embryonic fibroblasts as feeder cells providing the leukemia inhibitory factor that supports ES cell survival and prevents differentiation (Nichols and Smith, 2009). Pluripotency may be artificially acquired by a differentiated cell through nuclear reprogramming. Pioneer studies in amphibians have elegantly demonstrated that a nucleus from a differentiated somatic cell could be reprogrammed after transfer into an enucleated egg (Gurdon and Laskey, 1970). More recently, Shinya Yamanaka and collaborators have shown that mouse and human adult fibroblasts may return to an ES-like state and give rise to induced pluripotent stem (iPS) cells after ectopic expression of pluripotent genes such as *Oct-4*, *Sox-2*, *Klf4,* and *c-myc* (Takahashi et al., 2007; Takahashi and Yamanaka, 2006). Alternatively, reprogramming of a differentiated somatic cell can occur after fusion with ES cells (Cowan et al., 2005). Interestingly, although the two parental genomes coexist in the hybrid cells, the ES cell genetic profile is predominant and the hybrid cells acquire the morphology and differentiation potential of ES cells. Experimentally, pluripotency is evaluated with appropriate and rigorous *ex vivo* and *in vivo* assays. *Ex vivo*, cells are tested for their ability to produce derivatives of the three germ layers—ectoderm, mesoderm, and endoderm. *In vivo*, pluripotent stem cells generate teratomas after subcutaneous transplantation in immune-deficient mice. The gold standard assay to prove pluripotency consists in injecting manipulated cells in the inner cell mass of a host blastocyst and evaluating their contribution to ectodermal, mesodermal, and endodermal tissues as well as their germline transmission.

ES and iPS cells constitute an exceptional tool to study the processes of cell differentiation, to model human diseases, and to test the efficiency and toxicity of novel drugs. ES and iPS cells also represent a promising source for regenerative medicine. Among the questions and obstacles that face stem cell biologists, one concerns our capability to recapitulate in a petri dish the complex cellular events that occur naturally during early development. This means that researchers have to precisely define the optimal culture conditions to efficiently differentiate pluripotent stem cells into a specific cell lineage. Thus, the choice of concentration and timing of the growth factors, morphogens, hormones, and extracellular matrix (ECM) components that have to be supplemented in the culture medium as well as the setting of physical parameters (e.g., 3D culture, level of oxygen, rigidity and pattern of the substrate) are critical steps. Establishing such protocols is strongly dependent on our current knowledge on how a tissue is formed during ontogenesis. Since pluripotent stem cells are tumorigenic when placed in an adult cellular environment *in vivo*, conditions are absolutely required to produce at purity, or selectively isolate from the cultures, the differentiated cell population of interest without any remaining contamination with ES or iPS cells. A sequel of the derivation of ES or iPS cells into lines at a very precise developmental time point is the ability of these cells to self-renew indefinitely and express specific and immutable genetic and epigenetic profiles. In short, ES and iPS cell lines bear uniquely specific attributes (pluripotency, indefinite self-renewal capacity, and unchanged phenotype for a given line in given experimental conditions) that discriminate them starkly from adult stem cells.

Adult stem cells are found in many, if not all tissues (e.g., blood, bone marrow, connective tissue, adipose tissue, dental pulp, muscle, lung, brain, retina, skin, mammary gland, pancreas, liver, and intestine) and are responsible for tissue development during ontogeny, tissue homeostasis in the adult, and tissue regeneration after injury or stress. Adult tissue stem cells are very rare (the frequency of HSCs in the bone marrow is about 0,005%) and some are not easily accessible (e.g., neural stem cells that are located in the hippocampus and lateral ventricles). The self-renewal capacity of adult stem cells is dependent on the tissue renewal, permanent in tissues with high turn-over rate (blood, intestine, and skin), sporadic in tissues with low turn-over rate unless under stress, after injury or during growth (brain, skeletal muscle, and bone). Under steady-state condition, the blood, intestine, and skin compartments are completely renewed in a matter of days. Under stress (hemorrhage, irradiation, burns, etc.), the daily demand of differentiated cells continues to increase. One understands therefore the requirement of self-renewal capacity for the stem cells at the origin of these tissues and the necessity for the stem cells to be able to shift from lineage to population mechanism. On the contrary, in other tissues, tissue renewal is low or nil, except during development or after injury. For example, bone is renewed in $\approx$ 10 years (Bianco and Gehron Robey, 2000) and muscle satellite cells are activated only after traumatism. In these instances, there is in steady state, no requirement for stem cell activation. Although in these tissues stem cells may self-renew, the requirement for self-renewal is less absolute than in tissues with high cell turnover rate.

The differentiation potential of adult stem cells depends on the tissue in which they reside. Thus, they are multipotent, bipotent, or unipotent, according to the ability to differentiate into several, two, or one cell lineage(s). For example, neural stem cells are multipotential, giving rise to neurons, oligodendrocytes, and astrocytes, whereas satellite cells and germ stem cells are unipotent, producing only skeletal muscle fibers and gametes, respectively. Tissue adult stem cells do not express specific markers. Thus, procedures to enrich a cell population with stem cell activity usually require a large combination of cell surface markers. The hematopoietic system has served as a model to phenotypically distinguish populations of stem cells from those of progenitors still able to proliferate and differentiate in one or several lineages, but unable to self-renew. Irving Weissman and collaborators have shown that murine HSCs with the ability to repopulate the hematopoietic system (long-term and multilineage reconstitution) of irradiated adult recipients express the c-kit and Sca-1 antigens, but are negative for blood lineage markers (Lin) such as CD3, B220, Gr-1, Mac1, and Ter119 that are specifically expressed by T and B lymphocytes, granulocytes, macrophages, and erythrocytes, respectively (Spangrude et al., 1988). It has been further shown that injection of single c-kit$^+$Sca1$^+$CD34$^{low/neg}$Lin^{neg} cells into irradiated mice allowed reconstituting 21% of the recipients (Osawa et al., 1996). Thus, although the enrichment protocol had used at least 8 specific monoclonal antibodies, the isolation of HSCs was still not pure. More recently, it has been demonstrated that two members of the SLAM family (CD150 and CD48) were also differentially expressed by HSCs (Kiel et al., 2005). *In vivo* transplantations revealed that 1 out of 2 c-kit$^+$Sca1$^+$Lin^{neg}CD150$^+$CD48^{neg} cells harbored HSC activity. Alternatively, since HSCs in the adult are known to be relatively quiescent at the mitotic and metabolic levels (Hodgson and Bradley, 1979; Ploemacher and Brons, 1989), HSCs may be isolated on the basis of their DNA and RNA contents (Gothot et al., 1997). Finally, due to the expression of membrane transporters that exclude vital dyes such as the Hoescht 33342, it has been shown that stem cells could be enriched by flow cytometry as a side population (Goodell et al., 1996).

Thus, a retrospective analysis based on appropriate *ex vivo* and *in vivo* assays has to be performed, at a clonal level, to demonstrate that a population contains stem cells that are able to self-renew and differentiate in a balancing act maintaining the stem cell number through a lineage or population mechanism. Culture systems with specific cytokines, growth factors, hormones, ECM molecules, and stromal cells are used by stem cell biologists to test the potential of stem and progenitor cells to proliferate and differentiate. Clonogenic assays (such as culture of colonies in semi-solid medium) that allow single cells to grow and differentiate as clusters containing differentiated cell types are useful to enumerate and to identify stem cells and/or progenitors (Fauser and Messner, 1979; Leary and Ogawa, 1987; Metcalf, 1971). *Ex vivo* tests may also exist to study the potential of stem cells to self-renew. In principle, these assays rely on the ability of stem cells to eventually form colonies or spheres such as neurospheres or mesenspheres (Méndez-Ferrer et al., 2010; Reynolds and Weiss, 1992). The aim consists in demonstrating that a colony/sphere derives from a single cell and that, after dissociation and replating, it will give rise to novel secondary colonies/spheres with the same characteristics (such as phenotype or differentiation potential) as the primary colonies/spheres. However, experiments performed *in vivo* are still essential to definitively prove that a cell population harbors stem cell activity. Classical procedures consist in grafting cells of a defined phenotype into recipients with a different phenotype under normal conditions or after injury and evaluating the potential of stem cells to contribute to tissue homeostasis or regeneration (Harrison, 1980; Hodgson and Bradley, 1979; Ploemacher and Brons, 1989). Manipulated cells can be injected in the blood circulation or directly grafted in the tissue. Different approaches such as flow cytometry, immunohistochemistry, gene expression profiling, and measurement of physiological parameters can then be used to study the proliferative and differentiation potential of the grafted stem cells. Difference in phenotype between the graft and the recipient allows discriminating transplanted stem cells from the remaining endogenous ones. Showing that the grafted stem cells participate in the maintenance or restoration of the stem cell pool for a long period of time leading to the generation of chimeric recipients argue for their capability to self-renew. Serial transplantations from primary to secondary or tertiary recipients are required to reinforce this demonstration (Hellman et al., 1978). However, even for the prototypic

HSC, whose self-renewal capacity can be stringently tested via consecutive transplantations from one recipient mouse to the other, self-renewal has only been evidenced for only a few rounds of transplantations (Harrison and Astle, 1982; Wilson et al., 2008). This may be related to cellular alterations secondary to the transplantation stress that alters homing and engraftment without affecting stem cell function per se (Purton and Scadden, 2007), but may also be due to other mechanisms, such as reduction of telomere length (Allsopp et al., 2003).

The recently generalized use of genetic lineage tracing has considerably helped define adult tissue stem cells in contexts where transplantation models are not amenable or are difficult to set up. Lineage tracing indicates not only which, but also for how long a specific stem cell and its progeny are labeled (Grompe, 2012; Kretzschmar and Watt, 2012). Stem cells should give rise to a progeny permanently labeled in long-term follow-up, contrarily to progenitors whose progeny is labeled only in short-term. This method therefore provides information on the *in vivo* behavior of stem cells and progenitors via study of their clonal progeny in a non-stress situation. This method has proven essential to validate the existence of adult stem cells in many solid tissues and to characterize their phenotype and function.

Quiescence has been usually considered an essential property of adult stem cells. It has been extensively studied in bone marrow HSCs. The more recent data obtained via computational modeling suggest that murine bone marrow HSCs are dormant, dividing about every 145 days (Wilson et al., 2008). Stem cell exit from dormancy and active entry into the cell cycle appears to be proportional to the severity of the bone marrow depletion. Such recent data are reminiscent of the hypothesis laid down long ago that HSCs would not contribute to the daily turnover of hematopoiesis, being solicited only in cases of stress or injury, while the daily production of hematopoietic cells would be maintained by the proliferative activity of the progenitors, some of which would have self-renewing capacity (Metcalf, 1984). It has been proposed that the distinction between quiescent (able to retain DNA labels and thus qualified as label-retaining cells) and active stem cells be generalized to most types of adult stem cells (Li and Clevers, 2010). Quiescent stem cells would constitute a reserve called into service in case of emergency.

A last note concerning adult stem cell attributes is their modulation during development and aging. On this point also, HSCs have been the subject of numerous studies. In mammals, HSCs emerge from the endothelium of the ventral wall of the embryo dorsal aorta, then are considerably amplified in the fetal liver, and finally home to the bone marrow where they are maintained throughout adult life (Dzierzak and Bigas, 2018). With age, the number of HSCs tends to decrease, but their proliferation potential increases; in terms of differentiation, myeloid-biased HSCs are progressively increased at the expense of lymphoid-biased ones (Van Zant and Liang, 2012). Telomere loss may account for the poor engraftment of HSCs from elderly patients. Similar studies have been done for other types of stem cells.

In addition to appropriate *in vivo* and *ex vivo* assays, high-throughput technologies and computational approaches are emerging as instrumental methods to investigate stem cell biology. This conceptual and methodological framework provides stem cell biologists with a unique opportunity to explore in depth the molecular identity of stem cells, to reconstruct biologically relevant gene networks, and decipher their lineage history. These aspects will be discussed in more details in the following chapters.

1.3 The Regulation of Stem Cells and the Stem Cell Niche

How stem cells are regulated is a fundamental issue in the field of stem cell biology and regenerative medicine. Stem cells are controlled by both cell autonomous and non-cell autonomous mechanisms (Figure 1.2). Intrinsic regulators include transcription factors, cell cycle regulators, and components of the cytoskeleton. In addition, accumulating evidence supports a major role for epigenetic regulation such as nuclear domain organization, chromatin remodeling, long non-coding RNAs, and microRNAs in the biology of embryonic and adult stem cells. Gene networks accounting for the attributes of both human and mouse ES cells have been unraveled (Boyer et al., 2005; Wang et al., 2006); their core consists in a minimal number of transcription factors including *Oct4* (*Pou5f1*), *Nanog*, and *Sox2*. This discovery has been a critical step to define the factors essential for the reprogramming of somatic cells into iPS cells. The interaction between the core transcription factors and the epigenetic factors is critical for the maintenance of stemness (Boyer et al., 2006; Lee et al., 2006; Lu et al., 2009; Marks et al., 2012). High-throughput technologies and experimental approaches have revealed that these three transcription factors cooperate to regulate the expression of numerous target genes including their own transcription (Li and Belmonte, 2017; Shanak and Helms, 2020). Importantly, this core network also recruits additional molecules such as epigenetic and other transcriptional regulators. The pluripotency regulatory network is thus sensitive to fluctuations

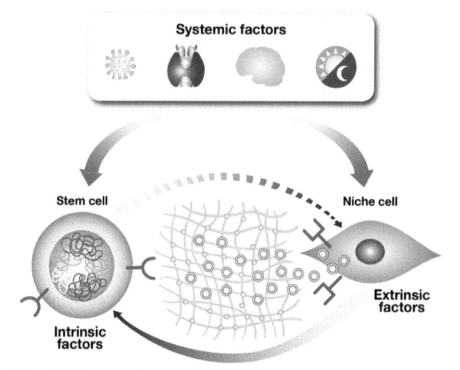

Figure 1.2   Factors involved in stem cell regulation. Intrinsic factors active in stem cells themselves (such as transcription and epigenetic factors) and extrinsic factors including cytokines, chemokines, growth factors (green circles), extracellular matrix components (light pink network), and extracellular vesicles (blue circles) released by niche cells cooperate to regulate stem cell functions. The dotted gray arrow illustrates the influence of stem cells on the phenotype, gene expression and functions of niche cells. On top of these signals, systemic factors including microbes, hormones, neurotransmitters, and the circadian clock, active at the level of the organism or provided by the environment, are emerging as crucial regulators of stem cell biology.

according to external stimuli and interactions with additional proteins, RNA molecules (such as micro-RNAs and long non-coding RNAs), and epigenetic factors (Li and Belmonte, 2017; Li and Izpisua Belmonte, 2018). These dynamic interactions confer to the pluripotency regulatory network a stable or unstable state, preventing or favoring ES cell differentiation, respectively. The ongoing description of gene networks in HSCs (Gazit et al., 2013; McKinney-Freeman et al., 2012; Novershtern et al., 2011; Tijssen et al., 2011) indicates which of the transcription factors may hold promise to convert somatic cells into this type of stem cells (Pereira et al., 2013).

Importantly, adult stem cells reside in specific microenvironments also called niches, firstly defined for HSCs. The concept of hematopoietic inductive microenvironment has been expounded by Norman Wolf and John Trentin in the 1960's (Trentin, 1989) and that of HSC niche has been subsequently proposed by Ray Schofield (Schofield, 1978). Remarkably, Alexander Friedenstein by studying bone marrow stem cells that give rise to connective tissue arrived also at the conclusion that some of the colonies derived from these stem cells were conducive for the homing of circulating HSCs (Friedenstein et al., 1970). Stem cell niches define the cellular contexts responsible for the molecular extrinsic regulations that keep in balance the commitment (determination) and self-renewal of stem cells not only through the production of factors such as cytokines, chemokines, morphogens, and cell adhesion molecules but also via mechanical changes such as the stiffness of the underlying ECM (Segel et al., 2019).

The notion of niches is nowadays generalized for most, if not all, adult tissue stem cells. Candidate niche cells are of diverse types. They might be differentiated cells belonging to a lineage distinct from that of stem cells (e.g., hub or cap cells of Drosophila reproductive organs), cells differentiated from the stem cells (e.g., Paneth cells of mouse intestine), or another type of stem cells (e.g., Mesenchymal Stem Cells (MSCs)) as a major component of the bone marrow HSC niche. Although the cells constituting these stem cell niches may have different identities in the blood, skeletal muscle, nervous and epithelial systems, canonical pathways such as members of the Wnt, Notch, Bone Morphogenetic Protein, and Hedgehog families are known to be active and to play a predominant role in the regulation of stem cell activity (Morrison and Spradling, 2008).

Study of germ stem cells (GSCs) in Drosophila has been particularly informative on how niche cells may control stem cell activity. The reason for this is the specific spatial organization of the testes and ovaries where differentiation proceeds along the anterior-posterior axis of the organ with GSCs at one end, the tip, in contact to specialized cap or hub cells constituting the niches (Tulina and Matunis, 2001; Xie and Spradling, 1998). A model has been established whereby when a GSC divides the daughter cell that is maintained in the niche retains the stem cell features. In contrast, the daughter cell moving away from the niche is not exposed anymore to the niche factors and thus progressively differentiates. The contacts between GSCs and their niches have been shown to directly influence the orientation of the mitotic spindle and the asymmetric division of GSCs. Indeed, when a GSC divides the daughter cell maintaining its physical contacts with the stromal cells stays undifferentiated. The other daughter cell, loosing its adhesion to the niche cells, is not exposed to the factors (such as decapentaplegic) delivered by the niche cells and starts to express genes involved in germ cell differentiation (Xie and Spradling, 1998). It must be underlined that such model relies on a lineage mechanism of stem cell renewal, i.e., asymmetrical division. However, since tissue stem cells have to be amplified during development or in conditions of stress, niche models must also account for a population mechanism of self-renewal. Such requirement implies the existence of a repertoire of niche cells or a repertoire of niche cell functions, allowing the adaptation to the varying tissue needs. Accumulating evidence suggests that the dialogue between stem cells and their niche cells should not be seen only as a unidirectional process with niche cells supporting stem cells, but more as a dynamic crosstalk between the niche and stem cells. For example, the transcriptome of stromal lines changes when these cells are exposed to HSCs and progenitor cells, very shortly after the initiation of the co-culture experiments (Charbord et al., 2014; Istvánffy et al., 2015). Also, leukemic myeloid cells favor the development of leukemia and myelofibrosis at the detriment of normal hematopoiesis by interfering with the osteoblastic HSC niche (Schepers et al., 2013). Thus, understanding how this crosstalk is regulated in normal conditions and deregulated after injury, stress, or cancer is a critical issue in the field of stem cell biology.

Bone marrow niches are not restricted to stem cells, but also associated to cells engaged in a differentiation pathway. Early on, the existence of stromal cells responsible for the self-renewal or expansion of HSCs have been hypothesized to be distinct from stromal cells responsible for the maturation of lymphoid lineages (Weissman, 1994). Cells in niches associated to B-lymphopoiesis have been well characterized and appear to be different at the different steps of differentiation (Mercier et al., 2012).

In addition to intrinsic, and extrinsic factors delivered locally by surrounding niche cells, systemic factors active at the level of the organism or provided by the environment emerge as critical regulators of adult stem cells (Figure 1.2). For example, Drosophila female infected by Wolbachia (intracellular bacteria maternally transmitted) produced more eggs than uninfected flies. This is due to an increase in GSC proliferation and a decrease of apoptosis in the germarium (Fast et al., 2011). Interestingly, Wolbachia strongly accumulate in niche cells in the ovaries as well as in neurosecretory cells in the brain secreting insulin-like peptides, suggesting that both local and systemic factors could influence GSC biology after Wolbachia infection

(Fast et al., 2011). In the mouse bone marrow, the release of HSCs into the blood flow has been shown to be under the control of the circadian clock through a molecular dialogue between bone marrow nerve cells and HSC niche cells (Méndez-Ferrer et al., 2008).

On top of these mechanisms, accumulating evidence highlights the important roles of extracellular vesicles (EVs) in cell-to-cell communications and in stem cell biology. Different types of EVs are identified depending on their size and ways of biogenesis (e.g., microparticules and exosomes) (Tkach and Théry, 2016). They are surrounded by a lipidic bilayer and contain RNA molecules and proteins within the cytosol. After delivery, EVs can act locally or at distance (via their circulation in body fluids) on target cells. EVs are released by neural stem and progenitor cells and have been shown to act on target cells by propagating cellular signaling (Cossetti et al., 2014). By taking advantage of two stromal lines of fetal origin with differing capacity to support HSCs *ex vivo*, we showed that EVs also play an important role on HSCs (Stik et al., 2017). Although both supportive and non-supportive stromal lines produced EVs, HSCs preferentially took up the EVs released by the supportive stromal cells through a mechanism not yet understood. This cell-to-cell communication process utilizing EVs strongly influences HSC gene expression profiling, their survival and hematopoietic potential, and appears to protect them from apoptosis (Durand et al., 2018; Stik et al., 2017).

1.4 Stem Cell Models

A number of models have been developed to account for the different properties of stem cells, including not only the defining properties of self-renewal and commitment, but also the proliferative capacity, the heterogeneity of the immature cell populations, and the plasticity of certain cell populations enriched in stem cells, i.e., their capacity to vary in phenotype or in behavior (fate switching) according to varying environmental conditions (Blau et al., 2001; Raff, 2003). Optimally, models should account not only for the cell functions, but also acknowledge the molecular mechanisms underlying these functions. This is a considerable task.

The major cellular model brought forth to account for hematopoiesis, i.e., the generation and maintenance of hematopoietic lineages from HSCs, has been hierarchical. The reader will find a considerable number of representations in the literature (Doulatov et al., 2012; Rossi et al., 2012). In this "tree" model, stem cells give rise to progenitors discriminated from stem cells by their lack of self-renewal ability, their commitment and their expression of lineage-specific markers. Progenitors will then produce precursor cells with distinctive differentiation potential. Thus, according to this model, a tissue is organized into different cell compartments (e.g., long-term HSCs, short-term HSCs, different types of progenitors and precursors). Cells within a compartment express homogeneously one or several feature(s) allowing the clear-cut discrimination from the adjacent compartments. Moreover, the cell flow from one compartment to the next is unidirectional (corresponding to loss of self-renewal capacity and irreversible differentiation). In such model, stem cells are intrinsically predetermined entities at the basis of tissue organization.

The hierarchical compartmentalized models are very attractive and easy to mathematically formalize. They have been used since a long time in hematology, for example to analyze the kinetics of hematopoietic lineages in humans in health and disease (Finch et al., 1970). However, even when they are made more complex to include a high number of compartments and hence of transition probabilities, they remain an approximation of reality, poorly fulfilling the criteria required for an encompassing model of hematopoiesis and other stem cell systems (Loeffler and Roeder, 2004). Two major road-blocks have been discerned (Graf, 2002; Metcalf, 2007; Quesenberry et al., 2002). Firstly, most data serving as basis for model elaboration were collected from transplantation studies, i.e., in conditions of extreme stress and hematopoietic damage. Secondly, hierarchical models did not account for the reversibility in differentiation and proliferation potential observed in many stem cell systems, including hematopoiesis. A number of recent experiments have recently challenged the hierarchical model of hematopoiesis both in mice and humans. Single cell barcoding tracing and refined functional assays have revealed the considerable degree of heterogeneity within cell compartments and unexpected cellular outputs. For example, a population of bone marrow $c\text{-}kit^+Sca\text{-}1^+Lin^-Flt3^+$ HSCs that exhibit lympho-myeloid but not erythro-megakaryocytic potential has been identified (Adolfsson et al., 2005). Single cell lineage tracing have highlighted the heterogeneity of common myeloid progenitors (CMP) and showed that most individual CMPs produce either erythroid or myeloid cells (Perié et al., 2015). In humans, erythro-megakaryocyte lineages may directly emerge from bone marrow HSCs and a shift in the architecture of the hematopoietic system has been proposed from the fetal to the adult stages (Notta et al., 2016).

Alternative models to the classical hierarchical model have been elaborated. Early on, a "screw" model has been proposed whereby a distinction was made between "actual" and "potential" stem cells corresponding to transit amplifying cells/progenitors recovering a self-renewing potential under regenerative conditions (Loeffler and Potten, 1997). More recent models include the phase-space model based on the hypothesis that the progeny of a given cell may possess greater stemness than that of the parent, although the possibility of such backward outcome declines with increasing differentiation (Kirkland, 2004). Similar dedifferentiation process is included as an essential variable in other models on HSC or MSC self-renewal and differentiation (Hoffmann et al., 2008; Krinner et al., 2010; Quesenberry et al., 2002). The study of hematopoiesis by a combination approach of lineage priming, single cell barcoding, and transcriptome analysis have recently generated many data advocating modified versions of the hierarchical tree (Jacobsen and Nerlov, 2019; Laurenti and Göttgens, 2018), classical Waddington's landscape of differentiation (Velten et al., 2017) or, as discussed below, a Markovian chain process (Wheat et al., 2020). Similar models are being developed for other stem cell systems and should be very useful to elucidate the reprogramming process and the development of cancers since the two concepts of stochasticity and determination have proven instrumental to understand the different phases of iPS cell generation (Yamanaka, 2009) and the origin and heterogeneity of tumors (Dick, 2008). The stochastic model is in keeping with the molecular, phenotypic, and functional heterogeneity observed in a tumor, which would result from random events operating at the single cell level. On the contrary, the hierarchical point of view suggests that tumors are organized like normal tissues and contain cancer stem cells with self-renewal and differentiation potential. Thus, according to the stochastic model, any cell of the tumor may have the ability to initiate a tumor whereas in the hierarchical model this potential is restricted to cancer stem cells. Interestingly, accumulating evidence suggests that tumor cells are highly plastic and that the tumor microenvironment may eventually confer cancer stem cell features in non-cancer stem cells (Batlle and Clevers, 2017). This has been illustrated for example in colorectal cancer where niche factors secreted by myofibroblasts restore the cancer stem cell phenotype in differentiated tumor cells (Vermeulen et al., 2010). The concept of cancer stem cells or tumor-initiating cells highlights that, together with gene mutation events, deregulation of the stem cell machinery or acquisition of stemness in inappropriate cells could be a major driver of tumorigenic processes. In addition, deregulations of stem cell activity during aging may also explain why tissue functions decline with age. This has been documented for several stem cell types including hematopoietic, intestinal, skeletal muscle, and skin stem cells (Schultz and Sinclair, 2016). For example, aged HSCs are lineage-biased and preferentially differentiate into the myeloid lineage, whereas hair follicle stem cells exhibit loss of function during aging affecting hair growth and regeneration.

At the molecular level, two models are used to describe stem cell differentiation (Zipori, 2004). In the first model, stem cells do not express any differentiation markers; they are said to be a blank slate. Commitment is then characterized by the appearance of the first differentiation markers followed by others as differentiation proceeds. In the second model, that of lineage priming, stem cells express lineage markers at low to moderate level. Differentiation is then characterized by the increase of markers specific for the differentiation pathway together with the progressive decline in markers characterizing alternative pathways. Transcriptional noise would account for lineage priming with stochastic oscillatory expression of certain transcripts (Graf and Stadtfeld, 2008). States with maximal expression of certain transcripts frozen by external conditions (microenvironment) would characterize committed cells.

Lineage priming was first described in HSCs (Hu et al., 1997) and confirmed since by many reports (Graf and Stadtfeld, 2008; Orkin and Zon, 2008). High-throughput population-based studies have suggested that lineage choice in stem cells was governed by transcriptional noise (Chang et al., 2008). However, a more recent study on individual cells indicates that bona fide stem cells, contrarily to progenitors, only sporadically express lineage regulators (Pina et al., 2012). Commitment would then be characterized by the expression of lineage regulators in a discrete and non-coordinated way, the precise identity of the first intervening regulator varying from one cell to the other. Association of single cell RNA-Seq and single cell cultures for the study of human bone marrow hematopoiesis has allowed describing the HSCs and their immediate progeny as a continuum of low-primed undifferentiated cells (Velten et al., 2017). From this central cloud of cells, discrete populations of restricted progenitors would emerge. A recent study using a very sensitive assay for detection and quantification of transcriptions factors has shown that single HSC can express transcription factors that are antagonistic in terms of differentiation lineage induction, a situation akin to lineage priming (Wheat et al., 2020). Analysis of the transcription factors along differentiation trajectories has led to a model of reversible transition states whereby the transcriptional state of a given cell is not fully predictive of the past or future states of parent or daughter cell. A Markov chain of transcriptional states would facilitate the maintenance of an uncommitted pool of stem cells; upon demand (for example, anemia resulting in high erythropoietin level) only cells in a certain transient transcriptional state (in this example cells with high expression of *Gata1* and *Gata2* and low expression of *PU.1*) would respond and commit to a distinct lineage (in this example erythropoiesis). In such a model, intrinsic stochasticity of gene expression can produce cellular behavior that may be incorrectly inferred to have arisen from deterministic dynamics.

These notions emphasize the complexity of stem cell systems, pointing out that, at the present stage of research, experimental data have to be confronted to theoretical modeling borrowing from the physical and mathematical fields. This approach is all the more necessary with the introduction of high-throughput data to provide a holistic view of global-scale genome organization accounting for stem cell functions and deregulations.

1.5 Cell Therapy Using Stem Cells

From a functional point of view, in the adult, the characteristic property of a population with stem cell activity (i.e., self-renewing and in most cases, multipotential) is its ability to regenerate (integrally repair) its tissue of origin after injury (Loeffler and Potten, 1997). Since the pioneering work of Donnall Thomas (Thomas et al., 1975) and the first successful bone marrow HSC transplantations in humans in the 1970s, HSCs have remained the major stem cell source used for cell therapy. Transplantation of these paradigmatic HSCs leads to total reconstitution of the blood system of an aplastic subject after a few days to weeks. The mechanism of reconstitution is well known. The HSC binds to specific sites on the endothelial lining of marrow sinuses and then crosses the endothelial barrier and homes in the bone marrow logettes where clones develop and differentiate. Finally the end-cells find their way back to the marrow sinuses and into the bloodstream. In short, HSCs reconstitute the blood system via proliferation and differentiation of its progeny and in concert with associated stromal niches that regulate the HSC activity.

This model of regeneration adequate for HSCs may hardly apply to MSCs. These cells are easily amplified *ex vivo*, contrasting in that respect with HSCs, the culture amplification of which remains a major challenge. In part because of the facility of *ex vivo* expansion leading to several million cells after one to two weeks, MSCs are nowadays used in preclinical models and in clinical trials. Some of the indications in the orthopedic field are expected due to MSC differentiation into osteoblasts and chondrocytes. But MSCs are also proposed for the treatment of other diseases. Regeneration by proliferation and differentiation of the progeny cannot explain the pre-clinical and clinical results obtained after transplantation of these cells inducing improvement of immunological disorders such as graft-versus-host disease or Crohn's disease. Secretion of cytokines and chemokines by cultured stromal cells and by MSCs is well known, suggesting that these molecules are implicated in the regenerative effect of transplanted MSCs. Of particular importance are the anti-apoptotic activity of the growth factors and the anti-inflammatory activity of certain interleukins (English et al., 2010; Singer and Caplan, 2011). Hence the emerging scenario for regeneration by MSCs implanted locally or injected systematically would be the secretion during a crucial, eventually short span of time of molecules active on tissue stem cells of different types or on differentiated cells such as macrophages.

Since the pioneering work of Howard Green (Green et al., 1979), Epidermal Stem Cells that can also be amplified by culture with relative ease are used for skin repair, primarily in extended burns. Large sheets of epidermis grown *in vitro* are implanted onto the burnt territories. In this case, one takes advantage of the properties of the diverse types of Epidermal Stem Cells, the attributes of which (self-renewal capacity, proliferative potential, and heterogeneity) are similar to those of HSCs.

It is theoretically possible to use other types of adult stem cells in the clinic. However, one may then face major problems. Are the tissues available? One may consider in that respect neural stem/progenitor cells that can be procured in fetal brain after therapeutic abortion; such exceptional procedure may only be envisioned as providing the proof of concept of the efficacy of a treatment, such has been done for Parkinson's disease (Peschanski et al., 1994). Can the stem cells be isolated in sufficient quantity or be amplified *ex vivo*? As already stated, purification of stem cells to homogeneity and their expansion *ex vivo* is not as yet feasible in most cases, including the best-studied case of HSCs. Finally, can stem cells be amplified *in vivo*, which would avoid the multiple difficulties of collection and manipulation *ex vivo*? In the case of HSCs, the expansion of the circulating blood pool is a common procedure, following administration of granulocyte colony-stimulating factor; the blood HSCs are then collected for secondary transplantation (Gianni et al., 1989). In the case of MSCs (considered both as stem cells and cells forming the HSC niche), treatment by parathormone (PTH) has been envisioned to expand the osteoblastic pool and increase HSC mobilization and engrafment, but such therapy has yet to prove its efficacy in the absence of deleterious side effects (Ohishi and Schipani, 2011; Wagers, 2012).

These difficulties underline the potential benefits that may be expected from reprogramming of somatic cells into iPS cells and the secondary differentiation of the latter into different somatic lineages. Transdifferentiation of somatic cells into another type of somatic cells may constitute an alternative. It is a more straightforward procedure than the generation of iPS cells since it results either from direct transition from a specific somatic cell to another (even without intervening cell divisions), or from transdetermination where a somatic cell of a given type reverts to a stem/progenitor cell, in turn converting to an somatic cell of a different type (Frisen, 2002). Transdifferentiation protocols have been described for many cell types, e.g., fibroblasts into myogenic cells (Weintraub et al., 1989), fibroblasts into cardiomyocytes (Ieda et al., 2010), fibroblasts into neurons (Vierbuchen et al., 2010), B-lymphocytes into macrophages (Xie et al., 2004) and macrophages into B-lymphocytes (Bussmann et al., 2009), adult pancreatic exocrine cells into endocrine ß-cells (Zhou et al., 2008), and human fibroblasts into hepatocytes (Zhu et al., 2014). Transdifferentiation is a natural process observed throughout the animal kingdom (Jopling et al., 2011; Slack, 2007). It is responsible for example for lens regeneration from retinal pigmental cells in newt or for transformations between imaginal disks in Drosophila. It occurs also in humans, resulting in the appearance of a cell population not expected in the adult in the tissue considered (metaplasia), which is either without functional consequence or harmful since predisposing to malignant transformation. It may also be beneficial in certain conditions, as an attempt by the organism to solve a pathological condition, e.g., the transdifferentiation of pancreatic endocrine α-cells into ß-cells in case of severe ß-cell loss (diabetes) in the mouse (Thorel et al., 2010). Moreover, epithelial-to-mesenchymal transition that may be considered a special case of transdifferentiation is an essential step in many developmental and repair processes and plays a major role in the development of many diseases such as cancer (Lamouille et al., 2014).

In conditions of regeneration, functional stem cells may be generated from somatic cells by dedifferentiation. The reader is referred to different reviews that have critically revisited the flurry of 2000's papers relating instances of stem cell plasticity following the description of the cloned Dolly sheep, thus culling bona fide examples of dedifferentiation and conversion of differentiated cells into stem cells (Blau et al., 2001; Graf, 2002; Raff, 2003). The somatic cells may be progenitors such as secretory Dll^{+} progenitors in a mouse's intestinal epithelium following extensive irradiation damage (van Es et al., 2012). They may be also differentiated cells such as airway epithelial (luminal secretory) cells converting into multipotent stem cells following basal stem cell ablation (Tata et al., 2013). These data have led to the concept that bona fide stem cells that insure tissue maintenance under homeostasis cannot always repopulate a damaged tissue. Dedifferentiated or transdifferentiated cells that insure the regeneration may be considered as facultative stem cells (Muñoz-Cánoves and Huch, 2018).

Although adult stem cells have been described in most tissues, there are instances where stem cells are not responsible for tissue regeneration after injury (Grompe, 2012). A typical example is that of the liver, where regeneration after partial surgical resection is due to proliferation of differentiated hepatocytes, contrarily to regeneration following some toxic injuries that result from the activity of stem cells (Grisham, 1997). Another example is that of endocrine pancreas where newborn beta cells are generated by division of insulin$^+$ pre-existing differentiated ß-cells (Dor et al., 2004).

Taken together these data underline the multiple mechanisms of tissue regeneration or repair after injury. Stem cells are not always implicated in the process and, when implicated, they can repair the tissues by other means than reconstitution following proliferation and differentiation. From a therapeutic standpoint, the different means of tissue reparation must be borne in mind. Auto- or allo-grafting of cell populations enriched in stem cells constitute only one of the options, depending, of course, on the clinical setting. From a fundamental point of view, the range of cells and of mechanisms implicated in the regeneration process has led to the proposition that a population of stem cells would constitute a specific but fleeting state, rather than a defined cell type with a precise molecular signature (Zipori, 2004). In line with the entity vs. state duality, investigations in philosophy are currently conducted to better characterize stemness (Laplane and Solary, 2019), as discussed in detail in a chapter of the present book. It appears presently reasonable to keep open the debate and consider that several models may account for the stem cell attributes.

1.6 Concluding Remarks

- Stem cells are cells that are able to self-renew and differentiate in a balancing act maintaining the stem cell number (symmetric vs. asymmetric division).
- Pluripotent ES cell lines derived from blastocyst embryos bear specific attributes (pluripotency, indefinite self-renewal capacity, and unchanged phenotype) that discriminate them from adult stem cells.

- Pluripotent stem cells constitute an exceptional tool to study cell differentiation, model human diseases, and for drug testing, and represent a promising source for regenerative medicine.
- Widely distributed adult stem cells are responsible for tissue development during ontogeny, tissue homeostasis in the adult and regeneration after injury or stress.
- The self-renewal capacity of adult stem cells is dependent on the tissue renewal; their differentiation potential (multipotent, bipotent, or unipotent) also depends on the tissue in which they reside.
- Stem cells are strictly regulated in time and space by a combination of cell autonomous intrinsic factors, non-cell autonomous extrinsic factors delivered locally by surrounding niche cells, and systemic factors.
- Different models borrowing from the physical and mathematical fields are developed to provide a holistic view of stem cell systems, including modified versions of the classical hierarchical tree model.
- Alterations of stem cell functions (e.g., during diseases or aging) impair tissue homeostasis and regeneration, and may lead to tumor formation.
- HSCs, the major stem cell source used for cell therapy, reconstitute the blood system via proliferation and differentiation of its progeny in concert with associated stromal niches that regulate HSC activity; this model of regeneration may not apply to other adult stem cells such as MSCs.
- In some instances dedifferentiated or transdifferentiated cells may function as facultative stem cells.

Acknowledgments

We thank Sophie Gournet (CNRS UMR7622) for excellent drawing assistance.

References

Adolfsson, J., Månsson, R., Buza-Vidas, N., Hultquist, A., Liuba, K., Jensen, C.T., Bryder, D., Yang, L., Borge, O.J., Thoren, L.A., *et al.* (2005). Identification of Flt3+ lympho-myeloid stem cells lacking erythro-megakaryocytic potential a revised road map for adult blood lineage commitment. Cell *121*, 295-306.

Allsopp, R.C., Morin, G.B., DePinho, R., Harley, C.B., and Weissman, I.L. (2003). Telomerase is required to slow telomere shortening and extend replicative lifespan of HSCs during serial transplantation. Blood *102*, 517-520.

Batlle, E., and Clevers, H. (2017). Cancer stem cells revisited. Nature medicine *23*, 1124-1134.

Becker, A.J., Mc, C.E., and Till, J.E. (1963). Cytological demonstration of the clonal nature of spleen colonies derived from transplanted mouse marrow cells. Nature *197*, 452-454.

Bianco, P., and Gehron Robey, P. (2000). Marrow stromal stem cells. J Clin Invest *105*, 1663-1668.

Blau, H.M., Brazelton, T.R., and Weimann, J.M. (2001). The evolving concept of a stem cell: entity or function? Cell *105*, 829-841.

Boyer, L.A., Lee, T.I., Cole, M.F., Johnstone, S.E., Levine, S.S., Zucker, J.P., Guenther, M.G., Kumar, R.M., Murray, H.L., Jenner, R.G., *et al.* (2005). Core transcriptional regulatory circuitry in human embryonic stem cells. Cell *122*, 947-956.

Boyer, L.A., Plath, K., Zeitlinger, J., Brambrink, T., Medeiros, L.A., Lee, T.I., Levine, S.S., Wernig, M., Tajonar, A., Ray, M.K., *et al.* (2006). Polycomb complexes repress developmental regulators in murine embryonic stem cells. Nature *441*, 349-353.

Bussmann, L.H., Schubert, A., Vu Manh, T.P., De Andres, L., Desbordes, S.C., Parra, M., Zimmermann, T., Rapino, F., Rodriguez-Ubreva, J., Ballestar, E., *et al.* (2009). A robust and highly efficient immune cell reprogramming system. Cell Stem Cell *5*, 554-566.

Cairns, J. (1975). Mutation selection and the natural history of cancer. Nature *255*, 197-200.

Chang, H.H., Hemberg, M., Barahona, M., Ingber, D.E., and Huang, S. (2008). Transcriptome-wide noise controls lineage choice in mammalian progenitor cells. Nature *453*, 544-547.

Charbord, P., Pouget, C., Binder, H., Dumont, F., Stik, G., Levy, P., Allain, F., Marchal, C., Richter, J., Uzan, B., *et al.* (2014). A systems biology approach for defining the molecular framework of the hematopoietic stem cell niche. Cell stem cell *15*, 376-391.

Cossetti, C., Iraci, N., Mercer, T.R., Leonardi, T., Alpi, E., Drago, D., Alfaro-Cervello, C., Saini, H.K., Davis, M.P., Schaeffer, J., *et al.* (2014). Extracellular vesicles from neural stem cells transfer IFN-γ via Ifngr1 to activate Stat1 signaling in target cells. Molecular cell *56*, 193-204.

Cowan, C.A., Atienza, J., Melton, D.A., and Eggan, K. (2005). Nuclear reprogramming of somatic cells after fusion with human embryonic stem cells. Science (New York, NY) *309*, 1369-1373.

Dick, J.E. (2008). Stem cell concepts renew cancer research. Blood *112*, 4793-4807.

Dor, Y., Brown, J., Martinez, O.I., and Melton, D.A. (2004). Adult pancreatic beta-cells are formed by self-duplication rather than stem-cell differentiation. Nature *429*, 41-46.

Doulatov, S., Notta, F., Laurenti, E., and Dick, J.E. (2012). Hematopoiesis: a human perspective. Cell Stem Cell *10*, 120-136.

Durand, C., Charbord, P., and Jaffredo, T. (2018). The crosstalk between hematopoietic stem cells and their niches. Curr Opin Hematol *25*, 285-289.

Dzierzak, E., and Bigas, A. (2018). Blood Development: Hematopoietic Stem Cell Dependence and Independence. Cell stem cell *22*, 639-651.

English, K., French, A., and Wood, K.J. (2010). Mesenchymal stromal cells: facilitators of successful transplantation? Cell Stem Cell *7*, 431-442.

Evans, M.J., and Kaufman, M.H. (1981). Establishment in culture of pluripotential cells from mouse embryos. Nature *292*, 154-156.

Fast, E.M., Toomey, M.E., Panaram, K., Desjardins, D., Kolaczyk, E.D., and Frydman, H.M. (2011). Wolbachia enhance Drosophila stem cell proliferation and target the germline stem cell niche. Science (New York, NY) *334*, 990-992.

Fauser, A.A., and Messner, H.A. (1979). Fetal hemoglobin in mixed hemopoietic colonies (CFU-GEMM), erythroid bursts (BFU-E) and erythroid colonies (CFU-E): assessment by radioimmune assay and immunofluorescence. Blood *54*, 1384-1394.

Finch, C.A., Deubelbeiss, K., Cook, J.D., Eschbach, J.W., Harker, L.A., Funk, D.D., Marsaglia, G., Hillman, R.S., Slichter, S., Adamson, J.W.*, et al.* (1970). Ferrokinetics in man. Medicine (Baltimore) *49*, 17-53.

Friedenstein, A.J., Chailakhjan, R.K., and Lalykina, K.S. (1970). The development of fibroblast colonies in monolayer cultures of guinea-pig bone marrow and spleen cells. Cell Tissue Kinet *3*, 393-403.

Frisen, J. (2002). Stem cell plasticity? Neuron *35*, 415-418.

Gazit, R., Garrison, B.S., Rao, T.N., Shay, T., Costello, J., Ericson, J., Kim, F., Collins, J.J., Regev, A., Wagers, A.J.*, et al.* (2013). Transcriptome analysis identifies regulators of hematopoietic stem and progenitor cells. Stem cell reports *1*, 266-280.

Gianni, A.M., Siena, S., Bregni, M., Tarella, C., Stern, A.C., Pileri, A., and Bonadonna, G. (1989). Granulocyte-macrophage colony-stimulating factor to harvest circulating haemopoietic stem cells for autotransplantation. Lancet *2*, 580-585.

Goodell, M.A., Brose, K., Paradis, G., Conner, A.S., and Mulligan, R.C. (1996). Isolation and functional properties of murine hematopoietic stem cells that are replicating in vivo. The Journal of experimental medicine *183*, 1797-1806.

Gothot, A., Pyatt, R., McMahel, J., Rice, S., and Srour, E.F. (1997). Functional heterogeneity of human CD34(+) cells isolated in subcompartments of the G0 /G1 phase of the cell cycle. Blood *90*, 4384-4393.

Graf, T. (2002). Differentiation plasticity of hematopoietic cells. Blood *99*, 3089-3101.

Graf, T., and Stadtfeld, M. (2008). Heterogeneity of embryonic and adult stem cells. Cell Stem Cell *3*, 480-483.

Green, H., Kehinde, O., and Thomas, J. (1979). Growth of cultured human epidermal cells into multiple epithelia suitable for grafting. Proc Natl Acad Sci U S A *76*, 5665-5668.

Grisham, J., Thorgeirsson, SS (1997). Liver stem cells. In Stem Cells, C. Potten, ed. (London: Academic Press), pp. 233-282.

Grompe, M. (2012). Tissue stem cells: new tools and functional diversity. Cell Stem Cell *10*, 685-689.

Gurdon, J.B., and Laskey, R.A. (1970). The transplantation of nuclei from single cultured cells into enucleate frogs' eggs. Journal of embryology and experimental morphology *24*, 227-248.

Harrison, D.E. (1980). Competitive repopulation: a new assay for long-term stem cell functional capacity. Blood *55*, 77-81.

Harrison, D.E., and Astle, C.M. (1982). Loss of stem cell repopulating ability upon transplantation. Effects of donor age, cell number, and transplantation procedure. The Journal of experimental medicine *156*, 1767-1779.

Hellman, S., Botnick, L.E., Hannon, E.C., and Vigneulle, R.M. (1978). Proliferative capacity of murine hematopoietic stem cells. Proceedings of the National Academy of Sciences of the United States of America *75*, 490-494.

Hodgson, G.S., and Bradley, T.R. (1979). Properties of haematopoietic stem cells surviving 5-fluorouracil treatment: evidence for a pre-CFU-S cell? Nature *281*, 381-382.

Hoffmann, M., Chang, H.H., Huang, S., Ingber, D.E., Loeffler, M., and Galle, J. (2008). Noise-driven stem cell and progenitor population dynamics. PLoS One *3*, e2922.

Hu, M., Krause, D., Greaves, M., Sharkis, S., Dexter, M., Heyworth, C., and Enver, T. (1997). Multilineage gene expression precedes commitment in the hemopoietic system. Genes Dev *11*, 774-785.

Ieda, M., Fu, J.D., Delgado-Olguin, P., Vedantham, V., Hayashi, Y., Bruneau, B.G., and Srivastava, D. (2010). Direct reprogramming of fibroblasts into functional cardiomyocytes by defined factors. Cell *142*, 375-386.

Istvánffy, R., Vilne, B., Schreck, C., Ruf, F., Pagel, C., Grziwok, S., Henkel, L., Prazeres da Costa, O., Berndt, J., Stümpflen, V., *et al.* (2015). Stroma-Derived Connective Tissue Growth Factor Maintains Cell Cycle Progression and Repopulation Activity of Hematopoietic Stem Cells In Vitro. Stem cell reports *5*, 702-715.

Jacobsen, S.E.W., and Nerlov, C. (2019). Haematopoiesis in the era of advanced single-cell technologies. Nature cell biology *21*, 2-8.

Jopling, C., Boue, S., and Izpisua Belmonte, J.C. (2011). Dedifferentiation, transdifferentiation and reprogramming: three routes to regeneration. Nat Rev Mol Cell Biol *12*, 79-89.

Karpowicz, P., Morshead, C., Kam, A., Jervis, E., Ramunas, J., Cheng, V., and van der Kooy, D. (2005). Support for the immortal strand hypothesis: neural stem cells partition DNA asymmetrically in vitro. J Cell Biol *170*, 721-732.

Kiel, M.J., He, S., Ashkenazi, R., Gentry, S.N., Teta, M., Kushner, J.A., Jackson, T.L., and Morrison, S.J. (2007). Haematopoietic stem cells do not asymmetrically segregate chromosomes or retain BrdU. Nature *449*, 238-242.

Kiel, M.J., Yilmaz, O.H., Iwashita, T., Yilmaz, O.H., Terhorst, C., and Morrison, S.J. (2005). SLAM family receptors distinguish hematopoietic stem and progenitor cells k and reveal endothelial niches for stem cells. Cell *121*, 1109-1121.

Kirkland, M.A. (2004). A phase space model of hemopoiesis and the concept of stem cell renewal. Exp Hematol *32*, 511-519.

Kretzschmar, K., and Watt, F.M. (2012). Lineage tracing. Cell *148*, 33-45.

Krinner, A., Hoffmann, M., Loeffler, M., Drasdo, D., and Galle, J. (2010). Individual fates of mesenchymal stem cells in vitro. BMC systems biology *4*, 73.

Lamouille, S., Xu, J., and Derynck, R. (2014). Molecular mechanisms of epithelial-mesenchymal transition. Nat Rev Mol Cell Biol *15*, 178-196.

Laplane, L., and Solary, E. (2019). Towards a classification of stem cells. eLife *8*.

Laurenti, E., and Göttgens, B. (2018). From haematopoietic stem cells to complex differentiation landscapes. Nature *553*, 418-426.

Leary, A.G., and Ogawa, M. (1987). Blast cell colony assay for umbilical cord blood and adult bone marrow progenitors. Blood *69*, 953-956.

Lee, T.I., Jenner, R.G., Boyer, L.A., Guenther, M.G., Levine, S.S., Kumar, R.M., Chevalier, B., Johnstone, S.E., Cole, M.F., Isono, K., *et al.* (2006). Control of developmental regulators by Polycomb in human embryonic stem cells. Cell *125*, 301-313.

Li, L., and Clevers, H. (2010). Coexistence of quiescent and active adult stem cells in mammals. Science *327*, 542-545.

Li, M., and Belmonte, J.C. (2017). Ground rules of the pluripotency gene regulatory network. Nature reviews Genetics *18*, 180-191.

Li, M., and Izpisua Belmonte, J.C. (2018). Deconstructing the pluripotency gene regulatory network. Nature cell biology *20*, 382-392.

Loeffler, M., and Potten, C. (1997). Stem cells and cellular pedigrees- a conceptual introduction. In Stem Cells, C. Potten, ed. (San Diego: Academic Press), pp. 1-28.

Loeffler, M., and Roeder, I. (2004). Conceptual models to understand tissue stem cell organization. Curr Opin Hematol *11*, 81-87.

Lu, R., Markowetz, F., Unwin, R.D., Leek, J.T., Airoldi, E.M., MacArthur, B.D., Lachmann, A., Rozov, R., Ma'ayan, A., Boyer, L.A., *et al.* (2009). Systems-level dynamic analyses of fate change in murine embryonic stem cells. Nature *462*, 358-362.

Marks, H., Kalkan, T., Menafra, R., Denissov, S., Jones, K., Hofemeister, H., Nichols, J., Kranz, A., Stewart, A.F., Smith, A., *et al.* (2012). The transcriptional and epigenomic foundations of ground state pluripotency. Cell *149*, 590-604.

Martin, G.R. (1981). Isolation of a pluripotent cell line from early mouse embryos cultured in medium conditioned by teratocarcinoma stem cells. Proceedings of the National Academy of Sciences of the United States of America *78*, 7634-7638.

McKinney-Freeman, S., Cahan, P., Li, H., Lacadie, S.A., Huang, H.T., Curran, M., Loewer, S., Naveiras, O., Kathrein, K.L., Konantz, M., *et al.* (2012). The transcriptional landscape of hematopoietic stem cell ontogeny. Cell Stem Cell *11*, 701-714.

Méndez-Ferrer, S., Lucas, D., Battista, M., and Frenette, P.S. (2008). Haematopoietic stem cell release is regulated by circadian oscillations. Nature *452*, 442-447.

Méndez-Ferrer, S., Michurina, T.V., Ferraro, F., Mazloom, A.R., Macarthur, B.D., Lira, S.A., Scadden, D.T., Ma'ayan, A., Enikolopov, G.N., and Frenette, P.S. (2010). Mesenchymal and haematopoietic stem cells form a unique bone marrow niche. Nature *466*, 829-834.

Mercier, F.E., Ragu, C., and Scadden, D.T. (2012). The bone marrow at the crossroads of blood and immunity. Nature reviews Immunology *12*, 49-60.

Metcalf, D. (1984). The hemopoietic colony stimulating factors (Amsterdam: Elsevier).

Metcalf, D. (2007). Concise review: hematopoietic stem cells and tissue stem cells: current concepts and unanswered questions. Stem Cells *25*, 2390-2395.

Metcalf, D., Moore, MAS (1971). Haemopoietic cells (Amsterdam: North Holland Publishing Company).

Morrison, S.J., and Spradling, A.C. (2008). Stem cells and niches: mechanisms that promote stem cell maintenance throughout life. Cell *132*, 598-611.

Muñoz-Cánoves, P., and Huch, M. (2018). Definitions for adult stem cells debated. Nature *563*, 328-329.

Nichols, J., and Smith, A. (2009). Naive and primed pluripotent states. Cell stem cell *4*, 487-492.

Notta, F., Zandi, S., Takayama, N., Dobson, S., Gan, O.I., Wilson, G., Kaufmann, K.B., McLeod, J., Laurenti, E., Dunant, C.F., *et al.* (2016). Distinct routes of lineage development reshape the human blood hierarchy across ontogeny. Science (New York, NY) *351*, aab2116.

Novershtern, N., Subramanian, A., Lawton, L.N., Mak, R.H., Haining, W.N., McConkey, M.E., Habib, N., Yosef, N., Chang, C.Y., Shay, T., *et al.* (2011). Densely interconnected transcriptional circuits control cell states in human hematopoiesis. Cell *144*, 296-309.

Ohishi, M., and Schipani, E. (2011). PTH and stem cells. Journal of endocrinological investigation *34*, 552-556.

Orkin, S.H., and Zon, L.I. (2008). Hematopoiesis: an evolving paradigm for stem cell biology. Cell *132*, 631-644.

Osawa, M., Hanada, K., Hamada, H., and Nakauchi, H. (1996). Long-term lymphohematopoietic reconstitution by a single CD34-low/negative hematopoietic stem cell. Science (New York, NY) *273*, 242-245.

Pereira, C.F., Chang, B., Qiu, J., Niu, X., Papatsenko, D., Hendry, C.E., Clark, N.R., Nomura-Kitabayashi, A., Kovacic, J.C., Ma'ayan, A., *et al.* (2013). Induction of a hemogenic program in mouse fibroblasts. Cell Stem Cell *13*, 205-218.

Perié, L., Duffy, K.R., Kok, L., de Boer, R.J., and Schumacher, T.N. (2015). The Branching Point in Erythro-Myeloid Differentiation. Cell *163*, 1655-1662.

Peschanski, M., Defer, G., N'Guyen, J.P., Ricolfi, F., Monfort, J.C., Remy, P., Geny, C., Samson, Y., Hantraye, P., Jeny, R., *et al.* (1994). Bilateral motor improvement and alteration of L-dopa effect in two patients with Parkinson's disease following intrastriatal transplantation of foetal ventral mesencephalon. Brain : a journal of neurology *117 (Pt 3)*, 487-499.

Pina, C., Fugazza, C., Tipping, A.J., Brown, J., Soneji, S., Teles, J., Peterson, C., and Enver, T. (2012). Inferring rules of lineage commitment in haematopoiesis. Nature cell biology *14*, 287-294.

Ploemacher, R.E., and Brons, R.H. (1989). Separation of CFU-S from primitive cells responsible for reconstitution of the bone marrow hemopoietic stem cell compartment following irradiation: evidence for a pre-CFU-S cell. Experimental hematology *17*, 263-266.

Potten, C.S., Hume, W.J., Reid, P., and Cairns, J. (1978). The segregation of DNA in epithelial stem cells. Cell *15*, 899-906.

Purton, L.E., and Scadden, D.T. (2007). Limiting factors in murine hematopoietic stem cell assays. Cell stem cell *1*, 263-270.

Quesenberry, P.J., Colvin, G.A., and Lambert, J.F. (2002). The chiaroscuro stem cell: a unified stem cell theory. Blood *100*, 4266-4271.

Raff, M. (2003). Adult stem cell plasticity: fact or artifact? Annu Rev Cell Dev Biol *19*, 1-22.

Reynolds, B.A., and Weiss, S. (1992). Generation of neurons and astrocytes from isolated cells of the adult mammalian central nervous system. Science (New York, NY) *255*, 1707-1710.

Rossi, L., Lin, K.K., Boles, N.C., Yang, L., King, K.Y., Jeong, M., Mayle, A., and Goodell, M.A. (2012). Less is more: unveiling the functional core of hematopoietic stem cells through knockout mice. Cell Stem Cell *11*, 302-317.

Schepers, K., Pietras, E.M., Reynaud, D., Flach, J., Binnewies, M., Garg, T., Wagers, A.J., Hsiao, E.C., and Passegué, E. (2013). Myeloproliferative neoplasia remodels the endosteal bone marrow niche into a self-reinforcing leukemic niche. Cell stem cell *13*, 285-299.

Schofield, R. (1978). The relationship between the spleen colony-forming cell and the haemopoietic stem cell. Blood Cells *4*, 7-25.

Schultz, M.B., and Sinclair, D.A. (2016). When stem cells grow old: phenotypes and mechanisms of stem cell aging. Development (Cambridge, England) *143*, 3-14.

Segel, M., Neumann, B., Hill, M.F.E., Weber, I.P., Viscomi, C., Zhao, C., Young, A., Agley, C.C., Thompson, A.J., Gonzalez, G.A., *et al.* (2019).

Niche stiffness underlies the ageing of central nervous system progenitor cells. Nature *573*, 130-134.

Shanak, S., and Helms, V. (2020). DNA methylation and the core pluripotency network. Developmental biology *464*, 145-160.

Siminovitch, L., McCulloch, E.A., and Till, J.E. (1963). The Distribution of Colony-Forming Cells among Spleen Colonies. J Cell Physiol *62*, 327-336.

Simons, B.D., and Clevers, H. (2011). Strategies for homeostatic stem cell self-renewal in adult tissues. Cell *145*, 851-862.

Singer, N.G., and Caplan, A.I. (2011). Mesenchymal stem cells: mechanisms of inflammation. Annu Rev Pathol *6*, 457-478.

Slack, J.M. (2007). Metaplasia and transdifferentiation: from pure biology to the clinic. Nat Rev Mol Cell Biol *8*, 369-378.

Spangrude, G.J., Heimfeld, S., and Weissman, I.L. (1988). Purification and characterization of mouse hematopoietic stem cells. Science (New York, NY) *241*, 58-62.

Spradling, A., Drummond-Barbosa, D., and Kai, T. (2001). Stem cells find their niche. Nature *414*, 98-104.

Stik, G., Crequit, S., Petit, L., Durant, J., Charbord, P., Jaffredo, T., and Durand, C. (2017). Extracellular vesicles of stromal origin target and support hematopoietic stem and progenitor cells. The Journal of cell biology *216*, 2217-2230.

Takahashi, K., Tanabe, K., Ohnuki, M., Narita, M., Ichisaka, T., Tomoda, K., and Yamanaka, S. (2007). Induction of pluripotent stem cells from adult human fibroblasts by defined factors. Cell *131*, 861-872.

Takahashi, K., and Yamanaka, S. (2006). Induction of pluripotent stem cells from mouse embryonic and adult fibroblast cultures by defined factors. Cell *126*, 663-676.

Tata, P.R., Mou, H., Pardo-Saganta, A., Zhao, R., Prabhu, M., Law, B.M., Vinarsky, V., Cho, J.L., Breton, S., Sahay, A., *et al.* (2013). Dedifferentiation of committed epithelial cells into stem cells in vivo. Nature *503*, 218-223.

Thomas, E., Storb, R., Clift, R.A., Fefer, A., Johnson, F.L., Neiman, P.E., Lerner, K.G., Glucksberg, H., and Buckner, C.D. (1975). Bone-marrow transplantation (first of two parts). N Engl J Med *292*, 832-843.

Thomson, J.A., Itskovitz-Eldor, J., Shapiro, S.S., Waknitz, M.A., Swiergiel, J.J., Marshall, V.S., and Jones, J.M. (1998). Embryonic stem cell lines derived from human blastocysts. Science (New York, NY) *282*, 1145-1147.

Thorel, F., Nepote, V., Avril, I., Kohno, K., Desgraz, R., Chera, S., and Herrera, P.L. (2010). Conversion of adult pancreatic alpha-cells to beta-cells after extreme beta-cell loss. Nature *464*, 1149-1154.

Tijssen, M.R., Cvejic, A., Joshi, A., Hannah, R.L., Ferreira, R., Forrai, A., Bellissimo, D.C., Oram, S.H., Smethurst, P.A., Wilson, N.K., *et al.* (2011). Genome-wide analysis of simultaneous GATA1/2, RUNX1, FLI1, and SCL binding in megakaryocytes identifies hematopoietic regulators. Dev Cell *20*, 597-609.

Till, J.E., and McCulloch, E.A. (1961). A direct measurement of the radiation sensitivity of normal mouse bone marrow cells. Radiat Res *14*, 213-222.

Tkach, M., and Théry, C. (2016). Communication by Extracellular Vesicles: Where We Are and Where We Need to Go. Cell *164*, 1226-1232.

Trentin, J. (1989). Hematopoietic microenironments. Historical perspectives, status and projections. In Handbook of the Hemopoietic Microenvironment, M. Tavassoli, ed. (Clifton, NJ: Humana Press), pp. 1-87.

Tulina, N., and Matunis, E. (2001). Control of stem cell self-renewal in Drosophila spermatogenesis by JAK-STAT signaling. Science *294*, 2546-2549.

van Es, J.H., Sato, T., van de Wetering, M., Lyubimova, A., Nee, A.N., Gregorieff, A., Sasaki, N., Zeinstra, L., van den Born, M., Korving, J., *et al.* (2012). Dll1+ secretory progenitor cells revert to stem cells upon crypt damage. Nat Cell Biol *14*, 1099-1104.

Van Zant, G., and Liang, Y. (2012). Concise review: hematopoietic stem cell aging, life span, and transplantation. Stem cells translational medicine *1*, 651-657.

Velten, L., Haas, S.F., Raffel, S., Blaszkiewicz, S., Islam, S., Hennig, B.P., Hirche, C., Lutz, C., Buss, E.C., Nowak, D., *et al.* (2017). Human haematopoietic stem cell lineage commitment is a continuous process. Nature cell biology *19*, 271-281.

Vermeulen, L., De Sousa, E.M.F., van der Heijden, M., Cameron, K., de Jong, J.H., Borovski, T., Tuynman, J.B., Todaro, M., Merz, C., Rodermond, H., *et al.* (2010). Wnt activity defines colon cancer stem cells and is regulated by the microenvironment. Nature cell biology *12*, 468-476.

Vierbuchen, T., Ostermeier, A., Pang, Z.P., Kokubu, Y., Südhof, T.C., and Wernig, M. (2010). Direct conversion of fibroblasts to functional neurons by defined factors. Nature *463*, 1035-1041.

Wagers, A.J. (2012). The stem cell niche in regenerative medicine. Cell Stem Cell *10*, 362-369.

Wang, J., Rao, S., Chu, J., Shen, X., Levasseur, D.N., Theunissen, T.W., and Orkin, S.H. (2006). A protein interaction network for pluripotency of embryonic stem cells. Nature *444*, 364-368.

Weintraub, H., Tapscott, S.J., Davis, R.L., Thayer, M.J., Adam, M.A., Lassar, A.B., and Miller, A.D. (1989). Activation of muscle-specific genes in pigment, nerve, fat, liver, and fibroblast cell lines by forced expression of MyoD. Proc Natl Acad Sci U S A *86*, 5434-5438.

Weissman, I.L. (1994). Developmental switches in the immune system. Cell *76*, 207-218.

Wheat, J.C., Sella, Y., Willcockson, M., Skoultchi, A.I., Bergman, A., Singer, R.H., and Steidl, U. (2020). Single-molecule imaging of transcription dynamics in somatic stem cells. Nature *583*, 431-436.

Wilson, A., Laurenti, E., Oser, G., van der Wath, R.C., Blanco-Bose, W., Jaworski, M., Offner, S., Dunant, C.F., Eshkind, L., Bockamp, E., *et al.* (2008). Hematopoietic stem cells reversibly switch from dormancy to self-renewal during homeostasis and repair. Cell *135*, 1118-1129.

Xie, H., Ye, M., Feng, R., and Graf, T. (2004). Stepwise reprogramming of B cells into macrophages. Cell *117*, 663-676.

Xie, T., and Spradling, A.C. (1998). decapentaplegic is essential for the maintenance and division of germline stem cells in the Drosophila ovary. Cell *94*, 251-260.

Yamanaka, S. (2009). Elite and stochastic models for induced pluripotent stem cell generation. Nature *460*, 49-52.

Zhou, Q., Brown, J., Kanarek, A., Rajagopal, J., and Melton, D.A. (2008). In vivo reprogramming of adult pancreatic exocrine cells to beta-cells. Nature *455*, 627-632.

Zhu, S., Rezvani, M., Harbell, J., Mattis, A.N., Wolfe, A.R., Benet, L.Z., Willenbring, H., and Ding, S. (2014). Mouse liver repopulation with hepatocytes generated from human fibroblasts. Nature *508*, 93-97.

Zipori, D. (2004). The nature of stem cells: state rather than entity. Nature reviews Genetics *5*, 873-878.

2

Transcription Regulation by Distal Enhancers: Dynamics of the 3D Genome

Eric Soler[1,2] and Frank Grosveld[3]

[1] IGMM, Univ Montpellier, CNRS, Montpellier, France
[2] Laboratory of Excellence GR-Ex, Université de Paris, France
[3] Department of Cell Biology, Erasmus Medical Centre, Rotterdam,
The Netherlands
E-mail: eric.soler@igmm.cnrs.fr; f.grosveld@erasmusmc.nl

2.1 Introduction

The regulation of gene expression occurs at multiple levels. The initial step consists of transcription, which converts DNA sequences into RNA products, which may subsequently be processed (e.g., spliced), transported (e.g., exported from the nucleus, or stored into specific cellular locations), and/or translated into proteins. Deciphering the rules that govern how cells transcribe a given gene at a given time, and at what level have been the subject of intense investigation over the past decades, and remain a central question in biology. The models that emerge imply that multiple regulatory steps are integrated in a dynamic fashion within the context of the spatial (i.e., three-dimensional (3D)) organization of the genome. In this chapter, we will focus on the role of enhancers as critical regulatory elements, together with transcription factors and structural regulatory factors as key organizers of the 3D genomic regulatory landscape.

2.2 Spatio-temporal Control of Gene Activity: The Pivotal Role of Enhancers

Whereas it is clear that bacterial and lower eukaryotic genes are regulated by proximal regulatory elements such as promoter proximal sequences, gene expression in metazoans escapes this rule. This was discovered in the 80's, starting with the first description of enhancers as DNA sequences that were located at a distance from the gene. When coupled to a gene and transfected into mammalian cells, an enhancer could stimulate the expression of the gene at a distance independent of its orientation relative to the promoter of the gene (Banerji et al. 1981). The earliest discovered enhancers were viral enhancers that were active in many different cell types, but these were soon followed by the discovery of enhancers that are naturally present in the genome, many of which show tissue/cell type specificity (Banerji et al. 1983, Gillies et al. 1983, Queen and Baltimore 1983). With the advance of transgenic technologies, minimal transgenic constructs were designed, which encompassed endogenous proximal promoter sequences fused to gene coding regions. However, these failed to express properly after random and stable integration in the genome. The use of promoter sequences did not recapitulate the tissue specificity and expression levels of endogenous genes, and such constructs were often silenced, expressed at very low levels, or aberrantly expressed at ectopic locations (Magram et al. 1985, Kollias et al. 1986, Trudel and Costantini 1987). This definitely proved the need to include additional, yet to be identified, elements in the transgene constructs, which were enhancers. Even though some of these were included, the transgenes were still poorly expressed. This was resolved by the discovery of Locus Control Regions (Grosveld et al. 1987, Fraser et al. 1993). The missing elements were combinations of transcriptional enhancers, which synergize with each other. They are embedded in the genome as short DNA sequences containing regulatory information in the form of transcription factor (TF) binding sites and will express the gene fully and independently of its position of integration in the genome with the exception of the inactivated X chromosome. Through their ability to recruit sequence-specific TFs together with associated co-factors (including additional non DNA-binding TFs and chromatin-modifying enzymes), transcriptional (super)enhancers are able to connect with the transcription machinery and thereby relay regulatory information to the surrounding promoters (Figure 2.1A). Enhancers and TFs are considered as critical elements driving spatiotemporal activities and expression levels of genes in most biological settings, including during development (Le Poul et al. 2020), cellular differentiation (Lara-Astiaso et al. 2014), responses to extracellular stimuli (Soucie et al. 2016), and signaling (Mullen et al. 2011, Trompouki et al. 2011) (Figure 2.1B). The critical role played by TFs and enhancers to sustain gene expression and support organism and

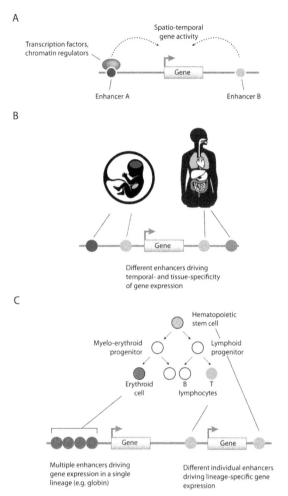

Figure 2.1 Enhancers driving complex gene expression patterns. (**A**) Schematic representation of enhancers and associated transcription factors (TFs) driving gene expression. Regulatory input is driven by DNA-binding and non DNA-binding TFs and their associated co-factors (chromatin modifiers, co-activators, and co-repressors) binding at enhancer sites. (**B**) Several enhancers drive composite gene expression during development and in adult tissues. The different enhancers are colored and the same color code is used to highlight the tissue-specific target gene expression pattern. (**C**) Examples of enhancers driving gene expression in select lineage within a given tissue (here, hematopoietic tissue). Some enhancers synergize to drive gene expression in one lineage, as it is the case for the globin genes loci (*left*), whereas for some other genes, various combinations of enhancers drive expression in specific cell types (e.g., here, hematopoietic stem cells, and T lymphocytes) (*right*).

tissue development is attested by the dramatic phenotypes associated with TF knock-out (KO) or enhancer deletions.

Inactivation of single TFs can broadly perturb gene expression patterns, impair cellular functions and/or differentiation capabilities. As TFs are usually able to bind hundreds to thousands of regulatory elements including promoters and enhancers, hence the disruptive phenotypes observed in TF KO are expected. The hematopoietic TFs GATA2, GATA1, or MYB for instance, control gene expression in hematopoietic stem and progenitor cells, mostly by binding enhancers. Knocking-out any of these factors leads to a severe hematopoietic phenotype, with a complete absence of blood formation, and embryonic lethality. In their absence, the regulatory networks necessary for blood stem cell formation, maintenance and differentiation fail to be established leading to severe hematopoietic failure, due to incapacity to activate proper transcriptional programs. Besides TFs KO, numerous studies have shown that deleting single enhancer sequences can also have dramatic consequences. For example, the *Sonic hedgehog* (*SHH*) gene encodes an essential morphogen required for a variety of patterning events during embryonic development. It is controlled by an array of at least 11 different enhancers, each driving SHH expression in a particular tissue (e.g., forebrain, hindbrain, epithelium or limb bud) at specific developmental time points (Anderson and Hill 2014, Williamson et al. 2019). Deletion of single individual *SHH* enhancers can abolish *SHH* expression in the target tissue without altering the expression dynamics in the other embryonic sites, underscoring the importance of enhancers for proper tissue patterning. For example, mutation of one of the enhancers that is located 1 Mb away from the gene leads to polydactyly (Lettice et al. 2002). Another example is the *Gata2* hematopoietic TF gene, which is regulated by at least 5 enhancers. Commitment of hematopoietic progenitors to the erythroid lineage is associated with a gradual decrease of GATA2 expression as the levels of the erythroid TF GATA1 raise. GATA1 targets *Gata2* enhancers where it provides repressive activity, leading to the switch from *Gata2* to *Gata1* expression, required for further erythroid differentiation. Interestingly, deletion of a single *Gata2* enhancer located -1.8 kb upstream of the *Gata2* transcription start site, affects the ability of cells to properly maintain *Gata2* repression, and shows a reactivation of *Gata2* in late committed erythroid cells (Snow et al. 2010), whereas deletion of the other enhancers does not phenocopy this situation and preferentially alter hematopoietic stem cell formation (Gao et al. 2013) or myeloid progenitors downstream of hematopoietic stem cells (Johnson et al. 2015). This shows that enhancers can play specific and non-redundant roles *in vivo*. Furthermore, it demonstrates that enhancers are not always associated with transcriptional stimulation, but can lead also lead to repression depending on the on the cellular context, the developmental window and on the TF occupancy (Stadhouders et al. 2015).

However, it should be noted that the above mentioned examples may not be representative of the majority of cases. Indeed, enhancers often function in an additive and overlapping fashion. The globin enhancers for instance drive strong globin gene expression for the efficient and massive hemoglobin production occurring in terminally differentiating erythroid cells (Figure 2.1C). As opposed to the different *SHH* enhancers, the globin enhancers drive gene expression in a single tissue type and in narrow windows of development and/or cellular differentiation (i.e., erythroid terminal maturation). Interestingly, the various enhancers of the globin genes enhancers function in a redundant fashion as deletion of individual enhancer sequences are not sufficient to abolish globin gene expression *in vivo*, they appear to be additive in their activity and for some of them their inactivation has only a moderate impact on total globin expression (Fraser et al. 1993, Hay et al. 2016). Deletion of the full set of enhancers leads to complete absence of globin expression and ineffective erythropoiesis. It is generally assumed that the presence of multiple enhancers with apparent redundant or overlapping function provides increased robustness to secure gene expression *in vivo* (Osterwalder et al. 2018). Enhancers evolve and perhaps the best example of how such evolution can occur was described in a recent paper on the regulatory sequences of the globin genes (Miyata et al. 2020 and references therein; Hardison 2020).

2.3 The Genome-wide Enhancer Landscape

Current evidence suggests that enhancers are the major determinants of cell identity and cell-type specificity in gene expression. The enhancer landscape is therefore informative to define a cell type, is characteristic of each tissue and reflects the cellular gene expression potential.

Genome-wide studies have revealed that the number of transcriptional enhancers greatly exceeds the number of genes present in mammalian genomes. Many enhancer sequences are poorly conserved across species, except at the actual very small TF-binding sites themselves, which makes it difficult to predict enhancers based solely on their DNA sequences (Schmidt et al. 2010, Villar et al. 2015). Candidate enhancers can be identified through a combination of distinct chromatin features including "open" accessible chromatin sites (e.g., nucleosome depleted regions hypersensitivity to DNAse I), and various epigenetic marks such as high H3K4me1, H3K4me2, and/or H3K27Ac histone modifications. In addition, enhancers are characterized by the binding of TFs, and the presence of the transcription machinery (RNA pol II). The key chromatin features correlating with putative enhancer sequences have been described in great details elsewhere (Cico et al. 2016), although it should be noted that these are only predictive features and do not imply functionality. One may directly assess enhancer function using conventional or high throughput reporter assays (Muerdter et al. 2015, Barakat et al. 2018, van Arensbergen et al. 2019) or (CRISPR-mediated) deletion to ascertain their activity. Furthermore, this epigenetic "definition" of enhancers may certainly not be universal as there are tens of other chromatin histone modifications (Tan et al. 2011, Sabari et al. 2017), and a unifying predictive model of enhancer chromatin features is still lacking as novel types of enhancer sites escaping the classical epigenetic definition are regularly uncovered (Pradeepa et al. 2016, Crespo et al. 2020). In any case, by using this epigenetic "definition" of putative enhancers, the genome-wide enhancer landscape has been defined in a variety of tissues and cell types. It showed that enhancers are frequently found within introns or intergenic regions, at both upstream and downstream of their target genes irrespective of the location. Most genes seem to be under the control of several enhancers. Interestingly, some dense clusters of enhancers were identified across the genome, spanning up to 50 kb, that were mostly associated with key lineage identity genes (Whyte et al. 2013). These regions, which were already observed on a handful of loci, including the globin genes loci in the early 80's, were recently renamed "Super-Enhancers" (SEs) (Hnisz et al. 2013, Whyte et al. 2013). Numerous SE structures have been identified and were shown to correlate with cell identity genes, or acquired near oncogenes in cancer cells (Hnisz et al. 2013). However, the term "Super-Enhancer" may be inappropriate as it implies that these structures may have added properties other than just the sum of their individual enhancer components. Nevertheless, it was shown by using the α-globin locus as a model, that the α-globin SE is no more than the sum of its components (Hay et al. 2016). A similar study conducted in ESCs carefully dissected SE functions to reach the conclusion that individual enhancers within SE structures act in an additive and partially redundant fashion (Moorthy et al. 2017). Taken altogether, these studies imply that, enhancers within SEs may only bring variable and additive inputs. Thus, SEs are better referred to as "enhancer clusters" to better reflect their biological features.

2.4 Enhancer Dynamics

The enhancer landscape molecularly defines cell-type specificity and differences across tissues. During the course of embryonic development or cellular differentiation, the enhancer profiles show dramatic differences and dynamic re-organization. They have been analyzed during hematopoietic differentiation in 16 cell stages ranging from immature hematopoietic stem cells (HSCs) to early committed precursors and mature cell types that form the myeloid, erythroid, and lymphoid lineages (Lara-Astiaso et al. 2014). It shows that enhancer establishment is initiated during early lineage commitment and often precedes the transcriptional activation of target genes. Interestingly, more than half of the dynamic enhancers detected in this study are already established in HSCs and uniquely maintained in the downstream relevant lineage, whereas they disappear from the other lineages (e.g., myeloid vs. lymphoid). For instance, the enhancers of the *Gata2* gene, which is important for HSC and myelo-erythroid lineage differentiation, are detectable in HSCs, early immature progenitors downstream of HSCs, and in myelo-erythroid progenitors, whereas they are extinguished in lymphoid cells and terminally differentiating erythroid cells. In addition, a large set of the dynamic enhancers (~40%) are established *de novo* during differentiation (e.g., this is the case for the *Gypa* gene enhancers which are *de novo* established in the erythroid lineage-committed cells) (Lara-Astiaso et al. 2014). Interestingly, within a single cell type, enhancers can show dramatic stimulus-induced dynamics to regulate cellular behavior (Ostuni et al. 2013) and cell plasticity (Soucie et al. 2016). Tissue-resident macrophages are fully differentiated long-lived cells that can patrol within tissues to monitor the presence of infectious contaminants. They can spontaneously enter into a state of self-renewal while maintaining their fully differentiated state. Whereas, the genome-wide position of enhancers does not significantly change between quiescent and self-renewing macrophages; many enhancers actually show a strong increase in their activity (Soucie et al. 2016). Macrophages have therefore, the ability to activate a set of poised enhancers, that become active in the self-renewing state, and that are associated with a network of genes important for the self-renewing process such as *Myc* and *Klf2*. Mechanistically, the MAF TFs (MafB and c-Maf) normally repress the self-renewal enhancers in macrophages, and in response to immune stimulation, proliferating resident macrophages transiently decrease the levels of MAF and therefore can access this gene-enhancer network to induce their self-renewing capacity (Soucie et al. 2016). Interestingly, this gene network is also used by embryonic stem cells (ESCs) for their self-renewal, however ESCs use a completely different set of enhancers to activate the same genes. Thus, a similar core of self-renewing genes can be used by different cells (ESCs or self-renewing macrophages) but through different cell type-specific enhancers. In macrophages, this is achieved in a transient and fully reversible manner. The fact that genes can be under the control of completely different sets of enhancers in different cell types has also been observed for broadly expressed genes. For example, the *Myc* proliferation gene is under the control of a completely different set of enhancers between ESCs and B-lymphocyte progenitors (Kieffer-Kwon et al. 2013). This probably reflects the use of cell type-specific TFs to control enhancer function and gene expression in different cells, with specific DNA-binding activities and chromatin occupancy. Taken altogether, these observations indicate that the enhancer landscape is highly dynamic and is a pivotal component of the gene regulatory networks in mammalian cells.

2.5 Enhancers and Promoters Are Distant Structures

A surprising discovery derived from the first genome-wide profiling of enhancers, was that mammalian enhancers are scattered throughout the genome, often at large distances from their target gene(s). Enhancers can be located far upstream of promoter sequences, within introns, downstream of genes, or within introns of unrelated genes. This greatly complicates the ability to comprehensively link enhancer–gene pairs and functionally ascribe the full complement of enhancers that regulate a single gene. For instance, and as referred to above, the *SHH* gene is essential for proper regulation of limb bud development, and the *SHH* limb bud distal enhancer lies within an intron of the *LMBR1* gene located approximately 1 mega-base (Mb) upstream of *SHH*. The *MYC* oncogene, which is a critical regulator of stem cells, cell cycle progression, metabolic control, and hematopoietic development, is another example of a complex distal regulation. In the intestine, it is controlled by an upstream enhancer located 300 kilobases (kb) apart in an intergenic region (Pomerantz et al. 2009, Tuupanen et al. 2009), whereas in T-lymphoid progenitors, it is under the control of a downstream intergenic enhancer lying 1.5 Mb away (Herranz et al. 2014, Kloetgen et al. 2020). Actually, most mammalian genes seem to be subjected to regulation by distal enhancers, with the median distance between enhancer–promoter pairs ranging from 80 to 120 kb according to different estimates, and based on comprehensive genome structural analyses (Sanyal et al. 2012, Jin et al. 2013, Kieffer-Kwon et al. 2013, Rao et al. 2014, Bonev et al. 2017). This particular organization of mammalian genomes leads to the question of how distal enhancers can control their target genes over such large distances, and how are the enhancer–gene communications organized *in vivo*? It also brings up the important issue how specificity is obtained as distant enhancers often do not regulate the gene(s) that are close-by, but regulate genes that are further away (see also below). In other words, why does an enhancer skip a gene(s) and regulate more distant ones, or why does it regulate a gene located in one orientation (e.g., upstream) but not a gene in the opposite orientation (downstream). And how can enhancers regulate more than one gene at the same time? The answers to these questions have only been answered partially but seem to involve transcription factor specificity and the 3D aspects of the genome.

2.6 Mechanisms of Enhancer Function: The 3D Genome

The prevalent model of transcription regulation by distal enhancers is "looping" whereby enhancers can be spatially positioned in close proximity to their target promoters as opposed to mechanisms that would involve a linear mechanism such as transcription itself. The flexibility of the chromatin fiber would allow looping to enable enhancers to communicate with their target genes despite the often very large linear distance that separates them. Looping could potentially be established in several ways. The enhancers may "scan" the surrounding environment and establish preferential contacts with the surrounding promoters that are stabilized by the interaction with transcription factors (Figure 2.2A). Alternatively, recent developments suggest that promoters and enhancers may be brought together through a mechanism of loop extrusion (see below, under TADs), where enhancers and promoters are brought in close proximity and would again be stabilized by transcription factor interaction.

The first evidence that chromatin looping actually occurs is derived from studies of the β-globin gene locus in erythroid cells. When the Locus Control Region of the β-globin locus was coupled to two genes in different orders, it showed that the proximal gene suppressed the distal gene (Hanscombe et al. 1991), that the level of expression/suppression was dependent on the relative distance of the genes (Dillon et al. 1997), and

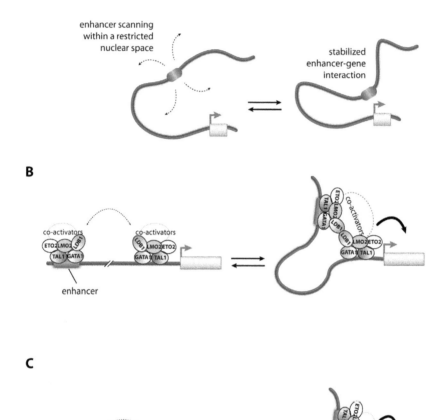

Figure 2.2   Chromatin looping mechanisms. (**A**) An enhancer "scans" the nuclear space until it partly stabilizes at a target gene promoter. Enhancer–promoter stability depends on enhancer-promoter pairs and on the TF content mediating the interaction. (**B,C**) Examples of TF-mediated chromatin looping. The LDB1–TF complex (containing the GATA1, TAL1, ETO2, LMO2 and LDB1 TFs) can mediate long-range chromatin interactions via LDB1 homodimerization. The resulting increased local density of transcriptional co-activators helps stimulating the transcription process. Alternatively, LDB1–CTCF heterodimerization leads to loop formation and transcriptional activity.

that multiple genes were expressed in an alternating fashion when regulated by one LCR/superenhancer (Wijgerde et al. 1995). These observations could only be explained by a looping mechanism. This was confirmed by direct biochemical evidence using 3C methodology (Tolhuis et al. 2002, Palstra et al. 2003, Andrieu-Soler and Soler 2020). It showed that the β-globin distal enhancers, known as the "Locus Control Region" (LCR), and located ~40 to 60 kb upstream of the globin genes cluster, are in close proximity to the globin gene promoters specifically in erythroid cells where the globin genes are highly expressed. Conversely, no such proximity could be observed in the brain where the globin genes are silent. Since then, enhancer–promoter looping structures have been reported for hundreds of genes across the genome, in multiple tissue- and cell-types and multicellular organisms, underscoring the widespread nature of this mode of enhancer function (van den Heuvel et al. 2015, Cico et al. 2016, Robson et al. 2019, Schoenfelder and Fraser 2019).

Several factors have been implicated in structuring and spatially organizing the genome. The ubiquitously expressed Lim-Domain Binding protein 1 (LDB1) TF was reported to act as a master regulator of chromatin looping (Soler et al. 2010, Deng et al. 2012, Stadhouders et al. 2012, Deng et al. 2014, Krivega et al. 2014) (Figure 2.2B). LDB1 is an adaptor protein that does not directly bind DNA but is able to assemble multiprotein complexes with TFs in several cell types or cellular contexts. It is an essential gene and its disruption leads to early embryonic lethality with multiple organ defects and complete absence of hematopoiesis (Mukhopadhyay et al. 2003). In hematopoietic cells, it plays a pivotal role in erythroid lineage differentiation (Li et al. 2010, Mylona et al. 2013, Love et al. 2014). LDB1 is able via its dimerization domain to establish self-interactions between two distally LDB1-bound elements, thereby creating a looped structure (Deng et al. 2012) (Figure 2.2B). This type of LDB1-mediated chromatin loop is present at the β-globin locus, where TF complexes containing LDB1 are both bound at the β-globin gene promoter and at the distal enhancers ~40 kb away. LDB1-mediated loops directly activate transcription of the globin gene in differentiating erythroid cells, and an artificially targeted LDB1 binding at this locus can force loop formation and transcriptional activity, confirming the causative role of LDB1 and chromatin loops in transcriptional activation (Deng et al. 2012, Deng et al. 2014). However, the LDB1 complexes tend to mostly (~90%) bind distal enhancer sites and rarely (less than 10%) promoters in erythroid cells (Soler et al. 2010). This leaves the question of how can LDB1 mediate self-dimerization enhancer–promoter looping if LDB1 is infrequently bound at the promoters? Recent evidence showed that in addition to LDB1–LDB1 dimerization, LDB1 is also able to heterodimerize with another key chromatin structural protein: the CTCF TF (Lee et al. 2017). It was shown that erythroid genes tend to frequently have CTCF binding at their promoter regions, offering the opportunity to bridge with LDB1-bound distal enhancers (Figure 2.2C). Besides LDB1, other factors have been implicated in regulating genome spatial organization such as YY1 (Weintraub et al. 2017), ZNF143 (Bailey et al. 2015), KLF1 (Drissen et al. 2004), the mediator complex (Kagey et al. 2010), the polcycomb repressive complex (Vieux-Rochas et al. 2015, Ogiyama et al. 2018), and more importantly the CTCF and cohesin complex (Dixon et al. 2016). With the exception of CTCF/cohesin (see below), the mechanisms of chromatin looping mediated by the above-mentioned factors remains poorly characterized, although it is worth mentioning that self-dimerization has been suggested to be responsible for YY1-mediated looping (Weintraub et al. 2017). Therefore, protein–protein interactions seem essential for chromatin looping and spatial organization of the genome.

2.7 Dynamic versus Stable Chromatin Looping

Increasing evidence indicates that chromatin loops are dynamic structures. At the β-globin locus, the different globin genes are arranged in tandem in the order of their developmental expression, where the embryonic β-globin genes (*Hbb-by* and *Hbb-bh1*) precede the adult-type β-globin genes (*Hbb-b1* and *Hbb-b2*). In mouse embryonic erythroid progenitors, the LCR containing the cluster of β-globin distal enhancers interacts with the embryonic globin genes (*Hbb-by* and *Hbb-bh1*), with this looping event correlating strongly with the activation of those genes. As development progresses, when definitive adult-type erythroid cells are formed, the LCR switches its interactions toward the downstream adult-type globin genes (*Hbb-b1* and *Hbb-b2*), and no longer contacts the embryonic genes (Palstra et al. 2003). This new loop formation correlates with the strong induction of the adult-type globin genes expression in definitive erythroid cells, and a loss of expression of the embryonic genes. Another example of dynamic chromatin looping correlating with gene expression changes is the regulation of the *Myb* proto-oncogene. In the highly proliferating immature erythroid progenitors, high levels of *Myb* expression sustain the rapid cell cycle progression. Upon entry into terminal differentiation, erythroid progenitors need to slow down their proliferation potential. This is achieved through a strong decrease in *Myb* expression, a necessary feature for proper erythroid maturation. It was shown that the *Myb* gene is controlled by an array of 5 distal upstream enhancers that physically interact with the *Myb* gene via LDB1-mediated chromatin looping (Stadhouders et al. 2012). Whereas the looped structure can be detected in proliferating progenitors expressing high levels of *Myb*, the global organization of the locus becomes dramatically destabilized upon induction of differentiation: differentiating cells show a remarkable loss of enhancer looping which directly correlates with loss of *Myb* expression. Mechanistically it was shown that LDB1 binding at *Myb* distal enhancers is severely destabilized when cells enter differentiation, leading to loss of chromatin looping and loss of *Myb* transcriptional stimulation, a necessary step for terminal erythroid differentiation (Stadhouders et al. 2012). Thus the *Myb* dynamic expression patterns are directly correlated with spatial reconfiguration of the locus *in vivo* in erythroid cells.

Whereas an abundant literature details the dynamics of chromatin loops, several examples of stable structures have been reported. During early embryonic development in drosophila enhancer–promoter loops seem to be pre-configured and stable (Ghavi-Helm et al. 2014). They are detected prior to gene activation and, surprisingly, show minimal changes between tissues and during development. They are frequently associated with paused RNA Pol II, which supports a model where pre-looped structures allow to establish a "poised" state where the transcriptional machinery is present but not yet fully active. Target gene activation is triggered via transcriptional pause–release by mechanisms that are not yet fully characterized but probably involve dynamic reorganization of TF binding and input from signaling molecules. Another study investigating the responses to hypoxia identified widespread pre-configured chromatin loops at hypoxia-inducible genes (Platt et al. 2016). During hypoxia, the HIF TFs are stabilized and activate hypoxia response genes to switch metabolic activities and help cells to deal with the lack of sufficient oxygen. The pre-established chromatin looping structures between HIF-binding enhancers and hypoxia-inducible genes probably allows a fast activation of the response genes under hypoxic conditions. More generally, it has been suggested that pre-established enhancer–promoter structures allow fast gene activation kinetics for inducible genes and for responses to signaling (Jin et al. 2013). Study of the primary human fibroblast cells (IMR90) chromatin interaction dynamics in response to TNFα treatment uncovered stable pre-existing enhancer–promoter loops. Such loops exist prior to TNFα stimulation and activation of TNFα-responsive enhancers does not associate with changes in long-range chromatin interactions. A similar observation has been made in other cell types (HUVEC, MCF7, LNCaP) upon stimulation with different signals (IFN-γ, β-oestradiol, and 5α-dihydrotestosterone) (Jin et al. 2013). This confirmed the native looped structure of multiple loci and led to the conclusion that by assembling specific chromatin pre-configurations, the different cell types can prioritize the target genes to be induced upon signaling. It may also provide a framework to explain how pleiotropic signals can elicit dramatically different transcriptional responses in different cell types.

Therefore, mammalian cells make use of both dynamic and "static" chromatin spatial organization for transcriptional control of gene expression. However, one should keep in mind that the looping model is probably not a universal model as some discrepancies between physical enhancer–promoter proximity and transcription activation have been reported. It was shown at the *SHH* locus that whereas the distal limb bud enhancer actually loops toward the SHH gene in developing limbs (Amano et al. 2009), the *SHH* brain enhancers do not show this typical behavior. By using ESCs differentiated into neural progenitors, it was shown that the spatial distance between *SHH* and the brain enhancers increases (as measured by 3D-FISH) during the course of *SHH* activation (Benabdallah et al. 2019). This finding was further confirmed *in vivo* in developing embryos. A proposed mechanism for this counterintuitive observation is that large molecular complexes (involving the PARP enzyme) are nucleated from the enhancer sites during *SHH* activation in neuronal lineages. The PARP-dependent PARylation initiated at the *SHH* brain enhancers may lead to large macromolecular assemblies or liquid–liquid phase-separated subnuclear compartments leading to the increased distance observed between *SHH* and its regulatory element during activation. Although the underlying mechanisms are not clear, the nucleation of transcription regulatory factors within the phase-separated subnuclear region could be compatible with gene activation even if the enhancer does not directly contact *SHH* as a transcriptionally permissive environment is created. Of note, increased nuclear volume of gene loci during activation have been observed in other systems such as the α-globin locus in erythroid cells, whereas in that case the distal enhancers strongly contact the α-globin genes (Brown et al. 2018). These interesting observations indicate that whereas long-range chromatin looping is a common feature, multiple evidence challenge the idea of stable enhancer–promoter loops as the basis for all enhancer–promoter communications.

2.8 Transcription Factories or Foci

An additional 3D transcriptional phenomenon is the observation that transcription (at least in part) takes place and is initiated in so-called transcription factories (also called foci), originally described in 1993 (Jackson et al. 1993, Wansink et al. 1993). These foci can be visualized in the nucleus by labeling or using antibodies against one of the components of the transcription machinery, e.g., RNA polymerase II itself (Sato et al. 2019) or Cdk9, which phosphorylates the polymerase for transcriptional elongation (e.g., (Ghamari et al. 2013)). These foci are thought to contain the interacting enhancers and genes of multiple co-regulated genes (Tolhuis et al. 2002, Drissen et al. 2004, Osborne et al. 2004, Osborne et al. 2007, Xu and Cook 2008, Schoenfelder et al. 2010, Noordermeer et al. 2011, de Wit et al. 2013, Fanucchi et al. 2013, Javierre et al. 2016, Beagrie et al. 2017, Oudelaar et al. 2019) and, interestingly, they occur in different combinations (Schoenfelder et al. 2010) in otherwise identical cells suggesting that genes and enhancers end up stably in these foci through some random walk type process creating a focus of high concentration of the relevant transcription factors and transcription machinery. However, it is not clear how these factories are formed. It has been suggested that they form at enhancers with a high number of transcription factor binding sites (Cho et al. 2018, Shrinivas et al. 2019, Sabari et al. 2020), which would nucleate to form a transcription factory, while it is known that enhancers are already activated before transcription takes place. Although suggested much earlier, the idea that such foci form through a mechanism of phase separation has become popular for a number of physiological processes (Sabari et al. 2020). It takes place when the concentration of a molecule reaches a critical point and it becomes energetically favorable for the molecule in question to partition into a well-defined low- and high-concentration phase. However, if phase separation underlies the mechanism of factory formation, it is likely to involve more than the concentration of a particular factor but a combination of different (transcription) factors. These factories appear to be very stable, but disappear during cell division at mitosis and then reform during late anaphase when the next cell cycle initiates (Imam A. *et al.* unpublished observations). It is less clear how stable the interactions are between individual enhancers and genes, although it has been known for some time that the process of transcription is periodic and that the level of an RNA transcript is largely determined by the frequency of such periods (Wijgerde et al. 1995).

2.9 Specificity of Enhancer–Promoter Contacts Within Higher Order Chromatin Structures

Whereas some enhancer–promoter pair specificity has been reported (Zabidi et al. 2015), other observations obtained from reporter assays, transient transgenic assays, and from experimental or disease situations where enhancers are translocated into a different chromatin environment suggest that enhancers are able to functionally establish contacts with promoters in a rather non-specific way. One intriguing question is therefore: how do enhancers specifically recognize their target promoters in the nuclear space? And how is specificity achieved? Part of the answers comes from physical constraints imposed by higher-order chromatin structures.

2.10 TADs: Topologically Associating Domains

Whole genome high-throughput chromatin interaction studies showed the existence of structural chromatin domains called Topologically Associating Domains or TADs (Dixon et al. 2012, Nora et al. 2012) (Figure 2.3A). TADs are domains of preferential self-interacting chromatin regions ranging from a hundred kb to 1 Mb in size. They partition mammalian chromosomes into domains where long-range chromatin interactions are favored within the domains rather than across domains. This spatial organization is believed to restrain enhancer contacts to genes present in the same TAD and to prevent ectopic interactions with genes from other TADs, because of the physical rule that two things tethered on a string have a much higher change to meet (and interact) than not being tethered on the same string. TADs and their borders were shown to be conserved across species, reinforcing the idea that they represent a fundamental aspect of chromosomal structure (Dixon et al. 2012, Rao et al. 2014). Furthermore, TADs appear surprisingly stable across cell types, despite extensive rewiring of transcriptome activity during cellular differentiation (Dixon et al. 2015) (Figure 2.3A). Comparing human embryonic stem cells (hESCs) with four hESC-derived lineages (mesendoderm, mesenchymal stem cells, neural progenitors, and trophoblast-like cells) highlighted conserved TAD boundary locations at the genome-wide scale. This suggests that cellular differentiation and the associated changes in gene expression do not modify overall TAD structure or borders, rather occur within these domains. Within TADs, looping proteins such as LDB1, CTCF, ZNF143, YY1, etc. participate in the establishment of specific enhancer–gene communications. Indeed, fine scale examination showed that intra-TAD interactions display a high degree of dynamic changes during differentiation, reflecting the dynamic establishment of enhancer–gene contacts in the different lineages (Dixon et al. 2015). Although most borders are invariant between cell types, some show cell type-specific activity. Comprehensive mapping of 3D chromatin organization at high resolution during ESC differentiation toward neuronal stem cells and cortical neurons has uncovered dynamic TAD borders (Bonev et al. 2017). A few hundred (~300) TAD borders show appearance or disappearance during ESC differentiation into neuronal lineages (these changes were missed in previous studies, probably due to the lower resolution of the assays). Among such borders, an ESC-specific TAD split into two consecutive TADs after the establishment of a novel border inside neural progenitors. This novel border is formed at the TSS of the Zfp608 gene during differentiation. Apparition of this border is concomitant with the transcriptional activation of Zfp608; however, artificial activation of this gene in ESC by introducing dCAS9 fused to transcription activation domains fails to create a border (despite activation of Zfp608 at levels similar to endogenous levels). This indicates that transcription is correlated with insulation at TAD borders, but promoter activation and transcription alone are not sufficient to generate a border (Bonev et al. 2017).

TAD-mediated "insulation" of chromatin interactions ensures some specificity to the enhancer–promoter contacts by limiting the choice of interactions. In this view, the TAD borders act as insulating elements by being able to "physically" impede enhancer contacts with irrelevant target genes. There is however additional as yet poorly understood specificity of promoter/enhancer interactions (see below). A large fraction of TAD borders is heavily bound by CTCF and the cohesin complex (Rao et al. 2014, Bonev et al. 2017). By being able to self-assemble, CTCF sites can establish and/or stabilize TAD borders. It was demonstrated that these CTCF–CTCF interactions are critically dependent on CTCF orientation, with converging motifs allowing interactions (Rao et al. 2014). Inverting CTCF motifs can rewire the interactions, leading to loss of looping (de Wit et al. 2015) or generation of novel looped structures with other properly oriented sites (Guo et al. 2015). The way that TADs are generated and stabilized by CTCF and cohesin is best explained by a loop

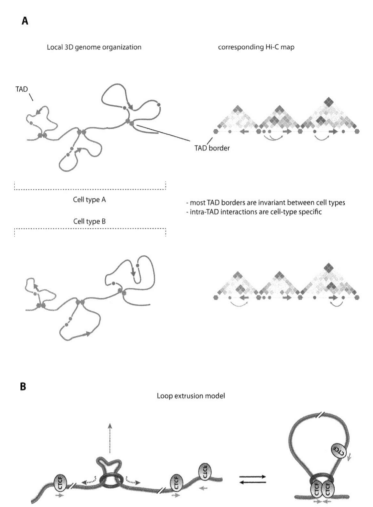

Figure 2.3 TADs and their formation by the loop extrusion model. (**A**) Schematic (*left*) and Hi-C (*right*) views of 3D genome organization. TADs are separated by borders, often enriched in CTCF/cohesin binding sites that are believed to prevent ectopic interactions between enhancers from one TAD with genes located in neighboring TADs, thereby favoring intra-TAD enhancer–gene interactions. Note that TAD borders are mostly invariant (with some exceptions) between cell types, but that intra-TAD chromatin looping is cell-type specific. (**B**) Schematic of the loop extrusion model. A cohesin ring travels along the chromatin fiber, initiating loop extrusion until it stops after encountering CTCF sites in a convergent orientation ("roadblock"). Note that this process is dynamic and that TADs and loop extrusion can be disassembled due to cohesion disengagement prior to new loop extrusion formation by another round of loop extrusion. CTCF motif orientation is shown as brown arrows underneath CTCF.

extrusion model, whereby a cohesin ring scans the DNA while extruding chromatin as it progresses until its stops when it encounters a CTCF site in the proper orientation (Dixon et al. 2016) (Figure 2.3B).

Interestingly, CTCF and cohesin despite showing extensive genome-wide overlap contribute differentially to chromatin organization (Zuin et al. 2014). Acute depletion of the CTCF protein using the auxin–degron system leads to a massive loss of insulation at most of the TAD boundaries (~80%) (Nora et al. 2017) and fewer intra-TAD contacts, whereas disruption of cohesin leads to loss of local chromatin interactions but TADs remain essentially intact (Zuin et al. 2014). The remaining ~20% TADs that escape CTCF depletion suggest that other mechanisms besides CTCF-mediated loops contribute to genome organization. The regulatory factors and mechanisms involved remain to be discovered. It should be noted that a significant fraction of TAD borders is not bound by CTCF (Bonev et al. 2017). Such borders often overlap with the promoter-associated H3K4me3 histone mark, active transcription, and are also bound by cohesin. How such CTCF-depleted borders form is currently unclear and is under extensive investigation.

2.11 Functional Relevance of TADs and Their Borders

The physical "rigid" insulating view of TAD borders should be taken with caution as single cell analyses have shown that, whereas TADs and borders are well observed at the cell population level, in individual cells these structure show a more flexible/dynamic behavior than previously thought (Cardozo Gizzi et al. 2019) and enhancer contacts across TAD borders can be readily detected (Luppino et al. 2020). This suggests that TADs are dynamic structures, forming continuously by loop extrusion, and that TAD borders are not fixed insulating structures but rather promote a certain degree of insulation by favoring intra-TAD interactions over inter-TAD contacts. Despite extensive data available on TADs and their borders, relatively few studies have shown the functional importance of such structures.

Careful dissection of the *Shh* TAD led to the surprising finding that TAD structure was not absolutely required for proper *Shh* regulation *in vivo* (Williamson et al. 2019). Deletion of single CTCF sites at the borders of the *Shh* TAD shifts the border to a site ~40 kb away and leads to decreased Shh contacts within its own TAD (where all its enhancers are located) and increased contacts with the neighboring TAD. However, all these CTCF deletions, despite shuffling the border, modifying long-range contacts, and leading to an apparent increased physical distance of *Shh* with its distal enhancers, do not lead to any observable phenotypic alterations *in vivo* (Williamson et al. 2019). *Shh* remains properly expressed in all embryonic tissues examined and retains proper tissue-specificity. This may be explained by remaining very transient contacts between the enhancers and the *Shh* promoter in the context of the modified TAD, that cannot be efficiently detected by imaging (3D-FISH) or by the chromatin looping assay used in this study. A similar observation was made at the Sox9 locus, where deletion of the CTCF sites at the TAD border did not lead to an altered TAD structure or fusion with the neighboring TAD (Despang et al. 2019). Fusions between the two adjacent TADs were only observed when internal CTCF sites were deleted in addition to the ones present at the border. But in all cases, *Sox9* developmental expression was only minimally altered and embryos developed normally. This suggests that, at least at some loci, TAD integrity is dispensable for proper developmental gene regulation. Indeed, at least at some loci, an individual TAD may not play a crucial role, but it is likely that the combination of TADs plays an essential structural role: the collection of TADs is probably essential, but losing a border does not matter much in general (as shown for at the above mentioned *loci*), but may result in aberrant expression (e.g., turning on a gene that should not be turned on (Lupianez et al. 2015)) in selected cases. Functional enhancer–gene contacts can be established even within altered TAD topologies (Despang et al. 2019, Williamson et al. 2019), which suggests that TF-mediated loops may be able to form independently of a CTCF-mediated organized scaffold. The formation of the recently proposed TF-mediated condensates through liquid–liquid phase separation (Hnisz et al. 2017, Boija et al. 2018) may be sufficient to ensure functional interactions. In that case, TADs may serve as spatially organized structures to optimize or favor such interactions in the nuclear space, without being absolutely required to establish them (Despang et al. 2019). These examples indicate that all TAD borders and overall structures are not functionally required *in vivo*. However, alterations of TAD borders have been observed in human diseases, associating with gene misregulation, indicating that at least some TADs are critically required for proper gene regulation *in vivo*. This will be detailed in the next paragraph. Taken altogether, these data suggest that in the genome, a variety of TADs and TAD borders exist, with apparently diverse functions. Whether this apparent functional diversity is due to the fact that some genes are more permissive to fluctuations in their expression (in the context for instance of an altered TAD structure) or is truly due to different types of TAD borders remains to be determined.

2.12 Alterations of Enhancers and 3D Genome Organization in Disease

Alterations of enhancer sequences by mutations or natural genetic variations have been documented (Lettice et al. 2012, Maurano et al. 2012, Ulirsch et al. 2016). Alterations linked to spatial chromatin organization are now accumulating, showing that disruption of enhancer long-range contacts or displacement of enhancers from their endogenous chromatin environment can have deleterious impacts. Genetic variants at the *HBS1L-MYB* intergenic interval were shown to alter LDB1 binding at a critical enhancer site located –84 kb upstream of *MYB* (Stadhouders et al. 2014). Decreased LDB1 binding leads to decreased looping of the –84 kb enhancer toward the *MYB* gene and decreased *MYB* expression in erythroid progenitors. Individuals bearing these variants show signs of accelerated erythroid differentiation and altered globin content. This represents an example of a subtle change in long-range enhancer–gene contacts causing an observable phenotype in humans. More generally, chromosomal translocations observed in cancer cells may reposition enhancers in ectopic locations where they can establish novel contacts with their surrounding environment, including with oncogenes. In inv(3)/t(3;3) acute myeloid leukemia (AML), an internal inversion within chromosome 3 associates with aberrant activation of the *EVI1* oncogene. In these AML, the inversion repositions a distal *GATA2* enhancer to the vicinity of the *EVI1* gene leading to both *GATA2* functional haploinsufficiency and ectopic *EVI1* activation, a driving event in leukemia initiation (Groschel et al. 2014, Yamazaki et al. 2014). This observation indicates that repositioning of a single enhancer can cause alteration of two unrelated distal genes leading to leukemic transformation. This type of structural variations leading to "enhancer hijacking" has been detected in other types of cancers (Northcott et al. 2014, Weischenfeldt et al. 2017), including medulloblastoma where the oncogenes *GFI1* and *GFI1B* get repositioned in the vicinity of enhancers, including enhancer clusters, and become aberrantly activated (Northcott et al. 2014).

In addition, a novel mechanism of cancer transformation involving aberrant enhancer activity was brought to light recently (Mansour et al. 2014, Navarro et al. 2015). The *TAL1* TF gene is a critical regulator of hematopoietic stem cells and erythroid cells. During hematopoiesis, commitment of HSCs to the lymphoid lineage is accompanied by TAL1 silencing. However, in a significant fraction (~20%) of T cell acute lymphoblastic leukemias (T-ALL), TAL1 becomes aberrantly activated following a typical internal chromosomal deletion (leading to the SIL-TAL1 oncogenic fusion), a feature considered as a major driving oncogenic event. Interestingly, a subset of TAL1-positive T-ALL show instead small genomic insertions in the vicinity of the TAL1 oncogene. These nucleotide insertions are able to create a "de novo" enhancer sequence, allowing the recruitment of critical TFs (e.g., MYB), that aberrantly activate TAL1 and maintain its expression in lymphoid cells (Mansour et al. 2014, Navarro et al. 2015). The concept of cancer-specific neo-enhancer has been validated in other types of cancer genomes (Hu et al. 2017, Rahman et al. 2017) and represents a previously unrecognized mode of cellular transformation.

More recently, series of more complex structural alterations were shown to directly alter TAD structures (Lupianez et al. 2015, Franke et al. 2016, Weischenfeldt et al. 2017). Recurrent alterations of the *EPHA4* locus have been detected in humans suffering from severe digit malformations. The alterations were diverse including large deletions, chromosomal inversions or duplications. However, each of these alterations disrupted one of the EPHA4 TAD border in such a way that 5' or 3' borders were deleted, duplicated, or repositioned at a different place within the locus after chromosomal inversions (Lupianez et al. 2015). The common feature among all these aberrant genetic events is that a strong limb-specific distal *EPHA4* enhancer establishes ectopic interactions with genes present in the TADs surrounding EPHA4 as a result of loss or displacement of EPHA4 TAD borders (Lupianez et al. 2015). This study was the first one demonstrating that TAD integrity is lost in a human disorder, and that intact TAD borders are required to prevent gene misregulation in disease. Other chromosomal duplications affecting the neighboring *SOX9* and *KCNJ* loci in humans, and their modeling in the mouse genome, show that duplications that encompass TAD borders can reshuffle the loci and lead to the generation of new domain structures with diverse phenotypic manifestations. Some of these duplications, despite affecting the normal TAD structure of the KCNJ and SOX9 TADs show no phenotypic effects as the duplication that encompasses the border between the KCNJ and SOX9 TADs generates a novel domain, called a "neo-TAD" that is insulated from the neighboring ones (Franke et al. 2016). Chromatin regions within the neo-TAD show interactions within their own domain, but do not contact regions from the KCNJ or SOX9 TADs, thereby preventing any gene misregulation, explaining the absence of phenotypic manifestations. Thus complex genomic copy number variations (CNVs) can have both deleterious phenotypes or be completely phenotypically silent. However, integration of genome 3D spatial organization is instrumental for the interpretation of complex genome rearrangements and CNVs that are routinely detected in the clinic.

Finally, cancer genomes were shown to be enriched for variations affecting CTCF sites and to have widespread boundary disruptions (Hnisz et al. 2016), and CTCF pathological dynamics may modify TAD boundaries and lead to aberrant oncogene activation by distal enhancers as was recently been shown at the *MYC* oncogene locus in T-ALL (Kloetgen et al. 2020).

2.13 Concluding Remarks

Alterations of the enhancer landscape and in the 3D genome organization have emerged as a critical component of genetic diseases and cancer initiation and progression. Better understanding of the rules governing long-range chromatin interactions, their confinement in 3D structures such as TADs, and how the different TADs are insulated from each other's will greatly enhance our ability to understand basic principles of gene regulations and unravel novel disease mechanisms and diagnostic tools.

Take-Home Message:

- The enhancer landscape is highly dynamic and is a pivotal component of the gene regulatory networks in mammalian cells;
- Mammalian enhancers are often distally located in respect to their target genes;
- Chromatin looping regulates enhancer–gene communications (although some exceptions do exist);
- Mammalian genomes are organized in spatial compartments called TADs;
- TADs and their borders confine gene expression and prevent ectopic enhancer contacts through spatial restriction;
- TADs and their borders are disrupted in genetic disorders and in various cancers.

Acknowledgements

ES is supported by the Labex EpiGenMed program "Investissements d'avenir", reference ANR-10-LABX-12-01; the Laboratory of Excellence GR-Ex, program "Investissements d'avenir" reference ANR-11-LABX-0051 (the labex GR-Ex is funded by the of the French National Research Agency, reference ANR-18-IDEX-0001); and by the Fondation pour la Recherche Médicale, Equipe FRM DEQ20180339221). FG is supported by Netherlands Organization of Scientific Research (NWO-BBOL) grant 737.016.014.

References

Amano, T., T. Sagai, H. Tanabe, Y. Mizushina, H. Nakazawa and T. Shiroishi (2009). "Chromosomal dynamics at the Shh locus: limb bud-specific differential regulation of competence and active transcription." *Dev Cell* 16(1): 47-57.

Anderson, E. and R. E. Hill (2014). "Long range regulation of the sonic hedgehog gene." *Curr Opin Genet Dev* 27: 54-9.

Andrieu-Soler, C. and E. Soler (2020). "When basic science reaches into rational therapeutic design: from historical to novel leads for the treatment of beta-globinopathies." *Curr Opin Hematol* 27(3): 141-8.

Bailey, S. D., X. Zhang, K. Desai, M. Aid, O. Corradin, R. Cowper-Sal Lari, B. Akhtar-Zaidi, P. C. Scacheri, B. Haibe-Kains and M. Lupien (2015). "ZNF143 provides sequence specificity to secure chromatin interactions at gene promoters. " *Nat Commun* 2: 6186.

Banerji, J., L. Olson and W. Schaffner (1983). "A lymphocyte-specific cellular enhancer is located downstream of the joining region in immunoglobulin heavy chain genes." *Cell* 33(3): 729-40.

Banerji, J., S. Rusconi and W. Schaffner (1981). "Expression of a beta-globin gene is enhanced by remote SV40 DNA sequences." *Cell* 27(2 Pt 1): 299-308.

Barakat, T. S., F. Halbritter, M. Zhang, A. F. Rendeiro, E. Perenthaler, C. Bock and I. Chambers (2018). "Functional Dissection of the Enhancer Repertoire in Human Embryonic Stem Cells." *Cell Stem Cell* 23(2): 276-88 e8.

Beagrie, R. A., A. Scialdone, M. Schueler, D. C. Kraemer, M. Chotalia, S. Q. Xie, M. Barbieri, I. de Santiago, L. M. Lavitas, M. R. Branco, J. Fraser, J. Dostie, L. Game, N. Dillon, P. A. Edwards, M. Nicodemi and A. Pombo (2017). "Complex multi-enhancer contacts captured by genome architecture mapping." *Nature* 543(7646): 519-24.

Benabdallah, N. S., I. Williamson, R. S. Illingworth, L. Kane, S. Boyle, D. Sengupta, G. R. Grimes, P. Therizols and W. A. Bickmore (2019). "Decreased Enhancer-Promoter Proximity Accompanying Enhancer Activation." *Mol Cell* 76(3): 473-84 e7.

Boija, A., I. A. Klein, B. R. Sabari, A. Dall'Agnese, E. L. Coffey, A. V. Zamudio, C. H. Li, K. Shrinivas, J. C. Manteiga, N. M. Hannett, B. J. Abraham, L. K. Afeyan, Y. E. Guo, J. K. Rimel, C. B. Fant, J. Schuijers, T. I. Lee, D. J. Taatjes and R. A. Young (2018). "Transcription Factors Activate Genes through the Phase-Separation Capacity of Their Activation Domains." *Cell* 175(7): 1842-55 e16.

Bonev, B., N. Mendelson Cohen, Q. Szabo, L. Fritsch, G. L. Papadopoulos, Y. Lubling, X. Xu, X. Lv, J. P. Hugnot, A. Tanay and G. Cavalli (2017). "Multiscale 3D Genome Rewiring during Mouse Neural Development." *Cell* 171(3): 557-72 e24.

Brown, J. M., N. A. Roberts, B. Graham, D. Waithe, C. Lagerholm, J. M. Telenius, S. De Ornellas, A. M. Oudelaar, C. Scott, I. Szczerbal, C. Babbs, M. T. Kassouf, J. R. Hughes, D. R. Higgs and V. J. Buckle (2018). "A tissue-specific self-interacting chromatin domain forms independently of enhancer-promoter interactions." *Nat Commns* 9(1): 3849.

Cardozo Gizzi, A. M., D. I. Cattoni, J. B. Fiche, S. M. Espinola, J. Gurgo, O. Messina, C. Houbron, Y. Ogiyama, G. L. Papadopoulos, G. Cavalli, M. Lagha and M. Nollmann (2019). "Microscopy-Based Chromosome Conformation Capture Enables Simultaneous Visualization of Genome Organization and Transcription in Intact Organisms." *Mol Cell* 74(1): 212-22 e5.

Cho, W. K., J. H. Spille, M. Hecht, C. Lee, C. Li, V. Grube and Cisse, II (2018). "Mediator and RNA polymerase II clusters associate in transcription-dependent condensates." *Science* 361(6400): 412-5.

Cico, A., C. Andrieu-Soler and E. Soler (2016). "Enhancers and their dynamics during hematopoietic differentiation and emerging strategies for therapeutic action." *FEBS Lett* 590(22): 4084-104.

Crespo, M., A. Damont, M. Blanco, E. Lastrucci, S. E. Kennani, C. Ialy-Radio, L. E. Khattabi, S. Terrier, M. Louwagie, S. Kieffer-Jaquinod, A. M. Hesse, C. Bruley, S. Chantalat, J. Govin, F. Fenaille, C. Battail, J. Cocquet and D. Pflieger (2020). "Multi-omic analysis of gametogenesis reveals a novel signature at the promoters and distal enhancers of active genes." *Nucleic Acids Res* 48(8): 4115-38.

de Wit, E., B. A. Bouwman, Y. Zhu, P. Klous, E. Splinter, M. J. Verstegen, P. H. Krijger, N. Festuccia, E. P. Nora, M. Welling, E. Heard, N. Geijsen, R. A. Poot, I. Chambers and W. de Laat (2013). "The pluripotent genome in three dimensions is shaped around pluripotency factors." *Nature* 501(7466): 227-31.

de Wit, E., E. S. Vos, S. J. Holwerda, C. Valdes-Quezada, M. J. Verstegen, H. Teunissen, E. Splinter, P. J. Wijchers, P. H. Krijger and W. de Laat (2015). "CTCF Binding Polarity Determines Chromatin Looping." *Mol Cell* 60(4): 676-84.

Deng, W., J. Lee, H. Wang, J. Miller, A. Reik, P. D. Gregory, A. Dean and G. A. Blobel (2012). "Controlling long-range genomic interactions at a native locus by targeted tethering of a looping factor." *Cell* 149(6): 1233-44.

Deng, W., J. W. Rupon, I. Krivega, L. Breda, I. Motta, K. S. Jahn, A. Reik, P. D. Gregory, S. Rivella, A. Dean and G. A. Blobel (2014). "Reactivation of developmentally silenced globin genes by forced chromatin looping." *Cell* 158(4): 849-60.

Despang, A., R. Schopflin, M. Franke, S. Ali, I. Jerkovic, C. Paliou, W. L. Chan, B. Timmermann, L. Wittler, M. Vingron, S. Mundlos and D. M. Ibrahim (2019). "Functional dissection of the Sox9-Kcnj2 locus identifies nonessential and instructive roles of TAD architecture." *Nat Genet* 51(8): 1263-71.

Dillon, N., T. Trimborn, J. Strouboulis, P. Fraser and F. Grosveld (1997). "The effect of distance on long-range chromatin interactions." *Mol Cell* 1(1): 131-9.

Dixon, J. R., D. U. Gorkin and B. Ren (2016). "Chromatin Domains: The Unit of Chromosome Organization." *Mol Cell* 62(5): 668-80.

Dixon, J. R., I. Jung, S. Selvaraj, Y. Shen, J. E. Antosiewicz-Bourget, A. Y. Lee, Z. Ye, A. Kim, N. Rajagopal, W. Xie, Y. Diao, J. Liang, H. Zhao, V. V. Lobanenkov, J. R. Ecker, J. A. Thomson and B. Ren (2015). "Chromatin architecture reorganization during stem cell differentiation." *Nature* 518(7539): 331-6.

Dixon, J. R., S. Selvaraj, F. Yue, A. Kim, Y. Li, Y. Shen, M. Hu, J. S. Liu and B. Ren (2012). "Topological domains in mammalian genomes identified by analysis of chromatin interactions." *Nature* 485(7398): 376-80.

Drissen, R., R. J. Palstra, N. Gillemans, E. Splinter, F. Grosveld, S. Philipsen and W. de Laat (2004). "The active spatial organization of the beta-globin locus requires the transcription factor EKLF." *Genes Dev* 18(20): 2485-90.

Fanucchi, S., Y. Shibayama, S. Burd, M. S. Weinberg and M. M. Mhlanga (2013). "Chromosomal contact permits transcription between coregulated genes." *Cell* 155(3): 606-20.

Franke, M., D. M. Ibrahim, G. Andrey, W. Schwarzer, V. Heinrich, R. Schopflin, K. Kraft, R. Kempfer, I. Jerkovic, W. L. Chan, M. Spielmann, B. Timmermann, L. Wittler, I. Kurth, P. Cambiaso, O. Zuffardi, G. Houge, L. Lambie, F. Brancati, A. Pombo, M. Vingron, F. Spitz and S. Mundlos

(2016). "Formation of new chromatin domains determines pathogenicity of genomic duplications." *Nature* 538(7624): 265-9.

Fraser, P., S. Pruzina, M. Antoniou and F. Grosveld (1993). "Each hypersensitive site of the human beta-globin locus control region confers a different developmental pattern of expression on the globin genes." *Genes Dev* 7(1): 106-13.

Gao, X., K. D. Johnson, Y. I. Chang, M. E. Boyer, C. N. Dewey, J. Zhang and E. H. Bresnick (2013). "Gata2 cis-element is required for hematopoietic stem cell generation in the mammalian embryo." *J Exp Med* 210(13): 2833-42.

Ghamari, A., M. P. van de Corput, S. Thongjuea, W. A. van Cappellen, W. van Ijcken, J. van Haren, E. Soler, D. Eick, B. Lenhard and F. G. Grosveld (2013). "In vivo live imaging of RNA polymerase II transcription factories in primary cells." *Genes Dev* 27(7): 767-77.

Ghavi-Helm, Y., F. A. Klein, T. Pakozdi, L. Ciglar, D. Noordermeer, W. Huber and E. E. Furlong (2014). "Enhancer loops appear stable during development and are associated with paused polymerase." *Nature* 512(7512): 96-100.

Gillies, S. D., S. L. Morrison, V. T. Oi and S. Tonegawa (1983). "A tissue-specific transcription enhancer element is located in the major intron of a rearranged immunoglobulin heavy chain gene." *Cell* 33(3): 717-28.

Groschel, S., M. A. Sanders, R. Hoogenboezem, E. de Wit, B. A. M. Bouwman, C. Erpelinck, V. H. J. van der Velden, M. Havermans, R. Avellino, K. van Lom, E. J. Rombouts, M. van Duin, K. Dohner, H. B. Beverloo, J. E. Bradner, H. Dohner, B. Lowenberg, P. J. M. Valk, E. M. J. Bindels, W. de Laat and R. Delwel (2014). "A single oncogenic enhancer rearrangement causes concomitant EVI1 and GATA2 deregulation in leukemia." *Cell* 157(2): 369-81.

Grosveld, F., G. B. van Assendelft, D. R. Greaves and G. Kollias (1987). "Position-independent, high-level expression of the human beta-globin gene in transgenic mice." *Cell* 51(6): 975-85.

Guo, Y., Q. Xu, D. Canzio, J. Shou, J. Li, D. U. Gorkin, I. Jung, H. Wu, Y. Zhai, Y. Tang, Y. Lu, Y. Wu, Z. Jia, W. Li, M. Q. Zhang, B. Ren, A. R. Krainer, T. Maniatis and Q. Wu (2015). "CRISPR Inversion of CTCF Sites Alters Genome Topology and Enhancer/Promoter Function." *Cell* 162(4): 900-10.

Hanscombe, O., D. Whyatt, P. Fraser, N. Yannoutsos, D. Greaves, N. Dillon and F. Grosveld (1991). "Importance of globin gene order for correct developmental expression." *Genes Dev* 5(8): 1387-94.

Hardison, R. C. (2020). "A Cambrian origin for globin gene regulation." *Blood* 136(3): 261-2.

Hay, D., J. R. Hughes, C. Babbs, J. O. J. Davies, B. J. Graham, L. Hanssen, M. T. Kassouf, A. M. Marieke Oudelaar, J. A. Sharpe, M. C. Suciu, J. Telenius, R. Williams, C. Rode, P. S. Li, L. A. Pennacchio, J. A. Sloane-Stanley, H. Ayyub, S. Butler, T. Sauka-Spengler, R. J. Gibbons, A. J. H. Smith, W. G. Wood and D. R. Higgs (2016). "Genetic dissection of the alpha-globin super-enhancer in vivo." *Nat Genet* 48(8): 895-903.

Herranz, D., A. Ambesi-Impiombato, T. Palomero, S. A. Schnell, L. Belver, A. A. Wendorff, L. Xu, M. Castillo-Martin, D. Llobet-Navas, C. Cordon-Cardo, E. Clappier, J. Soulier and A. A. Ferrando (2014). "A NOTCH1-driven MYC enhancer promotes T cell development, transformation and acute lymphoblastic leukemia." *Nat Med* 20(10): 1130-7.

Hnisz, D., B. J. Abraham, T. I. Lee, A. Lau, V. Saint-Andre, A. A. Sigova, H. A. Hoke and R. A. Young (2013). "Super-enhancers in the control of cell identity and disease." *Cell* 155(4): 934-47.

Hnisz, D., K. Shrinivas, R. A. Young, A. K. Chakraborty and P. A. Sharp (2017). "A Phase Separation Model for Transcriptional Control." *Cell* 169(1): 13-23.

Hnisz, D., A. S. Weintraub, D. S. Day, A. L. Valton, R. O. Bak, C. H. Li, J. Goldmann, B. R. Lajoie, Z. P. Fan, A. A. Sigova, J. Reddy, D. Borges-Rivera, T. I. Lee, R. Jaenisch, M. H. Porteus, J. Dekker and R. A. Young (2016). "Activation of proto-oncogenes by disruption of chromosome neighborhoods." *Science* 351(6280): 1454-8.

Hu, S., M. Qian, H. Zhang, Y. Guo, J. Yang, X. Zhao, H. He, J. Lu, J. Pan, M. Chang, G. Du, T. N. Lin, S. K. Kham, T. C. Quah, H. Ariffin, A. M. Tan, Y. Cheng, C. Li, A. E. Yeoh, C. H. Pui, A. J. Skanderup and J. J. Yang (2017). "Whole-genome noncoding sequence analysis in T-cell acute lymphoblastic leukemia identifies oncogene enhancer mutations." *Blood* 129(24): 3264-8.

Jackson, D. A., A. B. Hassan, R. J. Errington and P. R. Cook (1993). "Visualization of focal sites of transcription within human nuclei." *EMBO J* 12(3): 1059-65.

Javierre, B. M., O. S. Burren, S. P. Wilder, R. Kreuzhuber, S. M. Hill, S. Sewitz, J. Cairns, S. W. Wingett, C. Varnai, M. J. Thiecke, F. Burden, S. Farrow, A. J. Cutler, K. Rehnstrom, K. Downes, L. Grassi, M. Kostadima, P. Freire-Pritchett, F. Wang, B. Consortium, H. G. Stunnenberg, J. A. Todd, D. R. Zerbino, O. Stegle, W. H. Ouwehand, M. Frontini, C. Wallace,

M. Spivakov and P. Fraser (2016). "Lineage-Specific Genome Architecture Links Enhancers and Non-coding Disease Variants to Target Gene Promoters." *Cell* 167(5): 1369-84 e19.

Jin, F., Y. Li, J. R. Dixon, S. Selvaraj, Z. Ye, A. Y. Lee, C. A. Yen, A. D. Schmitt, C. A. Espinoza and B. Ren (2013). "A high-resolution map of the three-dimensional chromatin interactome in human cells." *Nature* 503(7475): 290-4.

Johnson, K. D., G. Kong, X. Gao, Y. I. Chang, K. J. Hewitt, R. Sanalkumar, R. Prathibha, E. A. Ranheim, C. N. Dewey, J. Zhang and E. H. Bresnick (2015). "Cis-regulatory mechanisms governing stem and progenitor cell transitions." *Sci Adv* 1(8): e1500503.

Kagey, M. H., J. J. Newman, S. Bilodeau, Y. Zhan, D. A. Orlando, N. L. van Berkum, C. C. Ebmeier, J. Goossens, P. B. Rahl, S. S. Levine, D. J. Taatjes, J. Dekker and R. A. Young (2010). "Mediator and cohesin connect gene expression and chromatin architecture." *Nature* 467(7314): 430-5.

Kieffer-Kwon, K. R., Z. Tang, E. Mathe, J. Qian, M. H. Sung, G. Li, W. Resch, S. Baek, N. Pruett, L. Grontved, L. Vian, S. Nelson, H. Zare, O. Hakim, D. Reyon, A. Yamane, H. Nakahashi, A. L. Kovalchuk, J. Zou, J. K. Joung, V. Sartorelli, C. L. Wei, X. Ruan, G. L. Hager, Y. Ruan and R. Casellas (2013). "Interactome maps of mouse gene regulatory domains reveal basic principles of transcriptional regulation." *Cell* 155(7): 1507-20.

Kloetgen, A., P. Thandapani, P. Ntziachristos, Y. Ghebrechristos, S. Nomikou, C. Lazaris, X. Chen, H. Hu, S. Bakogianni, J. Wang, Y. Fu, F. Boccalatte, H. Zhong, E. Paietta, T. Trimarchi, Y. Zhu, P. Van Vlierberghe, G. G. Inghirami, T. Lionnet, I. Aifantis and A. Tsirigos (2020). "Three-dimensional chromatin landscapes in T cell acute lymphoblastic leukemia." *Nat Genet* 52(4): 388-400.

Kollias, G., N. Wrighton, J. Hurst and F. Grosveld (1986). "Regulated expression of human A gamma-, beta-, and hybrid gamma beta-globin genes in transgenic mice: manipulation of the developmental expression patterns." *Cell* 46(1): 89-94.

Krivega, I., R. K. Dale and A. Dean (2014). "Role of LDB1 in the transition from chromatin looping to transcription activation." *Genes Dev* 28(12): 1278-90.

Lara-Astiaso, D., A. Weiner, E. Lorenzo-Vivas, I. Zaretsky, D. A. Jaitin, E. David, H. Keren-Shaul, A. Mildner, D. Winter, S. Jung, N. Friedman and I. Amit (2014). "Immunogenetics. Chromatin state dynamics during blood formation." *Science* 345(6199): 943-9.

Le Poul, Y., Y. Xin, L. Ling, B. Mühling, R. Jaenichen, D. Hörl, D. Bunk, H. Harz, H. Leonhardt, Y. Wang, E. Osipova, M. Museridze, D. Dharmadhikari, E. Murphy, R. Rohs, S. Preibisch, B. Prud'homme and N. Gompel (2020). "Deciphering the regulatory logic of a Drosophila enhancer through systematic sequence mutagenesis and quantitative image analysis." *bioRxiv:* 2020.06.24.169748.

Lee, J., I. Krivega, R. K. Dale and A. Dean (2017). "The LDB1 Complex Co-opts CTCF for Erythroid Lineage-Specific Long-Range Enhancer Interactions." *Cell Rep* 19(12): 2490-502.

Lettice, L. A., T. Horikoshi, S. J. Heaney, M. J. van Baren, H. C. van der Linde, G. J. Breedveld, M. Joosse, N. Akarsu, B. A. Oostra, N. Endo, M. Shibata, M. Suzuki, E. Takahashi, T. Shinka, Y. Nakahori, D. Ayusawa, K. Nakabayashi, S. W. Scherer, P. Heutink, R. E. Hill and S. Noji (2002). "Disruption of a long-range cis-acting regulator for Shh causes preaxial polydactyly." *Proc Natl Acad Sci U S A* 99(11): 7548-53.

Lettice, L. A., I. Williamson, J. H. Wiltshire, S. Peluso, P. S. Devenney, A. E. Hill, A. Essafi, J. Hagman, R. Mort, G. Grimes, C. L. DeAngelis and R. E. Hill (2012). "Opposing functions of the ETS factor family define Shh spatial expression in limb buds and underlie polydactyly." *Dev Cell* 22(2): 459-67.

Li, L., J. Y. Lee, J. Gross, S. H. Song, A. Dean and P. E. Love (2010). "A requirement for Lim domain binding protein 1 in erythropoiesis." *J Exp Med* 207(12): 2543-50.

Love, P. E., C. Warzecha and L. Li (2014). "Ldb1 complexes: the new master regulators of erythroid gene transcription." *Trends Genet* 30(1): 1-9.

Lupianez, D. G., K. Kraft, V. Heinrich, P. Krawitz, F. Brancati, E. Klopocki, D. Horn, H. Kayserili, J. M. Opitz, R. Laxova, F. Santos-Simarro, B. Gilbert-Dussardier, L. Wittler, M. Borschiwer, S. A. Haas, M. Osterwalder, M. Franke, B. Timmermann, J. Hecht, M. Spielmann, A. Visel and S. Mundlos (2015). "Disruptions of topological chromatin domains cause pathogenic rewiring of gene-enhancer interactions." *Cell* 161(5): 1012-25.

Luppino, J. M., D. S. Park, S. C. Nguyen, Y. Lan, Z. Xu, R. Yunker and E. F. Joyce (2020). "Cohesin promotes stochastic domain intermingling to ensure proper regulation of boundary-proximal genes." *Nat Genet.*

Magram, J., K. Chada and F. Costantini (1985). "Developmental regulation of a cloned adult beta-globin gene in transgenic mice." *Nature* 315(6017): 338-40.

Mansour, M. R., B. J. Abraham, L. Anders, A. Berezovskaya, A. Gutierrez, A. D. Durbin, J. Etchin, L. Lawton, S. E. Sallan, L. B. Silverman, M. L. Loh,

S. P. Hunger, T. Sanda, R. A. Young and A. T. Look (2014). "Oncogene regulation. An oncogenic super-enhancer formed through somatic mutation of a noncoding intergenic element." *Science* 346(6215): 1373-7.

Maurano, M. T., R. Humbert, E. Rynes, R. E. Thurman, E. Haugen, H. Wang, A. P. Reynolds, R. Sandstrom, H. Qu, J. Brody, A. Shafer, F. Neri, K. Lee, T. Kutyavin, S. Stehling-Sun, A. K. Johnson, T. K. Canfield, E. Giste, M. Diegel, D. Bates, R. S. Hansen, S. Neph, P. J. Sabo, S. Heimfeld, A. Raubitschek, S. Ziegler, C. Cotsapas, N. Sotoodehnia, I. Glass, S. R. Sunyaev, R. Kaul and J. A. Stamatoyannopoulos (2012). "Systematic localization of common disease-associated variation in regulatory DNA." *Science* 337(6099): 1190-5.

Miyata, M., N. Gillemans, D. Hockman, J. A. A. Demmers, J. F. Cheng, J. Hou, M. Salminen, C. A. Fisher, S. Taylor, R. J. Gibbons, J. J. Ganis, L. I. Zon, F. Grosveld, E. Mulugeta, T. Sauka-Spengler, D. R. Higgs and S. Philipsen (2020). "An evolutionarily ancient mechanism for regulation of hemoglobin expression in vertebrate red cells." *Blood* 136(3): 269-78.

Moorthy, S. D., S. Davidson, V. M. Shchuka, G. Singh, N. Malek-Gilani, L. Langroudi, A. Martchenko, V. So, N. N. Macpherson and J. A. Mitchell (2017). "Enhancers and super-enhancers have an equivalent regulatory role in embryonic stem cells through regulation of single or multiple genes." *Genome Res* 27(2): 246-58.

Muerdter, F., L. M. Boryn and C. D. Arnold (2015). "STARR-seq - principles and applications." *Genomics* 106(3): 145-50.

Mukhopadhyay, M., A. Teufel, T. Yamashita, A. D. Agulnick, L. Chen, K. M. Downs, A. Schindler, A. Grinberg, S. P. Huang, D. Dorward and H. Westphal (2003). "Functional ablation of the mouse Ldb1 gene results in severe patterning defects during gastrulation." *Development* 130(3): 495-505.

Mullen, A. C., D. A. Orlando, J. J. Newman, J. Loven, R. M. Kumar, S. Bilodeau, J. Reddy, M. G. Guenther, R. P. DeKoter and R. A. Young (2011). "Master transcription factors determine cell-type-specific responses to TGF-beta signaling." *Cell* 147(3): 565-76.

Mylona, A., C. Andrieu-Soler, S. Thongjuea, A. Martella, E. Soler, R. Jorna, J. Hou, C. Kockx, W. van Ijcken, B. Lenhard and F. Grosveld (2013). "Genome-wide analysis shows that Ldb1 controls essential hematopoietic genes/pathways in mouse early development and reveals novel players in hematopoiesis." *Blood* 121(15): 2902-13.

Navarro, J. M., A. Touzart, L. C. Pradel, M. Loosveld, M. Koubi, R. Fenouil, S. Le Noir, M. A. Maqbool, E. Morgado, C. Gregoire, S. Jaeger, E.

Mamessier, C. Pignon, S. Hacein-Bey-Abina, B. Malissen, M. Gut, I. G. Gut, H. Dombret, E. A. Macintyre, S. J. Howe, H. B. Gaspar, A. J. Thrasher, N. Ifrah, D. Payet-Bornet, E. Duprez, J. C. Andrau, V. Asnafi and B. Nadel (2015). "Site- and allele-specific polycomb dysregulation in T-cell leukaemia." *Nat Commun* 6: 6094.

Noordermeer, D., E. de Wit, P. Klous, H. van de Werken, M. Simonis, M. Lopez-Jones, B. Eussen, A. de Klein, R. H. Singer and W. de Laat (2011). "Variegated gene expression caused by cell-specific long-range DNA interactions." *Nat Cell Biol* 13(8): 944-51.

Nora, E. P., A. Goloborodko, A. L. Valton, J. H. Gibcus, A. Uebersohn, N. Abdennur, J. Dekker, L. A. Mirny and B. G. Bruneau (2017). "Targeted Degradation of CTCF Decouples Local Insulation of Chromosome Domains from Genomic Compartmentalization." *Cell* 169(5): 930-44 e22.

Nora, E. P., B. R. Lajoie, E. G. Schulz, L. Giorgetti, I. Okamoto, N. Servant, T. Piolot, N. L. van Berkum, J. Meisig, J. Sedat, J. Gribnau, E. Barillot, N. Bluthgen, J. Dekker and E. Heard (2012). "Spatial partitioning of the regulatory landscape of the X-inactivation centre." *Nature* 485(7398): 381-5.

Northcott, P. A., C. Lee, T. Zichner, A. M. Stutz, S. Erkek, D. Kawauchi, D. J. Shih, V. Hovestadt, M. Zapatka, D. Sturm, D. T. Jones, M. Kool, M. Remke, F. M. Cavalli, S. Zuyderduyn, G. D. Bader, S. VandenBerg, L. A. Esparza, M. Ryzhova, W. Wang, A. Wittmann, S. Stark, L. Sieber, H. Seker-Cin, L. Linke, F. Kratochwil, N. Jager, I. Buchhalter, C. D. Imbusch, G. Zipprich, B. Raeder, S. Schmidt, N. Diessl, S. Wolf, S. Wiemann, B. Brors, C. Lawerenz, J. Eils, H. J. Warnatz, T. Risch, M. L. Yaspo, U. D. Weber, C. C. Bartholomae, C. von Kalle, E. Turanyi, P. Hauser, E. Sanden, A. Darabi, P. Siesjo, J. Sterba, K. Zitterbart, D. Sumerauer, P. van Sluis, R. Versteeg, R. Volckmann, J. Koster, M. U. Schuhmann, M. Ebinger, H. L. Grimes, G. W. Robinson, A. Gajjar, M. Mynarek, K. von Hoff, S. Rutkowski, T. Pietsch, W. Scheurlen, J. Felsberg, G. Reifenberger, A. E. Kulozik, A. von Deimling, O. Witt, R. Eils, R. J. Gilbertson, A. Korshunov, M. D. Taylor, P. Lichter, J. O. Korbel, R. J. Wechsler-Reya and S. M. Pfister (2014). "Enhancer hijacking activates GFI1 family oncogenes in medulloblastoma." *Nature* 511(7510): 428-34.

Ogiyama, Y., B. Schuettengruber, G. L. Papadopoulos, J. M. Chang and G. Cavalli (2018). "Polycomb-Dependent Chromatin Looping Contributes to Gene Silencing during Drosophila Development." *Mol Cell* 71(1): 73-88 e5.

Osborne, C. S., L. Chakalova, K. E. Brown, D. Carter, A. Horton, E. Debrand, B. Goyenechea, J. A. Mitchell, S. Lopes, W. Reik and P. Fraser (2004). "Active genes dynamically colocalize to shared sites of ongoing transcription." *Nat Genet* 36(10): 1065-71.

Osborne, C. S., L. Chakalova, J. A. Mitchell, A. Horton, A. L. Wood, D. J. Bolland, A. E. Corcoran and P. Fraser (2007). "Myc dynamically and preferentially relocates to a transcription factory occupied by Igh." *PLoS Biol* 5(8): e192.

Osterwalder, M., I. Barozzi, V. Tissieres, Y. Fukuda-Yuzawa, B. J. Mannion, S. Y. Afzal, E. A. Lee, Y. Zhu, I. Plajzer-Frick, C. S. Pickle, M. Kato, T. H. Garvin, Q. T. Pham, A. N. Harrington, J. A. Akiyama, V. Afzal, J. Lopez-Rios, D. E. Dickel, A. Visel and L. A. Pennacchio (2018). "Enhancer redundancy provides phenotypic robustness in mammalian development." *Nature* 554(7691): 239-43.

Ostuni, R., V. Piccolo, I. Barozzi, S. Polletti, A. Termanini, S. Bonifacio, A. Curina, E. Prosperini, S. Ghisletti and G. Natoli (2013). "Latent enhancers activated by stimulation in differentiated cells." *Cell* 152(1-2): 157-71.

Oudelaar, A. M., C. L. Harrold, L. L. P. Hanssen, J. M. Telenius, D. R. Higgs and J. R. Hughes (2019). "A revised model for promoter competition based on multi-way chromatin interactions at the alpha-globin locus." *Nat Commun* 10(1): 5412.

Palstra, R. J., B. Tolhuis, E. Splinter, R. Nijmeijer, F. Grosveld and W. de Laat (2003). "The beta-globin nuclear compartment in development and erythroid differentiation." *Nat Genet* 35(2): 190-4.

Platt, J. L., R. Salama, J. Smythies, H. Choudhry, J. O. Davies, J. R. Hughes, P. J. Ratcliffe and D. R. Mole (2016). "Capture-C reveals preformed chromatin interactions between HIF-binding sites and distant promoters." *EMBO Rep* 17(10): 1410-21.

Pomerantz, M. M., N. Ahmadiyeh, L. Jia, P. Herman, M. P. Verzi, H. Doddapaneni, C. A. Beckwith, J. A. Chan, A. Hills, M. Davis, K. Yao, S. M. Kehoe, H. J. Lenz, C. A. Haiman, C. Yan, B. E. Henderson, B. Frenkel, J. Barretina, A. Bass, J. Tabernero, J. Baselga, M. M. Regan, J. R. Manak, R. Shivdasani, G. A. Coetzee and M. L. Freedman (2009). "The 8q24 cancer risk variant rs6983267 shows long-range interaction with MYC in colorectal cancer." *Nat Genet* 41(8): 882-4.

Pradeepa, M. M., G. R. Grimes, Y. Kumar, G. Olley, G. C. Taylor, R. Schneider and W. A. Bickmore (2016). "Histone H3 globular domain acetylation identifies a new class of enhancers." *Nat Genet* 48(6): 681-6.

Queen, C. and D. Baltimore (1983). "Immunoglobulin gene transcription is activated by downstream sequence elements." *Cell* 33(3): 741-8.

Rahman, S., M. Magnussen, T. E. Leon, N. Farah, Z. Li, B. J. Abraham, K. Z. Alapi, R. J. Mitchell, T. Naughton, A. K. Fielding, A. Pizzey, S. Bustraan, C. Allen, T. Popa, K. Pike-Overzet, L. Garcia-Perez, R. E. Gale, D. C. Linch, F. J. T. Staal, R. A. Young, A. T. Look and M. R. Mansour (2017). "Activation of the LMO2 oncogene through a somatically acquired neomorphic promoter in T-cell acute lymphoblastic leukemia." *Blood* 129(24): 3221-6.

Rao, S. S., M. H. Huntley, N. C. Durand, E. K. Stamenova, I. D. Bochkov, J. T. Robinson, A. L. Sanborn, I. Machol, A. D. Omer, E. S. Lander and E. L. Aiden (2014). "A 3D map of the human genome at kilobase resolution reveals principles of chromatin looping." *Cell* 159(7): 1665-80.

Robson, M. I., A. R. Ringel and S. Mundlos (2019). " Regulatory Landscaping: How Enhancer-Promoter Communication Is Sculpted in 3D." *Mol Cell* 74(6): 1110-22.

Sabari, B. R., A. Dall'Agnese and R. A. Young (2020). "Biomolecular Condensates in the Nucleus." *Trends Biochem Sci.*

Sabari, B. R., D. Zhang, C. D. Allis and Y. Zhao (2017). " Metabolic regulation of gene expression through histone acylations." *Nat Rev Mol Cell Biol* 18(2): 90-101.

Sanyal, A., B. R. Lajoie, G. Jain and J. Dekker (2012). " The long-range interaction landscape of gene promoters." *Nature* 489(7414): 109-13.

Sato, Y., L. Hilbert, H. Oda, Y. Wan, J. M. Heddleston, T. L. Chew, V. Zaburdaev, P. Keller, T. Lionnet, N. Vastenhouw and H. Kimura (2019). " Histone H3K27 acetylation precedes active transcription during zebrafish zygotic genome activation as revealed by live-cell analysis." *Development* 146(19).

Schmidt, D., M. D. Wilson, B. Ballester, P. C. Schwalie, G. D. Brown, A. Marshall, C. Kutter, S. Watt, C. P. Martinez-Jimenez, S. Mackay, I. Talianidis, P. Flicek and D. T. Odom (2010). " Five-vertebrate ChIP-seq reveals the evolutionary dynamics of transcription factor binding." *Science* 328(5981): 1036-40.

Schoenfelder, S. and P. Fraser (2019). "Long-range enhancer-promoter contacts in gene expression control." *Nat Rev Genet* 20(8): 437-55.

Schoenfelder, S., T. Sexton, L. Chakalova, N. F. Cope, A. Horton, S. Andrews, S. Kurukuti, J. A. Mitchell, D. Umlauf, D. S. Dimitrova, C. H. Eskiw, Y. Luo, C. L. Wei, Y. Ruan, J. J. Bieker and P. Fraser

(2010). "Preferential associations between co-regulated genes reveal a transcriptional interactome in erythroid cells." *Nat Genet* 42(1): 53-61.

Shrinivas, K., B. R. Sabari, E. L. Coffey, I. A. Klein, A. Boija, A. V. Zamudio, J. Schuijers, N. M. Hannett, P. A. Sharp, R. A. Young and A. K. Chakraborty (2019). "Enhancer Features that Drive Formation of Transcriptional Condensates." *Mol Cell* 75(3): 549-61 e7.

Snow, J. W., J. J. Trowbridge, T. Fujiwara, N. E. Emambokus, J. A. Grass, S. H. Orkin and E. H. Bresnick (2010). "A single cis element maintains repression of the key developmental regulator Gata2." *PLoS Gene* 6(9): e1001103.

Soler, E., C. Andrieu-Soler, E. de Boer, J. C. Bryne, S. Thongjuea, R. Stadhouders, R. J. Palstra, M. Stevens, C. Kockx, W. van Ijcken, J. Hou, C. Steinhoff, E. Rijkers, B. Lenhard and F. Grosveld (2010). "The genome-wide dynamics of the binding of Ldb1 complexes during erythroid differentiation." *Genes Dev* 24(3): 277-89.

Soucie, E. L., Z. Weng, L. Geirsdottir, K. Molawi, J. Maurizio, R. Fenouil, N. Mossadegh-Keller, G. Gimenez, L. VanHille, M. Beniazza, J. Favret, C. Berruyer, P. Perrin, N. Hacohen, J. C. Andrau, P. Ferrier, P. Dubreuil, A. Sidow and M. H. Sieweke (2016). "Lineage-specific enhancers activate self-renewal genes in macrophages and embryonic stem cells." *Science* 351(6274): aad5510.

Stadhouders, R., S. Aktuna, S. Thongjuea, A. Aghajanirefah, F. Pourfarzad, W. van Ijcken, B. Lenhard, H. Rooks, S. Best, S. Menzel, F. Grosveld, S. L. Thein and E. Soler (2014). "HBS1L-MYB intergenic variants modulate fetal hemoglobin via long-range MYB enhancers." *J Clin Invest* 124(4): 1699-710.

Stadhouders, R., A. Cico, T. Stephen, S. Thongjuea, P. Kolovos, H. I. Baymaz, X. Yu, J. Demmers, K. Bezstarosti, A. Maas, V. Barroca, C. Kockx, Z. Ozgur, W. van Ijcken, M. L. Arcangeli, C. Andrieu-Soler, B. Lenhard, F. Grosveld and E. Soler (2015). "Control of developmentally primed erythroid genes by combinatorial co-repressor actions." *Nat Commun* 6: 8893.

Stadhouders, R., S. Thongjuea, C. Andrieu-Soler, R. J. Palstra, J. C. Bryne, A. van den Heuvel, M. Stevens, E. de Boer, C. Kockx, A. van der Sloot, M. van den Hout, W. van Ijcken, D. Eick, B. Lenhard, F. Grosveld and E. Soler (2012). "Dynamic long-range chromatin interactions control Myb proto-oncogene transcription during erythroid development." *EMBO J* 31(4): 986-99.

Tan, M., H. Luo, S. Lee, F. Jin, J. S. Yang, E. Montellier, T. Buchou, Z. Cheng, S. Rousseaux, N. Rajagopal, Z. Lu, Z. Ye, Q. Zhu, J. Wysocka, Y. Ye, S. Khochbin, B. Ren and Y. Zhao (2011). "Identification of 67 histone marks and histone lysine crotonylation as a new type of histone modification." *Cell* 146(6): 1016-28.

Tolhuis, B., R. J. Palstra, E. Splinter, F. Grosveld and W. de Laat (2002). "Looping and interaction between hypersensitive sites in the active beta-globin locus." *Mol Cell* 10(6): 1453-65.

Trompouki, E., T. V. Bowman, L. N. Lawton, Z. P. Fan, D. C. Wu, A. DiBiase, C. S. Martin, J. N. Cech, A. K. Sessa, J. L. Leblanc, P. Li, E. M. Durand, C. Mosimann, G. C. Heffner, G. Q. Daley, R. F. Paulson, R. A. Young and L. I. Zon (2011). "Lineage regulators direct BMP and Wnt pathways to cell-specific programs during differentiation and regeneration." *Cell* 147(3): 577-89.

Trudel, M. and F. Costantini (1987). "A 3' enhancer contributes to the stage-specific expression of the human beta-globin gene." *Genes Dev* 1(9): 954-61.

Tuupanen, S., M. Turunen, R. Lehtonen, O. Hallikas, S. Vanharanta, T. Kivioja, M. Bjorklund, G. Wei, J. Yan, I. Niittymaki, J. P. Mecklin, H. Jarvinen, A. Ristimaki, M. Di-Bernardo, P. East, L. Carvajal-Carmona, R. S. Houlston, I. Tomlinson, K. Palin, E. Ukkonen, A. Karhu, J. Taipale and L. A. Aaltonen (2009). "The common colorectal cancer predisposition SNP rs6983267 at chromosome 8q24 confers potential to enhanced Wnt signaling." *Nat Genet* 41(8): 885-90.

Ulirsch, J. C., S. K. Nandakumar, L. Wang, F. C. Giani, X. Zhang, P. Rogov, A. Melnikov, P. McDonel, R. Do, T. S. Mikkelsen and V. G. Sankaran (2016). "Systematic Functional Dissection of Common Genetic Variation Affecting Red Blood Cell Traits." *Cell* 165(6): 1530-45.

van Arensbergen, J., L. Pagie, V. D. FitzPatrick, M. de Haas, M. P. Baltissen, F. Comoglio, R. H. van der Weide, H. Teunissen, U. Vosa, L. Franke, E. de Wit, M. Vermeulen, H. J. Bussemaker and B. van Steensel (2019). "High-throughput identification of human SNPs affecting regulatory element activity." *Nat Genet* 51(7): 1160-9.

van den Heuvel, A., R. Stadhouders, C. Andrieu-Soler, F. Grosveld and E. Soler (2015). "Long-range gene regulation and novel therapeutic applications." *Blood* 125(10): 1521-5.

Vieux-Rochas, M., P. J. Fabre, M. Leleu, D. Duboule and D. Noordermeer (2015). "Clustering of mammalian Hox genes with other H3K27me3 targets within an active nuclear domain." *Proc Natl Acad Sci U S A* 112(15): 4672-7.

Villar, D., C. Berthelot, S. Aldridge, T. F. Rayner, M. Lukk, M. Pignatelli, T. J. Park, R. Deaville, J. T. Erichsen, A. J. Jasinska, J. M. Turner, M. F. Bertelsen, E. P. Murchison, P. Flicek and D. T. Odom (2015). "Enhancer evolution across 20 mammalian species." *Cell* 160(3): 554-66.

Wansink, D. G., W. Schul, I. van der Kraan, B. van Steensel, R. van Driel and L. de Jong (1993). "Fluorescent labeling of nascent RNA reveals transcription by RNA polymerase II in domains scattered throughout the nucleus." *J Cell Biol* 122(2): 283-93.

Weintraub, A. S., C. H. Li, A. V. Zamudio, A. A. Sigova, N. M. Hannett, D. S. Day, B. J. Abraham, M. A. Cohen, B. Nabet, D. L. Buckley, Y. E. Guo, D. Hnisz, R. Jaenisch, J. E. Bradner, N. S. Gray and R. A. Young (2017). "YY1 Is a Structural Regulator of Enhancer-Promoter Loops." *Cell* 171(7): 1573-88 e28.

Weischenfeldt, J., T. Dubash, A. P. Drainas, B. R. Mardin, Y. Chen, A. M. Stutz, S. M. Waszak, G. Bosco, A. R. Halvorsen, B. Raeder, T. Efthymiopoulos, S. Erkek, C. Siegl, H. Brenner, O. T. Brustugun, S. M. Dieter, P. A. Northcott, I. Petersen, S. M. Pfister, M. Schneider, S. K. Solberg, E. Thunissen, W. Weichert, T. Zichner, R. Thomas, M. Peifer, A. Helland, C. R. Ball, M. Jechlinger, R. Sotillo, H. Glimm and J. O. Korbel (2017). "Pan-cancer analysis of somatic copy-number alterations implicates IRS4 and IGF2 in enhancer hijacking." *Nat Genet* 49(1): 65-74.

Whyte, W. A., D. A. Orlando, D. Hnisz, B. J. Abraham, C. Y. Lin, M. H. Kagey, P. B. Rahl, T. I. Lee and R. A. Young (2013). "Master transcription factors and mediator establish super-enhancers at key cell identity genes." *Cell* 153(2): 307-19.

Wijgerde, M., F. Grosveld and P. Fraser (1995). "Transcription complex stability and chromatin dynamics in vivo." *Nature* 377(6546): 209-13.

Williamson, I., L. Kane, P. S. Devenney, I. M. Flyamer, E. Anderson, F. Kilanowski, R. E. Hill, W. A. Bickmore and L. A. Lettice (2019). "Developmentally regulated Shh expression is robust to TAD perturbations." *Development* 146(19).

Xu, M. and P. R. Cook (2008). "Similar active genes cluster in specialized transcription factories." *J Cell Biol* 181(4): 615-23.

Yamazaki, H., M. Suzuki, A. Otsuki, R. Shimizu, E. H. Bresnick, J. D. Engel and M. Yamamoto (2014). "A remote GATA2 hematopoietic enhancer drives leukemogenesis in inv(3)(q21;q26) by activating EVI1 expression." *Cancer Cell* 25(4): 415-27.

Zabidi, M. A., C. D. Arnold, K. Schernhuber, M. Pagani, M. Rath, O. Frank and A. Stark (2015). "Enhancer-core-promoter specificity separates developmental and housekeeping gene regulation." *Nature* 518(7540): 556-9.

Zuin, J., J. R. Dixon, M. I. van der Reijden, Z. Ye, P. Kolovos, R. W. Brouwer, M. P. van de Corput, H. J. van de Werken, T. A. Knoch, I. W. F. van, F. G. Grosveld, B. Ren and K. S. Wendt (2014). "Cohesin and CTCF differentially affect chromatin architecture and gene expression in human cells." *Proc Natl Acad Sci U S A* 111(3): 996-1001.

3

Repair of DNA Double-Strand Breaks in Adult Stem Cells

Leyla Vahidi Ferdousi[‡], Haser Hasan Sutcu[‡#] and Miria Ricchetti*

Institut Pasteur, Team Stability of Nuclear and Mitochondrial DNA, Stem Cells and Development, Dept. of Developmental and Stem Cell Biology UMR3738 CNRS, 25, rue du Dr. Roux, 75724 Paris Cedex 15, France
[‡]Equal contribution
E-mail: miria.ricchetti@pasteur.fr
*Corresponding Author
[#]Present address: Institut de Radioprotection et de Sureté Nuclèaire (IRSN), Radiobiology of Accidental Exposure Laboratory (PSE-SANTE/ SERAMED/LRAcc), B.P. 17, F-92262 Fontenay-aux-Roses Cedex, France

3.1 Introduction

Living cells are challenged by DNA damage of exogenous and endogenous origin, which threaten cell survival and cell function if not repaired efficiently and accurately. Double-strand breaks (DSBs) are among the most fatal DNA lesions a cell can incur. DNA damage is especially dangerous for adult stem cells (ASCs), which are responsible for tissue homeostasis and regeneration in the adult. Indeed, in case of unrepaired or improperly repaired DSBs the fate of the stem cell progeny is at risk as well as the maintenance of the stem cell pool, which is indispensable throughout life, and relies on stem cell renewal.

Recent investigations indicate that in general ASCs repair DSBs more efficiently, and in some cases more accurately than their derived cells, suggesting that ASCs may have developed particularly performing mechanisms to maintain genome stability. However, ASCs seem to rely on the same repair mechanisms than other cells and their efficiency has not been mechanistically elucidated. In several cases, ASCs associate powerful DSB repair with reduced sensitivity to apoptosis.

We discuss here, the response to DNA damage of ASCs compared to engaged or more differentiated cells, including the possible implication of this process in tissue regeneration, ageing, and cancer development.

3.2 DNA Damage and Repair Mechanisms

DNA damage of exogenous or endogenous origin affects all cell types during the lifetime of the organism. In order to survive and function, cells should repair the DNA damage within the timing and under conditions that are compatible with the activity of the cell.

We distinguish single-strand breaks (SSBs), which consist in base or nucleotide alterations and affect one DNA strand whereas the second strand remains intact and serves as a template for repair, and double-strand breaks (DSBs) which consist in the rupture of both strands of the double-helix, thereby interrupting the chromosome continuity, and possibly leading to loss of information at the broken site. Importantly, multiple types of DNA damage can simultaneously be present, for instance genotoxic agents as X- and γ-irradiations (IR) generate both SSBs and DSBs, and DSBs may result from proximal SSBs.

According to the type of DNA alteration (base alkylation or oxidation, UV-induced thymidine dimers, formation of abasic sites, mispair, etc), SSBs are repaired by distinct repair pathways. SSB repair includes base excision repair (BER), nucleotide excision repair (NER), and mismatch repair (MMR), which require distinct protein machineries (Caldecott, 2008).

Here, we focus on DSBs, which are produced by endogenous processes, as DNA replication, reactive oxygen species (ROS) generated by cellular metabolism in particular mitochondrial respiration, or exogenous agents (e.g., irradiation, antitumor drugs). In addition to accidental DSBs, in certain cells DSBs are induced by specific nucleases to promote DNA rearrangements and genome reshuffling to increase global or *locus*-specific genomic variability. This is the case for V(D)J recombination in B- and T-lymphocytes, class-switch recombination, and meiosis (Alt et al., 2013; Bolcun-Filas & Schimenti, 2012). Accidental and programmed DSBs, of either endogenous or exogenous origin, are repaired by two major mechanisms, homologous recombination (HR) and non-homologous end-joining (NHEJ), and to a minor extent by alternative versions of these ones

(Chapman et al., 2012; Ciccia & Elledge, 2010). All repair mechanisms can potentially generate mutations (Ensminger & Löbrich, 2020; Malkova & Haber, 2012).

Differences in the maintenance of genome stability in different cells rely on the ability to repair DSBs as well as the load of DNA damage they are submitted to. Moreover, a variety of competing, supportive or mutually exclusive repair processes may operate simultaneously.

3.2.1 Homologous Recombination and Related Repair Mechanisms

HR requires the presence of an undamaged, homologous, and possibly prox-imally located copy of the broken DNA, normally the sister chromatid. Therefore, this repair pathway is prevalent during S and G_2/M phases of the cell cycle (Mao et al., 2008). HR also repairs DSBs induced during meiosis. Commitment to HR is initiated by 5'–3' resection of the broken ends by exonucleases or helicases coupled to endonucleases, with resected ends eventually invading the intact homologous strand. Prevention of end-resection inhibits HR and leads to the competing repair mechanism, NHEJ (see below). Therefore, interplay of distinct repair proteins at this stage (e.g., BRCA1 that promotes HR, and 53BP1 that promotes NHEJ) plays a major role in choice of the repair pathway (Bunting et al., 2010; Jachimowicz et al., 2019).

Initiation of end resection, normally requires the MRN complex (com-posed of the MRE11, RAD50, and NBS1 proteins), and the CtBP-interacting protein (CtIP) (Huertas, 2010). Extensive resection is completed by Exo1, which also recruits the coating replication protein A (RPA) on the newly formed single-strand DNA. RPA is then replaced by RAD51 in a BRCA2-dependent manner to form a nucleoprotein which starts homology search and strand invasion of the homologous region, generally on the sister chromatid. The invaded strand then serves as a template for DNA synthesis, and this step is followed by the physical exchange between chromosomes.

Although HR is considered an accurate repair process, as DNA synthesis is rather faithful, it can generate translocations if repair takes place on a homologous region located elsewhere than on the sister chromatid or the homologous chromosome (i.e., on a repeated sequence) (Guirouilh-Barbat et al., 2014). Moreover, in the presence of two alleles, repair may result in the duplicate of the donor (template) allele and loss of the recipient allele (broken end). This phenomenon of gene conversion results in loss of heterozygosity, which is associated with the occurrence of some cancers (Chapman et al., 2012; Pedersen & De, 2013).

Finally, other homology-directed repair mechanisms have been reported, which are associated with genome rearrangements (Maher et al., 2011; Malkova & Haber, 2012). These include single-strand annealing (SSA) and break-induced replication (BIR), which are initiated by resection of the broken DNA ends. In SSA, distinct homologous sequences (>30 nucleotides) located at different positions on complementary DNA strands become exposed after resection of the complementary strand, and are subsequently annealed generating deletion of the intervening sequence. The size of deletion can vary, as resection proceeds until a complementary region is found, that is able to synapse the DNA ends. The flanking sites are resected by endonucleases, followed by ligation and DNA synthesis performed by a DNA polymerase to fill the gap (Bhargava et al., 2016). Consequently, SSA more likely occurs at DSBs located in regions carrying repeated sequences (such as tandem repeats, interspersed repetitive DNA, etc.) (Currall et al., 2013).

In BIR, only one broken end is engaged in invasion of and recombination with the intact sequence. This mechanism, identified and dissected in yeast (Lydeard et al., 2010), has been evoked for non-reciprocal translocations, deletions, and copy number variations observed in human diseases and cancers (Hastings et al., 2009). BIR is a HR-related mechanism that induces mutations and affects genomic stability. As other HR pathways, BIR initiates with end-resection but only one (single-stranded) broken-end invades the template sequence forming a migrating D-loop (or replication bubble, as extended DNA synthesis occurs within the loop). In other HR pathways, both broken ends invade the template sequence, generating the Holliday junction, a structure that requires to be resolved to finalize the repair event. In BIR, DNA synthesis on the leading strand extends on the ssDNA. The lagging strand is synthesized through Okazaki fragments, as during DNA replication, although *via* an asynchronous mechanism that may lead to genomic instabilities. These events include misincorporations on the newly synthesized strand, or invasion (and then copy) on an homologous region located elsewhere in the genome (Kramara et al., 2018).

3.2.2 Classical and Alternative Non-Homologous End-Joining

NHEJ is the prevalent DSB repair pathway in mammalian cells. It takes place during the entire cell cycle, and consists in the joining of two DNA ends with minor or no modifications at the junction site, largely depending on the nature of the DNA ends. NHEJ can rejoin non-proximal DNA ends thereby generating chromosome translocations, and is therefore, a powerful mutagenic mechanism as well as a genome reshuffling tool. In this context, in the lymphoid tissue, NHEJ specifically repairs nuclease-induced DSBs during V(D)J recombination for diversification of the immunoglobulin and T-cell receptor repertoires (Alt et al., 2013). NHEJ is also responsible for evolutionary relevant but also potentially mutagenic events like the insertion of exogenous DNA sequences (e.g., viral DNA, transposons, mitochondrial DNA) (Daya et al., 2009; Ricchetti et al., 1999; Suzuki et al., 2009).

Classical NHEJ (C-NHEJ) requires the Ku complex (Ku70/Ku80 heterodimer) which binds DNA ends thereby preventing end-resection and inhibiting HR, and recruits DNA-dependent protein kinase catalytic subunit (DNA-PKcs) forming the activated DNA-PK complex, which is implicated in establishing end-synapsis (Chapman et al., 2012; Ciccia & Elledge, 2010), (Figure 3.1). If necessary, DNA ends are processed by specific enzymes resulting in loss or gain of nucleotides at the junction site. In addition to nucleases, Artemis opens the hairpins formed during V(D)J recombination and possibly other DNA ends, DNA polymerases λ and μ (Polμ) synthezise DNA. Interestingly, Polμ participates in NHEJ independently from its DNA synthesis activity, probably to reinforce end-synapsis (Chayot et al., 2012). Ligation of juxtaposed ends requires DNA ligase IV and its XRCC4 cofactor, together with Cernunnos/XLF.

The complexity of the NHEJ apparatus, and to some extent the relevance of this repair mechanism, increases with organism complexity (see next section). In this context a novel factor, PAXX, has been recently shown to operate in NHEJ in vertebrates. PAXX interacts with Ku proteins and stabilizes repair factors at the DSB site during the repair process. Although its exact role is still not clear, PAXX has a stimulatory effect on the gap filling activity of DNA polymerase λ during end-processing, and holds the role of bridging broken ends along with its paralogues XLF and XRCC4 (Craxton et al., 2018). Although essential for variability during V(D)J recombination, PAXX is suggested to be an accessory protein, as in the absence of PAXX alone DSB repair is not strongly ablated unless Cernunnos/XLF is also missing (Liu et al., 2017; Tadi et al., 2016). This is also the case for other NHEJ factors, which absence does not prevent this repair process to occur, although it results in a less efficient process that impacts on cell fitness and cell fate, including for stem cells (Chayot et al., 2010a; Lucas et al., 2009).

If there is no need of end-processing, C-NHEJ can join unaltered ends and thereby accurately repair the DNA break (Betermier et al., 2014). However,

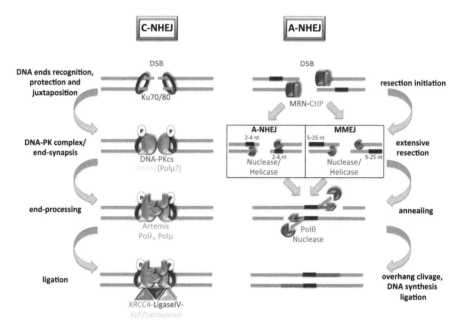

Figure 3.1   Schematic representation of classical and alternative NHEJ.

DSB repair by C-NHEJ (*left*) and A-NHEJ (*right*) are schematically represented. DNA ends (**blue traits**) are involved in successive interaction with C-NHEJ or A-NHEJ proteins, represented with different shapes, and named below. See text for details. The repair steps are described on the side. C-NHEJ results in accurate repair (sequence as before the DSB) if DNA ends were not modified during the repair process. Minor modifications may result from nucleotide(s) addition or loss by the action of DNA polymerases or nucleases, respectively. A-NHEJ is shown in parallel with MMEJ, the two mechanisms sharing several phases and relying on pairing of short, unrelated homologous sequences (**black traits**) of 2-4 nucleotides in A-NHEJ *versus* 5-25 nucleotides for MMEJ. Both mechanisms result in sequence loss, and therefore inaccurate repair.

the necessity of processing unligatable ends, and the possibility of joining unrelated DNA breaks generate C-NHEJ-dependent errors.

According to the prevalent view, when core C-NHEJ is impaired because of deficiency of one of its components, the error-prone and slower alternative-NHEJ (A-NHEJ, also called Alt-NHEJ) becomes operative (Dueva & Iliakis, 2013; Nussenzweig & Nussenzweig, 2007). A-NHEJ is initiated by poly (ADP-ribose) polymerase (PARP) activity that competes with Ku for binding DNA ends to direct repair through A-NHEJ rather than C-NHEJ (Wang et al., 2006). If A-NHEJ is engaged, DNA ends are processed by the MRN complex

(Rass et al., 2009) and the resection activity of nuclease CtIP. This step is followed by recruitment of the WRN helicase, DNA polymerase beta (Pol ß) (Ray et al., 2018), and polymerase theta (Pol θ) (Mateos-Gomez et al., 2017a). Pol θ is particularly important for A-NHEJ, for its interaction with 3' overhanging strand of the DNA and because of its interstrand activity (this DNA polymerase has also a helicase-like domain) (Black et al., 2019). Pol θ also plays an important role in stimulating microhomology-mediated end-joining (MMEJ, see next paragraph) of DSBs generated from replication stress (Wang et al., 2019) possibly counteracting the binding of the protein RPA with DNA, which normally stimulates HR (Mateos-Gomez et al., 2017b). Finally, A-NHEJ repair is completed by the scaffold protein XRCC1 and ligase I or ligase III (Soni et al., 2014) (Goodarzi & Jeggo, 2013) (Figure 3.1). Interestingly, repair proteins involved in A-NHEJ are also key players in other repair pathways, for instance HR (CtIP, WRN), BER and NER (XRCC1, ligase III), (McVey & Lee, 2008), and MMEJ (Pol θ, as mentioned above) (Figure 3.1).

A-NHEJ mainly relies on microhomologous sequences (2-4 nucleotides) at DNA breaks to promote ligation, and is therefore associated with mutations at the junction site. A-NHEJ has been also implicated in translocations associated with human blood malignancies, detected when C-NHEJ is not operative (Corneo et al., 2007; Yan et al., 2007). However, at least some V(D)J recombination events in B cells due to A-NHEJ seem to occur in the presence of functional C-NHEJ.

A special type of A-NHEJ is microhomology-mediated end-joining (MMEJ), which joins DNA ends through annealing of distant homologous (5-25 nucleotides) sequences exposed after 5'–3' resection of the broken ends (Figure 3.1). Because of the resection phase, which seems dependent on the MRN complex as well as CtIP, and BRCA1 (Chen et al., 2008), MMEJ has common features with SSA and can occur in the presence of functional NHEJ or HR. Since in A-NHEJ, DNA resection is necessary to ensure the annealing of broken ends, and this resection can be very extensive, A-NHEJ is necessarily error-prone, whereas this is not necessarily the case for C-NHEJ and HR (Betermier et al., 2014; Taty-Taty et al., 2016).

3.2.3 Context of Repair Processes

The repair machineries operate in the context of complex signaling to coordinate the repair process with cell survival and cell cycling (Goodarzi & Jeggo, 2013). The signaling pathways involved in the DNA damage response

(DDR), and generally orchestrated by protein p53 (Reinhardt & Schumacher, 2012), imply sensors that recognize the DNA damage, signal transducers that amplify the signal, mediators including those that activate DNA damage checkpoints at cell cycle transitions (G_1/S, intra-S, G_2/M), and effectors that actually repair the DNA damage. Engagement in DSB repair is associated with changes in the chromatin structure, primarily phosphorylation of histone H2AX. Enumeration of DSB markers like phosphorylated γH2AX and 53BP1 has been employed as one of the finest tools to measure the kinetics of DSB repair (Noon et al., 2010).

Differently from complex eukaryotes, prokaryotes essentially use HR to repair DSBs, although a few bacteria perform a rudimentary NHEJ (Aniukwu et al., 2008). Moreover, *Escherichia coli,* which do not possess NHEJ-related proteins, perform an alternative version of end-joining, which shares some mechanistic features with mammalian A-NHEJ (Chayot et al., 2010b). In the yeast *Saccharomyces cerevisiae*, HR is the predominant repair mechanism, in particular in haploid strains, although also NHEJ operates (Paques & Haber, 1999). Finally, the number of NHEJ components and the prevalence of this repair mechanism increases in organisms with larger genomes, including mammals (Lieber, 2010). It is tempting to speculate that increase in genome size is also linked to the end-joining-dependent ability to integrate, and eventually maintain, exogenous and unrelated sequences in the genome.

3.3 Cell Response to DSBs

Repairing DSBs is essential for the organism survival. Inefficient DSB repair is indeed associated with a variety of diseases and cancers (Ciccia & Elledge, 2010; O'Driscoll, 2012). In humans, impairment of NHEJ factors generally results in immunodeficiency due to the role of this repair mechanism in generating the diversity of B and T cells, and neurodegeneration due to the prevalent role of this repair mechanism in post-mitotic cells as neurons. Some NHEJ and HR factors are essential for cell function and their deletion results in embryonic lethality. There is increasing evidence that impaired DSB repair also impacts on ageing and tissue regeneration (Garinis et al., 2008; Vijg & Suh, 2013), largely due to defective maintenance of genome stability in adult stem cells (see below).

Cells react to DSBs with a variety of (generally p53-dependent) responses which are consequential for the cell fate (Reinhardt & Schumacher, 2012).

If DSBs are repaired efficiently and accurately, the original genomic information is preserved and the cell is expected to resume its original activity and function (Figure 3.2). Cells with misrepaired DSBs that generate point mutations may also resume the original activity if the mutation does not affect the gene function. However, large genome rearrangements or specific mutations which are incompatible with normal activity may result in proliferative and metabolic impairment or, alternatively, unregulated growth, which may lead to oncogenesis. This last outcome is of particular relevance for adult stem cells, which display characteristic of cancer stem cells (CSCs), namely self-renewal, differentiation potential, and high proliferation rate once activated. CSCs have been postulated at the origin of some cancers (Reya et al., 2001). Indeed, cancer can be considered as a malignant tissue with at its origin cells with aberrant self-renewal and differentiation capacities. These stem-like characteristics of CSCs may also be responsible for the occurrence of cancer cells with different genetic, epigenetic, metabolic, and functional characteristics that contribute to the heterogeneity of cancer tissues. Indeed, CSCs may produce progenitor cells of different differentiation levels. CSCs are considered to play at different levels in tumorigenesis: cancer initiation, cancer progression, metastatic growth, and importantly for the clinics, therapeutic resistance (Lytle et al., 2018). Indeed, quiescent and slow dividing "reserve" CSCs appear resistant to conventional therapies (as it is the case for the remarkable radio-resistance of some glioblastomas (Bao et al., 2006)) that target vigorously dividing cancer cells. Therapeutic developments to control CSCs underscore the need of understanding the origin of CSCs and their link with tissue-specific ASCs.

In the case of non-repaired DSBs, cells may undergo transient p21-mediated cell-cycle arrest, thereby favouring the repair of the DNA lesion before the cell resumes division, or p16- and p19-mediated senescence, an irreversible proliferative arrest that is believed to prevent oncogenic transformation while maintaining the cell alive, or else programmed cell death (apoptosis, through p53 targets like *Bax*, *Puma*, *Noxa*) which is also considered avoiding oncogenic transformation in cells carrying unrepaired DNA damage. In differentiated cells, these events generally result in the exhaustion of the cell carrying unrepaired DNA damage, whereas in ASCs the entire progeny risks to be affected. Moreover, exhaustion or reduction of the stem cell pool limits the regeneration of the tissue, and ultimately promotes ageing (Kenyon & Gerson, 2007; Ruzankina et al., 2008). In ASCs, DNA damage may also promote differentiation, which again results in the impoverishment of the stem cell reservoir and finally promotes ageing.

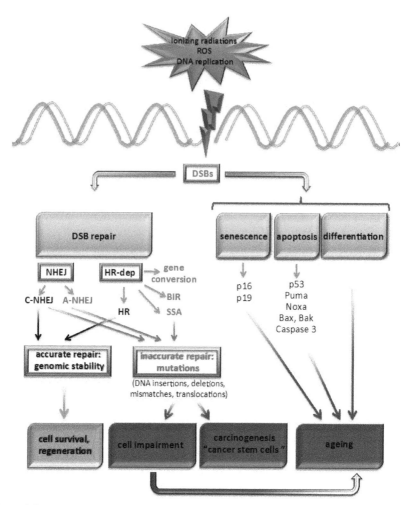

Figure 3.2 Schematic representation of DNA damage response in general and in adult stem cells.

Consequences of DSBs generated by exogenous or endogenous events. DSB repair by either NHEJ or HR-dependent (HR-dep) mechanisms can be accurate (**black arrows**) or inaccurate (**red arrows**). Accurate repair mechanisms are indicated in black, inaccurate repair mechanisms in red. Most consequences to DSBs are common to any cell, but some events are specific to ASCs. Accurate repair results in cell survival and function, which in case of ASCs promotes efficient tissue regeneration. Inaccurate repair generates mutations, which may result in cell impairment, and in case of ASCs lead to ageing for the reduction of the stem cell pool, or carcinogenesis because of the proliferation potential of these cells. Failure to repair DSBs results in ASC cell senescence, apoptosis, or unscheduled differentiation (for ASCs); all these processes ultimately lead to the reduction of the stem cells pool and thereby ageing.

Thus, the fine regulation of DDR is essential for the fate and survival of any cell, but the fitness of the entire organism depends on the DDR outcome of adult stem cells.

3.4 Stem Cells Resistance to Genotoxic stress

Cells and tissues have distinct sensitivities to DNA damage. Irradiation (IR) is the prevalent experimental system to induce DSBs, which is also used as localized anti-cancer treatment. The intensity of IR can be regulated and the extent of DNA lesions can be reasonably estimated (about 25–40 DSBs/Gy, in addition to a variety of single-stranded DNA lesions (Costes et al., 2010)). Highly proliferating tissues as the hematopoietic system and the intestine are generally highly radiosensitive. Acute radiation syndrome in humans appears at total body irradiation doses of 2–7.5 Gy in the hematopoietic system, 5.5 Gy in the gastrointestinal system, and >20 Gy in the neurovascular system, indicating that the hematopoietic system is the most radiosensitive tissue in the body (Shao et al., 2014). Tissue-specific sensitivity to IR depends on the response of stem and differentiated cells to DNA damage. Radioresistance of stem cells is a relevant factor in the clinics, as these cells are responsible for repopulation of the healthy tissue after radiotherapy.

The impact of DDR on ASCs has been investigated in multiple tissue and organismal contexts, despite limitations in isolating homogeneous populations of these cells and in large amounts as required for biochemical and molecular biology assessments. We describe below some of the most relevant responses.

3.5 Efficiency and Mechanisms of DSB Repair in Adult Stem Cells

ASCs are classed below according to the main tissues in the adult: epithelium, connective tissue, muscle, and nervous tissue. Germinal stem cells are considered separately. Induced pluripotent stem cells (iPSCs) are also reported, in comparison to somatic cells from which they derive.

3.5.1 Cells in the Epithelial Tissue: Epidermal, Mammary, and Intestinal Stem Cells

Multipotent bulge stem cells (BSCs), of epidermal origin, ensure hair follicle homeostasis and epidermal repair after wounding. These cells repair

IR-induced DSBs faster than non-bulge epidermal cells. Indeed, BSCs display more rapid disappearance of DSB markers (53BP1, γ-H2AX) and DNA damage (tested by semi-quantitative comet assay), as well as more efficient *in vitro* joining of DSB substrates (Sotiropoulou et al., 2010). BSCs carry larger nuclear amounts of the key NHEJ factor DNA-PKcs than non-bulge cells, and DNA-PKcs-impaired SCID mice fail to repair DSBs as efficiently as wild type mice, indicating that NHEJ plays a key role in repair of DSBs in BSCs. Interestingly, the expression of NHEJ as well as HR genes is not significantly different in BSCs compared to non-bulge cells. Fast NHEJ repair results in attenuated activation of p53 leading to reduced signals for apoptosis in BSCs than in non-bulge stem cells (see next section).

Upon activation during hair follicle regeneration, but not during quiescence, BSCs survival and differentiation depend on the HR factor BRCA1 (Sotiropoulou et al., 2013), as expected for proliferating cells. However, this is not the case for resident stem cells responsible for homeostasis of interfollicular epidermis and sebaceous glands. The reason for different requirement of BRCA1 in epidermis stem cells remains unknown, although distinct buffering of ROS-produced DNA damage or the occurrence of compensating repair mechanisms has been evoked.

Different responses to DNA damage in various types of skin ASCs were also demonstrated upon exposure to low doses of irradiation (LDI) that mimic levels emitted during medical investigations and interventions. BSCs of mice submitted to 50 mGy whole-body irradiation at weekly intervals for 10 weeks displayed comparable levels of DNA damage than sebaceous glands (SG) stem cells present in the same pilosebaceous unit. However, SGs underwent apoptosis whereas BSCs survived upon activation of a metabolic shift induced by the hypoxia-inducible factor 1α (HIF-1α) (Revenco et al., 2017) The author demonstrated that survival to LDI of epidermal stem cells already carrying a mutation results in cancer formation.

BRCA1 is also necessary for the differentiation of human mammary stem cells (MaSCs), as conditional knock-out of *Brca1* in these cells results in defects of mammary epithelial differentiation (Hakem, 2008). Irradiated mouse MaSCs that reside in the basal compartment of the mammary gland display less γ-H2AX foci than non-progenitor cells already 2 hours post-IR, indicating faster DSB repair in the stem cells (Woodward et al., 2007). MaSCs also display more NHEJ activity *ex vivo*, and higher levels of the repair protein 53BP1 (Chang et al., 2015)

The protective role of replication-dependent DSB repair, HR, is underscored in the small intestine, where two types of stem cells have been identified: cells at the bottom of the crypt (crypt base columnar cells, CBCs) are proliferating, thereby HR is operating, and radioresistant, whereas cells around the +4 position (intestinal stem cells, ISCs) are quiescent and radiosensitive (Hua et al., 2012; Li & Clevers, 2010). Upon irradiation, intestinal regeneration is promoted by quiescent ISCs that enhance NHEJ due to up-regulation of Ligase IV, and depending on Wnt signaling (Jun et al., 2016). However, high levels of Wnt would promote death of cycling ISCs (Blanpain et al., 2011; Metcalfe et al., 2014; Vitale et al., 2017). Although the dynamics and the heterogeneity of these cell populations keep revealing new details (Ritsma et al., 2014; Tajbakhsh, 2014), the response of stem cells to DNA damage can be distinct depending on their origin, cell cycle status, or both. CBCs cells repair DSBs more efficiently than transit amplifying (TA) daughter cells, and both proliferating cells repair more efficiently than differentiated gastrointestinal cells, as indicated by the kinetics of disappearance of γ-H2AX foci post-IR (Hua et al., 2012). HR appears more efficient in proliferating CBCs than in TA cells, as shown by the kinetic of disappearance of BRCA1 and RAD51 foci post-IR, whereas differentiated villus cells do not rely on these HR factors, as expected for non-proliferating cells. NHEJ is also essential for DSB repair in the intestine since SCID mice display radiosensitization of the gastrointestinal tract. Indeed, NHEJ, assessed by resolution of DNA-PKcs foci, appears slightly more efficient in CBCs than in TA cells whereas, surprisingly, villus cells fail to show these DSB markers. Thus, proliferating CBCs (and to a minor extent TA cells) display robust HR and classical-NHEJ repair pathways, whereas the poor repair efficiency of differentiated cells seem to be due to the lack of both mechanisms. It is unclear whether alternative, error-prone, NHEJ or other mechanisms are responsible for the residual DSB repair in these cells.

It should be noted that the response to DNA damage is not only cell-autonomous but may also depend on exogenous factors. A clear example is regulation of the DDR response and DSB repair in intestinal epithelial stem cells, i.e., ISCs, by IL22 (interleukin 22), a cytokine primarily produced by innate lymphoid cells (ILC3) (Gronke et al., 2019). These cells are primarily found in the gastrointestinal tract, and IL22 production is stimulated by the aryl hydrocarbon receptor (AhR), a key sensor that allows immune cells to adapt to the environment. AhR is stimulated by glucosinolate metabolites, and in mice, a diet free of glucosinolate metabolites negatively affects DDR (Gronke et al., 2019). IL22-deficient mouse epithelial stem cells fail to show signs of DDR, for instance, the DNA damage-dependent increase of *ATM* expression is ablated. Consequently, the ATM-dependent activation of proteins such as p53 and SMC (structural maintenance of chromosome), a protein that activates the S-phase checkpoint in response to irradiation, is lost. Furthermore, upon irradiation in the absence of IL22, the cell cycle arrest gene *p21* and the proapoptotic gene *Puma* are not induced in these cells. Altogether, these results show that IL22 is essential for the response to DNA damage and DNA repair in epithelial cells. As a consequence, in the absence of IL22, epithelial cells are impaired in repairing DSBs and undergoing apoptosis, providing a possible explanation for development of colon cancer (Gronke et al., 2019).

3.5.2 Cells in the Connective Tissue: Hematopoietic and Mesenchymal Stem Cells

Despite the hematopoietic tissue is very radiosensitive, highly enriched hematopoietic stem/progenitor cells (HSPCs) repair radiation-induced DSBs more efficiently than committed common myeloid (CMP) and granulo-cyte/macrophage (GMP) progenitors. Repair efficiency was demonstrated by faster kinetics of disappearance of DSB markers and residual DNA damage in HSPCs than in CMPs and GMPs post-IR (Mohrin et al., 2010). These data are in agreement with more ancient reports showing that radiore-sistance decreases during stem cell to progenitors transition (Ploemacher et al., 1992). Moreover, NHEJ factors are relatively more expressed in HSPCs than in progenitors upon irradiation. Quiescence by itself does not provide DNA damage protection, as *in vivo* and *in vitro* activated (proliferating) HSPCs display similar reduction in colony formation follow-ing IR than resting cells. As expected, resting HSPCs rely more on the NHEJ repair mechanism than proliferating HSPCs. Interestingly, NHEJ in quiescent HSPCs generated a dramatically high number of chromosome aberrations, which were not observed in proliferating cells. Mutations per-sisted even after a secondary transplantation of affected cells, but failed to result in leukemia or haematological abnormalities in engrafted mice (Mohrin et al., 2010).

Conversely, human umbilical cord blood-derived HSPCs delay DSB repair compared to progenitors (Milyavsky et al., 2010). The different response of mouse adult *versus* human cord-derived HSPCs may be species-specific or depend on the developmental stage. In this context, it has been evoked that during embryogenesis and until birth, the stem cell population expands, whereas in the adult it preserves blood homeostasis (Blanpain et al., 2011).

The role of NHEJ in the maintenance of mouse HSPCs is underscored by the observation of reduced stem cells pool in the absence of Ku80 (Rossi et al., 2007) and defective Ligase IV (Nijnik et al., 2007). Moreover, deficiency of the DSB repair signaling protein ATM results in depletion of HSPCs and impairment of their proliferation and self-renewal, which were initially attributed to increased ROS production in these cells (Ito et al., 2004). Notably, deficiency of the HR factor BRCA2 also leads to proliferation and self-renewal defects in mouse HSPCs (Navarro et al., 2006).

Importantly, the cytokine thrombopoietin (TPO) increases the efficiency of DNA-PKcs-dependent repair, "accurate" NHEJ, in HSPCs (de Laval et al., 2013b). TPO-dependent DNA-PK activation is regulated through direct interaction with the ERK signaling pathway factor pERK (de Laval et al., 2013a). These findings represent a further example of environmental signals (cytokines) affecting the DDR.

The capacity of HSPCs of repairing individual DSBs, as well as the possibility of experimentally directing the repair mechanism, have been recently assessed in the context of single or multiple DSBs induced by zinc finger and CRISPR/Cas9 nucleases (Schiroli et al., 2019). The aim of the editing process was to trigger HR in the presence of an exogenous donor sequence provided *ad hoc*, to generate a specifically designed sequence as a result of the repair process. This task is difficult to achieve since HSPCs, purified from human cord blood and cultured *ex vivo*, display low levels of the HR machinery as a consequence of being essentially quiescent. These cells nevertheless repaired one or more induced DSBs using HR thereby generating the desired sequence. Nuclease-induced DSB repair stimulated p53-dependent pathways that delayed cell proliferation, as a response to DNA damage, and constrained the yield and engraftments of edited HSPCs. These restrictions were lifted upon transient p53 inhibition, providing a possible strategy for accurate gene editing in human HSPCs. This process was operative in the context of a limited number of DSBs, since delayed cell proliferation could not be counteracted in the presence of massive DNA damage.

Whereas HSPCs in the bone marrow keep self-renewing and give rise to all types of hematopoietic cells, T lymphocytes in the thymus do not self-renew, and rely on bone marrow-derived progenitors that migrate through the blood and, once in the thymus, proliferate and differentiate into matured T cells. However, after a strong dose of IR that destroys HSPCs, the thymus auto-reconstitutes in a single wave due to proliferation and differentiation of highly radioresistant T-cell precursors (DN2 thymocytes). The reason for radioresistance of DN2 thymocytes has been recently identified in the rapid activation of the DDR, efficient repair of IR-induced DNA damage, and also induction of a protective G1/S checkpoint (Calvo-Asensio et al., 2018). Initially, the reason of the strong DDR of DN2 thymocytes appeared due to an efficient DNA repair machinery activated by V(D)J recombination. Indeed, developing T cells undergo V(D)J recombination to increase variability of the antigen receptor (TCR), and this process starts during the transition that leads to DN2 formation. Efficient NHEJ was indeed reported in DN2 thymocytes. However, V(D)J recombination occurs also in more advanced stages of T-cell development and these other cells are radiosensitive, indicating that additional reasons should be evoked to explain the elevated capacity of DN2 thymocytes to repair DNA damage.

Mesenchymal stem/stromal cells (MSCs), which are heterogeneous and are responsible for the support and regulation of HSPCs, notably play a role in the hematopoietic reconstitution after transplantation in irradiated bone marrow. Thus, MSCs play a critical role in the maintenance of HSPC niche, proliferation, and self-renewal (Dickhut et al., 2005), and are considered for therapeutic perspectives. MSCs are more radioresistant than HSPCs. Assessment of the repair mechanism in MSCs has been limited by the availability of only heterogeneous cultures, which, however, display robust NHEJ repair potential compared to more differentiated cells (Oliver et al., 2013) and also other stem cells. The response of MSCs to DNA damage has been recently reviewed (Banimohamad-Shotorbani et al., 2020).

3.5.3 Skeletal Muscle Stem Cells

Skeletal muscle stem cells, or satellite cells (SCs), have robust regenerative potential, and are quiescent in the adult unless activated for homeostasis or in response to muscle injury (Boldrin et al., 2012; Gross & Morgan, 1999). After a muscle injury, SCs enter the cell cycle and produce myoblasts which fuse to effect muscle regeneration (Tajbakhsh, 2009). In myoblasts, DNA damage inhibits MyoD-dependent differentiation activity through c-Abl-dependent cell-cycle arrest, allowing DNA repair before the onset of differentiation (Puri et al., 2002). The myogenic factor MyoD has been therefore proposed as DNA damage sensor.

Deficiency of the NHEJ factor Ku80 results in an accumulated DNA damage and SCs depletion, as well as alteration of myoblasts proliferation and differentiation, leading to precocious ageing (Didier et al., 2012). Interestingly, loss of a single Ku80 allele results in accelerated ageing, although without accumulation of DNA damage, but essentially by telomere shortening. Indeed, Ku80 also localizes to telomeres, where it participates to maintain telomere stability (Kazda et al., 2012).

During ageing, the muscle displays reduced regenerative potential after injury. However, repair of endogenous DNA damage in the proliferating myoblasts appears efficient in old as well as young mice (Cousin et al., 2013). Moreover, expression of key NHEJ and HR factors does not seem altered in SCs of young compared to that of old mice. The authors concluded that decreased regeneration potential is not accompanied by accumulation of DNA damage during ageing, thereby questioning the notion that DSBs, and perhaps DNA damage in general, play a key role in stem cell ageing (Garinis et al., 2008), at least in the muscle. However, these results are consistent with robust repair efficiency in myoblasts also during ageing, at least for endogenous DNA damage produced during DNA replication. This result is also consistent with the observation that although the number of SCs declines with age, the remaining SCs maintain their intrinsic capacity to regenerate muscle as efficiently as cells from younger mice (Collins et al., 2007).

Moreover, irradiation of quiescent SCs in young mice results in more extensive myogenic colony formation than old mice, indicating reduced repair activity of induced DNA damage in older cells. Conversely, irradiation of proliferating myoblasts does not display significant differences between young and old mice suggesting that, differently from NHEJ, proliferation-dependent HR repair is quite active in myoblasts in old as well as young mice (Cousin et al., 2013). NHEJ-impaired SCID mice display reduced myogenic colony formation compared to wild-type mice upon irradiation, as expected, suggesting that DDR plays a role in the expansion of myoblasts in culture. Strikingly, despite reduced myoblasts expansion, SCID mice are capable of robust muscle regeneration, leading the authors to conclude that repair of DSBs is not a key factor for muscle regeneration. However, SCID mice have been shown to improve muscle regeneration due to diminution of pro-inflammatory macrophages (Farini et al., 2012), independently of DNA damage repair, and the immune response has been shown to play a critical role in muscle regeneration (Saclier et al., 2013).

Thus, to clarify the role of DSB repair in muscle regeneration, DSB repair deficiency should be tested in the absence of immunological impairment. In this context, we reported that SCs repair IR-induced DSBs more efficiently than the committed progeny (Vahidi Ferdousi et al., 2014). Indeed, mouse SCs repair more efficiently than activated myoblasts and differentiated myotubes, with a decreasing gradient of repair efficiency toward more differentiated cells. This characteristic has been shown during differentiation *in vitro* and *in vivo*, in isolated SCs as well as SCs in the myofiber, suggesting that the high SC efficiency of DSB repair is cell autonomous. For instance, in cross-sections of the *tibialis anterior* muscle isolated at different time points after irradiation *in vivo*, SCs and differentiated myonuclei displayed a comparable number of unrepaired DSBs (evaluated by individual γ-H2AX foci) 30 minutes post-IR, but 90 minutes later the former had dramatically decreased the number of foci (i.e., repaired DSBs) compared to the latter.

Interestingly, myoblasts that are actively proliferating and therefore may account on active HR in addition to NHEJ, actually repair DSBs less efficiently than quiescent, non-dividing SCs. Moreover, post-mitotic myotubes repair much less efficiently than quiescent SCs (that are not cycling although they can re-enter the cell cycle upon activation), indicating that in this paradigm the cell cycle status affects the efficiency of IR-induced DSBs much less than the differentiation status. Importantly, SCs repair more efficiently than their committed progeny also after multiple cycles of irradiation, indicating that SCs may retain myogenic capacity also upon radiotherapy. This finding is relevant since in numerous cancers muscle regeneration around the tumor is important for recovery after radiotherapy. We also reported that SCs as well as proliferating myoblasts display low levels of NHEJ repair factors, and non-cycling SCs express also HR factors. Importantly, irradiation *per se* did not promote significant changes in the transcription of NHEJ factors, suggesting that SCs are basically equipped with an efficient DSB repair machinery.

Mechanistically, our study showed that repair in SCs largely depends on remarkably accurate C-NHEJ (Vahidi Ferdousi et al., 2014). The occurrence of C-NHEJ was assessed due to the involvement of DNA-PKcs. DSB repair in SCs that was monitored on exogenous episomes was extremely accurate, with an efficiency about 20-fold higher than standard MEFs (mouse embryonic fibroblasts) tested in the same set of experiments or in other contexts (for instance in (Chayot et al., 2012)). The reason by which SCs efficiently and accurately repair DSBs has not been elucidated yet.

To note, SCs do not undergo significant apoptosis after induced genotoxic stress. More than 90% SCs globally survive to high doses [20 Gy] of whole-body IR, whereas two-thirds of other muscle cells do not. Surviving SCs globally retain their differentiation potential. Thus, skeletal muscle stem cells display robust and accurate repair capacity of IR-induced DNA damage, and cells that cannot cope with these lesions may undergo senescence but do not elicit a strong apoptotic response. As for the other cases of ASCs discussed here, the entity and the type of DDR depends on the extent of DNA damage.

3.5.4 Neural Stem Cells

Neural stem cells (NSCs) generate glial cells like astrocytes, the predominant cells in mammalian brain, and also neurons. NSCs display robust NHEJ response following irradiation, observed by activation of ATM, and by 53BP1 and γ-H2AX proteins (Schneider et al., 2012). Intriguingly, the derived terminally differentiated astrocytes stably repress transcription of key NHEJ genes, thereby downregulating the DDR. It is suspected that astrocytes do so to avoid apoptosis upon genotoxic stress. This inhibition is not directly associated with terminal differentiation as post-mitotic neurons display detectable DDR.

Experiments conducted in the neural subventricular zone (SVZ) that provides neurogenesis in the adult, reveal that quiescent NSCs (qNSCs) have different response to DSBs than the downstream progenitors. Indeed, the latter undergo proliferation arrest, premature differentiation, and apoptosis upon IR, whereas the former do not activate apoptosis (Barazzuol et al., 2017). Consequently, upon IR damaged progenitor cells fail to give rise to progeny and conversely activate qNSCs that subsequently generate post-IR neuroblasts. It is though that this response serves to remove damaged progenitors. These processes are controlled in an ATM-dependent manner. Accordingly, *Atm*-deficient mice, do not activate these responses in the neural lineage, thereby qNSCs are not activated, and cells become remarkably radiosensitive. Other responses like cell senescence, permanent ATR-dependent checkpoint arrest, as well as other forms of cell death (*e.g.* necrosis) constitute alternative forms of cell loss that may contribute to the failure of qNSCs activation in $Atm^{-/-}$ mice.

As a comparison, embryonic neural stem and progenitor cells (NSPCs) display a more efficient NHEJ and thereby higher genome stability than MEFs. Although these are embryonic and not ASCs, it is interesting to note that during long-term culture NSPCs maintain higher DSB repair capacity and genome stability compared to MEFs. Robust maintenance of genome stability was associated with efficient apoptotic elimination of cells with damage (Mokrani et al., 2020). It remains to be assessed whether NSCs maintain the same capacity of preservation of genome stability in the adult.

3.5.5 Germinal Stem Cells

Quiescent spermatogonial stem cells (SSCs) are more radioresistant than proliferating and differentiated progeny (Moreno et al., 2001). Importantly, SSCs display different chromatin organization from somatic cells, lacking compacted heterochromatin. As a result, DSB repair is independent of the γ-H2AX signaling (Rube et al., 2011). Moreover, differently from differentiated cells and most tested ASCs, DSB repair is predominantly DNA-PKcs-independent, suggesting that male germ cells rely on alternative forms of end-joining to restore genome integrity. In agreement with other stem cells, however, ATM deficiency and accumulation of DNA damage result in loss of SSCs and self-renewal capacity, and finally in infertility in mice (Takubo et al., 2008). A hypomorphic mutation of RAD50 also leads to p53-dependent loss of spermatogonia (Bender et al., 2002), while RAD51 mutation renders female gonadal stem cells (GSCs) sensitive to cell death (Kujjo et al., 2010).

3.5.6 Induced Pluripotent Stem Cells

Induced pluripotent stem cells (iPSCs) are generated by genetically reprogramming and de-differentiating somatic cells into an embryonic-like pluripotent state, and are mostly generated from skin fibroblasts or peripheral blood mononuclear cells (PBMCs). As for other stem cells, genomic stability and maintenance is vital also for iPSCs which, however, may undergo additional genotoxic-producing processes than other stem cells (Yoshihara et al., 2017). Consequences are higher for iPSCs as they can contribute to all tissues (in animal models, e.g., mice), or be in turn differentiated into a variety of cells that can be potentially used for investigations and therapeutic interventions. These underlying processes include genetic alterations present in the original somatic cells from which they are generated, as well as the process of genetic re-programming itself. Moreover, the efficiency of DDR may also vary depending on the type and age of the cell they are derived from or the method of re-programming (Fan et al., 2011; Nagaria et al., 2016). Components of HR, NHEJ, and some components of other repair pathways are upregulated upon generation of iPSCs compared to the parental cells (Fan et al., 2011). Interestingly, iPSCs derived from human MSCs repair DSBs less efficiently but more accurately than their parental cells and rather share these characteristics with human embryonic stem cells (hESCs). Nevertheless iPSCs display a higher resistance to apoptosis than hESCs (Fan et al., 2011).

Of particular interest is the above mentioned dependence of DNA repair efficiency and DDR according to the age of the somatic cell donor. Multiple factors may be involved in this process. For instance, the expression of the deacetylase Sirt6, which plays a role in multiple biological processes and impacts on longevity as well as maintenance of DNA integrity, declines with aging (Tian et al., 2019). This process may negatively affect the NHEJ repair efficiency of iPSCs. Indeed, direct interaction of Sirt6 with Ku80 stimulates the interaction of the latter with DNA-PKcs, forming the DNA-PK complex (Ku70-Ku80-DNA-PKcs), triggering auto-phosphorylation of the kinase and thereby efficient NHEJ. The presence of higher levels of DNA damage in cells from aged organisms may result from less efficient NHEJ, in turn depending on reduced levels of Sirt6. This condition would translate into higher levels of DNA damage in somatic cell-derived iPSCs. Accordingly, iPSCs derived from old mice carry more DNA damage than those derived from young mice, and the low efficiency of NHEJ is improved upon induced expression of Sirt6 (Chen et al., 2017). Moreover, human-derived iPSCs with prolonged *in vitro* culturing and high number of passages display lower DSB repair efficiency, assessed by declined detection of DSB factors, compared to low passage iPSCs and hESCs (Simara et al., 2017).

3.6 Other Responses to DNA Damage

3.6.1 Apoptosis

We have seen in the previous paragraphs that inefficiently repaired DSBs may trigger different responses that ultimately decide the cell fate. We have also seen that several ASCs are relatively resistant to apoptosis (compared to more differentiated cells under the same genotoxic insult). Quiescent hair follicle bulge stem cells display more robust DSB repair, and are also more apoptosis-resistant than non-bulge cells upon IR (Sotiropoulou et al., 2010). Resistance to apoptosis is largely due to high expression of the anti-apoptotic factor *Bcl2*. This seems to be also the case for an enriched population of human keratinocyte stem cells, which is more resistant to 2 Gy IR-induced cell death than enriched population of progenitors, due to high expression of anti-apoptotic *Bcl2* and fast DSB repair (Rachidi et al., 2007). It should be noted that strong resistance to apoptosis is likely a successful strategy if accompanied, as in the case of bulge stem cells, by efficient DSB repair. Conversely, enhanced survival to apoptosis in cells with poor DNA repair capacity would likely accumulate potentially dangerous (stem) cells with genomic instability.

HSPCs also rely on more robust anti-apoptotic response than myeloid progenitors, which is mediated by lower expression of pro-apoptotic genes and higher expression of anti-apoptotic genes, although not *Bcl2* (Mohrin et al., 2010). In contrast, human umbilical cord blood-derived HSPCs undergo increased apoptosis, which could be reversed by high expression of anti-apoptotic *Bcl2*, and p53 inactivation (Milyavsky et al., 2010). Intriguingly, survival of mouse HSPCs after IR is limited by *Puma*, the p53 upregulated mediator of apoptosis, and deletion of *Puma* results in higher resistance to IR without increased malignancy in exposed animals (Yu et al., 2010).

Quiescent, radiosensitive ISCs at the +4 location undergo robust apoptosis in response to irradiation, i.e., at low radiation doses ($\leq$ 1 Gy) essentially all ISCs undergo rapid apoptotic death within 3–6 hours (Hua et al., 2012). Conversely, proliferating and radioresistant crypt base columnar cells in the small intestine require 12 Gy IR to undergo apoptosis, despite high pro-apoptotic *Puma* levels. This is not the case in the crypt of the large intestine, where columnar epithelial stem cells display strong resistance to IR-induced apoptosis up to 30 Gy (Hua et al., 2017). Under these conditions, cell death depends on the proliferative status of the cell, with CBCs in interphase dying by apoptosis, and CBCs in re-initiation of the cell cycle by mitotic death (representing about 1/3 and 2/3 of the CBC population, respectively).

In differentiated astrocytes, the expression of pro-apoptotic genes (*Puma*, *Bax*) increases weakly after high-dose irradiation (i.e., 20 Gy), resulting in high resistance to apoptosis, whereas *Puma* is strongly expressed in neural stem cells, leading to distinct apoptosis at 10 Gy (Schneider et al., 2012). Thus, differently from other stem cells, NSCs are more sensitive to apoptosis than the differentiated progeny, although in absolute values NSCs are more resistant to IR-induced apoptosis than stem cells from other tissues (e.g., hematopoietic or intestine). However, at low IR doses NSCs elicit a different response than progenitors. Indeed, already 6 hours post-IR (2 Gy) about half of the neural progenitor population undergoes apoptosis, with a peak at 48 hours post-IR. Conversely, under these conditions NSCs, regardless to their cell cycle state (quiescent or active), display resistance to apoptosis (Barazzuol et al., 2017).

IPSCs are hypersensitive to low doses of irradiation (1–2 Gy) and show high levels of apoptosis within 24 hours (Nagaria et al., 2016), which is not the case for their respective parental cells (Fan et al., 2011).

3.6.2 Senescence

Cellular senescence is revealed by senescence-associated beta-galactosidase activity, and increased expression of p16$^{\text{INK4a}}$ and p19^{ARF} (Reinhardt & Schumacher, 2012). Upon IR, quiescent hair follicle bulge stem cells, which display fast DSB repair and reduced apoptosis, do not senesce (Sotiropoulou et al., 2010), and this is also the case for melanocyte stem cells (MeSCs), which do not activate the p16$^{\text{INK4a}}$ or p19^{ARF}-dependent response after

5 Gy IR (Inomata et al., 2009). In the bone marrow, no senescence has been observed in wild type or $Mpl^{-/-}$ (Mpl is the TPO receptor) HSPCs. (de Laval et al., 2013b). Only a minor fraction of irradiated muscle satellite cells undergo senescence upon irradiation (Vahidi Ferdousi et al., 2014). Conversely, human MSCs have high risk to undergo senescence after several days in culture post-IR (Wang & Jang, 2009). Thus, to date cellular senescence does not seem relevant in response to DNA damage for most ASCs.

3.6.3 Differentiation

In addition to repair, apoptosis, and senescence, stem cells can also respond to DNA damage by undergoing differentiation. Whereas the differentiation ability of quiescent hair follicle bulge stem cells is not affected upon irradiation, damaged MeSCs, which have not efficiently repaired the DNA damage, differentiate prematurely without self-renewing, leading to loss of MeSCs and thereby pigmented melanocytes, and finally hair depigmentation (Inomata et al., 2009). IR-induced melanocyte differentiation has also been observed *in vitro*, but to a lesser extent than *in vivo*, suggesting that the niche, which is implicated in MeSC self-renewal, is also determinant for DNA damage-related differentiation. Premature hair graying, a sign of premature aging, has been observed after IR. Moreover, MeSCs are lost in an age-dependent manner (Nishimura et al., 2005). It was shown that ATM acts as a "stemness checkpoint," and is implicated in protecting MeSCs from differentiation by activating DDR. Indeed, *ATM* deficiency leads to hair graying due to IR-mediated MeSC differentiation, proposing a role for ATM in the maintenance of stemness and self-renewal, and in MeSC fate determination (Hakem, 2008; Inomata et al., 2009). Thus, premature differentiation is a relevant outcome for impaired DDR in MeSCs. This is not the case for other stem cells of the epithelial tissue, as differentiation ability upon IR is not altered in enriched populations of human keratinocyte stem cells, which keep proliferating (Tiberio et al., 2002) as the progenitors do (Harfouche et al., 2010).

Neuronal stem cell differentiation can be regulated by IR. Indeed, exposure to IR accelerates astrocyte differentiation of surviving NSCs (Ozeki et al., 2012). However, other studies report that IR inhibits NSC differentiation *in vitro* and *in vivo* (Kanzawa et al., 2006; Monje et al., 2002). Differentiation was observed in response to DNA damage in hematopoietic stem cells (Wang et al., 2012) but not in muscle stem cells. (Vahidi Ferdousi et al., 2014).

Mesoangioblasts are multipotent perivascular stem/progenitor cells that differentiate into multiple mesodermal cell types (although their myogenic role is still controversial). Mesoangioblasts can undergo osteogenic differentiation forming osteoblasts triggered by BMP2 (Bone Morphogenic Protein 2). An alternative, BMP2-independent, non-canonical way to trigger osteogenesis is through incorporation of the nucleotide analogue, idoxuridine (IdU) (used as antiviral drug) into the mesoangioblast DNA, which induces a mild DDR response. Mesoangioblasts display a mild increase of histone ATM-dependent H2AX phosphorylation. ATM also activates p38, which possibly delays the G2/M phase of the cell cycle, and also stimulates the osteogenic regulator RUNX2 (Runt-related transcription factor 2), which is essential to osteoblast differentiation (Rosina et al., 2019).

3.7 Concluding Remarks

- ASCs generally, but not exclusively, display more robust DSB repair ability and apoptosis resistance than their respective progeny.
- The pathways leading to DNA damage repair and apoptosis resistance start to be unveiled but mechanistic information is missing (different repair pathways, or a different combination of repair pathways in distinct ASC?).
- Only in limited cases the response of ASCs to DNA damage leads to senescence and precocious differentiation.
- The efficiency of DSB repair and resistance to apoptosis greatly vary among stem cells from different tissues and is possibly responsible for the distinct radioresistance and survival properties of these tissues.
- The duration of quiescence, the length of cell cycle, the number of cell divisions, the occurrence of distinct differentiation pathways, and the extent of endogenous metabolism may be responsible for the different resistance of ASC types to genotoxic stress/radiations.
- ASCs' response to genotoxic stress is relevant for diseases, including cancer for its possible origin through CSCs.
- ASCs' response to genotoxic stress is relevant for cancer treatment since repopulation of the healthy tissues after radiotherapy occurs through ASCs, and treatment-resistant cells may have properties of stem cells.
- The link between survival and maintenance of genome stability in ASCs is a key factor in tissue homeostasis and tissue regeneration, which have implications for the organism fitness and ageing.
- Developing novel approaches for the maintenance of genome stability in ASCs may select more robust stem cell populations with implications for cell therapies for myopathies and other tissue regeneration deficiencies.

References

Alt FW, Zhang Y, Meng FL, Guo C, Schwer B (2013) Mechanisms of programmed DNA lesions and genomic instability in the immune system. *Cell* 152: 417-429

Aniukwu J, Glickman MS, Shuman S (2008) The pathways and outcomes of mycobacterial NHEJ depend on the structure of the broken DNA ends. *Genes & development* 22: 512-527

Banimohamad-Shotorbani B, Kahroba H, Sadeghzadeh H, Wilson Iii DM, Maadi H, Samadi N, Hejazi MS, Farajpour H, Onari BN, Sadeghi MR (2020) DNA damage repair response in mesenchymal stromal cells: From cellular senescence and aging to apoptosis and differentiation ability. *Ageing research reviews* 62: 101125

Bao S, Wu Q, McLendon RE, Hao Y, Shi Q, Hjelmeland AB, Dewhirst MW, Bigner DD, Rich JN (2006) Glioma stem cells promote radioresistance by preferential activation of the DNA damage response. *Nature* 444: 756-760

Barazzuol L, Ju L, Jeggo PA (2017) A coordinated DNA damage response promotes adult quiescent neural stem cell activation. *PLoS biology* 15: e2001264

Bender CF, Sikes ML, Sullivan R, Huye LE, Le Beau MM, Roth DB, Mirzoeva OK, Oltz EM, Petrini JH (2002) Cancer predisposition and hematopoietic failure in Rad50(S/S) mice. *Genes & development* 16: 2237-2251

Betermier M, Bertrand P, Lopez BS (2014) Is non-homologous end-joining really an inherently error-prone process? *PLoS genetics* 10: e1004086

Bhargava R, Onyango DO, Stark JM (2016) Regulation of Single-Strand Annealing and its Role in Genome Maintenance. *Trends in genetics : TIG* 32: 566-575

Black SJ, Ozdemir AY, Kashkina E, Kent T, Rusanov T, Ristic D, Shin Y, Suma A, Hoang T, Chandramouly G, Siddique LA, Borisonnik N, Sullivan-Reed K, Mallon JS, Skorski T, Carnevale V, Murakami KS, Wyman C, Pomerantz RT (2019) Molecular basis of microhomology-mediated end-joining by purified full-length Polθ. *Nature communications* 10: 4423

Blanpain C, Mohrin M, Sotiropoulou PA, Passegue E (2011) DNA-damage response in tissue-specific and cancer stem cells. *Cell stem cell* 8: 16-29

Bolcun-Filas E, Schimenti JC (2012) Genetics of meiosis and recombination in mice. *International review of cell and molecular biology* 298: 179-227

Boldrin L, Neal A, Zammit PS, Muntoni F, Morgan JE (2012) Donor satellite cell engraftment is significantly augmented when the host niche is preserved and endogenous satellite cells are incapacitated. *Stem cells* 30: 1971-1984

Bunting SF, Callen E, Wong N, Chen HT, Polato F, Gunn A, Bothmer A, Feldhahn N, Fernandez-Capetillo O, Cao L, Xu X, Deng CX, Finkel T, Nussenzweig M, Stark JM, Nussenzweig A (2010) 53BP1 inhibits homologous recombination in Brca1-deficient cells by blocking resection of DNA breaks. *Cell* 141: 243-254

Caldecott KW (2008) Single-strand break repair and genetic disease. *Nature reviews Genetics* 9: 619-631

Calvo-Asensio I, Sugrue T, Bosco N, Rolink A, Ceredig R (2018) DN2 Thymocytes Activate a Specific Robust DNA Damage Response to Ionizing Radiation-Induced DNA Double-Strand Breaks. *Frontiers in immunology* 9: 1312

Chang CH, Zhang M, Rajapakshe K, Coarfa C, Edwards D, Huang S, Rosen JM (2015) Mammary Stem Cells and Tumor-Initiating Cells Are More Resistant to Apoptosis and Exhibit Increased DNA Repair Activity in Response to DNA Damage. *Stem cell reports* 5: 378-391

Chapman JR, Taylor MR, Boulton SJ (2012) Playing the end game: DNA double-strand break repair pathway choice. *Molecular cell* 47: 497-510

Chayot R, Danckaert A, Montagne B, Ricchetti M (2010a) Lack of DNA polymerase μ affects the kinetics of DNA double-strand break repair and impacts on cellular senescence. *DNA repair* 9: 1187-1199

Chayot R, Montagne B, Mazel D, Ricchetti M (2010b) An end-joining repair mechanism in Eschcrichia coli. *Proceedings of the National Academy of Sciences of the United States of America* 107: 2141-2146

Chayot R, Montagne B, Ricchetti M (2012) DNA polymerase mu is a global player in the repair of non-homologous end-joining substrates. *DNA repair* 11: 22-34

Chen L, Nievera CJ, Lee AY, Wu X (2008) Cell cycle-dependent complex formation of BRCA1.CtIP.MRN is important for DNA double-strand break repair. *The Journal of biological chemistry* 283: 7713-7720

Chen W, Liu N, Zhang H, Zhang H, Qiao J, Jia W, Zhu S, Mao Z, Kang J (2017) Sirt6 Promotes DNA End Joining in iPSCs Derived from Old Mice. *Cell reports* 18: 2880-2892

Ciccia A, Elledge SJ (2010) The DNA damage response: making it safe to play with knives. *Molecular cell* 40: 179-204

Collins CA, Zammit PS, Ruiz AP, Morgan JE, Partridge TA (2007) A population of myogenic stem cells that survives skeletal muscle aging. *Stem cells* 25: 885-894

Corneo B, Wendland RL, Deriano L, Cui X, Klein IA, Wong SY, Arnal S, Holub AJ, Weller GR, Pancake BA, Shah S, Brandt VL, Meek K, Roth DB (2007) Rag mutations reveal robust alternative end joining. *Nature* 449: 483-486

Costes SV, Chiolo I, Pluth JM, Barcellos-Hoff MH, Jakob B (2010) Spatiotemporal characterization of ionizing radiation induced DNA damage foci and their relation to chromatin organization. *Mutation research* 704: 78-87

Cousin W, Ho ML, Desai R, Tham A, Chen RY, Kung S, Elabd C, Conboy IM (2013) Regenerative capacity of old muscle stem cells declines without significant accumulation of DNA damage. *PloS one* 8: e63528

Craxton A, Munnur D, Jukes-Jones R, Skalka G, Langlais C, Cain K, Malewicz M (2018) PAXX and its paralogs synergistically direct DNA polymerase λ activity in DNA repair. *Nature communications* 9: 3877

Currall BB, Chiang C, Talkowski ME, Morton CC (2013) Mechanisms for Structural Variation in the Human Genome. *Current genetic medicine reports* 1: 81-90

Daya S, Cortez N, Berns KI (2009) Adeno-associated virus site-specific integration is mediated by proteins of the nonhomologous end-joining pathway. *Journal of virology* 83: 11655-11664

de Laval B, Pawlikowska P, Barbieri D, Besnard-Guerin C, Cico A, Kumar R, Gaudry M, Baud V, Porteu F (2013a) Thrombopoietin promotes NHEJ DNA repair in hematopoietic stem cells through specific activation of Erk and NF-kappaB pathways and their target IEX-1. *Blood*

de Laval B, Pawlikowska P, Petit-Cocault L, Bilhou-Nabera C, Aubin-Houzelstein G, Souyri M, Pouzoulet F, Gaudry M, Porteu F (2013b) Thrombopoietin-increased DNA-PK-dependent DNA repair limits hematopoietic stem and progenitor cell mutagenesis in response to DNA damage. *Cell stem cell* 12: 37-48

Dickhut A, Schwerdtfeger R, Kuklick L, Ritter M, Thiede C, Neubauer A, Brendel C (2005) Mesenchymal stem cells obtained after bone marrow transplantation or peripheral blood stem cell transplantation originate from host tissue. *Annals of hematology* 84: 722-727

Didier N, Hourde C, Amthor H, Marazzi G, Sassoon D (2012) Loss of a single allele for Ku80 leads to progenitor dysfunction and accelerated aging in skeletal muscle. *EMBO molecular medicine* 4: 910-923

Dueva R, Iliakis G (2013) Alternative pathways of non-homologous end joining (NHEJ) in genomic instability and cancer. *Translational Cancer Research* 2: 163-177

Ensminger M, Löbrich M (2020) One end to rule them all: Non-homologous end-joining and homologous recombination at DNA double-strand breaks. *The British journal of radiology* 93: 20191054

Fan J, Robert C, Jang YY, Liu H, Sharkis S, Baylin SB, Rassool FV (2011) Human induced pluripotent cells resemble embryonic stem cells demonstrating enhanced levels of DNA repair and efficacy of nonhomologous end-joining. *Mutation research* 713: 8-17

Farini A, Sitzia C, Navarro C, D'Antona G, Belicchi M, Parolini D, Del Fraro G, Razini P, Bottinelli R, Meregalli M, Torrente Y (2012) Absence of T and B lymphocytes modulates dystrophic features in dysferlin deficient animal model. *Experimental cell research* 318: 1160-1174

Garinis GA, van der Horst GT, Vijg J, Hoeijmakers JH (2008) DNA damage and ageing: new-age ideas for an age-old problem. *Nature cell biology* 10: 1241-1247

Goodarzi AA, Jeggo PA (2013) The repair and signaling responses to DNA double-strand breaks. *Advances in genetics* 82: 1-45

Gronke K, Hernández PP, Zimmermann J, Klose CSN, Kofoed-Branzk M, Guendel F, Witkowski M, Tizian C, Amann L, Schumacher F, Glatt H, Triantafyllopoulou A, Diefenbach A (2019) Interleukin-22 protects intestinal stem cells against genotoxic stress. *Nature* 566: 249-253

Gross JG, Morgan JE (1999) Muscle precursor cells injected into irradiated mdx mouse muscle persist after serial injury. *Muscle & nerve* 22: 174-185

Guirouilh-Barbat J, Lambert S, Bertrand P, Lopez BS (2014) Is homologous recombination really an error-free process? *Frontiers in genetics* 5: 175

Hakem R (2008) DNA-damage repair; the good, the bad, and the ugly. *The EMBO journal* 27: 589-605

Hastings PJ, Ira G, Lupski JR (2009) A microhomology-mediated break-induced replication model for the origin of human copy number variation. *PLoS genetics* 5: e1000327

Hua G, Thin TH, Feldman R, Haimovitz-Friedman A, Clevers H, Fuks Z, Kolesnick R (2012) Crypt base columnar stem cells in small intestines of mice are radioresistant. *Gastroenterology* 143: 1266-1276

Hua G, Wang C, Pan Y, Zeng Z, Lee SG, Martin ML, Haimovitz-Friedman A, Fuks Z, Paty PB, Kolesnick R (2017) Distinct Levels of Radioresistance in Lgr5(+) Colonic Epithelial Stem Cells versus Lgr5(+) Small Intestinal Stem Cells. *Cancer research* 77: 2124-2133

Huertas P (2010) DNA resection in eukaryotes: deciding how to fix the break. *Nature structural & molecular biology* 17: 11-16

Inomata K, Aoto T, Binh NT, Okamoto N, Tanimura S, Wakayama T, Iseki S, Hara E, Masunaga T, Shimizu H, Nishimura EK (2009) Genotoxic stress abrogates renewal of melanocyte stem cells by triggering their differentiation. *Cell* 137: 1088-1099

Ito K, Hirao A, Arai F, Matsuoka S, Takubo K, Hamaguchi I, Nomiyama K, Hosokawa K, Sakurada K, Nakagata N, Ikeda Y, Mak TW, Suda T (2004) Regulation of oxidative stress by ATM is required for self-renewal of haematopoietic stem cells. *Nature* 431: 997-1002

Jachimowicz RD, Goergens J, Reinhardt HC (2019) DNA double-strand break repair pathway choice - from basic biology to clinical exploitation. *Cell Cycle* 18: 1423-1434

Jun S, Jung YS, Suh HN, Wang W, Kim MJ, Oh YS, Lien EM, Shen X, Matsumoto Y, McCrea PD, Li L, Chen J, Park JI (2016) LIG4 mediates Wnt signalling-induced radioresistance. *Nature communications* 7: 10994

Kanzawa T, Iwado E, Aoki H, Iwamaru A, Hollingsworth EF, Sawaya R, Kondo S, Kondo Y (2006) Ionizing radiation induces apoptosis and inhibits neuronal differentiation in rat neural stem cells via the c-Jun NH2-terminal kinase (JNK) pathway. *Oncogene* 25: 3638-3648

Kazda A, Zellinger B, Rossler M, Derboven E, Kusenda B, Riha K (2012) Chromosome end protection by blunt-ended telomeres. *Genes & development* 26: 1703-1713

Kenyon J, Gerson SL (2007) The role of DNA damage repair in aging of adult stem cells. *Nucleic acids research* 35: 7557-7565

Kramara J, Osia B, Malkova A (2018) Break-Induced Replication: The Where, The Why, and The How. *Trends in genetics : TIG* 34: 518-531

Kujjo LL, Laine T, Pereira RJ, Kagawa W, Kurumizaka H, Yokoyama S, Perez GI (2010) Enhancing survival of mouse oocytes following chemotherapy or aging by targeting Bax and Rad51. *PloS one* 5: e9204

Li L, Clevers H (2010) Coexistence of quiescent and active adult stem cells in mammals. *Science* 327: 542-545

Lieber MR (2010) The mechanism of double-strand DNA break repair by the nonhomologous DNA end-joining pathway. *Annual review of biochemistry* 79: 181-211

Liu X, Shao Z, Jiang W, Lee BJ, Zha S (2017) PAXX promotes KU accumulation at DNA breaks and is essential for end-joining in XLF-deficient mice. *Nature communications* 8: 13816

Lucas D, Escudero B, Ligos JM, Segovia JC, Estrada JC, Terrados G, Blanco L, Samper E, Bernad A (2009) Altered hematopoiesis in mice lacking DNA polymerase mu is due to inefficient double-strand break repair. *PLoS genetics* 5: e1000389

Lydeard JR, Lipkin-Moore Z, Sheu YJ, Stillman B, Burgers PM, Haber JE (2010) Break-induced replication requires all essential DNA replication factors except those specific for pre-RC assembly. *Genes & development* 24: 1133-1144

Lytle NK, Barber AG, Reya T (2018) Stem cell fate in cancer growth, progression and therapy resistance. *Nature reviews Cancer* 18: 669-680

Maher RL, Branagan AM, Morrical SW (2011) Coordination of DNA replication and recombination activities in the maintenance of genome stability. *Journal of cellular biochemistry* 112: 2672-2682

Malkova A, Haber JE (2012) Mutations arising during repair of chromosome breaks. *Annual review of genetics* 46: 455-473

Mao Z, Bozzella M, Seluanov A, Gorbunova V (2008) DNA repair by nonhomologous end joining and homologous recombination during cell cycle in human cells. *Cell cycle (Georgetown, Tex)* 7: 2902 %U http://www.ncbi.nlm.nih.gov/pmc/articles/PMC2754209/

Mateos-Gomez PA, Kent T, Deng SK, McDevitt S, Kashkina E, Hoang TM, Pomerantz RT, Sfeir A (2017a) The helicase domain of Poltheta counteracts RPA to promote alt-NHEJ. *Nature structural & molecular biology* 24: 1116-1123

Mateos-Gomez PA, Kent T, Deng SK, McDevitt S, Kashkina E, Hoang TM, Pomerantz RT, Sfeir A (2017b) The helicase domain of Polθ counteracts RPA to promote alt-NHEJ. *Nature structural & molecular biology* 24: 1116-1123

McVey M, Lee SE (2008) MMEJ repair of double-strand breaks (director's cut): deleted sequences and alternative endings. *Trends in genetics : TIG* 24: 529-538

Metcalfe C, Kljavin NM, Ybarra R, de Sauvage FJ (2014) Lgr5+ stem cells are indispensable for radiation-induced intestinal regeneration. *Cell stem cell* 14: 149-159

Milyavsky M, Gan OI, Trottier M, Komosa M, Tabach O, Notta F, Lechman E, Hermans KG, Eppert K, Konovalova Z, Ornatsky O, Domany E, Meyn MS, Dick JE (2010) A distinctive DNA damage response in human hematopoietic stem cells reveals an apoptosis-independent role for p53 in self-renewal. *Cell stem cell* 7: 186-197

Mohrin M, Bourke E, Alexander D, Warr MR, Barry-Holson K, Le Beau MM, Morrison CG, Passegue E (2010) Hematopoietic stem cell quiescence promotes error-prone DNA repair and mutagenesis. *Cell stem cell* 7: 174-185

Mokrani S, Granotier-Beckers C, Etienne O, Kortulewski T, Grisolia C, de Villartay JP, Boussin FD (2020) Higher chromosome stability in embryonic neural stem and progenitor cells than in fibroblasts in response to acute or chronic genotoxic stress. *DNA repair* 88: 102801

Monje ML, Mizumatsu S, Fike JR, Palmer TD (2002) Irradiation induces neural precursor-cell dysfunction. *Nature medicine* 8: 955-962

Moreno SG, Dutrillaux B, Coffigny H (2001) Status of p53, p21, mdm2, pRb proteins, and DNA methylation in gonocytes of control and gamma-irradiated rats during testicular development. *Biology of reproduction* 64: 1422-1431

Nagaria PK, Robert C, Park TS, Huo JS, Zambidis ET, Rassool FV (2016) High-Fidelity Reprogrammed Human IPSCs Have a High Efficacy of DNA Repair and Resemble hESCs in Their MYC Transcriptional Signature. *Stem cells international* 2016: 3826249

Navarro S, Meza NW, Quintana-Bustamante O, Casado JA, Jacome A, McAllister K, Puerto S, Surralles J, Segovia JC, Bueren JA (2006) Hematopoietic dysfunction in a mouse model for Fanconi anemia group D1. *Molecular therapy : the journal of the American Society of Gene Therapy* 14: 525-535

Nijnik A, Woodbine L, Marchetti C, Dawson S, Lambe T, Liu C, Rodrigues NP, Crockford TL, Cabuy E, Vindigni A, Enver T, Bell JI, Slijepcevic P, Goodnow CC, Jeggo PA, Cornall RJ (2007) DNA repair is limiting for haematopoietic stem cells during ageing. *Nature* 447: 686-690

Nishimura EK, Granter SR, Fisher DE (2005) Mechanisms of hair graying: incomplete melanocyte stem cell maintenance in the niche. *Science* 307: 720-724

Noon AT, Shibata A, Rief N, Lobrich M, Stewart GS, Jeggo PA, Goodarzi AA (2010) 53BP1-dependent robust localized KAP-1 phosphorylation is essential for heterochromatic DNA double-strand break repair. *Nature cell biology* 12: 177-184

Nussenzweig A, Nussenzweig MC (2007) A backup DNA repair pathway moves to the forefront. *Cell* 131: 223-225

O'Driscoll M (2012) Diseases associated with defective responses to DNA damage. *Cold Spring Harbor perspectives in biology* 4

Oliver L, Hue E, Sery Q, Lafargue A, Pecqueur C, Paris F, Vallette FM (2013) Differentiation-related response to DNA breaks in human mesenchymal stem cells. *Stem cells* 31: 800-807

Ozeki A, Suzuki K, Suzuki M, Ozawa H, Yamashita S (2012) Acceleration of astrocytic differentiation in neural stem cells surviving X-irradiation. *Neuroreport* 23: 290-293

Paques F, Haber JE (1999) Multiple pathways of recombination induced by double-strand breaks in Saccharomyces cerevisiae. *Microbiology and molecular biology reviews : MMBR* 63: 349-404

Pedersen BS, De S (2013) Loss of heterozygosity preferentially occurs in early replicating regions in cancer genomes. *Nucleic acids research* 41: 7615-7624

Ploemacher RE, van Os R, van Beurden CA, Down JD (1992) Murine haemopoietic stem cells with long-term engraftment and marrow repopulating ability are more resistant to gamma-radiation than are spleen colony forming cells. *International journal of radiation biology* 61: 489-499

Puri PL, Bhakta K, Wood LD, Costanzo A, Zhu J, Wang JY (2002) A myogenic differentiation checkpoint activated by genotoxic stress. *Nature genetics* 32: 585-593

Rachidi W, Harfourche G, Lemaitre G, Amiot F, Vaigot P, Martin MT (2007) Sensing radiosensitivity of human epidermal stem cells. *Radiotherapy and oncology : journal of the European Society for Therapeutic Radiology and Oncology* 83: 267-276

Rass E, Grabarz A, Plo I, Gautier J, Bertrand P, Lopez BS (2009) Role of Mre11 in chromosomal nonhomologous end joining in mammalian cells. *Nature structural & molecular biology* 16: 819-824

Ray S, Breuer G, DeVeaux M, Zelterman D, Bindra R, Sweasy JB (2018) DNA polymerase beta participates in DNA End-joining. *Nucleic acids research* 46: 242-255

Reinhardt HC, Schumacher B (2012) The p53 network: cellular and systemic DNA damage responses in aging and cancer. *Trends in genetics : TIG* 28: 128-136

Revenco T, Lapouge G, Moers V, Brohée S, Sotiropoulou PA (2017) Low Dose Radiation Causes Skin Cancer in Mice and Has a Differential Effect on Distinct Epidermal Stem Cells. *Stem cells* 35: 1355-1364

Reya T, Morrison SJ, Clarke MF, Weissman IL (2001) Stem cells, cancer, and cancer stem cells. *Nature* 414: 105-111

Ricchetti M, Fairhead C, Dujon B (1999) Mitochondrial DNA repairs double-strand breaks in yeast chromosomes. *Nature* 402: 96-100

Ritsma L, Ellenbroek SI, Zomer A, Snippert HJ, de Sauvage FJ, Simons BD, Clevers H, van Rheenen J (2014) Intestinal crypt homeostasis revealed at single-stem-cell level by in vivo live imaging. *Nature* 507: 362-365

Rosina M, Langone F, Giuliani G, Cerquone Perpetuini A, Reggio A, Calderone A, Fuoco C, Castagnoli L, Gargioli C, Cesareni G (2019) Osteogenic differentiation of skeletal muscle progenitor cells is activated by the DNA damage response. *Scientific reports* 9: 5447

Rossi DJ, Bryder D, Seita J, Nussenzweig A, Hoeijmakers J, Weissman IL (2007) Deficiencies in DNA damage repair limit the function of haematopoietic stem cells with age. *Nature* 447: 725-729

Rube CE, Zhang S, Miebach N, Fricke A, Rube C (2011) Protecting the heritable genome: DNA damage response mechanisms in spermatogonial stem cells. *DNA repair* 10: 159-168

Ruzankina Y, Asare A, Brown EJ (2008) Replicative stress, stem cells and aging. *Mechanisms of ageing and development* 129: 460-466

Saclier M, Cuvellier S, Magnan M, Mounier R, Chazaud B (2013) Monocyte/macrophage interactions with myogenic precursor cells during skeletal muscle regeneration. *The FEBS journal* 280: 4118-4130

Schiroli G, Conti A, Ferrari S, Della Volpe L, Jacob A, Albano L, Beretta S, Calabria A, Vavassori V, Gasparini P, Salataj E, Ndiaye-Lobry D, Brombin C, Chaumeil J, Montini E, Merelli I, Genovese P, Naldini L, Di Micco R (2019) Precise Gene Editing Preserves Hematopoietic Stem Cell Function following Transient p53-Mediated DNA Damage Response. *Cell stem cell* 24: 551-565 e558

Schneider L, Fumagalli M, d'Adda di Fagagna F (2012) Terminally differentiated astrocytes lack DNA damage response signaling and are radioresistant but retain DNA repair proficiency. *Cell death and differentiation* 19: 582-591

Shao L, Luo Y, Zhou D (2014) Hematopoietic stem cell injury induced by ionizing radiation. *Antioxidants & redox signaling* 20: 1447-1462

Simara P, Tesarova L, Rehakova D, Matula P, Stejskal S, Hampl A, Koutna I (2017) DNA double-strand breaks in human induced pluripotent stem cell reprogramming and long-term in vitro culturing. *Stem cell research & therapy* 8: 73

Soni A, Siemann M, Grabos M, Murmann T, Pantelias GE, Iliakis G (2014) Requirement for Parp-1 and DNA ligases 1 or 3 but not of Xrcc1 in chromosomal translocation formation by backup end joining. *Nucleic acids research* 42: 6380-6392

Sotiropoulou PA, Candi A, Mascre G, De Clercq S, Youssef KK, Lapouge G, Dahl E, Semeraro C, Denecker G, Marine JC, Blanpain C (2010) Bcl-2 and accelerated DNA repair mediates resistance of hair follicle bulge stem cells to DNA-damage-induced cell death. *Nature cell biology* 12: 572-582

Sotiropoulou PA, Karambelas AE, Debaugnies M, Candi A, Bouwman P, Moers V, Revenco T, Rocha AS, Sekiguchi K, Jonkers J, Blanpain C (2013) BRCA1 deficiency in skin epidermis leads to selective loss of hair follicle stem cells and their progeny. *Genes & development* 27: 39-51

Suzuki J, Yamaguchi K, Kajikawa M, Ichiyanagi K, Adachi N, Koyama H, Takeda S, Okada N (2009) Genetic evidence that the non-homologous end-joining repair pathway is involved in LINE retrotransposition. *PLoS genetics* 5: e1000461

Tadi SK, Tellier-Lebegue C, Nemoz C, Drevet P, Audebert S, Roy S, Meek K, Charbonnier JB, Modesti M (2016) PAXX Is an Accessory c-NHEJ Factor that Associates with Ku70 and Has Overlapping Functions with XLF. *Cell reports* 17: 541-555

Tajbakhsh S (2009) Skeletal muscle stem cells in developmental versus regenerative myogenesis. *Journal of internal medicine* 266: 372-389

Tajbakhsh S (2014) Ballroom dancing with stem cells: placement and displacement in the intestinal crypt. *Cell stem cell* 14: 271-273

Takubo K, Ohmura M, Azuma M, Nagamatsu G, Yamada W, Arai F, Hirao A, Suda T (2008) Stem cell defects in ATM-deficient undifferentiated spermatogonia through DNA damage-induced cell-cycle arrest. *Cell stem cell* 2: 170-182

Taty-Taty GC, Chailleux C, Quaranta M, So A, Guirouilh-Barbat J, Lopez BS, Bertrand P, Trouche D, Canitrot Y (2016) Control of alternative end joining by the chromatin remodeler p400 ATPase. *Nucleic acids research* 44: 1657-1668

Tian X, Firsanov D, Zhang Z, Cheng Y, Luo L, Tombline G, Tan R, Simon M, Henderson S, Steffan J, Goldfarb A, Tam J, Zheng K, Cornwell A, Johnson A, Yang JN, Mao Z, Manta B, Dang W, Zhang Z, Vijg J, Wolfe A, Moody K, Kennedy BK, Bohmann D, Gladyshev VN, Seluanov A, Gorbunova V (2019) SIRT6 Is Responsible for More Efficient DNA Double-Strand Break Repair in Long-Lived Species. *Cell* 177: 622-638 e622

Tiberio R, Marconi A, Fila C, Fumelli C, Pignatti M, Krajewski S, Giannetti A, Reed JC, Pincelli C (2002) Keratinocytes enriched for stem cells are protected from anoikis via an integrin signaling pathway in a Bcl-2 dependent manner. *FEBS letters* 524: 139-144

Vahidi Ferdousi L, Rocheteau P, Chayot R, Montagne B, Chaker Z, Flamant P, Tajbakhsh S, Ricchetti M (2014) More efficient repair of DNA double-strand breaks in skeletal muscle stem cells compared to their committed progeny. *Stem cell research* 13: 492-507

Vijg J, Suh Y (2013) Genome instability and aging. *Annual review of physiology* 75: 645-668

Vitale I, Manic G, De Maria R, Kroemer G, Galluzzi L (2017) DNA Damage in Stem Cells. *Molecular cell* 66: 306-319

Wang D, Jang DJ (2009) Protein kinase CK2 regulates cytoskeletal reorganization during ionizing radiation-induced senescence of human mesenchymal stem cells. *Cancer research* 69: 8200-8207

Wang J, Sun Q, Morita Y, Jiang H, Gross A, Lechel A, Hildner K, Guachalla LM, Gompf A, Hartmann D, Schambach A, Wuestefeld T, Dauch D, Schrezenmeier H, Hofmann WK, Nakauchi H, Ju Z, Kestler HA, Zender L, Rudolph KL (2012) A differentiation checkpoint limits hematopoietic stem cell self-renewal in response to DNA damage. *Cell* 148: 1001-1014

Wang M, Wu W, Wu W, Rosidi B, Zhang L, Wang H, Iliakis G (2006) PARP-1 and Ku compete for repair of DNA double strand breaks by distinct NHEJ pathways. *Nucleic acids research* 34: 6170-6182

Wang Z, Song Y, Li S, Kurian S, Xiang R, Chiba T, Wu X (2019) DNA polymerase θ (POLQ) is important for repair of DNA double-strand breaks caused by fork collapse. *The Journal of biological chemistry* 294: 3909-3919

Woodward WA, Chen MS, Behbod F, Alfaro MP, Buchholz TA, Rosen JM (2007) WNT/beta-catenin mediates radiation resistance of mouse mammary progenitor cells. *Proceedings of the National Academy of Sciences of the United States of America* 104: 618-623

Yan CT, Boboila C, Souza EK, Franco S, Hickernell TR, Murphy M, Gumaste S, Geyer M, Zarrin AA, Manis JP, Rajewsky K, Alt FW (2007) IgH class switching and translocations use a robust non-classical end-joining pathway. *Nature* 449: 478-482

Yoshihara M, Hayashizaki Y, Murakawa Y (2017) Genomic Instability of iPSCs: Challenges Towards Their Clinical Applications. *Stem cell reviews and reports* 13: 7-16

Yu H, Shen H, Yuan Y, XuFeng R, Hu X, Garrison SP, Zhang L, Yu J, Zambetti GP, Cheng T (2010) Deletion of Puma protects hematopoietic stem cells and confers long-term survival in response to high-dose gamma-irradiation. *Blood* 115: 3472-3480

4

Haematopoietic Stem Cell Niches in the Bone Marrow

Claire Fielding and Simón Méndez-Ferrer*

Haematology Department, University of Cambridge, Wellcome-MRC
Cambridge Stem Cell Institute and National Health Service Blood and
Transplant, Cambridge, UK
E-mail: sm2116@cam.ac.uk
*Corresponding Author

4.1 Haematopoietic Stem Cell Niches

During mammalian postnatal life, the bone marrow (BM) is the primary
site of haematopoiesis and haematopoietic stem cell (HSC) maintenance.
Although the generation of HSCs from endothelial cells (through a process
called endothelial-to-haematopoietic transition (EHT)) mainly occurs in large
embryo arteries, recent studies have suggested that EHT can occur in the
BM of avians during the perinatal period [1]. Haematopoiesis is a continuous
process resulting in blood cell production via controlled self-renewal, proliferation, and differentiation of HSCs in the BM, followed by egress of mature
progeny into the peripheral circulation [2].

The haematopoietic stem cell niche concept was introduced by Schofield
in 1978, where he first proposed the regulation of stem cell self-renewal by
the surrounding microenvironmental niche [3]. The HSC niche resides in the
BM, which is a complex organ consisting of a medullary cavity protected by
a shell of bone. Bone is further divided into trabecular bone located at the
ends of long bone and cortical bone, which forms the largest part of the long
bones.

The BM HSC niche is an essential microenvironment, which evolves and
responds to the physiological demands of HSCs. The inner bone surface is
termed "endosteum" and it was the first anatomical niche of the BM to be

functionally described [4–6]. Since then, numerous cellular components that contribute to endosteal and central BM HSC niches have been described. The central marrow emerged as another important niche, which is highly vascularized and where a large proportion of HSCs reside by the venous sinusoids [7]. Subsequently, progress has been made to confirm that distinct coexisting BM niches are important in orchestrating the fate of HSCs and tightly regulate the processes that occur in the BM including self-renewal, quiescence, engraftment, and lineage differentiation. The HSC niches thus respond and adapt to the demands of HSCs in both physiological and pathological conditions [8].

4.2 The Contribution of Structural Components to the Bone Marrow HSC Niche

4.2.1 Vascular Niche Components

The BM is highly vascularized to provide nutrients and oxygen. The nutrient artery and vein infiltrate the compact bone and subsequently branch to form small arterioles. These arterioles connect via transition zone vessels to the venous sinusoids near the endosteum. BM sinusoids form a complex network and are mainly found in the central marrow, operating as the site where HSCs and leukocytes enter systemic circulation due to their permeability [9, 10]. However, transcortical vessels crossing the bone have been recently suggested to carry up to 80% of arterial blood and 60% of venous blood and might therefore be important routes for cell trafficking in and out of the BM [11].

The different types of blood vessels in the BM allow to distinguish two main vascular niches: arteriolar and sinusoidal. Each niche is accompanied by operating a unique function and harbor differing cellular niche components. Sinusoids [12], arteries, arterioles [13], and transitional capillaries [14] are formed by different types of endothelial cells, which can be distinguished utilizing molecular markers. The majority of HSCs are located within <5 μm of a vessel and endothelial cells secrete important cytokines that regulate HSCs [15]. Conditional deletion of either stem cell factor or the chemokine *Cxcl12* from endothelial cells can reduce HSC numbers [16, 17]. E-selectin is an adhesion molecule mainly expressed by endothelial cells in the BM. It has been demonstrated through knockout mice or administration of E-selectin antagonists where E-selectin regulates HSC proliferation and self-renewal capacity [18]. Additionally, E-selectin binding can render acute myeloid leukemia (AML) cells resistant to chemotherapy in perivascular BM niches [19].

4.2.2 Mesenchymal Stromal Cells

A key cellular niche component is composed by the stromal cells that produce key niche factors that directly act on HSCs. Perivascular cells containing BM mesenchymal stem cell (MSC) activity can be identified by the expression of nestin [20] and are divided into two subsets according to their GFP expression in *Nestin-Gfp* transgenic mice: Nes-GFP[bright] and Nes-GFP.[dim] Nes-GFP[dim] cells are subendothelial cells in BM sinusoids and Nes-GFP[bright] cells comprise endothelial and perivascular cells in the arterioles [13] and the transition zone vessels [10, 21]. MSCs can be further divided to neuron-glial antigen (NG2) + cells [13], Cxcl12 abundant reticular (CAR) cells [22, 23], and leptin receptor (LepR)–expressing cells [16], all of which overlap with Nes-GFP+ cells to varying degrees [24, 25].

Ng2+ cells ensheathing the arterioles have been postulated as important for regulating the quiescence of HSCs via the secretion of Cxcl12 [13]. Depletion of Ng2-Cre[ERT2] labeled cells or conditional deletion of *Cxcl12* in Ng2+ cells led to a redistribution of HSCs away from the arterioles along with decreased quiescence [13, 26]. LEPR+ cells, marked by the receptor for leptin, which is secreted by adipocytes, express high levels of kit ligand (stem cell factor) and Cxcl12 and are found in BM sinusoids [16]. Some studies have demonstrated that LEPR+ cells also carry out a similar function as NG2+ cells and NG2 has been suggested to label nestin+ sinusoidal cells in another study [27]. Therefore, these divergent interpretations might be due to disagreements in the recombination pattern and distribution of the cells targeted by the Cre lines utilized in these studies [25]. Additionally, other non-haematopoietic cells, such as BM adipocytes, have been shown to be an important source of stem cell factor to regulate haematopoiesis [28]. Similarly, Cxcl12-abundant reticular (CAR) cells defined by Cxcl12 expression likely comprise a heterogeneous cell population located throughout the BM [23]. Recent studies have cross-compared the gene expression signatures of murine HSC-supporting and -non supporting stromal cell lines and have identified a modular network of paracrine signaling pathways including the majority of known regulators, and previously unrecognized potential niche factors [29, 30]. These algorithms and other studies have revealed a large overlap between LEPR+ cells, CAR cells, and Nes-GFP[dim] cells [24]. Nestin+ cell ablation *in vivo* results in reduced HSC homing [20] and BM content [31] without affecting other hematopoietic cells, demonstrating that nestin+ cells are an important component of the HSC niche. In addition, conditional deletion of either Cxcl12 or Kitl in perivascular stromal cells similarly reduces HSC numbers [16, 32]. However, conditional deletion of *Cxcl12* in LEPR+ cells did not alter HSC numbers or function in a separate study [26]. Collectively, the results from studies comparing the effect of niche factor deletion from the candidate perivascular niche cells, support the notion that different specialized HSC niches might co-exist. Although further clarification of HSC niche composition and heterogeneity is required, it is possible that distinct vascular BM niches can orchestrate the balance between quiescence and proliferation, which is necessary for homeostasis but also regeneration of the BM following injury.

The BM HSC niche exhibits dynamic properties to ensure that the physiological demands of HSCs are met. Single-cell studies have demonstrated that both HSCs and MSCs are heterogeneous cell populations [33–35]. Therefore, due to the presence of functionally and molecularly distinct populations of HSCs, an emerging concept suggests that distinct subpopulations of normal and malignant HSCs [2] are regulated by specialized niches [36].

HSC maintenance requires that a subset of HSCs remains in a quiescent and a low metabolic state, allowing for protection of genomic integrity. However, HSCs are required to become activated to support haematopoiesis and replenish lost or damaged cells. Therefore, one school of thought suggests that these two different HSC states (dormant/activated) could be maintained and supported by distinctive niches. The distribution of HSCs in the BM is associated with different metabolic activities, measured for instance via their reactive oxygen species (ROS) levels. Itkin et al. found that HSCs exhibiting a lower metabolic state and reactive oxygen species (ROS^{low}) are more quiescent and localized to arterioles near the endosteum, while ROS^{high} HSCs are more active and located near the sinusoids [10]. However, direct measurement of oxygen concentration has found the lowest levels in perisinusoidal BM niches [37], and cells often located near sinusoids (such as mature megakaryocytes, discussed below) have been shown to regulate HSC quiescence. Therefore, different cell types distributed in endosteal and central BM niches, including osteoblasts, MSCs, neuroglial cells, and megakaryocytes may impinge on normal or malignant HSC fate [36, 38].

Interestingly, a bidirectional exchange of mitochondria between BM-MSCs and HSCs or leukemic stem cells (LSCs) has been described to reprogram the metabolism of HSCs/LSCs and their niches during chemotherapy and hematopoietic recovery following irradiation. On the one hand, chemotherapy causes excessive mitochondrial-derived ROS levels in LSC and triggers mitochondrial donation from MSCs, fostering LSC chemoresistance and relapse, [31, 39–41]. On the other hand, irradiation (used as a conditioning regimen for HSC transplantation) reduces mitochondrial function and ROS levels in BMSCs; however, high ATP levels and low AMPK activity in transplanted HSCs promote Cx43- and cell-contact-dependent mitochondrial transfer from HSPCs to BMSCs, favoring niche recovery and subsequently hematopoietic regeneration [42, 43].

4.2.3 Mature Haematopoietic Cells

Initial research focused on the contributions of HSC niche cells of non-haematopoietic origin. However, as tissue homeostasis demands regulatory feedback, it has been suggested that HSC behavior could be regulated in part by the HSC progeny.

One of the first examples includes macrophages, cells that are located throughout the BM, where they act on perivascular stromal cells to modulate their Cxcl12 production, in turn influencing HSC release from the niche [44–46]. Depletion of CD169+ macrophages and subsequent reduction in *Cxcl12* expression in perivascular niche cells, leads to increased HSC mobilization [44, 46]. Furthermore, aged neutrophils that return to the BM on a daily basis to be cleared by macrophages are able to modulate this process by activating these macrophages [47]. Additionally, histamine production by myeloid cells has been shown to regulate HSC quiescence and myeloid differentiation [48]. HSCs isolated from histamine receptor (H_2R)-deficient mice exhibited reduced engraftment capacity following transplantation [48]. Collectively, these results suggest that myeloid-biased HSCs are regulated by myeloid cells directly via histamine-H_2R signaling [48].

An additional haematopoietic cell of myeloid lineage (the megakaryocyte) has been shown to regulate HSC quiescence. Imaging studies indicate that around 35% of HSCs are located adjacent to megakaryocytes and exhibit a ROS^{low} quiescent state [10]. Along this line, other studies have indicated that megakaryocytes are able to maintain HSC quiescence via transforming growth factor (TGF)β, the chemokine Cxcl4/platelet factor 4, and the C-type lectin domain family 2 (CLEC-2) [49–51].

4.2.4 Adipocytes

In humans, adipocytes progressively replace active haematopoietic tissues in the BM with ageing. However, the role of adipocytes as a HSC niche cell has been debated. Naveiras et al. negatively correlated the abundance of adipose tissue with the number of HSCs [52]. This process was accelerated in fatless A-ZIP/F1 mice or mice treated with peroxisome proliferator–activated receptor γ (PPAR$_\gamma$) inhibitor bisphenol A diglycidyl ether (BADGE), which is a known inhibitor of adipogenesis [52]. However, BM adipocytes promote myelo-erythroid differentiation [53]. Since different adipocyte subtypes exist in the BM [54], it will be important to dissect their roles in the regulation of haematopoiesis.

4.2.5 Bone-associated Cells

During development, osteoclast-deficient mice exhibit a greatly diminished BM cavity and overall haematopoietic cell numbers [55]. However, their role in steady-state haematopoiesis remains controversial, with mixed reports regarding the direct role of osteoclasts in HSC regulation [56, 57]. In addition, it has been demonstrated that mature osteoblasts negatively regulate the HSC pool via osteopontin production [58, 59]. However, subsequent studies deleting major niche factors from osteoblasts did not affect HSC numbers in the BM, indicating that they may not be critical HSC niche cells. However, the variable penetrance and recombination efficiency of compared Cre lines and the different abundance of cell types producing the same HSC niche factors should be taken into consideration when interpreting these results. Subsequent studies identified perivascular stromal cells including osteoprogenitors as essential HSC niche-forming cells [20, 60]. Osteolineage cells that line the bone surface express higher levels of mRNA for embigin and angiogenin when located near HSCs, and deletion of these molecules caused loss of HSC quiescence [61], further highlighting the relevance of the endosteal HSC niche in keeping HSC quiescent under proliferative stress.

4.2.6 Innervation

The BM is highly innervated by various types of nerves, of which the autonomic branch is predominant [62]. Sympathetic nerve fibers enter the BM through the nutrient foramen and are closely associated with the blood vessels, before sprouting and innervating different BM regions [63], although some nerves may reach the BM associated with transcortical vessels in bone. Sympathetic nerves are ensheathed by non-myelinating Schwann cells, which have been reported to maintain HSC quiescence via secreting TGF-β activator molecules and inducing TGF-β-SMAD signaling in HSCs [64]. This signaling contributes toward HSC quiescence through increased phosphorylation of Smad2 and Smad3, hence supporting the maintenance and self-renewal of HSCs [64]. The sympathetic nervous system (SNS) regulates haematopoiesis directly and indirectly predominantly via the stromal cells, mediated by neurotransmitters binding to adrenergic receptors. Adrenergic receptors are classified into two groups: α and β and further classified into $\alpha 1$, $\alpha 2$, $\beta 1$, $\beta 2$, and $\beta 3$ [65].

Most notably, the SNS plays a major role in regulating the proliferation and differentiation of HSPCs, and the migration of HSPCs and leukocytes between the BM and extramedullary sites [66]. This was initially suggested due to catecholamine levels in the blood and HSPCs egress from the BM to the circulation both adhering to circadian rhythms [67]. A neurally-driven circadian release of HSCs occurs during the resting period of mice [66]. The chemoattractant molecule Cxcl12 along with its receptor Cxcr4 is a key gatekeeper of HSPCs and leukocytes in the BM [68]. Circulating HSPCs peak 5 hours after the initiation of light in mice, orchestrated via circadian nora-drenaline secretion by the SNS, targeting $\beta 3$-adrenergic receptor expressed on stromal cells [66]. This binding causes a decrease in the nuclear content of Sp1 transcription factor and finally downregulation of Cxcl12 [66]. It has also been demonstrated that norepinephrine binding to $\beta 2$-adrenergic receptors on osteoblasts causes a reduction in Cxcl12 expression and subsequent egress of HSCs from the BM after G-CSF–enforced HSC mobilization [69]. Additionally, catecholamines can directly affect human CD34+ cell proliferation and migration, affecting their BM engraftment of NOD-SCID mice, via canonical Wnt signaling activation [70].

More recently, it has been observed that a morning peak of norepinephrine and TNF induces vascular permeability, leading to a temporal increase in ROS levels, altogether causing HSPC proliferation, differentiation, and migration [71]. Instead, at night a second peak of TNF increases melatonin secretion and is associated with reduced vascular permeability and HSPC ROS levels, ensuring HSPC maintenance [71].

On the other hand, the parasympathetic nervous system (PNS) uses acetylcholine (ACh) as the main postsynaptic neurotransmitter, which binds to either muscarinic or nicotinic receptors. One study suggested that the PNS may innervate the distal femoral metaphysis [72] and another study similarly supported the presence of cholinergic innervation within the BM of rats [73]. However, the bone anabolic effect of the PNS was suggested to be indirectly mediated through the inhibition of central sympathetic tone via muscarinic receptors in the brain [74], which also regulate G-CSF–induced

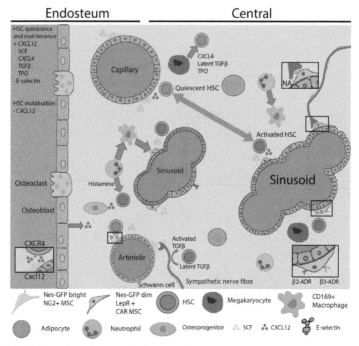

Figure 4.1 Schematic of anatomical BM HSC niches. Numerous cell types have been identified as key components of the HSC niche. In particular, the stromal populations are primarily associated with the vasculature throughout the BM and contribute key cytokines to aid HSC maintenance. Mature blood cell types: megakaryocytes, macrophages, adipocytes, and neutrophils also contribute to HSC maintenance or mobilization.

HSC mobilization [75]. Therefore, additional research is warranted to investigate the local and long-range effects of cholinergic signals.

A recent study has shown that parasympathetic cholinergic signals coordinate with sympathetic signals to regulate the egress and homing of HSPCs and leukocytes. At night, the PNS acts to dampen the noradrenergic sympathetic branch and decrease BM egress of HSPCs and leukocytes [76]. In contrast, epinephrine released at night in circulation can stimulate β2-adrenergic receptor and increase vascular adhesion and subsequent BM homing [77]. During the day, sympathetic noradrenergic fibers decrease Cxcl12-mediated BM retention of HSCs and leukocytes, whereas sympathetic cholinergic signals inhibit vascular adhesion to allow cells migrate into circulation [76].

4.3 Concluding Remarks

The dissection of the BM HSC niches has greatly improved in the recent years, although most of this knowledge comes from mouse and fish models, and our understanding of the human HSC niches is much more limited. Recent advances in imaging, transgenic models, and single-cell technologies [33, 35, 78] have revealed an unexpected degree of complexity, suggesting the possibility that numerous co-existing niches might regulate HSCs and their progeny. Future studies are needed to test this possibility and unravel the potential of niche-targeting strategies to further improve our capacity to modulate haematopoiesis under physiology and disease.

Take Home Messages

- There is heterogeneity in the anatomical location and the cellular and molecular composition of BM HSC niches.
- Both endosteal and central BM niches are highly vascularized but are enriched in different types of blood vessels and associated MSCs that regulate HSC function.
- Central and endosteal BM niches might have different functions in daily/emergency hematopoiesis and lympho/myeloid skewing.
- Neural and endocrine signals regulate these different niches to adjust hematopoiesis to organismal demands.

Acknowledgments

Original work discussed in this chapter was supported by core support grants from the Wellcome Trust (203151/Z/16/Z) and the MRC to the Cambridge Stem Cell Institute, MRC-AMED grant MR/V005421/1, National Health Service Blood and Transplant (United Kingdom), European Union's Horizon 2020 research (ERC-2014-CoG-648765) and a Programme Foundation Award (C61367/A26670) from Cancer Research UK to S.M.-F.

References

[1] Yvernogeau, L. et al. In vivo generation of haematopoietic stem/progenitor cells from bone marrow-derived haemogenic endothelium. *Nat Cell Biol* 21, 1334-1345, doi:10.1038/s41556-019-0410-6 (2019).

[2] Laurenti, E.& Gottgens, B. From haematopoietic stem cells to complex differentiation landscapes. *Nature* 553, 418-426, doi:10.1038/nature25022 (2018).

[3] Schofield, R. The relationship between the spleen colony-forming cell and the haemopoietic stem cell. *Blood cells* 4, 7-25 (1978).

[4] Calvi, L. M. et al. Osteoblastic cells regulate the haematopoietic stem cell niche. *Nature* 425, 841-846 (2003).

[5] Zhang, J. et al. Identification of the haematopoietic stem cell niche and control of the niche size. *Nature* 425, 836-841 (2003).

[6] Arai, F. et al. Tie2/angiopoietin-1 signaling regulates hematopoietic stem cell quiescence in the bone marrow niche. *Cell* 118, 149-161 (2004).

[7] Acar, M. et al. Deep imaging of bone marrow shows non-dividing stem cells are mainly perisinusoidal. *Nature* 526, 126-130, doi:10.1038/nature15250 (2015).

[8] Sanchez-Aguilera, A. & Mendez-Ferrer, S. The hematopoietic stem-cell niche in health and leukemia. *Cell Mol Life Sci* 74, 579-590, doi:10.1007/s00018-016-2306-y (2017).

[9] Kopp, H. G., Avecilla, S. T., Hooper, A. T. & Rafii, S. The bone marrow vascular niche: Home of HSC differentiation and mobilization. *Physiology* 20, 349-356, doi:10.1152/physiol.00025.2005 (2005).

[10] Itkin, T. et al. Distinct bone marrow blood vessels differentially regulate haematopoiesis. *Nature* 532, 323-328, doi:10.1038/nature17624 (2016).

[11] Gruneboom, A. et al. A network of trans-cortical capillaries as mainstay for blood circulation in long bones. *Nat Metab* 1, 236-250, doi:10.1038/s42255-018-0016-5 (2019).

[12] Hooper, A. T. et al. Engraftment and reconstitution of hematopoiesis is dependent on VEGFR2-mediated regeneration of sinusoidal endothelial cells. *Cell stem cell* 4, 263-274, doi:10.1016/j.stem.2009.01.006 (2009).

[13] Kunisaki, Y. et al. Arteriolar niches maintain haematopoietic stem cell quiescence. *Nature* 502, 637-643, doi:10.1038/nature12612 (2013).

[14] Kusumbe, A. P., Ramasamy, S. K. & Adams, R. H. Coupling of angiogenesis and osteogenesis by a specific vessel subtype in bone. *Nature* 507, 323-328, doi:10.1038/nature13145 (2014).

[15] Kiel, M. J., Yilmaz, O. H., Iwashita, T., Terhorst, C. & Morrison, S. J. SLAM family receptors distinguish hematopoietic stem and progenitor cells and reveal endothelial niches for stem cells. *Cell* 121, 1109-1121 (2005).

[16] Ding, L., Saunders, T. L., Enikolopov, G. & Morrison, S. J. Endothelial and perivascular cells maintain haematopoietic stem cells. *Nature* 481, 457-462, doi:10.1038/nature10783 (2012).

[17] Greenbaum, A. et al. CXCL12 in early mesenchymal progenitors is required for haematopoietic stem-cell maintenance. *Nature* 495, 227-230, doi:10.1038/nature11926 (2013).

[18] Winkler, I. G. et al. Vascular niche E-selectin regulates hematopoietic stem cell dormancy, self renewal and chemoresistance. *Nat Med* 18, 1651-1657, doi:10.1038/nm.2969 (2012).

[19] Barbier, V. et al. Endothelial E-selectin inhibition improves acute myeloid leukaemia therapy by disrupting vascular niche-mediated chemoresistance. *Nature*Nature communications 11, 2042, doi:10.1038/s41467-020-15817-5 (2020).

[20] Mendez-Ferrer, S. et al. Mesenchymal and haematopoietic stem cells form a unique bone marrow niche. *Nature* 466, 829-834, doi:10.1038/nature09262 (2010).

[21] Isern, J. et al. The neural crest is a source of mesenchymal stem cells with specialized hematopoietic stem-cell-niche function. *eLife* 3, doi:10.7554/eLife.03696 (2014).

[22] Sugiyama, T., Kohara, H., Noda, M. & Nagasawa, T. Maintenance of the hematopoietic stem cell pool by CXCL12-CXCR4 chemokine signaling in bone marrow stromal cell niches. *Immunity* 25, 977-988 (2006).

[23] Omatsu, Y. et al. The essential functions of adipo-osteogenic progenitors as the hematopoietic stem and progenitor cell niche. *Immunity* 33, 387-399, doi:S1074-7613(10)00322-5 [pii]10.1016/j.immuni.2010.08.017 (2010).

[24] Mende, N. et al. Prospective isolation of non-hematopoietic cells of the niche and their differential molecular interactions with HSCs. *Blood* 134, 1214-1226, doi:10.1182/blood.2019000176 (2019).

[25] Mendez-Ferrer, S. Molecular interactome between HSCs and their niches. *Blood* 134, 1197-1198, doi:10.1182/blood.2019002615 (2019).

[26] Asada, N. et al. Differential cytokine contributions of perivascular haematopoietic stem cell niches. *Nat Cell Biol* 19, 214-223, doi:10.1038/ncb3475 (2017).

[27] Wang, L. et al. Identification of a clonally expanding haematopoietic compartment in bone marrow. *Embo J* 32, 219-230, doi:10.1038/emboj.2012.308 (2013).

[28] Zhang, Z. et al. Bone marrow adipose tissue-derived stem cell factor mediates metabolic regulation of hematopoiesis. *Haematologica* 104, 1731-1743, doi:10.3324/haematol.2018.205856 (2019).

[29] Charbord, P. et al. A systems biology approach for defining the molecular framework of the hematopoietic stem cell niche. *Cell stem cell* 15, 376-391, doi:10.1016/j.stem.2014.06.005 (2014).

[30] Desterke, C. et al. Inferring Gene Networks in Bone Marrow Hematopoietic Stem Cell-Supporting Stromal Niche Populations. *iScience* 23, 101222, doi:10.1016/j.isci.2020.101222 (2020).

[31] Forte, D. et al. Bone Marrow Mesenchymal Stem Cells Support Acute Myeloid Leukemia Bioenergetics and Enhance Antioxidant Defense and Escape from Chemotherapy. *Cell metabolism* 32, 829-843 e829, doi:10.1016/j.cmet.2020.09.001 (2020).

[32] Greenbaum, A. et al. CXCL12 in early mesenchymal progenitors is required for haematopoietic stem-cell maintenance. *Nature*, doi:10.1038/nature11926 (2013).

[33] Baryawno, N. et al. A Cellular Taxonomy of the Bone Marrow Stroma in Homeostasis and Leukemia. *Cell* 177, 1915-1932 e1916, doi:10.1016/j.cell.2019.04.040 (2019).

[34] Tikhonova, A. N. et al. The bone marrow microenvironment at single-cell resolution. *Nature*, doi:10.1038/s41586-019-1104-8 (2019).

[35] Baccin, C. et al. Combined single-cell and spatial transcriptomics reveal the molecular, cellular and spatial bone marrow niche organization. *Nat Cell Biol* 22, 38-48, doi:10.1038/s41556-019-0439-6 (2020).

[36] Mendez-Ferrer, S. et al. Bone marrow niches in haematological malignancies. *Nat Rev Cancer* 20, 285-298, doi:10.1038/s41568-020-0245-2 (2020).

[37] Spencer, J. A. et al. Direct measurement of local oxygen concentration in the bone marrow of live animals. *Nature* 508, 269-273, doi:10.1038/nature13034 (2014).

[38] Morrison, S. J. & Scadden, D. T. The bone marrow niche for haematopoietic stem cells. *Nature* 505, 327-334, doi:10.1038/nature12984 (2014).

[39] Moschoi, R. et al. Protective mitochondrial transfer from bone marrow stromal cells to acute myeloid leukemic cells during chemotherapy. *Blood* 128, 253-264, doi:10.1182/blood-2015-07-655860 (2016).

[40] Marlein, C. R. et al. NADPH oxidase-2 derived superoxide drives mitochondrial transfer from bone marrow stromal cells to leukemic blasts. *Blood* 130, 1649-1660, doi:10.1182/blood-2017-03-772939 (2017).

[41] Mistry, J. J. et al. ROS-mediated PI3K activation drives mitochondrial transfer from stromal cells to hematopoietic stem cells in response to infection. *Proc Natl Acad Sci U S A* 116, 24610-24619, doi:10.1073/pnas.1913278116 (2019).

[42] Golan, K. et al. Bone marrow regeneration requires mitochondrial transfer from donor Cx43-expressing hematopoietic progenitors to stroma. *Blood* 136, 2607-2619, doi:10.1182/blood.2020005399 (2020).

[43] Mendez-Ferrer, S. HSCs revive their niche after transplantation. *Blood* 136, 2597-2598, doi:10.1182/blood.2020008923 (2020).

[44] Chow, A. et al. CD169(+) macrophages provide a niche promoting erythropoiesis under homeostasis and stress. *Nat Med* 19, 429-436, doi:10.1038/nm.3057 (2013).

[45] Christopher, M. J., Rao, M., Liu, F., Woloszynek, J. R. & Link, D. C. Expression of the G-CSF receptor in monocytic cells is sufficient to mediate hematopoietic progenitor mobilization by G-CSF in mice. *J Exp Med* 208, 251-260, doi:jem.20101700 [pii]10.1084/jem.20101700 (2011).

[46] Winkler, I. G. et al. Bone marrow macrophages maintain hematopoietic stem cell (HSC) niches and their depletion mobilizes HSCs. *Blood*, doi:10.1182/blood-2009-11-253534 (2010).

[47] Casanova-Acebes, M. et al. Rhythmic modulation of the hematopoietic niche through neutrophil clearance. *Cell* 153, 1025-1035, doi:10.1016/j.cell.2013.04.040 (2013).

[48] Chen, X. et al. Bone Marrow Myeloid Cells Regulate Myeloid-Biased Hematopoietic Stem Cells via a Histamine-Dependent Feedback Loop. *Cell stem cell*, doi:10.1016/j.stem.2017.11.003 (2017).

[49] Bruns, I. et al. Megakaryocytes regulate hematopoietic stem cell quiescence through CXCL4 secretion. *Nat Med* 20, 1315-1320, doi:10.1038/nm.3707 (2014).

[50] Zhao, M. et al. Megakaryocytes maintain homeostatic quiescence and promote post-injury regeneration of hematopoietic stem cells. *Nat Med* 20, 1321-1326, doi:10.1038/nm.3706 (2014).

[51] Nakamura-Ishizu, A., Takubo, K., Kobayashi, H., Suzuki-Inoue, K. & Suda, T. CLEC-2 in megakaryocytes is critical for maintenance of hematopoietic stem cells in the bone marrow. *J Exp Med* 212, 2133-2146, doi:10.1084/jem.20150057 (2015).

[52] Naveiras, O. et al. Bone-marrow adipocytes as negative regulators of the haematopoietic microenvironment. *Nature* 460, 259-263, doi:10.1038/nature08099 (2009).

[53] Boyd, A. L. et al. Acute myeloid leukaemia disrupts endogenous myelo-erythropoiesis by compromising the adipocyte bone marrow niche. *Nat Cell Biol* 19, 1336-1347, doi:10.1038/ncb3625 (2017).

[54] Cawthorn, W. P. et al. Bone marrow adipose tissue is an endocrine organ that contributes to increased circulating adiponectin during caloric restriction. *Cell metabolism* 20, 368-375, doi:10.1016/j.cmet.2014.06. 003 (2014).

[55] Mansour, A. et al. Osteoclasts promote the formation of hematopoietic stem cell niches in the bone marrow. *Journal of Experimental Medicine*, doi:10.1084/jem.20110994 (2012).

[56] Kollet, O. et al. Osteoclasts degrade endosteal components and promote mobilization of hematopoietic progenitor cells. *Nat Med* 12, 657-664 (2006).

[57] Takamatsu, Y. et al. Osteoclast-mediated bone resorption is stimulated during short-term administration of granulocyte colony-stimulating factor but is not responsible for hematopoietic progenitor cell mobilization. *Blood* 92, 3465-3473 (1998).

[58] Stier, S. et al. Osteopontin is a hematopoietic stem cell niche component that negatively regulates stem cell pool size. *Journal of Experimental Medicine*, doi:10.1084/jem.20041992 (2005).

[59] Nilsson, S. K. et al. Osteopontin, a key component of the hematopoietic stem cell niche and regulator of primitive hematopoietic progenitor cells. *Blood* 106, 1232-1239 (2005).

[60] Sacchetti, B. et al. Self-renewing osteoprogenitors in bone marrow sinusoids can organize a hematopoietic microenvironment. *Cell* 131, 324-336 (2007).

[61] Silberstein, L. et al. Proximity-Based Differential Single-Cell Analysis of the Niche to Identify Stem/Progenitor Cell Regulators. *Cell stem cell*, doi:10.1016/j.stem.2016.07.004 (2016).

[62] del Toro, R. & Mendez-Ferrer, S. Autonomic regulation of hematopoiesis and cancer. *Haematologica* 98, 1663-1666, doi:10.3324/ haematol.2013.084764 (2013).

[63] Calvo, W. The innervation of the bone marrow in laboratory animals. *The American journal of anatomy* 123, 315-328 (1968).

[64] Yamazaki, S. et al. Nonmyelinating Schwann cells maintain hematopoietic stem cell hibernation in the bone marrow niche. *Cell* 147, 1146-1158, doi:10.1016/j.cell.2011.09.053 (2011).

[65] Rosenbaum, D. M., Rasmussen, S. G. & Kobilka, B. K. The structure and function of G-protein-coupled receptors. *Nature* 459, 356-363, doi:10.1038/nature08144 (2009).

[66] Mendez-Ferrer, S., Lucas, D., Battista, M. & Frenette, P. S. Haematopoietic stem cell release is regulated by circadian oscillations. *Nature* 452, 442-447, doi:10.1038/nature06685 (2008).

[67] Maestroni, G. J. M. et al. Neural and endogenous catecholamines in the bone marrow. Circadian association of norepinephrine with hematopoiesis? *Experimental Hematology* (1998).

[68] Nagasawa, T. et al. Defects of B-cell lymphopoiesis and bone-marrow myelopoiesis in mice lacking the CXC chemokine PBSF/SDF-1. *Nature*, doi:10.1038/382635a0 (1996).

[69] Katayama, Y. et al. Signals from the sympathetic nervous system regulate hematopoietic stem cell egress from bone marrow. *Cell* 124, 407-421 (2006).

[70] Spiegel, A. et al. Catecholaminergic neurotransmitters regulate migration and repopulation of immature human CD34+ cells through Wnt signaling. *Nat Immunol* 8, 1123-1131, doi:ni1509 [pii]10.1038/ni1509 (2007).

[71] Golan, K. et al. Daily Onset of Light and Darkness Differentially Controls Hematopoietic Stem Cell Differentiation and Maintenance. *Cell stem cell*, doi:10.1016/j.stem.2018.08.002 (2018).

[72] Bajayo, A. et al. Skeletal parasympathetic innervation communicates central IL-1 signals regulating bone mass accrual. *Proceedings of the National Academy of Sciences of the United States of America*, doi:10.1073/pnas.1206061109 (2012).

[73] Artico, M. et al. Noradrenergic and cholinergic innervation of the bone marrow. *International journal of molecular medicine* 10, 77-80 (2002).

[74] Shi, Y. et al. Signaling through the M3 muscarinic receptor favors bone mass accrual by decreasing the sympathetic activity. *Cell metabolism* 11, 231-238, doi:10.1016/j.cmet.2010.01.005.Signaling (2010).

[75] Pierce, H. et al. Cholinergic Signals from the CNS Regulate G-CSF-Mediated HSC Mobilization from Bone Marrow via a Glucocorticoid Signaling Relay. *Cell stem cell* 20, 648-658 e644, doi:10.1016/j.stem.2017.01.002 (2017).

[76] Garcia-Garcia, A. et al. Dual cholinergic signals regulate daily migration of hematopoietic stem cells and leukocytes. *Blood* 133, 224-236, doi:10.1182/blood-2018-08-867648 (2019).

[77] Scheiermann, C. et al. Adrenergic nerves govern circadian leukocyte recruitment to tissues. *Immunity* 37, 290-301, doi:10.1016/
j.immuni.2012.05.021 (2012).

[78] Tikhonova, A. N. et al. The bone marrow microenvironment at single-
cell resolution. *Nature* 569, 222-228, doi:10.1038/s41586-019-1104-8
(2019).

5

Computational Models of Spatio-temporal Stem Cell Organization

Torsten Thalheim and Joerg Galle*

Interdisciplinary Center for Bioinformatics, University Leipzig, Haertelstr.
16–18, 04107 Leipzig, Germany
E-mail: galle@izbi.uni-leipzig.de
*Corresponding Author

5.1 Introduction

From a conceptual point of view, a "**stem cell**" (SC) is a cell occupying one
of a set of regulatory states that renders it capable of (i) self-maintaining,
(ii) specifying into different lineages, and accordingly (iii) contributing to
the regeneration of the tissue of its origin (Loeffler and Roeder, 2002).
Cells with such potential, i.e., in such a state, are found in virtually any
kind of mammalian tissue and can be induced from many if not all other
cell states, although with different efficiency (Liu et al. 2019). To ensure
tissue function and homeostasis, one usually assumes that the number of
SCs, their heterogeneity and accordingly the accessibility of the respective
regulatory states have to be conserved over the entire lifetime of an individual.
Changes of the numbers of SCs and of their capability of interconversion are
observed during ageing and in disease (reviewed in: Schultz and Sinclair,
2016). These changes can be interpreted as a changed accessibility of one or
more SC states, i.e., as changed "SC plasticity." They often come along with
impaired tissue function. Thus, understanding the mechanisms that control
"SC organization," i.e., the dynamics of reaching and leaving SC states is
a major challenge of present systems' medicine. Different mechanisms of
SC organization are still under debate. There is evidence that they vary from
tissue to tissue (Simons and Clevers, 2011).

A growing number of experimental studies aim at understanding how observable phenomena on the level of tissue dynamics are governed by the regulatory mechanisms controlling SC organization. In our opinion, such an insight is a prerequisite for providing quantitative predictions of tissue behavior under perturbations and thus for designing effective interventions. Here, **computational tissue models** can contribute to generate and test hypotheses. Spatio-temporal models of several growing and regenerative tissues have been developed in the last decade. Among them are the compartment models of blood (Roeder et al., 2005), cellular Potts models of blood vessels (Merks et al., 2008), and individual cell-based models (IBMs) of the liver (Hoehme et al., 2010). They all consider a cell state (or type) that carries SC properties. Accordingly, these models enable to describe perturbations of tissue homeostasis during malignant transformation based on changed SC organization (liver: Hoehme et al. 2018, vessel: Szabo& Merks 2017, blood: Roeder and Glauche, 2006).

In our group, we develop three-dimensional (3D) IBMs, where cells are described as physical objects that can move, grow, and divide and can form contacts to other cells or substrates (Galle et al., 2005). Combining this kind of biomechanical cell model with a model of environmentally controlled SC organization, we provided the first 3D IBM of the intestinal crypt that allows simulation of the organization of intestinal stem cells (ISCs) under homeostatic and several pathological conditions (Buske et al., 2011). Subsequently, we applied it to simulate clonal development in the intestine (Thalheim et al. 2016).

In order to enable simulation of SC differentiation in homogeneous compartments, we developed the model of noise-driven SC organization (Hoffmann et al., 2008) and applied it in studies of the environmental dependence of growth and differentiation of mesenchymal stem cells (Krinner et al., 2010). In these studies, the source of the assumed stochasticity (noise) remained undefined and it became clear that a mechanistic understanding of SC organization requires a multi-scale approach including the molecular, cellular, and tissue level of organization. There is increasing evidence that epigenetic regulation is a major component of this regulation.

Here, we provide an insight into the state-of-the-art computational modeling of spatio-temporal SC organization. We start providing an overview about general concepts of SC organization (Section 5.2). Afterwards, we discuss models of the intestinal crypt, developed in our group, as examples of extrinsically regulated SC organization (Section 5.3) and explain model extensions to intrinsic, epigenetic regulation (Section 5.4), which allows studying phenomena of SC aging (Section 5.5). We discuss variants of these models that have been applied to simulate *in vitro* SC systems (Section 5.6) and address interrelations of tissue shape and SC organization.

5.2 Concepts of Stem Cell Organization

A basic question regarding SC organization is whether it is regulated by external signals or by cell intrinsic processes. Extrinsic control of SCs implicates a one-to-one relation between the environment and the regulatory state of the cells (Figure 5.1A). Thus, in a homogenous environment, cells divide symmetrically and their daughters will overtake the same fate until changes of the environment induce state changes. Accordingly, conservation of the SC state requires a defined environment: a "SC niche" (Lander et al., 2012). If SCs leave their niche, they become specified. This can happen already during the process of division and mimic an asymmetric division, in particular, if a heterogeneous environment is considered as niche as e.g., in the intestine (Sato et al., 2011). In contrast, intrinsic regulation enables the cells to occupy

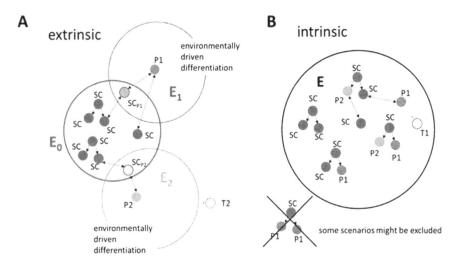

Figure 5.1 Concepts of stem cell organization. (**A**) Extrinsic regulation: The cell's environment $E_{0,1,2}$ defines their functional state. SCs divide symmetrically. Changing the state requires changing the environment (e.g. $E_0 \rightarrow E_1$). (**B**) Intrinsic regulation: SCs self-renew or differentiate in a homogenous environment E. Depending on the actual mechanism, some scenarios might be excluded (cross). In both scenarios, state changes are reversible as long as cells are not terminally differentiated (hollow circles).

different states in a homogenous environment (Figure 5.1B). Cells under intrinsic control are actually capable of dividing asymmetrically and the daughter cells can acquire different states over time without changes in the environment. As an example, hematopoietic stem cells have been suggested to follow basically intrinsic regulation (Copley et al., 2012). However, both types of regulation represent simplified models and both might contribute to SC organization.

This is evident considering that the environment controls cell metabolism and thus impacts epigenetics. For example, modification of DNA and histones requires specific metabolites (Shyh-Chang and Ng, 2017). Changed availability of these metabolites can thus change the epigenetic state of the cells and may disturb their function. This is well documented in the case of tissue transformation where nutrients and oxygen deficiency contribute to the so-called "cancer stem cell" formation and expansion (Thankamony et al. 2020).

Beside the question whether SCs organize according to intrinsic or extrinsic mechanisms, there is a second important question regarding their organization, that is, whether or not SC differentiation is reversible. In the past, SC organization was generally assumed to follow the "pedigree concept." This concept assumes that differentiation of SCs is irreversible and the SC state cannot be reached from other states. Accordingly, cells never gain "stemness" and losing all SCs of a tissue will definitely result in loss of the tissue. The pedigree concept has been very successful in describing tissue homeostasis and enabled estimates of tissue specific fractions of asymmetrically and symmetrically dividing SCs and of related cell cycle times (Doupe et al., 2012; Loeffler et al., 1986). However, in the case of tissue damage or transformation, its predictive power is limited. Here, de-differentiation events can frequently occur requiring a more flexible model approach. The "concept of flexibility" assumes that SC regulatory states can potentially be reached from all other states. State conversion may result from intrinsic fluctuations of the regulatory states of the cells or from fluctuations in environmental signaling. In case of very small conversion rates, the concept suggests a similar cell behavior as the pedigree concept. However, as long as there is a finite rate of gaining stemness, a tissue, in contrast to what is implicated by the pedigree concept, will be capable of regeneration even after elimination of all SCs. In the case of ISCs, this kind of behavior was predicted in 2011 (Buske et al., 2011). Subsequently, several studies explicitly validated this prediction (Roth et al., 2012; van Es et al., 2012).

The concept of flexibility represents the basis of all computational tissue models that have been developed in our group. In our models of the intestinal crypt, this kind of regulation has been introduced as a set of deterministic rules defining how the cells change their regulatory state in a direct relation to the environment (Buske et al., 2011; Thalheim et al. 2016). In the following, we present our intestinal crypt model as an example of models with flexible stem cell fate decisions under extrinsic control.

5.3 Intestinal Stem Cell Organization as a Paradigm

The adult intestinal epithelium is a rapidly self-renewing tissue. Cell production in the epithelium starts near the base of small tissue invaginations called crypts. At the base of the crypts, SCs produce numerous progeny, which move up along the crypt axis. These cells continue proliferating while in parallel becoming committed either to an absorptive or to a secretory fate. Reaching the crypt orifice, they become post-mitotic and terminally differentiated and are shed from the tissue a few days after. However, as an exception in the small intestine, cells that become committed to the secretory Paneth lineage move down the crypt while differentiating until they occupy their final position at the very bottom of the crypt, where they stay for several weeks. Due to this permanent regeneration process, the intestine of mice and men does self-renew within a few days.

Due to its simple structure, the intestinal epithelium has been one of the early studied spatially organized SC systems and several model approaches have been applied to the system (reviewed in: Almet et al. 2020). In 2011, we introduced a 3D IBM of the mouse intestinal crypt under homeostatic conditions (Buske et al., 2011). Assuming a fixed crypt shape, this model describes the turnover of the major cell types of the tissue (Figure 5.2A). It considers specification of SCs into enterocytes (ECs) and two types of secretory cells: Paneth cells (PCs) and Goblet cells (GCs). Fate decisions are assumed to be regulated by extrinsic stimulation of Wnt and Notch signaling. The activities of these pathways control: (i) whether a cell is capable of proliferating, and (ii) into which lineage a cell will specify and eventually differentiate (Figure 5.2B). There are two sources of Wnt: PCs and non-epithelial cells. Wnt-activity of the cells changes with their position along the crypt-villus axis, decreasing with the distance from the bottom of the crypt. Notch activity depends on secretory cells. These cells induce Notch signaling in neighboring cells by providing the required ligands. Fate decisions are reversible in the model as long as the cells are not terminally differentiated.

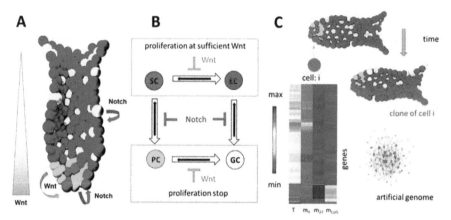

Figure 5.2 Model of the intestinal crypt. (**A**) A longitudinal cross-section of a crypt shows the distribution of stem cells (SC), Paneth cells (PC), Goblet cells (GC), and enterocytes (EC). (**B**) In the basic model, the activity of the Wnt and Notch pathway control lineage specification and proliferation activity. (**C**) In the extended model, an intrinsic regulatory network, based on an artificial genome, enables simulation of transcriptional activity (T) of genes in dependence to level of H3K4me3 (m_4), H3K27me3 (m_{27}), and DNA methylation (m_{CpG}) of gene promoters. The regulatory states (T, m_4, m_{27}, m_{CpG}) are affected by extrinsic regulation and proliferation activity. They can be specified for individual cells, or averaged over cell types and clones (Thalheim et al., 2018a)

Terminal differentiation is reached if Wnt-signaling falls below a threshold value.

Achievements: The model is capable of consistently describing a plethora of experimental data in crypt steady state and after perturbations of the Notch and Wnt pathways. Moreover, the model recapitulates monoclonal conversion within the crypt on the same time scale as observed *in vivo*. The described SC system is robust to manipulation as e.g., the removal of entire cell populations including the SCs.

Limitations: Experimental results in mice suggest that many intestinal crypt processes depend on the interaction between epithelial and stromal cells (fibroblasts: Roulis and Flavell, 2016, telocytes: Shoshkes-Carmel et al. 2018). Such interactions are not included in our model so far, limiting its predictive power. Moreover, the model does not include a detailed model of Wnt regulation. Alternative models of the epithelium describe the signaling e.g., by a system of coupled differential equations (van Leeuwen et al., 2009).

A further problem with the model represents its strictly external regulation. Consequently, the system fully adapts to any change in the environment. This is in contrast to experimental findings showing that SCs from the small intestine conserve their identity after transplantation into the large intestine (Fukuda et al. 2014). Thus, some regulatory memory must be encoded. Here, model extensions considering epigenetic regulation comes into focus (Figure 5.2C).

5.4 Models of Epigenetic Regulation

Computational models of this kind of regulation have been developed in the last decade, starting with (Dodd et al. 2007), in order to answer the questions: (i) how SC states are established during development and (ii) how stable they are during homeostasis and aging. These models provide a straightforward explanation of cell intrinsic regulation and the capability of asymmetric cell division. While asymmetric cell division might be explained by asymmetric partition of proteins in the two daughters (Beckmann et al. 2007), it can also be explained by limited inheritance of epigenetic states (Binder et al. 2013).

In our group, we focused on changes of the tri-methylation of lysine 4 and 27 of histone 3 (H3K4me3, H3K27me3) and associated changes in promoter DNA methylation. We developed a model of epigenetic regulation of transcription taking into account these modifications. We explicitly modeled binding of methyl-transferases to chromatin associated with a gene thereby considering the actual regulatory state of this gene (Binder et al. 2013). This state emerges as a consequence of interactions between histone and DNA methylation and transcription (Figure 5.3A). The model enables to simulate lineage specification based on changes of histone modifications. A potential regulation is based on H3K4me3 and H3K27me3 bivalent modified promoters of major transcription factors. Thereby, limited inheritance of both marks at their promoters leads to robust SC specification (Figure 5.3B,C). We applied this kind of model to mesenchymal (Hamidouche et al. 2017) and hematopoietic SC organization (Thalheim et al. 2017).

5.5 Stem Cells During Development and Aging

Similar to stem cell fate decisions, also developmental and age-related phenomena of stem cell organization depend on epigenetic regulation. We here focus again on the intestinal tissue. At birth, the small intestine of mice has

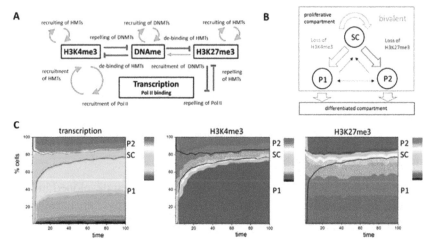

Figure 5.3   Epigenetic regulation of stem cell fate. (**A**) The simulation model considers interactions between transcription, histone modifications, and DNA methylation. (**B**) A potential mechanism of lineage specification is loss of H3K4me3–H3K27me3 bivalent modification of gene promoters. Dependent on the inheritance of the histone modifications after cell division, SCs either self-renew or specify into progenitors. Progenitors leave the proliferative compartment and differentiate terminally. (**C**) Starting simulations with a pure SC population, over time a quasi-steady state develops with stable fractions of the different cell types. Each type is characterized by its transcription and epigenetic profile (Thalheim et al., 2017)

narrow finger-like villi and the crypts just start developing. A specific type of cells, localized in the so-called "intervillus junctions," has been suggested to initiate crypt formation and thereby maturation of adult ISCs. In an excellent study, Kazakevych ct al. (2017) demonstrated that this maturation depends on both changes in DNA and histone methylation. However, accounting for these changes might not be sufficient to model this developmental stage because shape change of the tissue potentially feeds back on SC regulation (Guiu et al. 2019). Thus, an IBM would have to take into account self-organizing tissue shape. This is a strong challenge for mathematical modeling. Currently, a 3D IBM of ISC maturation is, to our best knowledge, missing.

In contrast, in modeling SC aging one can ignore shape changes of the tissue at first approximation. Accordingly, models of this process are available, aiming at explaining associated aberrant transcription (Przybilla et al. 2014) or clonal de-regulation (Bast et al. 2018, Klose et al. 2019). In the intestine, an age-related functional decline is manifest e.g., resulting in a reduced capability of tissue regeneration after radiation (Martin et al., 1998). This effect has been linked to changes in DNA methylation and was proposed to render the tissue more susceptible to chronic inflammation and transformation (Maegawa et al., 2010).

In order to describe such changes in DNA methylation, we combined our model of epigenetic regulation of transcription with the 3D IBM of the intestinal crypt and further extended our DNA methylation model (Figure 5.4A). We assumed a local increase of the *de novo* DNA methylation activity following DNA damage such as double strand breaks (Thalheim et al. 2018a). Thus, we linked hyper-methylation of gene promoters to repeated DNA damage even in the case of successful repair (Figure 5.4B). Experimental findings indicate that hyper-methylation most frequently occurs for H3K4me3–H3K27me3 modified genes (Rakyan et al. 2010). These genes are enriched in gene sets linked to developmental processes and tissue regeneration, which might explain the functional decline of hyper-methylated cells with age. Notably, to

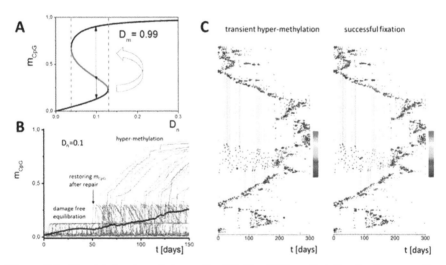

Figure 5.4 Simulated aging of stem cells. (**A**) The DNA methylation level m_{CpG} of a promoter (fraction of methylated CpGs) depends on the *de novo* DNA methylation activity D_n. Starting with low methylation, increasing D_n above a threshold can result in a switch to much higher methylation (*arrow*). Assuming that such increase happens during DNA repair, the mechanism can describe ongoing hyper-methylation. (**B**) Shown is the methylation level m_{CpG} of a selected gene promoter for 100 different cells (thin lines) in response to damage. Thick line: average value of m_{CpG}. Note that during repair the *de novo* DNA methylation activity D_n switches from 0.1 to 0.3. (**C**) Fixation of a hyper-methylation event requires that the affected SC wins clonal competition. Shown are simulated SC genealogies with unsuccessful (**left**) and successful (**right**) fixation. Color code: methylation level (red: m_{CpG} = 1, blue: m_{CpG} = 0). (Thalheim et al., 2018a)

become fixed in the crypt, hyper-methylation of a gene has to occur in clones generated by those ISCs that win the clonal competition. Simulation of such competition is straightforward applying IBMs (Figure 5.4C).

Experimental studies on long-term epigenetic changes following moderate radiation and/or loss of DNA repair function demonstrated that DNA stress induces also global epigenetic changes in histone modification profiles (Herberg et al. 2019). However, it remains open whether these epigenetic changes are observed in most ISCs or are due to the expansion of a particular subpopulation of cells that are stress-resistant (Thalheim et al. 2020). Here, one touches the problem of SC heterogeneity and interconversion. While recent experimental studies provide deep insight into these processes (reviewed by: Goodell et al. 2015), IBMs of the related dynamics in the intestine are missing.

5.6 Stem Cells in Artificial Environments

Following the outlined SC concept, cultivated SCs are quasi-stem cells and their regulatory states extend the set of *in vivo* SC states. Transcriptional profiles of such artificial SCs largely differ from normal tissue. This has been shown, e.g., for mesenchymal SCs (Ghazanfari et al. 2017). While the cells are capable of inducing spheroid and organoid growth, or of producing engineered tissue, the question remains whether they can turn back *in vivo* into bona fide SCs. This problem has become a central problem of tissue engineering (Liu et al. 2019).

In our studies of *in vitro* SC behavior in intestinal organoid culture (Sato et al., 2009), this problem was ignored. The focus was on the interrelations between tissue shape and SC organization. Based on the crypt model, a model of growing intestinal tissue was introduced (Buske et al., 2012). In the model, the basal membrane surrounding the organoid is represented by a network of semi-flexible polymers. This structure assigns a bending modulus to the cell monolayer and can be reorganized according to cell–polymer interactions. Springs connecting neighboring cells model the properties of the cell's intercalated cytoskeletons. All cell fate decisions are reversible and the cells fully adapt to the environment.

Experimental studies have demonstrated that in organoid culture PCs are essential for ISC self-renewal (Sato et al., 2011), and that ISCs with a particular high Wnt activity preferentially specify into the PC type (Farin et al. 2012). Intestinal organoid cultures contain R-spondin that enables such particular high Wnt activation in SCs. In the organoid model, we assume

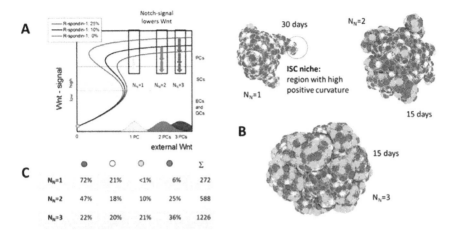

Figure 5.5 Tissue shape feeds back on stem cell organization. (**A**) In simulated organoid culture, PC neighbors control Wnt signaling in SCs. PCs provide Wnt (external) but in parallel repress Wnt activity (internal) by inducing Notch activity. For a number N_N of PCs neighbors, Wnt is balanced at SC levels (*arrows*). N_N depends on the R-spondin concentration in the culture. (**B**) Depending on N_N, SC niches of different size and shape form, containing different numbers of SCs. Gray network: basal membrane. (**C**) Cell type composition of the organoid for different N_N (Thalheim et al., 2018b)

that, if SCs specify into PCs, they stabilize Notch and thereby suppress Wnt signaling in the neighboring SCs (Thalheim et al. 2018b). If a sufficient number N_N of PCs is present in the SC neighborhood, Wnt activation (Figure 2A) and suppression by PCs are balanced at the SC level (Figure 5.5A). The number N_N depends on the R-spondin concentration. Without a PC limitation, a pure SC–PC population would expand. The PC specification was limited by assuming that it occurs only if local organoid curvature exceeds a certain positive threshold. This links organoid shape and SC specification (Figure 5.5B). In order to maintain SCs, their niche has to incorporate a defined number of PCs per SC to balance Wnt signaling. This leads to a specific shape of the niche and defines the number of SCs and PCs that can locate inside. Accordingly, R-spondin controls SC numbers in the organoid via SC niche shape (Figure 5.5C).

5.7 Summary

Computational models enable simulation of SC organization based on different hypotheses and thereby allow for testing the consistency of these hypotheses with experimental data. This is particularly useful if experiments are time consuming and costly. Here, we provided examples of modeling approaches to SC specification and ageing. We demonstrated that choosing an IBM approach enables a straightforward simulation of spatio-temporal distribution of specific lineages and of clonal competition in the tissue. Moreover, it opens perspectives for modeling self-organizing tissue shape based on biomechanics principles and thus for a description of 3D tissue development. Modeling transcriptional activity of major SC pathways, either in a rule-based manner or applying sophisticated TF-network models, is essential to cover SC behavior. Nevertheless, there is growing evidence that many phenotypic switches are controlled by epigenetic regulation, which has therefore to be considered in the computer models.

5.8 Concluding Remarks

3D IBMs have been established for various tissues including blood vessels, liver, and the small intestine.

This type of model:

- enables to generate hypotheses in order to guide experimental design and to reduce experimental costs
- enables to causally link succeeding observations by simulation of tissue dynamics
- can be extended, e.g., including epigenetic regulation etc.
- can be developed with a long-term horizon, i.e., if computational capacities are improved, further questions can be addressed based on the same model.

According to these potentials, we expect IBMs to support stem cell research also in the future. However, as any other computational model, they are limited, i.e., never provide a one-to-one description of the tissue under consideration. Thus, the hypotheses generated need always to be validated experimentally.

Acknowledgement

This study was supported by the Bundesministerium für Bildung und Forschung (grant: INDRA, grant number: BMBF 031A312).

References

Almet AA, Maini PK, Moulton DE, Byrne HM. Modeling perspectives on the intestinal crypt, a canonical system for growth, mechanics, and remodeling. Current Opinion in Biomed. Engineering 2020; 15: 32-39

Bast L, Calzolari F, Strasser MK, et al. Increasing Neural Stem Cell Division Asymmetry and Quiescence Are Predicted to Contribute to the Age-Related Decline in Neurogenesis. Cell Rep. 2018; 25(12): 3231-3240.e8.

Beckmann J, Scheitza S, Wernet P, Fischer JC, Giebel B. Asymmetric cell division within the human hematopoietic stem and progenitor cell compartment: identification of asymmetrically segregating proteins. Blood. 2007; 109(12): 5494-5501.

Binder H, Steiner L, Przybilla J, Rohlf T, Prohaska S, Galle J. Transcriptional regulation by histone modifications: towards a theory of chromatin re-organization during stem cell differentiation. Phys Biol. 2013; 10(2): 026006.

Buske P, Przybilla J, Loeffler M, et al. On the biomechanics of stem cell niche formation in the gut–modelling growing organoids. FEBS J. 2012; 279(18): 3475-3487.

Buske P, Galle J, Barker N, Aust G, Clevers H, Loeffler M. A comprehensive model of the spatio-temporal stem cell and tissue organisation in the intestinal crypt. PLoS Comput Biol. 2011; 7(1):e1001045.

Copley MR, Beer PA, Eaves CJ. Hematopoietic stem cell heterogeneity takes center stage. Cell Stem Cell. 2012; 10(6): 690-697.

Dodd IB, Micheelsen MA, Sneppen K, Thon G. Theoretical analysis of epigenetic cell memory by nucleosome modification. Cell. 2007; 129(4): 813-22.

Doupé DP, Jones PH. Interfollicular epidermal homeostasis: dicing with differentiation. Exp Dermatol. 2012; 21(4): 249-253.

Farin HF, Van Es JH, Clevers H. Redundant sources of Wnt regulate intestinal stem cells and promote formation of Paneth cells. Gastroenterology. 2012; 143(6): 1518-1529.e7.

Fukuda M, Mizutani T, Mochizuki W, Matsumoto T, Nozaki K, Sakamaki Y, Ichinose S, Okada Y, Tanaka T, Watanabe M, Nakamura T. Small intestinal stem cell identity is maintained with functional Paneth cells in heterotopically grafted epithelium onto the colon. Genes Dev. 2014; 28(16): 1752-7.

Galle J, Loeffler M, Drasdo D. Modeling the effect of deregulated proliferation and apoptosis on the growth dynamics of epithelial cell populations in vitro. Biophys J. 2005; 88(1): 62-75.

Ghazanfari R, Zacharaki D, Li H, Ching Lim H, Soneji S, Scheding S. Human Primary Bone Marrow Mesenchymal Stromal Cells and Their in vitro Progenies Display Distinct Transcriptional Profile Signatures. Sci Rep. 2017; 7(1): 10338.

Goodell MA, Nguyen H, Shroyer N. Somatic stem cell heterogeneity: diversity in the blood, skin and intestinal stem cell compartments. Nat Rev Mol Cell Biol. 2015; 16(5): 299-309

Guiu J, Hannezo E, Yui S, Demharter S, Ulyanchenko S, Maimets M, Jørgensen A, Perlman S, Lundvall L, Mamsen LS, Larsen A, Olesen RH, Andersen CY, Thuesen LL, Hare KJ, Pers TH, Khodosevich K, Simons BD, Jensen KB. Tracing the origin of adult intestinal stem cells. Nature. 2019; 570(7759): 107-111

Hamidouche Z, Rother K, Przybilla J, et al. Bistable Epigenetic States Explain Age-Dependent Decline in Mesenchymal Stem Cell Heterogeneity. Stem Cells. 2017; 35(3): 694-704.

Herberg M, Siebert S, Quaas M, et al. Loss of Msh2 and a single-radiation hit induce common, genome-wide, and persistent epigenetic changes in the intestine. Clin Epigenetics. 2019; 11(1): 65. Published 2019 Apr 27.

Hoehme S, Bertaux F, Weens W, Grasl-Kraupp B, Hengstler JG, Drasdo D. Model Prediction and Validation of an Order Mechanism Controlling the Spatiotemporal Phenotype of Early Hepatocellular Carcinoma. Bull Math Biol. 2018; 80(5): 1134-1171

Hoehme S, Brulport M, Bauer A, et al. Prediction and validation of cell alignment along microvessels as order principle to restore tissue architecture in liver regeneration. Proc Natl Acad Sci U S A. 2010; 107(23): 10371-10376.

Hoffmann M, Chang HH, Huang S, Ingber DE, Loeffler M, Galle J. Noisc-driven stem cell and progenitor population dynamics. PLoS One. 2008; 3(8):e2922.

Kazakevych J, Sayols S, Messner B, Krienke C, Soshnikova N. Dynamic changes in chromatin states during specification and differentiation of adult intestinal stem cells. Nucleic Acids Res. 2017; 45(10): 5770-5784.

Klose M, Florian MC, Gerbaulet A, Geiger H, Glauche I. Hematopoietic Stem Cell Dynamics Are Regulated by Progenitor Demand: Lessons from a Quantitative Modeling Approach. Stem Cells. 2019; 37(7): 948-957.

Krinner A, Hoffmann M, Loeffler M, Drasdo D, Galle J. Individual fates of mesenchymal stem cells in vitro. BMC Syst Biol. 2010; 4: 73.

Lander AD, Kimble J, Clevers H, et al. What does the concept of the stem cell niche really mean today? BMC Biol. 2012; 10: 19.

Liu T, Chen L, Zhao Z, Zhang S. Toward a Reconceptualization of Stem Cells from Cellular Plasticity. Int J Stem Cells. 2019; 12(1): 1-7.

Loeffler, M., and Roeder, I. Tissue stem cells: definition, plasticity, heterogeneity, self-organization and models–a conceptual approach. Cells Tissues Organs 2002; 171, 8–26.

Loeffler M, Stein R, Wichmann HE, Potten CS, Kaur P, Chwalinski S. Intestinal cell proliferation. I. A comprehensive model of steady-state proliferation in the crypt. Cell Tissue Kinet. 1986; 19(6): 627-645.

Maegawa S, Hinkal G, Kim HS, et al. Widespread and tissue specific age-related DNA methylation changes in mice. Genome Res. 2010; 20(3): 332-340.

Martin K, Kirkwood TB, Potten CS. Age changes in stem cells of murine small intestinal crypts. Exp Cell Res. 1998; 241(2): 316-323.

Merks RM, Perryn ED, Shirinifard A, Glazier JA. Contact-inhibited chemotaxis in de novo and sprouting blood-vessel growth PLoS Comput Biol. 2008; 4(9):e1000163.

Przybilla J, Rohlf T, Loeffler M, Galle J. Understanding epigenetic changes in aging stem cells–a computational model approach. Aging Cell. 2014; 13(2): 320-8.

Rakyan VK, Down TA, Maslau S, Andrew T, Yang TP, Beyan H, Whittaker P, McCann OT, Finer S, Valdes AM, Leslie RD, Deloukas P, Spector TD. Human aging-associated DNA hypermethylation occurs preferentially at bivalent chromatin domains. Genome Res. 2010; 20(4): 434-9

Roeder I, Horn M, Glauche I, Hochhaus A, Mueller MC, Loeffler M. Dynamic modeling of imatinib-treated chronic myeloid leukemia: functional insights and clinical implications. Nat Med. 2006; 12(10): 1181-1184.

Roeder I, Kamminga LM, Braesel K, Dontje B, de Haan G, Loeffler M. Competitive clonal hematopoiesis in mouse chimeras explained by a stochastic model of stem cell organization. Blood. 2005; 105(2): 609-616.

Roth S, Franken P, Sacchetti A, et al. Paneth cells in intestinal homeostasis and tissue injury. PLoS One. 2012; 7(6):e38965.

Roulis M, Flavell RA. Fibroblasts and myofibroblasts of the intestinal lamina propria in physiology and disease. Differentiation. 2016; 92(3): 116-131.

Sato T, van Es JH, Snippert HJ, et al. Paneth cells constitute the niche for Lgr5 stem cells in intestinal crypts. Nature. 2011; 469(7330): 415-418.

Sato T, Vries RG, Snippert HJ, et al. Single Lgr5 stem cells build crypt-villus structures in vitro without a mesenchymal niche. Nature. 2009; 459(7244): 262-265.

Schultz MB, Sinclair DA. When stem cells grow old: phenotypes and mechanisms of stem cell aging. Development. 2016; 143(1): 3-14.

Shoshkes-Carmel M, Wang YJ, Wangensteen KJ, Tóth B, Kondo A, Massasa EE, Itzkovitz S, Kaestner KH. Subepithelial telocytes are an important source of Wnts that supports intestinal crypts. Nature. 2018; 557(7704): 242-246

Shyh-Chang N, Ng HH. The metabolic programming of stem cells. Genes Dev. 2017; 31(4): 336-346

Simons BD, Clevers H. Strategies for homeostatic stem cell self-renewal in adult tissues. Cell. 2011; 145(6): 851-862.

Szabó A, Merks RMH. Blood vessel tortuosity selects against evolution of aggressive tumor cells in confined tissue environments: A modeling approach. PLoS Comput Biol. 2017; 13(7):e1005635.

Thankamony AP, Saxena K, Murali R, Jolly MK, Nair R. Cancer SCPlasticity - A Deadly Deal. Front Mol Biosci. 2020; 7: 79.

Thalheim T, Hopp L, Herberg M, et al. Fighting Against Promoter DNA Hyper-Methylation: Protective Histone Modification Profiles of Stress-Resistant Intestinal Stem Cells. Int J Mol Sci. 2020; 21(6): 1941.

Thalheim T, Herberg M, Galle J. Linking DNA Damage and Age-Related Promoter DNA Hyper-Methylation in the Intestine. Genes (Basel). 2018a;9(1).

Thalheim T, Quaas M, Herberg M, Braumann UD, Kerner C, Loeffler M, Aust G, Galle J. Linking stem cell function and growth pattern of intestinal organoids. Dev Biol. 2018b; 433(2): 254-261.

Thalheim T, Herberg M, Loeffler M, Galle J. The Regulatory Capacity of Bivalent Genes-A Theoretical Approach. Int J Mol Sci. 2017; 18(5): 1069.

Thalheim T, Buske P, Przybilla J, Rother K, Loeffler M, Galle J. Stem cell competition in the gut: insights from multi-scale computational modelling. J R Soc Interface. 2016; 13(121): 20160218

van Es JH, Sato T, van de Wetering M, et al. Dll1+ secretory progenitor cells revert to stem cells upon crypt damage. Nat Cell Biol. 2012; 14(10): 1099-1104.

van Leeuwen IM, Mirams GR, Walter A, et al. An integrative computational model for intestinal tissue renewal. Cell Prolif. 2009; 42(5): 617-636.

6

Transcriptomics Investigations

Christophe Desterke[1] and Pierre Charbord[2]

[1]Université Paris Sud, Hôpital Paul Brousse, Villejuif, France
[2]Sorbonne Université, CNRS, Inserm U1156, Institut de Biologie Paris
Seine, Laboratoire de Biologie du Développement/UMR7622, 9 Quai
St-Bernard, 75005 Paris, France
Email: christophe.desterke@inserm.fr; pierre.charbord@sorbonne-universite.fr

6.1 From Omics to Data

Biological experiments to investigate cellular processes now involve analyses
at Omics level. "Omics" is derived from the Sanskrit prefix "om" meaning
"globally," so Omics studies suggest studies at the global cell or tissue levels.
For example, transcriptome analysis comprises the study of all transcripts
present in the cell. In recent years, with the development of new technologies,
new fields of Omics have been developed to characterize molecules present in
a cell. Genomics have allowed studying genomic DNA not only at large level
using Whole Genome Sequencing (WGS), but also at more restricted level,
such as gene coding sequencing using Whole Exome Sequencing (WES).
With the help of algorithm workflows applied to these DNA studies, it is
now possible to characterize several events in the genome such as insertions,
deletions, single nucleotide variations, and large structural variations. For
genomics big data studies, Linux appears to be an indispensable operating
system to carry out reproducible analyses due to the possibility to build
algorithm workflows and to tune their computing. Scripting with "Bourne
against shell" (BASH) commands inside this operating system allows performing operations on text files including several billions of lines resulting
from Next Generation Sequencing (NGS). Debian Linux, a large community

139

of developers, maintains a stable operating system with several medical applications (Möller et al., 2010).

In this chapter, we will first indicate the major programming languages and platforms necessary for bioinformatics studies, without being exhaustive in such a rapidly evolving field of research. We will then focus on transcriptomics, in "bulk" studies in which each sample includes a large number of cells of different identities, and in single-cell studies in which each cell constitutes a sample by itself. The last sections will be dedicated to network biology and machine learning classifiers.

6.2 Data Analysis and Standardization in Omics

The major problem in bioinformatics is the reproducibility of analyses due to the great diversity of tools and methods used to obtain results. One of the major solutions currently employed is the development of virtual subsystems containing the algorithm pipeline for analyses and being able to be deployed at distance or on the cloud. Data analysis standardization is based on the efficiency of different programming languages. BASH UNIX language implemented in Linux open-source operating system is a starting point to build algorithm pipelines with verbose in log files for NGS data. In Linux operating system BASH terminal, it is possible to write a log file during the algorithm pipeline execution to check if the execution of the algorithm takes place correctly and within a reasonable time. Data science in the field of Omics is well developed by different open-source communities such as those using R or Python languages. For a greater efficiency in bioinformatics, it is recommended to understand the UNIX architecture and be aware of a minimum of three programming languages.

Python: Python language helps develop many skills easily compatible with the structured data matrices used in Omics biology. Matrices can be easily processed with "Numpy" and "Pandas" Python libraries. Using Python language, intensive efforts are made to control all dependencies including mathematical libraries necessary to run the bioinformatics algorithms necessary for genomic analysis.

Conda: Conda environment allows building and activating environments that virtualize the management of dependencies between libraries. A virtual environment is a kind of subsystem that contains all the libraries necessary for the execution of the algorithms installed in the system, and these are independent of the operating system of the user. The main interest of the virtual environment is to control the version of the libraries employed to execute the pipeline of algorithms. The majority of open-source softwares used in bioinformatics are easy to install via Bioconda. For example, snakePipes epigenetics workflows were developed to analyze epigenetic data in different kinds of settings (ChIP-seq, RNA-seq, Bisulfite-seq, ATAC-seq, Hi-C, and single-cell RNA-seq) through Conda repositories accessed via simple command lines (Bhardwaj et al., 2019).

Frameworks: Galaxy is a Python framework realized initially to help researchers extract information from genomic sequences obtained from NGS data. This online framework is an interactive graphical interface that allows sending to clusters some jobs composed of command line workflows. In addition, some public data were uploaded in the galaxy framework by an international consortium with guidelines for training (Batut et al., 2018). Snakemake is another useful language of workflows adapted to Python language. Similarly to the old GNU language MakeFile that describes steps of software compilation, Snakemake language describes all the steps of a workflow: for each command at each step (input, output, type of execution), a corresponding command-line is included in the file. This workflow language has been already implemented in NGS frameworks such as Sequanix (Desvillechabrol et al., 2018) or VIPER for RNA-sequencing (Cornwell et al., 2018).

Containers: Docker containers allow developing and virtualizing environments that contain complex frameworks of analysis with all of their dependencies. Docker container can contain framework for metagenome analysis in a stable Virtual machine used for both research in and teaching of genomics, and the study of the molecular evolution of genomes assembled from small single-cell projects or large-scale complex environmental metagenomes (Murovec et al., 2019). Some biology-oriented containers are collected in the biocontainers (biocontainers.pro) repository, an open-source and community-driven framework that provides platform-independent executable environments for bioinformatics software (da Veiga Leprevost et al., 2017). Cloud-oriented platforms with microservices are developed for metabolomics: PhenoMeNal consortium maintains the web portal (https://portal.phenomenal-h2020.eu) using a cloud platform implemented with open-source Openstack solution (Emami Khoonsari et al., 2019). This cloud-oriented metabolism analysis platform implements standard scientific open-source web environments such as Galaxy and Jupyter notebook.

Other languages: JAVA language can be valorized through the Biojava project aimed at simplifying bioinformatics analyses by implementing parsers, data structures, and algorithms for common tasks in genomics, structural biology, ontologies, and phylogenetics (Lafita et al., 2019). The Perl language that was very useful to manage genomic sequences starts to be less used and therefore less well-maintained by the general community of programmers. The R statistical language is very important due to the diversity of its libraries oriented for the treatment of Omics data and for the maintenance of the Bioconductor Repository (Sepulveda, 2019); it includes special Omics tutorials that are published in an open-source journal (https://f1000research.com/gateways/bioconductor).

6.3 Study of Transcriptomes Using Microarrays

Expression analyses applied to genes at global cell level was initially investigated by microarray technologies. Different principles of microarray were developed. A first approach was the dual color DNA array technology, which allowed the dual hybridization of two samples marked with different fluorochromes: generally cyanine 3 for control and cyanine 5 for tested samples. After normalization, the ratios of the two signals are calculated to obtain a relative RNA quantification between the two samples. In a second approach, simple color microarray technology was developed, especially the well-known "inject printing" technology developed by Affymetrix. In a third approach, DNA-beads microarray technology was developed by Illumina. The mathematical normalization step is different for each type of technology. Scans from GeneCHIP Affymetrix technology display raw data in *.CEL format which can be normalized by using Robust-Multichip Average (RMA) (Irizarry et al., 2003). After normalization, the gene expression matrix is built and can be processed in the R environment for Differentially Expressed Gene (DEG) analysis or unsupervised analysis such as principal component analysis through FactoMineR R-package (Lê et al., 2008). Since log-transformed data distribution is close to a Gaussian curve, downstream supervised identification of DEGs can be performed by studying their regulation using linear regression models. Implementation of Bayesian probability rules on differential gene discovery improves the efficiency of these algorithms. Implementation of p-value corrections is necessary to reduce the False positive Discovery Rate (FDR) linked to the high dimensionality of the data. LIMMA (linear model for microarray analysis) R-package robustly implements the different steps of analysis for DEG discovery

(Diboun et al., 2006). After supervised analysis, gene expression heatmap with clustering can visualize principal gene deregulations: made4 R-package from Bioconductor repository can easily draw this type of graph starting from the matrix of gene quantifications (Culhane et al., 2005).

Regardless of the software for analysis it is imperative to check the quality of the samples according to Quality Control (QC) criteria. The total number of expressed genes must be similar from one sample to another, i.e., similar means and variances, which is best assessed by comparing the sample quartiles. Likewise, the expressed gene frequency distributions must be relatively comparable from one sample to another, which is best assessed by superimposing the frequency distribution curves. The samples corresponding to different identities (cell populations, functions, response to different drugs or other biological variables) are usually distributed randomly in the PCA space, contrasting with the samples of a given identity that should be clustered together. Samples not fulfilling the QC criteria must be discarded. In addition, if ANOVA is used for the statistical analysis, the gene distribution over all samples should fit, at least approximately after logarithmic transformation, to a Gaussian. If not, the statistical analysis should rely on non-parametric test such as Wilcoxon's rank sum test or on ANOVA with Welch's correction. Finally, there should be enough replicates per sample: in general biological triplicates (each corresponding to a different experiment), but not technical replicates (corresponding to the splitting of one experiment into different microarrays), are recommended. However, a larger number of replicates might be preferable, taking into account the possible outliers that will have to be discarded, and the subsequent analyses. In any case, the experimental procedure should be carefully devised to avoid the exclusion of samples resulting in loss of statistical power.

For DEG determination, one has to choose the threshold for p-values (corrected or not for FDR) and the fold difference for the two-group comparison. P-values <0,05 and |fold difference| ≤ 1,5 are common choices, but one must keep in mind that these are arbitrary values: a reasonable choice has to be made according to the expression set and to the subsequent analyses that will be carried out since, as will be discussed, for some analyses there should be minimal discrimination or no discrimination at all. Using the DEGs as new variables, a new PCA should be made to confirm that the first component corresponds to the contrast between the two identities under comparison and to evaluate how other identities included in the study are distributed in the PCA space.

In many studies, it is desirable to include new datasets using the same identities in order to increase the statistical power and the biological relevance of the results. A rich resource to find datasets of interest is the Gene Omnibus repository (GEO) maintained by the National Center for Biotechnology Information (ncbi.nlm.nih.gov). The new datasets might result from studies made using different platforms or even different species, usually leading to a limited loss of genes not present on certain microarrays or in certain species. The major drawback of the merging of different datasets is that the primary source of differential expression is almost always across batches (i.e., the different datasets) rather than across biological groups, which can be visualized by PCA in which the first component corresponds to the contrast between the different batches. One has therefore to remove this "technical" batch effect in order to unravel the biological factors of interest. There is a wide range of tools available to remove the batch effect (Lazar et al., 2013), including the easy-to-use software developed by Partek©. However, whatever the software available, one must bear in mind that it is not possible to remove a batch effect if the batch factor coincides with a biological factor of interest. For example, if one series of results consists of control and the other of drug-treated samples, removal of the batch technical effect will result in removal of the biological effect of the drug. These simple considerations emphasize again the necessity of careful design of the experimental procedure.

After having characterized DEGs, it is essential to identify which functions, pathways, cell compartments, cell populations, and other biological information are implicated in the gene signature. To provide this complex information, databases have been developed such as: Gene Ontology with three different sub-domains—biological process, molecular function, and cellular compartments (The Gene Ontology Consortium, 2017), KEGG for pathways and metabolism (Ogata et al., 1999), Reactome and Biocarta for pathways (Joshi-Tope et al., 2005), and CellPhone for ligand-receptor analysis (Efremova et al., 2020). Some online interfaces of these databases have been developed such as: DAVID, The Database for Annotation Visualization and Integrated Discovery (Huang et al., 2009), Toppgene (Chen et al., 2009), and Enrichr (Kuleshov et al., 2016).

The identification of DEGs and the subsequent functional enrichment analysis as delineated above has two major disadvantages: (i) fixing a more or less arbitrary threshold for statistical significance may miss relevant, but modest relative to noise, biological differences, and (ii) analysis of a gene list, even when using the different databases cited above, may miss small sets of genes acting in concert. To overcome these difficulties the Broad

Institute has developed the user-friendly application Gene Set Enrichment Analysis (GSEA) (Subramanian et al., 2005). GSEA evaluates microarray data at the level of gene sets deposited in the Molecular Signatures Database (MSigDB). This approach considers expression sets consisting in samples belonging to two phenotypes, and ranks the genes according to their differential expression. The distribution of genes of the ranked gene list is then examined in each of the gene sets, evaluating whether the genes are located at random or concentrated at the top or bottom of the distribution. The gene sets with random distribution are discarded, while for the retained gene sets the statistical significance of the preferential shift at top or bottom is evaluated by iterative phenotype- or gene set-based permutations (≥ 1000 permutations are advised). It is advised to retain gene sets for which the FDR q-value is less than 0.25, but, of course, the lower the value, the higher the biological relevance of the gene set. A considerable number of gene sets ($>10,000$) are deposited in MSigDB, including not only functional sets (canonical pathways and experimental signatures curated from publications), but also genes sharing cis-regulatory motifs up or down-stream of their coding sequences, oncogenic or immunologic signatures, cytogenetic sets, and neighborhood sets. MSigDB also includes a collection of 50 "hallmark" gene sets representing well-defined biological states or processes (Liberzon et al., 2015). An increasing number of publications use only GSEA to evaluate the functional enrichment according to two compared phenotypes. However, in studies in which the final relevant gene sets are obtained after intersecting more than two pairwise identities, the DEG-based strategy remains invaluable (Charbord et al., 2014; Desterke et al., 2020). In this case, it is not recommended to use stringent statistical thresholds that may result after intersection in genes too few for significant enrichment analysis.

6.4 Study of Transcriptomes Using Next Generation Sequencing (NGS)

At first view, NGS data are a bit scaring due to their dimension and their structure consisting in expression matrices organized in specific genomic format. Knowledge of the specific formats and understanding of their respective structures are important efforts to make to employ correctly the specific workflows for data interpretation. Briefly, sequence data from samples are contained in sequence files with FASTQ format comprising for each read of sequence: the identifier of the sequence, the raw sequence letters, the identifier of the quality and the quality at each base position. Genomic alignment requires using a genome of reference in traditional FASTA format (nucleotide sequence in which nucleotides are represented using single-letter codes). After alignment, data are stocked in SAM format consisting in a pile of reads with genomic coordinates (chromosome name, start and end positions). SAM files can be compressed in BAM files rendered binary with SAMtools utilities (Li et al., 2009). Transcriptomes of reference used in RNA sequencing are in GTF files including genomic coordinates and transcript annotations with splicing information between each exon. The Picard and Genome Analysis Toolkits (GATK) developed by the Broad Institute are important softwares to manipulate SAM/BAM files and for variant discovery. Files in BED format can be also of interest for annotation of sequence files using BEDTOOLS algorithm (Quinlan, 2014).

Steps in Transcriptome Studies Using RNA-sequencing

Alignment: Transcriptome investigation by RNA sequencing for a species with known genome is usually done by paired-end short-read sequencing. The major complexity of transcriptome analysis using NGS data resides in the alignment of reads taking into account the splicing heterogeneity of all known transcripts. Some of the tools dedicated to align the reads are well-known: the Tophat2 aligner was developed to perform interpolation between genome and transcriptome information during the alignment (Trapnell et al., 2009); the STAR ultrafast aligner was developed to treat splicing information with a minimum of 32 Gigabytes of RAM (Dobin et al., 2013); and the Salmon algorithm uses a pseudo-alignment step for fast alignment by building a transcriptome index (Patro et al., 2017).

Quantification of transcripts: Depending on the workflow used and the downstream analyses, there are different ways to quantify transcripts in RNA-sequencing. Cufflinks algorithm following Tophat2 quantifies the transcripts as Fragments Per Kilobase of transcript per Million (FPKM) in paired-end sequencing, or Reads Per Kilobase of transcript per Million (RPKM) in single-end sequencing (Trapnell et al., 2012). Data transformation of FPKM into logarithm (X+1), with X as input matrix, can be a solution to approach a Gaussian distribution. It is also possible to quantify transcripts by absolute counts using HTSeq Python algorithm adapted to paired-end sequencing (Anders et al., 2015). Matrix of counts obtained with HTSeq can be scarce and zero-inflated conferring a negative binomial distribution to the data. VOOM function in EdgeR R-package is an interesting data transformation of count matrix insuring flexibility in downstream analyses (Law et al., 2014).

Testing differential expression: Differential expression analysis can be done in different ways depending on the data distribution. EdgeR, an adaptation of the LIMMA R-package to FPKM/RPKM or TPM (Transcript Per Million obtained with the Salmon algorithm) quantification of data, identifies a scaling factor adjusting each of the sample variance to the overall experimental variance (Robinson et al., 2010). DeSeq2 algorithm allows obtaining differentially expressed genes using count matrices, in which data are distributed following a negative binomial. All the information given in the section on microarrays concerning QC, DEG determination, removal of batch effect, enrichment analysis, and GSEA remains valid for NGS transcriptomes.

6.5 Study of Single Cell Transcriptomes

Analysis of the transcriptome of single cells is a major recent technology aimed primarily at characterizing the cell heterogeneity within a tissue. It makes use of two kinds of technology: in plate-based techniques cells are isolated into wells on a plate, while in droplet-based methods each cell is captured in its own microfluidic droplet. Regardless of the methods, errors can occur: multiple cells can be collected in a well or captured in a drop (doublets or multiplets), while in some wells or drops there might be no viable cells or no cells at all. The latter case is most frequent in droplet-based methods relying on a low flow of input cells to control the doublet rate. Cells from drops or wells are chemically broken down to isolate mRNA. The mRNA of each cell is labeled with a well- or droplet-specific barcode. Molecules of mRNA are then reverse-transcribed into cDNAs, which are linearly amplified to increase their detection. In many protocols, the addition to DNA fragments of molecular tags (Unique Molecular Identifiers: UMIs) before amplification allows identifying reads from distinct mRNA molecules transcribed from the same gene. After library construction, data are sequenced using NGS technology. Because of the cellular barcodes, libraries can be pooled together (i.e., multiplexed) for sequencing. The resulting reads are grouped according to their barcodes (i.e., demultiplexed), and in UMI-based protocols further demultiplexed to produce counts of mRNA molecules (count data).

From this point, the analysis of the data is a complex procedure. The interested reader will find an excellent recent review detailing the different steps and giving best-practice recommendations (Luecken and Theis, 2019). The most relevant steps will be discussed here. Live, non-damaged, single cells should be selected according to at least three major QC criteria: the count depth (frequency of counts per cell barcode), the number of detected genes, and the fraction of mitochondrial counts. Cells with excessively high counts and detected genes may represent doublets or multiplets. On the contrary, a subset of barcodes with low count depth, few detected genes, and high fraction of mitochondrial genes may correspond to wells or drops containing too few cells or damaged cells. QC steps must also be carried out at the level of transcripts by filtering out genes that are expressed in only a few cells and are therefore not informative on the cell heterogeneity. Application of the QC criteria may result in a substantial reduction of cells and genes. For example, in a study on mouse bone marrow microenvironmental cell populations, 1064 cells were removed out of 20,097 cells and the mean value of detected genes was 2227,2 (Tikhonova et al., 2019).

Data generated in the matrix of genes versus live single cells are then processed for normalization and scaling. As indicated above, obtaining single cell transcriptomes include several steps (cell capture, reverse-transcription, and sequencing), each with its own inherent variability. A gene may therefore be detected in one cell, but not in another of the same type merely because of the sampling. It is then necessary to try and correct for these technical dropouts (zero counts due to sampling), assuming that all cells initially contained a similar number of mRNA molecules. Several approaches exist to normalize the data, in particular Scran for non-full-length datasets. The normalization and scaling step allows transforming raw count data into data with a more Gaussian distribution. As described for microarrays, in many studies it is desirable to compare single cell data sets across different conditions, technologies, or species. The remove batch effect method implemented in Seurat is based on the identification of gene–gene correlation patterns that are conserved across data sets.

After filtering, normalization, and scaling, the Seurat R-package (Butler et al., 2018), among others, analyzes the cell heterogeneity by focusing on the most variable genes, considered to be the more "informative." Genes are binned according to their mean expression and the genes with the highest coefficient of variation are selected in each bin. Typically, a range of 500–2000 genes are selected for further downstream analyses. The dimensions of the data are then reduced based on the fact that the essential biological factors accounting for the gene expression are far fewer than the number of genes. The primary method for dimension reduction is PCA that identifies a small number of relevant components explaining the cell heterogeneity. PCA has the great advantage of ranking the components according to the percentage of variance explained, each component being uncorrelated, orthogonal, to another. The number of selected components, corresponding to an elbow on the representative "scree" plot, is usually in the range 30–50. As single cell data are not linear, some nonlinear dimension reduction methods have been developed for data visualization, such as the t-distributed stochastic neighbor embedding (t-SNE), the Manifold Approximation and Projection (UMAP), ForceAtlas 2, and diffusion maps. Diffusion maps are principally used for the study of continuous processes such as differentiation.

The next step is to find out clusters of cells with similar expression profile. First, one identifies the nearest neighbors of a given cell using Euclidian distances in a PCA-reduced expression space. In this k-Nearest-Neighbors (kNN) approach the value for k is usually in the range of 5–100 depending on the size of the dataset. In the resulting graph, each cell is represented as a node connected to its k-most similar cells. The Louvain algorithm then detects communities as groups of cells having more links between them than expected from the number of links the cells have in total. A resolution parameter and the choice of k help optimize the scale of the cluster partition. The cell identity of a cluster is assessed by considering the genes top-ranked by p-values that usually include known marker genes of a given population. It is also possible to search for up-regulated marker genes by comparing the genes of a cluster to all others using Student or Wilcoxon's sum rank tests. Finally, all methods described above for enrichment analysis using microarrays can also be applied here, although in many cases clusters are small and genes too few for significant enrichment in gene categories.

Clustering is but the first step to characterize the cell diversity. The second step, trajectory analysis, tries to capture the dynamics of a system, i.e., the transitions between cell identities, the branching points in a differentiation process, or the gradual changes in biological functions. Single cell data are considered as a snapshot in a continuous process. Trajectory inference aims at finding paths minimizing the transcriptional changes between neighboring cells. The pseudotime variable describes the ordering of cells along the paths. A number of algorithms are available, and the selection of a specific model should depend on the expected complexity of the paths (connected or disconnected trajectories, tree with two or more bifurcations, or presence of cycles) (Saelens et al., 2019). The recent algorithm, PArtition-based Graph Abstraction (PAGA), establishes a PAGA graph in which single cell clusters (corresponding to the Louvain clusters) are nodes and edges correspond to trajectories (Wolf et al., 2019). High-confidence paths can be identified, taking into account the connectivity scores indicative of the strength of connection between the nodes, and a random-walk–based distance defining the pseudotime. This algorithm, contrary to others such as Monocle, is not restricted to data with a tree-like topology, but provides with several possible trajectories depending on the connectivity scores. At present, there are a number of articles using PAGA that the interested reader can peruse. For example, trajectories from Hematopoietic Stem Cells (HSCs) to different lineages in developing human liver are reported in Popescu et al. (Popescu et al., 2019).

One problem with the establishment of trajectories is fixing the starting point for the pseudotemporal ordering. When dealing with stem cell–like populations and their differentiation into different lineages, one may calculate the entropy of the different populations (see for example (Grün et al., 2016)). In a highly differentiated cell population, expressing mostly one type of transcripts (e.g., endocrine cells), the entropy value tends to zero. On the contrary, the value reaches a maximum in a cell population, such as stem cells, expressing a large range of transcripts, each at low level. A more general way of finding the flux of cells along a trajectory is to measure the RNA velocity of single cells (La Manno et al., 2018). The RNA velocity is estimated by the balance between production of spliced from unspliced mRNA, and the mRNA degradation. Unspliced mRNA counts, estimated by counting reads that include intronic sequences, represent 15%–25% of the total reads depending on the single cell protocol (full-length–based SMART-Seq2 versus 3' end–based 10X Chromium or inDrop). In case of gene induction, unspliced mRNAs are in excess as compared to spliced mRNAs; the converse is true in case of gene repression. Phase portraits representing for a given gene the unspliced/spliced mRNAs ratio over the whole range of cells indicate whether the gene is induced, repressed, or in steady state. Velocity estimates can be made with or without cell or gene pooling. Velocity vectors are represented in low-dimensional PCA space (according to the highest-ranked principal components) or non-linear embeddings (e.g., t-SNE, UMAP). The size and direction of the vectors inform on the cell trajectories, directly indicating the starting and end points without any assumption about the root or calculation of a minimal spanning tree. However, the method has some limitations, namely that a single experimental observation (a snapshot) captures the full mRNA cycle (induction, repression, and steady state), and that a common splicing rate applies to all genes. These assumptions are not verified when the studied cell population comprises subsets with different RNA kinetics. scVelo is a refined method developed to overcome the limitations (Bergen et al., 2020). It enables inferring a latent time that, contrarily to pseudo-time methods, is based only on the transcriptional dynamics. It also enables uncovering putative driver genes of the transcriptional changes. The study of RNA velocity should not be considered as a substitute for trajectory inference methods, but rather as complementary, both approaches aiming at solving the difficult problem of the dynamics of the process under study.

6.6 Network Biology

Genes do not function independently but in associations whose outputs do not merely result from the additive effect of each gene but reflect a complex intracellular network of molecular interactions. It is therefore essential to move beyond single-gene comparisons and characterize the system-level features of the dataset under study; in other words to understand how genes are related and find out the gene organization into sub-networks (i.e., modules) that may be characteristic of a tissue-specific pattern or a biological process (e.g., response to drug or genetic modification).

Weighted Gene Correlation Analysis (WGCNA) is one such method developed by Steve Horvath and co-workers (Horvath, 2011; Langfelder and Horvath, 2008). The analysis proceeds through different steps briefly summarized here. As its name indicates the starting point is the measure of the correlation between two genes. The correlation factor is modified into a signed adjacency by taking the correlation absolute value, or modifying the correlation range from [-1, 1] to [0, 1]. A power value ß $>$ 0 of the adjacency is then selected to amplify the strong correlations and punish the weak ones. Making the same operation for all genes results in an adjacency matrix. The gene connectivity is defined by the sum of pairwise adjacencies of the given gene with all other genes. The gene topological overlap measure (TOM) is calculated from the connectivity of a given gene with its one-step neighbors. A gene co-expression network is then constructed from the TOM matrix and using a hierarchical clustering procedure. The branches of the hierarchical cluster tree define the different modules. Finally, to find the biological significance of a module one considers the most connected "hub" genes in the module, and calculates the correlation between the hub exemplar gene and one or several external trait(s) of biological interest. Importantly, this analysis is valid only when the frequency distribution of the connectivity follows a power law characterized by a few highly connected nodes contrasting with a majority of poorly connected ones. The network is then said to exhibit a scale-free topology. The ß power is chosen so as to select the network with the optimal scale-free topology. Scale-free networks are robust against random suppression of a node since most nodes are poorly connected; however, it is vulnerable to the attack of hub nodes whose removal will result in more or less complete module disintegration. WGCNA is widely used to find out modules and within the modules the hub genes that are critical to the biological process and/or cellular entity characterizing the module. It works well using microarrays either on unselected datasets or after minimal selection since a stringent discrimination might result in a non–scale-free topology. Of note, the number of samples should be in excess of 15, which requires careful experimental design. We find WGCNA most useful when several cell populations are well segregated according to the major principal components (Charbord et al., 2014; Desterke et al., 2020); in this case the population-specific modules usually correspond to well-defined biological process. When studying NGS and single cell transcriptomes, the zero-inflated sparse matrix can be a problem, solved after exclusion of the poorly expressed genes.

Being based on gene correlations, the links found between genes using WGCNA cannot be taken as causal relationships. In addition, most of the gene-to-gene links may correspond to indirect relationships implicating at least a third partner. To find out links corresponding to direct relationships, some of which causative, one has to look for other algorithms. One such is the multivariate information-based inductive causation (miic) recently developed by Hervé Isambert and co-workers (Verny et al., 2017). This algorithm takes for association measures the mutual information. Contrarily to the correlation, based on the covariance of two genes related to the variance of each, the mutual information is based on the gene entropy that takes into account the whole gene distribution visualized by a histogram plot after discretization of the gene frequency values. The mutual information between two genes can be estimated by the sum of the entropy of each gene distribution minus the entropy of their joint distribution. Mutual information aims at measuring general dependence-relationships, while correlation measures only linear relationships. Starting from a pre-existing network miic calculates mutual information involving more than two points, searching for "v-structures" in which one variable is dependent on two independent ones. A generic decomposition of mutual information enables distinguishing direct interactions from indirect ones. The end result is a graph modified from the pre-existing network by pruning all of the indirect interactions and outlining some directed links that may be taken as causative. It may also include information about latent "hidden" variables responsive for some of the causative links. The hubs in the miic graph usually differ from those in the pre-existing network. This algorithm is very promising, but requires a large number of observations as compared to variables. We use it for the analysis of single cell dataset, taking a WGCNA module as the pre-existing network (Desterke et al., 2020).

In the methods described so far, hubs need not be transcription factors: they may be soluble mediators acting on diverse targets, enzymes implicated in multiple reactions, cell receptors or adhesive molecules activating different signaling pathways. However, if the pre-existing network contains only genes encoding transcription factors and their downstream targets, miic is appropriate to display Gene Regulatory Networks (GRN) in which hubs represent transcription factors with causative oriented links to their downstream targets that may be other transcription factors. GRN are in theory completely oriented graphs. The construction of a causal model from purely observational data is a very difficult task, considering that there are $4^{n(n-1)/2}$ models using n variables. In other words, there are already more than 4000 models considering only 4 variables. Consequently, the construction of GRNs cannot be done without selection of transcription factors according to previous information collected from e.g., Chip-Seq analysis or perturbation analysis. A global approach for GRN construction has been adopted first for *Escherichia coli* and more recently for mammals (CellNet) (Cahan et al., 2014; Faith et al., 2007). CellNet has been used primarily to evaluate to what extent tissue-engineered samples differ from their *in vivo* counterparts. The first step in CellNet is to assemble a large number of publicly available expression sets representing a large range of murine and human cells and tissue types, ensuring that each cell or tissue is represented with a minimum of 10 conditions (controls and perturbations, i.e., deletion or overexpression of activators or repressors). The expression sets are then integrated in a single dataset. GRNs are reconstructed using a modified version of the Context Likelihood of Relatedness (CRL), an algorithm based on mutual information. CellNet is available as an R-package. It will indicate to what extent a cell or tissue-specific GRN is established in the query expression set. Query expression sets must be microarray or RNA-Seq datasets

In order to identify GRN from single cell transcriptomes boolean modeling appears to be a promising approach, and has been used for the study of early stages of blood development in the mouse (Moignard et al., 2015). The expression of 33 transcription factors and 9 markers has been analyzed in 3934 single cell profiles. First, each gene was associated with a boolean binary variable (1 = activated, 0 = inactive) according to a predefined threshold. This was done at the cell level for each of the 33 transcription factors, resulting in a 33-letter sentence for each transcription factor. Each gene was also given a boolean update rule specifying how its expression value changes over time according to the other 32 genes. Of the 3934 sentences, 3070 were observed as unique binary states, and 1448 as states that can be connected by single gene transitions. These 1448 transition states were then assigned to six embryonic stages, generating a directed state graph. The application for each transcription factor of boolean update functions consistent with expression changes across the entire graph allowed establishing a core network of 20 transcription factors. This work illustrates how complex is the discovery of a GRN from single cell data after careful selection of the transcription factors to be studied.

6.7 Machine Learning Classifiers

In many instances one is interested in finding set(s) of genes that may predict a known quantitative outcome or fit to a known classification. In this case the best approach is to divide the dataset by randomly assigning the genes into a training set and a validation set. Gene expression values (inputs) and the measurable output(s) are assembled in the training set. The input values are fed into a computer program (learning algorithm) to produce outputs in response to the inputs. The learning algorithm is such that it can modify the input/output relationship in response between original and generated outputs. The validation set is used to estimate the prediction error for best model selection. One example would be to find a set of genes that would predict the frequency of relapse or the resistance to chemotherapy in cases of acute myeloid leukemia (AML) or other cancer types. This kind of problem is called supervised since it includes a measurable outcome (e.g., the frequency of relapse) that should be predicted by the predictor set from the transcriptomic data. By contrast, unsupervised analysis aims at understanding only how the data are organized: in our example how leukemia samples can be clustered according to the expression dataset (a problem discussed at length in the previous sections). Algorithms to solve supervised learning problems can be schematically divided into two major categories: least squares methods assume that the prediction function is well approximated by a globally linear function, while k-nearest neighbors methods rely on local approximations. Regardless of the methods, obtaining useful results implies to restrict the potential solutions by imposing constraints. In the Lasso method, the regression coefficients are shrunk by imposing a penalty on their sizes. This algorithm has been used to relate gene expression to the survival of AML patients (Ng et al., 2016). Using a large dataset of 495 patients the method has allowed identifying a 17-gene signature predictive of the outcome. Lasso has also been used for the correct classification of AML versus other types of leukemia and healthy donors (Warnat-Herresthal et al., 2020). Of note, provided that the training and validation set are sampled in a similar fashion, heterogeneity is not a problem and removal of batch effect is not necessary when merging sets from different platforms. Importantly, many of the genes found according to Lasso were not differentially expressed, which underlines the power of the algorithm. Random Forest is a decision tree algorithm that builds a large collection of de-correlated trees before averaging them. We have used a shrunken centroid and the Random Forest algorithms to discover the genes most discriminant of a population of HSC-supportive bone marrow stromal cells (Desterke et al., 2020). The interested reader will find an in-depth description of the algorithms used for supervised learning in Hastie et al. (2009).

6.8 Conclusion

Nowadays, the analysis of transcriptomes is essential to the study of stem cells, providing with a range of tools useful for in-depth understanding of the stem cell attributes (self-renewal, differentiation, proliferation, and apoptosis) in different biological contexts, as well as for obtaining optimal procedures for *ex vivo* engineering of tissues from stem cells. "Computational stem cell biology" development depends on the tight collaboration between computationalists and experimentalists (Cahan et al., 2021). This chapter indicates in a non-exhaustive way the major methods used to study cell transcriptomes either in "bulk" (microarrays and NGS RNA-Seq) or in single cell studies. There is a wide range of outputs: identification of differentially expressed genes, gene pathways, cell compartments, cell populations, cell trajectories, characterization of gene networks in undirected graphs or directed ones, and identification of predictor genes. These outputs can be obtained using several approaches and a number of algorithms. In our experience, the association of bulk and single cell methods is very informative. Due to space constraint, it has not been possible to give in this chapter the algorithm details; however, we have tried to provide with the concepts underlying the algorithms. (Figure 6.1) illustrates some of the methodological problems that have been discussed.

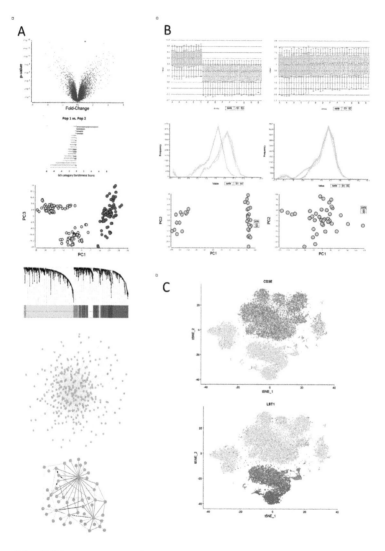

Figure 6.1 **(A)**From gene expression set to oriented gene network. Schematic flow chart: full details are in (Desterke et al., 2020). The dataset included a large number of samples studied in bulk using microarrays. The samples corresponded to three cell populations: turquoise, blue, and gray. From top to bottom:

—Differentially expressed genes (DEGs), identified by comparing using ANOVA turquoise to blue populations, are shown on the Volcano plot

—Enrichment analysis using DAVID. Each bar corresponds to one Gene Ontology category.

—Principal Component Analysis (PCA) using a matrix including DEGs and all samples: The first component corresponds to the contrast between the extreme right blue and the extreme

Figure 6.1 (*Continued*) left turquoise cell populations. The gray population in the middle has little weight in the analysis.

—Weighted Gene Correlation Analysis (WGCNA) made using the same matrix as for PCA. The gene dendogram given by hierarchical clustering reveals two major branches: one on the left corresponding to the turquoise cell population, and the other on the right corresponding to the blue cell population.

—WGCNA plot showing the organization of genes of the turquoise module. Each dot corresponds to one gene; the most connected genes are in the center.

—Multivariate information–based inductive causation (miic) plot of the most connected genes of the turquoise module. Some links are oriented (*black arrows*). Red and black links correspond to positive and negative correlations, respectively.

(**B**)Removal of batch effect (RBE) using a microarray dataset. The dataset included two distinct batches S1 an S2. RBE was done using Partek©. Without RBE (**left panels**) the two series are distinct considering the total number of genes in the different samples (**upper panel**), the gene frequency distribution in the different samples (**middle panel**), and the sample distribution in the PCA space (**lower panel**). This distinction is no longer apparent after RBE (**right panels**).

(**C**) Analysis of the transcriptome of 33,000 human peripheral blood mononuclear cells. The nonlinear dimension reduction method, t-distributed stochastic neighbor embedding (t-SNE) was applied to the single cell transcriptome. Four major clusters were visualized. High expression of CD3E and LST1 markers characterizes the lymphoid and myeloid clusters, respectively (**upper and lower panels**, respectively). Gray dots correspond to marker-negative cells.

Take Home Messages:

- Experiments must be carefully designed ahead of any transcriptional analysis, keeping in mind the overall outputs one wants to obtain according to the studied biological issue.
- The quality of the samples must be checked according to more or less numerous criteria depending on the methods. Samples failing the Quality Control analysis must be discarded. The possible exclusion of such samples must be taken into account when planning the experiment.
- The flowchart of the analysis must be clearly indicated in a way ensuring reproducibility by other teams.
- Even the most refined algorithms cannot provide with "real" results. In other terms, in all cases an experimental assessment must be made, confirming at least some of the predictions suggested by the analysis (the relevance of e.g., some genes, pathways, trajectories, networks, predictors).

References

Anders, S., Pyl, P.T., and Huber, W. (2015). HTSeq-a Python framework to work with high-throughput sequencing data. Bioinformatics. 31, 166–169.

Batut, B., Hiltemann, S., Bagnacani, A., Baker, D., Bhardwaj, V., Blank, C., Bretaudeau, A., Brillet-Guéguen, L., Čech, M., Chilton, J., et al. (2018). Community-Driven Data Analysis Training for Biology. Cell Syst. 6, 752-758.e1.

Bergen, V., Lange, M., Peidli, S., Wolf, F.A., and Theis, F.J. (2020). Generalizing RNA velocity to transient cell states through dynamical modeling. Nat. Biotechnol. 38, 1408–1414.

Bhardwaj, V., Heyne, S., Sikora, K., Rabbani, L., Rauer, M., Kilpert, F., Richter, A.S., Ryan, D.P., and Manke, T. (2019). snakePipes: facilitating flexible, scalable and integrative epigenomic analysis. Bioinformatics. 35, 4757–4759.

Butler, A., Hoffman, P., Smibert, P., Papalexi, E., and Satija, R. (2018). Integrating single-cell transcriptomic data across different conditions, technologies, and species. Nat. Biotechnol. 36, 411–420.

Cahan, P., Li, H., Morris, S.A., Lummertz da Rocha, E., Daley, G.Q., and Collins, J.J. (2014). CellNet: network biology applied to stem cell engineering. Cell 158, 903–915.

Cahan, P., Cacchiarelli, D., Dunn, S.-J., Hemberg, M., de Sousa Lopes, S.M.C., Morris, S.A., Rackham, O.J.L., Del Sol, A., and Wells, C.A. (2021). Computational Stem Cell Biology: Open Questions and Guiding Principles. Cell Stem Cell 28, 20–32.

Charbord, P., Pouget, C., Binder, H., Dumont, F., Stik, G., Levy, P., Allain, F., Marchal, C., Richter, J., Uzan, B., et al. (2014). A systems biology approach for defining the molecular framework of the hematopoietic stem cell niche. Cell Stem Cell 15, 376–391.

Chen, J., Bardes, E.E., Aronow, B.J., and Jegga, A.G. (2009). ToppGene Suite for gene list enrichment analysis and candidate gene prioritization. Nucleic Acids Res. 37, W305-311.

Cornwell, M., Vangala, M., Taing, L., Herbert, Z., Köster, J., Li, B., Sun, H., Li, T., Zhang, J., Qiu, X., et al. (2018). VIPER: Visualization Pipeline for RNA-seq, a Snakemake workflow for efficient and complete RNA-seq analysis. BMC Bioinformatics 19, 135.

Culhane, A.C., Thioulouse, J., Perrire, G., and Higgins, D.G. (2005). MADE4: an R package for multivariate analysis of gene expression data. Bioinformatics. 21, 2789–2790.

Desterke, C., Petit, L., Sella, N., Chevallier, N., Cabeli, V., Coquelin, L., Durand, C., Oostendorp, R.A.J., Isambert, H., Jaffredo, T., et al. (2020). Inferring Gene Networks in Bone Marrow Hematopoietic Stem Cell-Supporting Stromal Niche Populations. iScience 23, 101222.

Desvillechabrol, D., Legendre, R., Rioualen, C., Bouchier, C., van Helden, J., Kennedy, S., and Cokelaer, T. (2018). Sequanix: a dynamic graphical interface for Snakemake workflows. Bioinformatics. 34, 1934–1936.

Diboun, I., Wernisch, L., Orengo, C.A., and Koltzenburg, M. (2006). Microarray analysis after RNA amplification can detect pronounced differences in gene expression using limma. BMC Genomics 7, 252.

Dobin, A., Davis, C.A., Schlesinger, F., Drenkow, J., Zaleski, C., Jha, S., Batut, P., Chaisson, M., and Gingeras, T.R. (2013). STAR: ultrafast universal RNA-seq aligner. Bioinformatics. 29, 15–21.

Efremova, M., Vento-Tormo, M., Teichmann, S.A., and Vento-Tormo, R. (2020). CellPhoneDB: inferring cell-cell communication from combined expression of multi-subunit ligand-receptor complexes. Nat. Protoc. 15, 1484–1506.

Emami Khoonsari, P., Moreno, P., Bergmann, S., Burman, J., Capuccini, M., Carone, M., Cascante, M., de Atauri, P., Foguet, C., Gonzalez-Beltran, A.N., et al. (2019). Interoperable and scalable data analysis with microservices: applications in metabolomics. Bioinformatics. 35, 3752–3760.

Faith, J.J., Hayete, B., Thaden, J.T., Mogno, I., Wierzbowski, J., Cottarel, G., Kasif, S., Collins, J.J., and Gardner, T.S. (2007). Large-scale mapping and validation of Escherichia coli transcriptional regulation from a compendium of expression profiles. PLoS Biol. 5, e8.

The Gene Ontology Consortium (2017). Expansion of the Gene Ontology knowledgebase and resources. Nucleic Acids Res. 45, D331–D338.

Grün, D., Muraro, M.J., Boisset, J.-C., Wiebrands, K., Lyubimova, A., Dharmadhikari, G., van den Born, M., van Es, J., Jansen, E., Clevers, H., et al. (2016). De Novo Prediction of Stem Cell Identity using Single-Cell Transcriptome Data. Cell Stem Cell 19, 266–277.

Hastie, T., Tibshirani, R., and Friedman, J. (2009). The Elements of Statistical Learning, Data Mining, Inference, and Prediction (Springer).

Horvath, S. (2011). Weighted Network Analysis (Springer).

Huang, D.W., Sherman, B.T., and Lempicki, R.A. (2009). Systematic and integrative analysis of large gene lists using DAVID bioinformatics resources. Nat. Protoc. 4, 44–57.

Irizarry, R.A., Bolstad, B.M., Collin, F., Cope, L.M., Hobbs, B., and Speed, T.P. (2003). Summaries of Affymetrix GeneChip probe level data. Nucleic Acids Res. 31, e15.

Joshi-Tope, G., Gillespie, M., Vastrik, I., D'Eustachio, P., Schmidt, E., de Bono, B., Jassal, B., Gopinath, G.R., Wu, G.R., Matthews, L., et al. (2005). Reactome: a knowledgebase of biological pathways. Nucleic Acids Res. 33, D428-432.

Kuleshov, M.V., Jones, M.R., Rouillard, A.D., Fernandez, N.F., Duan, Q., Wang, Z., Koplev, S., Jenkins, S.L., Jagodnik, K.M., Lachmann, A., et al. (2016). Enrichr: a comprehensive gene set enrichment analysis web server 2016 update. Nucleic Acids Res. 44, W90-97.

La Manno, G., Soldatov, R., Zeisel, A., Braun, E., Hochgerner, H., Petukhov, V., Lidschreiber, K., Kastriti, M.E., Lönnerberg, P., Furlan, A., et al. (2018). RNA velocity of single cells. Nature 560, 494–498.

Lafita, A., Bliven, S., Prlić, A., Guzenko, D., Rose, P.W., Bradley, A., Pavan, P., Myers-Turnbull, D., Valasatava, Y., Heuer, M., et al. (2019). BioJava 5: A community driven open-source bioinformatics library. PLoS Comput. Biol. 15, e1006791.

Langfelder, P., and Horvath, S. (2008). WGCNA: an R package for weighted correlation network analysis. BMC Bioinformatics 9, 559.

Law, C.W., Chen, Y., Shi, W., and Smyth, G.K. (2014). voom: Precision weights unlock linear model analysis tools for RNA-seq read counts. Genome Biol. 15, R29.

Lazar, C., Meganck, S., Taminau, J., Steenhoff, D., Coletta, A., Molter, C., Weiss-Solís, D.Y., Duque, R., Bersini, H., and Nowé, A. (2013). Batch effect removal methods for microarray gene expression data integration: a survey. Brief. Bioinform. 14, 469–490.

Lê, S., Josse, J., and Husson, F. (2008). FactoMineR: An R Package for Multivariate Analysis. J. Stat. Softw. 25, 1–18.

Li, H., Handsaker, B., Wysoker, A., Fennell, T., Ruan, J., Homer, N., Marth, G., Abecasis, G., Durbin, R., and 1000 Genome Project Data Processing Subgroup (2009). The Sequence Alignment/Map format and SAMtools. Bioinformatics. 25, 2078–2079.

Liberzon, A., Birger, C., Thorvaldsdóttir, H., Ghandi, M., Mesirov, J.P., and Tamayo, P. (2015). The Molecular Signatures Database (MSigDB) hallmark gene set collection. Cell Syst. 1, 417–425.

Luecken, M.D., and Theis, F.J. (2019). Current best practices in single-cell RNA-seq analysis: a tutorial. Mol. Syst. Biol. 15, e8746.

Moignard, V., Woodhouse, S., Haghverdi, L., Lilly, A.J., Tanaka, Y., Wilkinson, A.C., Buettner, F., Macaulay, I.C., Jawaid, W., Diamanti, E., et al. (2015). Decoding the regulatory network of early blood development from single-cell gene expression measurements. Nat. Biotechnol. 33, 269–276.

Möller, S., Krabbenhöft, H.N., Tille, A., Paleino, D., Williams, A., Wolstencroft, K., Goble, C., Holland, R., Belhachemi, D., and Plessy, C. (2010). Community-driven computational biology with Debian Linux. BMC Bioinformatics 11 Suppl 12, S5.

Murovec, B., Deutsch, L., and Stres, B. (2020). Computational framework for high-quality production and large-scale evolutionary analysis of metagenome assembled genomes. Mol Biol Evol.37, 593–598.

Ng, S.W.K., Mitchell, A., Kennedy, J.A., Chen, W.C., McLeod, J., Ibrahimova, N., Arruda, A., Popescu, A., Gupta, V., Schimmer, A.D., et al. (2016). A 17-gene stemness score for rapid determination of risk in acute leukaemia. Nature 540, 433–437.

Ogata, H., Goto, S., Sato, K., Fujibuchi, W., Bono, H., and Kanehisa, M. (1999). KEGG: Kyoto Encyclopedia of Genes and Genomes. Nucleic Acids Res. 27, 29–34.

Patro, R., Duggal, G., Love, M.I., Irizarry, R.A., and Kingsford, C. (2017). Salmon provides fast and bias-aware quantification of transcript expression. Nat. Methods 14, 417–419.

Popescu, D.-M., Botting, R.A., Stephenson, E., Green, K., Webb, S., Jardine, L., Calderbank, E.F., Polanski, K., Goh, I., Efremova, M., et al. (2019). Decoding human fetal liver haematopoiesis. Nature 574, 365–371.

Quinlan, A.R. (2014). BEDTools: The Swiss-Army Tool for Genome Feature Analysis. Curr Protoc Bioinformatics. 47, 11.12.1–34.

Robinson, M.D., McCarthy, D.J., and Smyth, G.K. (2010). edgeR: a Bioconductor package for differential expression analysis of digital gene expression data. Bioinformatics. 26, 139–140.

Saelens, W., Cannoodt, R., Todorov, H., and Saeys, Y. (2019). A comparison of single-cell trajectory inference methods. Nat. Biotechnol. 37, 547–554.

Sepulveda, J.L. (2020). Using R and Bioconductor in Clinical Genomics and Transcriptomics. J Mol Diagn. 22, 3–20.

Subramanian, A., Tamayo, P., Mootha, V.K., Mukherjee, S., Ebert, B.L., Gillette, M.A., Paulovich, A., Pomeroy, S.L., Golub, T.R., Lander, E.S., et al. (2005). Gene set enrichment analysis: a knowledge-based approach for interpreting genome-wide expression profiles. Proc. Natl. Acad. Sci. U. S. A. 102, 15545–15550.

Tikhonova, A.N., Dolgalev, I., Hu, H., Sivaraj, K.K., Hoxha, E., Cuesta-Domínguez, Á., Pinho, S., Akhmetzyanova, I., Gao, J., Witkowski, M., et al. (2019). The bone marrow microenvironment at single-cell resolution. Nature 569, 222–228.

Trapnell, C., Pachter, L., and Salzberg, S.L. (2009). TopHat: discovering splice junctions with RNA-Seq. Bioinformatics. 25, 1105–1111.

Trapnell, C., Roberts, A., Goff, L., Pertea, G., Kim, D., Kelley, D.R., Pimentel, H., Salzberg, S.L., Rinn, J.L., and Pachter, L. (2012). Differential gene and transcript expression analysis of RNA-seq experiments with TopHat and Cufflinks. Nat. Protoc. 7, 562–578.

da Veiga Leprevost, F., Grüning, B.A., Alves Aflitos, S., Röst, H.L., Uszkoreit, J., Barsnes, H., Vaudel, M., Moreno, P., Gatto, L., Weber, J., et al. (2017). BioContainers: an open-source and community-driven framework for software standardization. Bioinforma. Oxf. Engl. 33, 2580–2582.

Verny, L., Sella, N., Affeldt, S., Singh, P.P., and Isambert, H. (2017). Learning causal networks with latent variables from multivariate information in genomic data. PLoS Comput. Biol. 13, e1005662.

Warnat-Herresthal, S., Perrakis, K., Taschler, B., Becker, M., Baler, K., Beyer, M., Günther, P., Schulte-Schrepping, J., Seep, L., Klee, K., et al. (2020). Scalable Prediction of Acute Myeloid Leukemia Using High-Dimensional Machine Learning and Blood Transcriptomics. iScience 23, 100780.

Wolf, F.A., Hamey, F.K., Plass, M., Solana, J., Dahlin, J.S., Göttgens, B., Rajewsky, N., Simon, L., and Theis, F.J. (2019). PAGA: graph abstraction reconciles clustering with trajectory inference through a topology preserving map of single cells. Genome Biol. 20, 59.

7

The Regulatory Network of X Chromosome Inactivation

Jeffrey Boeren, Sabina Jašarević and Joost Gribnau*

Department of Developmental Biology, Oncode Institute, Erasmus MC,
The Netherlands
E-mail: j.gribnau@erasmusmc.nl
*Corresponding Author

7.1 Introduction

Males and females differ by their sex chromosome composition but neverthe-less display balanced expression of X-linked genes. This is accomplished by a complex dosage compensation process, necessary for correct functioning of a cell, which evolved early during eutherian evolution (Gribnau and Groote-goed 2012). Throughout evolution, the fate of the X and Y chromosomes that evolved from one and the same autosome has changed. The Y chromo-some degenerated to a gene-poor chromosome whereas the X chromosome has retained most of its original genes. Susumu Ohno stated that due to degeneration of the Y chromosome, genes on the X chromosome became haploinsufficient, triggering upregulation of dose-sensitive X chromosomal genes in males (Ohno, 1967). However, he also hypothesized that this upregu-lation was not restricted to males, but it also occurred on both X chromosomes in females. To overcome this imbalance, female-specific downregulation co-evolved. Mary Lyon postulated that in placental mammals downregulation of X-linked genes in female cells is accomplished by inactivation of one of the two X chromosomes during early female development (Lyon 1961). Subsequent studies have shown that in mice and several other mammals (not including human) there are two waves of X chromosome inactivation. Imprinted XCI (iXCI) is initiated in all cells of the pre-implantation embryo, in which the paternal X chromosome is exclusively inactivated (Takagi & Sasaki, 1975). At a later stage of development, in the inner cell mass of the blastocyst, the paternal X chromosome is reactivated, followed by a wave of random XCI (rXCI), with either the maternal or paternal X chromosome being inactivated, initiated just after implantation.

The key regulatory region for XCI is the X inactivation centre (Xic) (Figure 7.1). The Xic is the region on the X chromosome required for initiation of XCI. Within this region, both protein-coding and lncRNA genes were identified to be required for the proper initiation of XCI. The best studied element in the Xic is the X-inactive specific transcript (*Xist*). *Xist* encodes a functional lncRNA that is upregulated at the onset of XCI, is transcribed from and coats the entire future inactive X chromosome (Xi), thereby recruiting the silencing machinery to the future inactive X chromosome (Brockdorff et al., 2020). *Xist* is required for XCI, as an X chromosome that carries a deletion of *Xist* is never inactivated (Brockdorff et al., 1992; Brown et al., 1992; Penny et al., 1996). Before XCI is initiated, *Xist* expression is kept at low levels, pointing to the involvement of a repressor of XCI. The antisense transcribed *Xist* overlapping gene *Tsix* was identified as the repressor of *Xist* (Lee & Lu, 1999). *Tsix* also encodes a lncRNA that before XCI is expressed biallelically. After XCI, *Tsix* remains transiently expressed monoallelically from the active X chromosome (Xa). When *Tsix* activity is fully abolished on one of the two X chromosomes, XCI is skewed toward initiation of XCI on the X carrying the mutation (Lee & Lu, 1999).

Decades of research led to the identification of more genes and elements in the Xic that positively (*Jpx, Ftx, Rnf12*) or negatively regulate *Xist* (*Xite, Linx, LinxP*) (Figure 7.1). More genes/elements have been found on the Xic, such as *Slc16a2* and *Cdx4*, but to this day, the function of these remains unknown. Repressors of *Xist* were mostly found upstream of *Tsix* and activators often located downstream of *Xist*. Experiments mapping chromatin interactions within the Xic, revealed that the Xic is separated over two Topologically Associated Domains (TADs) and that the overlapping open reading frames of *Tsix/Xist* are located on the border of the TADs, separating the *Tsix* and *Xist* promoter in two distinct TADs (Nora et al., 2012). TADs represent *cis* regions with a higher frequency of chromatin interactions and diverse chromatin conformations of which the borders are usually marked by CTCF and Cohesin binding (Dixon et al., 2012; Giorgetti et al., 2014). TADs play a role in shaping regulatory environments in which expression of genes that cluster together can be coordinated (Nora et al., 2012). This chromatin

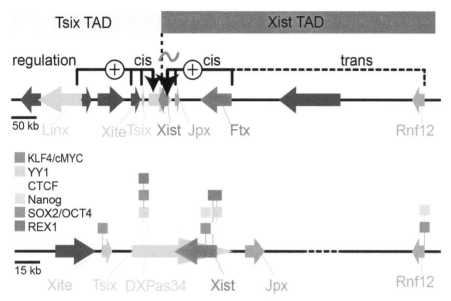

Figure 7.1 The X inactivation centre (Xic). A schematic representation of the Xic in the mouse genome. **Top panel:** Genes colored in green represent Xist repressors and genes colored in red represent Xist activators acting in cis to regulate Xist; Rnf12 acts in trans as a dose-dependent activator of XCI. **Bottom panel:** the transcription factor network in the regulation of XCI.

organization thus physically separates *Tsix* and its regulators from *Xist* and its regulators by placing them in two different regulatory environments.

The elements that regulate XCI are thus located within the Xic and lie in two different TADs. In this chapter, we will discuss how the regulatory network directing gene activity in the Xic leads to female exclusive XCI, and highlights the important role of pluripotency factors in the proper initiation of XCI.

7.2 The X:Autosome Ratio Dictates Initiation of XCI

XCI involves many levels of regulation, and the Xic is the crucial region for the ability of an X chromosome to inactivate. It has been thought for a long time that Xic-mediated pairing and nuclear positioning of the X chromosomes as a factor directing XCI counting and initiation. In this model, two X chromosomes pair up together, facilitating the cell in counting the number of X chromosomes and decide how many and which chromosomes need to be inactivated (Augui et al., 2007; Xu et al., 2006). The pairing process was therefore thought to exchange factors to establish asymmetry of the two X chromosomes. This would explain why male cells do not initiate XCI, since the sole X cannot pair with another X. Depletion of *Oct4* in female embryonic stem cells (ESCs) was found to block pairing resulting in the inactivation of both X chromosomes, arguing that pairing is important for counting (Donohoe et al., 2009). However, recent advances studying multinucleated cells and tethering the X chromosome to the nuclear lamina revealed that pairing is not the primary determinant of random XCI, but may be the mere consequence of transcriptional activation of genes in the Xic (Barakat et al., 2014; Pollex & Heard., 2019).

Studies with multinucleated heterokaryons also indicated that XCI was dependent on trans-acting factors that diffuse across the membranes of the two nuclei (Barakat et al., 2014). These trans-acting factors are involved in setting up the probability for an X chromosome to be inactivated (Monkhorst et al., 2008). When each chromosome has a probability to be inactivated, there must be a mechanism to sense how many X chromosomes are present and how many X chromosomes should be inactivated. Experiments with tetraploid ESCs have shown that these cells preferably keep one active X per diploid genome (Monkhorst et al., 2008). This and other observations indicated that the X:autosome ratio dictates the probability to initiate XCI, predicting the presence of X-encoded activators (numerators) and autosomally encoded inhibitors (denominators) of XCI (Monkhorst et al., 2008; Monkhorst et al., 2009). Both activators and inhibitors of XCI act in *trans* to regulate *Xist* expression through interaction with several elements and genes located within the Xic. Feedback mechanisms involving rapid downregulation of X-encoded activators upon XCI, and a close link between XCI and loss of pluripotency prevents XCI of too many or too less X chromosomes respectively.

7.3 Initiating X-Inactivation, the Regulatory Environment of the Xic

The Xic harbors several coding and non-coding elements involved in regulating *Xist* and *Tsix* expression in *cis*. One of these—*Jpx*—is a non-coding locus located upstream of *Xist*. *Jpx* is co-expressed with *Xist* during ESC differentiation, partially escapes XCI, and is known as an activator of *Xist*. Initial studies indicated that both copies of *Jpx* are required for XCI and proper cell differentiation, and that both copies of *Jpx* are required for proper initiation of XCI in female ESCs (Tian, Sun & Lee., 2010). Increasing the copy numbers of *Jpx* transgenes positively correlated with *Xist* expression, suggesting a *trans* mechanisms in the activation of *Xist* by *Jpx* through *Jpx*-mediated eviction of CTCF at the *Xist* promoter (Karner et al., 2020; Carmona et al., 2018; Sun et al., 2013). A recent study also provided evidence that the function of *Jpx* RNA is conserved between mouse and human since human *Jpx* can rescue the phenotype of a heterozygous *Jpx* deletion in mice, despite the great divergence in structure and sequence of the *Jpx* RNA (Karner et al., 2020).

A possible role for *Jpx* in *cis* regulation of *Xist* came from experiments indicating that *Jpx* might influence *Tsix*, as truncation of *Tsix* in *cis* of the *Jpx* deletion restored viability of differentiating ESCs (Tian, Sun & Lee., 2010). In addition, a heterozygous deletion spanning 500 kb from *Jpx* up to *Rnf12* only resulted in skewed *Xist* expression from the wild-type chromosome, but no further phenotype, suggesting that *Jpx* and its neighboring elements have a *cis*-preference (Barakat et al., 2014). In addition, introduction of a transgene containing *Jpx*, *Ftx*, and *Slc16a2* did not affect *Xist* expression, which argues against a *trans*-acting role for *Jpx* in XCI (Barakat et al., 2014). These findings indicate that *Jpx*'s action in XCI might involve both *cis*- and *trans*-regulatory mechanisms that need further investigation to discriminate between both pathways.

Another non-coding locus found upstream of *Xist*, is *Ftx*. The *Ftx* locus produces a lncRNA and is, as described for *Jpx*, co-expressed with *Xist* during differentiation, and similar to *Jpx* partially escapes XCI (Chureau et al., 2011). Various *Ftx* isoforms are produced and the locus contains a microRNA cluster with unknown function. Experiments with male ESCs harboring a deletion of *Ftx* showed reduced transcription levels of many genes located in the entire Xic with *Xist* expression levels being affected most (Chureau et al., 2011). Subsequent studies with female *Ftx* transgenic and knockout ESC lines indicated that *Ftx* acts in *cis* to regulate *Xist*, and that positive regulation of *Xist* does not involve the *Ftx* RNA but is mediated by *Ftx* transcriptional activity (Furlan et al., 2018, Barakat et al., 2014). The role of *Ftx* thus might therefore to create an environment for active transcription across its flanking genes and especially *Xist*. Indeed, transcription of *Ftx* prevents the accumulation of heterochromatin marks on histones in the region of *Ftx* (Chureau et al., 2011). *Ftx*-mediated activation of *Xist* might also involve mass action (recruitment of shared transcription factors) or mediated through shared enhancers that might be limited to the TAD *Xist* and *Ftx* are embedded in, but also this question needs further investigation.

Tsix acts as the key negative regulator of *Xist*. Loss of *Tsix* expression is a characteristic for the Xi, and when *Tsix* activity was affected on one chromosome, this led to a complete skewing of *Xist* upregulation from the affected chromosome (Lee & Lu., 1999). XCI also occurred faster on the affected allele than on the wild-type allele (Lee & Lu., 1999; Lee, 2005; Monkhorst et al., 2008). Studies involving a *Tsix* truncation, showed this to result in elevated H3K4 di-methylation at the *Xist* promoter, indicating *Tsix* to act through chromatin remodeling (Navarro et al., 2005). Whether this involves *Tsix* RNA-mediated deposition of chromatin remodeling proteins or complexes, or the act of transcription leading to chromatin changes at the reciprocal promoter needs further investigation. Other studies have shown that overlapping transcription of *Xist* and *Tsix* leads to mutual repression involving transcriptional interference (Mutzel et al., 2019). These findings indicate that *Tsix* negatively affects *Xist* transcription through multiple mechanisms.

Another potential negative regulator of *Xist* is the non-coding locus Large INtervening transcript in the Xic, *Linx*. *Linx* is co-expressed with *Tsix* in male and female ESCs (Nora et al., 2012). This study also reported *Linx* often to be expressed monoallelically prior to XCI, uncovering a transcriptional asymmetry of the two X chromosomes prior to XCI (Nora et al., 2012). Furthermore, the probability of expressing *Tsix* on one X chromosome increased when *Linx* was expressed in cis (Nora et al., 2012). Recent studies indicate that expression of *Linx* is involved in shaping the transcriptional landscape of the Xic independent of its RNA product and *Tsix* (Galupa et al., 2020). *Linx* acts as a repressor of *Xist* through inter TAD promoter interaction, and intriguingly introduction of the *Linx* promoter (*LinxP*) in the *Xist* TAD domain resulted in enhanced *Xist* expression, highlighting an important instructive role for higher order chromatin structure in *Xist* regulation (Galupa et al., 2020). The importance of the *cis*-regulatory environment was further addressed by inversion of *Tsix–Xist* in male and female ESCs, resulting in a switch in expression dynamics with aberrant *Xist* expression in undifferentiated ESCs (Van Bemmel et al., 2019). In addition, another lncRNA gene *Xite*, located upstream of *Tsix*, plays a crucial role in maintaining the *Xist/Tsix* boundary. These findings highlight the complexity of gene regulation in XCI and shows that the TAD-environment provides important instructions for the function of a particular element and that this function might depend on other cis-regulatory elements.

7.4 The Pluripotency Network and Other Trans-acting Factors of XCI

XCI initiation occurs in a specific developmental time window in female embryos and is beautifully recapitulated in differentiating female ESCs. It is associated with downregulation of pluripotency factors and exit of the pluripotent state in which both X chromosomes are active in female cells. Pluripotency factors have been found to act in *trans* as inhibitors of XCI, either through repressing *Xist* or activating *Tsix* directly, or independently by regulating factors that act on *Tsix* or *Xist*. ChIP seq studies revealed OCT4, SOX2 and Nanog binding sites within *Xist* intron 1 and in the *Xite/Tsix* promoters (Figure 7.1) (Donohoe et al., 2009; Navarro et al., 2008; Navarro et al., 2010). Interfering with expression of these factors affects the timing and level of *Xist* expression. For instance, *Oct4* silencing in male ESCs led to upregulation of *Xist* expression, almost compatible with female levels of *Xist* (Navarro et al., 2008). Depletion of *Oct4* in female ESCs resulted in biallelic upregulation of *Xist* (Donohoe et al., 2009). Mutating the binding motifs of OCT4 and SOX2 in *Tsix* resulted in abolished stimulation of *Tsix* transcription and silencing *Oct4* led to downregulation of *Tsix, Nanog,* and *Sox2*, but also upregulation of *Xist* (Donohoe et al., 2009; Navarro et al., 2008). Together, these results imply that *Oct4* can inhibit *Xist* expression by stimulating *Tsix* expression and thus keeping an X chromosome in its active state.

Pluripotency factors *Rex1*, *Klf4,* and *c-Myc* seem to have a direct effect on *Tsix* expression, as KLF4 and c-MYC bind to a region between the *Tsix* regulatory element DXPas34 and the *Tsix* promoter and REX1 binds DXPas34 directly (Figure 7.1) (Navarro et al., 2010). In addition, *Rex1* was found to be a positive regulator of *Tsix* expression (Navarro et al., 2010). DXPas34 is a CG-rich repeat downstream of the *Tsix* promoter, with a unique DNA methylation profile on the active X chromosome (Prissette et al., 2001). Deletion of DXPas34 in male ESCs led to a strong reduction of *Tsix* transcription and ectopic XCI (Vigneau et al., 2006), and resulted in lowered recruitment of basal transcription factors to the *Tsix* promoter and strongly reduced *Tsix* RNA levels (Navarro et al., 2010). Knockdown experiments with *Rex1* shRNA affected *Tsix* RNA levels by reducing it by half, but KLF4 and c-MYC binding was unaffected, suggesting that KLF4 and c-MYC bind independently of REX1 to the DXPas34-Tsix promoter region. In addition, RNA polymerase II levels and H3K36 tri-methylation (associated with transcriptional elongation) in the 3' end of *Tsix* around the

Xist promoter were reduced, implying reduced transcriptional activity of *Tsix* (Navarro et al., 2010). Furthermore, *c-Myc* is known as a global regulator of transcriptional elongation and may aid in *Tsix*-mediated repression of *Xist* (Rahl et al., 2010).

REX1 is an autosomal factor that evolved from the retro-transposition of autosomal factor *Yy1* and therefore has similar binding motifs (Makhlouf et al., 2014). YY1 has been shown to bind promoter regions of both *Xist* and *Tsix*, as described for REX1 (Figure 7.1) (Gontan et al., 2018; Makhlouf et al., 2014; Navarro et al., 2010). *Yy1* is considered an activator of *Xist*, since loss of *Yy1* prevents *Xist* upregulation. ESCs knocked down for *Yy1* do not show *Xist* upregulation (Makhlouf et al., 2014). In addition, YY1 and REX1 compete for binding at the *Xist* 5' region. In wild-type ESCs, YY1 occupies this region but in ESCs overexpressing REX1 the YY1 occupancy is reduced and REX1 occupancy increased (Makhlouf et al., 2014). Yet, the overall levels of YY1 were not reduced in REX1 overexpressing ESCs, showing a competition for the binding site in *Xist* (Makhlouf et al., 2014). REX1 dosage therefore seems important for the action of YY1 as a trans-acting activator of *Xist*. Recently, YY1 has been established as a global mediator of enhancer–promoter loops (Weintraub et al., 2017). YY1 could therefore contribute to *Xist* activation in XCI by establishing enhancer–promoter loops. Further experiments, for instance targeted mutagenesis of Xist–YY1 binding sites could uncover the exact function of YY1 in regulating XCI.

REX1 was identified as a target of the X-encoded activator of XCI, *Rnf12* (Gontan et al., 2012). *Rnf12* is located in the Xist-TAD and encodes an E3-ubiquitin ligase that is upregulated in the early stages of differentiation and targets different proteins including REX1 for proteasomal degradation. Overexpression of *Rnf12* leads to upregulation of *Xist* and ectopic XCI (both in male and female ESCs), whereas disrupting *Rnf12* results in an XCI phenotype in female mouse ESCs (Barakat et al., 2011; Jonkers et al., 2009). *Rnf12* therefore acts as dose-dependent and *trans*-acting factor to activate XCI. The concentration of X-encoded RNF12 is higher in female cells with two Xas, and therefore REX1 levels lower compared to male cells, directly linking RNF12:REX1 to female exclusive initiation of XCI.

Knockout of *Rnf12 in vivo* leads to failure of iXCI and embryonic lethality (Shin et al., 2010). This phenotype was rescued by an additional knockout of *Rex1* (Gontan et al., 2018). In *Rnf12* knockout mice, REX1 is not targeted for degradation and its repressing activity on *Xist* prevents iXCI to be initiated. Removal of *Rex1* in this *Rnf12* negative background thus allows *Xist* upregulation and iXCI to be initiated. REX1 therefore is a strong inhibitor of imprinted XCI and degradation of REX1 is required for normal development. These results also suggest a more prominent role of *Rnf12–Rex1* action in iXCI than rXCI as double KO mice are born. Although XCI is skewed in *Rnf12* heterozygous knockout mice and differentiating ESCs, *Rnf12* null phenotype was only observed in differentiating ESCs (Barakat et al., 2011; Shin et al., 2014; Wang et al., 2017). These results suggest that other activators of random XCI must act in parallel with *Rnf12–Rex1* axis to induce female exclusive XCI.

Besides REX1-mediated regulation of *Xist*, Nanog, OCT4, and SOX2 also retain binding sites within *Xist* intron 1 (Figure 7.1). Release of these factors from intron 1 coincides with *Xist* upregulation (Navarro et al., 2008). Female ESCs carrying a heterozygous deletion of *Xist* intron 1 display a very mild XCI phenotype, but this study also indicated this region to act together with *Tsix* in the repression of *Xist* (Minkovsky et al., 2013). Nanog expression seems to anti-correlate with *Xist* expression in ESCs as *Xist* was upregulated after inducing depletion of Nanog (Navarro et al., 2008). Furthermore, no downregulation of *Tsix* was observed upon deletion of Nanog, in line with the observation that *Tsix* lacks Nanog-binding sites. Therefore, it seems that Nanog exerts its function on *Xist* only. In addition, *Nanog* null cells showed a slight upregulation of *Rnf12*. Nanog, OCT4, and SOX2 were found to have binding sites upstream of *Rnf12* (Figure 7.1) (Navarro et al., 2011). Inhibition of *Oct4* in male ESCs resulted in slightly increased *Rnf12* levels not enough to induce XCI. However, when *Nanog* and *Sox2* were silenced additionally to *Oct4*, *Rnf12* expression increased to a level compatible with XCI in female ESCs (Navarro et al., 2011). Therefore, *Oct4*, *Sox2*, and *Nanog* act in concert to repress the XCI-activator *Rnf12* and thus inhibit XCI.

The loss of most individual pluripotency factors and pluripotency factor–binding sites has limited effects on *Xist* expression. We can therefore only speculate that the pluripotency factors act in concert to inhibit XCI as well as ensure proper timing in development and differentiation. For most of the regulators of XCI, such as *Tsix, Jpx, Ftx, Rnf12*, and even *Xist* itself, it is not fully known yet what regulates their activity. Pluripotency factors might also play a role in regulating other factors involved in the regulation of XCI.

7.5 The Balancing Act of Activators and Inhibitors in Initiation of XCI

Dosage-sensing by X-encoded activators specifies the process of XCI, where autosomally encoded inhibitors of XCI provide the threshold for XCI initiation. Several of the described XCI-inhibitors appear to be pluripotency factors providing a tight link between loss of pluripotency and initiation of XCI that happens just after implantation of the embryo. Upon development or during ESC differentiation the levels of XCI-activators rise whereas the levels of XCI-inhibitors decrease. In female cells, the level of XCI-activators is twofold compared to male cells and sufficient to pass the threshold and initiate XCI. So far, *Rnf12* and *Jpx* are the only factors described as *trans*-acting activators, and although the robustness of XCI is affected in $Rnf12^{+/-}$ and $Jpx^{+/-}$ cells, XCI is still initiated in a significant percentage of differentiating mutant female ESCs. This and several other findings indicate the presence of additional activators of XCI which need to be identified in future studies.

Exclusive inactivation of one X chromosome can only be accomplished by robust feedback (Figure 7.2). *Rnf12* is indeed among the first genes silenced after *Xist* upregulation, and the half-life of both RNF12 and its target REX1 are relatively short facilitating a quick response to XCI (Gontan et al., 2012). In addition, at least one active copy of *Rnf12* is required for

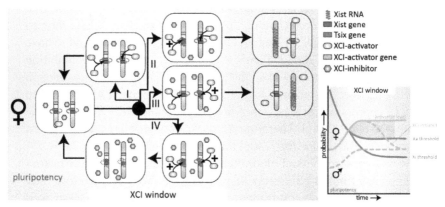

Figure 7.2   Upon differentiation or development the concentration of the XCI-activator increases (*green*); and the concentration of the XCI-inhibitors drops (*blue*). As a result, in female cells and not in male cells the threshold for XCI initiation is reached, leading to accumulation of Xist in cis (+), initiated with a specific probability (*arrows*, all four possible outcomes are shown). Cells that do not initiate XCI cannot differentiate (I) allowing the XCI initiation process to proceed. Cells that initiate XCI on both X chromosomes (IV) will silence XCI-activators on both X chromosomes, leading to loss of Xist and a restart of the XCI initiation process. Finally, in cells that initiate XCI on one X chromosome (II and III) Xist mediated silencing of the XCI-activator gene in cis and a subsequent drop in the XCI-activator concentration prevents initiation of XCI on the second X. Xist expression is maintained on the Xi as a consequence of a lower threshold for Xist expression.

sustained XCI in female ESCs, indicating that when cells initiate XCI on both X chromosomes this leads to repression of *Xist*, allowing the process to start over again (Barakat et al., 2014). In contrast to *Rnf12, Jpx* escapes XCI, and therefore cannot attribute to negative feedback preventing one X chromosome too many. Following downregulation of XCI-activators on one X chromosome, facilitated by initiation of XCI, *Xist* expression needs to be maintained, requiring additional mechanisms, possibly involving shutdown of the *Tsix* TAD relieving repression on *Xist*. As XCI activators are involved in determining the probability for an X chromosome to initiate XCI, in a certain timespan no XCI may be initiated. Interestingly, studies with female cells that cannot initiate XCI have shown that ESC differentiation is blocked through inhibition of the MAPK and GSK3 pathways and stimulation of the Akt pathway (Schulz et al., 2014). This indicates that cells that do not initiate XCI retain their pluripotent state until XCI is initiated, providing a properly timed, high fidelity and robust XCI process.

7.6 Concluding Remarks

The studies discussed in this chapter indicate that *trans*-acting factors act in concert with *cis*-regulatory elements on the X chromosome to ensure the proper regulation of *Xist* and XCI. The timing of XCI is tightly linked to loss of pluripotency, and female exclusive initiation of XCI is mediated by X-linked activators of XCI. Although a lot of progress has been made and important regulatory factors and elements have been identified, others still need identification. This includes additional activators of XCI, and a better view on the role of individual and combinations of factors and elements (in the context of the chromatin organization) in orchestrating *Xist* expression.

So far, most studies aimed at identification and determination of the role of several factors in XCI were performed in mouse, providing a detailed but not yet complete picture of the complexity of this process. However, studies examining XCI in other mammalian species highlight important differences in kinetics of the XCI process. For instance, human pre-implantation female embryos show biallelic *XIST* accumulation, whereas human female somatic as well as ESCs and induced pluripotent stem cells (iPSCs) mostly retain a single *XIST* coated X chromosome. This suggests that other feedback mechanisms might be involved in establishment of the inactive X, and that the present ESC and iPSC culturing protocols do not result in cells that represent the human epiblast. Also, the function of *TSIX* which only partially overlaps *XIST* is unclear, suggesting that other factors and mechanisms might be in play in the regulation of human XCI. Fortunately, this field is moving forward very quickly, and in combination with CRISPR/Cas9 targeting, these "naive" ESCs or iPSCs will serve as powerful model systems to study the regulation of human XCI.

Take home messages:

- XCI is regulated by trans-acting X-encoded activators and autosomally encoded inhibitors of XCI that regulate Xist expression acting through several cis-acting elements located within the XIC.
- Initiation of XCI is tightly linked to development by implementing the pluripotency transcription factor network as inhibitors of XCI.
- The XIC contains several lncRNA genes with a very diverse function in XCI, ranging from silencing or activation through lncRNA-mediated enzyme recruitment, through transcriptional interference, or by promoter co-activation mechanisms.
- Female exclusive XCI of one single X chromosome requires tight, fast, and robust feedback systems.

References

Augui, S., Filion, G. J., Huart, S., Nora, E., Guggiari, M., Maresca, M., … Heard, E. (2007). Sensing X Chromosome Pairs Before X Inactivation via a Novel X-Pairing Region of the Xic. *Science*, *318*(5856), 1632–1636.

Barakat, T. S., Gunhanlar, N., Pardo, C. G., Achame, E. M., Ghazvini, M., Boers, R., … Gribnau, J. (2011). RNF12 activates Xist and is essential for X chromosome inactivation. *PLoS Genetics*, *7*(1), 1–12. https://doi.org/10.1371/journal.pgen.1002001

Barakat, T. S., Loos, F., Van Staveren, S., Myronova, E., Ghazvini, M., Grootegoed, J. A., & Gribnau, J. (2014). The trans-activator RNF12 and cis-acting elements effectuate X chromosome inactivation independent of X-pairing. *Molecular Cell*, *53*(6), 965–978. https://doi.org/10.1016/j.molcel.2014.02.006

Brockdorff, N., Ashworth, A., Kay, G. F., McCabe, V. M., Norris, D. P., Cooper, P. J., … Rastan, S. (1992). The product of the mouse Xist gene is a 15 kb inactive X-specific transcript containing no conserved ORF and located in the nucleus. *Cell*, *71*(3), 515–526. https://doi.org/10.1016/0092-8674(92)90519-I

Brockdorff, N., Bowness, J. S., & Wei, G. (2020). Progress toward under-standing chromosome silencing by Xist RNA. *Genes & Development*, *34*(11–12), 733–744. https://doi.org/10.1101/gad.337196.120

Brown, C. J., Hendrich, B. D., Rupert, J. L., Lafrenière, R. G., Xing, Y., Lawrence, J., & Willard, H. F. (1992). The human XIST gene: Analysis of a 17 kb inactive X-specific RNA that contains conserved repeats and is highly localized within the nucleus. *Cell*, *71*(3), 527–542. https://doi.org/10.1016/0092-8674(92)90520-M

Carmona, S., Lin, B., Chou, T., Arroyo, K., & Sun, S. (2018). LncRNA Jpx induces Xist expression in mice using both trans and cis mechanisms. *PLoS Genetics*, *14*(5), 1–21. https://doi.org/10.1371/journal.pgen.1007378

Chureau, C., Chantalat, S., Romito, A., Galvani, A., Duret, L., Avner, P., & Rougeulle, C. (2011). Ftx is a non-coding RNA which affects Xist expression and chromatin structure within the X-inactivation center region. *Human Molecular Genetics*, *20*(4), 705–718. https://doi.org/10.1093/hmg/ddq516

Dixon, J. R., Selvaraj, S., Yue, F., Kim, A., Li, Y., Shen, Y., … Ren, B. (2012). Topological domains in mammalian genomes identi-fied by analysis of chromatin interactions. *Nature*, *485*(7398), 376–380. https://doi.org/10.1038/nature11082

Donohoe, M. E., Silva, S. S., Pinter, S. F., Xu, N., & Lee, J. T. (2009). The pluripotency factor, Oct4, interacts with Ctcf and also con-trols X-chromosome pairing and counting. *Nature*, *460*(7251), 128–132. https://doi.org/10.1038/nature08098

Furlan, G., Gutierrez Hernandez, N., Huret, C., Galupa, R., van Bemmel, J. G., Romito, A., … Rougeulle, C. (2018). The Ftx Noncoding Locus Controls X Chromosome Inactivation Indepen-dently of Its RNA Products. *Molecular Cell*, *70*(3), 462-472.e8. https://doi.org/10.1016/j.molcel.2018.03.024

Galupa, R., Nora, E. P., Worsley-Hunt, R., Picard, C., Gard, C., van Bemmel, J. G., … Heard, E. (2020). A Conserved Noncoding Locus Regulates Ran-dom Monoallelic Xist Expression across a Topological Boundary. *Molec-ular Cell*, *77*(2), 352-367.e8. https://doi.org/10.1016/j.molcel.2019.10.030

Giorgetti, L., Galupa, R., Nora, E. P., Piolot, T., Lam, F., Dekker, J., … Heard, E. (2014). Predictive polymer modeling reveals coupled fluc-tuations in chromosome conformation and transcription. *Cell*, *157*(4), 950–963. https://doi.org/10.1016/j.cell.2014.03.025

Gontan, C., Achame, E. M., Demmers, J., Barakat, T. S., Rentmeester, E., Van Ijcken, W., … Gribnau, J. (2012). RNF12 initiates X-chromosome inactivation by targeting REX1 for degradation. *Nature*, *485*(7398), 386–390. https://doi.org/10.1038/nature11070

Gontan, C., Mira-Bontenbal, H., Magaraki, A., Dupont, C., Barakat, T. S., Rentmeester, E., … Gribnau, J. (2018). REX1 is the critical target of RNF12 in imprinted X chromosome inactivation in mice. *Nature Communications*, *9*(1), 1–12. https://doi.org/10.1038/s41467-018-07060-w

Gribnau, J., & Grootegoed, J. A. (2012). Origin and evolution of X chromosome inactivation. *Current Opinion in Cell Biology*, *24*(3), 397–404. https://doi.org/10.1016/j.ceb.2012.02.004

Jonkers, I., Barakat, T. S., Achame, E. M., Monkhorst, K., Kenter, A., Rentmeester, E., … Gribnau, J. (2009). RNF12 Is an X-Encoded Dose-Dependent Activator of X Chromosome Inactivation. *Cell*, *139*(5), 999–1011. https://doi.org/10.1016/j.cell.2009.10.034

Karner, H., Webb, C. H., Carmona, S., Liu, Y., Lin, B., Erhard, M., … Sun, S. (2020). Functional Conservation of LncRNA JPX Despite Sequence and Structural Divergence. *Journal of Molecular Biology*, *432*(2), 283–300. https://doi.org/10.1016/j.jmb.2019.09.002

Lee, J. T. (2005). Molecular biology: Regulation of X-chromosome counting by Tsix and Xite sequences. *Science*, *309*(5735), 768–771. https://doi.org/10.1126/science.1113673

Lee, J. T., & Lu, N. (1999). Targeted mutagenesis of Tsix leads to nonrandom X inactivation. *Cell*, *99*(1), 47–57. https://doi.org/10.1016/S0092-8674(00)80061-6

Lyon, M. F. (1961). Gene action in the X chromosome of the mouse (Mus muscu/us L). *Nature*, *190*, 372–373.

Makhlouf, M., Ouimette, J. F., Oldfield, A., Navarro, P., Neuillet, D., & Rougeulle, C. (2014). A prominent and conserved role for YY1 in Xist transcriptional activation. *Nature Communications*, *5*, 1–12. https://doi.org/10.1038/ncomms5878

Minkovsky, A., Barakat, T. S., Sellami, N., Chin, M. H., Gunhanlar, N., Gribnau, J., & Plath, K. (2013). The pluripotency factor-bound intron 1 of xist is dispensable for X chromosome inactivation and reactivation In Vitro and In Vivo. *Cell Reports*, *3*(3), 905–918. https://doi.org/10.1016/j.celrep.2013.02.018

Monkhorst, K., de Hoon, B., Jonkers, I., Achame, E. M., Monkhorst, W., Hoogerbrugge, J., … Gribnau, J. (2009). The probability to initiate X chromosome inactivation is determined by the X to autosomal ratio and X chromosome specific allelic properties. *PLoS ONE*, *4*(5). https://doi.org/10.1371/journal.pone.0005616

Monkhorst, K., Jonkers, I., Rentmeester, E., Grosveld, F., & Gribnau, J. (2008). X Inactivation Counting and Choice Is a Stochastic Process: Evidence for Involvement of an X-Linked Activator. *Cell*, *132*(3), 410–421. https://doi.org/10.1016/j.cell.2007.12.036

Mutzel, V., & Schulz, E. G. (2020). Dosage Sensing, Threshold Responses, and Epigenetic Memory: A Systems Biology Perspective on Random X-Chromosome Inactivation. *BioEssays*, *42*(4), 1–14. https://doi.org/10.1002/bies.201900163

Mutzel, V., Okamoto, I., Dunkel, I., Saitou, M., Giorgetti, L., Heard, E., & Schulz, E. G. (2019). A symmetric toggle switch explains the onset of random X inactivation in different mammals. *Nature Structural and Molecular Biology*, *26*(May). https://doi.org/10.1038/s41594-019-0214-1

Navarro, P., Moffat, M., Mullin, N. P., & Chambers, I. (2011). The X-inactivation trans-activator Rnf12 is negatively regulated by pluripotency factors in embryonic stem cells. *Human Genetics*, *130*(2), 255–264. https://doi.org/10.1007/s00439-011-0998-5

Navarro, P., Oldfield, A., Legoupi, J., Festuccia, N., Dubois, A. S., Attia, M., … Avner, P. (2010). Molecular coupling of Tsix regulation and pluripotency. *Nature*, *468*(7322), 457–460. https://doi.org/10.1038/nature09496

Navarro, P., Oldfield, A., Legoupi, J., Festuccia, N., Dubois, A. S., Attia, M., … Avner, P. (2008). Molecular coupling of Xist regulation and pluripotency. *Science*, *321*(5896), 1693–1695. https://doi.org/10.1126/science.1160952 ARTICLE

Navarro, P., Pichard, S., Ciaudo, C., Avner, P., & Rougeulle, C. (2005). Tsix transcription across the Xist gene alters chromatin conformation without affecting Xist transcription: Implications for X-chromosome inactivation. *Genes and Development*, *19*(12), 1474–1484. https://doi.org/10.1101/gad.341105

Nora, E. P., Lajoie, B. R., Schulz, E. G., Giorgetti, L., Servant, N., Piolot, T., … Dekker, J. (2012). Spatial partitioning of the regulatory landscape of the X- inactivation center. *Nature*, *485*(7398), 381–385. https://doi.org/10.1038/nature11049.Spatial

Ohno, S. (1967). *Sex Chromosomes and Sex-Linked Genes*. Berlin: Springer.

Penny, G. D., Kay, G. F., Sheardown, S. A., Rastan, S., & Brockdorff, N. (1996). Requirement for Xist in X chromosome inactivation. *Nature*, *379*(6561), 131–137. https://doi.org/10.1038/379131a0

Pollex, T., & Heard, E. (2019). Nuclear positioning and pairing of X-chromosome inactivation centers are not primary determinants during initiation of random X-inactivation. *Nature Genetics*, *51*(2), 285–295. https://doi.org/10.1038/s41588-018-0305-7

Prissette, M., El-Maarri, O., Arnaud, D., Walter, J., & Avner, P. (2001). Methylation profiles of DXPas34 during the onset of X-inactivation. *Human Molecular Genetics*, *10*(1), 31–38. https://doi.org/10.1093/hmg/10.1.31

Rahl, P. B., Lin, C. Y., Seila, A. C., Flynn, R. A., McCuine, S., Burge, C. B., ... Young, R. A. (2010). c-Myc regulates transcriptional pause release. *Cell*, *141*(3), 432–445. https://doi.org/10.1016/j.cell.2010.03.030

Schulz, E. G., Meisig, J., Nakamura, T., Okamoto, I., Sieber, A., Picard, C., ... Heard, E. (2014). The two active X chromosomes in female ESCs block exit from the pluripotent state by modulating the ESC signaling network. *Cell Stem Cell*, *14*(2), 203–216. https://doi.org/10.1016/j.stem.2013.11.022

Shin, J., Bossenz, M., Chung, Y., Ma, H., Byron, M., Zhu, X., ... Bach, I. (2010). Maternal Rnf12/RLIM is required for imprinted X chromosome inactivation in mice JongDae. *Nature*, *467*(7318), 977–981. https://doi.org/10.1038/nature09457.Maternal

Sun, S., Del Rosario, B. C., Szanto, A., Ogawa, Y., Jeon, Y., & Lee, J. T. (2013). Jpx RNA activates Xist by evicting CTCF. *Cell*, *153*(7), 1537. https://doi.org/10.1016/j.cell.2013.05.028

Takagi, N., & Sasaki, M. (1975). Preferential Inactivation of the Paternally derived X Chromosome in the Extraembryonic Membranes of the Mouse. *Nature*, *256*(5519), 640–642. https://doi.org/10.1038/256640a0

Tian, D., Sun, S., & Lee, J. T. (2010). The long noncoding RNA, Jpx, Is a molecular switch for X chromosome inactivation. *Cell*, *143*(3), 390–403. https://doi.org/10.1016/j.cell.2010.09.049

van Bemmel, J. G., Mira-Bontenbal, H., & Gribnau, J. (2016). Cis- and trans-regulation in X inactivation. *Chromosoma*, *125*(1), 41–50. https://doi.org/10.1007/s00412-015-0525-x

Vigneau, S., Augui, S., Navarro, P., Avner, P., & Clerc, P. (2006). An essential role for the DXPas34 tandem repeat and Tsix transcription in the counting process of X chromosome inactivation. *Proceedings of the National Academy of Sciences of the United States of America*, *103*(19), 7390–7395. https://doi.org/10.1073/pnas.0602381103

Wang, F., McCannell, K. N., Bošković, A., Zhu, X., Shin, J. D., Yu, J., ... Bach, I. (2017). Rlim-Dependent and -Independent Pathways for X Chromosome Inactivation in Female ESCs. *Cell Reports*, *21*(13), 3691–3699. https://doi.org/10.1016/j.celrep.2017.12.004

Weintraub, A. S., Li, C. H., Zamudio, A. V., Sigova, A. A., Hannett, N. M., Day, D. S., ... Young, R. A. (2017). YY1 Is a Structural Regulator of Enhancer-Promoter Loops. *Cell*, *171*(7), 1573-1588.e28. https://doi.org/10.1016/j.cell.2017.11.008

Xu, N., Tsai, C., & Lee, J. (2006). Transient homologous chromosome pairing marks the onset of X inactivation. *Science*, *311*(5764), 1149–1152.

8

Directed Differentiation of Human-induced Pluripotent Stem Cells into Hepatic Cells: A Transposable Example of Disease Modeling and Regenerative Medicine Applications

Antonietta Messina[1,2,3,*], **Eléanor Luce**[1,2,3,*] **and Anne Dubart-Kupperschmitt**[1,2,3,*]

[1]INSERM unité mixte de recherche (UMR_S) 1193, Villejuif, F-94800, France
[2]UMR_S 1193, Université Paris-Saclay, Villejuif, F-94800, France
[3]Fédération Hospitalo-Universitaire (FHU) Hépatinov, Villejuif, F-94800, France
Email: antonietta.messina@inserm.fr; eleanor.luce@inserm.fr; anne.dubart@inserm.fr
*Corresponding Author(s)

8.1 Introduction

Human embryonic stem cells (hESCs) have first been considered as the best source providing unlimited numbers of cells for medical research, drug development, or screening, as well as toxicology studies. Isolated for the first time in 1998 (Thomson, 1998), their self-renewal ability and pluripotency, regulated by transcription factors such as OCT4, SOX2, and NANOG, allowed the in vitro generation of more than 200 types of specialized cells that may theoretically be used for various research purposes and medical applications (Doğan, 2018). However, the ethical and moral concerns arisen with their use in research led to the exploration of adult stem cells (ASCs) as an alternative. Found in various tissues and organs such as the brain, bone marrow, blood, teeth, heart, and gut (Caglayan et al., 2019), human

ASCs are in charge of generating specialized cells that replace dying or damaged cells, acting in the maintenance of tissues and organs of which they are more or less specific. While ASCs hold great promise for research and therapy, some important limitations must be overcome before their use in clinics. Indeed, ASCs represent a minor cell population in native tissues and the extraction, concentration, and administration processes of these cells are very complex and invasive for both the donors and the patients (Dulak et al., 2015). In addition, they have limited self-renewal ability and rapidly become senescent after several cell divisions, impairing extensive amplification. In 2007, two independent research teams succeeded in reprogramming human somatic cells to an undifferentiated state, opening a brand-new perspective in the world of biology and medicine (Takahashi et al., 2007; Yu et al., 2007). Defined as human induced pluripotent stem cells (hiPSCs), these rejuvenated cells shared many common characteristics with hESCs, in terms of transcriptome profile or epigenetics, showing the same self-renewal and pluripotency abilities. Moreover, because hiPSCs only require somatic cells from a skin biopsy, blood or even urine cells to be generated, they avoid the ethical concerns brought by the destruction of human embryos during the hESC isolation process (Lo and Parham, 2009). Since 2007, hiPSCs proved to provide: (1) a model for studying the early stages of human embryonic development; (2) an in vitro model of the physiological development to better understand tissue repair and regeneration; (3) models of human diseases to screen new drugs or to study their pathogenesis mechanisms and toxicology; (4) cells to be used in therapy as an alternative to organ transplantation.

The potential use of hiPSCs mainly relies on their pluripotency. Indeed, stem cells, including hiPSCs, can generate various cell types of the human body. This process, called differentiation, is the result of a very complex mechanism in which multiple biological pathways induce the generation of new specialized cells (Oh and Jang, 2019). Physiologically, the differentiation process begins early during embryonic development. Consequently, two questions arise: (1) How much do we know about the molecular and cellular mechanisms that regulate cell differentiation in the developing embryo? (2) Can we recapitulate such a complicated process in vitro? Indeed, we still do not know in detail every single step of human embryonic development. Comparative medicine, which is based on the concept that other animal species share physiological or behavioural characteristics with humans, allowed significant advances in our understanding of human embryo development (Ericsson et al., 2013). Many of the involved pathways can now be replicated with reproducibility to generate human differentiated cells in vitro.

This chapter aims to describe how the guided differentiation of pluripotent stem cells allows generating human specialized cells that can be used as models for fundamental research and clinical approaches. In this chapter, we will discuss the developmental origins of the liver and how this knowledge could be translated to efficiently produce hepatic cells in vitro, a critical issue in fundamental biology and regenerative medicine. Using the hepatic cells as an illustrative example, this conceptual and methodological framework could be then transposed and adapted to the differentiation of most other types of tissues and organs.

8.2 Developmental Origins of the Liver

Although the development of a new organism from the fertilized egg is a continuous process, it is classically divided into three major steps: (i) the segmentation (or cleavage) that converts a unicellular state (the fertilized egg) into a multicellular state (the blastocyst in mammals), (ii) the gastrulation allowing the rearrangements of the founding cells of the three germ layers (ectoderm, mesoderm, and endoderm), (iii) the histogenesis and organogenesis during which the different tissues and organs of the developing organism are formed (Alberts, 2015; Vaillancourt and Lafond, 2009).

The development of the germ layers provides the foundation for the development of all the organs and tissues of organisms. The ectoderm is giving rise to the neuroderm (brain, spinal cord, and neural crest cells), the epidermis, and sensory placodes. The skeletal muscles, dermis, tendons, various internal organs such as the heart, kidneys, and gonads, as well as the vascular and hematopoietic systems and different types of connective tissues derive from the mesoderm. Finally, the endoderm gives rise to the epithelial tissue lining the digestive or respiratory tract, and organs such as the pancreas and liver (Figure 8.1A).

At the centre of numerous physiological processes, the liver is the result of a complex differentiation program that includes exogenous signal gradients and a vast cohort of transcription factors. Macronutrient metabolism, lipid and cholesterol homeostasis, blood-volume regulation, immune system support, endocrine pathways, and the metabolism of xenobiotic compounds, are only some of the vital processes carried out by this organ (Trefts et al., 2017).

The cells giving rise to the liver are located in the ventral part of the anterior endoderm, near the cardiac mesoderm and the septum transversum, a transient mesenchymal structure. This spatial organization enables them to

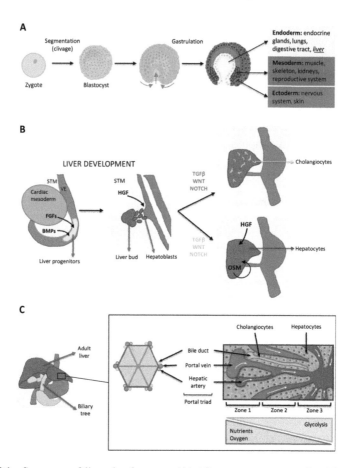

Figure 8.1   Summary of liver development. (**A**) After zygote segmentation (cleavage) the generated blastocyst undergoes gastrulation and gives rise to three germ layers: ectoderm, mesoderm, and endoderm. (**B**) Liver progenitors located in the ventral part of the anterior endoderm, near the cardiac mesoderm and the septum transversum (STM) receive FGFs and BMPs signals from the cardiac mesoderm and the STM respectively. After formation of the liver bud due to the hepatoblast proliferation and migration into the STM, the activation or inhibition of TGFB, NOTCH, and WNT pathways leads to the hepatoblast differentiation into cholangiocytes or hepatocytes respectively. Pathways activated during human liver embryogenesis are indicated in green and pathways inactivated are indicated in red. (**C**) Architecture of the adult liver organized in lobules. Cross section of the liver lobule showing the metabolic zonation and the resulting gradients. White arrows show the blood flow direction.
STM: septum transversum; VE: ventral endoderm; FGF: Fibroblast growth factor; BMP: bone morphological protein; HGF: Hepatocyte Growth Factor; OSM: Oncostatin M.
Figure inspired from: Zorn, 2008 in StemBook and Stenvall et al., 2014 in Physics in Medicine & Biology

receive FGF (Fibroblast Growth Factor) 1 and 2 signals from the cardiac mesoderm. Moreover, the septum transversum located near the endoderm secretes proteins of the BMP family (Bone Morphological Protein 2 and 4) (Figure 8.1B). These signals will induce the differentiation and the morphological changes required for the liver bud formation. Other important signaling pathways such as TGF-β and WNT/βcat induce the hepatic specification of definitive endoderm leading the hepatic progenitors to arise. Called hepatoblasts, these progenitors migrate from the liver bud to the septum transversum under the influence of transcription factors such as GATA4, GATA6, PROX-1, HNF6, and FOXA (Forkhead Box Proteins A). At this stage, hepatoblasts are non-polarized bipotential cells able to further differentiate into the two main liver cells: the hepatocytes and the cholangiocytes. The respective intensities of TGF-β, NOTCH, WNT3a, and βCAT signaling will determine hepatic progenitors' fate toward hepatocytes or cholangiocytes.

The activin/TGF-β pathway strongly activated around the portal vein, induces hepatoblasts to re-arrange in a monolayer configuration, generating the ductal plate. Here, mainly under Notch and Wnt3a signaling pathways, they differentiate into cholangiocyte precursors expressing Sox9, one of the most specific and early markers of cholangiocytes in the liver. Small luminal spheres will then appear, lined on the portal side by biliary cells that express markers such as Sox9, osteopontin, and Krt19 (cytokeratin 19), and by hepatoblasts on the parenchymal side (Antoniou et al., 2009; Ober and Lemaigre, 2018). These structures will then grow along the portal vein and mature, becoming the bile duct entirely lined by polarized cholangiocytes (Lemaigre, 2020). The signaling pathways controlling the bile duct morphogenesis are mostly constituted by TGF-β, NOTCH, and WNT. Cholangiocytes can be easily distinguished from the other hepatic populations because of their cuboidal shape, the strong expression of tight junction protein ZO-1 and the adherent junction protein E-cadherin, as well as the presence of both the nuclear factor Sox9 and a primary cilium on their apical surface which extends toward the duct lumen and modulates the secretory and proliferative functions of these cells.

However, hepatoblasts located further away from the portal vein differentiate into hepatocytes. In this region, the CCAAT/enhancer-binding protein (C/EBP) α is triggered and inhibits TGF-β signals while activating members of the HNF (hepatocyte nuclear factor) family such as HNF1α and HNF4α, co-activators of most of the hepatocyte-specific genes. Hepatocyte differentiation is a gradual process prompted by HGF (hepatocyte growth factor), glucocorticoid hormones, and oncostatin M (OSM), which is secreted by hematopoietic cells since the liver is the main hematopoietic organ at this stage of the development. These factors promote hepatoblast proliferation and upregulation of the proteins involved in cell adhesion and polarity such as the tight junction protein claudin-2 and E-cadherin. Newly formed hepatocytes start to express specific proteins such as albumin and ZO-1 (tight junction protein 1) while endothelial cells (ECs), migrate between the hepatocytes, secreting a matrix and regulatory factors (Ober and Lemaigre, 2018; Trefts et al., 2017). This mechanism leads to the formation of the hepatic cords and contributes to the generation of the so-called Disse space, which enables the exchanges between blood plasma and hepatocytes. Hepatic cells then, begin to acquire their typical morphology and polarization, express specific proteins such as the MRP-2 (Multidrug resistance-associated protein 2) while bile canaliculi are formed and start draining the bile toward the bile ducts lined by cholangiocytes. The last step in the human hepatocyte differentiation is their functional maturation that is very progressive lasting up to late childhood. It has been demonstrated that hepatocytes show a remarkable heterogeneity in the biochemical and physiological functions they perform according to their position to the portal vein. The WNT/β-catenin pathway, as well as HGF signaling, and the gradient of oxygen, have been demonstrated to play a key role in this phenomenon known as metabolic zonation (Benhamouche et al., 2006; Gebhardt, 2014). At the microscopic scale, the liver is made of lobules in which the cords of hepatocytes radiate from a central vein toward a portal triad consisting of a hepatic artery that supplies oxygen, a hepatic portal vein where blood is rich in nutrients absorbed from the small intestine but poorly oxygenated and, a bile duct that carries bile away from the hepatocytes and the liver lobule (Figure 8.1C). Fully mature hepatocytes define three distinct zones distributed along the axis connecting the portal vein to the central vein. Hepatocytes in closest proximity to the portal vein (zone 1) receive the greatest amount of nutrients and oxygen content from blood perfusion but are also first in line to be affected by toxins transported from the gut through the portal vein. Here, the WNT/βcat signaling is inhibited resulting in the activation of the oxidative liver functions such as gluconeogenesis, β-oxidation of fatty acids, and cholesterol synthesis. The zone 3 is located near the central vein, where WNT/βcat activity is higher and processes as glycolysis, lipogenesis, cytochrome P-450–based drug detoxification, glutamine synthesis, and heme synthesis are upregulated (Gebhardt, 2014). Finally, zone 2 hepatocytes are interspersed between zone 1 and zone 3, showing a mixed profile of functions.

8.3 hiPSCs and Directed Differentiation In Vitro

8.3.1 Hepatoblast Differentiation

In the absence of the self-renewal factor FGF2 (Fibroblast Growth Factor 2), human pluripotent stem cells (hPSCs) spontaneously differentiate in suspension and form aggregates called "embryoid bodies" composed of differentiated cells from the three germ layers. In 2002, Jones et al. took advantage of this phenomenon and showed that the culture of embryoid bodies formed from mouse embryonic stem cells allowed the differentiation into hepatocytes after 12 days (Jones et al., 2002). A year later, the first hESC differentiation protocol into functional hepatocytes was published (Rambhatla et al., 2003). However, because embryoid bodies are composed of a heterogeneous population of endoderm, ectoderm, and mesoderm cells, the yield of hepatocyte differentiation was very low. The addition of specific growth factors can direct pluripotent stem cell differentiation toward the cell type of interest. Most of the differentiation protocols developed aim at mimicking the first steps of embryonic development and liver organogenesis in vitro (Figure 8.2). At each step of the protocol, RT-PCR or quantitative RT-PCR analyses allow to monitor specific marker gene expression and flow cytometry, immunofluorescence, or Western Blot analyses to study the appearance of specific protein synthesis. The first step of all the protocols developed so far consists in inducing the endoderm germ layer formation by activation of the Activin/Nodal pathway, using mostly a high concentration of Activin A, a member of the TGF-β family. Other morphogens such as WNT3a or chemical compounds like Ly2904002, an inhibitor of the PI3K (phosphatidylinositol 3-kinase), are generally used to improve the efficiency of the endoderm differentiation (Hay et al., 2008). To specify the definitive endoderm cells to the hepatic lineage, members of the BMP and FGF families are often used, sometimes in combination with other molecules such as retinoic acid, which appears to improve differentiation both in vitro and in vivo (Negishi et al., 2010; Touboul et al., 2010), or inhibitors of the TGF-β pathway such as SB431542 (Touboul et al., 2016). The most robust protocols lead to up to 90% of cells expressing both AFP (Alpha-Fetoprotein), the foetal form of albumin, and KRT19, a feature of the hepatoblasts.

8.3.2 Hepatocyte Differentiation

All differentiation protocols for the generation of hiPSC-derived hepatocyte (iHeps) use growth factor cocktails in which the most important components are the cytokines HGF and OSM, mimicking the organogenesis signaling

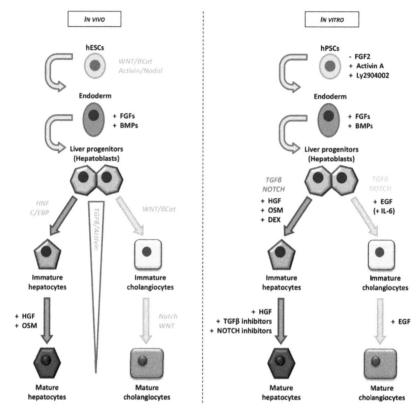

Figure 8.2 Principal steps of human embryogenesis and liver organogenesis (**left panel**) mimicked and recapitulated in protocols for human pluripotent stem cell differentiation into hepatocytes and cholangiocytes (**right panel**). BMP: Bone morphogenic proteins; EGF: Epithelial Growth Factor; FGF: Fibroblast Growth Factor; HGF: Hepatocyte Growth Factor; HNF: Hepatocyte Nuclear Factor; IL-6: Interleukin-6; OSM: Oncostatin M. Pathways activated during human liver embryogenesis or cell differentiation *in vitro* are indicated in green and pathways inactivated are indicated in red.

pathways (Figure 8.2). Moreover, the use of inhibitors of the NOTCH pathway, such as the compound E or inhibitors of the TGF-β pathway such as the SB431542 is also used in several protocols (Lv et al., 2015; Touboul et al., 2016). Indeed, as already described, the activation of the NOTCH and TGF-β pathways directs the hepatoblast differentiation into cholangiocytes, whereas their inhibition will lead to a hepatocytic fate. Approximately two weeks of treatment result in differentiated cells expressing hepatocyte markers such as HNF4α, albumin, A1AT (Alpha-1-anti-trypsin), and CYP3A7 (a foetal isoform of the CYP3A4). Moreover, functional studies confirm that the iHeps obtained can synthesize albumin and can fulfil functions related to energy metabolisms such as the glycogen storage assessed by PAS (Periodic Acid Schiff) staining, or to detoxification functions carried out by the P450 cytochrome activity (Ma et al., 2013; Ulvestad et al., 2013).

However, many teams are still currently working to improve the conditions of liver cell differentiation because everybody agrees that cells obtained after PSC differentiation better mimic foetal or newborn rather than adult hepatocytes (Baxter et al., 2015; Luce et al., 2021). This lack of cell differentiation and/or maturation in culture can be explained by the absence of other cell types that physiologically bring extracellular matrix (ECM) support and secrete specific factors, and/or because of low cell–cell interactions due to the absence of a defined organization in space. Indeed, co-culture systems constituted of multiple cell types, such as non-parenchymal human hepatic stellate cell line or human umbilical vein endothelial cells, proved to induce the generation of hepatocytes showing closer morphologic and phenotypic features than primary human hepatocytes (Freyer et al., 2017; Javed et al., 2014). Furthermore, since the monolayer culture does not mimic physiological conditions, three-dimensional (3D) approaches have been developed. These enable the generation of more complex structures called "organoids" which better mimic the tissue organization in vivo. These 3D structures can be considered as a miniature reproduction of an organ and may provide suitable alternatives to generate hepatocytes showing higher activity and long-term functions in vitro (Lou and Leung, 2018; Sakabe et al., 2020). Indeed, by providing an increased cell density, cell–cell interactions, and signaling, as well as a gradient of oxygen, 3D models seem to significantly influence the acquisition of liver-specific functions. These new systems have proven useful in developmental and toxicological studies, drug discovery, disease modeling, and new bioengineering systems are under investigation for the development of new liver therapy approaches (Augustyniak et al., 2019; Messina et al., 2020; Wu et al., 2019).

8.3.3 Cholangiocyte Differentiation

As previously mentioned, the differentiation of hepatoblasts into cholangiocytes relies on the activation of both NOTCH and TFG-β signaling pathways (Figure 8.2). In the literature, the activation of the NOTCH pathway has been mainly performed using recombinant Jagged1 (De Assuncao et al.,

2015) or co-culture with OP9 stromal cells (Ogawa et al., 2015). The activation of the TFG-β signaling pathway has been reported by the addition of recombinant TGF-β1 (Ogawa et al., 2015) or activin A (a member of the TGF-β superfamily). Furthermore, it has been shown that the addition of retinoic acid with the activin A allowed suppressing hepatoblast markers and that the further use of FGF10 permitted to induce early biliary markers (Sampaziotis et al., 2015). Finally, in addition to Notch and TGF-β, members of the Wnt family also regulate differentiation of hepatoblasts to biliary cells in mice, like Wnt3A, and improve the proliferation and survival capacity of bile duct cells (Hussain et al., 2004). Other cytokines have been also tested such as interleukin 6 (IL-6), present in foetal liver, or the growth hormone (GH), a regulator of the insulin-like growth factor-1 (IGF1) pathway whose receptor is expressed in periportal hepatoblasts (Dianat et al., 2014; Luce and Dubart-Kupperschmitt, 2020).

The combined activation of these signaling pathways leads to the differentiation of hepatoblasts into immature cholangiocytes expressing early biliary markers, such as KRT19 and SOX9. However, during the development, the maturation of cholangiocytes happens with the re-arrangement of the ductal plate into tubular structures, and most of the functions of the biliary epithelium are associated with several absorption and secretion processes, requiring apico-basal polarity of the cells. As this polarization cannot be accurately reproduced by cells cultured in monolayer, almost all the differentiation protocol published so far end by a stage of cell maturation in 3D culture systems using mostly a mix of Collagen I and Matrigel (Dianat et al., 2014; Ogawa et al., 2015) or Matrigel and EGF that promotes spontaneous differentiation of progenitor cells into cystic structures (Luce and Dubart-Kupperschmitt, 2020; Sampaziotis et al., 2015). Prolonged culture in these conditions results in polarized cyst formation with a central lumen demonstrating characteristic functional properties such as γ-Glutamyl Transferase (GGT) activity and active transport of molecules regulated by specific apical membrane proteins.

8.4 hiPSC-derived Liver Cell Applications: Toward the Personalized Medicine

Because of their ability to self-renew and differentiate into almost all cell types, the use of hiPSCs is becoming an integral part of modern medicine (Suman et al., 2019). The possibility to culture cells generated from patients with diverse pathologies allows the development of in vitro models of pathologies including complex multifactorial diseases of unknown genetic identity such as NAFLD (Non-Alcoholic fatty liver disease), cancer, or inherited liver disorders (Bailey et al., 2018; Jarrett et al., 2017). Pathological pathways can be closely studied and effective pharmacological approaches, in terms of drug-response and control of recorded side effects, can be developed (Vogenberg et al., 2010). Moreover, the ability to culture human liver organoids will permit to accurately predict the metabolism or the toxicity of a compound before its use on patients in clinical trials. Furthermore, with the advent of regenerative medicine, new treatments have been developed such as cell or cell/gene therapy. Due to its innate ability to regenerate and self-repair, the liver is particularly responsive to these new approaches; pre-clinical trials of cell therapy already showed a liver regeneration in animal models affected by liver failure or genetic diseases. Primary human hepatocytes (PHHs) were the first candidates for cell therapy. However, PHHs are most of the time harvested from livers of poor quality that are not suitable for transplantation; they are not easily cultured in vitro and are susceptible to freeze-thaw damage (Elaut et al., 2006). Moreover, the number of cells required for each application is huge and PHHs suffer from the same donor shortage than liver transplantation. For this reason, hiPSC-derived cells appeared as an alternative source of liver cells for cell therapy (Figure 8.3). Indeed, in addition to providing a suitable platform for drug toxicity testing and disease modeling, they would also provide the number of cells required for transplantation and any other regenerative approaches developed so far. The next sections will briefly describe the progress reported during the last decades in the use of hiPSC technology for clinical applications.

8.4.1 Disease Modeling

The differentiation of hiPSCs obtained from patient biopsies into liver cells allowed the development of many liver disease models so far. Pathologies such as familial hypercholesterolemia, Crigler Najjar, Wilson's disease, $\alpha 1$ antitrypsin deficiency, liver fibrosis, NASH, NAFLD, and acquired diseases due to toxin products, have been modeled and characterized, leading to a better knowledge of pathology mechanisms and the individual drug responses (Ouchi et al., 2019; Wang et al., 2018; Wu et al., 2019). In this context, liver organoids have demonstrated to maintain some of the hepatocytes' most important functions, such as albumin secretion, cytochrome P450 metabolism, bile acid production, ammonia elimination, low-density lipoprotein uptake, and glycogen storage, for over 3 months (Goulart et al., 2019).

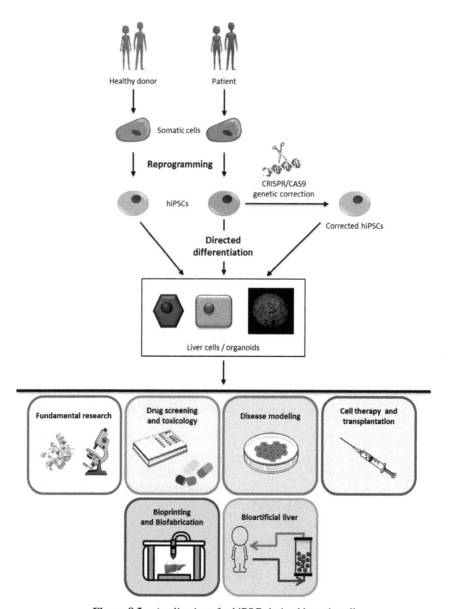

Figure 8.3   Applications for hiPSC-derived hepatic cells.

Although most of the current researches and clinical trials use unmodified hiPSCs and derivatives, the development of the CRISPR/Cas9 technology enables the genetic modification of hiPSCs to correct genetic disorders in vitro (Alves-Bezerra et al., 2019), such as hereditary tyrosinemia type I, Wilson's disease, and familial hypercholesterolemia (Bailey et al., 2018; Caron et al., 2019; Jarrett et al., 2017). This gene-editing technology is based on the direct correction of the mutation or the insertion of a customized DNA sequence at a precise location to phenotypically correct genetic deleterious variants (Komor et al., 2017). The proof of concept of the potential use of CRISPR/Cas9-modified hiPSC-derived cells to treat some human diseases have been done in a mouse model of haemophilia A, by transplanting endothelial cells differentiated from genetically corrected hiPSCs (i.e., with a non-mutated factor VIII gene) (Park et al., 2015).

8.4.2 Drug Screening and Toxicology Studies

The drug response consists of a fine balance between the therapeutic benefits and the toxic and side effects related to its use. Preclinical animal models have always been used to screen drug candidates through the investigation of the absorption, distribution, metabolism, excretion, and toxicity mechanisms (ADMET) (Waring et al., 2015). However, predicting human-specific liver toxicity using animal models is particularly challenging because drugs do not always have the same toxicity in animals and humans due to the differences in physiology, metabolism, and disease adaptations. Most of the currently available drugs are metabolized by the enzymes of the cytochrome P450 (CYP450) system. These enzymes are responsible for individual variation in drug ADMET and the genetic variation between animals and humans are the main causes of failure during preclinical drug testing (Li et al., 2019). The advantage of using hiPSC technology in this field, besides being human cells, is the possibility to generate a library that reflects the human genetic diversity and polymorphism. Therefore, the selection of drugs, their mechanisms of action, their related toxicity, and the therapeutic cell responses will be investigated with better precision, minimizing harmful side effects and ensuring more successful outcomes for each patient compared to animal model studies (Underhill and Khetani, 2018). The hiPSC-derived liver organoids are nowadays considered the most appropriate tool to evaluate drug efficacy and to screen drugs before whole organism studies on animal models and clinical trials (Augustyniak et al., 2019). Indeed, these 3D structures showed to be able to correctly metabolize molecules such as rifampicin, omeprazole, phenobarbital, and paracetamol, allowing also the discovery of new drug-adverse effects on the human liver that have never been previously found in 2D cultures or animal models (Du et al., 2014; Takayama et al., 2013).

To further improve these models, devices called organ-on-a-chip have also been developed. Using microfluidic technology, these devices can mimic molecules and oxygen flows in vitro, predicting with high accuracy drug clearances, toxicity, and mechanisms of action (Mittal et al., 2019; Wang et al., 2018). Indeed, hiPSC-derived hepatocytes differentiated in such devices reach a high cytochrome P450 activity, so that the response to drug treatment can be recorded with higher accuracy than in previously reported systems or animal models (Banaeiyan et al., 2017; Kang et al., 2020).

8.4.3 Bio Artificial Liver Devices

Liver transplantation remains the only curative treatment for end-stage liver diseases. However, the limited number of available organs highly affects this procedure and often requires the use of external devices to sustain the liver functions of the more seriously affected patients (Larsen, 2019). The available artificial liver (AL) support systems detoxify waste molecules relying on a purely mechanical/chemical/physical method. However, bioartificial liver (BAL) devices, containing cell bioreactors, are currently investigated to also fulfil other functions such as the enzymatic detoxification, the biotransformation, and the protein synthesis (Sakiyama et al., 2017). PHHs are the ideal cells as the biological component for a BAL, but as already mentioned, they are not readily available. For this reason, the hiPSCs represent a good alternative for large-scale production of functional hepatocytes to be incorporated in a BAL system. To ensure the device's ability to sustain a patient liver function, about 10^{10}–20^{10} hepatocytes are needed (Yamashita et al., 2018) and the differentiation process of stem cells is currently the only way to obtain such a cell mass. Despite the fact that no stem cell-based BAL system has undergone human trial yet, human hepatoblastoma cells-derived organoids cultured in a fluidized bed bioreactor showed to improve the BAL performances in a porcine model of severe liver failure, improving the animal general conditions and paving the way to the development of BALs that could host human liver organoids in the near future (Selden et al., 2017).

8.4.4 Cell Transplantation

Cell transplantation is a technique that has been already tested in clinics to treat patients affected by genetic disorders, mostly children, and that led to clinical improvement and/or partial correction of the underlying metabolic defect. Indeed, this therapeutic approach aims to infuse cells through intrasplenic or portal vein injection to replace the damaged cells of an organ and to ensure the recovery of the organ-missing functions with a procedure far less invasive than organ transplantation (Lee et al., 2018). Moreover, the transplantation of patient-derived liver cells would avoid the need for a life-long immunosuppressive treatment and thus their strong side effects. However, even if the ongoing clinical trials are showing encouraging results, this approach requires that the architecture of the liver is intact for cells engraftment. Indeed, when the ECM is altered due to the disease (cancer, cirrhosis, NASH, etc.), even by performing multiple injections, no therapeutic effect is obtained. Current efforts are thus made to recapitulate the structural liver organization by using scaffolds on which cells or organoids can be cultured before transplantation (Li et al., 2013; Lou and Leung, 2018). Indeed, the prospect of using engineered livers from hiPSC-derived cells for organ transplantation has become a major goal of regenerative medicine (Brody, 2016). In 2013, human liver tissue was generated coculturing hiPSC-derived hepatic endodermal cells with human umbilical vein endothelial cells (HUVECs) and human mesenchymal stem cells (MSCs). These liver organoids, once engrafted in mice, were able to secrete human alpha-1-antitrypsin and albumin, displayed cytochrome P450 activity, and improved the survival of mice in a toxin-induced liver injury model, as a proof of concept of the great promise of hiPSC technology to alleviate the critical shortage of donor organs (Takebe et al., 2013). Though organoids with more complex structure, composed of cholangiocytes or biliary structures, for example, have not yet been generated, the organoid model represents a major breakthrough in cell biology that has revolutionized biomedical research.

8.4.5 Bioprinting of 3D Liver Tissues and Whole Organ Assembly

A strategy to improve liver cell therapy is to develop patterned 2D cultures and 3D tissue structures in which multiple cell types can be organized in space, enabling intercellular junctions, cell polarity, and the extension of lifespan and function in vitro (Lou and Leung, 2018). It has been demonstrated that providing pre-defined extrinsic forces to enhance cell–cell interactions by employing matrices and/or supports with engineered geometry (micropatterning, microwells, and micro-fluid dynamics) enhanced cell and organoid viability (Brassard and Lutolf, 2019).

In the past 10 years, ink-jet and 3D printers have been developed to allow the manipulation of living cells by using special bio-inks purposely made of biocompatible materials such as gelatin, alginate, fibrin, hyaluronan, laminin, and collagen (Matai et al., 2020; Nakamura et al., 2005) which provide biological, mechanical, and structural support that sustains over time the intercellular interactions. hiPSC-derived hepatocytes and organoids in combination with mesenchymal and endothelial cells have been bio-printed to build organoids that showed hepatic function and protein/amino acid metabolism for more than 18 days in culture (Goulart et al., 2019). Bioprinted liver tissues were first commercialized in 2014 (Organovo, San Diego, CA) and are used today for long-term and tissue-level assessment of toxicity, liver biology, and disease modeling, as well as many other bioengineered constructs recently generated through several bioprinting techniques (Faulkner-Jones et al., 2015; Goulart et al., 2019; Ma et al., 2016).

Implantable organs could also be obtained by the decellularization of whole organs and subsequent reseeding with appropriate cell types. This process consists of the chemical removal of all cells from an organ to retrieve its cellular matrix skeleton. By reseeding new cells on the obtained scaffold, a reproduction of the original organ can be obtained since the most important ECM molecules, which are required for site-specific engraftment and cell differentiation, are still in place. In this case, the presence of an intact native ECM is of central importance, as it not only provides a platform for cell growth but also mediates biochemical and molecular signaling (Brovold et al., 2018). Liver constructs were obtained in 2011 by perfusing cells such as hepatic progenitors and endothelial cells through the vasculature of decellularized animal livers. Cells were able to repopulate the scaffold and to differentiate into hepatocytes and cholangiocytes (Baptista et al., 2011). Many other attempts have been carried out, but the engineered grafts generated until now failed to survive in pre-clinical trials mostly due to their inability to maintain the vascular and bile duct network after transplantation and to enable the in vivo regeneration of hepatocytes inside the scaffold. Further researches are currently ongoing aiming at the creation of readily available and sustainable organs for transplantation.

8.5 Concluding Remarks

- The directed differentiation of hiPSCs is a reproducible method for the generation of most of the cell types of the human body, as illustrated with hepatic cells, making them a concrete renewable cell source for medical research (Shi et al., 2017).
- Generated from somatic cells of both healthy individuals and patients, hiPSC-derived hepatic cells have been already used for: the development and screening of new drugs (Augustyniak et al., 2019; Mittal et al., 2019), disease modeling (Caron et al., 2019; Dianat et al., 2013; Hannoun et al., 2016), and the design and fabrication of mini-liver constructs.
- iPSC-derived hepatic cells could also be used in clinics as an unlimited source of cells which could in the future (i) improve the patient quality of life before and after surgery as hiPSC-derived hepatocytes could be used in BAL systems either enabling the regeneration of the native liver or supporting the patient liver functions until the transplantation; (ii) limit the need of invasive surgery, eventually decreasing the overall need for transplantation, using the cell therapy approach to treat various forms of liver disease as the metabolic disorders; (iii) create bioengineered transplantable organs (Matai et al., 2020; Sakiyama et al., 2017).
- However, before their use in therapy an extensive validation of the functionality of iPSC-derived cells needs to be performed along with a deep study of their genetic stability (Volarevic et al., 2018). Indeed, even if our laboratory showed no appearance of de novo mutation during the differentiation process (Steichen et al., 2014) and highlighted that the safety mainly relied on the selection of a high-quality hiPSC clone (Steichen et al., 2019), additional studies are still needed.
- The absence of remaining undifferentiated cells in the graft must also be verified as teratoma development could be a risk after transplantation. Straightforward procedures that are able to remove non-differentiated cells must be set up before moving toward transplantation (Ronen and Benvenisty, 2012; Steichen et al., 2019).
- In addition to the tremendous number of cells required for transplantation and BAL systems, building 3D structures of appreciable size ensuring a therapeutic effect is very complex; a full characterization and preservation of the cell phenotype are mandatory.

References

Alberts, B., 2015. Molecular biology of the cell, Sixth edition. ed. Garland Science, Taylor and Francis Group, New York, NY.

Alves-Bezerra, M., Furey, N., Johnson, C.G., Bissig, K.-D., 2019. Using CRISPR/Cas9 to model human liver disease. JHEP Reports 1, 392–402.

Antoniou, A., Raynaud, P., Cordi, S., Zong, Y., Tronche, F., Stanger, B.Z., Jacquemin, P., Pierreux, C.E., Clotman, F., Lemaigre, F.P., 2009. Intrahepatic Bile Ducts Develop According to a New Mode of Tubulogenesis Regulated by the Transcription Factor SOX9. Gastroenterology 136, 2325–2333.

Augustyniak, J., Bertero, A., Coccini, T., Baderna, D., Buzanska, L., Caloni, F., 2019. Organoids are promising tools for species-specific in vitro toxicological studies. J Appl Toxicol 39, 1610–1622.

Bailey, M.H., Tokheim, C., Porta-Pardo, E., Sengupta, S., Bertrand, D., Weerasinghe, A., Colaprico, A., Wendl, M.C., Kim, J., Reardon, B., et al., 2018. Comprehensive Characterization of Cancer Driver Genes and Mutations. Cell 174, 1034–1035.

Banaeiyan, A.A., Theobald, J., Paukštyte, J., Wölfl, S., Adiels, C.B., Goksör, M., 2017. Design and fabrication of a scalable liver-lobule-on-a-chip microphysiological platform. Biofabrication 9, 015014.

Baptista, P.M., Siddiqui, M.M., Lozier, G., Rodriguez, S.R., Atala, A., Soker, S., 2011. The use of whole organ decellularization for the generation of a vascularized liver organoid. Hepatology 53, 604–617.

Baxter, M., Withey, S., Harrison, S., Segeritz, C.-P., Zhang, F., Atkinson-Dell, R., Rowe, C., Gerrard, D.T., Sison-Young, R., Jenkins, R., Henry, J., Berry, A.A., Mohamet, L., Best, M., Fenwick, S.W., Malik, H., Kitteringham, N.R., Goldring, C.E., Piper Hanley, K., Vallier, L., Hanley, N.A., 2015. Phenotypic and functional analyses show stem cell-derived hepatocyte-like cells better mimic fetal rather than adult hepatocytes. Journal of Hepatology 62, 581–589.

Benhamouche, S., Decaens, T., Godard, C., Chambrey, R., Rickman, D.S., Moinard, C., Vasseur-Cognet, M., Kuo, C.J., Kahn, A., Perret, C., Colnot, S., 2006. Apc Tumor Suppressor Gene Is the "Zonation-Keeper" of Mouse Liver. Developmental Cell 10, 759–770. https://doi.org/10.1016/j.devcel.2006.03.015

Brassard, J.A., Lutolf, M.P., 2019. Engineering Stem Cell Self-organization to Build Better Organoids. Cell Stem Cell 24, 860–876.

Brody, H., 2016. Regenerative medicine. Nature 540, S49–S49.

Brovold, M., Almeida, J.I., Pla-Palacín, I., Sainz-Arnal, P., Sánchez-Romero, N., Rivas, J.J., Almeida, H., Dachary, P.R., Serrano-Aulló, T., Soker, S., Baptista, P.M., 2018. Naturally-Derived Biomaterials for Tissue Engineering Applications, in: Chun, H.J., Park, K., Kim, C.-H., Khang, G. (Eds.), Novel Biomaterials for Regenerative Medicine, Advances in Experimental Medicine and Biology. Springer Singapore, Singapore, pp. 421–449.

Caglayan, S., Ahrens, T.D., Cieślar-Pobuda, A., Staerk, J., 2019. Modern Ways of Obtaining Stem Cells, in: Stem Cells and Biomaterials for Regenerative Medicine. Elsevier, pp. 17–36.

Caron, J., Pène, V., Tolosa, L., Villaret, M., Luce, E., Fourrier, A., Heslan, J.-M., Saheb, S., Bruckert, E., Gómez-Lechón, M.J., Nguyen, T.H., Rosenberg, A.R., Weber, A., Dubart-Kupperschmitt, A., 2019. Low-density lipoprotein receptor-deficient hepatocytes differentiated from induced pluripotent stem cells allow familial hypercholesterolemia modeling, CRISPR/Cas-mediated genetic correction, and productive hepatitis C virus infection. Stem Cell Res Ther 10, 221.

De Assuncao, T.M., Sun, Y., Jalan-Sakrikar, N., Drinane, M.C., Huang, B.Q., Li, Y., Davila, J.I., Wang, R., O'Hara, S.P., Lomberk, G.A., Urrutia, R.A., Ikeda, Y., Huebert, R.C., 2015. Development and characterization of human-induced pluripotent stem cell-derived cholangiocytes. Lab Invest 95, 684–696.

Dianat, N., Dubois-Pot-Schneider, H., Steichen, C., Desterke, C., Leclerc, P., Raveux, A., Combettes, L., Weber, A., Corlu, A., Dubart-Kupperschmitt, A., 2014. Generation of functional cholangiocyte-like cells from human pluripotent stem cells and HepaRG cells. Hepatology 60, 700–714.

Dianat, N., Steichen, C., Vallier, L., Weber, A., Dubart-Kupperschmitt, A., 2013. Human Pluripotent Stem Cells for Modelling Human Liver Diseases and Cell Therapy. CGT 13, 120–132.

Doğan, A., 2018. Embryonic Stem Cells in Development and Regenerative Medicine, in: Turksen, K. (Ed.), Cell Biology and Translational Medicine, Volume 1, Advances in Experimental Medicine and Biology. Springer International Publishing, Cham, pp. 1–15.

Du, Y., Wang, J., Jia, J., Song, N., Xiang, C., Xu, J., Hou, Z., Su, X., Liu, B., Jiang, T., Zhao, D., Sun, Y., Shu, J., Guo, Q., Yin, M., Sun, D., Lu, S., Shi, Y., Deng, H., 2014. Human Hepatocytes with Drug Metabolic Function Induced from Fibroblasts by Lineage Reprogramming. Cell Stem Cell 14, 394–403.

Dulak, J., Szade, K., Szade, A., Nowak, W., Józkowicz, A., 2015. Adult stem cells: hopes and hypes of regenerative medicine. Acta Biochim Pol 62, 329–337.

Ericsson, A.C., Crim, M.J., Franklin, C.L., 2013. A brief history of animal modeling. Mo Med 110, 201–205.

Faulkner-Jones, A., Fyfe, C., Cornelissen, D.-J., Gardner, J., King, J., Courtney, A., Shu, W., 2015. Bioprinting of human pluripotent stem cells and their directed differentiation into hepatocyte-like cells for the generation of mini-livers in 3D. Biofabrication 7, 044102.

Freyer, N., Greuel, S., Knöspel, F., Strahl, N., Amini, L., Jacobs, F., Monshouwer, M., Zeilinger, K., 2017. Effects of Co-Culture Media on Hepatic Differentiation of hiPSC with or without HUVEC Co-Culture. IJMS 18, 1724.

Gebhardt, R., 2014. Liver zonation: Novel aspects of its regulation and its impact on homeostasis. WJG 20, 8491.

Goulart, E., de Caires-Junior, L.C., Telles-Silva, K.A., Araujo, B.H.S., Rocco, S.A., Sforca, M., de Sousa, I.L., Kobayashi, G.S., Musso, C.M., Assoni, A.F., Oliveira, D., Caldini, E., Raia, S., Lelkes, P.I., Zatz, M., 2019. 3D bioprinting of liver spheroids derived from human induced pluripotent stem cells sustain liver function and viability in vitro. Biofabrication 12, 015010.

Hannoun, Z., Steichen, C., Dianat, N., Weber, A., Dubart-Kupperschmitt, A., 2016. The potential of induced pluripotent stem cell derived hepatocytes. Journal of Hepatology 65, 182–199.

Hay, D.C., Zhao, D., Fletcher, J., Hewitt, Z.A., McLean, D., Urruticoechea-Uriguen, A., Black, J.R., Elcombe, C., Ross, J.A., Wolf, R., Cui, W., 2008. Efficient Differentiation of Hepatocytes from Human Embryonic Stem Cells Exhibiting Markers Recapitulating Liver Development In Vivo. Stem Cells 26, 894–902.

Hussain, S.Z., Sneddon, T., Tan, X., Micsenyi, A., Michalopoulos, G.K., Monga, S.P.S., 2004. Wnt impacts growth and differentiation in ex vivo liver development. Experimental Cell Research 292, 157–169.

Jarrett, K.E., Lee, C.M., Yeh, Y.-H., Hsu, R.H., Gupta, R., Zhang, M., Rodriguez, P.J., Lee, C.S., Gillard, B.K., Bissig, K.-D., Pownall, H.J., Martin, J.F., Bao, G., Lagor, W.R., 2017. Somatic genome editing with CRISPR/Cas9 generates and corrects a metabolic disease. Sci Rep 7, 44624.

Javed, M.S., Yaqoob, N., Iwamuro, M., Kobayashi, N., Fujiwara, T., 2014. Generation of hepatocyte-like cells from human induced pluripotent stem (iPS) cells by co-culturing embryoid body cells with liver non-parenchymal cell line TWNT-1. J Coll Physicians Surg Pak 24, 91–96.

Jones, E.A., Tosh, D., Wilson, D.I., Lindsay, S., Forrester, L.M., 2002. Hepatic Differentiation of Murine Embryonic Stem Cells. Experimental Cell Research 272, 15–22.

Kang, Y.B. (Abraham), Eo, J., Bulutoglu, B., Yarmush, M.L., Usta, O.B., 2020. Progressive hypoxia-on-a-chip: An in vitro oxygen gradient model for capturing the effects of hypoxia on primary hepatocytes in health and disease. Biotechnology and Bioengineering 117, 763–775.

Komor, A.C., Badran, A.H., Liu, D.R., 2017. CRISPR-Based Technologies for the Manipulation of Eukaryotic Genomes. Cell 169, 559.

Larsen, F.S., 2019. Artificial liver support in acute and acute-on-chronic liver failure: Current Opinion in Critical Care 25, 187–191.

Lee, C.A., Sinha, S., Fitzpatrick, E., Dhawan, A., 2018. Hepatocyte transplantation and advancements in alternative cell sources for liver-based regenerative medicine. J Mol Med 96, 469–481.

Lemaigre, F.P., 2020. Development of the Intrahepatic and Extrahepatic Biliary Tract: A Framework for Understanding Congenital Diseases. Annu. Rev. Pathol. Mech. Dis. 15, 1–22.

Li, Y., Meng, Q., Yang, M., Liu, D., Hou, X., Tang, L., Wang, X., Lyu, Y., Chen, X., Liu, K., Yu, A.-M., Zuo, Z., Bi, H., 2019. Current trends in drug metabolism and pharmacokinetics. Acta Pharmaceutica Sinica B 9, 1113–1144.

Li, Y.-S., Harn, H.-J., Hsieh, D.-K., Wen, T.-C., Subeq, Y.-M., Sun, L.-Y., Lin, S.-Z., Chiou, T.-W., 2013. Cells and Materials for Liver Tissue Engineering. Cell Transplant 22, 685–700.

Lo, B., Parham, L., 2009. Ethical Issues in Stem Cell Research. Endocrine Reviews 30, 204–213.

Lou, Y.-R., Leung, A.W., 2018. Next generation organoids for biomedical research and applications. Biotechnology Advances 36, 132–149.

Luce, E., Dubart-Kupperschmitt, A., 2020. Pluripotent stem cell-derived cholangiocytes and cholangiocyte organoids, in: Methods in Cell Biology. Elsevier, pp. 69–93.

Luce, E., Messina, A., Duclos-Vallée, J-C. and Dubart-Kupperschmitt, A., 2021. Advanced techniques and awaited clinical applications for human pluripotent stem cell differentiation into hepatocytes, in: Hepatology. https://doi.org/10.1002/hep.31705.

Lv, L., Han, Q., Chu, Y., Zhang, M., Sun, L., Wei, W., Jin, C., Li, W., 2015. Self-renewal of hepatoblasts under chemically defined conditions by iterative growth factor and chemical screening. Hepatology 61, 337–347.

Ma, X., Duan, Y., Tschudy-Seney, B., Roll, G., Behbahan, I.S., Ahuja, T.P., Tolstikov, V., Wang, C., McGee, J., Khoobyari, S., Nolta, J.A., Willenbring, H., Zern, M.A., 2013. Highly Efficient Differentiation of Functional Hepatocytes from Human Induced Pluripotent Stem Cells. STEM CELLS Translational Medicine 2, 409–419.

Ma, X., Qu, X., Zhu, W., Li, Y.-S., Yuan, S., Zhang, H., Liu, J., Wang, P., Lai, C.S.E., Zanella, F., Feng, G.-S., Sheikh, F., Chien, S., Chen, S., 2016. Deterministically patterned biomimetic human iPSC-derived hepatic model via rapid 3D bioprinting. Proc Natl Acad Sci USA 113, 2206–2211.

Matai, I., Kaur, G., Seyedsalehi, A., McClinton, A., Laurencin, C.T., 2020. Progress in 3D bioprinting technology for tissue/organ regenerative engineering. Biomaterials 226, 119536.

Messina, A., Luce, E., Hussein, M., Dubart-Kupperschmitt, A., 2020. Pluripotent-Stem-Cell-Derived Hepatic Cells: Hepatocytes and Organoids for Liver Therapy and Regeneration. Cells 9, 420.

Mittal, R., Woo, F.W., Castro, C.S., Cohen, M.A., Karanxha, J., Mittal, J., Chhibber, T., Jhaveri, V.M., 2019. Organ-on-chip models: Implications in drug discovery and clinical applications. J Cell Physiol 234, 8352–8380.

Nakamura, M., Kobayashi, A., Takagi, F., Watanabe, A., Hiruma, Y., Ohuchi, K., Iwasaki, Y., Horie, M., Morita, I., Takatani, S., 2005. Biocompatible Inkjet Printing Technique for Designed Seeding of Individual Living Cells. Tissue Engineering 11, 1658–1666.

Negishi, T., Nagai, Y., Asaoka, Y., Ohno, M., Namae, M., Mitani, H., Sasaki, T., Shimizu, N., Terai, S., Sakaida, I., Kondoh, H., Katada, T., Furutani-Seiki, M., Nishina, H., 2010. Retinoic acid signaling positively regulates liver specification by inducing wnt2bb gene expression in medaka. Hepatology 51, 1037–1045.

Ober, E.A., Lemaigre, F.P., 2018. Development of the liver: Insights into organ and tissue morphogenesis. Journal of Hepatology 68, 1049–1062.

Ogawa, M., Ogawa, S., Bear, C.E., Ahmadi, S., Chin, S., Li, B., Grompe, M., Keller, G., Kamath, B.M., Ghanekar, A., 2015. Directed differentiation of cholangiocytes from human pluripotent stem cells. Nat Biotechnol 33, 853–861.

Oh, Y., Jang, J., 2019. Directed Differentiation of Pluripotent Stem Cells by Transcription Factors. Molecules and Cells 42.

Ouchi, R., Togo, S., Kimura, M., Shinozawa, T., Koido, M., Koike, H., Thompson, W., Karns, R.A., Mayhew, C.N., McGrath, P.S., McCauley,

H.A., Zhang, R.-R., Lewis, K., Hakozaki, S., Ferguson, A., Saiki, N., Yoneyama, Y., Takeuchi, I., Mabuchi, Y., Akazawa, C., Yoshikawa, H.Y., Wells, J.M., Takebe, T., 2019. Modeling Steatohepatitis in Humans with Pluripotent Stem Cell-Derived Organoids. Cell Metabolism 30, 374-384.e6.

Park, H.-J., Choi, Y.-J., Kim, J.W., Chun, H.-S., Im, I., Yoon, S., Han, Y.-M., Song, C.-W., Kim, H., 2015. Differences in the Epigenetic Regulation of Cytochrome P450 Genes between Human Embryonic Stem Cell-Derived Hepatocytes and Primary Hepatocytes. PLoS ONE 10, e0132992.

Rambhatla, L., Chiu, C.-P., Kundu, P., Peng, Y., Carpenter, M.K., 2003. Generation of Hepatocyte-Like Cells from Human Embryonic Stem Cells. Cell Transplant 12, 1–11.

Ronen, D., Benvenisty, N., 2012. Genomic stability in reprogramming. Current Opinion in Genetics & Development 22, 444–449.

Sakabe, K., Takebe, T., Asai, A., 2020. Organoid Medicine in Hepatology. Clinical Liver Disease 15, 3–8.

Sakiyama, R., Blau, B.J., Miki, T., 2017. Clinical translation of bioartificial liver support systems with human pluripotent stem cell-derived hepatic cells. WJG 23, 1974.

Sampaziotis, F., Cardoso de Brito, M., Madrigal, P., Bertero, A., Saeb-Parsy, K., Soares, F.A.C., Schrumpf, E., Melum, E., Karlsen, T.H., Bradley, J.A., Gelson, W.T.H., Davies, S., Baker, A., Kaser, A., Alexander, G.J., Hannan, N.R.F., Vallier, L., 2015. Cholangiocytes derived from human induced pluripotent stem cells for disease modeling and drug validation. Nat Biotechnol 33, 845–852.

Selden, C., Bundy, J., Erro, E., Puschmann, E., Miller, M., Kahn, D., Hodgson, H., Fuller, B., Gonzalez-Molina, J., Le Lay, A., Gibbons, S., Chalmers, S., Modi, S., Thomas, A., Kilbride, P., Isaacs, A., Ginsburg, R., Ilsley, H., Thomson, D., Chinnery, G., Mankahla, N., Loo, L., Spearman, C.W., 2017. A clinical-scale BioArtificial Liver, developed for GMP, improved clinical parameters of liver function in porcine liver failure. Sci Rep 7, 14518.

Shi, Y., Inoue, H., Wu, J.C., Yamanaka, S., 2017. Induced pluripotent stem cell technology: a decade of progress. Nat Rev Drug Discov 16, 115–130.

Steichen, C., Hannoun, Z., Luce, E., Hauet, T., Dubart-Kupperschmitt, A., 2019. Genomic integrity of human induced pluripotent stem cells: Reprogramming, differentiation and applications. WJSC 11, 729–747.

Steichen, C., Luce, E., Maluenda, J., Tosca, L., Moreno-Gimeno, I., Desterke, C., Dianat, N., Goulinet-Mainot, S., Awan-Toor, S., Burks, D., Marie, J., Weber, A., Tachdjian, G., Melki, J., Dubart-Kupperschmitt, A., 2014.

Messenger RNA- Versus Retrovirus-Based Induced Pluripotent Stem Cell Reprogramming Strategies: Analysis of Genomic Integrity. STEM CELLS Translational Medicine 3, 686–691.

Suman, S., Domingues, A., Ratajczak, J., Ratajczak, M.Z., 2019. Potential Clinical Applications of Stem Cells in Regenerative Medicine, in: Ratajczak, M.Z. (Ed.), Stem Cells, Advances in Experimental Medicine and Biology. Springer International Publishing, Cham, pp. 1–22.

Takahashi, K., Tanabe, K., Ohnuki, M., Narita, M., Ichisaka, T., Tomoda, K., Yamanaka, S., 2007. Induction of Pluripotent Stem Cells from Adult Human Fibroblasts by Defined Factors. Cell 131, 861–872.

Takayama, K., Kawabata, K., Nagamoto, Y., Kishimoto, K., Tashiro, K., Sakurai, F., Tachibana, M., Kanda, K., Hayakawa, T., Furue, M.K., Mizuguchi, H., 2013. 3D spheroid culture of hESC/hiPSC-derived hepatocyte-like cells for drug toxicity testing. Biomaterials 34, 1781–1789.

Takebe, T., Sekine, K., Enomura, M., Koike, H., Kimura, M., Ogaeri, T., Zhang, R.-R., Ueno, Y., Zheng, Y.-W., Koike, N., Aoyama, S., Adachi, Y., Taniguchi, H., 2013. Vascularized and functional human liver from an iPSC-derived organ bud transplant. Nature 499, 481–484.

Thomson, J.A., 1998. Embryonic Stem Cell Lines Derived from Human Blastocysts. Science 282, 1145–1147.

Touboul, T., Chen, S., To, C.C., Mora-Castilla, S., Sabatini, K., Tukey, R.H., Laurent, L.C., 2016. Stage-specific regulation of the WNT/β-catenin pathway enhances differentiation of hESCs into hepatocytes. Journal of Hepatology 64, 1315–1326.

Touboul, T., Hannan, N.R.F., Corbineau, S., Martinez, A., Martinet, C., Branchereau, S., Mainot, S., Strick-Marchand, H., Pedersen, R., Di Santo, J., Weber, A., Vallier, L., 2010. Generation of functional hepatocytes from human embryonic stem cells under chemically defined conditions that recapitulate liver development. Hepatology 51, 1754–1765.

Trefts, E., Gannon, M., Wasserman, D.H., 2017. The liver. Current Biology 27, R1147–R1151.

Ulvestad, M., Nordell, P., Asplund, A., Rehnström, M., Jacobsson, S., Holmgren, G., Davidson, L., Brolén, G., Edsbagge, J., Björquist, P., Küppers-Munther, B., Andersson, T.B., 2013. Drug metabolizing enzyme and transporter protein profiles of hepatocytes derived from human embryonic and induced pluripotent stem cells. Biochemical Pharmacology 86, 691–702.

Underhill, G.H., Khetani, S.R., 2018. Advances in Engineered Human Liver Platforms for Drug Metabolism Studies. Drug Metab Dispos 46, 1626–1637.

Vaillancourt, C., Lafond, J. (Eds.), 2009. Human Embryogenesis, Methods in Molecular Biology. Humana Press, Totowa, NJ.

Vogenberg, F.R., Isaacson Barash, C., Pursel, M., 2010. Personalized medicine: part 1: evolution and development into theranostics. P T 35, 560–576.

Volarevic, V., Markovic, B.S., Gazdic, M., Volarevic, A., Jovicic, N., Arsenijevic, N., Armstrong, L., Djonov, V., Lako, M., Stojkovic, M., 2018. Ethical and Safety Issues of Stem Cell-Based Therapy. Int. J. Med. Sci. 15, 36–45.

Wang, Y., Wang, H., Deng, P., Chen, W., Guo, Y., Tao, T., Qin, J., 2018. In situ differentiation and generation of functional liver organoids from human iPSCs in a 3D perfusable chip system. Lab Chip 18, 3606–3616.

Waring, M.J., Arrowsmith, J., Leach, A.R., Leeson, P.D., Mandrell, S., Owen, R.M., Pairaudeau, G., Pennie, W.D., Pickett, S.D., Wang, J., Wallace, O., Weir, A., 2015. An analysis of the attrition of drug candidates from four major pharmaceutical companies. Nat Rev Drug Discov 14, 475–486.

Wu, L.-J., Chen, Z.-Y., Wang, Y., Zhao, J.-G., Xie, X.-Z., Chen, G., 2019. Organoids of liver diseases: From bench to bedside. World J. Gastroenterol. 25, 1913–1927.

Yamashita, T., Takayama, K., Sakurai, F., Mizuguchi, H., 2018. Billion-scale production of hepatocyte-like cells from human induced pluripotent stem cells. Biochemical and Biophysical Research Communications 496, 1269–1275.

Yu, J., Vodyanik, M.A., Smuga-Otto, K., Antosiewicz-Bourget, J., Frane, J.L., Tian, S., Nie, J., Jonsdottir, G.A., Ruotti, V., Stewart, R., Slukvin, I.I., Thomson, J.A., 2007. Induced Pluripotent Stem Cell Lines Derived from Human Somatic Cells. Science 318, 1917–1920.

9

Hydra and the Evolution of Stem Cells

Alexander Klimovich and Thomas C.G. Bosch*

Zoological Institute, University Kiel, Germany
E-mail: tbosch@zoologie.uni-kiel.de
Address for correspondence: Thomas C.G. Bosch
Zoological Institute, Christian-Albrechts-University Kiel
Am Botanischen Garten 7, 24118 Kiel, Germany
Tel +49-431-880-4169, Fax +49-431-880-4747

9.1 Introduction

Hydra's stem cells are exceptional for two reasons. First, they represent one of the evolutionarily most ancient stem cell systems in the animal kingdom, and thus can shed light on the evolution of stem cells and their ancient molecular signatures and functions. Second, *Hydra*'s stem cells indefinitely maintain their self-renewal capacity establishing eternal lineages, which are the basis for *Hydra*'s remarkable non-senescence and resilience. Much of this potential can be traced back to the asexual mode of reproduction by budding, which requires a tissue consisting of stem cells with continuous self-renewal capacity and a dynamic adjustment of stem cell behavior to ever-changing environment. Emerging novel technologies and the availability of genomic resources reveal that: (i) stem cell self-renewal and differentiation in *Hydra* are governed by the coordinated actions of conserved signaling pathways and regulatory mechanisms that also operate in stem cells in vertebrates, strongly suggesting a common evolutionary origin of these cell types; (ii) stem cell behavior, developmental programs, environment sensing and immunity are integrated at the molecular level; and (iii) the environment, and particularly-the microbiota, has a profound effect on both normal functioning of the stem cells and their abnormal behavior such as tumorigenesis. Here we present *Hydra* as a strategic model system for stem cell research with great fundamental and high translational relevance.

9.2 The "Mystery" of *Hydra*'s Life Cycle: A Non-senescent Organism with Three Eternal Stem Cell Lineages

To many researchers, *Hydra* is an enigma in evolution. It represents a rare case of an animal with negligible senescence and extreme longevity (Jones et al., 2014). The animal model broke onto the scene as early as in the Eighteenth century when a Swiss naturalist and private scholar, Abraham Trembley, used *Hydra*'s extensive regenerative capacity to demonstrate (Trembley, 1744) to his disciples that–against prevailing belief–not everything in nature is "preformed" but that living creatures can form *de novo* by simply following nature's laws. About 230 years later, another feature of *Hydra* has been uncovered: the asexual mode of budding and constantly active axial patterning processes are based on the presence of continuously dividing cells (Gierer et al., 1972). More recently, these findings were complemented by an experimental evidence that *Hydra* demonstrates neither age-associated increase in mortality nor decline in fertility–the typical signs of senescence (Martínez, 1998; Schaible et al., 2015), and hence is a truly non-senescent animal.

Hydra belongs to the phylum Cnidaria which is a sister group of the Bilateria (Martindale et al., 2002; Schwentner and Bosch, 2015; Telford and Copley, 2011). Cnidarians were members of the pre-Ediacaran fauna between 1200 and 600 Myr ago and diverged from the main line of metazoan evolution long before the prc-Cambrian radiation. Interestingly, Cnidarians not only possess most of the gene families found in bilaterians (Chapman et al., 2010; Kusserow et al., 2005; Putnam et al., 2007) but also retain many genes that have been lost in other animal lineages, *Drosophila* and *Caenorhabditis elegans* (Kortschak et al., 2003; Miller et al., 2005; Technau et al., 2005). Cnidarians have all eumetazoan features and are characterized by true tissues connected by tight junctions, sensory, nerve and muscle cells, a gastric cavity, and a blastopore (Nielsen et al., 1996). In contrast to the triploblastic Bilateria, Cnidaria are diploblastic and possess one oral-aboral axis and a radially symmetrical body.

Hydra's simple body plan is made of a single oral-aboral body axis consisting of the head with a hypostome and tentacles, the tube-like body column, the foot region with peduncle cells, and the basal disc for the attachment to the substrate (Figure 9.1). *Hydra*'s body consists of two monolayered epithelia, the ectoderm at the outside, and the endoderm surrounding the gastric cavity. Both epithelial layers are separated by the extracellular matrix (ECM) called mesoglea, which provides stability and elasticity to the polyp (Naik et al., 2020; Sarras, 2012).

Hydra possesses about 20 different cell types originating from three independent stem cell lineages: the ectodermal epithelial cells, the endodermal epithelial cells, and the interstitial cells (Bosch, 2009; Bosch et al., 2010). Presence of only three cell lineages and the low diversity of cell types results in unusual multifunctionality of the cells (Hemmrich et al., 2012). Endodermal epitheliomuscular cells, while accomplishing self-renewal as stem cells, also contribute to digestion and contraction at the same time. Ectodermal epithelial cells, besides self-renewing, produce glycocalyx, mucus for attachment to a substrate, and assert contractility to the polyp (Szymanski and Yuste, 2019). The body shape is predetermined by the epitheliomuscular cells of both tissue layers. Further cell types include nerve cells, forming a diffuse net within the entire body (Bode et al., 1973), gland cells secreting mucus and digestive enzymes to the gastric cavity, different types of nematocytes, and germ cells, all descendants of the interstitial cell lineage. Nematocytes originate from proliferating and differentiating nests of nematoblasts and are taken up in the tentacles by battery cells, a specially differentiated type of ectodermal epithelial cells, and harboring up to 24 nematocytes (Campbell, 1987; Slautterback, 1967; Wood and Novak, 1982). Diverse molecular and cell biological methods as well as the availability of mutants resulted in a remarkably deep understanding of the cell lineages and their cell cycle characteristics (Campbell, 1967a; Campbell, 1967b; Campbell and David, 1974; David and Campbell, 1972; David and Challoner, 1974; David and Gierer, 1974; Siebert et al., 2019; Takano and Sugiyama, 1984). Statistical cloning experiments (Bosch and David, 1990; David and Murphy, 1977) demonstrated that *Hydra*'s cells have an astonishing ever-lasting potential to self-renew and differentiate following strict spatio-temporal rules; and that interstitial stem cells can do both: to differentiate into somatic cells and–under appropriate environmental stimuli–into germ cells. Thus, while ectodermal and endodermal epithelial cells are unipotent stem cells, interstitial cells are truly multipotent stem cells, able to give rise to somatic and germ-line cells (Figure 9.2). This lineage, therefore, represents an extremely rare case in the animal world, where multipotent stem cells, typically restricted to early embryogenesis stages, are present in an adult organism (Juliano et al., 2010; Tanaka and Reddien, 2011).

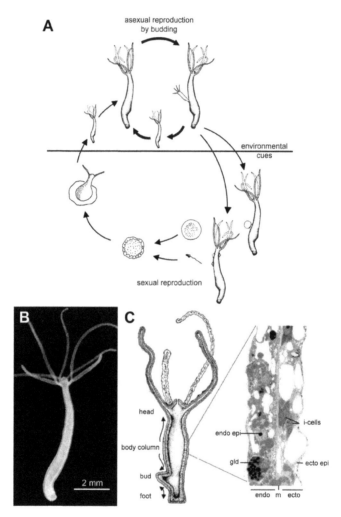

Figure 9.1   Life cycle and anatomy of *Hydra*. (A) The freshwater polyp reproduces primarily asexually by budding and yet is able to produce gametes and reproduce sexually upon induction by environmental cues. Image provided by S. Fraune. (B) A polyp of *Hydra vulgaris* AEP strain. (C) Anatomy of a *Hydra* polyp. Mitotic stem cells in the body column self-renew, terminally differentiate and are displaced toward the head, tentacles, or foot (arrows), resulting in a constant flux of cells toward the extremities, with older cells being lost. A detailed view on a histological longitudinal section of *Hydra* tissue demonstrates two monolayered epithelia, the ectoderm (ecto) and the endoderm (endo) separated by the extracellular matrix (m). While interstitial stem cells (i-cells) are interspersed between the ectodermal cells (ecto epi), the gland cells (gld) are found in between the endodermal epithelial cells (endo epi). (Modified from Bosch, 2007).

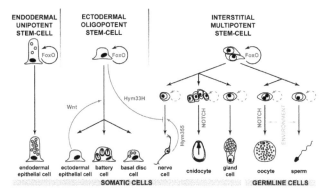

Figure 9.2   Three stem-cell lineages give rise to all the diversity of cell types in *Hydra*. Stem cells proliferate indefinitely (reversionary arrows), while transient amplifying cells have limited proliferation capacity (dashed reversionary arrows), and after 1-2 division rounds give rise to clones of differentiated cells. Factors controlling stemness (FoxO), commitment, and differentiation (diffusible molecules Wnt, Hym33H and Hym355, contact signaling Notch) are outlined in red. Environmental factors (temperature and metabolic state) affect gametogenesis.

How do the stem cell lineages contribute to the *Hydra*'s unique life cycle? In fact, the unlimited proliferation of stem cells is fundamental for *Hydra*'s asexual reproduction by budding and its non-senescence (Figure 9.1). Stem cells located in the middle body column efficiently do both, tissue construction and tissue maintenance, throughout the lifetime of a polyp. Stem cells of all three lineages continuously proliferate and undergo terminal differentiation (Bosch, 2009; Bosch et al., 2010). Consequently, there is a continuous displacement of cells from the body toward the upper (hypostome and tentacles) and lower (the basal disc) extremities of a polyp, as well as into the bud, which is forming in the lower part of the body column (Campbell, 1967a; Campbell, 1967b). In spite of such an intensive tissue turnover, the number of cells in the *Hydra* polyp as well as proportions between cells of different types and lineages stays remarkably stable (Bode et al., 1973). This suggests that a delicate balance between the self-renewal and differentiation behavior of stem cells is tightly controlled. In addition, activity of stem cells in *Hydra* must be tightly adjusted to the environment, allowing for a stable maintenance of tissue homeostasis for extended periods. The functionality of *Hydra*'s stem cells evidently never declines, which is in sharp contrast to most other animal species. Decline in stem cell number and functionality is recognized as one of the hallmarks of aging (López-Otín et al., 2013) and as a chief cellular mechanism underlying organismal senescence and limited lifespan (van Deursen, 2014). Owing to its unique biology, the non-senescent *Hydra* provides extremely valuable insights into the mechanisms that are at work to escape, delay, or suppress senescence (Mortzfeld and Bosch, 2017; Schenkelaars et al., 2018).

In sum, tissue function and maintenance in *Hydra* are based on three tissue-specific stem cells: ectodermal and endodermal epitheliomuscular cells and interstitial stem cells. All three stem cell types have unlimited self-renewal capacity and can differentiate in one or more cell types. What molecular and cellular mechanisms control the behavior of the stem cells in *Hydra*?

9.3 Revisiting Stem Cells in *Hydra* with Emerging Methodologies

Thanks to technological breakthroughs of the last decades, *Hydra* has become a highly informative model system in developmental biology and stem cell research, providing the advantages of a simple body plan, high regeneration capacity, clonal propagation, short reproduction time, perfectly understood cell biology, and easy cultivability. Given these attributes, the model organism *Hydra* has been used to study a variety of biological questions in an evolutionary context, including morphogenesis and pattern formation (Gee et al., 2010; Gierer et al., 1972; Meinhardt and Gierer, 1974; Meinhardt and Gierer, 2000; Watanabe et al., 2014), stem cell behavior and aging (Anton-Erxleben et al., 2009; Boehm and Bosch, 2012; Boehm et al., 2012; Bosch and David, 1986; Bosch et al., 2010; Buzgariu et al., 2014; Hemmrich et al., 2012; Klimovich et al., 2018; Nishimiya-Fujisawa and Kobayashi, 2018; Siebert et al., 2008), tumor formation (Domazet-Lošo et al., 2014; Rathje et al., 2020), and regeneration (Bosch, 2007; Buzgariu et al., 2018; Chera et al., 2006; Govindasamy et al., 2014; Reiter et al., 2012). More recently, *Hydra* became an important model organism to investigate the fundamental principles of immunity and gene-environment interactions (Augustin et al., 2012; Bosch, 2013; McFall-Ngai and Bosch, 2020; Mortzfeld et al., 2018; Mortzfeld et al., 2019), as well as for addressing emerging neurobiological questions (Augustin et al., 2017; Bosch et al., 2017; Dupre and Yuste, 2017; Han et al., 2018; Klimovich and Bosch, 2018; Klimovich et al., 2020; Koizumi, 2016; Murillo-Rincon et al., 2017).

The studies of the *Hydra*'s stem cells were particularly fueled in the last years thanks to the advances of three technologies: (i) the development of the transgenesis technology (Klimovich et al., 2019; Wittlieb et al., 2006);

(ii) the emergence of large scale transcriptomic resources and, particularly, the application of cell-sorting and single-cell analysis (Hemmrich et al., 2012; Klimovich et al., 2020; Siebert et al., 2019); (iii) 16S rDNA profiling and metagenomic sequencing (Franzenburg et al., 2013a; Fraune and Bosch, 2007; Mortzfeld et al., 2018; Rathje et al., 2020).

The genetic manipulation via embryo microinjection opened new possibilities for differential labeling and *in vivo* imaging and tracing of cells, the functional characterization of genes and molecular mechanisms by constitutive or conditional gain- and loss-of-function analyses (reviewed in (Klimovich et al., 2019)). Transgenesis not only provided ultimate evidence that stem cells in *Hydra* maintain eternal lineages (Wittlieb et al., 2006), but also showed a remarkable phenotypic plasticity of individual cells in response to positional signals (Khalturin et al., 2007). Combining *in vivo* tracing of genetically-labeled interstitial stem cells and tissue transplantation showed that multipotent interstitial stem cells are stationary in homeostatic conditions. However, when exposed to tissue depleted of the interstitial cell lineage, these cells migrate and repopulate emptied niches, pointing to existence of some instructive microenvironment cues emanating from the differentiated products of the interstitial cells or derived from the epithelial cells (Boehm and Bosch, 2012). In addition, by *in vivo* monitoring of transgenic gland cells, we observed to our great surprise, that zymogen cells continuously transdifferentiate into granular mucous cells in the head region (Siebert et al., 2008). These findings demonstrate that, in addition to stem cell-based mechanisms, transdifferentiation is essential for normal development and maintenance of cell type complexity in *Hydra*. This might indicate that in organisms which diverged before the origin of bilaterian animals, commitment and differentiation might be less stable events than in more complex metazoans.

Studies using transcriptome analysis at single-cell resolution provided deeper insights into the molecular logic which operates within cell lineages and cell types of *Hydra*. While Siebert and colleagues collected and sequenced ~25,000 single-cell transcriptomes to reveal the dynamics of gene expression that accompany cell specification and differentiation in all three *Hydra* lineages (Siebert et al., 2019), we have focused specifically on the interstitial cell lineage (Klimovich et al., 2020). Together, the two studies provided a comprehensive view of the bifurcating trajectories of progenitor states that lead to the differentiating cell types. Intriguingly, novel transient states, that previously escaped detection by conventional methods, have been identified. For instance, bipotential progenitors derived from interstitial stem cells in the ectodermal layer cross the ECM to supply the endodermal layer with neurons and gland cells (Siebert et al., 2019). Furthermore, single-cell transcriptomics combined with whole-mount *in situ* hybridization and transgenic reporter lines uncovered a remarkable diversity of neuronal cell types in *Hydra*. The interstitial stem cells give rise to multiple spatially restricted populations of molecularly distinct and functionally diverse neurons that build up an astonishingly complex nerve net in *Hydra* (Klimovich et al., 2020; Siebert et al., 2019). Beyond that, the single-cell transcriptomics revealed transcription factor signatures, cis-regulatory elements and gene regulatory networks that specify each cell type within a *Hydra* polyp. This allowed, for the first time, a direct comparison of the *Hydra*'s cell type repertoire to the cell types and tissues of other cnidarians and bilaterians. While the molecular fingerprint of the *Hydra* ectodermal cells provided a clear evidence for a highly conserved ectoderm program shared by all Metazoa, and the endodermal signature revealed a clear homology to the pharyngeal ectoderm of the sea anemone and to bilaterian endoderm, the single-cell data provided no support for the origin of mesoderm within the endoderm (Arendt, 2019).

Together, these findings provide truly novel insights into how *Hydra* builds its body with only a limited number of different cell types. These achievements also brought *Hydra*'s stem cells into a key position in comparative stem cell research by delivering answers to questions such as: Are certain features common to all stem cells in all Metazoa? How do stem cells in basal metazoan differ from stem cells in Bilateria? How do stem cells evolve? Since the advent of next-generation sequencing and 16S rRNA profiling (Weisburg et al., 1991), it is getting increasingly recognized that epithelia of all animals including humans, are colonized by microbes. In each species, a specific and diverse microbial community, the microbiome, together with its multicellular hosts form one functional unit-the holobiont (Bosch and McFall-Ngai, 2021; McFall-Ngai et al., 2013). *Hydra* is not an exception: the polyp's epithelial surface is densely colonized by a stable species-specific bacterial community (Fraune and Bosch, 2007). The *Hydra* host shapes its microbiome using a rich repertoire of antimicrobial peptides (AMPs) (Augustin et al., 2017; Bosch et al., 2009; Franzenburg et al., 2013b). By combining the 16S rDNA sequencing, experimental manipulation of both the host and its microbiome, we have recently uncovered that the presence and structure of the microbiome is critical for the proper stem cell activity, tissue homeostasis, and health of *Hydra*. To our surprise, an invasion of the normal *Hydra* microbiome by a single environmental bacterium, a spirochete, causes substantial alterations in stem cell behavior and triggers tumor formation (Rathje et al., 2020).

Recognition of the microbiome's impact on stem cell activity has contributed to a paradigm shift in developmental biology: stem cells are now being embedded into the metaorganism concept and are engaged in a molecular dialogue within a holobiont (Hadfield and Bosch, 2020).

9.4 Stem Cells in *Hydra* Are Controlled by Conserved Transcription Factors and Effectors

Tight control of stem cell activity and cell-cell communication are critical for the generation and maintenance of multicellularity in animals; therefore, one would expect that the molecular toolkits involved in these processes are highly conserved across metazoans. Transcriptome sequencing corroborated these expectations and uncovered the molecular factors that enable *Hydra* to indefinitely maintain stem cells self-renewal capacity. First, *Hydra*'s stem cell specific transcriptomes contain signatures of genes coding for highly conserved transcription factors (Hemmrich et al., 2012; Klimovich et al., 2020; Siebert et al., 2019). For instance, transcripts of two regulators, FoxO and Myc, were found overrepresented in stem cells of *Hydra*, and further studies using transgenesis provided experimental evidence that both are indispensable for stem cell homeostasis in *Hydra* (Ambrosone et al., 2012; Boehm et al., 2012; Hartl et al., 2010; Hartl et al., 2014; Hartl et al., 2019). Interestingly, recent work has implicated FoxO orthologues in mammals in playing central role in maintenance of neural (Paik et al., 2009; Renault et al., 2009), adult hematopoietic (Miyamoto et al., 2007; Tothova and Gilliland, 2007), and embryonic stem cells (Zhang et al., 2011). Moreover, FoxO homologues consistently confer increased life span and stress resistance in flies and worms, while FoxO3a is known as an important component of the genetic signature of human exceptional longevity (Flachsbart et al., 2009; Jünger et al., 2003; Kenyon, 2010; Li et al., 2009). Similarly, the bilaterian Myc protooncogene is a master regulator of such fundamental cellular processes as growth, proliferation, differentiation, metabolism, and apoptosis (Conacci-Sorrell et al. , 2014; Dang, 2012; Eilers and Eisenman, 2008). In addition, HMG-B3b, a member of the high-mobility group (HMG) super family, and KLF13, a representative of the Krüppel-like factors, both were overrepresented in the *Hydra* interstitial stem cell transcriptome (Hemmrich et al., 2012). Their orthologues are known to regulate the balance between self-renewal and differentiation in mouse hematopoietic stem cells (Nemeth et al., 2003). CUX1 transcription factor, also predominantly expressed in interstitial stem cells, is involved in the control of many cellular processes, including determination of cell identity, cell cycle progression, and cell-cell communication in bilaterians (Sansregret and Nepveu, 2008). Further investigation will shed light onto the functional role of these conserved transcription factors in maintaining stemness of three stem cell lineages and in governing lineage-specific transcriptional profiles in *Hydra*. Intriguingly, a number of transcription factors commonly associated with stemness and pluripotency in vertebrates, such as Nanog and Oct3/4 (Boiani and Schöler, 2005; Noggle et al., 2005; Pan and Thomson, 2007), were found to be missing in *Hydra*. Thus, the vertebrate-specific core transcription factor circuitry (Oct3/4, Nanog, and Sox2) appears to be a late emerging invention in bilaterian evolution (Hemmrich et al., 2012). This is prompting the question, whether the function of these regulators is taken over by other, non-related and perhaps non-conserved transcription factors in *Hydra*.

Beyond the vast diversity of ancient transcription factors, highly conserved proteins of PIWI family and their molecular targets, piRNAs, were shown indispensable for maintenance of stem cell lineages in *Hydra* (Juliano et al., 2014; Lim et al., 2014). Additionally, RNA-interacting proteins Nanos and Vasa are specifically enriched in the interstitial stem cells of *Hydra* and particularly, in the germline cells (Hemmrich et al., 2012; Mochizuki et al., 2000; Mochizuki et al., 2001). These observations support the emerging view that a set of RNA-binding proteins found ubiquitously in the animal kingdom, from sponges to mammals, was in fact the first circuitry evolved in the common ancestor of all Metazoa to regulate stem cell proliferation and differentiation (Alié et al., 2015; Boehm et al., 2012; Fierro-Constaín et al., 2017; Hemmrich et al., 2012).

Finally, several highly conserved effector molecules were recently shown to play crucial role in stem cell homeostasis in *Hydra*. We have demonstrated that transcripts coding for the Lamin nuclear protein are present in all stem cells in *Hydra* (Klimovich et al., 2018). In bilaterian species, Lamins are known to play a central role in maintaining stem cell activity and tissue homeostasis, as well as in controlling cellular and organismal senescence (Gruenbaum et al., 2005; Lans and Hoeijmakers, 2006; Mattout et al., 2006). To our surprise, we discovered that the proliferation of *Hydra*'s stem cells is remarkably robust against the disturbances in Lamin expression levels and in the nuclear envelope structure, in contrast to other animals (Klimovich et al., 2018). We argued that this extraordinary robustness and low complexity of the nuclear envelope of *Hydra* may allow for its ever-lasting stem cells renewal and non-senescence. We also performed a functional analysis of *Hydra* orthologues of three apoptosis-related proteins: Bcl-2, Bax-Inhibitor, and Lifeguard. Using transgenesis, we unveiled that, similar to their function in bilaterians, these effectors protect the *Hydra* epithelial cells from apoptosis, pointing to their ancestral anti-apoptotic role in stem cell maintenance (Motamedi et al., 2019). Finally, we investigated the role of the planar cell polarity in *Hydra*. By generating and characterizing in detail a series of transgenic lines, we provided evidence that homologues of Fat-like and Dachsous cadherins are indispensable for cell adhesion, spindle orientation, and tissue organization in *Hydra* (Brooun et al., 2020).

Taken together, these observations unveil a surprisingly complex ancient molecular landscape of *Hydra* stem cells. They suggest that the last common ancestor of Eumetazoa has already evolved a sophisticated toolkit of transcription factors, RNA-binding proteins, and effector molecules that enabled the emergence of stem cells and transition to stable tissue-level complexity.

9.5 Epigenetic Control of the Stem Cells in *Hydra*

One of the most intriguing questions in *Hydra* biology is how a stable axial patterning is maintained in a polyp while its tissue is constantly growing and being displaced (Figure 9.1). Axis patterning during development is accompanied by large-scale gene expression changes, commonly attributed to differential activity of transcription factors. This hypothesis was supported by Siebert and co-authors, who not only uncovered differential expression of transcription factors in cells from distinct body compartments of *Hydra* (body column, head, foot), but also matched this differential expression to a profile of enriched TF-binding motifs in the gene regulatory regions (Siebert et al., 2019). Using ATAC-seq technology, they demonstrated that, for example, the paired box (Pax) motif is enriched in regulatory regions of genes expressed during early and mid-stages of nematocyte differentiation, while the POU motif is enriched only in late stages. The ETS domain binding motif was particularly enriched in genes expressed in endodermal and ectodermal epithelial cells in the extremities (tentacles and foot). These observations indicate that the cells drastically change their gene expression programs once they cross an invisible border between the body column and head or foot. What mechanisms establish and maintain this remarkably sharp border between the domains of proliferation and differentiation?

We recently showed that the sharp boundary separating tissue in the body column from head and foot tissue depends on histone acetylation (López-Quintero et al., 2020). Histone deacetylation disrupts this boundary by affecting numerous developmental pathways including the Wnt components and prevents stem cells from entering the position-dependent differentiation program. These findings were corroborated and extended by Reddy and co-authors, who used high-throughput transcriptome sequencing, pharmacological perturbation, and chromatin profiling to uncover a remarkably complex landscape of epigenetic modifications in *Hydra* genome (Reddy et al., 2019). Diverse activation- and repression-associated epigenetic marks, such as histone trimethylation (H3K4me3) and acetylation (H3K27ac and H3K9ac) differentially occupy gene bodies, promoters, and intergenic regions that act as proximal cis-regulatory elements. Reddy and co-authors provided evidence that this complex code of epigenetic marks controls the occupancy of specific DNA motifs by zinc-finger, T-box, and ETS-related transcription factors, that in turn, regulate the pattern-dependent activity of stem cells.

Together, these findings provide evidence that a dynamic histone acetylation/methylation switch of distinct gene regulatory elements combined with differential activity of multiple transcription factors regulate chromatin architecture, gene expression, and balance between self-renewal and differentiation in the *Hydra* cells. These results uncover a remarkably complex ancient regulatory mechanism of position-dependent cell fate control, which was present in the common ancestor of cnidarians and bilaterians, at least 600 million years ago.

9.6 Taxonomically Restricted Genes (TRGs) Have Their Saying

All animals, from sponges to humans, share a highly conserved developmental molecular toolkit (Carroll, 2008; Miller et al., 2005; Putnam et al., 2007; Sebé-Pedrós et al., 2018). However, up to 40% of genes in the genome of every animal species are represented by non-conserved, taxonomically restricted genes (TRGs) (Nelson and Buggs, 2016; Tautz and Domazet-Lošo, 2011). These genes show no homology to other genes outside that species or clade, may drive morphological specification and allow organisms to adapt to changing environment (Babonis et al., 2016; Belcaid et al., 2019; Chen et al., 2013; Hwang et al., 2007; Johnson, 2018; Johnson and

Tsutsui, 2011; Milde et al., 2009; Santos et al., 2017). Although the role of TRGs has been underestimated for a long time, it is becoming increasingly recognized that they have important implications in evolutionary processes (Khalturin et al., 2009; Tautz and Domazet-Lošo, 2011). Notably, in *Hydra*, cells of all three stem cell lineages contain a considerable number of *Hydra*-specific gene products, which lack recognizable homologs in other, even closely related, cnidarian species. First, nematocytes were found particularly rich in transcripts encoded by TRGs, which is not surprising considering that they represent a highly-specialized cell type found exclusively in Cnidaria (Hwang et al., 2007; Milde et al., 2009). What is truly surprizing, TRGs are even more abundant in the transcriptomes of *Hydra* neurons (Klimovich et al., 2020)– a cell type commonly considered as synapomorphic for Eumetazoa. Accumulating observations provide evidence that neuron-specific TRGs in *Hydra* are implicated in neuronal communication (Takahashi et al., 2008; Yum et al., 1998), pattern formation (Khalturin et al., 2008; Takahashi et al., 2005), and interactions with the commensal microbiota (Augustin et al., 2009; Augustin et al., 2017; Franzenburg et al., 2013b; Fraune et al., 2010; Klimovich et al., 2020; Taubenheim et al., 2020). Moreover, small peptides encoded in TRGs may form intricate regulatory circuits of interlineage communication and contribute to tissue homeostasis in *Hydra* (Fujisawa, 2008). For instance, while a *Hydra*-specific peptide Hym-355 secreted by neurons promotes neuronal commitment of interstitial cells and their maturation to neurons (Takahashi et al., 2000), a secreted peptide Hym-33H produced by the epithelial cells counteracts this and inhibits the neuronal differentiation (Figure 9.2). Another example is a pair of closely-related peptides, named Eco1 and Eco2, encoded in taxonomically restricted genes and produced by epithelial cells in *Hydra*. We have recently demonstrated that the expression of Eco1 and Eco2 depends on temperature and bacterial colonization (Taubenheim et al., 2020). Loss-of-function experiments revealed that Eco peptides are involved in the regulation of pattern formation and have an antagonistic function to Wnt signaling in *Hydra*. Hence, our results demonstrate how environmental cues can be linked to developmental processes by the regulation of non-conserved genes that modulate conserved signaling pathways. Recent studies open even more exciting perspectives: Reddy and co-authors have recently demonstrated that not only small effector molecules, but also transcription factors, essential for polyp's patterning, are encoded in *Hydra*-specific genes (Reddy et al., 2019).

These observations provide an absolutely novel view: many important processes, fundamental for Metazoa, such as stem cell maintenance, differentiation, and interlineage communication might be regulated via non-conserved molecules. However, only a miniscule number of TRGs encoded in *Hydra* genome are identified and functionally characterized so far, and the questions how are TRGs embedded into the gene regulatory networks and how do they interact with conserved transcriptional regulators and signaling pathways are among the most exciting topics of *Hydra* research.

9.7 Stem Cells Interact with the Environment

It is becoming increasingly recognized that the core genetic machineries of stem cell control and development do not act in isolation. In fact, the development of virtually all metazoans is sensitive to environmental abiotic (such as temperature, salinity, or photoperiod) or biotic cues (such as those emanating from predators, conspecifics, or food) (Gilbert, 2012). Moreover, accumulating evidence shows that symbiotic microorganisms greatly impact the host development as well as tissue maintenance (Hadfield and Bosch, 2020). However, it remains poorly understood, how the developmental regulatory networks integrate sensing the instructive environmental cues, stem cell behavior, tissue homeostasis, and development.

Previous studies on *Hydra* provided evidence for a strict dependence of tissue homeostasis and development on the environment. Cell proliferation and budding frequency are tightly linked to the feeding conditions (Bosch and David, 1984; Holstein et al., 1991). Environmental cues, such as reduced temperature or starvation, also induce the proliferation of germ line precursors, formation of gametes and sexual reproduction (Bosch and David, 1986; Littlefield et al., 1991; Sugiyama and Fujisawa, 1978). Finally, the polyp body size is strictly temperature-dependent (McGill, 1908; Stiven, 1965). These observations point to existence of a mechanistic link between environmental sensing and the molecular circuits of stem cell control.

The protein FoxO was one of few transcription factors strongly expressed in all the three stem cell lineages in *Hydra* and absent from terminally differentiated cells (Boehm et al., 2012; Hemmrich et al., 2012). This was very exciting, since FoxO is known in other systems to maintain stemness, to delay the aging process, and to confer longevity (Giannakou et al., 2004; Kenyon et al., 1993; Kimura et al., 1997). In human, specific polymorphisms in the *foxO3a* gene have been shown positively associated with longevity in centenarians, thus assigning the longevity factor role to FoxO (Anselmi et al., 2009; Flachsbart et al., 2009; Kojima et al., 2004; Li et al., 2009; Pawlikowska et al., 2009). In addition, in many organisms including

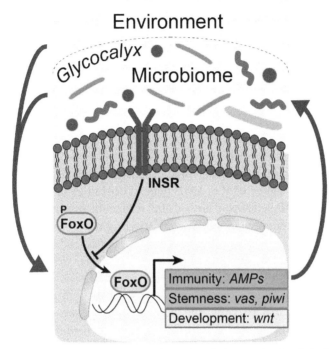

Figure 9.3 The conserved transcription factor FoxO is a molecular hub that mediates gene-environment interactions by integrating environmental signals with stem cell control, developmental pathways, and immune homeostasis.

C. elegans, *Drosophila*, and mouse, FoxO transcription factor has been shown to prevent from aging and confer increased life span (Becker et al., 2010; Flachsbart et al., 2009; Kenyon, 2010; Kimura et al., 1997). Together, these observations prompted us to test the hypothesis, whether FoxO might be a molecular link between sensing the environmental cues, stem cell maintenance, and tissue homeostasis in *Hydra*.

We have assessed the role of FoxO in stem cell maintenance of *Hydra* in gain- and loss-of-function experiments (Boehm et al., 2012). Overexpression of FoxO in the interstitial cell lineage impairs the balance between stem cell renewal and differentiation: the proliferation of interstitial stem cells and nematoblasts increases. Moreover, excess of FoxO activity causes a strong upregulation of stemness genes, such as *vasa* and *piwi* (Figure 9.3). Conversely, a functional FoxO knock-down results in a significant down-regulation of stemness genes, in accumulation of terminally differentiated cells, substantial reduction of polyp size, lower number of epithelial cells per polyp, and is accompanied by a considerable slowdown of population growth rate (Boehm et al., 2012; Mortzfeld et al., 2019). These observations point toward an implication of FoxO in the maintenance of stem cell self-renewal and thereby continuous growth in *Hydra*. Intriguingly, the downregulation of FoxO also caused considerable changes in the expression of numerous genes coding for AMPs (Figure 9.3). In fact, most genes coding for epithelially expressed AMP families including the Hydramacin, Kazal, and Arminin were strongly downregulated in the FoxO-deficient animals, indicating that FoxO positively regulates AMP expression (Mortzfeld et al., 2018). This was supported by *in silico* analysis that predicted multiple FoxO-binding sites in the promoter sequences of the AMP genes (Boehm et al., 2012). Since FoxO-dependent AMP genes play an essential role in the assembly of species-specific microbial communities of *Hydra* (Bosch et al., 2009; Franzenburg et al., 2013b) and loss of tissue homeostasis in *Hydra* leads to changes in the symbiotic bacterial community (Fraune et al., 2009), we next hypothesized that FoxO, in addition to its role in stem cell self-renewal, may affect bacterial colonization. Our experiments uncovered that, in contrast to the control *Hydra* polyps that assemble and maintain their species-specific microbiome for decades (Fraune and Bosch, 2007), FoxO loss-of-function polyps were impaired in selection for bacteria resembling the native microbiome and were more susceptible to colonization by foreign bacteria (Figure 9.3). Therefore, reduction in AMP expression in FoxO-deficient polyps correlates with impaired microbial colonization, highlights the central role of AMP in shaping microbial communities and embeds the conserved tissue regulator and aging antagonist FoxO into the metaorganism concept.

Exploring epithelial FoxO loss-of-function mutants, we also observed that multiple genes involved in developmental signaling were dysregulated in FoxO-deficient *Hydra*. Specifically, the expression of genes involved in axial patterning, such as Wnt11, its receptor Fzd1/7 (Guder et al., 2006), and thrombospondin-family regulators (Hamaguchi-Hamada et al., 2016; Lommel et al., 2018) were altered by epithelial FoxO inactivation (Mortzfeld et al., 2019). Canonical Wnt/β-Catenin signaling is a central player in the head organiser and in pattern formation along the oral-aboral body axis (Broun et al., 2005; Gee et al., 2010; Hobmayer et al., 2000; Nakamura et al., 2011). Furthermore, expression of several members of the developmentally relevant TGF-β pathway such as ACV, Cer-1, TGF-β2, and DAN, were affected by FoxO deficiency (Mortzfeld et al., 2019). These findings were fully corroborated in experiments using pharmacological interference approach and qRT-PCR, and allowed us to conclude that FoxO transcription factor directly regulates the Wnt signaling pathway, which in turn, controls the expression and activity of the TGF-β pathway (Mortzfeld et al., 2019). Binding of FoxO to conserved regulatory genomic elements, therefore, modulates the expression of genes involved in stem cell homeostasis, immune function, and development. Since stem cell activity, developmental programs and immune homeostasis in *Hydra* are sensitive to the environmental inputs (McFall-Ngai and Bosch, 2020), we hypothesized that FoxO might integrate the environmental signals and translate them into specific transcriptional activity. What molecular pathways might relay environmental cues onto FoxO activity? In a search for such a candidate molecular mechanism, we noticed that the insulin-like growth factor (IGF)-like signaling (IIS) pathway is an important environmental sensor up-stream of FoxO in bilaterians (Greer and Brunet, 2005; Kenyon, 2010; Salih and Brunet, 2008). Does insulin receptor (INSR), the key component of the IIS, function as the environmental sensor in *Hydra*? Using INSR knock-down, we disrupted the IIS signaling in *Hydra* and uncovered, to our surprise, that the number of cells per polyp was substantially elevated in IIS-deficient hydras–opposite to the FoxO loss-of-function phenotype. Additionally, several components of the TGF-β pathway, such as the ACV protein, were consistently up-regulated upon INSR knock-down. These findings provide evidence that IIS acts up-stream from FoxO and negatively regulates its activity (Mortzfeld et al., 2019).

It is becoming increasingly recognized that stem cells are exposed to very heterogeneous environments (Belkaid and Naik, 2013; Brown et al., 2013; Hadfield and Bosch, 2020), and that the phenotypic characteristics of an organism are determined by both genetic and environmental factors, and the interactions between the two (Gillespie and Turelli, 1989). Our findings provide novel insights into the molecular logics of these gene-environment interactions. We unveiled that the stem cell regulatory machinery and developmental programs are intertwined at the molecular level with the environment, including the resident microbiota. The conserved transcriptional regulator FoxO acts as a molecular hub, integrates diverse upstream signals, and controls activity of conserved stemness factors, developmental pathways, and immune system components.

What are the implications of the molecular coupling of stem cell behavior, developmental programs, environment sensing, and the immune homeostasis? In many organisms, age-related decline in stem cell activity has been observed to be associated with drastic changes in the immune system (López-Otín et al., 2013). While adaptive immunity significantly declines,

a phenomenon called immuno-aging, innate immunity seems to be activated, inducing a pro-inflammatory profile, also referred to as inflamm-aging
(Salminen et al., 2008), which causes age-related inflammatory diseases and
is a major contributor to age-associated frailty, morbidity, and mortality (Dato
et al., 2017). The appreciation that stem cell maintenance, development, and
immunity are part of a global program, the "Eco-Aging" (Mortzfeld and
Bosch, 2017), acknowledges the multi-organismic nature of all living beings
and presents a new conceptual framework for explaining the process of aging.

9.8 Microbiome Contributes to Tumor Formation in *Hydra*

A plethora of molecular mechanisms ensure a proper spatial and temporal
control of stem cell activity in *Hydra* on multiple levels–transcription, epigenetics, nuclear organization, interlineage communication, and adjustment
to the environment. The fact that *Hydra*'s lifespan is virtually unlimited,
poses an intriguing question: Is such an elaborated system of stem cell
control indeed exceptionally robust or is it still prone to errors? A tumor
is an abnormal growth of body tissue, which is due to a disturbed tissue
homeostasis–cell proliferation, differentiation, or apoptosis (Hanahan and
Weinberg, 2011). The evolutionary origin of tumor long remained unknown.
The phylostratigraphic analysis of genes involved in tumor formation in
mammals has uncovered their ancient evolutionary origin and has predicted
that most metazoans might be prone to develop tumors (Domazet-Lošo et al.,
2010). We provided the first evidence for naturally occurring tumors in *Hydra*
(Figure 9.4). Using histological and immunochemical methods as well as *in
situ* hybridization, we demonstrated that tumors in two *Hydra* species–*Hydra
oligactis* and *Pelmatohydra robusta* are composed of female germline precursor cells that are not able to terminally differentiate into oocytes and nurse
cells. Moreover, these abnormal female precursor cells acquire resistance to
apoptosis, insensitivity to environmental stimuli, and an increased migratory
potential. Numerous genes related to cell cycle, apoptosis, genomic stability,
and metabolism regulation were differentially expressed in tumorous cells.
Together, these acquired features of tumor cells in *Hydra* greatly resemble the
hallmarks of invasive mammalian tumors (Hanahan and Weinberg, 2011).

To our surprise, the microscopic and molecular analysis revealed that
the microbiota of tumorous polyps was also drastically different from that
of the healthy polyps. Using extensive manipulation with both, the host and
its microbiome, we provided evidence that the microbes in fact drive tumor
formation in *Hydra*. We demonstrated that an environmental stressor has

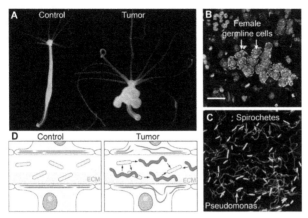

Figure 9.4 Microbiome drives tumor formation in *Hydra*. (A) Tumorous polyps have conspicuous tissue budges in the body column. (B) Uncontrolled accumulation of interstitial cells in *Hydra* tumors is due to a differentiation arrest of committed female precursor cells. (C) Two bacteria, an environmental Spriochete and a commensal Pseusomonas, are found in the mesoglea of tumorous polyps. (D) The interplay between these two bacteria and some virulence factors produced by them affect the *H. oligactis* cells, their morphology (including the actin cytoskeleton depicted as red curved lines), disturb the tissue homeostasis and cause tumor formation.

likely disrupted the normal bacterial community within *Hydra* and, hence, allowed an environment bacterium to colonize the polyps. This foreign bacterium from the phylum Spirochaetes became increasingly abundant in the mesoglea of *Hydra* (Figure 9.4). The contact of this intruder with the already present commensal bacterium Pseudomonas completely shifted gene expression and behavior of both microbes and lead to production of multiple virulence factors, such as the flagellum, proteases, and toxins. These molecules cause structural changes in *Hydra* epithelial cells, disturb the tissue homeostasis and developmental pathways, ultimately cause tumor formation, and gravely affect fitness of the polyps. What is particularly interesting– the foreign spirochete exerts its harmful influence only in the presence of the commensal Pseudomonas, pointing to a surprizing hidden pathogenic potential within the normal microbiome. How these interactions between the two bacteria occur at the molecular level and which biochemical mechanisms are involved in the cancerous transformations of the host cells are a subject of currently ongoing studies.

This work has contributed to a paradigm shift in developmental biology. First, it demonstrated that cancers can be found not only in bilaterians, but in pre-bilaterian animals as well. It became accepted that tumor formation is a legacy of transition to multicellularity in the early animal evolution (Aktipis et al., 2015). Second, our findings highlighted the role of the environment as a source of opportunistic tumorigenic microbes. These findings complement the mounting evidence that cancers in mammals and humans have a unique tumor-associated bacterial community (Atreya and Turnbaughe, 2020; Nejman et al., 2020) and are frequently triggered by environmental microbes (Chang and Parsonnet, 2010; Cho and Blaser, 2012; Cummins and Tangney, 2013). Finally, our study highlights the protective role of the resident microbiome in guarding the tissue homeostasis. This is consistent with observations on mice (Rosshart et al., 2017), implicating that the natural microbiota provides resistance against harmful, cancerogenic influences. Future research will show whether this ability of the microbiome may also be harnessed for prevention of cancer.

9.9 Open Questions and Future Perspectives

In the recent decades, *Hydra* has become a strategically important model organism for understanding stem cell biology and gene-environment interactions. Most importantly, due to its experimental accessibility, the *Hydra* model allows deep mechanistic insights into the fundamental questions of how stem cells function and evolve. Modern technologies allow addressing causation and, in turn, fuel conceptual shifts in metaorganism function. Beyond this great basic importance, *Hydra* has a high translational relevance: insights gained from the *Hydra* model can be applied directly to understanding human development and disease. Our work has shown that non-aging, long-lived organisms such as *Hydra* can provide important insights into the mechanisms of aging and longevity. Understanding these processes can help developing therapeutic approaches to delay the decline of stem cell and immune functionality and therefore has significant implications for the extension of "health span" by prevention of age-related diseases and the tuning of heathy aging (Nebel and Bosch, 2012). The *Hydra* model allows uncovering the fundamental principles of oncogenesis and the organism's natural resources to fight the cancer. There are three major unresolved issues concerning *Hydra*'s stem cells. First, what are the cellular and molecular mechanisms that sense the diverse environment and instruct the stem cells in *Hydra*? We hypothesize that the nervous system might play the leading role in providing these instructive cues (Klimovich and Bosch, 2018). Future studies will uncover how neurons may perceive the environmental signals and adjust the animal's internal vital processes: development, physiology, tissue homeostasis, and behavior.

Second, accumulating evidence indicates that microbes directly impact the stem cell behavior in *Hydra*. What are the microbial products, receptors, and signaling cascades within *Hydra* cells that enable this crosstalk? A technological platform, including high-throughput meta-genomic and tran-scriptomic sequencing, metabolomic profiling, and microbiome manipulation is already established and promises unique insights into the biology of *Hydra* holobiont. Moreover, identifying these mechanisms may also shed light onto how millions of years of co-evolution between the microbes and their hosts have shaped the animal's developmental pathways (McFall-Ngai and Bosch, 2020).

Finally, what molecular mechanisms underlie the divergence of animal species? If the structure and function of transcription factors, RNA-binding proteins and developmental signaling pathways is so highly conserved across the animal kingdom, how did the diversity of animal species emerge? The extensive conservation of the developmental toolkit stays is in astounding contrast to the diversity of animal cell type repertoires, life histories, life spans, and adaptations. We hypothesize that, in addition to a rapid evolution of transcriptional control elements (enhancers) regulating conserved genes expression (Jeong et al., 2008; Rebeiz et al., 2009), non-conserved protein-coding genes, the TRGs, also may play a pivotal role in animal evolution (Khalturin et al., 2009). Further studies will uncover how TRGs interact with conserved transcriptional regulators and signaling pathways to control stem cells and development.

The imminent availability of methods for functional analyses of genes and the massive advances in molecular technology that are presently taking place, make *Hydra* a powerful and also intellectually attractive system for studying stem cells. Organisms become models when they support sustainable opportunities with uncompromising experimental rigor and ease of use. With the molecular dissecting of the components controlling stem cell behavior in *Hydra*, the stage is now set to uncover the mystery of "stemness" and deciphering the fundamental components controlling pluripotency and lineage commitment that underlie all stem cell systems.

9.10 Concluding Remarks

In the recent decades, emerging technologies have transformed *Hydra* from a curiosity to a remarkably informative model for evolutionary developmental biology. The ongoing development of new molecular tools opened new perspectives for studies of both, the *Hydra* host and its associated microbiome, and provided deep mechanistic insights into biological phenomena of fundamental importance. Three of them are particularly worth mentioning:

- From the very beginning of Eumatazoa evolution, the animal stem cells were governed by surprisingly complex mechanisms of position-dependent cell fate control, that employ conserved transcription factors, effector molecules, and epigenetic mechanisms;
- Non-conserved phylum- and species-specific genes play an essential role in stem cell maintenance, differentiation, and interlineage communication within multicellular animals;
- The environment, and particularly-the microbiota, has a profound effect on normal functioning of the stem cells, owing to a tight integration of stem cell homeostasis, development, environment sensing, and immunity at the molecular level.

These findings have important implications for understanding the stem cell biology and aging, and are indispensable for harnessing the stem cell potential in health span extension strategies and regenerative medicine.

References

Aktipis, A. C., Boddy, A. M., Jansen, G., Hibner, U., Hochberg, M. E., Maley, C. C. and Wilkinson, G. S. (2015). Cancer across the tree of life: Cooperation and cheating in multicellularity. *Philos. Trans. R. Soc. B Biol. Sci* 370, 20140219.

Alié, A., Hayashi, T., Sugimura, I., Manuel, M., Sugano, W., Mano, A., Satoh, N., Agata, K. and Funayama, N. (2015). The ancestral gene repertoire of animal stem cells. *Proc. Natl. Acad. Sci. U. S. A.* 112, E7093-100.

Ambrosone, A., Marchesano, V., Tino, A., Hobmayer, B. and Tortiglione, C. (2012). Hymyc1 downregulation promotes stem cell proliferation in *Hydra* vulgaris. *PLoS One* 7, e30660.

Anselmi, C. V., Malovini, A., Roncarati, R., Novelli, V., Villa, F., Condorelli, G., Bellazzi, R. and Puca, A. A. (2009). Association of the FOXO3A locus with extreme longevity in a Southern Italian centenarian study. *Rejuvenation Res.* 12, 95-104.

Anton-Erxleben, F., Thomas, A., Wittlieb, J., Fraune, S. and Bosch, T. C. G. (2009). Plasticity of epithelial cell shape in response to upstream signals: a whole-organism study using transgenic *Hydra*. *Zoology (Jena)*. 112, 185-94.

Arendt, D. (2019). Many ways to build a polyp. Trends Genet. 35, 885-887.

Atreya, C. E. and Turnbaugh, P. J. (2020). Probing the tumor micro(b)environment. *Science* 368, 938-939.

Augustin, R., Anton-Erxleben, F., Jungnickel, S., Hemmrich, G., Spudy, B., Podschun, R. and Bosch, T. C. G. (2009). Activity of the novel peptide arminin against multiresistant human pathogens shows the considerable potential of phylogenetically ancient organisms as drug sources. *Antimicrob. Agents Chemother.* 53, 5245-5250.

Augustin, R., Fraune, S., Franzenburg, S. and Bosch, T. C. G. (2012). Where simplicity meets complexity: hydra, a model for host-microbe interactions. In *Recent Advances on Model Hosts,* pp. 71-81. Springer.

Augustin, R., Schröder, K., Murillo Rincón, A. P., Fraune, S., Anton-Erxleben, F., Herbst, E.-M., Wittlieb, J., Schwentner, M., Grötzinger, J., Wassenaar, T. M., et al. (2017). A secreted antibacterial neuropeptide shapes the microbiome of *Hydra*. *Nat. Commun.* 8, 698.

Babonis, L. S., Martindale, M. Q. and Ryan, J. F. (2016). Do novel genes drive morphological novelty? An investigation of the nematosomes in the sea anemone Nematostella vectensis. *BMC Evol. Biol.* 16, 114.

Becker, T., Loch, G., Beyer, M., Zinke, I., Aschenbrenner, A. C., Carrera, P., Inhester, T., Schultze, J. L. and Hoch, M. (2010). FOXO-dependent regulation of innate immune homeostasis. *Nature.* 463, 369-73.

Belcaid, M., Casaburi, G., McAnulty, S. J., Schmidbaur, H., Suria, A. M., Moriano-Gutierrez, S., Pankey, M. S., Oakley, T. H., Kremer, N., Koch, E. J., et al. (2019). Symbiotic organs shaped by distinct modes of genome evolution in cephalopods. *Proc. Natl. Acad. Sci.* 116, 3030-3035.

Belkaid, Y. and Naik, S. (2013). Compartmentalized and systemic control of tissue immunity by commensals. *Nat. Immunol.* 14, 646-53.

Bode, H., Berking, S., David, C. N., Gierer, A., Schaller, H. and Trenkner, E. (1973). Quantitative analysis of cell types during growth and morphogenesis in *Hydra. Wilhelm Roux'Archiv für Entwicklungsmechanik der Org.* 171, 269-285.

Boehm, A.-M. and Bosch, T. C. G. (2012). Migration of multipotent interstitial stem cells in *Hydra*. *Zoology (Jena)*. 115, 275-82.

Boehm, A.-M., Khalturin, K., Anton-Erxleben, F., Hemmrich, G., Klostermeier, U. C., Lopez-Quintero, J., Oberg, H.-H., Puchert, M., Rosenstiel,

P., Wittlieb, J., et al. (2012). FoxO is a critical regulator of stem cell maintenance in immortal *Hydra*. *Proc. Natl. Acad. Sci. U. S. A.* 109, 19697-702.

Boiani, M. and Schöler, H. R. (2005). Regulatory networks in embryo-derived pluripotent stem cells. *Nat. Rev. Mol. Cell Biol.* 6, 872-84.

Bosch, T. C. G. (2007). Why polyps regenerate and we don't: towards a cellular and molecular framework for *Hydra* regeneration. *Dev. Biol.* 303, 421-33.

Bosch, T. (2009). *Hydra* and the evolution of stem cells. *Bioessays* 31, 478-86.

Bosch, T. C. G. (2013). Cnidarian-microbe interactions and the origin of innate immunity in metazoans. *Annu. Rev. Microbiol.* 67, 499-518.

Bosch, T. C. G. and David, C. N. (1984). Growth regulation in *Hydra*: Relationship between epithelial cell cycle length and growth rate. *Dev. Biol.* 104, 161-171.

Bosch, T. C. G. and David, C. N. (1986). Male and female stem cells and sex reversal in *Hydra* polyps. *Proc. Natl. Acad. Sci. U. S. A.* 83, 9478-82.

Bosch, T. C. and David, C. N. (1990). Cloned interstitial stem cells grow as contiguous patches in hydra. *Dev. Biol.* 138, 513-5.

Bosch, T. C. G. and McFall-Ngai, M. (2021). Animal development in the microbial world: Re-thinking the conceptual framework. *Cur. Top. Dev. Biol.* 141, 399-427.

Bosch, T. C. G., Augustin, R., Anton-Erxleben, F., Fraune, S., Hemmrich, G., Zill, H., Rosenstiel, P., Jacobs, G., Schreiber, S. and Leippe, M. (2009). Uncovering the evolutionary history of innate immunity: the simple metazoan *Hydra* uses epithelial cells for host defence. *Dev. Comp. Immunol.* 33, 559-569.

Bosch, T., Anton-Erxleben, F., Hemmrich, G. and Khalturin, K. (2010). The *Hydra* polyp: nothing but an active stem cell community. *Dev. Growth Differ.* 52, 15-25.

Bosch, T. C. G., Klimovich, A., Domazet-Lošo, T., Gründer, S., Holstein, T. W., Jékely, G., Miller, D. J., Murillo-Rincon, A. P., Rentzsch, F., Richards, G. S., et al. (2017). Back to the basics: cnidarians start to fire. *Trends Neurosci.* 40, 92-105.

Brooun, M., Klimovich, A., Bashkurov, M., Pearson, B. J., Steele, R. E. and McNeill, H. (2020). Ancestral roles of atypical cadherins. *Proc. Natl. Acad. Sci.* 117, 19310-19320.

Broun, M., Gee, L., Reinhardt, B. and Bode, H. R. (2005). Formation of the head organizer in hydra involves the canonical Wnt pathway. *Development.* 132, 2907-16.

Brown, E. M., Sadarangani, M. and Finlay, B. B. (2013). The role of the immune system in governing host-microbe interactions in the intestine. *Nat. Immunol.* 14, 660-7.

Buzgariu, W., Crescenzi, M. and Galliot, B. (2014). Robust G2 pausing of adult stem cells in *Hydra. Differentiation* 87, 83-99.

Buzgariu, W., Wenger, Y., Tcaciuc, N., Catunda-Lemos, A.-P. and Galliot, B. (2018). Impact of cycling cells and cell cycle regulation on *Hydra* regeneration. *Dev. Biol.* 433, 240-253.

Campbell, R. D. (1967a). Tissue dynamics of steady state growth in *Hydra* littoralis. I. Patterns of cell division. *Dev. Biol.* 15, 487-502.

Campbell, R. D. (1967b). Tissue dynamics of steady state growth in *Hydra* littoralis. II. Patterns of tissue movement. *J. Morphol.* 121, 19-28.

Campbell, R. D. (1987). Organization of the nematocyst battery in the tentacle of hydra: Arrangement of the complex anchoring junctions between nematocytes, epithelial cells, and basement membrane. *Cell Tissue Res.* 249, 647-655.

Campbell, R. D. and David, C. N. (1974). Cell cycle kinetics and development of *Hydra* attenuata. II. Interstitial cells. *J. Cell Sci.* 16, 349-58.

Carroll, S. B. (2008). Evo-Devo and an Expanding Evolutionary Synthesis: A Genetic Theory of Morphological Evolution. *Cell* 134, 25-36.

Chang, A. H. and Parsonnet, J. (2010). Role of bacteria in oncogenesis. *Clin. Microbiol. Rev.* 23, 837-57.

Chapman, J. A., Kirkness, E. F., Simakov, O., Hampson, S. E., Mitros, T., Weinmaier, T., Rattei, T., Balasubramanian, P. G., Borman, J., Busam, D., et al. (2010). The dynamic genome of *Hydra. Nature* 464, 592-596.

Chen, S., Krinsky, B. H. and Long, M. (2013). New genes as drivers of phenotypic evolution. *Nat. Rev. Genet.* 14, 645.

Chera, S., de Rosa, R., Miljkovic-Licina, M., Dobretz, K., Ghila, L., Kaloulis, K. and Galliot, B. (2006). Silencing of the hydra serine protease inhibitor Kazal1 gene mimics the human;SPINK1 pancreatic phenotype. *J. Cell Sci.* 119, 846-857.

Cho, I. and Blaser, M. J. (2012). The human microbiome: At the interface of health and disease. *Nat. Rev. Genet.* 13, 260-70.

Conacci-Sorrell, M., McFerrin, L. and Eisenman, R. N. (2014). An overview of MYC and its interactome. *Cold Spring Harb. Perspect. Med.* 4, a014357.

Cummins, J. and Tangney, M. (2013). Bacteria and tumours: Causative agents or opportunistic inhabitants? *Infect. Agent. Cancer.* 8, 11.

Dang, C. V. (2012). MYC on the path to cancer. *Cell* 149, 22-35.

Dato, S., Rose, G., Crocco, P., Monti, D., Garagnani, P., Franceschi, C. and Passarino, G. (2017). The genetics of human longevity: an intricacy of genes, environment, culture and microbiome. *Mech. Ageing Dev.* 65, 147-155.

David, C. N. and Campbell, R. D. (1972). Cell cycle kinetics and development of *Hydra* attenuata. I. Epithelial cells. *J. Cell Sci.* 11, 557-68.

David, C. N. and Challoner, D. (1974). Distribution of interstitial cells and differentiating nematocytes in nests in *Hydra* attenuata. *Integr. Comp. Biol.* 14, 537-54.

David, C. N. and Gierer, A. (1974). Cell cycle kinetics and development of *Hydra* attenuata. III. Nerve and nematocyte differentiation. *J. Cell Sci.* 16, 359-75.

David, C. N. and Murphy, S. (1977). Characterization of interstitial stem cells in hydra by cloning. *Dev. Biol.* 58, 372-383.

Domazet-Lošo, T., Tautz, D., Domazet-Loso, T. and Tautz, D. (2010). Phylostratigraphic tracking of cancer genes suggests a link to the emergence of multicellularity in metazoa. *BMC Biol.* 8, 66.

Domazet-Lošo, T., Klimovich, A., Anokhin, B., Anton-Erxleben, F., Hamm, M. J., Lange, C. and Bosch, T. C. (2014). Naturally occurring tumours in the basal metazoan *Hydra*. *Nat. Commun.* 5, 4222.

Dupre, C. and Yuste, R. (2017). Non-overlapping neural networks in *Hydra* vulgaris. *Curr. Biol.* 27, 1085-1097.

Eilers, M. and Eisenman, R. N. (2008). Myc's broad reach. *Genes Dev.* 22, 2755-66.

Fierro-Constaín, L., Schenkelaars, Q., Gazave, E., Haguenauer, A., Rocher, C., Ereskovsky, A., Borchiellini, C. and Renard, E. (2017). The conservation of the germline multipotency program, from sponges to vertebrates: A stepping stone to understanding the somatic and germline origins. *Genome Biol. Evol.* 9, 474-488.

Flachsbart, F., Caliebe, A., Kleindorp, R., Blanché, H., Von Eller-Eberstein, H., Nikolaus, S., Schreiber, S. and Nebel, A. (2009). Association of FOXO3A variation with human longevity confirmed in German centenarians. *Proc. Natl. Acad. Sci. U. S. A.* 106, 2700-5.

Franzenburg, S., Fraune, S., Altrock, P. M., Künzel, S., Baines, J. F., Traulsen, A. and Bosch, T. C. G. (2013a). Bacterial colonization of *Hydra* hatchlings follows a robust temporal pattern. *ISME J.* 7, 781.

Franzenburg, S., Walter, J., Künzel, S., Wang, J., Baines, J. F., Bosch, T. C. G. and Fraune, S. (2013b). Distinct antimicrobial peptide expression determines host species-specific bacterial associations. *Proc. Natl. Acad. Sci.* 110, E3730-E3738.

Fraune, S. and Bosch, T. C. G. (2007). Long-term maintenance of species-specific bacterial microbiota in the basal metazoan *Hydra*. *Proc. Natl. Acad. Sci.* 104, 13146-13151.

Fraune, S., Abe, Y. and Bosch, T. C. G. (2009). Disturbing epithelial homeostasis in the metazoan *Hydra* leads to drastic changes in associated microbiota. *Environ. Microbiol.* 11, 2361-2369.

Fraune, S., Augustin, R., Anton-Erxleben, F., Wittlieb, J., Gelhaus, C., Klimovich, V. B., Samoilovich, M. P. and Bosch, T. C. G. (2010). In an early branching metazoan, bacterial colonization of the embryo is controlled by maternal antimicrobial peptides. *Proc. Natl. Acad. Sci. U. S. A.* 107, 18067-72.

Fujisawa, T. (2008). *Hydra* peptide project 1993-2007. *Dev. Growth Differ.* 50 Suppl 1, S257-68.

Gee, L., Hartig, J., Law, L., Wittlieb, J., Khalturin, K., Bosch, T. C. G. and Bode, H. R. (2010). β-catenin plays a central role in setting up the head organizer in hydra. *Dev. Biol.* 340, 116-124.

Giannakou, M. E., Goss, M., Jünger, M. A., Hafen, E., Leevers, S. J. and Partridge, L. (2004). Long-lived Drosophila with overexpressed dFOXO in adult fat body. *Science* 305, 361.

Gierer, A., Berking, S., Bode, H., David, C. N., Flick, K., Hansmann, G., Schaller, H. and Trenkner, E. (1972). Regeneration of hydra from reaggregated cells. *Nat. New Biol.* 39, 98-101.

Gilbert, S. F. (2012). Ecological developmental biology: Environmental signals for normal animal development. *Evol. Dev.* 14, 20-8.

Gillespie, J. H. and Turelli, M. (1989). Genotype-environment interactions and the maintenance of polygenic variation. *Genetics* 121, 129-38.

Govindasamy, N., Murthy, S. and Ghanekar, Y. (2014). Slow-cycling stem cells in hydra contribute to head regeneration. *Biol. Open* 3, 1236-44.

Greer, E. L. and Brunet, A. (2005). FOXO transcription factors at the interface between longevity and tumor suppression. *Oncogene* 24, 7410-25.

Gruenbaum, Y., Margalit, A., Goldman, R. D., Shumaker, D. K. and Wilson, K. L. (2005). The nuclear lamina comes of age. *Nat Rev Mol Cell Biol* 6, 21-31.

Guder, C., Philipp, I., Lengfeld, T., Watanabe, H., Hobmayer, B. and Holstein, T. W. (2006). The Wnt code: Cnidarians signal the way. *Oncogene* 25, 7450-60.

Hadfield, M. G. and Bosch, T. C. G. (2020). Cellular dialogues between hosts and microbial symbionts. In *Cellular Dialogues in the Holobiont.*

Hamaguchi-Hamada, K., Kurumata-Shigeto, M., Minobe, S., Fukuoka, N., Sato, M., Matsufuji, M., Koizumi, O. and Hamada, S. (2016). Thrombospondin type-1 repeat domain-containing proteins are strongly expressed in the head region of hydra. *PLoS One.* 11: e0151823.

Han, S., Taralova, E., Dupre, C. and Yuste, R. (2018). Comprehensive machine learning analysis of *Hydra* behavior reveals a stable basal behavioral repertoire. *Elife* 7, e32605.

Hanahan, D. and Weinberg, R. a (2011). Hallmarks of cancer: the next generation. *Cell* 144, 646-74.

Hartl, M., Mitterstiller, A.-M., Valovka, T., Breuker, K., Hobmayer, B. and Bister, K. (2010). Stem cell-specific activation of an ancestral myc pro-tooncogene with conserved basic functions in the early metazoan *Hydra*. *Proc. Natl. Acad. Sci. U. S. A.* 107, 4051-6.

Hartl, M., Glasauer, S. G., Valovka, T., Breuker, K., Hobmayer, B. and Bister, K. (2014). *Hydra* myc2, a unique pre-bilaterian member of the myc gene family, is activated in cell proliferation and gametogenesis. *Biol. Open* 3, 397-407.

Hartl, M., Glasauer, S., Gufler, S., Raffeiner, A., Puglisi, K., Breuker, K., Bister, K. and Hobmayer, B. (2019). Differential regulation of myc homologs by Wnt/β-Catenin signaling in the early metazoan *Hydra*. *FEBS J.* 286, 2295-2310.

Hemmrich, G., Khalturin, K., Boehm, A.-M., Puchert, M., Anton-Erxleben, F., Wittlieb, J., Klostermeier, U. C., Rosenstiel, P., Oberg, H.-H., Domazet-Loso, T., et al. (2012). Molecular signatures of the three stem cell lineages in hydra and the emergence of stem cell function at the base of multicellularity. *Mol. Biol. Evol.* 29, 3267-80.

Hobmayer, B., Rentzsch, F., Kuhn, K., Happel, C. M., Von Laue, C. C., Snyder, P., Rothbächer, U. and Holstein, T. W. (2000). WNT signalling molecules act in axis formation in the diploblastic metazoan *Hydra*. *Nature* 407, 186-9.

Holstein, T. W., Hobmayer, E. and David, C. N. (1991). Pattern of epithelial cell cycling in hydra. *Dev. Biol.* 148, 602-11.

Hwang, J. S., Ohyanagi, H., Hayakawa, S., Osato, N., Nishimiya-Fujisawa, C., Ikeo, K., David, C. N., Fujisawa, T. and Gojobori, T. (2007). The evolutionary emergence of cell type-specific genes inferred from the gene expression analysis of *Hydra*. *Proc. Natl. Acad. Sci.* 104, 14735-14740.

Jeong, S., Rebeiz, M., Andolfatto, P., Werner, T., True, J. and Carroll, S. B. (2008). The Evolution of Gene Regulation Underlies a Morphological Difference between Two Drosophila Sister Species. *Cell* 132, 783-793.

Johnson, B. R. (2018). Taxonomically Restricted Genes Are Fundamental to Biology and Evolution. Front. *Genet.* 9, 407.

Johnson, B. R. and Tsutsui, N. D. (2011). Taxonomically restricted genes are associated with the evolution of sociality in the honey bee. *BMC Genomics* 12, 164.

Jones, O. R., Scheuerlein, A., Salguero-Gómez, R., Camarda, C. G., Schaible, R., Casper, B. B., Dahlgren, J. P., Ehrlén, J., García, M. B., Menges, E. S., et al. (2014). Diversity of ageing across the tree of life. *Nature* 505, 169-73.

Juliano, C. E., Swartz, S. Z. and Wessel, G. M. (2010). A conserved germline multipotency program. *Development* 137, 4113-4126.

Juliano, C. E., Reich, A., Liu, N., Götzfried, J., Zhong, M., Uman, S., Reenan, R. A., Wessel, G. M., Steele, R. E. and Lin, H. (2014). PIWI proteins and PIWI-interacting RNAs function in *Hydra* somatic stem cells. *Proc. Natl. Acad. Sci.* 111, 337-342.

Jünger, M. A., Rintelen, F., Stocker, H., Wasserman, J. D., Végh, M., Radimerski, T., Greenberg, M. E. and Hafen, E. (2003). The Drosophila Forkhead transcription factor FOXO mediates the reduction in cell number associated with reduced insulin signaling. *J. Biol.* 2, 20.

Kenyon, C. (2010). A pathway that links reproductive status to lifespan in Caenorhabditis elegans. In *Annals of the New York Academy of Sciences,* pp. 156-162.

Kenyon, C., Chang, J., Gensch, E., Rudner, A. and Tabtiang, R. (1993). A C. elegans mutant that lives twice as long as wild type. *Nature* 366, 461-4.

Khalturin, K., Anton-Erxleben, F., Milde, S., Plötz, C., Wittlieb, J., Hemmrich, G. and Bosch, T. C. G. (2007). Transgenic stem cells in *Hydra* reveal an early evolutionary origin for key elements controlling self-renewal and differentiation. *Dev. Biol.* 309, 32-44.

Khalturin, K., Anton-Erxleben, F., Sassmann, S., Wittlieb, J., Hemmrich, G. and Bosch, T. C. G. (2008). A novel gene family controls species-specific morphological traits in *Hydra*. *PLoS Biol.* 6, e278.

Khalturin, K., Hemmrich, G., Fraune, S., Augustin, R. and Bosch, T. C. G. (2009). More than just orphans: are taxonomically-restricted genes important in evolution? *Trends Genet.* 25, 404-413.

Kimura, K. D., Tissenbaum, H. A., Liu, Y. and Ruvkun, G. (1997). Daf-2, an insulin receptor-like gene that regulates longevity and diapause in Caenorhabditis elegans. *Science* 277, 942-6.

Klimovich, A. V. and Bosch, T. C. G. (2018). Rethinking the role of the nervous system: Lessons from the *Hydra* holobiont. *BioEssays* 40, 1800060.

Klimovich, A., Rehm, A., Wittlieb, J., Herbst, E.-M., Benavente, R. and Bosch, T. C. G. (2018). Non-senescent *Hydra* tolerates severe disturbances in the nuclear lamina. *Aging (Albany NY)* 10, 951.

Klimovich, A., Wittlieb, J. and Bosch, T. C. G. (2019). Transgenesis in *Hydra* to characterize gene function and visualize cell behavior. *Nat. Protoc.* 14, 2069-2090.

Klimovich, A., Giacomello, S., Björklund, Å., Faure, L., Kaucka, M., Giez, C., Murillo-Rincon, A. P., Matt, A.-S., Willoweit-Ohl, D., Crupi, G., et al. (2020). Prototypical pacemaker neurons interact with the resident microbiota. *Proc Natl Acad Sci U S A* 117, 17854-17863.

Koizumi, O. (2016). Origin and Evolution of the Nervous System Considered from the Diffuse Nervous System of Cnidarians. In *The Cnidaria, Past, Present and Future,* pp. 73-91. Springer.

Kojima, T., Kamei, H., Aizu, T., Arai, Y., Takayama, M., Nakazawa, S., Ebihara, Y., Inagaki, H., Masui, Y., Gondo, Y., et al. (2004). Association analysis between longevity in the Japanese population and polymorphic variants of genes involved in insulin and insulin-like growth factor 1 signaling pathways. *Exp. Gerontol.* 39, 1595-8.

Kortschak, R. D., Samuel, G., Saint, R. and Miller, D. J. (2003). EST analysis of the cnidarian Acropora millepora reveals extensive gene loss and rapid sequence divergence in the model invertebrates. *Curr. Biol.* 13, 2190-2195.

Kusserow, A., Pang, K., Sturm, C., Hrouda, M., Lentfer, J., Schmidt, H. A., Technau, U., Von Haeseler, A., Hobmayer, B., Martindale, M. Q., et al. (2005). Unexpected complexity of the Wnt gene family in a sea anemone. *Nature* 433, 156-160.

Lans, H. and Hoeijmakers, J. H. J. (2006). Cell biology: Ageing nucleus gets out of shape. *Nature* 440, 32-34.

Li, Y., Wang, W. J., Cao, H., Lu, J., Wu, C., Hu, F. Y., Guo, J., Zhao, L., Yang, F., Zhang, Y. X., et al. (2009). Genetic association of FOXO1A and FOXO3A with longevity trait in Han Chinese populations. *Hum. Mol. Genet.* 18, 4897-904.

Lim, R. S. M., Anand, A., Nishimiya-Fujisawa, C., Kobayashi, S. and Kai, T. (2014). Analysis of *Hydra* PIWI proteins and piRNAs uncover early evolutionary origins of the piRNA pathway. *Dev. Biol.* 386, 237-251.

Littlefield, C. L., Finkemeier, C. and Bode, H. R. (1991). Spermatogenesis in *Hydra* oligactis. II. How temperature controls the reciprocity of sexual and asexual reproduction. *Dev. Biol.* 146, 292-300.

Lommel, M., Strompen, J., Hellewell, A. L., Balasubramanian, G. P., Christofidou, E. D., Thomson, A. R., Boyle, A. L., Woolfson, D. N., Puglisi, K., Hartl, M., et al. (2018). *Hydra* mesoglea proteome identifies Thrombospondin as a conserved component active in head organizer restriction. *Sci. Rep.* 8, 11753.

López-Otín, C., Blasco, M. A., Partridge, L., Serrano, M. and Kroemer, G. (2013). The hallmarks of aging. *Cell* 153, 1194-1217.

López-Quintero, J. A., Torres, G. G., Neme, R. and Bosch, T. C. G. (2020). Boundary maintenance in the ancestral metazoan *Hydra* depends on histone acetylation. *Dev. Biol.* 458, 200-214.

Martindale, M. Q., Finnerty, J. R. and Henry, J. Q. (2002). The radiata and the evolutionary origins of the bilaterian body plan. *Mol. Phylogenet. Evol.*

Martínez, D. E. (1998). Mortality patterns suggest lack of senescence in hydra. *Exp. Gerontol.* 33, 217-25.

Mattout, A., Dechat, T., Adam, S. A., Goldman, R. D. and Gruenbaum, Y. (2006). Nuclear lamins, diseases and aging. Curr. Opin. *Cell Biol.* 18, 335-341.

McFall-Ngai, M. and Bosch, T. C. G. (2020). Animal development in the microbial world: The power of experimental model systems. Cur. Top. *Dev. Biol.* 141, 371-97.

McFall-Ngai, M., Hadfield, M. G. G., Bosch, T. C. G., Carey, H. V., Domazet-Lošo, T., Douglas, A. E., Dubilier, N., Eberl, G., Fukami, T., Gilbert, S. F., et al. (2013). Animals in a bacterial world, a new imperative for the life sciences. *Proc. Natl. Acad. Sci.* 110, 3229-3236.

McGill, C. (1908). The effect of low temperatures on *Hydra. Biol. Bull.* 14, 78-88.

Meinhardt, H. and Gierer, A. (1974). Applications of a theory of biological pattern formation based on lateral inhibition. J. *Cell Sci.* 15, 321-346.

Meinhardt, H. and Gierer, A. (2000). Pattern formation by local self-activation and lateral inhibition. *Bioessays* 22, 753-760.

Milde, S., Hemmrich, G., Anton-Erxleben, F., Khalturin, K., Wittlieb, J. and Bosch, T. C. G. (2009). Characterization of taxonomically restricted genes in a phylum-restricted cell type. *Genome Biol.* 10, R8.

Miller, D. J., Ball, E. E. and Technau, U. (2005). Cnidarians and ancestral genetic complexity in the animal kingdom. *Trends Genet.* 21, 536-539.

Miyamoto, K., Araki, K. Y., Naka, K., Arai, F., Takubo, K., Yamazaki, S., Matsuoka, S., Miyamoto, T., Ito, K., Ohmura, M., et al. (2007). Foxo3a is essential for maintenance of the hematopoietic stem cell pool. *Cell Stem Cell* 1, 101-12.

Mochizuki, K., Sano, H., Kobayashi, S., Nishimiya-Fujisawa, C. and Fujisawa, T. (2000). Expression and evolutionary conservation of nanos-related genes in *Hydra*. *Dev. Genes Evol.* 210, 591-602.

Mochizuki, K., Nishimiya-Fujisawa, C. and Fujisawa, T. (2001). Universal occurrence of the vasa-related genes among metazoans and their germline expression in *Hydra*. *Dev. Genes Evol.* 211, 299-308.

Mortzfeld, B. M. and Bosch, T. C. G. (2017). Eco-Aging: stem cells and microbes are controlled by aging antagonist FoxO. *Curr. Opin. Microbiol.* 38, 181-187.

Mortzfeld, B. M., Taubenheim, J., Fraune, S., Klimovich, A. V and Bosch, T. C. G. (2018). Stem cell transcription factor FoxO controls microbiome resilience in *Hydra*. *Front. Microbiol.* 9, 629.

Mortzfeld, B., Taubenheim, J., Klimovich, A., Fraune, S., Rosenstiel, P. and Bosch, T. C. (2019). Temperature and insulin signaling regulate body size in *Hydra* by the Wnt and TGF-beta pathways. *Nat. Commun.* 10, 3257.

Motamedi, M., Lindenthal, L., Wagner, A., Kemper, M., Moneer, J., Steichele, M., Klimovich, A., Wittlieb, J., Jenewein, M. and Böttger, A. (2019). Apoptosis in hydra: Function of hybcl-2 like 4 and proteins of the transmembrane bax inhibitor motif (tmbim) containing family. *Int. J. Dev. Biol.* 63, 259-270.

Murillo-Rincon, A. P., Klimovich, A., Pemöller, E., Taubenheim, J., Mortzfeld, B., Augustin, R. and Bosch, T. C. G. (2017). Spontaneous body contractions are modulated by the microbiome of *Hydra*. *Sci. Rep.* 7, 15937.

Naik, S., Unni, M., Sinha, D., Rajput, S. S., Reddy, P. C., Kartvelishvily, E., Solomonov, I., Sagi, I., Chatterji, A., Patil, S., et al. (2020). Differential tissue stiffness of body column facilitates locomotion of *Hydra* on solid substrates. *J. Exp. Biol.* 223, jeb232702.

Nakamura, Y., Tsiairis, C. D., Özbek, S. and Holstein, T. W. (2011). Autoregulatory and repressive inputs localize *Hydra* Wnt3 to the head organizer. *Proc. Natl. Acad. Sci.* 108, 9137-9142.

Nebel, A. and Bosch, T. C. G. (2012). Evolution of human longevity: Lessons from hydra. *Aging (Albany. NY)* 4, 730-1.

Nejman, D., Livyatan, I., Fuks, G., Gavert, N., Zwang, Y., Geller, L. T., Rotter-Maskowitz, A., Weiser, R., Mallel, G., Gigi, E., et al. (2020). The human tumor microbiome is composed of tumor type-specific intracellular bacteria. *Science* 368, 973-980.

Nelson, P. A. and Buggs, R. J. A. (2016). Next generation apomorphy: the ubiquity of taxonomically restricted genes. *Next Gener. Syst.* 85, 237.

Nemeth, M. J., Curtis, D. J., Kirby, M. R., Garrett-Beal, L. J., Seidel, N. E., Cline, A. P. and Bodine, D. M. (2003). Hmgb3: An HMG-box family member expressed in primitive hematopoietic cells that inhibits myeloid and B-cell differentiation. *Blood* 102, 1298-306.

Nielsen, C., Scharff, N. and Eibye-Jacobsen, D. (1996). Cladistic analyses of the animal kingdom. *Biol. J. Linn. Soc.* 57, 385-410.

Nishimiya-Fujisawa, C. and Kobayashi, S. (2018). Roles of germline stem cells and somatic multipotent stem cells in *Hydra* sexual reproduction. In: Kobayashi K., Kitano T., Iwao Y., Kondo M. (eds) Reproductive and Developmental Strategies. Diversity and Commonality in Animals. Springer, Tokyo, pp. 123-155.

Noggle, S. A., James, D. and Brivanlou, A. H. (2005). A molecular basis for human embryonic stem cell pluripotency. *Stem Cell Rev.* 1, 111-8.

Paik, J. hye, Ding, Z., Narurkar, R., Ramkissoon, S., Muller, F., Kamoun, W. S., Chae, S. S., Zheng, H., Ying, H., Mahoney, J., et al. (2009). FoxOs cooperatively regulate diverse pathways governing neural stem cell homeostasis. *Cell Stem Cell.* 5, 540-53. e

Pan, G. and Thomson, J. A. (2007). Nanog and transcriptional networks in embryonic stem cell pluripotency. *Cell Res.* 17, 42-49.

Pawlikowska, L., Hu, D., Huntsman, S., Sung, A., Chu, C., Chen, J., Joyner, A. H., Schork, N. J., Hsueh, W. C., Reiner, A. P., et al. (2009). Association of common genetic variation in the insulin/IGF1 signaling pathway with human longevity. *Aging Cell* 8, 460-72.

Putnam, N. H., Srivastava, M., Hellsten, U., Dirks, B., Chapman, J., Salamov, A., Terry, A., Shapiro, H., Lindquist, E., Kapitonov, V. V, et al. (2007). Sea Anemone Genome Reveals Ancestral Eumetazoan Gene Repertoire and Genomic Organization. *Science* 317, 86-94.

Rathje, K., Mortzfeld, B., Hoeppner, M. P., Taubenheim, J., Bosch, T. C. G. and Klimovich, A. (2020). Dynamic interactions within the host-associated microbiota cause tumor formation in the basal metazoan *Hydra*. *PLoS Pathog.* 16, e1008375.

Rebeiz, M., Pool, J. E., Kassner, V. A., Aquadro, C. F. and Carroll, S. B. (2009). Stepwise modification of a modular enhancer underlies adaptation in a drosophila population. *Science* 326, 1663-1667.

Reddy, P. C., Gungi, A., Ubhe, S., Pradhan, S. J., Kolte, A. and Galande, S. (2019). Molecular signature of an ancient organizer regulated by Wnt/β-catenin signalling during primary body axis patterning in *Hydra*. *Commun. Biol.* 2, 434.

Reiter, S., Crescenzi, M., Galliot, B. and Buzgariu, W. (2012). *Hydra*, a versatile model to study the homeostatic and developmental functions of cell death. *Int. J. Dev. Biol.* 56, 593-604.

Renault, V. M., Rafalski, V. A., Morgan, A. A., Salih, D. A. M., Brett, J. O., Webb, A. E., Villeda, S. A., Thekkat, P. U., Guillerey, C., Denko, N. C., et al. (2009). FoxO3 regulates neural stem cell homeostasis. *Cell Stem Cell* 5, 527-39.

Rosshart, S. P., Vassallo, B. G., Angeletti, D., Hutchinson, D. S., Morgan, A. P., Takeda, K., Hickman, H. D., McCulloch, J. A., Badger, J. H., Ajami, N. J., et al. (2017). Wild mouse gut microbiota promotes host fitness and improves disease resistance. *Cell* 171, 1015-1028.

Salih, D. A. and Brunet, A. (2008). FoxO transcription factors in the maintenance of cellular homeostasis during aging. *Curr. Opin. Cell Biol.* 20, 126-36.

Salminen, A., Huuskonen, J., Ojala, J., Kauppinen, A., Kaarniranta, K. and Suuronen, T. (2008). Activation of innate immunity system during aging: NF-kB signaling is the molecular culprit of inflamm-aging. *Ageing Res. Rev.* 7, 83-105.

Sansregret, L. and Nepveu, A. (2008). The multiple roles of CUX1: Insights from mouse models and cell-based assays. *Gene* 412, 84-94.

Santos, M. E., Le Bouquin, A., Crumière, A. J. J. and Khila, A. (2017). Taxon-restricted genes at the origin of a novel trait allowing access to a new environment. *Science* 358, 386-390.

Sarras, M. P. (2012). Components, structure, biogenesis and function of the *Hydra* extracellular matrix in regeneration, pattern formation and cell differentiation. *Int. J. Dev. Biol.* 56, 567-76.

Schaible, R., Scheuerlein, A., Dańko, M. J., Gampe, J., Martínez, D. E. and Vaupel, J. W. (2015). Constant mortality and fertility over age in *Hydra*. *Proc. Natl. Acad. Sci.* 112, 15701-15706.

Schenkelaars, Q., Tomczyk, S., Wenger, Y., Ekundayo, K., Girard, V., Buzgariu, W., Austad, S. and Galliot, B. (2018). *Hydra*, a model system for deciphering the mechanisms of aging and resistance to aging. In *Conn's Handbook of Models for Human Aging,* Eds. Ram, J.L,Conn, P.M., pp. 507-520.

Schwentner, M. and Bosch, T. C. G. (2015). Revisiting the age, evolutionary history and species level diversity of the genus *Hydra* (Cnidaria: Hydrozoa). *Mol. Phylogenet. Evol.* 91, 41-55.

Sebé-Pedrós, A., Chomsky, E., Pang, K., Lara-Astiaso, D., Gaiti, F., Mukamel, Z., Amit, I., Hejnol, A., Degnan, B. M. and Tanay, A. (2018). Early metazoan cell type diversity and the evolution of multicellular gene regulation. *Nat. Ecol. Evol.* 2, 1176-1188.

Siebert, S., Anton-Erxleben, F. and Bosch, T. C. G. (2008). Cell type complexity in the basal metazoan *Hydra* is maintained by both stem cell based mechanisms and transdifferentiation. *Dev. Biol.* 313, 13-24.

Siebert, S., Farrell, J. A., Cazet, J. F., Abeykoon, Y., Primack, A. S., Schnitzler, C. E. and Juliano, C. E. (2019). Stem cell differentiation trajectories in *Hydra* resolved at single-cell resolution. *Science* 365, eaav9314.

Slautterback, D. B. (1967). The cnidoblast-musculoepithelial cell complex in the tentacles of hydra. *Zeitschrift für Zellforsch. und Mikroskopische Anat.* 79, 296-318.

Stiven, A. E. (1965). The relationship between size, budding rate, and growth efficiency in three species of hydra. *Res. Popul. Ecol.* 7, 1-15.

Sugiyama, T. and Fujisawa, T. (1978). Genetic analysis of developmental mechanisms in *Hydra*. II. Isolation and characterization of an interstitial cell-deficient strain. *J. Cell Sci.* 29, 35-52.

Szymanski, J. R. and Yuste, R. (2019). Mapping the whole-body muscle activity of *Hydra* vulgaris. *Curr. Biol.* 29, 1807-1817.

Takahashi, T., Koizumi, O., Ariura, Y., Romanovitch, A., Bosch, T. C., Kobayakawa, Y., Mohri, S., Bode, H. R., Yum, S., Hatta, M., et al. (2000). A novel neuropeptide, Hym-355, positively regulates neuron differentiation in *Hydra*. *Development* 127, 997-1005.

Takahashi, T., Hatta, M., Yum, S., Gee, L., Ohtani, M., Fujisawa, T. and Bode, H. R. (2005). Hym-301, a novel peptide, regulates the number of tentacles formed in hydra. *Development* 132, 2225-2234.

Takahashi, T., Hayakawa, E., Koizumi, O. and Fujisawa, T. (2008). Neuropeptides and their functions in *Hydra*. *Acta Biol. Hung.* 59, 227-235.

Takano, J. and Sugiyama, T. (1984). Genetic analysis of developmental mechanisms in hydra. XII. Analysis of chimaeric hydra produced from a normal and a slow-budding strain (L4). *J. Embryol. Exp. Morphol.* 80, 155-173.

Tanaka, E. M. and Reddien, P. W. (2011). The cellular basis for animal regeneration. *Dev. Cell* 21, 172-185.

Taubenheim, J., Willoweit-Ohl, D., Knop, M., Franzenburg, S., He, J., Bosch, T. C. G. and Fraune, S. (2020). Bacteria- and temperature-regulated peptides modulate β-catenin signaling in hydra. *Proc. Natl. Acad. Sci. U. S. A.* 117, 21459-21468.

Tautz, D. and Domazet-Lošo, T. (2011). The evolutionary origin of orphan genes. *Nat. Rev. Genet.* 12, 692.

Technau, U., Rudd, S., Maxwell, P., Gordon, P. M. K., Saina, M., Grasso, L. C., Hayward, D. C., Sensen, C. W., Saint, R. and Holstein, T. W. (2005). Maintenance of ancestral complexity and non-metazoan genes in two basal cnidarians. *Trends Genet.* 21, 633-639.

Telford, M. J. and Copley, R. R. (2011). Improving animal phylogenies with genomic data. *Trends Genet.* 27, 186-195.

Tothova, Z. and Gilliland, D. G. (2007). FoxO transcription factors and stem cell homeostasis: Insights from the hematopoietic system. *Cell Stem Cell* 1, 14-152.

Trembley, A. (1744). *Mémoires pour servir à l'histoire d'un genre de polypes d'eau douce, à bras en forme de cornes.* Jean and Herman Verbeek, Leiden.

van Deursen, J. M. (2014). The role of senescent cells in ageing. *Nature* 509, 439-446.

Watanabe, H., Schmidt, H. A., Kuhn, A., Höger, S. K., Kocagöz, Y., Laumann-Lipp, N., Özbek, S. and Holstein, T. W. (2014). Nodal signalling determines biradial asymmetry in *Hydra*. *Nature* 515, 112-115.

Weisburg, W. G., Barns, S. M., Pelletier, D. A. and Lane, D. J. (1991). 16S ribosomal DNA amplification for phylogenetic study. *J. Bacteriol.* 173, 697-703.

Wittlieb, J., Khalturin, K., Lohmann, J. U., Anton-Erxleben, F. and Bosch, T. C. G. (2006). Transgenic *Hydra* allow in vivo tracking of individual stem cells during morphogenesis. *Proc. Natl. Acad. Sci. U. S. A.* 103, 6208-11.

Wood, R. L. and Novak, P. L. (1982). The anchoring of nematocysts and nematocytes in the tentacles of hydra. *J. Ultrasructure Res.* 81, 104-116.

Yum, S., Takahashi, T., Koizumi, O., Ariura, Y., Kobayakawa, Y., Mohri, S. and Fujisawa, T. (1998). A novel neuropeptide, Hym-176, induces contraction of the ectodermal muscle in *Hydra*. *Biochem. Biophys. Res. Commun.* 248, 584-590.

Zhang, X., Yalcin, S., Lee, D. F., Yeh, T. Y. J., Lee, S. M., Su, J., Mungamuri, S. K., Rimmelé, P., Kennedy, M., Sellers, R., et al. (2011). FOXO1 is an essential regulator of pluripotency in human embryonic stem cells. *Nat. Cell Biol.* 13, 1092-9.

10

Regeneration in Anamniotic Vertebrates

Prayag Murawala[1,2,*] and Dunja Knapp[3,*]

[1]Mount Desert Island Biological Laboratory (MDIBL), PO Box 35,
Salisbury Cove, ME-04609-7250, USA
[2]Clinic for Kidney and Hypertension Diseases, Hannover Medical School,
Hannover – 30625, Germany
[3]DFG-Center for Regenerative Therapies Dresden/TU Dresden,
Fetscherstraße 105, 01307 Dresden, Germany
E-mail: pmurawala@mdibl.org; dunja.knapp@tu-dresden.de
*Corresponding Authors

10.1 Introduction

The ability to repair lost or damaged tissue after injury varies greatly across species, organs, developmental stage, age or even gender (Kang et al., 2013; Seifert et al., 2012). Among vertebrates, the most remarkable examples of regeneration are found in the clade of anamniotes, a group that includes fish and amphibians. For example, teleost fish (zebrafish) (Figure 10.1A) can regenerate fins, heart, and parts of the brain. In urodele amphibians (salamanders) (Figure 10.1B), the limb, heart, spinal cord, brain, and parts of the eye are examples of tissues and organs that can regenerate faithful replicas. In tail-less amphibians (anuran frogs and toads) (Figure 10.1C), regenerative capacity is typically limited to pre-metamorphic stages. In *Xenopus laevis*, the larval tail, limb buds, lens, and skin can regenerate; however, this regenerative capacity gradually declines as metamorphosis proceeds (Dent, 1962; Muneoka et al., 1986b). This makes *Xenopus* a unique model to study regeneration, because it allows for comparison between regenerative and non-regenerative conditions within the same species.

Regarding the extent and mode of restoration, one can make a general distinction between tissue repair and regeneration. Tissue repair refers to mending a localized damage by relatively little tissue growth. Such repair

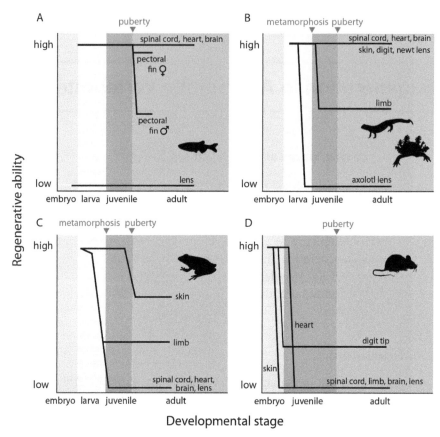

Figure 10.1 Regenerative capacity as a function of developmental stage among vertebrate models of regeneration. Modified with permission from (Seifert and Voss, 2013). In general, anamniotes (**A**: fish, **B**: salamanders and **C**: frogs) have a much larger regenerative capacity than mammals (**D**).

may or may not restore some structure and functionality and it often results in the formation of a scar. Tissue repair is an integral part of homeostasis in all animals, and it is essential for survival. Regeneration refers to *de novo* formation of new structures with more or less original tissue complexity, three-dimensional form, and functionality.

Mechanistically, different body structures take distinct routes to regenerate. Regeneration of appendages (limbs or tail) in vertebrates occurs via a process called epimorphosis, which involves formation of a regeneration bud called a blastema. The blastema is a mass of undifferentiated proliferating cells that forms at the amputation site and undergoes an embryonic-like developmental process to give rise to all tissues in the new, patterned structure. Regeneration of organs, such as heart, brain, or lung, does not involve formation of an obvious blastema (Jensen et al., 2021). Rather, proliferation happens broadly through the organ or in specialized niches that are also involved in adult homeostasis. Finally, some organs such as liver regenerate via compensatory hypertrophy, which does not replace the lost portion, but the remaining tissue expands to compensate functionally for the lost tissue.

Despite the long history of regeneration research, an understanding of the cellular and molecular mechanisms of regeneration has been limited by the lack of molecular and genetic techniques, particularly in the model species that regenerate well. Recent advances in transcriptomics, genomics, and gene editing in the main vertebrate models of regeneration have created important new tools. Some of the key questions at the heart of regeneration that can now be addressed are: How do injury signals activate normally quiescent progenitors? What are the progenitor cells for individual tissues and what is the developmental potential of individual progenitors? How does new tissue integrate with the old? How do cells know what structures they need to rebuild i.e., how is the information of their position in the body molecularly encoded?

Experiments in various animal models have converged toward surprisingly uniform results. First, it appears that vertebrates typically utilize lineage-restricted progenitor cells to regenerate lost structures (a few exceptions will be discussed), although the mechanisms by which such progenitors arise vary. Secondly, the molecular and cellular events that follow blastema formation are similar to those that govern the formation of the corresponding structure during embryonic development. In this chapter, we will cover limb regeneration in detail and provide an overview of the few other important organs that regenerate across these species.

10.2 Appendage Regeneration in Amphibians

Regrowing a fully functional organ requires restoring its full tissue diversity and spatial organization (Figure 10.2A). This process has to incorporate several aspects including: conveying the information that a structure was lost or damaged; acquiring progenitors for all missing tissues; controlled cell proliferation; patterning and differentiation. Appendages such as limb or tail have been widely used to study tissue regeneration as they are easily accessible and not vital for the organism.

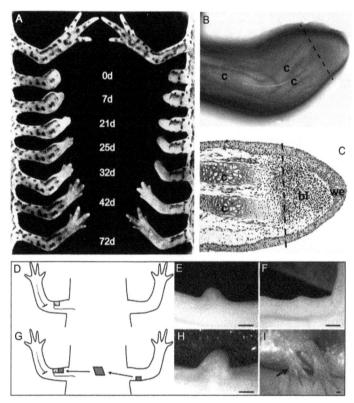

Figure 10.2   Limb regeneration in the urodeles. **A**. Regeneration of a limb amputated through the lower arm (**left**) and upper arm (**right**) of an adult newt *Notophthalmus viridescens*. By 7 days after amputation (7d) blastema is clearly visible distal to the amputation plane. At 72 days after amputation (72d), the regenerated limb has restored all the segments although it still did not reach the full size of the original limb (**top**). Note that the regenerated part of the limb corresponds exactly to the segment that was removed, irrespective of the plane of the amputation. Modified from (Goss, 1969). **B**. Regenerating limb of a larval salamander *Ambystoma mexicanum*. In the blastema, distal to the amputation plane marked by a dashed line, differentiated structures such as bone are not discernible. **C**. A section through the midbud (cone) stage blastema of a related salamander *Ambystoma maculatum*. Note the thickened wound epidermis (we) at the tip. Blastema cells (bl) have an undifferentiated mesenchymal phenotype. Differentiated cartilage (c) is visible in the mature part of the stump. Modified from (Edwards, 2008). **D–I**. Accessory limb model (ALM) demonstrated that deviation of nerve endings to an anterior lateral injury site combined with transplantation of posterior limb skin to that site are sufficient to elicit ectopic limb outgrowth in the axolotl. Deviating a nerve to a lateral wound (**D**) induces formation of a transient blastema–like bump (**E**). Such a bump regresses (**F**) unless cells from the opposite side of the limb (**red**) are provided (**G**), in which case the bump continues to grow (**H**) and eventually gives rise to an ectopic limb (**I**). Reproduced from (Endo et al., 2004).

Limbs are complex structures that consist of multiple tissues such as skin, muscle, bones, nerves, and vasculature. Furthermore, limbs are patterned along three axes, proximodistal (from shoulder to fingers), anteroposterior (from thumb to last finger), and dorsoventral (from back of the hand to the palm). Following amputation, limb regeneration proceeds through three morphologically and molecularly distinguishable phases (Bryant et al., 2002; Knapp et al., 2013). Regeneration begins with the wound-healing phase. During this phase, epidermal cells from the edges of the wound crawl over the cut surface to cover the wound and form a specialized wound epidermis. Over the next few days, during the blastema formation phase, the wound epidermis thickens into a multi-layered regeneration-specific epidermis which starts to secrete many signaling molecules and undifferentiated cells with mesenchymal appearance start accumulating beneath it, collectively forming a blastema (Figure 10.2B, C). These mesenchymal cells are the progenitors that give rise to all inner tissues of there generate, such as dermis, bones, muscles, vasculature, and innervation, and in tail regeneration also spinal cord. Once the blastema has grown to a critical size, it does not require any further context information from the stump. If transplanted to an ectopic "neutral" place, such as body trunk or the eye socket, such a blastema regenerates the same structure as it would produce in its original place (Stocum, 1968).

Blastema formation is followed by a re-development phase, which includes differentiation of blastema cells into mature tissue types and morphogenesis. During this phase many signaling pathways that were used in the limb development are redeployed, and the blastema resembles an embryonic developing limb bud morphologically and in the gene expression profile (Gerber et al., 2018; Knapp et al., 2013). However, the circumstances that lead to activation and accumulation of prospective blastema cells at the tip of the stump differ greatly from those involved in forming a limb bud. In addition, the size of the field, as well as the cellular and extra-cellular environment of a blastema and a limb bud are very different. Limb bud development starts with specification of the limb field within the lateral plate mesoderm (LPM) by the action of retinoic acid and certain homeobox (Hox) genes. A series of epithelial–mesenchymal inductions between LPM of the prospective limb bud area and overlying ectoderm result in the formation of an important organizing structure called the apical ectodermal ridge (AER) (Capdevila and Izpisua Belmonte, 2001). In contrast, the blastema is formed from a mature limb and in response to wounding and tissue loss; therefore, it is likely that the injury-specific signals are crucial for activating blastema progenitors.

10.2.1 Launching Regeneration via Wounding

The initial response to injury is migration of epidermal cells from the edges of the wound to quickly re-establish epithelial integrity and thus seal off the wound from outside influences (Hay, 1961). The wound epidermis, as well as dermal fibroblasts that accumulated beneath the wound epidermis, up-regulate specific extracellular matrix components and matrix metallo-proteases (Calve, 2010; Gerber, 2018; Kato, 2003) that help remodel the damaged tissue. It is likely that this response is initiated by physical changes related to tissue damage, such as oxidative stress (Gauron et al., 2013; Love et al., 2013; Niethammer et al., 2009), ion flux (Adams, 2007; Tseng, 2010), as well as inflammation (Godwin et al., 2013), apoptosis (Tseng et al., 2007) or presence of a blood clot (Godwin et al., 2010). In various animal models, it has been observed that these changes play important roles in wound healing and in regeneration, yet how these inputs integrate into downstream cellular response is still unknown. Interestingly, several studies suggest that the intensity or duration of such inputs, as well as the dependence on them, differ between lateral injury healing and regeneration (Adams et al., 2007; Gauron et al., 2013).

The early response to limb amputation is morphologically and in gene expression profile similar to a lateral limb injury, which does not lead to the formation of a new limb. However, amputations but not lateral wounds result in generation of a stable and highly proliferative blastema. Therefore, tissue damage starts the initial steps of regeneration, but it is not sufficient to launch a full regenerative response. Which additional conditions are needed to start the series of events leading to the formation of a new limb?

10.2.2 Role of the Nerves in Inducing Blastema

A large body of evidence demonstrated the nerve dependence of tetrapod appendage regeneration. This includes fin regeneration in teleost fish and limb regeneration in salamanders and frogs (Kumar and Brockes, 2012). If the nerves that innervate the salamander limbs are severed prior to amputa-tion, wound epidermis does not transform into secretory epithelium called apical epidermal cap (AEC). Several factors supplied by the nerve, such as fibroblast growth factor (FGF2) and keratinocyte growth factor (KGF), are thought to act upon basal keratinocytes of the wound epidermis and induce them to transform into AEC. AEC is analogous to the AER—an organizer structure present in developing tetrapod limbs, which is necessary for limb development (Christensen and Tassava, 2000; Satoh et al., 2008). AEC seems essential for the blastema growth because replacing AEC with mature skin results in a small blastema-like structure which fails to grow (Mescher, 1976; Wigmore and Holder, 1986).

Apart from its role in inducing/maintaining AEC, nerves are required to support blastema cell proliferation. In denervated limbs, blastema cells do not accumulate underneath the wound epidermis (Satoh et al., 2010; Satoh et al., 2007). Conversely, in an experimental model called the accessory limb model (ALM), an ectopic blastema can be induced by directing severed nerve bundles to the injuries on the side of the limb, which normally just heal (Figure 10.2D, E) (Endo et al., 2004). This suggested an inductive role of nerves in blastema formation. Nerve factors that supported blastema cell growth *in vivo* or *in vitro* include FGF2 (Mullen et al., 1996), neuregulin/glial growth factor (GGF2) (Farkas et al., 2016; Wang et al., 2000), Anterior Gradient protein (nAG) (Kumar et al., 2007). However, none of these factors suffice for the complete limb regeneration leaving it open whether this is due to a limited factor supply or to a requirement for additional factors.

In contrast to AEC, the function of the AER in salamander limb development does not depend on nerves and complete aneurogenic limbs can develop. This can be achieved by parabiosing two early embryos and removing the spinal cord, and thus preventing limb innervation, from one of them (Yntema, 1959a, b). Curiously, such aneurogenic limbs can regenerate even in the absence of nerves. How does limb innervation make regeneration dependent on the presence of the nerves? This can be explained by the observation that the expression of nAG—one of the aforementioned nerve factors, is down-regulated in developing limb buds upon innervation. Such innervated limbs show reactivation of nAG during regeneration, first in the Schwann cells and subsequently in the wound epidermis, in a nerve-dependent manner. In contrast, nAG expression persists in the epidermis even in the fully developed aneurogenic limbs, and thus its availability in these limbs does not depend on nerves (Kumar et al., 2011).

10.2.3 Maintenance of Blastema Outgrowth by the Shh–Fgf Activation Loop

Directing nerves to the site of injury in the accessory limb model induces ectopic proliferation and the accumulation of a blastema-like cell mass (Figure 10.2E), but it is not sufficient to promote development of a complete limb. After a period of growth, such ectopic blastemas regress (Figure 10.2F). These ectopic blastemas can progress to form ectopic limbs only if cells from different sides of the limb come into contact (Figure 10.2G–I) (Endo et al., 2004). This and other grafting experiments have shown that the juxtaposition of anterior and posterior limb tissue plus innervation is necessary and sufficient to induce complete limb regeneration in salamanders. What is the molecular and cellular mechanism behind this phenomenon?

Molecular basis of the requirement for both anterior and posterior tissue during limb regeneration has recently been clarified in axolotls and it is analogous to the molecular mechanism that controls limb bud outgrowth during development. During limb development, two signaling centers that secrete diffusible molecules are necessary for the limb bud outgrowth: AER at the distal tip of the bud and at the posterior side of the bud a region called the zone of polarizing activity (ZPA). ZPA cells secrete the morphogen Sonic Hedgehog (SHH), which is involved in maintaining a positive feedback loop between FGF8 in the AER and FGF10 in the mesenchymal cells of the limb bud. SHH is restricted to the posterior part of the bud by the presence of its transcriptional activator dHAND at the posterior side, and the transcriptional repressor GLI3 at the anterior side. This loop is necessary to maintain limb bud outgrowth (reviewed in (Zeller et al., 2009)). Later on, during limb bud patterning, SHH is involved in specifying the posterior part of the limb.

In the salamander blastema, similar to the limb buds, *Shh* is expressed in the posterior part of the blastema, *Fgf8* in the blastema cells just below the AEC, and *Fgf10* in the mesenchymal cells of the blastema (Christensen et al., 2002; Han et al., 2001; Imokawa and Yoshizato, 1997; Nacu et al., 2016). It was recently shown in the accessory limb model that pharmacological activation of SHH signaling on the anterior side can substitute for posterior skin graft and promote growth of nerve-induced blastema into complete accessory limbs on the anterior part of the limb, whereas activation of FGF8 signaling in the nerve-induced blastema at the posterior part can substitute an anterior skin graft leading to the formation of the accessory limbs at the posterior surface. These results showed that cross-activation between anteriorly restricted FGF8 and posteriorly restricted SHH is an essential part of normal limb regeneration (Nacu et al., 2016).

10.2.4 Cellular Sources and Differentiation Potential of Blastema Cells

Blastema formation requires a permissive and inductive environment as well as competent cells. Previously, we discussed the sources of signals that are needed to promote and maintain blastema growth, but what is the origin of the blastema cells? Which tissues contribute to the progenitor pool and what path do they take to transform into blastema cells?

The majority of the blastema cells are locally derived; approximately 0.5 mm of limb tissue proximal to the amputation participates in making of blastema (Currie et al., 2016). Furthermore, series of histological observations and cell tracking experiments have established that most of the tissues present in the limb contribute to the blastema (Chalkley, 1954; Hay and Fischman, 1961; Kragl et al., 2009; Muneoka et al., 1986a). Multiple studies also investigated the differentiation potential of urodele blastema cells. In such approaches, various partially purified tissue implants were tested for the ability to rescue regeneration of X-ray irradiated limbs (Dunis and Namenwirth, 1977; Wallace and Wallace, 1973). Alternatively, tissues were traced after labeling with dyes (Echeverri and Tanaka, 2002; Morrison et al., 2006). These experiments suggested that urodele cells can contribute to a broad set of tissues during regeneration. Although such techniques suffered from inherent difficulties to separate limb tissues from each other, they led to speculations that blastema consists of pluripotent stem cells and that their pluripotent state may be necessary to achieve self-organizing properties of blastema.

More recent reinvestigation of the differentiation potential of the blastema cells combined a ubiquitously labeled transgenic axolotl (Caggs:GFP) with embryonic grafting to indelibly label individual cell populations (Kragl et al., 2009). This technique allowed distinct labeling of epidermis, muscle, connective tissue, nervous system, and blood vessels with a GFP-fluorescence in a mature limb. By tracing labeled cells, it was shown that each of these populations provide the progenitors that remain restricted to contribute only to their specific lineage. Limited flexibility was observed only within connective tissue (CT) lineages, where soft CT gave rise to the bone in addition to soft CT (Gerber et al., 2018). Similarly, lineages remain restricted to their tissue of origin during regeneration of the *Xenopus* tail (Gargioli and Slack, 2004), zebrafish fin (Knopf et al., 2011; Stewart and Stankunas, 2012; Tu and Johnson, 2011), as well as in the mouse digit tip (Rinkevich et al., 2011). Based on these findings, it appears that a blastema consisting of heterogeneous lineage-restricted progenitors, may be a universal feature of vertebrate appendage regeneration.

The previously described transplantation and genetic-labeling methods demonstrated that during regeneration, tissues generally remain within broad embryonically defined lineages. But, are there multipotent progenitors for more closely related lineages? For example, all limb CT subtypes arise from embryonically related source (lateral plate mesoderm) but are phenotypically very diverse. They include soft CT such as various fibroblasts and hard CT such as tenocytes, skeletal, and periskeletal cells. CT lineage is the most abundant lineage contributing to the blastema but due to its heterogeneity, traditional grafting experiments could not answer how individual CT cell types participate during limb regeneration. Development of single cell tracing and single cell transcriptomics finally allowed dissecting cell transitions and lineage hierarchies during regeneration. Similar to mammals, skeletal cells in axolotl do not participate in regeneration, and periskeletal cells only contribute to bone repair near the amputation plane. Fibroblasts are responsible for bulk of regenerated CT including distal skeletal elements (Currie et al., 2016; Gerber et al., 2018). Single cell clonal tracing showed that a single fibroblast cell can contribute to diverse CT types including tenocytes, skeletal, and periskeletal cells. Hence, fibroblast-derived blastema cells mirror an embryonic limb bud-like multipotent phenotype. Behavior of fibroblasts in anamniotes is in stark contrast to mammals where injuries typically promote scarring instead of forming blastema.

10.2.4.1 Becoming blastema cell—dedifferentiation versus stem cell activation

How do highly proliferative blastema cells arise from the more or less quiescent homeostatic cells present within uninjured tissues? There are two major routes how this could happen; by recruitment of pre-existing stem cells or via dedifferentiation of differentiated cells. The mechanisms of recruiting progenitors differ for different cell types. Looking deeper into two examples: connective tissue and muscle, we will see that both mechanisms can be used.

10.2.4.1.1 Muscle

Skeletal muscle tissue in salamanders consists of muscle fibers and their associated satellite-stem cells, which are characterized by the expression of the paired box transcription factor PAX7. Mature muscle fibers are formed by fusion of myoblasts. Thus, they constitute multi-nucleated syncytia with a striated pattern of organized sarcomeres. In multiple experiments, primarily in newts, fragmentation of muscle cells was observed in cell culture (Kumar et al., 2000; Lo et al., 1993), in the animals using histological methods (Calve and Simon, 2011; Echeverri et al., 2001; Hay, 1959) and *in vivo* by cell labeling with dyes (Echeverri et al., 2001). Although these observations supported the hypothesis that myotube fragmentation and proliferation of resulting myoblasts occur during regeneration, due to limitations of the used methodologies they could not prove that myotube fragmentation represents a genuine mechanism with a significant contribution to regenerating muscle. This question was conclusively answered by permanent genetic labeling of multinucleated myofibers separately from muscle satellite cells, in two different salamander species: the newt *Notophthalmus viridescens* and axolotl. Muscle-specific expression of Cre-recombinase mediated permanent genetic labeling specifically in multinucleated muscle fibers. Following limb amputation through labeled muscle, some of the regenerated muscles in the newt limbs were also labeled, implying that mature myofibers gave rise to muscle progenitors. In contrast, there was no sign of contribution from labeled myofibers to the regenerated muscle (nor to any other cell type) in the axolotl limbs (Sandoval-Guzman et al., 2014). Recently, development of knock-in technology in axolotl allowed permanent labeling of muscle satellite cells in the *Pax7: ER-Cre-ER* transgenic line (Fei et al., 2017). Tracing of labeled cells showed that newly formed muscles in regenerated limb are derived from *Pax7+* satellite-like cells in axolotl. PAX7+ satellite cells also appear to be the major progenitors for muscle regeneration in zebrafish (Berberoglu et al., 2017), *Xenopus* (Gargioli and Slack, 2004), and for the muscle repair in amniotes (Lepper et al., 2009). Therefore, it seems that two alternative strategies evolved to regrow muscle: via PAX7+ stem cells, which appears to be a universal strategy for muscle repair in vertebrates, and myofiber dedifferentiation, which might be a trait derived within the newt branch.

10.2.4.1.2 Connective Tissue (CT)

In the axolotl, single cell transcriptome analysis of regenerating CT showed that mature uninjured limbs lack cells resembling blastema cells, showing that blastema does not arise by quick amplification of pre-existing blastema-like cells. Instead, all major specialized CT subtypes close to the amputation surface (excluding bone and cartilage) lose their differentiated tissue markers and converge toward their multipotent common progenitor—the mesenchymal cells of the developing limb bud (Gerber et al., 2018). Interestingly, as described in Section 2.4, fibroblast-derived blastema cells are somehow privileged and they over-proportionately contribute to all CT subtypes in the regenerated part of the limb.

Dedifferentiation of CT subtypes (osteoblasts) has also been demonstrated in zebrafish fin regeneration. Like other appendages, zebrafish fins regenerate via formation of a blastema. The zebrafish tail fin consists of a series of segmented bony rays connected by soft tissue with nerves and blood vessels, all covered by the epidermis. Each bony ray consists of two concave hemi-rays with an inner space filled with intra-ray mesenchymal cells. Ray bones, the major tissue in the fin, consist of differentiated bone cells, called osteoblasts, which secrete bone matrix. Upon fin amputation, osteoblasts downregulate expression of bone-differentiation markers and induce genes expressed by bone progenitors (Knopf et al., 2011; Sousa et al., 2011; Stewart and Stankunas, 2012). In contrast to axolotls, these dedifferentiated osteoblasts appear to be a major if not the only source of newly formed bone, during normal regeneration. However, genetic ablation of osteoblasts (by activating production of a lethal metabolite by the enzyme nitro-reductase), did not prevent bone regeneration (Singh et al., 2012). This finding revealed an unexpected plasticity of the fin to activate alternative mechanisms in order to generate *de novo* osteoblasts. Further cell lineage tracing and ablation experiments identified two populations: $sp7^-/mmp9^+$, osteoblast progenitor cells reserved in ray joint niches, and mesenchymal $sp7^-/mmp9^-$ cells that can replenish osteoblasts. Thus, regeneration of osteoblasts in the fin employs both mechanisms; dedifferentiation of mature osteoblasts and recruitment of undifferentiated progenitors from the same embryonic source (Ando et al., 2017).

10.2.5 Specification and Re-specification of Positional Information

The limb blastema precisely regenerates the missing portion of the limb. This implies that cells in the mature limb can access the information about their position in the limb. This is called positional identity. Positional identity is the characteristic of the connective tissue, while other cell types appear to be positionally naive (Nacu et al., 2013). Positional information is acquired during limb bud development, but in which form it is conserved in the adult limb remains an open question. Our knowledge about acquiring positional information during limb development is based mostly on work in chick and mouse. It is hypothesized that during embryonic development, retinoic acid (RA) emanating from the trunk forms a proximodistal gradient in the developing limb bud (reviewed in Zeller et al., 2009). In the proximal-most limb bud regions, RA activates expression of the homeobox transcription factor MEIS1, which specifies proximal cell identity (upper arm or thigh). FGF secreted from the AER restricts the zone of RA activity, thus promoting distalization (Mercader et al., 2000). These two opposing gradients result in the expression of Hox genes, which specify limb segments: upper arm (*HoxA9*), lower arm (*HoxA9* and *HoxA11*), and hand (*HoxA9*, *HoxA11*, and

HoxA13). In axolotl limb development as well as in amniotes, proximal Hox genes (*HoxA9*) are activated first and then progressively more distal Hox genes (Gardiner et al., 1995; Roensch et al., 2013). Consequently, limb segment specification proceeds from proximal toward distal via mechanism called "Progressive Specification."

CT cells in a mature limb maintain their fixed positional identity, but during regeneration, positional information has to be re-specified in order to regenerate more distal limb elements. A temporal analysis of Hox gene activation during regeneration following amputation through the upper arm, showed that the upper arm marker *HoxA9* is activated first, followed by the lower arm marker *HoxA11*, and finally the hand marker *HoxA13* (Roensch et al., 2013). This order of events is consistent with the progressive specification as it happens during limb bud development.

Re-specification of positional identity is normally only possible in one direction; toward the more distal fate—the phenomenon described as the "Rule of Distal Transformation." However, treatment with RA acting via MEIS1 can override the rule of distal transformation and proximalize the identity of the blastema cells, resulting in serial duplications (Maden, 1980; Mercader et al., 2005). For example, limbs amputated at the wrist may grow a complete limb, if exposed to RA. Due to the proximalizing effect of RA and distal localization of the FGF-signaling components in the blastema, it is thought that antagonism between these factors patterns the proximodistal axis of blastema, similar to the limb bud. Puzzlingly, it appears that in the blastema, location of the most active RA, signaling is not at the proximal base of the blastema but in the most distal tip—in the AEC (Monaghan and Maden, 2012). Therefore, although RA can impinge on the positional identity, the role of endogenous RA in the proximo-distal patterning of the blastema remains ambiguous. The simplest scenario is that during regeneration, blastema cells' default positional identity is determined by epigenetic memory appropriate to the position of the amputation, and over time cells at the distal end become distalized by the signals from the wound epidermis. In this scenario epigenetic memory substitutes the role of RA in patterning a developing limb bud.

10.2.6 Frogs—Regenerative Capacity Depends on Developmental Stage

Xenopus tadpoles can regenerate their developing limbs and tails in a fashion very similar to that described in the urodeles. However, *Xenopus* tadpoles regenerate an amputated hind limb only during larval stages.

With the progression of metamorphosis, the ability to regenerate progressively declines resulting in more and more hypomorphic regenerates. Eventually, in post-metamorphic frogs, limb amputation results at best in a single spike (Dent, 1962; Muneoka et al., 1986b). The important question that ensues is; what causes the loss of regenerative capability.

Transplantation of limb cells between tadpoles and froglets suggested that loss of the regenerative ability is not related to the physiological state of the host, but rather to the intrinsic properties of the limb cells. When limb buds from tadpoles were transplanted onto stumps of the post-metamorphic froglets, almost normal regenerates are formed. Conversely, cells of the froglet blastema grafted onto tadpole stump either didn't contribute to the regenerate or only formed a spike typical of the froglet stage (Sessions and Bryant, 1988). Further transplantation experiments have shown that the regenerative decline is not due to the loss of competence in the wound epidermis but in the mesenchymal cells, which lose the ability to express some patterning genes such as *Fgf10* or *Shh* (Carlson, 1982; Yokoyama et al., 2000). It is thought that this loss of competence is mediated by methylation of some limb-specific enhancers (Yakushiji et al., 2007).

Interestingly, in recent heterochronic transplantation experiments, the mesenchymal cells of the old host also made a substantial contribution to the resulting regenerates when treated with a growth factor cocktail consisting of WNT/β-catenin, SHH, FGF10, and thymosin β4. However, the cocktail rendered those old cells regeneration-competent only if young cells were also provided, highlighting the need for still unknown factors produced by the young cells (Lin et al., 2013). Remarkably, old host cells started producing a subset of blastema cell markers in these circumstances, implying that a partial reprograming of the old cells was achieved. It will be very important to correlate these findings with the situation in mammals, which also regenerate their limb buds embryonically, but no longer once the limb buds differentiate.

10.3 Other Examples of Regeneration

Besides appendages, which were the focus of the previous chapter, in anamniotic species various other organs have the capacity to replace missing parts. Brain, heart, and lens regeneration proceed without an apparent blastema. They were subject to intensive research because of their obvious medical relevance and will be discussed in some detail below.

10.3.1 Brain Regeneration

While in teleost fish and amphibians, CNS injury leads to neurogenesis (Lust and Tanaka, 2019), the mammalian central nervous system (CNS) has little capacity for repair. Although, in mammals, injuries induce proliferation of astrocytes and oligodendrocytes, this leads to a glial scar. In general, neural regeneration relies on the neural stem/progenitor cells that are retained in neurogenic niches. These cells retain some embryonic features such as contact to the ventricle and a cytoplasmic process that connects them to the pial surface. In salamanders, regeneration of brain tissue is attributed to the ependymoglia cells that line the ventricle and express mixed glia and neuroepithelial markers such as GFAP and Sox2, respectively (Amamoto et al., 2016; Berg et al., 2011). In zebrafish, neural progenitors differ across brain niches. Ventricular GFAP+, Sox2+ radial glia cells that correspond to ependymal cells of salamanders act as neural progenitors in the dorsal telencephalon (Kroehne et al., 2011), whereas Nestin$^+$ neuroepithelial-like cells, which represent a developmentally earlier stage, act as major neural progenitors in the optic tectum and cerebellum (Kaslin et al., 2017).

To what extent does repair recover the original neuronal diversity, structural organization, and functionality? Diverse neuronal subtypes seem to be produced after stab injury in zebrafish telencephalon (Kroehne et al., 2011; Lange et al., 2020). However, potential for perfect repair in some areas is lost during ontogeny, such as the competence to regenerate Purkinje cells in the cerebellum as a consequence of the loss of the stem cell niche during juvenile development (Kaslin et al., 2017). Also, a shift toward symmetric neurogenic divisions that produce neurons quickly but deplete the stem cell population over time has been observed (Barbosa et al., 2015). In the red spotted newt, pharmacological ablation of dopaminergic neuron in the mesencephalon or cholinergic neurons in telencephalon results in restoration of the ablated neuron type (Berg et al., 2011). Similarly, in the axolotl, a diverse array of neurons is produced after removal of a portion of the pallium (dorsal part of the telencephalon) (Amamoto et al., 2016). These new neurons acquire functional electrophysiological traits and respond to afferent inputs. However, these neurons fail to produce long-distance connections even after 42 weeks post injury, showing limitations to integrate into the already patterned brain (Amamoto et al., 2016).

Similar to limb regeneration, the immune system also seems to be essential for brain regeneration, at least in zebrafish (Kyritsis et al., 2012). Additionally, neuronal inputs are essential for brain regeneration in these diverse species. For example, both in *Xenopus* and in axolotl, regeneration of telencephalic regions, such as the olfactory bulb, depends on inputs from the olfactory nerves (Maden et al., 2013; Yoshino and Tochinai, 2006). In general, there is a correlation between the extent of homeostatic adult neurogenesis and the regenerative potential. In mammals, adult neurogenic niches are only found in two restricted areas of the telencephalon: the dentate gyrus of the hippocampus and the olfactory bulb, while in fish and salamanders ventricular neuroepithelial and/or radial glia-like cells, which have been linked to neurogenesis, persist in most areas of the adult brain (Grandel and Brand, 2011; Maden et al., 2013). In anurans (*Xenopus*), the number of undifferentiated and proliferating cells in the brains of metamorphosing animals declines alongside their regenerative abilities (Yoshino and Tochinai, 2004). Nevertheless, there is no direct causal connection between homeostatic and regenerative neurogenesis. For example, in red spotted newt, homeostatic neurogenesis is restricted to the forebrain, but quiescent ependymoglia cells can also proliferate in response to injury and regenerate lost tissue in other brain regions (Berg et al., 2010). It remains to be seen how newly formed connectome of the neurons in a regenerated brain translates to its functional recovery.

10.3.2 Heart Regeneration

Contrary to mammals where cardiac injury typically leads to scarring, with minimal regeneration of the heart muscle (myocardium), zebrafish and salamanders can replace cardiomyocytes (the specialized muscle cells of the vertebrate myocardium). Zebrafish can regenerate their hearts after losing 20% of the ventricle muscledue to amputation or cryoinjury. Cryoinjury, which more closely resembles degeneration of the human heart after myocardial infarction, causes cardiomyocyte and vasculature death in the area. This is followed by proliferation in all layers of the heart: epicardium, myocardium, and endocardium. In the next weeks, cell debris is cleared and initially replaced by a fibrotic scar. Such a scar in mammalian hearts remains, but in zebrafish it is replaced by cardiomyocytes and new blood vessels. Eventually, the heart returns to nearly its original form. Interestingly, this ability is not shared by medaka—another teleost fish model, which seems at least partly attributable to the differences in immune response between these models (Lai, 2017).

Which cells give rise to the new cardiomyocytes? Genetic cell-tracking in zebrafish demonstrated that regenerated myocardium was completely or almost completely derived from the pre-existing cardiomyocytes. Mature cardiomyocytes loosen contacts with each other, their sarcomeric structure becomes disorganized and they re-express some of the early muscle lineage markers (Jopling et al., 2010; Kikuchi et al., 2010).

Myocardium in vertebrates comprises a mixture of diploid and polyploid cardiomyocytes. Polyploidization occurs naturally, and in adult mammals, the majority of cardiomyocytes are polyploid and often binucleated. In contrast, in adult zebrafish, only 1% of cardiomyocytes are polyploid. It seems that mononucleated diploid cardiomyocytes are better poised for heart regeneration as compared to the multinucleated and polyploid cardiomyocytes (Gonzalez-Rosa et al., 2018). Recent work in zebrafish, *Xenopus*, and mouse all point to evidence that thyroid pathway promotes formation of multinucleated cardiomyocytes. Overactivation of the thyroid pathway in the zebrafish heart, led to an increase in binucleated cardiomyocytes, which perform poorly in heart regeneration. Correspondingly, inhibition of thyroid pathway signaling in mouse cardiomyocytes led to an increase in mononucleated cardiomyocytes and improved cardiac regenerative potential in adults (Hirose et al., 2019). Thus, it appears that several roadblocks stand in the way of efficient heart regeneration and both, modulation of immune response, as well as activation of those remaining diploid cardiomyocytes, might be needed to aptly mend human hearts.

10.3.3 Lens Regeneration in Amphibians via Transdifferentiation

Lens regeneration is limited to some salamanders and only a few species offish and frogs. Newts can regenerate their lens independently of age, *Xenopus* only at pre-metamorphic stages, while axolotl can regenerate lens only during a limited time after hatching. In *Xenopus*, the new lens regenerates from corneal epithelium, while in axolotls and newts it regenerates from the iris (Figure 10.3) (Suetsugu-Maki et al., 2012).

During lens regeneration in newts, pigmented epithelial cells (PECs) lose the pigment, reorganize the extracellular matrix, and re-enter the cell cycle. The process of de-differentiation is accompanied by an upregulation of certain pluripotency markers such as *Sox2, Klf4*, as well as an accumulation of the stem cell marker Nucleostemin in the nucleoli (Maki et al., 2009; Maki et al., 2007), but it is not known if these factors play a role in allowing lineage switching. In any case, newt PECs do not become multipotent because upon implantation into limb blastema, they only form the lens in this foreign environment, indicating that they can only switch into a defined phenotype (Ito et al., 1999; Reyer et al., 1973). The expression of two early markers of lens development, *Pax6* and *Prox1*, during lens regeneration from the

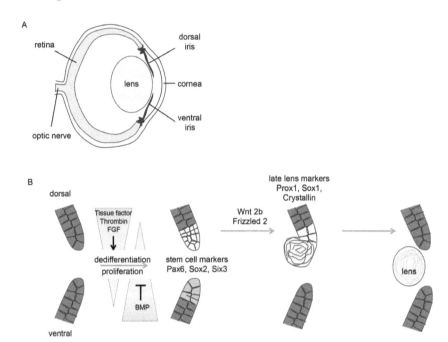

Figure 10.3   Transdifferentiation during lens regeneration in newts. **A.** Schematic representation of the eye. In *Xenopus*, a new lens regenerates from the corneal epithelium, which, like the lens, develops from the embryonic surface ectoderm. Newts regenerate the lens from the dorsal iris, which develops from the embryonic neuroectoderm, thus representing a real example of transdifferentiation. **B.** Transdifferentiation of the pigmented epithelium of the dorsal iris upon lens removal in the newt. The process of PEC dedifferentiation is accompanied by an upregulation of stem cell markers such as Sox2 and Klf4, and accumulation of Nucleostemin in the nucleoli. Transcription factors Six3, Pax6, and Prox1 mediate differentiation into lens; cells start expressing late lens markers (Sox1, Prox1, and Crystallins) and the structure transforms into functional lens. Tissue Factor and BMP limit the process to only the dorsal part of the iris. Figure modified from (Knapp and Tanaka, 2012).

newt iris and *Xenopus* cornea suggests activation of a program similar to embryonic lens development (Mizuno et al., 1999). Newt lens regeneration is the classic example of regeneration where terminally differentiated cells transform into a different mature phenotype—the phenomena called transdifferentiation. Nevertheless, it will be important to revisit that question in the future using modern transgenic techniques to examine if transdifferentiation is exclusively responsible for regenerating the newt lens or stem cell–like cells exist and play a role also in this system.

The comparison between permissive and non-permissive conditions is the key entry point for understanding the regulation of lens regeneration. Interestingly, in newts new lens formation can be elicited only from the dorsal side of the iris. An intrinsic difference between dorsal and ventral iris in the newt is the expression of the transmembrane Tissue Factor (TF) only in a patch of cells on the dorsal side (Godwin et al., 2010). TF is suggested to act as a scaffold for the blood clot components and leukocytes, which may act as a source of FGF2 signal that induces S-phase reentry. Additional differences must exist between dorsal and ventral side, because ectopic ventral expression of TF was not sufficient to promote transdifferentiation of the ventral iris. Observed differences include preferential expression of BMP on the ventral side and Wnt on dorsal (Grogg et al., 2005; Hayashi et al., 2006). Indeed, combined addition of FGF2 and WNT3a, as well as inhibition of BMP signaling was able to induce lens transdifferentiation of ventral explants. Notably, iris PECs of other vertebrates that cannot regenerate lens also have the potential to transdifferentiate *in vitro* and form lentoids (Asami et al., 2007; Kosaka et al., 1998). This may suggest that regeneration of lens by the iris may be the basal form of lens regeneration while regeneration from the cornea may be a result of convergent evolution for restoring the lens.

10.4 Regeneration and Metabolism

One notable trait that distinguishes anamniotes from mammals is their slow metabolism coupled with exothermy. The acquisition of the ability to maintain a constant body temperature (endothermy) requires a fast metabolism that is regulated by the thyroid hormone. Although thyroid hormone signaling is intact in anamniotes, they usually maintain low level of plasma thyroid hormone. It is noteworthy that thyroid hormone which causes metamorphosis has been associated with a loss of regeneration in *Xenopus*. Hence, it is plausible that the regenerative ability has been traded off for the ability to regulate body temperature (Hirose et al., 2019). Consistent with this argument, among poikilotherm amniotes (lizards), there are also examples of regeneration (even if imperfect) of complex body parts such as tail including the spinal cord (Sun et al., 2018). In evolutionary terms, it is plausible that the ability to regenerate may not offer a selective advantage to the species with fast metabolism, as their need to feed frequently and move fast, does not grant them the time needed to complete the regeneration process. Investigating the effects of metabolism in a broader range of species and cell populations will help elucidate its impact on the regenerative capabilities.

10.5 Concluding Remarks

- Different vertebrate taxa show huge diversity in their capacity to regenerate in regard to which tissues and organs, as well as developmental stages are capable of regeneration. Such differences exist even between related groups, such as the ability of heart and retina regeneration in zebrafish but its absence or impairment in Japanese medaka (Lai et al., 2017; Lust and Wittbrodt, 2018) or lifelong regeneration of lens in newts but stage-dependent in axolotls (Suetsugu-Maki et al., 2012).

- A key step in regeneration is the acquisition of specific progenitor cells. The cellular mechanisms that provide progenitor cells for regeneration include dedifferentiation, stem cell activation, and transdifferentiation (lens). Interestingly, even related species can apply different strategies as exemplified by the ability of myotubes to dedifferentiate in newts but not in axolotls (Sandoval-Guzman et al., 2014)). These differences highlight the importance of studying the regeneration in diverse models.

- In most vertebrate regenerative systems, tissue damage-signals, wound epithelium, nerves, and immune system play critical roles in stimulating regenerative response.

- Progenitor cells largely maintain the tissue lineages that were specified in development. Even in the cases when lineage boundaries are crossed, fate switching is limited to related lineages (like in the examples of the inter-conversion between connective tissue types or conversion of cornea or iris into lens). In cases when de-differentiation happens, it seems that de-differentiation proceeds only to a recent common progenitor stage (connective tissue de-differentiate into progenitors that resemble mesenchymal cells of the developing limb bud; iris retains its neurocctoderm lineage by expressing Pax6). Attainment of a pluripotent state during regeneration has not been observed in any vertebrate system.

- Finally, once the progenitor state is achieved, redeployment of developmental pathways in reconstructing the pattern seems to be a general strategy. Notable exceptions to the rule are the function of the nerve in limb regeneration, which was not necessary in developing limb bud, and the still questioned role of retinoic acid in proximo-distal patterning of the salamander limb blastema. These and other examples of adapting mature tissue to substitute for distinct conditions in development will be interesting to consider in terms of the evolutionary origin, maintenance, and loss of the ability to regenerate.

Acknowledgments

We thank Attila Tóth and Keith Gunapala for advice on the manuscript. PM is supported by grants from NIH – COBRE (5P20GM104318-08) and DFG (429469366). DK is supported by DFG (SA 3349/4-1)

References

Adams, D.S., Masi, A., Levin, M., 2007. H+ pump-dependent changes in membrane voltage are an early mechanism necessary and sufficient to induce Xenopus tail regeneration. Development 134, 1323–1335.

Amamoto, R., Huerta, V.G., Takahashi, E., Dai, G., Grant, A.K., Fu, Z., Arlotta, P., 2016. Adult axolotls can regenerate original neuronal diversity in response to brain injury. Elife 5.

Ando, K., Shibata, E., Hans, S., Brand, M., Kawakami, A., 2017. Osteoblast Production by Reserved Progenitor Cells in Zebrafish Bone Regeneration and Maintenance. Dev Cell 43, 643–650 e643.

Asami, M., Sun, G., Yamaguchi, M., Kosaka, M., 2007. Multipotent cells from mammalian iris pigment epithelium. Dev Biol 304, 433–446.

Barbosa, J.S., Sanchez-Gonzalez, R., Di Giaimo, R., Baumgart, E.V., Theis, F.J., Gotz, M., Ninkovic, J., 2015. Neurodevelopment. Live imaging of adult neural stem cell behavior in the intact and injured zebrafish brain. Science 348, 789–793.

Berberoglu, M.A., Gallagher, T.L., Morrow, Z.T., Talbot, J.C., Hromowyk, K.J., Tenente, I.M., Langenau, D.M., Amacher, S.L., 2017. Satellite-like cells contribute to pax7-dependent skeletal muscle repair in adult zebrafish. Dev Biol 424, 162–180.

Berg, D.A., Kirkham, M., Beljajeva, A., Knapp, D., Habermann, B., Ryge, J., Tanaka, E.M., Simon, A., 2010. Efficient regeneration by activation of neurogenesis in homeostatically quiescent regions of the adult vertebrate brain. Development 137, 4127–4134.

Berg, D.A., Kirkham, M., Wang, H., Frisen, J., Simon, A., 2011. Dopamine controls neurogenesis in the adult salamander midbrain in homeostasis and during regeneration of dopamine neurons. Cell Stem Cell 8, 426–433.

Bryant, S.V., Endo, T., Gardiner, D.M., 2002. Vertebrate limb regeneration and the origin of limb stem cells. Int J Dev Biol 46, 887–896.

Calve, S., Simon, H.G., 2011. High resolution three-dimensional imaging: Evidence for cell cycle reentry in regenerating skeletal muscle. Dev Dyn 240, 1233–1239.

Capdevila, J., Izpisua Belmonte, J.C., 2001. Patterning mechanisms controlling vertebrate limb development. Annu Rev Cell Dev Biol 17, 87–132.

Carlson, B.M., 1982. The regeneration of axolotl limbs covered by frog skin. Dev Biol 90, 435–440.

Chalkley, D.T., 1954. A quantitative histological analysis of forelimb regeneration in triturus viridescens. Journal of Morphology. 94, 21–70.

Christensen, R.N., Tassava, R.A., 2000. Apical epithelial cap morphology and fibronectin gene expression in regenerating axolotl limbs. Dev Dyn 217, 216–224.

Christensen, R.N., Weinstein, M., Tassava, R.A., 2002. Expression of fibroblast growth factors 4, 8, and 10 in limbs, flanks, and blastemas of Ambystoma. Dev Dyn 223, 193–203.

Currie, J.D., Kawaguchi, A., Traspas, R.M., Schuez, M., Chara, O., Tanaka, E.M., 2016. Live Imaging of Axolotl Digit Regeneration Reveals Spatiotemporal Choreography of Diverse Connective Tissue Progenitor Pools. Dev Cell 39, 411–423.

Dent, J.N., 1962. Limb regeneration in larvae and metamorphosing individuals of the South African clawed toad. J Morphol 110, 61–77.

Dunis, D.A., Namenwirth, M., 1977. The role of grafted skin in the regeneration of x-irradiated axolotl limbs. Dev Biol 56, 97–109.

Echeverri, K., Clarke, J.D., Tanaka, E.M., 2001. In vivo imaging indicates muscle fiber dedifferentiation is a major contributor to the regenerating tail blastema. Dev Biol 236, 151–164.

Echeverri, K., Tanaka, E.M., 2002. Ectoderm to mesoderm lineage switching during axolotl tail regeneration. Science 298, 1993–1996.

Edwards, R.G., 2008. From embryonic stem cells to blastema and MRL mice. Reprod Biomed Online 16, 425–461.

Endo, T., Bryant, S.V., Gardiner, D.M., 2004. A stepwise model system for limb regeneration. Dev Biol 270, 135–145.

Farkas, J.E., Freitas, P.D., Bryant, D.M., Whited, J.L., Monaghan, J.R., 2016. Neuregulin-1 signaling is essential for nerve-dependent axolotl limb regeneration. Development 143, 2724–2731.

Fei, J.F., Schuez, M., Knapp, D., Taniguchi, Y., Drechsel, D.N., Tanaka, E.M., 2017. Efficient gene knockin in axolotl and its use to test the role of satellite cells in limb regeneration. Proc Natl Acad Sci U S A 114, 12501–12506.

Gardiner, D.M., Blumberg, B., Komine, Y., Bryant, S.V., 1995. Regulation of HoxA expression in developing and regenerating axolotl limbs. Development 121, 1731–1741.

Gargioli, C., Slack, J.M., 2004. Cell lineage tracing during Xenopus tail regeneration. Development 131, 2669–2679.

Gauron, C., Rampon, C., Bouzaffour, M., Ipendey, E., Teillon, J., Volovitch, M., Vriz, S., 2013. Sustained production of ROS triggers compensatory proliferation and is required for regeneration to proceed. Sci Rep 3, 2084.

Gerber, T., Murawala, P., Knapp, D., Masselink, W., Schuez, M., Hermann, S., Gac-Santel, M., Nowoshilow, S., Kageyama, J., Khattak, S., Currie, J.D., Camp, J.G., Tanaka, E.M., Treutlein, B., 2018. Single-cell analysis uncovers convergence of cell identities during axolotl limb regeneration. Science 362.

Godwin, J.W., Liem, K.F., Jr., Brockes, J.P., 2010. Tissue factor expression in newt iris coincides with thrombin activation and lens regeneration. Mech Dev 127, 321–328.

Godwin, J.W., Pinto, A.R., Rosenthal, N.A., 2013. Macrophages are required for adult salamander limb regeneration. Proc Natl Acad Sci U S A 110, 9415–9420.

Gonzalez-Rosa, J.M., Sharpe, M., Field, D., Soonpaa, M.H., Field, L.J., Burns, C.E., Burns, C.G., 2018. Myocardial Polyploidization Creates a Barrier to Heart Regeneration in Zebrafish. Dev Cell 44, 433–446 e437.

Goss, R.J., 1969. Principles of Regeneration. Academic Press, New York.

Grandel, H., Brand, M., 2011. Zebrafish limb development is triggered by a retinoic acid signal during gastrulation. Dev Dyn 240, 1116–1126.

Grogg, M.W., Call, M.K., Okamoto, M., Vergara, M.N., Del Rio-Tsonis, K., Tsonis, P.A., 2005. BMP inhibition-driven regulation of six-3 underlies induction of newt lens regeneration. Nature 438, 858–862.

Han, M.J., An, J.Y., Kim, W.S., 2001. Expression patterns of Fgf-8 during development and limb regeneration of the axolotl. Dev Dyn 220, 40–48.

Hay, E.D., 1959. Electron microscopic observations of muscle dedifferentiation in regenerating Amblystoma limbs. Dev Biol 1, 555–585.

Hay, E.D., Fischman, D.A., 1961. Origin of the blastema in regenerating limbs of the newt Triturus viridescens. An auto radiographic study using tritiated thymidine to follow cell proliferation and migration. Dev Biol 3, 26–59.

Hayashi, T., Mizuno, N., Takada, R., Takada, S., Kondoh, H., 2006. Determinative role of Wnt signals in dorsal iris-derived lens regeneration in newt eye. Mech Dev 123, 793–800.

Hirose, K., Payumo, A.Y., Cutie, S., Hoang, A., Zhang, H., Guyot, R., Lunn, D., Bigley, R.B., Yu, H., Wang, J., Smith, M., Gillett, E., Muroy, S.E., Schmid, T., Wilson, E., Field, K.A., Reeder, D.M., Maden, M., Yartsev,

M.M., Wolfgang, M.J., Grutzner, F., Scanlan, T.S., Szweda, L.I., Buffen-stein, R., Hu, G., Flamant, F., Olgin, J.E., Huang, G.N., 2019. Evidence for hormonal control of heart regenerative capacity during endothermy acquisition. Science 364, 184–188.

Imokawa, Y., Yoshizato, K., 1997. Expression of Sonic hedgehog gene in regenerating newt limb blastemas recapitulates that in developing limb buds. Proc Natl Acad Sci U S A 94, 9159–9164.

Ito, M., Hayashi, T., Kuroiwa, A., Okamoto, M., 1999. Lens formation by pigmented epithelial cell reaggregate from dorsal iris implanted into limb blastema in the adult newt. Dev Growth Differ 41, 429–440.

Jensen TB, Giunta P, Schultz NG, Griffiths J, Duerr TJ, Kyeremateng Y, Wong H, Adesina A, Monaghan JR., 2021. Lung injury in axolotl sala-manders induces an organ-wide proliferation response. Dev Dyn. doi: 10.1002/dvdy.315.

Jopling, C., Sleep, E., Raya, M., Marti, M., Raya, A., Izpisua Belmonte, J.C., 2010. Zebrafish heart regeneration occurs by cardiomyocyte dedifferenti-ation and proliferation. Nature 464, 606–609.

Kang, J., Nachtrab, G., Poss, K.D., 2013. Local Dkk1 crosstalk from breeding ornaments impedes regeneration of injured male zebrafish fins. Dev Cell 27, 19–31.

Kaslin, J., Kroehne, V., Ganz, J., Hans, S., Brand, M., 2017. Distinct roles of neuroepithelial-like and radial glia-like progenitor cells in cerebellar regeneration. Development 144, 1462–1471.

Kikuchi, K., Holdway, J.E., Werdich, A.A., Anderson, R.M., Fang, Y., Egnaczyk, G.F., Evans, T., Macrae, C.A., Stainier, D.Y., Poss, K.D., 2010. Primary contribution to zebrafish heart regeneration by gata4(+) cardiomyocytes. Nature 464, 601–605.

Knapp, D., Schulz, H., Rascon, C.A., Volkmer, M., Scholz, J., Nacu, E., Le, M., Novozhilov, S., Tazaki, A., Protze, S., Jacob, T., Hubner, N., Haber-mann, B., Tanaka, E.M., 2013. Comparative transcriptional profiling of the axolotl limb identifies a tripartite regeneration-specific gene program. PLoS One 8, e61352.

Knapp, D., Tanaka, E.M., 2012. Regeneration and reprogramming. Curr Opin Genet Dev 22, 485–493.

Knopf, F., Hammond, C., Chekuru, A., Kurth, T., Hans, S., Weber, C.W., Mahatma, G., Fisher, S., Brand, M., Schulte-Merker, S., Weidinger, G., 2011. Bone regenerates via dedifferentiation of osteoblasts in the zebrafish fin. Dev Cell 20, 713–724.

Kosaka, M., Kodama, R., Eguchi, G., 1998. In vitro culture system for iris-pigmented epithelial cells for molecular analysis of transdifferentiation. Exp Cell Res 245, 245–251.

Kragl, M., Knapp, D., Nacu, E., Khattak, S., Maden, M., Epperlein, H.H., Tanaka, E.M., 2009. Cells keep a memory of their tissue origin during axolotl limb regeneration. Nature 460, 60–65.

Kroehne, V., Freudenreich, D., Hans, S., Kaslin, J., Brand, M., 2011. Regeneration of the adult zebrafish brain from neurogenic radial glia-type progenitors. Development 138, 4831–4841.

Kumar, A., Brockes, J.P., 2012. Nerve dependence in tissue, organ, and appendage regeneration. Trends Neurosci 35, 691–699.

Kumar, A., Delgado, J.P., Gates, P.B., Neville, G., Forge, A., Brockes, J.P., 2011. The aneurogenic limb identifies developmental cell interactions underlying vertebrate limb regeneration. Proc Natl Acad Sci U S A 108, 13588–13593.

Kumar, A., Godwin, J.W., Gates, P.B., Garza-Garcia, A.A., Brockes, J.P., 2007. Molecular basis for the nerve dependence of limb regeneration in an adult vertebrate. Science 318, 772–777.

Kumar, A., Velloso, C.P., Imokawa, Y., Brockes, J.P., 2000. Plasticity of retrovirus-labelled myotubes in the newt limb regeneration blastema. Dev Biol 218, 125–136.

Kyritsis, N., Kizil, C., Zocher, S., Kroehne, V., Kaslin, J., Freudenreich, D., Iltzsche, A., Brand, M., 2012. Acute inflammation initiates the regenerative response in the adult zebrafish brain. Science 338, 1353–1356.

Lai, S.L., Marin-Juez, R., Moura, P.L., Kuenne, C., Lai, J.K.H., Tsedeke, A.T., Guenther, S., Looso, M., Stainier, D.Y., 2017. Reciprocal analyses in zebrafish and medaka reveal that harnessing the immune response promotes cardiac regeneration. Elife 6.

Lange, C., Rost, F., Machate, A., Reinhardt, S., Lesche, M., Weber, A., Kuscha, V., Dahl, A., Rulands, S., Brand, M., 2020. Single cell sequencing of radial glia progeny reveals the diversity of newborn neurons in the adult zebrafish brain. Development 147.

Lepper, C., Conway, S.J., Fan, C.M., 2009. Adult satellite cells and embryonic muscle progenitors have distinct genetic requirements. Nature 460, 627–631.

Lin, G., Chen, Y., Slack, J.M., 2013. Imparting regenerative capacity to limbs by progenitor cell transplantation. Dev Cell 24, 41–51.

Lo, D.C., Allen, F., Brockes, J.P., 1993. Reversal of muscle differentiation during urodele limb regeneration. Proc Natl Acad Sci U S A 90, 7230–7234.

Love, N.R., Chen, Y., Ishibashi, S., Kritsiligkou, P., Lea, R., Koh, Y., Gallop, J.L., Dorey, K., Amaya, E., 2013. Amputation-induced reactive oxygen species are required for successful Xenopus tadpole tail regeneration. Nat Cell Biol 15, 222–228.

Lust, K., Tanaka, E.M., 2019. A Comparative Perspective on Brain Regeneration in Amphibians and Teleost Fish. Dev Neurobiol 79, 424–436.

Lust, K., Wittbrodt, J., 2018. Activating the regenerative potential of Muller glia cells in a regeneration-deficient retina. Elife 7.

Maden, M., 1980. Intercalary regeneration in the amphibian limb and the rule of distal transformation. J Embryol Exp Morphol 56, 201–209.

Maden, M., Manwell, L.A., Ormerod, B.K., 2013. Proliferation zones in the axolotl brain and regeneration of the telencephalon. Neural Dev 8, 1.

Maki, N., Suetsugu-Maki, R., Tarui, H., Agata, K., Del Rio-Tsonis, K., Tsonis, P.A., 2009. Expression of stem cell pluripotency factors during regeneration in newts. Dev Dyn 238, 1613–1616.

Maki, N., Takechi, K., Sano, S., Tarui, H., Sasai, Y., Agata, K., 2007. Rapid accumulation of nucleostemin in nucleolus during newt regeneration. Dev Dyn 236, 941–950.

Marshall, L.N., Vivien, C.J., Girardot, F., Pericard, L., Scerbo, P., Palmier, K., Demeneix, B.A., Coen, L., 2019. Stage-dependent cardiac regeneration in Xenopus is regulated by thyroid hormone availability. Proc Natl Acad Sci U S A 116, 3614–3623.

Mercader, N., Leonardo, E., Piedra, M.E., Martinez, A.C., Ros, M.A., Torres, M., 2000. Opposing RA and FGF signals control proximodistal vertebrate limb development through regulation of Meis genes. Development 127, 3961–3970.

Mercader, N., Tanaka, E.M., Torres, M., 2005. Proximodistal identity during vertebrate limb regeneration is regulated by Meis homeodomain proteins. Development 132, 4131–4142.

Mescher, A.L., 1976. Effects on adult newt limb regeneration of partial and complete skin flaps over the amputation surface. J Exp Zool 195, 117–128.

Mizuno, N., Mochii, M., Yamamoto, T.S., Takahashi, T.C., Eguchi, G., Okada, T.S., 1999. Pax-6 and Prox 1 expression during lens regeneration from Cynops iris and Xenopus cornea: evidence for a genetic program common to embryonic lens development. Differentiation 65, 141–149.

Monaghan, J.R., Maden, M., 2012. Visualization of retinoic acid signaling in transgenic axolotls during limb development and regeneration. Dev Biol 368, 63–75.

Morrison, J.I., Loof, S., He, P., Simon, A., 2006. Salamander limb regeneration involves the activation of a multipotent skeletal muscle satellite cell population. J Cell Biol 172, 433–440.

Mullen, L.M., Bryant, S.V., Torok, M.A., Blumberg, B., Gardiner, D.M., 1996. Nerve dependency of regeneration: the role of Distal-less and FGF signaling in amphibian limb regeneration. Development 122, 3487–3497.

Muneoka, K., Fox, W.F., Bryant, S.V., 1986a. Cellular contribution from dermis and cartilage to the regenerating limb blastema in axolotls. Dev Biol 116, 256–260.

Muneoka, K., Holler-Dinsmore, G., Bryant, S.V., 1986b. Intrinsic control of regenerative loss in Xenopus laevis limbs. J Exp Zool 240, 47–54.

Nacu, E., Glausch, M., Le, H.Q., Damanik, F.F., Schuez, M., Knapp, D., Khattak, S., Richter, T., Tanaka, E.M., 2013. Connective tissue cells, but not muscle cells, are involved in establishing the proximo-distal outcome of limb regeneration in the axolotl. Development 140, 513–518.

Nacu, E., Gromberg, E., Oliveira, C.R., Drechsel, D., Tanaka, E.M., 2016. FGF8 and SHH substitute for anterior-posterior tissue interactions to induce limb regeneration. Nature 533, 407–410.

Niethammer, P., Grabher, C., Look, A.T., Mitchison, T.J., 2009. A tissue-scale gradient of hydrogen peroxide mediates rapid wound detection in zebrafish. Nature 459, 996–999.

Reyer, R.W., Woolfitt, R.A., Withersty, L.T., 1973. Stimulation of lens regeneration from the newt dorsal iris when implanted into the blastema of the regenerating limb. Dev Biol 32, 258–281.

Rinkevich, Y., Lindau, P., Ueno, H., Longaker, M.T., Weissman, I.L., 2011. Germ-layer and lineage-restricted stem/progenitors regenerate the mouse digit tip. Nature 476, 409–413.

Roensch, K., Tazaki, A., Chara, O., Tanaka, E.M., 2013. Progressive specification rather than intercalation of segments during limb regeneration. Science 342, 1375–1379.

Sandoval-Guzman, T., Wang, H., Khattak, S., Schuez, M., Roensch, K., Nacu, E., Tazaki, A., Joven, A., Tanaka, E.M., Simon, A., 2014. Fundamental differences in dedifferentiation and stem cell recruitment during skeletal muscle regeneration in two salamander species. Cell Stem Cell 14, 174–187.

Satoh, A., Cummings, G.M., Bryant, S.V., Gardiner, D.M., 2010. Neurotrophic regulation of fibroblast dedifferentiation during limb skeletal regeneration in the axolotl (Ambystoma mexicanum). Dev Biol 337, 444–457.

Satoh, A., Gardiner, D.M., Bryant, S.V., Endo, T., 2007. Nerve-induced ectopic limb blastemas in the Axolotl are equivalent to amputation-induced blastemas. Dev Biol 312, 231–244.

Satoh, A., Graham, G.M., Bryant, S.V., Gardiner, D.M., 2008. Neurotrophic regulation of epidermal dedifferentiation during wound healing and limb regeneration in the axolotl (Ambystoma mexicanum). Dev Biol 319, 321–335.

Seifert, A.W., Monaghan, J.R., Smith, M.D., Pasch, B., Stier, A.C., Michonneau, F., Maden, M., 2012. The influence of fundamental traits on mechanisms controlling appendage regeneration. Biol Rev Camb Philos Soc 87, 330–345.

Seifert, A.W., Voss, S.R., 2013. Revisiting the relationship between regenerative ability and aging. BMC Biol 11, 2.

Sessions, S.K., Bryant, S.V., 1988. Evidence that regenerative ability is an intrinsic property of limb cells in Xenopus. J Exp Zool 247, 39–44.

Singh, S.P., Holdway, J.E., Poss, K.D., 2012. Regeneration of amputated zebrafish fin rays from de novo osteoblasts. Dev Cell 22, 879–886.

Sousa, S., Afonso, N., Bensimon-Brito, A., Fonseca, M., Simoes, M., Leon, J., Roehl, H., Cancela, M.L., Jacinto, A., 2011. Differentiated skeletal cells contribute to blastema formation during zebrafish fin regeneration. Development 138, 3897–3905.

Stewart, S., Stankunas, K., 2012. Limited dedifferentiation provides replacement tissue during zebrafish fin regeneration. Dev Biol 365, 339–349.

Stocum, D.L., 1968. The urodele limb regeneration blastema: a self-organizing system. I. Morphogenesis and differentiation of autografted whole and fractional blastemas. Dev Biol 18, 457–480.

Suetsugu-Maki, R., Maki, N., Nakamura, K., Sumanas, S., Zhu, J., Del Rio-Tsonis, K., Tsonis, P.A., 2012. Lens regeneration in axolotl: new evidence of developmental plasticity. BMC Biol 10, 103.

Sun, A.X., Londono, R., Hudnall, M.L., Tuan, R.S., Lozito, T.P., 2018. Differences in neural stem cell identity and differentiation capacity drive divergent regenerative outcomes in lizards and salamanders. Proc Natl Acad Sci U S A 115, E8256-E8265.

Tseng, A.S., Adams, D.S., Qiu, D., Koustubhan, P., Levin, M., 2007. Apoptosis is required during early stages of tail regeneration in Xenopus laevis. Dev Biol 301, 62–69.

Tu, S., Johnson, S.L., 2011. Fate restriction in the growing and regenerating zebrafish fin. Dev Cell 20, 725–732.

Wallace, B.M., Wallace, H., 1973. Participation of grafted nerves in amphibian limb regeneration. J Embryol Exp Morphol 29, 559–570.

Wang, L., Marchionni, M.A., Tassava, R.A., 2000. Cloning and neuronal expression of a type III newt neuregulin and rescue of denervated, nerve-dependent newt limb blastemas by rhGGF2. J Neurobiol 43, 150–158.

Wigmore, P., Holder, N., 1986. The effect of replacing different regions of limb skin with head skin on regeneration in the axolotl. J Embryol Exp Morphol 98, 237–249.

Yakushiji, N., Suzuki, M., Satoh, A., Sagai, T., Shiroishi, T., Kobayashi, H., Sasaki, H., Ide, H., Tamura, K., 2007. Correlation between Shh expression and DNA methylation status of the limb-specific Shh enhancer region during limb regeneration in amphibians. Dev Biol 312, 171–182.

Yntema, C.L., 1959a. Blastema formation in sparsely innervated and aneurogenic forelimbs of amblystoma larvae. J Exp Zool 142, 423–439.

Yntema, C.L., 1959b. Regeneration in sparsely innervated and aneurogenic forelimbs of Amblystoma larvae. J Exp Zool 140, 101–123.

Yokoyama, H., Yonei-Tamura, S., Endo, T., Izpisua Belmonte, J.C., Tamura, K., Ide, H., 2000. Mesenchyme with fgf-10 expression is responsible for regenerative capacity in Xenopus limb buds. Dev Biol 219, 18–29.

Yoshino, J., Tochinai, S., 2004. Successful reconstitution of the non-regenerating adult telencephalon by cell transplantation in Xenopus laevis. Dev Growth Differ 46, 523–534.

Yoshino, J., Tochinai, S., 2006. Functional regeneration of the olfactory bulb requires reconnection to the olfactory nerve in Xenopus larvae. Dev Growth Differ 48, 15–24.

Zeller, R., Lopez-Rios, J., Zuniga, A., 2009. Vertebrate limb bud development: moving towards integrative analysis of organogenesis. Nat Rev Genet 10, 845–858.

11

Stem Cells and Regeneration in Plants

Marco Da Costa[1,2,*] and Philippe Rech[1,3,*]

[1]Institut Jean-Pierre Bourgin, INRAE, AgroParisTech, Université
Paris-Saclay, 78000, Versailles, France
[2]Sorbonne Université, Centre National de la Recherche Scientifique (CNRS),
Institut de Biologie Paris-Seine (IBPS), Laboratoire de Biologie du
Développement (LBD), UMR7622, Paris, France
[3]Institut de Systématique, Evolution, Biodiversité (ISYEB – UMR 7205 –
CNRS, MNHN, SU, EPHE), Muséum national d'Histoire naturelle,
57 rue Cuvier, 75005 Paris, France
E-mail: marco.da_costa@sorbonne-universite.fr;
philippe.rech@sorbonne-universite.fr
*Corresponding Authors

11.1 Introduction

Terrestrial plants have colonized a remarkably wide range of different ecosystems despite being sessile. One key to this evolutionary success is their tremendous ability to adapt to the local environment. This developmental and morphological plasticity is based on the life-long production of new repetitive tissue modules, whose shape and physiology can be modulated efficiently by developmental and environmental signals (de Jong and Leyser, 2012).

Unlike animals in which organogenesis occurs mainly during embryogenesis, plants have to produce most of their organs post-embryonically and can grow and give rise to new organs throughout their life (Heidstra and Sabatini, 2014). This unique developmental ability is due to the existence of small groups of pluripotent and undifferentiated stem cells located in proliferative growth centers at the growing apices of plants, the meristems. There are two main meristems in plants: the shoot apical meristem (SAM), located at the shoot apex, and the root apical meristem (RAM), located at the root tip, giving birth to above-ground tissues and organs and the underground root system, respectively (Figure 11.1) (Greb and Lohmann, 2016). Both the SAM and the RAM are primary meristems and are established early in embryogenesis and maintained throughout postembryonic life.

Although animal and plant stem cells differ greatly in terms of cell morphology and differentiation patterns, they share several common features including similarities in their maintenance mechanisms (Heidstra and Sabatini, 2014; Shi et al., 2017). The maintenance of stem cell activity requires the existence of a specialized microenvironment known as the stem cell niche (SCN) (Heidstra and Sabatini, 2014). A stem cell niche is composed of a group of stem cells and regulators cells. Stem cells are maintained in an undifferentiated state by short-range signals sent by regulators cells. The rate of division of a stem cell is very low, but each time one stem cell divides, one of the daughter cells escape from the zone of influence of the regulator cells. The escaping cell will then divide a lot and enter into differentiation to form new tissue or organs. The defining attributes of self-renewal and tissue generation limit stem cells to those positions that remain in place to produce long-term lineages (Laux, 2003).

11.2 SAM Organization and Regulation

Similarly to the well-characterized SCN found in Drosophila, for example in the model plant *Arabidopsis thaliana*, the SAM is a multicellular hemispherical dome-shaped structure organized into several clonally distinct cell layers and, based on gene expression patterns and signaling activities, subdivided into three different zones or functional domains (Bustamante et al., 2016).

The SAM harbors and maintains a reservoir of stem cells in the central zone (CZ), constituted by three well-defined cell layers, L1 to L3. Underneath the CZ lies the organizing center (OC), a small group of cells that regulates the stem cell population above it and forms with the CZ, the SCN (Carles and Fletcher, 2003). The homeostasis of this population of cells is essential to maintain a stable meristem. Cells located in the CZ divide at a slow rate and with their division, the daughter cells are moved outward and displaced laterally into the peripheral zone (PZ), a ring of cells that surrounds the CZ and that is characterized by a high mitotic activity, or underneath into the rib zone (RZ). The RZ provides multipotent cells responsible for the development of the internal tissues of the stem whereas cells in the PZ undergo sequential differentiation gaining distinct identities and giving rise to organ *primordia* such as leaves and flowers (Figure 11.1A) (Shi and Vernoux, 2019).

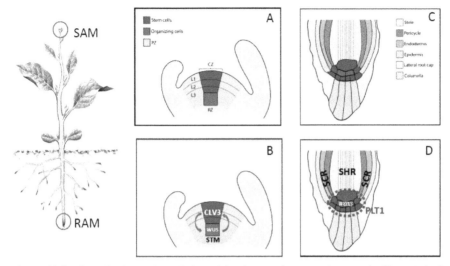

Figure 11.1 Organization and regulation of the shoot and root meristems. (**A**) Shoot Apical Meristem. The central zone (CZ) of the shoot apical meristem contains the organizing center (**red**) and the stem cells (**blue**), which generate daughter cells that divide while traversing the meristem peripheral zone (PZ) before differentiation and incorporation into lateral organs such as leaves. The rib zone (RZ) will give rise to the stem of the plant. Lines indicate lineage relationship in the different tissues with anticlinally dividing stem cells in layer 1 (L1) and L2. Stem cells in L3 divide into all directions and their progeny are incorporated into the growing stem. (**B**) Maintenance of the stem cell niche in the SAM. WUS, CLV3, and STM are transcription factors described in section SAM organization and regulation, Sections 11.2.1 and 11.2.3, hereafter. (**C**) Root Apical Meristem. The root meristem is comprised of well-arranged tissue layers. The quiescent center (**red**) is surrounded by tissue-specific stem cells (**blue**), the progeny of which sustain root growth. Lines indicate the lineage relationship of stem cells and their progeny. (**D**) Maintenance of the stem cell niche in the RAM. WOX5, PLT1, SCR, and SHR are transcription factors described in section RAM organization and regulation, Section 11.2.1 hereafter.

Whereas a majority of our current knowledge is derived from studies in Arabidopsis, genes related to the maintenance and organization of the SAM seem to be highly conserved among different plant species (Fletcher, 2018). However, our knowledge remains partial, mainly because of the complex dynamic nature of the SAM. As cells are displaced through the different functional domains, cells have to adjust their fate or identity accordingly to their position relative to the neighboring cells. Inversely, acquisition of specific identities will also participate to the maintenance and structuration of the meristem (Shi and Vernoux, 2019). Consistently, the dynamic nature of the SAM depends strongly on intercellular communications that coordinate the development and maintenance of each functional domain. As for animal, morphogenic substances that form gradients, such as hormones, intercellular movement of proteins/transcription factors (TF) and peptides or microRNAs, are the key pilots instructing qualitatively the transcriptional responses and *per se* the different cellular state (Hofhuis and Heidstra, 2018; Pierre-Jerome et al., 2018; Shi and Vernoux, 2019; Uchida and Torii, 2019).

11.2.1 The Core WUS/CLV3 Loop

Within the established SAM, the stem cell population is stably maintained throughout a negative feedback loop between the homeodomain TF WUSCHEL (WUS) and a founding member of the CLAVATA3/ESR-RELATED (CLE) family of small peptides, CLV3 and a mechanism that involves communication between the OC and the CZ. *WUS* mRNA is exclusively expressed and translated in the OC of the SCN. From the OC, the WUS protein migrates *via* cytoplasmic bridges, the plasmodesmata, to regulate non-cell autonomously the transcription of downstream target genes involved in biological processes related to meristem growth, cell division, and hormonal signaling (Aichinger et al., 2012; Hofhuis and Heidstra, 2018).

In particular, WUS is required to maintain the undifferentiated identity and the pluripotency state of stem cells of all three layers of the CZ by preventing notably their differentiation (Yadav et al., 2013).

WUS movement from the OC into the above stem cells forms a concentration gradient in which a low concentration of WUS proteins directly activates expression of *CLV3* gene, while high concentration in the OC acts as a transcriptional repressor of *CLV3* transcription (Perales et al., 2016). This dual function of WUS protein could allow formation of sharp threshold dependent "on-off" boundaries (Hofhuis and Heidstra, 2018). Additionally, it was revealed recently that WUS cooperatively interacts with the GRAS-domain HAIRY MERISTEM (HAM) TFs to inhibit *CLV3* expression in the lower part of the OC explaining the exclusive expression pattern of *CLV3* in the uppermost layer of the SAM despite the presence of WUS protein in cells all around the OC (Figure 11.1B) (Zhou et al., 2018).

In stem cells, CLV3 peptide is then secreted to the extracellular space where it binds in part to the leucine-rich repeat receptor kinases complexes CLV1/CLV2/CORYNE that restrict the spatial expression domain of *WUS* and therefore the stem cell fate (Biedermann and Laux, 2018; Pierre-Jerome et al., 2018). This negative feedback loop between stem cells and the OC ensures stem cell homeostasis in the SAM and provides a model of how the number of stem cells is dynamically controlled.

11.2.2 Role of Hormones in SAM Regulation

Beside this fundamental duo, the coordinated behavior of stem cells, layers, and domains of the SAM is due to the integration of a multitude of interconnected signaling pathways and involves a multilayered regulation by multiple factors (Janocha and Lohmann, 2018; Pierre-Jerome et al., 2018; Uchida and Torii, 2019). All of these molecular pathways, even if their connection to the core WUS/CLV3 circuit continue to be highly elusive, lead to a robust homeostatic mechanism that balances stem cell numbers and differentiation rates in a dynamic cellular environment (Shi and Vernoux, 2019).

Plant hormone signaling pathways and most notably cytokinin (CK) and auxin, are the main driving actors of a strong interregional communications required for the coordination of the fine-tuning of the meristem maintenance and for the developmental decisions with appropriate cell behavior (Janocha and Lohmann, 2018). Cytokinin inhibits differentiation throughout the SAM and promotes stem cell proliferation in the center, while high concentration of auxin promotes organ formation, growth, and differentiation in the periphery. Consequently, stem cells in the center of the SAM present weak concentration of auxin (Vernoux et al., 2011) and inversely, cytokinin-dependent gene activity is mostly confined to the OC (Gordon et al., 2009).

11.2.2.1 Cytokinin

CK is thought to be important for SAM development as it activates *WUS* expression and leads indirectly to the specification of stem cell fate (Gordon et al., 2009). The production site of active cytokinin in the SAM is presumed to be the epidermal L1 layer because the expression of *LONELY GUY* gene which encodes an enzyme catalyzing the final step of cytokinin biosynthesis, is restricted to the L1 (Chickarmane et al., 2012). But the expression of the Cytokinin ARABIDOPSIS HISTIDINE KINASE (AHK) receptors genes is in inner tissues including the OC, but excluded from the L1 and L2 layers (Chickarmane et al., 2012). Thus, the active cytokinin molecules produced in the L1 layer move basipetally and activate primary cytokinin responses in the OC. Upon sensing cytokinin, AHK receptors trigger a multistep phosphorelay to ultimately activate Type-B Arabidopsis Response Regulators (ARRs) TFs. These transcriptional activators directly bind to *WUS* promoter to induce *WUS* transcription in the deeper layer of the SAM (Xie et al., 2018). Furthermore, it has been shown that the WUS protein directly represses expression of type-*A ARRs* genes encoding CK signaling repressors, resulting in enhanced cytokinin responses sensitivity in the OC and in enhanced WUS function in the SAM (Zhao et al., 2010). Beside their role in transcriptional activation of *WUS*, cytokinin signaling is also known to stabilize the WUS protein. As activation of cytokinin signaling occurs in the OC but is excluded from the L1 and L2 layers, WUS protein level decreases by protein destabilization according to the distance from the OC (Snipes et al., 2018). In addition to the CLV3 pathway, CK appears to be another feedback loop controlling the spatial distribution pattern of WUS protein and the stem cell activity in the SAM (Snipes et al., 2018).

11.2.2.2 Auxin

In contrast to CK, a classic role of auxin is to specify organ *primordium* fate in the PZ during SAM development. Although these two hormones seem to be antagonist, recent studies have revealed their synergistic interactions in the regulation of the stem cell activity and SAM development (Lee et al., 2019; Zhao et al., 2010).

Among the 22 Auxin Response TFs (ARF) present in Arabidopsis, ARF5, also named Monopteros (MP) plays a dominant role in specifying meristematic and *primordium* fate by orchestrating gene expression (Lee et al., 2019). In the absence of auxin, ARF5/MP dimerizes with AUXIN/INDOLE-3-ACETIC ACID (Aux/IAA) proteins that repress auxin signaling. Upon sensing auxin on TRANSPORT INHIBITOR RESISTANT1 (TIR1) receptors localized in the nucleus, Aux/IAA repressors are degraded through the action of SCFTIR1 ubiquitin ligase. ARF5/MP activity is subsequently derepressed and triggers the expression of many auxin response genes. ARF5/MP proteins are present at low levels in the CZ where auxin concentration is weak and at high levels in the PZ of the SAM with a higher auxin concentration (Zhao et al., 2010). Noteworthy, the maintenance of a weak concentration of auxin in the SCN depends also on lateral organs that provide a positive feedback to the stem cells fate through an auxin transport mechanism (Shi et al., 2018). As in animals, stem cell proliferation and meristem homeostasis seem to also be controlled by a feedback regulation from differentiating progeny cells (Shi and Vernoux, 2019).

Consistent with this accumulation pattern, ARF5/MP modulates different downstream target genes in the CZ and PZ. In the PZ, ARF5/MP activates AHP6, a dominant negative component of the CK signal transduction pathway, thereby establishing inhibitory fields of cytokinin that allow organ initiation (Besnard et al., 2014). This cellular fate transition is also modulated by the activity of the basic helix-loop-helix (bHLH) transcription factor HECATE1 (HEC1) which integrates environmental signals to control the auxin–cytokinin hormonal balance (Gaillochet et al., 2017).

In the CZ, ARF5/MP directly represses type-*A ARRs* genes encoding CK signaling inhibitors, reinforcing cytokinin signaling, WUS activity and the meristematic fate (Zhao et al., 2010). ARF5/MP is also known to repress the expression of DORNRÖSCHEN (DRN/ESR1), a CLV3 expression activator, strengthening the role of auxin in the regulation of stem cells (Luo et al., 2018). Interestingly, WUS itself behave as a rheostat allowing instructive low levels of auxin signaling in the stem cells required for their maintenance and in the same time render them unresponsive to high levels needed for cell differentiation (Ma et al., 2019).

11.2.3 Other SAM Regulators

Beside the essential role of hormones, multiple other signaling pathways are integrated to adjust the proliferative activity of the stem cells in the SAM. Among them, microRNAs, such as miR394 and miR165/166, act as mobile molecules to form gradient through the SAM layers and tissues to potentiate respectively the role of WUS and HD-ZIPIII TFs, which are known to regulate stem cell activity (Uchida and Torii, 2019). Additionally, there is now strong evidence that the mechanical forces generated by the acquisition of the complex and dynamic three-dimensional shape of the SAM in turn provide cues for the patterning of specific domains of cell identities (Landrein and Ingram, 2019). These cues pattern the expression domain of a key regulator of the meristematic cell identity, SHOOT MERISTEMLESS TF (STM) in the center of the SAM and at the same time control the expression of a specific set of TFs, cell division orientation and hormone distribution needed to allow the separation of the emerging organ from the rest of the meristem.

In addition to these few signaling pathways described above, the proliferative activity of the stem cells and finally the developmental decisions are also adjusted in perfect accordance with the local environment of the plant. Consistently, multiple environmental inputs and feedbacks, such as changes in light and response to nutritional status, are now known to impact the level of cytokinin signaling in the OC and stem cell activity (Janocha and Lohmann, 2018). The evolutionary conserved core regulator, TARGET OF RAPAMYCIN (TOR) kinase complex plays also an important role in integrating environmental signaling and energy-rich metabolites signaling to the transcriptional regulation of *WUS* and stem cell function (Janocha and Lohmann, 2018). Finally, reactive oxygen species (ROS) might also be a link between environmental cues and the balance between cell proliferation and differentiation in the SAM (Wany et al., 2018). Enzymes of the ROS pathway are differently expressed in the spatial domains within the SAM, generating a high accumulation of O_2 in the central zone to maintain stemness by regulating *WUS* and *CLV3* expression whereas H_2O_2 is inversely accumulated in the peripheral zone to repress *WUS* and promotes stem cell differentiation (Yang et al., 2018).

11.3 RAM Organization and Regulation

Higher vascular land plants have successfully colonized the terrestrial environment *via* the evolution of root systems that mainly penetrate substrates, anchor plants, and uptake water and nutrients necessary for plant growth (Du and Scheres, 2018). Root systems consist of roots derived from the embryo but also roots arising from post-embryonic existing roots, the lateral roots (LR), or from non-root tissues, the adventitious roots (Du and Scheres, 2018). LR branching from the parent root is crucial for maximizing the function of the root system.

The RAM can be divided into three main zones: (a) the meristematic zone at the root tip has a high mitotic activity and contains the SCN, (b) the elongation zone, containing the cells that after cell divisions have left the meristematic zone and are now elongating, and (c) the differentiation zone, containing cells that have acquired their destined cell fates (Drisch and Stahl, 2015). In the meristematic zone, stem cells are arranged around the quiescent center (QC), a group of less mitotically active cells, that may serve as a central signaling organizer that maintains and positions the stem cells (Figure 11.1C) (Dinneny and Benfey, 2008). The rarely dividing QC cells are thought to be less stress sensitive and protected from DNA damage, two essential properties that can therefore maintain their longevity. Thus, the QC could act as a long-term reservoir to replace the surrounding stem cells if stress is occurring (Fulcher and Sablowski, 2009). The layer arrangement of stem cell niches within the RAM appears to be in sharp contrast with the situation in the SAM, because only cells in contact with the QC act as stem cells (Uchida and Torii, 2019).

The stem cells continuously divide asymmetrically, generating new stem cells still in contact with the QC and daughter cells that undergo several transit-amplifying divisions before rapid elongation and differentiation (Rahni et al., 2016). The position of the stem cells remains the same throughout development and each stem cell contributes to the production of only one or two specific cell types, resulting in the formation of columns of specialized cell files. Thus, stem cells in the RAM act as "lineage-specific" stem cells and consists of four types, namely, columella stem cell (CSC), epidermal and lateral root cap stem cell, cortex/endodermis stem cell, and stele stem cell generating for the last ones the vascular tissues (Scheres et al., 1994). The columella stem cells (CSCs) give rise to the differentiated columella cells which contain starch granules needed for graviperception (Drisch and Stahl, 2015).

11.3.1 RAM Master Regulators

Transcription factors combined with phytohormones, small signaling molecules, and miRNAs play also an essential role in regulating the stem cell fate and RAM maintenance during root development (Shimotohno and Scheres, 2019).

The central role that *WUS* plays in the maintenance of stem cell fate and SAM functioning is mirrored in root meristems by *WUSCHEL-RELATED HOMEOBOX5 (WOX5)* gene. *WOX5* is expressed in the quiescent center and contribute to quiescence in the root by preventing cell divisions in the QC (Figure 11.1D) (Forzani et al., 2014). Like WUS, WOX5 appears to move certainly *via* plasmodesmata from the QC domain to the neighboring stem cell and particularly in the CSC to maintain them in an undifferentiated state *via* a chromatin-mediated repression of their differentiation (Pi et al., 2015). Interestingly, WOX5 and WUS are capable of functioning interchangeably in the control of shoot and root SCNs (Sarkar et al., 2007). In the CSCs, WOX5 recruits protein members including HISTONE DEACETYLASE 19 (HDA19) to induce histone deacetylation and repress consequently the differentiation factor CYCLING DOF FACTOR 4 (CDF4). As for its paralog in SAM, the expression pattern of *WOX5* depends on short-range signals mediated by CLE peptides (Stahl et al., 2013). The small peptide *CLE40* is expressed from differentiated columella cells and regulates CSC fate *via* the receptor-like kinases ARABIDOPSIS CRINKLY4 (ACR4) and CLV1 by restricting the expression domain of *WOX5* to the QC. As WOX5 protein content decreased from the QC to the differentiated columella cells, an opposite gradient of CDF4 protein is formed allowing the exit of stem cell descendants from the stem cell state (Pi et al., 2015). Similar to their roles in the SAM, the HAM members were shown to interact with WOX5 to act as mediators to maintain the indeterminate activity of the root meristem (Zhou et al., 2015).

Aside from these minor similarities, the SAM and the RAM differ broadly in terms of molecular mechanisms. The AP2/ERF TF family, PLETHORA (PLT) proteins, are distributed in a concentration gradient that encompass the meristem with highest levels in the SCN where they regulate the specification of QC and the activity of surrounding stem cells (Figure 11.1D) (Santuari et al., 2016). The PLT protein gradient is also distributed according to the distance from the QC to form a developmentally instructive guide to regulate the transitions from stem cell to differentiated cell (Santuari et al., 2016). The CLE-like small peptides (CLEL) (also named RGFs) positively regulate and define *PLT* expression and protein stability (Matsuzaki et al., 2010). In parallel to PLT function, the GRAS-transcription factors SHORTROOT (SHR) and SCARECROW (SCR) are also required for the SCN specification and homeostasis (Figure 11.1D) (Aida et al., 2004). Besides WOX5 and PLT, SHR is a prominent example of a mobile TF that is expressed in the stele of the *Arabidopsis* root, but moves *via* plasmodesmata, one layer further into the endodermis, into the cortex/endodermis stem cells, and into the QC cells where it activates *SCR* expression (Gallagher et al., 2004). These TFs are known to co-operate in the regulation of *WOX5* (Sarkar et al., 2007), and have also been found to physically bind among others to cell cycle regulators to define the position of the asymmetric cell divisions in the cortex/endodermis stem cell necessary for the formation of distinct cortex and endodermal cell layers (Long et al., 2015). The QC activity and the stem cell behaviors are then tightly controlled by complex regulatory networks that depend on signaling gradients with feedback control.

As for SAM development, root development is also dependent on the local environment. The nitrate is an essential nutrient resource for growth and development, and its availability is directly connected to the QC regulatory PLT–SCR complex to adjust root foraging (Shimotohno and Scheres, 2019). The distribution of ROS gradient follows a complex pattern of redox potentials in the RAM. The QC and the adjacent cells show the most reduced potentials, while the redox potentials in the differentiation zone are being more oxidized. This differential ROS distribution controls root SCN maintenance by regulating PLTs and SCR stem cell regulators and establishes the boundary between cell proliferation and differentiation (Yang et al., 2018).

11.3.2 Role of Hormones in RAM Regulation

Considerable crosstalks between hormonal pathways are necessary for integrating external and internal cues into the dynamic developmental processes of the root stem cell maintenance, proliferation, and differentiation. Several TFs have been shown to be regulated by and act in concert with them. Auxin plays a dominant role in root initiation and development by acting as a dose-dependent signal. An auxin gradient, build up from the QC and the surrounding root stem cells by local biosynthesis and auxin transport, regulates zonation and differentiation of root cells in a complex interplay with the PLT proteins (Drisch and Stahl, 2015). In line with this, the expression of *PLT* is auxin inducible (Mähönen et al., 2014) and in turn PLT has a positive feedback action on auxin signaling through transcriptional regulation of major auxin response factors, such as ARF5/MP, but also on auxin biosynthesis enzymes and on auxin transporters (Santuari et al., 2016). *WOX5* expression was also reported to be auxin inducible and responsible for the establishment of an auxin maximum in the root tip (Gonzali et al., 2005).

Inversely, cytokinins play a pivotal role in root meristem balance and act antagonistically to auxin. CKs control the switch from meristematic to differentiated cell fates by suppressing auxin signaling and transport where cells leave the meristematic zone. This is mediated by the repressor AUX/IAA SHORTHYPOCOTYL2 (SHY2), which is activated by cytokinin *via* type-B ARR1 (Dello Ioio et al., 2008). In the QC, SCR directly suppresses *ARR1* expression and indirectly titrates auxin production, which also contributes to auxin accumulation in QC (Moubayidin et al., 2013). Furthermore, AHK-mediated cytokinin signaling negatively regulate *WOX5* expression possibly by modulating the auxin flux in the root and *per se* promotes QC differentiation (Zhang et al., 2013).

11.3.3 LR Formation

LR branching from the parent root is crucial for maximizing the function of the root system. LRs are usually derived from the single-layered pericycle cells that surrounds the central vasculature tissues of parent roots, notably from xylem pole-pericycle (XPP) cells (Du and Scheres, 2018). Because of this characteristic, only XPP cells are thought to be "semi-meristematic" (Parizot et al., 2008). LR organogenesis starts with the specification of a subset of competent XPP cells (Du and Scheres, 2018). In specified XPP cells, an auxin maximum is created and maintained which activates, through the auxin signaling pathway, a de-repression of ARFs transcription factors that are essential for LR initiation. Then, from the XPP founder cells, subsequent rounds of cell divisions are launched to establish a dome-shaped *primordium* that grows through the overlaying tissues of the parent root, including the endodermis, cortex, and epidermis and finally emerge as a LR (Du and Scheres, 2018). These emerged LRs possess a fully functional meristem that is highly reminiscent of the primary root meristem, in which the QC and the stem cells surrounding it form the core of the LR meristem. However, a functional meristem is already formed in sub-emergent *primordia* constituted by a minimum of 3 to 5 cell layers, a stage which precedes the recognizable meristem architecture (Laskowski et al., 2008). At this developmental phase, *WOX5* expression is induced and marks the occurrence of a new SCN (Goh et al., 2016; Rosspopoff et al., 2017). Its establishment is tightly associated with a proper formation of an auxin maximum in the *primordium* and with the activity of PLTs and SCR, whose expression precede *WOX5* (Du and Scheres, 2017).

11.4 SAM Regeneration

Since their divergence with animals 1.6 billion years ago, plants live a sessile lifestyle and are exposed to severe environmental threats (Meyerowitz, 2002). They acquired several abilities to survive by developing regeneration capacities to a high degree of complexity to minimize the loss from damages (Birnbaum and Sánchez Alvarado, 2008). Thus, plants are able to regenerate cells, tissues, or even a complete organ (Sugimoto et al., 2019). Upon hormonal induction through *in vitro* culture, plants may also reconstitute whole organs from explants (*de novo* organ regeneration) or fully restore individuals from a set of few highly regenerative cells (Sugimoto et al., 2019). Even though the ability to regenerate varies from one species to the other, in these cases, the regeneration process relies largely on the ability for *de novo* specification of new stem cell niches and establishment of *de novo* apical meristems.

11.4.1 Tip Regeneration

To illustrate the extremely plastic cellular state of meristematic cells and their ability to molecularly interchange cell fates to rebuild or to form a new SCN (Perez-Garcia and Moreno-Risueno, 2018), ablation experiments were realized *in planta* on functional meristems.

The complete removal of the QC and all surrounding stem cells led spontaneously to the re-construction from the remaining cells of a functional root tip containing re-established stem cell niches through a process largely piloted by the progressive spatial patterning of auxin and cytokinin signals (Figure 11.2A) (Efroni et al., 2016). Despite the fact that they have been specified or have a pre-assigned identity, the remaining cells within the meristem are reprogrammed to other cell fates, by using a regeneration process

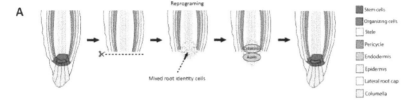

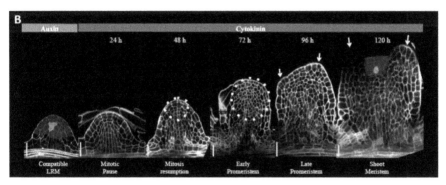

Figure 11.2   Schematic representation of regeneration systems in plants. **A.** Root tip regeneration in Arabidopsis. Sequential cellular events involved in root-to-shoot direct conversion. In an Arabidopsis compatible lateral root meristem exposed to cytokinin, cells first stop dividing (24 hours), then resume mitotic activity (48 hours and on; white-dashed circles) to form a SM from which leaf *primordia* are quickly bulging out (96 hours and 120 hours; *white arrows*). **B.** Cells highlighted in red and in yellow mark the meristems and clusters prefiguring the quiescent or organizing centers respectively.

that recapitulates the embryonic developmental pathway (Efroni et al., 2016). This competency shows their pluri- or multipotency characters and their wide potential to adopt stem cell behavior (Sugimoto et al., 2019). However, not all switches in cell fate are possible because once meristematic cells undergo differentiation, cell fate transitions are not possible (Efroni et al., 2016).

When localized damages or stresses are applied to the root stem cells, the QC divides rapidly to replace dead cells, an inverse situation to postembryonic development during which QC is maintained in a slow division rate. During the regeneration of the damaged stem cells, the brassinosteroid hormone signalization pathway, another class of phytohormone, promotes expression of ETHYLENE RESPONSE FACTOR 115 (ERF115) TF which transduces the BR-positive effect on QC cell divisions (Heyman et al., 2013). In contrast, upon damage of the entire QC, a new auxin maximum is recreated in a few cells above the original QC. The new positional information provided by auxin guides the expression of the root stem cells maintenance genes *PLT*, *SHR*, and *SCR* (Xu et al., 2006) that will re-specifies cell fate by establishing a new QC and a new SCN (Xu et al., 2006).

Similarly to the root meristem, SAM integrity is very well preserved and regenerative capacities are also observed from this self-organizing tissue assembly in the adult plant. Specific ablation of the stem cells in the SAM clearly showed re-specification into stem cell identity from cells located in the PZ. Cells in the PZ are then able to retain a potency to be re-specified, to rapidly re-establish a new SCN and finally to regenerate a functional meristem (Reinhardt et al., 2003). Consistent with the re-specification, *CLV3*-expressing cells appeared in the PZ after the ablation, following the re-establishment of *WUS* expression (Reinhardt et al., 2003). However, while in the RAM a whole new meristem may regenerate after removal, complete removal of SAM activates axillary meristems (from where new branches will develop) instead of activating a regenerative process (Reinhardt et al., 2003). So far, there is no clear explanation to this divergence (Perez-Garcia and Moreno-Risueno, 2018).

11.4.2 *De novo* Regeneration

In the last century, the pioneering works of Skoog and Miller paved the foundation of *in vitro* culture of *de novo* organogenesis in various plant species, which led to a breakthrough in facilitating its application in the agricultural industry and biological research (Sang et al., 2018). In this generation of *de novo* structures that are not present in the original explant, the *ratio* between exogenous auxin and cytokinin in culture media determines the type of regenerated organs. Although there are many different types of regeneration in plants that rely on certain tissues or types of cells and implying distinctive biological mechanisms, *de novo* shoot regeneration in Arabidopsis is certainly one of the most investigated and best understood mode of regeneration in plants (Radhakrishnan et al., 2018).

Explants derived from somatic tissues are first incubated in an auxin-rich medium to produce a mass of proliferating cells followed by shoot organogenesis on a second cytokinin-rich medium (Figure 11.2B) (Radhakrishnan et al., 2018). It has been acknowledged that the auxin-induced cells acquired always a pluripotent state through a dedifferentiation process, a mechanism involved in other plant regeneration events (Sugimoto et al., 2019). However, this hypothesis has been extensively revised by studies, including ours, which demonstrate the involvement of mechanisms required for lateral root formation, regardless of the origin of the tissue (Atta et al., 2009; Chatfield et al., 2013; Rosspopoff et al., 2017; Sugimoto et al., 2010). Interestingly, the pericycle-like cells from which the pluripotent lateral roots are formed, are found in most plant organs outside of the root, and have been considered as a reservoir of reprogrammable "adult" stem cells that are broadly located throughout the plant (Kareem et al., 2016).

Then, instead of a dedifferentiation state, a root meristem identity appears to be a prerequisite for the regenerative competency of established lateral root meristem (LRM) or LRM-like *primordia* to generate the SCN of the *de novo* shoot meristem (Atta et al., 2009; Sugimoto et al., 2010). Despite being structures differently organized and requiring different set of regulators, the conversion from a LRM to a SAM appears to be a relatively easy process which occurs after a few rounds of cell division (Figure 11.2C) (Rosspopoff et al., 2017). However, this organ identity switch is only possible in a narrow developmental window in which the LRM is sub-emergent and nearly constituted of root meristem (Rosspopoff et al., 2017). The competency of these sub-emergent LRM (-like) is closely tied to the presence of a root SCN and the SCN maintenance genes (Kareem et al., 2015; Rosspopoff et al., 2017). Consistently, PLTs proteins have a fundamental role to form the pluripotent founder cells during LRM development but are also required for cellular reprogramming toward shoot fate after the cytokinin treatment by activating downstream TF genes known to regulate SAM activity (Kareem et al., 2015). Although it is fairly certain that the root meristem identity is critical for the regenerative competency of LRM cells, the molecular mechanism behind the complex process of acquisition of root identity and the fate conversion of root (-like) to shoot tissue is still largely unknown (Sugimoto et al., 2019).

11.4.3 Key Players During DNSO (*de novo* Shoot Organogenesis)

During the *de novo* shoot formation from LRM (-like), endogenous auxin/CK balance and their crosstalk are crucial in re-specifying the pluripotent root SCNs by providing the positional information that will progressively re-pattern cell identities into a functional shoot meristem (Atta et al., 2009; Cheng et al., 2013; Pierre-Jerome et al., 2018; Rosspopoff et al., 2017).

Firstly, DNSO starts by transferring the competent LRM (-like) into an environment constituted of a high concentration of CKs, known to promote shoot cellular identity and the subsequent shoot development (Skoog and Miller, 1957). Consistently with their role into the SAM maintenance and development, many components of CK biosynthesis, perception and signaling pathways play positive roles in DNSO, by promoting expression of shoot stem cell regulator factors like *WUS* but also by inhibiting expression of the root ones, as *WOX5* (Shin et al., 2019). Then, concomitantly to the transfer on the CK-rich medium, the root stem cell regulators such as *PLTs*, *SCR*, and *WOX5* are rapidly downregulated and the root meristem identity is lost while the shoot fate is simultaneously initiated (Atta et al., 2009; Radhakrishnan et al., 2018; Rosspopoff et al., 2017). Combined with other evidences, the SM initiation in this conversion process may occur through a transdifferentiation process, i.e., the direct transition of a root SCN into a shoot one (Rosspopoff et al., 2017), a case never observed in animals.

During the initial stages of the transition, the expression of shoot-specific genes is not confined to specific domains (Atta et al., 2009; Kareem et al., 2015; Meng et al., 2017; Rosspopoff et al., 2017), leaving a narrow developmental window through which a defined root or shoot identity cannot be assigned (Kareem et al., 2015; Radhakrishnan et al., 2018; Rosspopoff et al., 2017). In this transient phase, the organogenetic programs are remarkably plastic as the same SCN can be reversed repeatedly to grow as root- or shoot-specific tissues by alternating cytokinin and auxin treatments (Rosspopoff et al., 2017).

During the progressive establishment of the future shoot meristem, auxin and cytokinin response signals gradually form a mutually exclusive pattern providing positional information for specifying shoot SCN (Cheng et al., 2013; Meng et al., 2017; Pierre-Jerome et al., 2018; Rosspopoff et al., 2017). Influenced by these informative signals, the shoot stem cell regulator factors progressively display a restricted expression in the confined domain of a classical SAM (Chatfield et al., 2013; Radhakrishnan et al., 2018; Rosspopoff et al., 2017). Notably, in the cytokinin signaling region, type-B ARRs transcription activators act in concert with miRNA-regulated HD-ZIP III transcription factors to directly activate and maintain *WUS* expression in order to promote fate transition of LRM (-like) cells (Zhang et al., 2017). In parallel, type-B ARRs further reinforce *WUS* expression in its domain by suppressing YUCCA-mediated auxin biosynthesis and consequently the level of auxin (Meng et al., 2017). Besides CK, auxin hormone and their key perception and signaling mediators, TIR1 receptor and ARFs TF respectively, are also central players affecting the regeneration of a *de novo* shoot meristem from LRM and LRM-like (Qiao et al., 2012; Wang et al., 2018). MONOPTEROS (MP)/ARF5 and ARF10 positively participate in the conversion process, possibly through the transcriptional activation of the shoot cell identity gene, *STM*, and also cytokinin signaling activators (Ckurshumova et al., 2014; Qiao et al., 2012).

11.5 Concluding Remarks

The emergence of cutting-edge techniques such as single cell RNA sequencing, time lapse imaging, and cell lineage analysis has led to considerable advances in understanding the complex meristems functioning and the underlying molecular mechanisms. However, many discoveries have still to be achieved for the comprehension of the pluri- or multipotency character that rely on certain tissues or cell types and the processes involved in regenerative mechanisms. These ones are fundamental to answer questions addressing the stem cell specification, cell fate transitions, and cellular reprogramming. Future studies on *de novo* shoot regeneration processes, such as the conversion process, should drive to the discovery of new regulators that are able to confer regenerative potential to recalcitrant plant species or enhancing regeneration efficiency, and to prosper their applications in agriculture and biotechnology.

11.6 Take on Messages

SAM and RAM have a well-known organization in zones, layers, and domains and their regulation is finely tuned by transcription factors and hormonal gradients.

In addition, small peptides, microRNAs, and epigenetic marks are also implied in the transcriptional responses and *per se* the different cellular states necessary for the maintenance of stem cell niches.

Plant regeneration can occur from different tissues, for example *via* meristems neoformation and/or stem cell niche reorganization.

References

Aichinger, E., Kornet, N., Friedrich, T., and Laux, T. (2012). Plant stem cell niches. Annu. Rev. Plant Biol. *63*, 615–636.

Aida, M., Beis, D., Heidstra, R., Willemsen, V., Blilou, I., Galinha, C., Nussaume, L., Noh, Y.-S., Amasino, R., and Scheres, B. (2004). The PLETHORA genes mediate patterning of the Arabidopsis root stem cell niche. Cell *119*, 109–120.

Atta, R., Laurens, L., Boucheron-Dubuisson, E., Guivarc'h, A., Carnero, E., Giraudat-Pautot, V., Rech, P., and Chriqui, D. (2009). Pluripotency of Arabidopsis xylem pericycle underlies shoot regeneration from root and hypocotyl explants grown in vitro. Plant J. *57*, 626–644.

Besnard, F., Refahi, Y., Morin, V., Marteaux, B., Brunoud, G., Chambrier, P., Rozier, F., Mirabet, V., Legrand, J., Lainé, S., et al. (2014). Cytokinin signalling inhibitory fields provide robustness to phyllotaxis. Nature *505*, 417–421.

Biedermann, S., and Laux, T. (2018). Plant Development: Adding HAM to Stem Cell Control. Curr. Biol. *28*, R1261–R1263.

Birnbaum, K.D., and Sánchez Alvarado, A. (2008). Slicing across kingdoms: regeneration in plants and animals. Cell *132*, 697–710.

Bustamante, M., Matus, J.T., and Riechmann, J.L. (2016). Genome-wide analyses for dissecting gene regulatory networks in the shoot apical meristem. J. Exp. Bot. *67*, 1639–1648.

Carles, C.C., and Fletcher, J.C. (2003). Shoot apical meristem maintenance: the art of a dynamic balance. Trends Plant Sci. *8*, 394–401.

Chatfield, S.P., Capron, R., Severino, A., Penttila, P.-A., Alfred, S., Nahal, H., and Provart, N.J. (2013). Incipient stem cell niche conversion in tissue culture: using a systems approach to probe early events in WUSCHEL-dependent conversion of lateral root primordia into shoot meristems. Plant J. *73*, 798–813.

Cheng, Z.J., Wang, L., Sun, W., Zhang, Y., Zhou, C., Su, Y.H., Li, W., Sun, T.T., Zhao, X.Y., Li, X.G., et al. (2013). Pattern of auxin and cytokinin responses for shoot meristem induction results from the regulation of cytokinin biosynthesis by AUXIN RESPONSE FACTOR3. Plant Physiol. *161*, 240–251.

Chickarmane, V.S., Gordon, S.P., Tarr, P.T., Heisler, M.G., and Meyerowitz, E.M. (2012). Cytokinin signaling as a positional cue for patterning the apical-basal axis of the growing Arabidopsis shoot meristem. Proc. Natl. Acad. Sci. *109*, 4002–4007.

Ckurshumova, W., Smirnova, T., Marcos, D., Zayed, Y., and Berleth, T. (2014). Irrepressible *MONOPTEROS/ARF5* promotes *de novo* shoot formation. New Phytol. *204*, 556–566.

Dello Ioio, R., Nakamura, K., Moubayidin, L., Perilli, S., Taniguchi, M., Morita, M.T., Aoyama, T., Costantino, P., and Sabatini, S. (2008). A genetic framework for the control of cell division and differentiation in the root meristem. Science *322*, 1380–1384.

Dinneny, J.R., and Benfey, P.N. (2008). Plant stem cell niches: standing the test of time. Cell *132*, 553–557.

Drisch, R.C., and Stahl, Y. (2015). Function and regulation of transcription factors involved in root apical meristem and stem cell maintenance. Front. Plant Sci. *6*, 505.

Du, Y., and Scheres, B. (2017). PLETHORA transcription factors orchestrate de novo organ patterning during *Arabidopsis* lateral root outgrowth. Proc. Natl. Acad. Sci. *114*, 11709–11714.

Du, Y., and Scheres, B. (2018). Lateral root formation and the multiple roles of auxin. J. Exp. Bot. *69*, 155–167.

Efroni, I., Mello, A., Nawy, T., Ip, P.-L., Rahni, R., DelRose, N., Powers, A., Satija, R., and Birnbaum, K.D. (2016). Root Regeneration Triggers an Embryo-like Sequence Guided by Hormonal Interactions. Cell *165*, 1721–1733.

Fletcher, J.C. (2018). The CLV-WUS Stem Cell Signaling Pathway: A Roadmap to Crop Yield Optimization. Plants Basel Switz. *7*.

Forzani, C., Aichinger, E., Sornay, E., Willemsen, V., Laux, T., Dewitte, W., and Murray, J.A.H. (2014). WOX5 suppresses CYCLIN D activity to establish quiescence at the center of the root stem cell niche. Curr. Biol. *24*, 1939–1944.

Fulcher, N., and Sablowski, R. (2009). Hypersensitivity to DNA damage in plant stem cell niches. Proc. Natl. Acad. Sci. *106*, 20984–20988.

Gaillochet, C., Stiehl, T., Wenzl, C., Ripoll, J.-J., Bailey-Steinitz, L.J., Li, L., Pfeiffer, A., Miotk, A., Hakenjos, J.P., Forner, J., et al. (2017). Control of plant cell fate transitions by transcriptional and hormonal signals. ELife *6*.

Gallagher, K.L., Paquette, A.J., Nakajima, K., and Benfey, P.N. (2004). Mechanisms regulating SHORT-ROOT intercellular movement. Curr. Biol. *14*, 1847–1851.

Goh, T., Toyokura, K., Wells, D.M., Swarup, K., Yamamoto, M., Mimura, T., Weijers, D., Fukaki, H., Laplaze, L., Bennett, M.J., et al. (2016). Quiescent center initiation in the Arabidopsis lateral root primordia is dependent on the SCARECROW transcription factor. Development *143*, 3363–3371.

Gonzali, S., Novi, G., Loreti, E., Paolicchi, F., Poggi, A., Alpi, A., and Perata, P. (2005). A turanose-insensitive mutant suggests a role for WOX5 in auxin homeostasis in Arabidopsis thaliana: WOX5 modulates auxin homeostasis in Arabidopsis. Plant J. *44*, 633–645.

Gordon, S.P., Chickarmane, V.S., Ohno, C., and Meyerowitz, E.M. (2009). Multiple feedback loops through cytokinin signaling control stem cell number within the Arabidopsis shoot meristem. Proc. Natl. Acad. Sci. U. S. A. *106*, 16529–16534.

Greb, T., and Lohmann, J.U. (2016). Plant Stem Cells. Curr. Biol. *26*, R816–R821.

Heidstra, R., and Sabatini, S. (2014). Plant and animal stem cells: similar yet different. Nat. Rev. Mol. Cell Biol. *15*, 301–312.

Heyman, J., Cools, T., Vandenbussche, F., Heyndrickx, K.S., Van Leene, J., Vercauteren, I., Vanderauwera, S., Vandepoele, K., De Jaeger, G., Van Der Straeten, D., et al. (2013). ERF115 Controls Root Quiescent Center Cell Division and Stem Cell Replenishment. Science *342*, 860–863.

Hofhuis, H.F., and Heidstra, R. (2018). Transcription factor dosage: more or less sufficient for growth. Curr. Opin. Plant Biol. *45*, 50–58.

Janocha, D., and Lohmann, J.U. (2018). From signals to stem cells and back again. Curr. Opin. Plant Biol. *45*, 136–142.

de Jong, M., and Leyser, O. (2012). Developmental Plasticity in Plants. Cold Spring Harb. Symp. Quant. Biol. *77*, 63–73.

Kareem, A., Durgaprasad, K., Sugimoto, K., Du, Y., Pulianmackal, A.J., Trivedi, Z.B., Abhayadev, P.V., Pinon, V., Meyerowitz, E.M., Scheres, B., et al. (2015). PLETHORA Genes Control Regeneration by a Two-Step Mechanism. Curr. Biol. *25*, 1017–1030.

Kareem, A., Radhakrishnan, D., Sondhi, Y., Aiyaz, M., Roy, M.V., Sugimoto, K., and Prasad, K. (2016). De novo assembly of plant body plan: a step ahead of Deadpool. Regeneration *3*, 182–197.

Landrein, B., and Ingram, G. (2019). Connected through the force: mechanical signals in plant development. J. Exp. Bot.

Laskowski, M., Grieneisen, V.A., Hofhuis, H., Hove, C.A.T., Hogeweg, P., Marée, A.F.M., and Scheres, B. (2008). Root system architecture from coupling cell shape to auxin transport. PLoS Biol. *6*, e307.

Laux, T. (2003). The stem cell concept in plants: a matter of debate. Cell *113*, 281–283.

Lee, Z.H., Hirakawa, T., Yamaguchi, N., and Ito, T. (2019). The Roles of Plant Hormones and Their Interactions with Regulatory Genes in Determining Meristem Activity. Int. J. Mol. Sci. *20*.

Long, Y., Smet, W., Cruz-Ramírez, A., Castelijns, B., de Jonge, W., Mähönen, A.P., Bouchet, B.P., Perez, G.S., Akhmanova, A., Scheres, B., et al. (2015). Arabidopsis BIRD Zinc Finger Proteins Jointly Stabilize Tissue Boundaries by Confining the Cell Fate Regulator SHORT-ROOT and Contributing to Fate Specification. Plant Cell *27*, 1185–1199.

Luo, L., Zeng, J., Wu, H., Tian, Z., and Zhao, Z. (2018). A Molecular Framework for Auxin-Controlled Homeostasis of Shoot Stem Cells in Arabidopsis. Mol. Plant.

Ma, Y., Miotk, A., Łutikoviæ, Z., Ermakova, O., Wenzl, C., Medzihradszky, A., Gaillochet, C., Forner, J., Utan, G., Brackmann, K., et al. (2019). WUSCHEL acts as an auxin response rheostat to maintain apical stem cells in Arabidopsis. Nat. Commun. *10*, 5093.

Mähönen, A.P., Ten Tusscher, K., Siligato, R., Smetana, O., Díaz-Triviño, S., Salojärvi, J., Wachsman, G., Prasad, K., Heidstra, R., and Scheres, B. (2014). PLETHORA gradient formation mechanism separates auxin responses. Nature *515*, 125–129.

Matsuzaki, Y., Ogawa-Ohnishi, M., Mori, A., and Matsubayashi, Y. (2010). Secreted Peptide Signals Required for Maintenance of Root Stem Cell Niche in Arabidopsis. Science *329*, 1065–1067.

Meng, W.J., Cheng, Z.J., Sang, Y.L., Zhang, M.M., Rong, X.F., Wang, Z.W., Tang, Y.Y., and Zhang, X.S. (2017). Type-B ARABIDOPSIS RESPONSE REGULATORs Specify the Shoot Stem Cell Niche by Dual Regulation of WUSCHEL. Plant Cell *29*, 1357–1372.

Meyerowitz, E.M. (2002). Plants compared to animals: the broadest comparative study of development. Science *295*, 1482–1485.

Moubayidin, L., Di Mambro, R., Sozzani, R., Pacifici, E., Salvi, E., Terpstra, I., Bao, D., van Dijken, A., Dello Ioio, R., Perilli, S., et al. (2013). Spatial Coordination between Stem Cell Activity and Cell Differentiation in the Root Meristem. Dev. Cell *26*, 405–415.

Parizot, B., Laplaze, L., Ricaud, L., Boucheron-Dubuisson, E., Bayle, V., Bonke, M., De Smet, I., Poethig, S.R., Helariutta, Y., Haseloff, J., et al. (2008). Diarch Symmetry of the Vascular Bundle in Arabidopsis Root Encompasses the Pericycle and Is Reflected in Distich Lateral Root Initiation. Plant Physiol. *146*, 140–148.

Perales, M., Rodriguez, K., Snipes, S., Yadav, R.K., Diaz-Mendoza, M., and Reddy, G.V. (2016). Threshold-dependent transcriptional discrimination underlies stem cell homeostasis. Proc. Natl. Acad. Sci. *113*, E6298–E6306.

Perez-Garcia, P., and Moreno-Risueno, M.A. (2018). Stem cells and plant regeneration. Dev. Biol. *442*, 3–12.

Pi, L., Aichinger, E., van der Graaff, E., Llavata-Peris, C.I., Weijers, D., Hennig, L., Groot, E., and Laux, T. (2015). Organizer-Derived WOX5 Signal Maintains Root Columella Stem Cells through Chromatin-Mediated Repression of CDF4 Expression. Dev. Cell *33*, 576–588.

Pierre-Jerome, E., Drapek, C., and Benfey, P.N. (2018). Regulation of Division and Differentiation of Plant Stem Cells. Annu. Rev. Cell Dev. Biol. *34*, 289–310.

Qiao, M., Zhao, Z., Song, Y., Liu, Z., Cao, L., Yu, Y., Li, S., and Xiang, F. (2012). Proper regeneration from in vitro cultured Arabidopsis thaliana requires the microRNA-directed action of an auxin response factor. Plant J. *71*, 14–22.

Radhakrishnan, D., Kareem, A., Durgaprasad, K., Sreeraj, E., Sugimoto, K., and Prasad, K. (2018). Shoot regeneration: a journey from acquisition of competence to completion. Curr. Opin. Plant Biol. *41*, 23–31.

Rahni, R., Efroni, I., and Birnbaum, K.D. (2016). A Case for Distributed Control of Local Stem Cell Behavior in Plants. Dev. Cell *38*, 635–642.

Reinhardt, D., Frenz, M., Mandel, T., and Kuhlemeier, C. (2003). Microsurgical and laser ablation analysis of interactions between the zones and layers of the tomato shoot apical meristem. Development *130*, 4073–4083.

Rosspopoff, O., Chelysheva, L., Saffar, J., Lecorgne, L., Gey, D., Caillieux, E., Colot, V., Roudier, F., Hilson, P., Berthomé, R., et al. (2017). Direct conversion of root primordium into shoot meristem relies on timing of stem cell niche development. Development *144*, 1187–1200.

Sang, Y.L., Cheng, Z.J., and Zhang, X.S. (2018). Plant stem cells and de novo organogenesis. New Phytol. *218*, 1334–1339.

Santuari, L., Sanchez-Perez, G.F., Luijten, M., Rutjens, B., Terpstra, I., Berke, L., Gorte, M., Prasad, K., Bao, D., Timmermans-Hereijgers, J.L.P.M., et al. (2016). The PLETHORA Gene Regulatory Network Guides Growth and Cell Differentiation in Arabidopsis Roots. Plant Cell *28*, 2937–2951.

Sarkar, A.K., Luijten, M., Miyashima, S., Lenhard, M., Hashimoto, T., Nakajima, K., Scheres, B., Heidstra, R., and Laux, T. (2007). Conserved factors regulate signalling in Arabidopsis thaliana shoot and root stem cell organizers. Nature *446*, 811–814.

Scheres, B., Wolkenfelt, H., Willemsen, V.A., Terlouw, M., Lawson, E., Dean, C., and Weisbeek, P. (1994). Embryonic Origin of the Arabidopsis Primary Root and Root-Meristem Initials. Development *120*.

Shi, B., and Vernoux, T. (2019). Patterning at the shoot apical meristem and phyllotaxis. In Current Topics in Developmental Biology, (Elsevier), pp. 81–107.

Shi, B., Guo, X., Wang, Y., Xiong, Y., Wang, J., Hayashi, K.-I., Lei, J., Zhang, L., and Jiao, Y. (2018). Feedback from Lateral Organs Controls Shoot Apical Meristem Growth by Modulating Auxin Transport. Dev. Cell *44*, 204–216.e6.

Shi, D., Tavhelidse, T., Thumberger, T., Wittbrodt, J., and Greb, T. (2017). Bifacial stem cell niches in fish and plants. Curr. Opin. Genet. Dev. *45*, 28–33.

Shimotohno, A., and Scheres, B. (2019). Topology of regulatory networks that guide plant meristem activity: similarities and differences. Curr. Opin. Plant Biol. *51*, 74–80.

Shin, J., Bae, S., and Seo, P.J. (2019). De novo shoot organogenesis during plant regeneration. J. Exp. Bot. *71*, 63–72.

Skoog, F., and Miller, C.O. (1957). Chemical regulation of growth and organ formation in plant tissues cultured in vitro. Symp. Soc. Exp. Biol. *11*, 118–130.

Snipes, S.A., Rodriguez, K., DeVries, A.E., Miyawaki, K.N., Perales, M., Xie, M., and Reddy, G.V. (2018). Cytokinin stabilizes WUSCHEL by acting on the protein domains required for nuclear enrichment and transcription. PLoS Genet. *14*, e1007351.

Stahl, Y., Grabowski, S., Bleckmann, A., Kühnemuth, R., Weidtkamp-Peters, S., Pinto, K.G., Kirschner, G.K., Schmid, J.B., Wink, R.H., Hülsewede, A., et al. (2013). Moderation of Arabidopsis Root Stemness by CLAVATA1 and ARABIDOPSIS CRINKLY4 Receptor Kinase Complexes. Curr. Biol. *23*, 362–371.

Sugimoto, K., Jiao, Y., and Meyerowitz, E.M. (2010). Arabidopsis regeneration from multiple tissues occurs via a root development pathway. Dev. Cell *18*, 463–471.

Sugimoto, K., Temman, H., Kadokura, S., and Matsunaga, S. (2019). To regenerate or not to regenerate: factors that drive plant regeneration. Curr. Opin. Plant Biol. *47*, 138–150.

Uchida, N., and Torii, K.U. (2019). Stem cells within the shoot apical meristem: identity, arrangement and communication. Cell. Mol. Life Sci. *76*, 1067–1080.

Vernoux, T., Brunoud, G., Farcot, E., Morin, V., Van den Daele, H., Legrand, J., Oliva, M., Das, P., Larrieu, A., Wells, D., et al. (2011). The auxin signalling network translates dynamic input into robust patterning at the shoot apex. Mol. Syst. Biol. *7*, 508.

Wang, L., Liu, Z., Qiao, M., and Xiang, F. (2018). miR393 inhibits in vitro shoot regeneration in Arabidopsis thaliana via repressing TIR1. Plant Sci. *266*, 1–8.

Wany, A., Foyer, C.H., and Gupta, K.J. (2018). Nitrate, NO and ROS Signaling in Stem Cell Homeostasis. Trends Plant Sci. *23*, 1041–1044.

Xie, M., Chen, H., Huang, L., O'Neil, R.C., Shokhirev, M.N., and Ecker, J.R. (2018). A B-ARR-mediated cytokinin transcriptional network directs hormone cross-regulation and shoot development. Nat. Commun. *9*, 1604.

Xu, J., Hofhuis, H., Heidstra, R., Sauer, M., Friml, J., and Scheres, B. (2006). A molecular framework for plant regeneration. Science *311*, 385–388.

Yadav, R.K., Perales, M., Gruel, J., Ohno, C., Heisler, M., Girke, T., Jönsson, H., and Reddy, G.V. (2013). Plant stem cell maintenance involves direct transcriptional repression of differentiation program. Mol. Syst. Biol. *9*, 654.

Yang, S., Yu, Q., Zhang, Y., Jia, Y., Wan, S., Kong, X., and Ding, Z. (2018). ROS: The Fine-Tuner of Plant Stem Cell Fate. Trends Plant Sci. *23*, 850–853.

Zhang, T.-Q., Lian, H., Zhou, C.-M., Xu, L., Jiao, Y., and Wang, J.-W. (2017). A Two-Step Model for de Novo Activation of WUSCHEL during Plant Shoot Regeneration. Plant Cell *29*, 1073–1087.

Zhang, W., Swarup, R., Bennett, M., Schaller, G.E., and Kieber, J.J. (2013). Cytokinin induces cell division in the quiescent center of the Arabidopsis root apical meristem. Curr. Biol. *23*, 1979–1989.

Zhao, Z., Andersen, S.U., Ljung, K., Dolezal, K., Miotk, A., Schultheiss, S.J., and Lohmann, J.U. (2010). Hormonal control of the shoot stem-cell niche. Nature *465*, 1089–1092.

Zhou, Y., Liu, X., Engstrom, E.M., Nimchuk, Z.L., Pruneda-Paz, J.L., Tarr, P.T., Yan, A., Kay, S.A., and Meyerowitz, E.M. (2015). Control of plant stem cell function by conserved interacting transcriptional regulators. Nature *517*, 377–380.

Zhou, Y., Yan, A., Han, H., Li, T., Geng, Y., Liu, X., and Meyerowitz, E.M. (2018). HAIRY MERISTEM with WUSCHEL confines CLAVATA3 expression to the outer apical meristem layers. Science *361*, 502–506.

12

Hematopoietic Development in Vertebrates

Hanane Khoury[1,2], Thierry Jaffredo[1,*], and Laurent Yvernogeau[1,*]

[1]Sorbonne Université, IBPS, CNRS UMR7622, Inserm U1156,
Laboratoire de Biologie du Développement; 75005 Paris
[2]Present address: Department of Hematology, St. Jude Children's Research
Hospital; 262 Danny Thomas, Memphis 38105 TN; USA
E-mail : thierry.jaffredo@sorbonne-universite.fr;
laurent.yvernogeau@sorbonne-universite.fr
*Corresponding Authors

12.1 History of the Concept of HSCs

In 1908, Alexander Maximov, a Russian histologist, put forward a unitarian theory according to which all blood cells derive from a unique precursor that he designated "Stammzelle" (stem cell). However, the first studies of the hematopoietic system really started in 1914, when Franz Ernst Christian Neumann (1834–1918) described the bone marrow (BM) as the organ of blood cell formation.

In 1945, the United States detonated two nuclear bombs over Hiroshima and Nagasaki. Many people died afterward from hematopoietic failure due to a lethal dose of irradiation. This was the starting point for seminal discoveries in the field of hematopoietic cell transplantation. In the early 1950's, Jacobson and Lorenz showed that adult mouse spleen and BM cells were able to fully reconstitute the hematopoietic system of a lethally irradiated animal, proving the existence of cells responsible for replenishing hematopoiesis (Jacobson, et al., 1951; Lorenz, et al., 1951).

In 1960, Till & McCulloch revealed that the injection of BM cells into lethally irradiated mice led to the formation of hematopoietic nodules within the spleen. These nodules were designated as Colony Forming Unit in the Spleen, CFU-S in short (McCulloch and Till, 1960). These nodules are essentially formed of myeloid, erythroid, and a few megakaryocytic cells but also contained less differentiated hematopoietic cells (HCs), hence demonstrating a hematopoietic hierarchy within the nodules. Further experiments showed that CFU-S are not able to maintain long-term hematopoiesis hence do not contain hematopoietic stem cells (HSCs) but rather hematopoietic progenitors endowed with an important multiplication and differentiation potential (Jones, et al., 1990; Hodgson and Bradley, 1979).

In the adult amniotes and amphibians, HSCs reside within the BM whereas in fish, they are found in the kidney. Both organs provide micro-environments for either HSC maintenance/quiescence or differentiation/proliferation. However, it is now well accepted that HSCs emerge early during embryonic development mainly in the aorta, but also in associated umbilical and vitelline arteries for mammalian species. After emergence, HSCs undergo a massive expansion in the fetal liver or in the caudal hematopoietic tissue in fish before reaching their final destination i.e., the BM or the kidney marrow in fish. Whether HSCs can be generated in other organs (i.e., fetal liver, BM) was under debate until it was shown that the fetal/young adult BM contributes to the generation of HSCs (Yvernogeau, et al., 2019).

In both embryos and adults, HSCs are rare and difficult to characterize. This considerably limits their fundamental study and their use in clinical therapies. The paradigm of HSCs is that it is still impossible to isolate them at purity since no specific marker has been identified to date. Using a combination of membrane and intracellular markers, cell fractions enriched in HSCs can be isolated, and their respective hematopoietic potentials can be evaluated retrospectively through *in vivo* and/or *ex vivo* assays.

12.2 On the Origin of Blood

The origin of HSCs has been the subject of intense debates. Development of the hematopoietic system is characterized by the existence of multiple and successive waves, occurring in different anatomical locations. There are also species-specificities. The chicken embryo was instrumental in decoding many rules of developmental hematopoiesis. To investigate how the blood system forms during development, Moore and Owen used the chicken embryo, a model already utilized for many descriptive studies since the end of the 19th century because of its direct visibility and accessibility after shell opening and its development independent from a maternal circulation. Moore and Owen applied an array of experimental approaches: irradiation/restoration, parabiosis, grafting of hematopoietic rudiments (thymus, spleen, and bursa of Fabricius) on the chorioallantoic membrane (CAM) ((Moore and Metcalf, 1970) for review). To follow the cell origins in these different designs, they used the pair of sexual chromosomes as a marker (in birds ZW for the female, ZZ for the male). This labeling system has some drawbacks: (1) the chromosomes can be observed only in 5–10% of the cells arrested at the metaphase stage of mitosis after colcemid treatment, which elicits high mortality of the embryos; (2) the adequate sex combination is left to chance. Despite these inherent drawbacks, investigators discovered an important trait that characterizes the development of the blood system: HSCs colonize all hematopoietic rudiments apart from the yolk sac (YS). Similar type of experiments using mouse thymic rudiments allowed Moore and Owen to conclude in 1970 that, also in mammalian developmental hematopoiesis, HSCs emerge in one location and differentiate in another one (Moore and Metcalf, 1970).

Benefiting from accurate functional tests, the mouse embryo took the lead and was instrumental in building a precise, comprehensive picture of the hematopoietic development in mammals. Hematopoietic development is organized into three separated waves. The first one takes place in the YS at embryonic day (E)7.5 in the mouse embryo (Palis, et al., 1999), 19–22 hours in the chicken embryo (Minko, et al., 2003), 21 days in the human embryo, and 5-somite-stage in the zebrafish embryo (Figure 12.1). This wave gives rise to primitive erythroid cells, macrophages, and megakaryocytes (Palis, et al., 1999). Shortly after the onset of this primitive hematopoiesis, the first definitive erythro-myeloid progenitors and immune-restricted progenitors emerge from the YS at E8.5 in the mouse and E2.5 in the chicken embryo. The third wave occurs in the embryo proper in a specific region named aorta–gonad–mesonephros (AGM), and is characterized by the production of HSCs and adult-type lympho-myeloid progenitors (Gritz and Hirschi, 2016; Drevon and Jaffredo, 2014). HSCs, which are detectable by direct transplantation into adult recipients, are not present until E10.5 in the AGM but hematopoietic emergence is detected from E9.5 closely associated with vitelline and umbilical arteries (Tavian, et al., 2001; de Bruijn, et al., 2000). Subsequently, transplantable HSCs are found in the YS, placenta, and head ((Dzierzak and Bigas, 2018) for review), but it is not clear whether HSCs emerge *de novo* in these sites. Shortly after their emergence, HSCs migrate and either colonize the fetal liver in mammals or reach the sub-aortic mesenchyme in birds. In both situations, they undergo multiplication and amplification and eventually migrate to the definitive hematopoietic organs (BM, thymus, spleen...) where they will reside during the lifetime of the organism.

12.3 An Intra-embryonic Source of HSCs

At the dawn of embryology, scientists described the seemingly coincident emergence of endothelial and hematopoietic cells in the YS giving rise to the concept of hemangioblast (Murray, 1932). It is not difficult to imagine why the YS was the first hematopoietic site to be investigated. Indeed, it shows the first signs of hematopoiesis testified by the presence of the red blood islands and was naturally proposed to be the site of HSC production. In the mid and late 60s, Moore and Owen proposed that the YS was the site where HSCs emerge in a unique, early, developmental event and subsequently colonize the hematopoietic organs (Moore and Owen, 1967). The experimental design used a series of graft and parabiosis experiments relying on the identification of sex chromosomes. In addition, they also showed that hematopoietic stem/progenitor cells (HSPCs) were of extrinsic origin (see above). These two pillars made the core of the hematogenous theory. During the 70s, the relevance of this hypothesis was investigated in mammalian species by culturing either E7 whole mouse embryo or the separated YS and embryo. The conclusion was that the YS was the only site where HSCs were formed reinforcing the central role of this tissue in HSC generation (Moore and Metcalf, 1970). Based on this and on other reports (Weissman, et al., 1977), the hematogenous theory became a dogma and was widely accepted among the scientific community.

In 1972, a surgical technique to construct a unique type of quail/chicken chimera was devised by Martin (Martin, 1972). The flatness of the avian embryo as it lies on the surface of the spherical YS facilitates its dissection, allowing it to be replaced with the embryo of another bird species *in ovo*. The quail/chicken chimera consisted in grafting a quail embryo on a chicken YS at a very early stage of development, before the onset of heart beating. When such manipulations were carried out, there were a rapid reconstitution and joining of blood vessels between the grafted embryo body and the host YS and the free circulation of blood cells was established between the YS and the body, without any disruption of normal development in the resulting chimeras. The chimeric embryos possessed hematopoietic organs in which the cells belonging to blood lineages in the thymus, spleen, bursa of Fabricius, and BM were all of quail origin while stromal cells were of chicken (Martin, et al., 1978; Dieterlen-Lièvre, 1975). Circulating blood was chicken-derived (derived from YS progenitors) until E5, became mixed and then richer and richer in quail-derived erythrocytes (Beaupain, et al., 1979). The replacement of YS-derived hematopoiesis by intra-embryonic hematopoiesis was

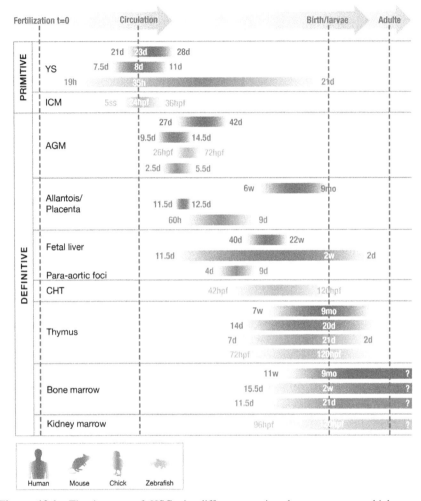

Figure 12.1 The journey of HSCs in different species: human, mouse, chicken, and zebrafish. Starting with fertilization, blood circulation, birth and larvae, and finishing with adult life.

The four species share two major waves of hematopoiesis: primitive and definitive waves.

Human, mouse, and chicken have the same hematopoietic organs, except that in the chicken the role of the fetal liver is replaced by the para-aortic foci. In the zebrafish, the AGM and thymus have the same role as in the other species. The ICM is the equivalent of the YS and the CHT is the equivalent of the fetal liver. The kidney marrow replaced the role of the bone marrow.

Abbreviations: YS, yolk sac; ICM: inner cell mass; CHT: caudal hematopoietic tissue; d: day; hpf: hours post-fertilization; mo: months.

Adapted from Rowe et al., 2016

strikingly confirmed when the chimeras were built between congenic strains of chickens, differing in their immunoglobulin allotypes or in their major histocompatibility antigens (Lassila, et al., 1978). These results led to the indisputable conclusion that YS progenitors give rise to a short-lived progeny and that an intra-embryonic source was responsible for the production of both hematopoietic progenitors and HSCs that seed the hematopoietic organs. Intra-aortic hematopoietic clusters (IAHCs) had been described from the beginning of the 20th century in both avian and mammalian species as groups of hematopoietic cells, located on the aortic floor, protruding into the aortic lumen (Dieterlen-Lièvre, et al., 2006). This aortic region thus appeared as an interesting candidate to produce the cells that seeded the definitive hematopoietic organs. In quail/chicken chimeras, these intra-aortic clusters are quail-derived, i.e., they have an *in situ* intra-embryonic origin (Dieterlen-Lievre and Martin, 1981). Moreover, their affinity for anti-ITAG2b integrin (also known as CD41) antibody as well as for CD45 authenticates their hematopoietic progenitor nature (Corbel, 2002; Jaffredo, et al., 1998). Furthermore, the peri-aortic region of the chicken embryo, dissociated into single cells and seeded in a semi-solid medium, gave rise to 3–4 times more colonies (erythroid or macrophages) than the BM from a newly hatched animal (Cormier and Dieterlen-Lievre, 1988).

a. Mouse Aorta

Based on results obtained in birds, two groups identified the aorta as a major site of HSC emergence in mice (Godin, et al., 1993; Medvinsky, et al., 1993). Prior to formation of the single aorta harboring HSCs, the presumptive aortic territory is contained within the splanchnopleural mesoderm located in the caudal region of the embryo (Cumano, et al., 1996). From E8.5, the splanchnopleural mesoderm gives rise to a structure containing the paired aortas, the sub-aortic mesenchyme and the endoderm of the digestive tract.This structure is called the para-aortic splanchnopleura (P-Sp) (Godin, et al., 1995). From E9.5 to E11.5, the P-Sp evolves to form the AGM region comprising the dorsal aorta and its underlying mesenchyme and the urogenital system (Medvinsky, et al., 1993). It is within the AGM that IAHCs can be observed from E9.5 until E14.5 in the aortic lumen (Yokomizo and Dzierzak, 2010).

Mouse IAHCs are similar to those observed in chickens and are mainly present in the aortic floor but in this species, IAHCs can also be found in the aortic roof marking here a difference with birds (de Bruijn, et al., 2002). Evidence that definitive hematopoiesis emerge in the aorta was provided through experiments of P-Sp or YS culture isolated prior to the establishment of circulation between the YS and the embryo (Cumano, et al., 1996). By grafting P-Sp or YS cells into adult immunodeficient mice, the authors concluded that only the aortic region allowed reconstructing the lymphoid lineage (Godin, et al., 1993). At the same time, another group showed that the AGM carries a CFU-S activity initiated from E9, which peaks at E10, and then declines. Concurrently, CFU-S activity increases in the fetal liver, suggesting that hematopoietic progenitors emerging in the aorta subsequently colonize the fetal liver (Medvinsky, et al., 1993).

The first hematopoietic progenitors are detected in the P-Sp while the circulation between the YS and the embryo is already established. This situation precludes any conclusion on the site of emergence of the aortic HSCs. However, Cumano et al. (Cumano, et al., 1996) tackled this question by testing the hematopoietic potential of the YS and P-Sp isolated before the establishment of blood circulation. By placing YS or P-Sp tissues in organotypic culture for 2 to 4 days, the authors demonstrated that the P-Sp contains progenitors capable of producing B- and T-lymphocytes before they were detected in the YS, suggesting that the P-Sp was the primary source of definitive hematopoiesis. In parallel, it was shown that cells from E11 AGM could reconstitute the different hematopoietic blood lineages when injected into primary and secondary irradiated adult mice, hence testifying the presence of *bona fide* HSCs (Medvinsky and Dzierzak, 1996).

b. Amphibian Aorta

In amphibians, removal of the ventral blood island (VBI), the equivalent of the YS, resulted in the absence of circulating red blood cells. The origin of adult progenitors was hence first exclusively attributed to the VBI (Federici, 1926). This hypothesis was then invalidated in the amphibian *Rana pipiens* by transplanting various territories subsequently identified according to their ploidy difference (Hollyfield, 1966). These experiments first showed the existence of an alternative source of progenitors. When the cytogenetically labeled VBI was transplanted to the neurula stage, no larval erythrocyte came from the graft. This approach then made it possible to locate an alternative source of progenitors: if the dorsal mesodermal territory containing the pronephros is transplanted, erythrocytes from the graft are detected in the larva and adult indicating that this region could be at the origin of the adult hematopoietic system. In agreement with this hypothesis, the lymphoid population of the larva is also derived from the dorsal region (Turpen, et al., 1983; Turpen, et al., 1981). The emergence of HSCs from the dorsal region was then demonstrated in Xenopus (Tompkins, 1980). The activity of the hematopoietic region is characterized by the presence of HCs in the pronephric tubules, the aorta, and the cardinal veins (Turpen, et al., 1981). The origin of these cells was precisely located in the dorsal lateral plate (DLP) (Kau and Turpen, 1983; Turpen and Knudson, 1982). Data indicate that the DLP produces HCs contributing to larval and adult hematopoiesis. In particular, this territory participates in definitive erythropoiesis, thymic and hepatic hematopoietic activities (Chen and Turpen, 1995; Bechtold, et al., 1992; Maeno, et al., 1985; Kau and Turpen, 1983). Using dye injection into blastomeres and the follow up of the blastomere progeny, R. Patient's group was able to map the origin of the different hematopoietic progenitors in stage 32 blastula cells. This study highlighted that DLP and VBI lineages are already separated at a very early stage: only the C3 dorsal blastomere contributes to the formation of the DLP and to the emergence of HCs in the aortic region of the embryo (Ciau-Uitz, et al., 2000).

12.4 Construction of the Aorta and Establishment of the Dorso-ventral Polarity

During gastrulation, the mesoderm is laid down between the ectoderm and the endoderm. It subsequently regionalizes according to the embryonic axes. Medio-lateral regionalization of the mesoderm results in the formation of different compartments: notochord, somites, intermediate mesoderm, and lateral plate mesoderm. The latter, therefore comprises the embryonic and extra-embryonic mesoderm. The coelom splits the lateral mesoderm into two layers: a dorsal one, the somatopleural mesoderm, and a ventral one designated as the splanchnopleural mesoderm. These layers associate respectively with the ectoderm to give rise to the somatopleura, or the endoderm to give rise to the splanchnopleura. HCs and a cohort of endothelial cells (ECs) arise from the splanchnopleura. The blood islands are formed from the extra-embryonic splanchnopleural mesoderm and associated endothelial and hematopoietic progenitors. In the embryo proper, the splanchnopleural mesoderm gives rise to angioblasts that form capillaries, which will merge to give a vascular network closely associated with the endoderm. The vascular system of the viscera and the aorta are also formed this way.

Another source of ECs is the somite (Ambler, et al., 2001; Noden, 1989; Wilting, et al., 1995; Pardanaud and Dieterlen-Lievre, 1993), the probable source of angioblasts colonizing the somatopleura. To substantiate this hypothesis, Pardanaud and co-workers orthotopically transplanted either the last-formed or two somites from the quail into the chicken embryo, and followed the distribution of QH1$^+$ ECs (QH1 is an antibody recognizing quail endothelial and hematopoietic cells). The somite-derived QH1$^+$ ECs spread into the body wall, limbs, and kidney but also integrated into the roof and sides of the aortic endothelium (Pardanaud, et al., 1996). Notably, they never penetrated visceral organs nor integrated the aortic floor. In contrast, angioblasts derived from quail splanchnopleural mesoderm, grafted on top of the host splanchnopleura, also migrated out but invaded all embryonic territories, that is, not only limb bud and body wall, but also the floor of the aorta and visceral organs. Thus, somite-derived and splanchnopleura-derived angioblasts have distinct homing potentials. Furthermore, when integrated in the aortic floor—and only there—cells of splanchnopleural origin proliferated into clusters. These results demonstrate that the embryo is vascularized by two distinct sources of endothelial precursors, splanchnopleural mesoderm and paraxial mesoderm, i.e., somites. EC precursors originating from the paraxial mesoderm vascularize the body wall, limbs, and kidney. ECs originating from the splanchnopleural mesoderm colonize the internal organs and contribute to hematopoiesis (Pardanaud, et al., 1996). Finally, the endothelial and hematopoietic potential of these two mesoderm tissues can be flip-flopped by prior culturing them with endoderm or ectoderm, or by treatment with several growth factors. The endoderm induced a hemangiopoietic potential in the associated mesoderm. Indeed, the association of somatopleural mesoderm with endoderm promoted the "ventral homing" and the production of HCs from mesoderm, which is not normally endowed with this potential. The hemangiopoietic induction by endoderm could be mimicked by Vascular Endothelial Growth Factor (VEGF), basic Fibroblast Growth Factor (bFGF), and Transforming Growth Factor $\beta 1$ (TGF-$\beta 1$). In contrast, contact with ectoderm or Epidermal Growth Factor (EGF)/Transforming Growth Factor α (TGF-α) treatment totally abrogated the hemangiopoietic capacity of the splanchnopleural mesoderm, which produced pure angioblasts with no "ventral homing" behavior (Pardanaud and Dieterlen-Lievre, 1999).

While Pardanaud et al. transplanted a single or two somites, Pouget et al. performed grafts of the pre-somitic mesoderm (PSM) (material corresponding to a length of 10 somites) and analyzed the contribution of the graft to the formation of the aortic endothelium at different stages (Pouget, et al., 2006). In sum, during their formation, the paired aortas are uniquely composed of ECs of splanchnopleural origin. From E2, while the aortas are still paired, somite-derived ECs migrate and replace the initial roof and sides of splanchnopleural origin. When the aortas fuse into a single vessel,

ECs derived from the splanchnopleura are restricted to the aortic floor, hence limiting the blood-forming ability to the ventral side of the aorta. The second phase of aortic remodeling occurs during and after aortic hematopoiesis. At the time of IAHC formation, ECs from the somite migrate and locate immediately underneath the IAHCs. Since IAHCs are produced at the expense of splanchnopleural ECs, ECs of somitic origin progressively replace the floor ensuring the integrity of the vessel during the phase of HSC emergence. At the completion of hematopoiesis, the aortic floor of splanchnopleural origin has been replaced by ECs originating from the somite. Hence, within a short period of time, the origin of the aortic endothelium has changed from splanchnopleura to somatopleura (somite). In addition to replacing the aortic endothelium, the somite was also shown to give rise to the aortic smooth muscle cells (Pouget, et al., 2008; Wiegreffe, et al., 2007). This somite-derived EC population is also present in the mouse (Ema et al., 2006). Orthotopic grafts of mouse somites in a chicken recipient showed a gradual and complete replacement of the chicken aortic endothelium by mouse-derived ECs (Yvernogeau, et al., 2012). Smooth muscle cells forming the tunica of the aorta are also established in two successive stages: cells originating from the lateral plate and placed under the aorta differentiate under the vessel, then a second generation of smooth muscle cells of sclerotomal origin forms the final tunica around the aorta concomitantly with the replacement of ECs (Wiegreffe, et al., 2009; Wasteson, et al., 2008) (Figure 12.2).

12.5 Role of the Sub-aortic Mesenchyme

Runx1 is a transcription factor, which is a key regulator of the so-called endothelial to hematopoietic transition (EHT) that occurs during aortic hematopoiesis (Swiers, et al., 2013; Chen, et al., 2009). EHT is the process by which ECs of the aorta transdifferentiate to become hematopoietic, generating IAHCs. *Runx1* inactivation in the mouse embryo results in the absence of IAHCs (Chen, et al., 2009; North, et al., 1999) and fetal liver HSCs (North, et al., 2002) leading to embryonic death at E12.5, albeit primitive hematopoiesis is preserved. Runx1 deficiency in embryonic stem cells also prevents the formation of HC from hemogenic EC (Lancrin, et al., 2009). In the chicken embryo, *RUNX1* expression is spatially and temporally controlled during the course of definitive hematopoiesis (Richard, et al., 2013). *RUNX1* expression, restricted to the EC layer, initiates around 50 hours of incubation in the lateral part of the paired aortas shortly followed by *c-MYB* and *PU1* expression. It then spreads latero-ventrally to become expressed by the whole

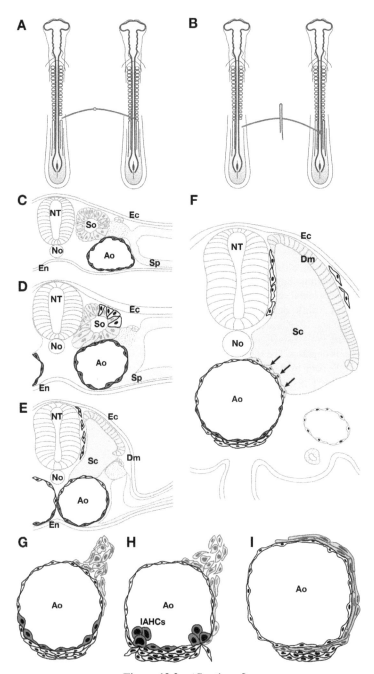

Figure 12.2 (*Continued*)

Figure 12.2 (***Continued***) Dynamics of aorta formation and role of the dorsal and ventral structures. Schematic drawings representing the role of the somite and the splanchnopleural mesoderm in the formation of the aorta. (**A, B**) Quail into chicken or chicken into chicken isotopic and isochronic grafts of a single somite (**A**) or PSM (**B**) performed at E2, at the level of the wing bud. (**C–F**) Cross-sections at the mid-trunk level. (**C**) Early-paired aorta stage. Aortic ECs (**red**) derive from the splanchnopleural mesoderm (**pink**) earlier on. The grafted somite(s) appear in green. (**D**) Mid-paired aorta stage. A small population of ECs (**yellow-green**) emerge from the dorso-lateral aspect of the somite. The splanchnopleural mesoderm has folded to come in close contact with the lateral part of the aortic rudiment. (**E**) Late-paired aorta stage. ECs from the somite have migrated to form the vascular network of the body wall and limbs and have replaced the initial aortic roof. Folding of the splanchnopleural mesoderm around the aorta is now conspicuous. Splanchnopleural mesoderm-derived cells insinuate between the ventral aspect of the aorta and the endoderm, separating the two tissues. (**F**) Early fusion stage. The aortic roof is entirely made of ECs originating from the somite. ECs of the sides and floor remain of splanchnopleural origin and are endowed with a hemogenic potential. Ventral-most cells of the sclerotome (*arrows*) migrate close to the aorta and begin to express smooth muscle cell–specific markers. Cells of the splanchnopleural mesoderm keep on separating the ventral aspect of the aorta from the endoderm. (**G**) Late fusion stage. Hemogenic ECs undergo EHT and are rounding. Smooth muscle cells from the sclerotome progressively wrap the aorta. Splanchnopleural mesoderm–derived cells have formed a new tissue immediately underneath the aorta (**pink**) designated as the sub-aortic mesenchyme. (**H**) IAHCs formation. EHT has occurred and IAHCs are formed in the ventral aspect of the aorta. As EHT occurs, hemogenic ECs are progressively replaced by ECs from the somite. Formation of the smooth muscle cell layer is ongoing. The sub-aortic mesenchyme has thickened. (**I**) Post-IAHC stage. IAHCs have left the aorta by the blood stream. The hemogenic EC has been totally replaced by ECs from the somite. The smooth muscle cell layer has formed and has moved the sub-aortic mesenchyme ventrally.

Abbreviations: Ao: aorta; Dm: dermomyotome; Ec: ectoderm; En: endoderm; IAHCs: intra-aortic hematopoietic cluster cells; No: notochord; NT: neural tube; So: somite; Sc: sclerotome; Sp: splanchnopleural mesoderm.

ventral endothelium and the IAHCs at 70 hours of incubation. This lateral-to-ventral progression is suggestive of *RUNX1* induction by adjacent cells or tissues. It also reveals that the aortic endothelium is primed to respond to this induction. The sub-aortic mesenchyme, closely associated to ECs of the aortic floor appeared as a likely candidate. By making a slit on one side of the embryo between the somites and the intermediate mesoderm to prevent mesoderm ingression underneath the aorta, it is possible to prevent the formation of the sub-aortic mesenchyme (Richard, et al., 2013). The slit does not impair vessel formation nor arterial identity, as shown by *VE-CADHERIN* and *DELTA LIKE4* expression, but blocked the initiation of *RUNX1* expression and the subsequent formation of IAHCs. These results clearly showed that the sub-aortic mesenchyme is a key element required for proper initiation of aortic hematopoiesis. To rule out the fact that the sub-aortic mesenchyme was carrying IAHC precursors, lateral plate meso-derm (but not the ECs of the vascular system) was specifically labeled by inoculation of CFDA-SE (carboxyfluorescein diacetate n-succinimidyl ester) into the coelom, one day before the formation of IAHCs. This compound enters the cells by diffusion and is cleaved by intracellular enzymes into a fluorescent, non-diffusible product, unable to leave the cell. When ana-lyzed at the time of aortic hematopoiesis, the sub-aortic mesenchyme was found labeled; however, the IAHCs were free of CFDA-SE, indicating that the sub-aortic mesenchyme did not harbor the forerunners of the IAHCs (Richard, et al., 2013).

12.6 IAHC Formation is a Conserved Mechanism in Vertebrates

It is now admitted that HSCs are generated via EHT, in the AGM region of vertebrate embryos and notably in the dorsal aorta (Ivanovs, et al., 2011; Bertrand, et al., 2010; Boisset, et al., 2010; Kissa and Herbomel, 2010; Lam, et al., 2010; Tavian, et al., 2001; Jaffredo, et al., 1998; Medvinsky and Dzierzak, 1996; Müller, et al., 1994). The process of HSC emergence is highly conserved and is regulated both in time and space albeit few species-specific differences exist (Klaus and Robin, 2017) (Figure 12.3). For instance, HSC generation is polarized and restricted to the ventral side of the aorta in the chicken, zebrafish, and human embryos (Ivanovs, et al., 2017; Yvernogeau and Robin, 2017; Bertrand, et al., 2010; Kissa and Her-bomel, 2010; Lam, et al., 2010; Tavian, et al., 1996), while in the mouse embryo, EHT is also occurring in the dorsal part of the aorta (Yokomizo and Dzierzak, 2010; Taoudi and Medvinsky, 2007; de Bruijn, et al., 2002). HSCs are produced via the formation of IAHCs in chicken, mouse, and human embryos but not in zebrafish where single cells emerge underneath the aorta, along the elongated YS, and are immediately released into the blood stream through the underlying cardinal vein (Figure 12.3) (Tavian, et al., 1996; Kissa and Herbomel, 2010; Yokomizo and Dzierzak, 2010; Yvernogeau and Robin, 2017).

IAHC/HSC generation is also regulated in time. In zebrafish embryos, this generation starts at around 26 hours post-fertilization (hpf) with a peak at 40hpf and totally ceases after 72 hpf (Bertrand, et al., 2010; Kissa and Herbomel, 2010). Using whole-mount immunostaining combined with 3D reconstruction confocal imaging system, the repartition and number

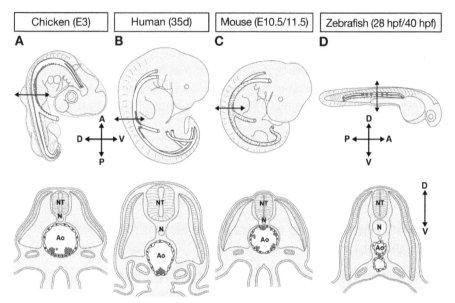

Figure 12.3 Spatial localization of IAHC cells in different species: chicken, human, mouse, and zebrafish embryos. (**A**) In chicken, IAHCs are restricted to the anterior part and to the ventral side of the dorsal aorta. (**B**) In human, IAHCs are present in the middle part of the aorta, around the umbilical artery, and restricted to the ventral side of the dorsal aorta. (**C**) In the mouse, IAHCs are mainly observed in the ventral part of the aorta but can also be found in the dorsal part. (**D**) In the zebrafish, HCs emerge as single cell in the aorta, along the elongated YS, and are immediately released through the underlying vein.

Abbreviations: Ao, aorta; N, notochord; NT; neural tube. D, dorsal; V, ventral; A, anterior; P, posterior.

Adapted from Yvernogeau et al., 2020.

of IAHCs generated in the mouse and chicken aorta had been precisely determined (Yvernogeau and Robin, 2017; Yokomizo and Dzierzak, 2010). In the mouse embryo, IAHC emergence starts at E9.5 to reach a peak of ≈ 700 IAHC cells at E10.5. Then, IAHC number progressively decreases and ceases after E14.5 (Yokomizo and Dzierzak, 2010). In the chicken embryo, IAHC emergence occurs only in the anterior portion of the aorta and starts at E2.25 to reach a peak of ≈ 1,500 IAHC cells at E3. IAHC number then decreases and are no longer observed after E5.5 (Yvernogeau and Robin, 2017). Finally, in the human embryo, IAHC/HSC emergence occurs in the middle part of the aorta from day 27 to day 42. This emergence is maximum at day 35 and generate an estimated ≈ 1,000 IAHC cells per embryo (Ivanovs, et al., 2017; Mascarenhas, et al., 2009; Tavian, et al., 1996) (Figures 12.1 and 12.3).

12.7 BM ECs Can Generate a Transient Wave of HSPCs

Whether EHT is occurring at late embryonic stages and for instance within the BM was under debate since decades. Indeed, how can one prove that ECs of the liver or the BM are capable to generate HSCs? Since ECs and HSCs share a lot of surface markers, the use of transgenic reporter lines to follow the progeny of ECs preclude any conclusion on the "spatial" origin of the tagged HSCs (YS? Aorta? BM?...). The only conclusion that can be made is that all HSCs are derived from ECs.

T. Jaffredo's group recently demonstrated that the late fetal/young adult BM ECs were capable to generate, *de novo*, a cohort of HSPCs through an EHT similar to the one observed in the aorta (Yvernogeau, et al., 2019). By transplanting PSM from GFP$^+$ transgenic embryo into wild-type chicken at the level of the limb, they could follow the formation of the vascularization of limb (see chapter above on the origin of ECs of the limb). They showed that limb vascularization was entirely derived from the PSM and identified by FACS and time-lapse confocal imaging that ECs were generating HSPCs using the same EHT as the one observed in the aorta. The newly-emerged HSPCs are capable of colonizing secondary hematopoietic organs (reflecting their multi-potency) and are long-lived since they could identify these cells in adult transplanted chicken (>5-month-old), reflecting their self-renewal capacity. The hemogenic potential of the BM ECs is also found in mammals. Using an inducible *VE-Cadherin-Cre* transgenic mouse, Yvernogeau et al. (2019) could confirm the observations made in birds. Finally, using transcriptomic approach, they demonstrated that the hemogenic ECs of the BM are transcriptionally close to the one generated in the aorta. This breakthrough discovery reveals that the micro-environment of the BM can also trigger ECs to become hemogenic and to ultimately generate HSPCs. The identification of the molecular signals originating from the BM allowing EHT to occur is of major interest and may have therapeutic potential.

12.8 Systems Used to Study Hematopoietic Cell Commitment

So far, it is not possible to generate HSCs *de novo* or to amplify them *in vitro*, which could help to overcome the HSC transplant flaws. This lack weighs heavily on health budgets and calls for a better understanding of the molecular mechanisms underlying the EHT to be able to generate HSCs from ECs *in vitro*, and use them for therapeutic purposes.

Therefore, considering this challenge, efforts have been devoted in designing *in vitro* models that faithfully recapitulate the EHT. The first system developed relies on the embryonic stem cells' (ESCs) potential to give rise to almost all cell types. Indeed, by adding specific growth factors or by over expressing specific transcription factor genes, it is possible to orient ESC toward a chosen cell fate. Thus, thanks to the use of appropriate culture conditions, ESCs can undergo differentiation and give rise to cellular structures called blast-colony forming cells, which turns out to be the *in vitro* equivalent of the nascent mesoderm, that do contain progenitors able to give rise to both ECs and HCs (Choi, et al., 1998). These early progenitors, when cultured with specific growth factors, first differentiate into hemogenic ECs that subsequently give rise to blood cells through EHT (Eilken, et al., 2009; Lancrin, et al., 2009). However, the number of hemogenic ECs is reported to be not numerous enough to study the molecular mechanisms underlying EHT.

Different groups have focused on developing model systems that explicitly and reproducibly recapitulate *in vivo* hematopoiesis. The isolation of murine and human embryonic stem cells (ESCs), the reprogramming of induced pluripotent stem cells (iPSCs) and the direct conversion of somatic cells offer a novel and unique opportunity to study blood development. Here, we will summarize the most recent advances in generating HSCs from these various sources. The team of Kateri Moore showed that the overexpression of four transcription factors, *Gata2*, *Gfi1b*, *c-Fos*, and *Etv6* within murine fibroblasts induced a hemogenic EC fate that subsequently produced hematopoietic precursor cells that express *CD34*, *Sca1*, and *Prominin1* within a global endothelial program (Pereira, et al., 2013). However, these cells displayed poor multilineage potential *in vitro*. This hemogenic phenotype completed by the absence of CD45 expression was later shown to be present in ECs of the placenta. When grafted into irradiated mice following co-culture with stromal cells, these hemogenic ECs engraft into primary and secondary recipient mice (Pereira, et al., 2016). Lacaud and co-workers showed that the ectopic expression of five transcription factors (*Erg*, *Gata2*, *Lmo2*, *Runx1C*, and *Scl*) reprogrammed fibroblasts to hematopoietic progenitors with erythrocyte, granulocyte, macrophage, and megakaryocyte potential. This reprogramming passes through a hemogenic endothelium state, that in contrast to Peirera et al., gives rise to robust clonogenic potential (Batta, et al., 2014). In recent reports, Daley and co-workers showed that seven transcription factors (*ERG, HOXA5, HOXA9, HOXA10, LCOR, RUNX1, and SPI1*) are sufficient to convert hemogenic endothelium into HSPCs that engraft primary and secondary mouse recipients with myeloid, B and T potentials. These factors are transferred into so-called CD34$^+$ hemogenic ECs and are likely to mimic molecular events occurring within the dorsal embryonic aorta (Sugimura, et al., 2017). In another study published in the same issue of Nature, Rafii and colleagues reported that constitutive expression of the transcription factors *FOSB, GFI1, RUNX1*, and *SPI1* within adult ECs co-cultured with vascular niche cells induce a hemogenic state characterized by the expression of RUNX1 and the subsequent expression of CD45. When transplanted into immuno-compromised recipients, these cells were capable of primary and secondary engraftment (Lis, et al., 2017). Interestingly, this protocol is a refinement of a previous study published in 2014 identifying the transcription factors and the requirement for a vascular niche (Sandler, et al., 2014).

The most promising approach into generating HSCs is to produce a hemogenic endothelium intermediate. Improving the differentiation protocols will also be of prime importance to improve (1) our knowledge on how ECs transdifferentiate to become hemogenic ECs then hematopoietic (stem) cells; (2) our ability to generate, at will, HSCs from naïve ECs.

To overcome this limitation, T. Jaffredo's team has developed an efficient *in vitro* model that recapitulates all the steps of mesoderm commitment toward hematopoietic production, including EHT (Yvernogeau, et al., 2016). This approach uses uncommitted quail PSM pieces put under specific culture conditions to orient the mesoderm cells toward an endothelial fate. In these conditions, hemogenic ECs are specified from the endothelium and are testified by the expression of the transcription factor *RUNX1*. Finally, hemogenic ECs undergo EHT and give birth to HCs. For each step, cells are characterized by the expression of specific genes or markers. Thereby, to study the molecular mechanisms underlying EHT, transcriptomes will be generated for each population. Specific gene expression signatures will be obtained using comprehensive meta-analysis and candidate genes will be studied using functional experiments.

12.9 Concluding Remarks

HSCs are probably one of the most studied and most well-known cell types among the stem cells. They benefit a number of phenotypic and functional assays that allowed approaching to purify the HSC population. However, despite efforts made in the last three decades, it is not possible yet to isolate HSCs at purity nor to create them *de novo* from non-hematopoietic sources thereby urgently requesting additional approaches to fill this gap.

One of the strategies to achieve this goal is to understand how HSCs are formed in the embryo. This is a long and winding path that passes through the fine-tuned dissection of embryonic hematopoiesis, the identification of the different types of progenitors produced by the different embryonic sites of hematopoiesis and by the discovery of the source of the definitive, adult type, HSCs. This source has been identified since the mid-seventies of the last century for the chicken and by the mid-nineties for the mouse embryo. It was quite at the same time that similar conclusions were drawn for the human embryo together with the idea that HSC emergence was an extremely conserved mechanism across all vertebrate species. Despite this strong conservation species-specificities complexify the painting. Several key cellular mechanisms of HSC formation, including their intra-embryonic origin, were obtained using the avian model. It is also this model that has untangled the question of the restriction of HSC production to the ventral side of the aorta and the endothelial origin of hematopoietic cells. However, does the chicken embryo represent a paradigm among vertebrates regarding the polarity of HSC production by the aorta since there are clearly various species specificities? The zebrafish uses an opposite gradient of Hedgehog (dorsal) and BMP (ventral) to regionalize the aorta; whereas, the mouse uses another, less clearly defined strategy to restrict HSC production to the aortic floor. More efforts are clearly needed to solve this important question. The endothelial origin of HSCs appears to be conserved as is the molecular control of HSC emergence i.e., the key role of Runx1 and the EHT. However, nothing is known about how HSCs become specified within the aorta since this vessel produces hundreds of hematopoietic cells among which only a few will be endowed with an HSC potential. Does it obey to a stochastic mechanism or is it something subtler that escapes our scrutiny? This is the very beginning of the fine dissection of this phenomenon. This is probably one of the most important questions to solve if one wants to mimic HSC production *ex vivo* from non-hematopoietic sources. This will pass through an in-depth knowledge of the intrinsic and extrinsic molecular mechanisms operating *in vivo* to produce HSCs but also to the adaptation and adjustment of the culture conditions to generate HSCs *ex vivo*. There is still a long way before we were able to produce HSCs for curative purposes but the different approaches that research groups worldwide are undertaking will undoubtedly feed our knowledge and certainly converge in the coming years to help producing this grail coveted for so long.

Bullet Points

- Not possible yet to isolate HSCs at purity nor to create them *de novo* from non-hematopoietic sources.
- Understanding how HSCs are formed in the embryo will certainly untangle key mechanisms toward *ex vivo* HSC production.
- HSCs first emerge in the embryo at the level of the aorta from hemogenic endothelial cells.
- The sub-aortic mesenchyme plays a critical role in triggering Runx1 expression and IAHC formation.
- The bone marrow of the late fetus/young adult contains hemogenic endothelial cells able to produce hematopoietic stem and progenitor cells.

References

Ambler, C.A., Nowicki, J.L., Burke, A.C., and Bautch, V.L. (2001). Assembly of trunk and limb blood vessels involves extensive migration and vasculogenesis of somite-derived angioblasts. Dev Biol 234, 352–64.

Batta, K., Florkowska, M., Kouskoff, V., and Lacaud, G. (2014). Direct reprogramming of murine fibroblasts to hematopoietic progenitor cells. Cell Rep 9, 1871–1884.

Beaupain, D., Martin, C., and Dieterlen-Lievre, F. (1979). Are developmental hemoglobin changes related to the origin of stem cells and site of erythropoiesis? Blood 53, 212–25.

Bechtold, T.E., Smith, P.B., and Turpen, J.B. (1992). Differential stem cell contributions to thymocyte succession during development of Xenopus laevis. J Immunol 148, 2975–82.

Bertrand, J.Y., Chi, N.C., Santoso, B., Teng, S., Stainier, D.Y., and Traver, D. (2010). Haematopoietic stem cells derive directly from aortic endothelium during development. Nature 464, 108–11.

Boisset, J.C., van Cappellen, W., Andrieu-Soler, C., Galjart, N., Dzierzak, E., and Robin, C. (2010). In vivo imaging of haematopoietic cells emerging from the mouse aortic endothelium. Nature 464, 116–20.

Chen, M.J., Yokomizo, T., Zeigler, B.M., Dzierzak, E., and Speck, N.A. (2009). Runx1 is required for the endothelial to haematopoietic cell transition but not thereafter. Nature 457, 887–91.

Chen, X.D., and Turpen, J.B. (1995). Intraembryonic origin of hepatic hematopoiesis in Xenopus laevis. J Immunol 154, 2557–67.

Choi, K., Kennedy, M., Kazarov, A., Papadimitriou, J.C., and Keller, G. (1998). A common precursor for hematopoietic and endothelial cells. Development 125, 725–32.

Ciau-Uitz, A., Walmsley, M., and Patient, R. (2000). Distinct origins of adult and embryonic blood in Xenopus. Cell 102, 787–96.

Corbel, C. (2002). Expression of alphaVbeta3 integrin in the chick embryo aortic endothelium. Int J Dev Biol 46, 827–30.

Cormier, F., and Dieterlen-Lievre, F. (1988). The wall of the chick embryo aorta harbours M-CFC, G-CFC, GM-CFC and BFU-E. Development 102, 279–85.

Cumano, A., Dieterlen-Lievre, F., and Godin, I. (1996). Lymphoid potential, probed before circulation in mouse, is restricted to caudal intraembryonic splanchnopleura. Cell 86, 907–16.

de Bruijn, M., Ma, X., Robin, C., Ottersbach, K., Sanchez, M.J., and Dzierzak, E. (2002). Hematopoietic stem cells localise to the endothelial cell layer in the midgestation mouse aorta. Immunity 16, 673–683.

de Bruijn, M.F., Speck, N.A., Peeters, M.C., and Dzierzak, E. (2000). Definitive hematopoietic stem cells first develop within the major arterial regions of the mouse embryo. Embo J 19, 2465–74.

Dieterlen-Lièvre, F. (1975). On the origin of haematopoietic stem cells in avian embryos: an experimental approach. J. Embryol. Exp. Morphol. 33, 609–619.

Dieterlen-Lièvre, F., and Martin, C. (1981). Diffuse intraembryonic hemopoiesis in normal and chimeric avian development. Dev Biol 88, 180–91.

Dieterlen-Lièvre, F., Pouget, C., Bollérot, K., and Jaffredo, T. (2006). Are Intra-Aortic Hemopoietic Cells Derived from Endothelial Cells During Ontogeny? Trends in Cardiovascular Medicine 16, 128–139.

Drevon, C., and Jaffredo, T. (2014). Cell interactions and cell signaling during hematopoietic development. Exp Cell Res 329, 200–6.

Dzierzak, E., and Bigas, A. (2018). Blood Development: Hematopoietic Stem Cell Dependence and Independence. Cell Stem Cell 22, 639–651.

Eilken, H.M., Nishikawa, S., and Schroeder, T. (2009). Continuous single-cell imaging of blood generation from haemogenic endothelium. Nature 457, 896–900.

Federici, H. (1926). Recherches experimentales sur les potentialités de l'îlot sanguin chez l'embryon Rana fusca. Arch. Biol. 36, 466–488.

Godin, I., Dieterlen-Lievre, F., and Cumano, A. (1995). Emergence of multipotent hemopoietic cells in the yolk sac and paraaortic splanchnopleura in mouse embryos, beginning at 8.5 days postcoitus. Proc Natl Acad Sci U S A 92, 773–7.

Godin, I.E., Garcia-Porrero, J.A., Coutinho, A., Dieterlen-Lievre, F., and Marcos, M.A. (1993). Para-aortic splanchnopleura from early mouse embryos contains B1a cell progenitors. Nature 364, 67–70.

Gritz, E., and Hirschi, K.K. (2016). Specification and function of hemogenic endothelium during embryogenesis. Cellular and molecular life sciences : CMLS 73, 1547–67.

Hodgson, G.S., and Bradley, T.R. (1979). Properties of haematopoietic stem cells surviving 5-fluorouracil treatment: evidence for a pre-CFU-S cell? Nature 281, 381–2.

Hollyfield, J.G. (1966). The origin of erythroblasts in Rana pipiens tadpoles. Developmental Biology 14, 461–480.

Ivanovs, A., Rybtsov, S., Ng, E.S., Stanley, E.G., Elefanty, A.G., and Medvinsky, A. (2017). Human haematopoietic stem cell development: from the embryo to the dish. Development 144, 2323–2337.

Ivanovs, A., Rybtsov, S., Welch, L., Anderson, R.A., Turner, M.L., and Medvinsky, A. (2011). Highly potent human hematopoietic stem cells first emerge in the intraembryonic aorta-gonad-mesonephros region. J Exp Med 208, 2417–27.

Jacobson, L.O., Simmons, E.L., Marks, E.K., and Eldredge, J.H. (1951). Recovery from radiation injury. Science 113, 510–11.

Jaffredo, T., Gautier, R., Eichmann, A., and Dieterlen-Lievre, F. (1998). Intraaortic hemopoietic cells are derived from endothelial cells during ontogeny. Development 125, 4575–83.

Jones, R.J., Wagner, J.E., Celano, P., Zicha, M.S., and Sharkis, S.J. (1990). Separation of pluripotent haematopoietic stem cells from spleen colony-forming cells. Nature 347, 188–9.

Kau, C.L., and Turpen, J.B. (1983). Dual contribution of embryonic ventral blood island and dorsal lateral plate mesoderm during ontogeny of hemopoietic cells in *Xenopus laevis*. J. Immunol. 131, 2262–2266.

Kissa, K., and Herbomel, P. (2010). Blood stem cells emerge from aortic endothelium by a novel type of cell transition. Nature 464, 112–5.

Klaus, A., and Robin, C. (2017). Embryonic hematopoiesis under microscopic observation. Dev Biol 428, 318–327.

Lam, E.Y., Hall, C.J., Crosier, P.S., Crosier, K.E., and Flores, M.V. (2010). Live imaging of Runx1 expression in the dorsal aorta tracks the emergence of blood progenitors from endothelial cells. Blood 116, 909–14.

Lancrin, C., Sroczynska, P., Stephenson, C., Allen, T., Kouskoff, V., and Lacaud, G. (2009). The haemangioblast generates haematopoietic cells through a haemogenic endothelium stage. Nature 457, 892–5.

Lassila, O., Eskola, J., Toivanen, P., Martin, C., and Dieterlen-Lievre, F. (1978). The origin of lymphoid stem cells studied in chick yolk sac-embryo chimaeras. Nature 272, 353–4.

Lis, R., Karrasch, C.C., Poulos, M.G., Kunar, B., Redmond, D., Duran, J.G.B., Badwe, C.R., Schachterle, W., Ginsberg, M., Xiang, J., et al. (2017). Conversion of adult endothelium to immunocompetent haematopoietic stem cells. Nature 545, 439–445.

Lorenz, E., Uphoff, D., Reid, T.R., and Shelton, E. (1951). Modification of irradiation injury in mice and guinea pigs by bone marrow injections. J Natl Cancer Inst 12, 197–201.

Maeno, M., Tochainai, S., and Katagiri, C. (1985). Differential participation of ventral and dorsolateral mesoderms in the hemopoiesis of *Xenopus*, as revealed in diploid-triploid or interspecific chimeras. Dev Biol. 110, 503–508.

Martin, C. (1972). Method of explantation in ovo of the blastoderm of bird embryos. C R Seances Soc Biol Fil 166, 283–5.

Martin, C., Beaupain, D., and Dieterlen-Lievre, F. (1978). Developmental relationships between vitelline and intra-embryonic haemopoiesis studied in avian 'yolk sac chimaeras'. Cell differentiation 7, 115–30.

Mascarenhas, M.I., Parker, A., Dzierzak, E., and Ottersbach, K. (2009). Identification of novel regulators of hematopoietic stem cell development through refinement of stem cell localization and expression profiling. Blood 114, 4645–53.

McCulloch, E.A., and Till, J.E. (1960). The radiation sensitivity of normal mouse bone marrow cells, determined by quantitative marrow transplantation into irradiated mice. Radiat Res 13, 115–25.

Medvinsky, A., and Dzierzak, E. (1996). Definitive hematopoiesis is autonomously initiated by the AGM region. Cell 86, 897–906.

Medvinsky, A.L., Samoylina, N.L., Muller, A.M., and Dzierzak, E.A. (1993). An early pre-liver intraembryonic source of CFU-S in the developing mouse. Nature 364, 64–7.

Minko, K., Bollerot, K., Drevon, C., Hallais, M.F., and Jaffredo, T. (2003). From mesoderm to blood islands: patterns of key molecules during yolk sac erythropoiesis. Gene Expr Patterns 3, 261–72.

Moore, M.A., and Metcalf, D. (1970). Ontogeny of the haemopoietic system: yolk sac origin of in vivo and in vitro colony forming cells in the developing mouse embryo. Br J Haematol 18, 279–96.

Moore, M.A., and Owen, J.J. (1967). Chromosome marker studies in the irradiated chick embryo. Nature 215, 1081–2.

Müller, A.M., Medvinsky, A., Strouboulis, J., Grosveld, F., and Dzierzak, E. (1994). Development of hematopoietic stem cell activity in the mouse embryo. Immunity 1, 291–301.

Murray, P.D.F. (1932). The development "in vitro" of blood of the early chick embryo. Strangeways Res. Lab. Cambridge., 497–521.

Noden, D.M. (1989). Embryonic origins and assembly of blood vessels. Am Rev Respir Dis 140, 1097–103.

North, T., Gu, T.L., Stacy, T., Wang, Q., Howard, L., Binder, M., Marin-Padilla, M., and Speck, N.A. (1999). Cbfa2 is required for the formation of intra-aortic hematopoietic clusters. Development 126, 2563–75.

North, T.E., de Bruijn, M.F., Stacy, T., Talebian, L., Lind, E., Robin, C., Binder, M., Dzierzak, E., and Speck, N.A. (2002). Runx1 expression marks long-term repopulating hematopoietic stem cells in the midgestation mouse embryo. Immunity 16, 661–72.

Palis, J., Robertson, S., Kennedy, M., Wall, C., and Keller, G. (1999). Development of erythroid and myeloid progenitors in the yolk sac and embryo proper of the mouse. Development 126, 5073–84.

Pardanaud, L., and Dieterlen-Lievre, F. (1993). Emergence of endothelial and hemopoietic cells in the avian embryo. Anat Embryol (Berl) 187, 107–14.

Pardanaud, L., and Dieterlen-Lievre, F. (1999). Manipulation of the angiopoietic/hemangiopoietic commitment in the avian embryo. Development 126, 617–27.

Pardanaud, L., Luton, D., Prigent, M., Bourcheix, L.-M., Catala, M., and Dieterlen-Lièvre, F. (1996). Two distinct endothelial lineages in ontogeny, one of them related to hemopoiesis. Development 122, 1363–1371.

Pereira, C.F., Chang, B., Gomes, A., Bernitz, J., Papatsenko, D., Niu, X., Swiers, G., Azzoni, E., de Bruijn, M.F., Schaniel, C., et al. (2016). Hematopoietic Reprogramming In Vitro Informs In Vivo Identification of Hemogenic Precursors to Definitive Hematopoietic Stem Cells. Dev Cell 36, 525–39.

Pereira, C.F., Chang, B., Qiu, J., Niu, X., Papatsenko, D., Hendry, C.E., Clark, N.R., Nomura-Kitabayashi, A., Kovacic, J.C., Ma'ayan, A., et al. (2013). Induction of a hemogenic program in mouse fibroblasts. Cell Stem Cell 13, 205–18.

Pouget, C., Gautier, R., Teillet, M.-A., and Jaffredo, T. (2006). Somite-derived cells replace ventral aortic hemangioblasts and provide aortic smooth muscle cells of the trunk. Development 133, 1013–1022.

Pouget, C., Pottin, K., and Jaffredo, T. (2008). Sclerotomal origin of vascular smooth muscle cells and pericytes in the embryo. Developmental Biology 315, 437–447.

Richard, C., Drevon, C., Canto, P.-Y., Villain, G., Bollérot, K., Lempereur, A., Teillet, M.-A., Vincent, C., Rosselló Castillo, C., Torres, M., et al. (2013). Endothelio-Mesenchymal Interaction Controls runx1 Expression and Modulates the notch Pathway to Initiate Aortic Hematopoiesis. Developmental Cell 24, 600–611.

Rowe, R.G., Mandelbaum, J., Zon, L.I., and Daley, G.Q. (2016). Engineering Hematopoietic Stem Cells: Lessons from Development. Cell Stem Cell 18, 707–20.

Sandler, V.M., Lis, R., Liu, Y., Kedem, A., James, D., Elemento, O., Butler, J.M., Scandura, J.M., and Rafii, S. (2014). Reprogramming human endothelial cells to haematopoietic cells requires vascular induction. Nature 511, 312–8.

Sugimura, R., Jha, D.K., Han, A., Soria-Valles, C., da Rocha, E.L., Lu, Y.F., Goettel, J.A., Serrao, E., Rowe, R.G., Malleshaiah, M., et al. (2017). Haematopoietic stem and progenitor cells from human pluripotent stem cells. Nature 545, 432–438.

Swiers, G., Rode, C., Azzoni, E., and de Bruijn, M.F. (2013). A short history of hemogenic endothelium. Blood Cells Mol Dis.

Taoudi, S., and Medvinsky, A. (2007). Functional identification of the hematopoietic stem cell niche in the ventral domain of the embryonic dorsal aorta. Proc Natl Acad Sci U S A 104, 9399–403.

Tavian, M., Coulombel, L., Luton, D., San Clemente, H., Dieterlen-Lièvre, F., and Peault, B. (1996). Aorta-associated CD34[+] hematopoietic cells in the early human embryo. Blood 87, 67–72.

Tavian, M., Robin, C., Coulombel, L., and Peault, B. (2001). The human embryo, but not its yolk sac, generates lympho-myeloid stem cells: mapping multipotent hematopoietic cell fate in intraembryonic mesoderm. Immunity 15, 487–95.

Tompkins, R.V., E. P; Reinschmidt, D. (1980). Origin of hemopoietic stem cells in amphibian ontogeny. In Developement and differentiation of Vertebrate Lymphocytes., Horton, J.D., ed. (Amsterdam: Elsevier/north Holland,), pp. 25–34.

Turpen, J.B., and Knudson, C.M. (1982). Ontogeny of hematopoietic cells in Rana pipiens: precursor cell migration during embryogenesis. Dev Biol 89, 138–51.

Turpen, J.B., Knudson, C.M., and Hoefen, P.S. (1981). The early ontogeny of hematopoietic cells studied by grafting cytogenetically labeled tissue anlagen: localization of a prospective stem cell compartment. Dev Biol 85, 99–112.

Turpen, J.B., Marrion, R.M., and Williams, K. (1983). Peritoneal exudate in larval Rana pipiens contains cells that are embryologically derived from dorsal lateral plate mesoderm. Dev Comp Immunol 7, 295–302.

Wasteson, P., Johansson, B.R., Jukkola, T., Breuer, S., Akyurek, L.M., Partanen, J., and Lindahl, P. (2008). Developmental origin of smooth muscle cells in the descending aorta in mice. Development 135, 1823–32.

Weissman, I.L., Papaioannou, V.E., and Gardner, R.L. (1977). Fetal hematopoietic origins of the adult hematolymphoid system. Cold Spring Harbor 5, 33–43.

Wiegreffe, C., Christ, B., Huang, R., and Scaal, M. (2007). Sclerotomal origin of smooth muscle cells in the wall of the avian dorsal aorta. Dev Dyn 236, 2578–85.

Wiegreffe, C., Christ, B., Huang, R., and Scaal, M. (2009). Remodeling of aortic smooth muscle during avian embryonic development. Dev Dyn 238, 624–31.

Wilting, J., Brand-Saberi, B., Huang, R., Zhi, Q., Kontges, G., Ordahl, C.P., and Christ, B. (1995). Angiogenic potential of the avian somite. Dev Dyn 202, 165–71.

Yokomizo, T., and Dzierzak, E. (2010). Three-dimensional cartography of hematopoietic clusters in the vasculature of whole mouse embryos. Development 137, 3651–61.

Yvernogeau, L., Auda-Boucher, G., and Fontaine-Perus, J. (2012). Limb bud colonization by somite-derived angioblasts is a crucial step for myoblast emigration. Development 139, 277–87.

Yvernogeau, L., Gautier, R., Khoury, H., Menegatti, S., Schmidt, M., Gilles, J.F., and Jaffredo, T. (2016). An *in vitro* model of hemogenic endothelium commitment and hematopoietic production. Development 143, 1302–12.

Yvernogeau, L., Gautier, R., Petit, L., Khoury, H., Relaix, F., Ribes, V., Sang, H., Charbord, P., Souyri, M., Robin, C., et al. (2019). In vivo generation of haematopoietic stem/progenitor cells from bone marrow-derived haemogenic endothelium. Nat Cell Biol 21, 1334–1345.

Yvernogeau, L., and Robin, C. (2017). Restricted intra-embryonic origin of bona fide hematopoietic stem cells in the chicken. Development 144, 2352–2363.

13

Developmental Biology of Hematopoietic Stem Cells: Non-cell Autonomous Mechanisms

Leslie Nitsche and Katrin Ottersbach[*]

Centre for Regenerative Medicine, Institute for Regeneration and Repair,
University of Edinburgh, Edinburgh, UK
E-mail: katrin.ottersbach@ed.ac.uk
*Corresponding Author

13.1 Introduction

Unlike pluripotent stem cells (PSCs), somatic stem cells are often difficult
to maintain and manipulate in culture. This is due to the fact that tissue-
specific stem cells reside in complex, multi-factorial, three-dimensional
environments, which control their generation, maintenance, proliferation,
self-renewal, differentiation, and migration and which are challenging to
recreate *in vitro*. The immediate surroundings of a stem cell are generally
referred to as its "niche", which is commonly composed of supporting cells
that regulate stem cell behavior via direct cell–cell interactions and through
the secretion of soluble factors that bind to receptors on the surface of stem
cells, thus triggering downstream intracellular signaling pathways. Other
important components of the niche are the extracellular matrix and the local
oxygen concentration.

The behavior of stem cells can also be controlled by more distant events
of what would then be referred to as their microenvironment, which could
also include systemic signals coming, for example, from the brain. The wider
context of their microenvironment is especially at play during development
as tissues constantly develop and are re-modelled in close proximity to each
other, often producing a functional interplay between neighboring tissues that
will be further discussed below.

Hematopoietic stem cells (HSCs) are self-renewing cells that have the potential to repopulate the entire hematopoietic system after transplantation and are therefore of paramount interest for treatment of blood and immune system disorders. Despite many attempts over the decades to develop robust protocols for the *ex vivo* expansion of HSCs, these clinically important cells are an example of a somatic stem cell that has proven to be difficult to maintain and expand in culture, although some remarkable progress has been made recently (Wilkinson et al., 2019). For this reason, the cellular composition of the bone marrow (BM) niche, in which HSCs mostly reside in adult life, as well as the signals that regulate HSC behavior have been the subject of intense research over many years (reviewed in (Gao et al., 2018; Morrison and Scadden, 2014)), with the advent of single-cell RNA sequencing technology allowing for a detailed dissection of the cellular heterogeneity within the BM (Baccin et al., 2020; Baryawno et al., 2019; Tikhonova et al., 2019; Wolock et al., 2019). Many of these different cell types have been linked to HSC support, including osteolineage cells, endothelial cells, perivascular cells, mesenchymal stromal cells (MSCs), non-myelinating Schwann cells, sympathetic nerves, and even mature blood cell types such as macrophages and megakaryocytes, many of which provide important HSC factors such as Cxcl12 and Kitl (Gao et al., 2018). A picture is starting to emerge in which different cell types/niches selectively maintain HSCs when they are in a quiescent or an active state, or more mature progenitors versus HSCs, which will be discussed in more detail elsewhere in this book.

In contrast to the wealth of knowledge that has been accumulated for the BM over the years, comparatively little is known about the niches that regulate HSC generation, maturation, expansion, and migration during embryonic and fetal development. One striking feature of developmental hematopoiesis is the employment of shifting locations. This may be due to the constant remodeling of tissues as part of normal morphogenesis, which may result in some niches no longer being supportive for HSCs, or it may be easier for HSCs to find new locations that better support the next step in their maturation, rather than for niches to be repurposed for the changing needs of the developing blood system.

13.2 Developmental Niches

Broadly speaking, developmental hematopoiesis can be divided into three overlapping waves. The first of these, known as the primitive wave, initiates at embryonic day (E)7.5 of mouse development in the extra-embryonic yolk sac (YS) and consists of primitive erythrocytes, macrophages, and megakaryocytes (Figure 13.1A). Its main purpose is to provide oxygen supply and aid during morphogenesis (reviewed in (McGrath et al., 2015)). This is followed by the first of two definitive waves, which also originates from the YS and is characterized by the appearance of two important blood progenitors, the erythro-myeloid progenitor (Palis et al., 2001), which has no lymphoid potential, and a complementary cell type in the form of an immune-restricted lymphoid-primed progenitor, that has no erythroid or megakaryocyte potential (Boiers et al., 2013) (Figure 13.1B,C). The third wave initiates through the emergence of the first directly transplantable HSCs, which are detected from E10.5 in the intra-embryonic aorta-gonads-mesonephros (AGM) region (Figure 13.1D) (Medvinsky and Dzierzak, 1996; Muller et al., 1994). Subsequently, HSCs are also detected in the YS, the embryonic head, and the placenta (Gekas et al., 2005; Li et al., 2012; Ottersbach and Dzierzak, 2005), although it is currently unclear whether they are generated independently in all of those additional sites. Shortly after their appearance, HSCs start colonizing the fetal liver (Gekas et al., 2005), where their numbers increase dramatically and where the hierarchy of different blood lineages is established, reminiscent of the adult hematopoietic tree (Figure 13.1E).

It is thus clear that the embryo utilizes different niches for the consecutive steps of HSC generation, maturation, expansion, and differentiation, which is an obligatory developmental pathway that cannot be completed entirely through the direct transplantation of nascent HSCs into the adult BM niche (Mascarenhas et al., 2016). The AGM is the first and most robust site of HSC emergence, and will therefore be the focus of the remaining sections of this chapter; however, it cannot support the expansion and differentiation of HSCs (Godin et al., 1999). This happens to some degree in the placenta, but primarily in the fetal liver (Gekas et al., 2005). Attention has therefore focused on the fetal liver microenvironment to find methods for the *ex vivo* expansion of HSCs, although detailed knowledge on the fetal liver niche remains scarce. There is evidence for the atypical Notch ligand Dlk1 and Dlk1-expressing hepatic progenitors as strong supporters of HSC maintenance and expansion in the fetal liver (Chou and Lodish, 2010; Moore et al., 1997), although, interestingly, Dlk1 appears to have the opposite effect on HSC numbers in the AGM (Mirshekar-Syahkal et al., 2013). In addition, and in analogy to the BM niche, HSCs in the fetal liver were recently observed to locate close to NG2+ perivascular cells, which were found to be essential for HSC support (Khan et al., 2016). Furthermore, the same study reported an expansion

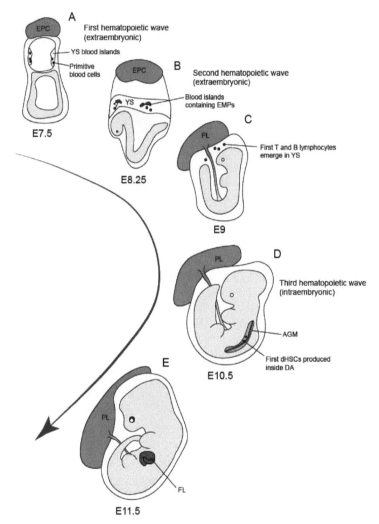

Figure 13.1   The three hematopoietic waves and their developmental niches. (**A**) The first (primitive) hematopoietic wave originates in the extra-embryonic yolk sac (YS) of the E7.5 mouse embryo and includes primitive erythrocytes, macrophages, and megakaryocytes. (**B**) The second (definitive) hematopoietic wave produces erythroid-myeloid progenitors (EMPs) within blood islands derived from the E8.25 YS. (**C**) The second wave continues at E9 to produce the first lymphoid-primed progenitors. (**D**) The third (definitive) hematopoietic wave stems from the aorta-gonad-mesonephros (AGM) region at E10.5, which produces the first definitive hematopoietic stem cells (dHSCs). (**E**) HSCs go on to seed the fetal liver (FL) where the cells will expand and proliferate. From this point on, the FL becomes the primary site of embryonic hematopoiesis. EPC-ectoplacental cone; DA—dorsal aorta; PL—placenta.

of the vascular network in the fetal liver coincident with an expanded HSC pool, while a remodeling of the hepatic vasculature around birth may offer an explanation why HSCs relocate to the BM at this time point. A novel way for HSCs to shape their own niche was also recently described for the zebrafish-equivalent of the fetal liver—the caudal hematopoietic tissue (Tamplin et al., 2015). Specifically, through *in vivo* imaging, HSCs were shown to extravasate from the circulation into the sub-vascular space, where they caused endothelial cells to form a protective pocket around them, which also included a stromal cell that determined the plane of HSC division. The same study also showed images that suggested that a similar "cuddling" process takes place in the mouse fetal liver. However, despite these numerous studies, robust protocols for the *ex vivo* expansion of HSCs based on data from the fetal liver, remain elusive.

13.3 Cell-intrinsic Factors of Hematopoietic Stem Cell Generation

13.3.1 The Endothelial-to-Hematopoietic Transition

As mentioned above, definitive HSCs (dHSCs) emerge during the third hematopoietic wave in the AGM. The process by which they first arise is termed the endothelial-to-hematopoietic transition (EHT), which occurs in a narrow developmental window beginning around E10.5 in mice. A subset of endothelial cells (ECs) in the wall of the dorsal aorta acquire a hematopoietic transcriptional program that transforms them into hemogenic endothelial cells (HECs) that can ultimately give rise to hematopoietic stem and progenitor cells (HSPCs) (Figure 13.2A–D). These transcriptional changes are brought on by numerous signals coming from both inside the cells as well as the surrounding environment, which further encourage the formation of intra-aortic hematopoietic clusters (IAHCs) by the HECs, which are aggregates of cells attached to the aortic wall co-expressing endothelial and hematopoietic markers. The nascent precursor cells mature within IAHCs until they bud off as single dHSCs (Figure 13.2C,D) (Boisset et al., 2015; de Bruijn et al., 2002; Jaffredo et al., 1998; Lancrin et al., 2009; Rybtsov et al., 2014). This process has notably been captured via live imaging in both zebrafish and mice (Bertrand et al., 2010; Boisset et al., 2010; Kissa and Herbomel, 2010). Strikingly, quantification of absolute dHSC numbers in the mouse embryo have shown that there are likely no more than ~1 dHSC present in the AGM by the end of embryonic day (E)10 (Kumaravelu et al., 2002). As soon as they have emerged, dHSCs will enter the circulation and carry on toward the fetal liver.

The maturation of dHSCs from VE-cadherin–expressing endothelial cells has been further divided into four major steps that are defined by differences in temporal emergence and by the sequential upregulation of hematopoietic markers, beginning with VE-cad+CD41+CD43-CD45- pro-HSCs, which go on to become VE-cad+CD41+CD43+CD45- type I pre-HSCs, followed by VE-cad+CD41+CD43+CD45+ type II pre-HSCs and finally dHSCs. Type II pre-HSCs are phenotypically the same as the matured dHSCs, but can be discerned using transplantation experiments, as only dHSCs are able to fully repopulate the blood system long term when transplanted directly into irradiated mice (Rybtsov et al., 2014; Rybtsov et al., 2011).

13.3.2 Transcription Factor Dynamics During the EHT

Due to the intricacy and complexity of the processes unfolding in the AGM at the time of dHSC emergence, the EHT must be tightly controlled by a network of transcription factors that are reliably up- or downregulated at the correct time. Runx1 and Gata2 are indispensable for HSC emergence and are widely accepted to be the most critical transcription factors of the EHT (Chen et al., 2009; de Pater et al., 2013; Okuda et al., 1996; Tsai et al., 1994; Wang et al., 1996). Runx1 knockout in mice, as well as knockout of the Runx1 cofactor CBFβ, results in embryonic lethality during midgestation, while Runx1 haploinsufficiency displays a dose-dependent effect on HSC activity in the AGM (Cai et al., 2000). Importantly, Runx1 expression in the aortic endothelium is thought to be the main driving force behind the transition from hemogenic endothelium to hematopoietic progenitor cell (Chen et al., 2009; Kissa and Herbomel, 2010), whereas it is no longer required after the EHT for dHSC maintenance (Tober et al., 2013). Genome-wide analysis of transcription factor interactions revealed that Runx1 interacts with other known hematopoietic factors such as Scl/Tal1 and Fli1 in HECs in order to initiate hematopoietic fate-associated transcription (Lichtinger et al., 2012). Other direct targets of Runx1 include Gfi1 and Gfi1b, which are responsible for the downregulation of the endothelial program and initiation of the hematopoietic program during the onset of hematopoiesis (Lancrin et al., 2012).

Along with Runx1, Gata2 is the other major conductor of hematopoietic fate. This transcription factor is expressed in the endothelium and in hematopoietic cells of the AGM, as well as the YS, fetal liver, and placenta. Tissue-specific deletions have notably shown that Gata2 is necessary for the

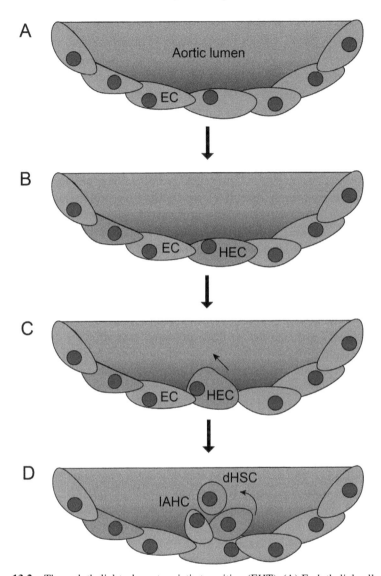

Figure 13.2 The endothelial-to-hematopoietic transition (EHT). (**A**) Endothelial cells (ECs, **pink**) line the wall of the dorsal aorta. (**B**) A subset of ECs switches to a hematopoietic transcriptional program forming hemogenic endothelial cells (HECs, **gray**). (**C**) HECs undergo morphological changes that cause them to round off and adhere to the endothelial wall inside the aortic lumen. (**D**) Intra-aortic hematopoietic clusters (IAHCs) containing HECs and HSC precursors are formed. Inside these clusters, single definitive hematopoietic stem cells (dHSCs, **green**) mature and bud off, from where they will travel through the circulation to the fetal liver.

generation and survival of HSCs specifically in the AGM (de Pater et al., 2013; Ling et al., 2004). The Notch pathway is understood to induce both Runx1 and Gata2 in the AGM (Burns et al., 2005; Robert-Moreno et al., 2008). Gata2 expression in particular appears to be entirely dependent on Notch1 and its ligand Jagged1 (Jag1), as Jag1 knockout leads to complete Gata2 loss and a lack of definitive hematopoiesis (Robert-Moreno et al., 2008). One mechanism linking Notch1 and Gata2 is through the Notch-dependent activation of Hes1, which functions to repress Gata2 after its expression is no longer required at high levels in the AGM (Guiu et al., 2013). If Gata2 levels are sustained for a prolonged period, hematopoietic progenitor formation becomes severely defective, highlighting the importance of timing during definitive hematopoiesis. Furthermore, investigation of expression dynamics using a Gata2 reporter line found that Gata2 is expressed in a pulsatile manner in single cells undergoing the EHT and that these cells are likely to be in an unstable genetic state during the transition process (Eich et al., 2018). Notch signaling itself becomes attenuated by E11.5 in the AGM, after which HSC emergence no longer relies on the Notch pathway (Souilhol et al., 2016b).

Another study showed that Notch1 becomes restricted by Jag1 in the hemogenic endothelium (HE), in order to induce a hematopoietic fate rather than an arterial identity (Gama-Norton et al., 2015). Arterial specification is likewise mediated by the transcription factor Sox17, which is expressed strongly in arterial compared to venous cells (Corada et al., 2013). Sox17 acts upstream of Notch signaling in both arterial fate specification as well as in the HE and the emerging HSCs, where Sox17 has also been shown to be highly expressed (Clarke et al., 2013). In the HE, Sox17 acts as a repressor of Runx1 and Gata2 while maintaining endothelial arterial fate. When Sox17 is downregulated, Runx1 and Gata2-mediated hematopoietic transcriptional programs are initiated. Runx1 notably reciprocally inhibits Sox17, prompting a rapid increase in Runx1 expression once Sox17 starts to become downregulated (Lizama et al., 2015).

The transcriptional network that enables the formation of dHSCs from ECs has in recent years been subject to detailed analyses using modern techniques such as single-cell RNA sequencing (scRNA-seq) technology, which allows researchers to dissect the transcriptome of each individual cell of interest. One of the first studies that investigated the EHT using PCR-based single-cell expression analysis employed a +23 Runx1 enhancer-reporter element (23GFP) to sort HECs and directly look at the initiation of the hematopoietic program. Swiers et al. found that HECs acquire hematopoietic properties already at E9.5, much earlier than previously thought (Swiers et al., 2013). A major breakthrough in the field was the first successful scRNA-seq of early and late-stage pre-HSCs. These experiments revealed complex transcriptional dynamics, a role for mTOR signaling in the pre-HSCs compared to ECs and increased proliferation for pre-HSCs in the AGM, which stands in contrast to slow-cycling adult HSCs (Zhou et al., 2016). Coupled with single-cell ATAC-sequencing (scATAC-seq), which allows insight into chromatin accessibility within single cells, scRNA-seq has been employed to track transcriptional landscapes across the EHT trajectory (Zhu et al., 2020). This has been particularly informative for transcription factor activity at the various EHT phases and has given more detailed insight into each respective pro/pre-HSC stage. As such, scRNA-seq has allowed the identification of novel EHT markers, which will facilitate future research in the field. Oatley et al. recently published data describing the cell surface molecule CD44 as a robust marker for arterial endothelial cells that give rise to HSPCs. In combination with the known markers VE-Cadherin and cKit, CD44 can be used to further investigate different steps of the EHT process (Oatley et al., 2020). Finally, a newly published study by Vink et al. applied iterative scRNA-seq rounds in order to obtain the transcriptome of the extremely rare newly formed dHSCs in the AGM. By using data generated in previous rounds of sequencing to inform the markers chosen for dHSC isolation, it was eventually possible to generate single-cell datasets of this minute cell population (Vink et al., 2020). Table 13.1 provides further details on the specific cell populations sequenced in the above-mentioned studies.

13.4 Supporting Nascent Hematopoietic Stem Cells

The AGM niche comprises numerous different cell types, including endothelial, mesenchymal, and nervous system cells, which together orchestrate a network of signals that enable the emergence of the first dHSCs. Researching non-cell autonomous support provided by the AGM microenvironment is a critical part of recreating HSC generation *in vitro* for medical purposes. Extensive progress has been made in the understanding of niche signaling, including the cell types that are involved, the pathways that become activated and the signals that are secreted. The rest of this chapter will focus on these different components of the niche and how together they ensure dHSC emergence (Figure 13.3). It will also give insight into the history of the field and the areas that still require further investigation.

Table 13.1 Cell populations analyzed in the outlined sequencing studies

Population Type	Markers	Embryonic Stage	Method	Reference
ECs	Runx1+23GFP⁻Cdh5⁺Ter119⁻CD45⁻CD41⁻	E8.5,	scqRT-PCR,	Swiers et al. 2013
HECs	Runx1+23GFP⁺Cdh5⁺Ter119⁻CD45⁻CD41⁻	E9.25,	microarrays	
HSPCs	Runx1+23GFP⁺Cdh5⁺Ter119⁻CD45⁻CD41⁺	E10.5		
ECs	CD31⁺Cdh5⁺CD41⁻CD43⁻CD45⁻Ter119⁻	E11.5		
Pre-HSC I	CD31⁺CD45⁻CD41^{lo}cKit⁺EPCR^{hi}	E11.5	10-cell/	Zhou et al. 2016
Pre-HSC II	CD31⁺CD45⁺CD41^{lo}	E11.5	scRNA-seq	
HSC (FL)	Lin⁻Sca1⁺Mac1^{lo}EPCR⁺	E12		
HSC (FL)	CD45⁺CD150⁺CD48⁻EPCR⁺	E14		
ECs	Ter119⁻CD41⁻CD45⁻CD31⁺CD144⁺ESAM⁺Kit⁻GFP⁻	E9.5, E10.5	scRNA-seq	Zhu et al. 2020
HECs	Ter119⁻CD41⁻CD45⁻CD31⁺CD144⁺ESAM⁺Kit$^{lo/-}$GFP⁺	E9.5, E10.5		
E+HE+IACs	Ter119⁻CD41$^{med/-}$CD31⁺CD144⁺ESAM⁺	E9.5, E10.5		
IACs	Ter119⁻CD41$^{med/-}$CD31⁺CD144⁺ESAM⁺Kit⁺	E10.5, E11.5		
Pre-HSC II	CD45⁺CD27⁺CD144⁺	E11.5	scRNA-seq,	
E+HE+IACs	Ter119⁻CD41$^{med/-}$CD31⁺CD144⁺ESAM⁺CD44⁺	E10.5	scATAC-seq	
ECs-HSPCs	Cdh5⁺	E10.5	scRNA-seq	Oatley et al. 2020
Non-HECs	Cdh5⁺CD44⁻	E10	scqRT-PCR	
HECs	Cdh5⁺CD44^{lo}ckit⁻			
Pre-HSPC I	Cdh5⁺CD44^{lo}ckit⁺			
Pre-HSPC II	Cdh5⁺CD44^{hi}			
Pro-HSCs	Cdh5⁺CD41⁺CD43⁻CD45⁻	E10	scqRT-PCR	
Pre-HSC I	Cdh5⁺CD41⁺CD43⁺CD45⁻			
Pre-HSC II	Cdh5⁺CD41⁺CD43⁺CD45⁺			
Non-HECs	Cdh5⁺CD44⁻	E9.5/10/11	bulk RNA-seq	
HECs	Cdh5⁺CD44^{lo}ckit⁻			
Pre-HSPC I	Cdh5⁺CD44^{lo}ckit⁺	E9.5/10		
Pre-HSPC II	Cdh5⁺CD44^{hi}	E11		
IAHCs	CD31⁺cKit^{hi}Gata2^{med}	E11		Vink et al. 2020
Multipotent HSPCs	CD31^{hi}SSC^{lo}cKit^{hi}Gata2^{med}CD27^{+N}	E11	scRNA-seq	

EC—endothelial cells; HEC—hemogenic endothelial cells; HSPC—hematopoietic stem and progenitor cell; HSC—hematopoietic stem cell; FL—fetal liver; IAC—intra-aortic cluster (cell).

E11.5 mouse embryo

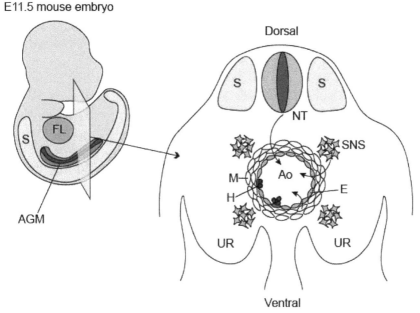

Figure 13.3   The aorta-gonad-mesonephros (AGM) niche. Diagram of a transverse section along the dorso-ventral axis depicting cell types present in the E11.5 AGM. Curved arrows indicate secretion of signals received by the developing HSCs. FL—fetal liver; S—somites; NT—neural tube; Ao—(dorsal) aorta; EC—endothelial cell (**orange**); MC—mesenchymal cell (**yellow**); HC—hematopoietic cell (**red**); SNS—sympathetic nervous system cell cluster (**purple**); UR—urogenital ridges.

13.4.1 Endothelial Cells

Endothelial cells clearly stand at the center of definitive hematopoiesis through direct transformation of endothelial cells into hematopoietic cells during the EHT. The conversion of endothelial cells into HSCs, thereby mimicking the EHT, has notably been successful *in vitro*. Adult endothelial cells were co-cultured with vascular niche endothelial cells and transiently induced with the transcription factors *FOSB*, *GFI1*, *RUNX1*, and *SPI1*, which produced functional HSCs (Lis et al., 2017). This is regarded as a major development in the field, as the cells generated were engraftable HSCs capable of self-renewal and the reconstitution of all hematopoietic lineages. The key to this successful conversion from endothelium to HSPC appears to be the signals provided by the vascular niche through co-culture, including Notch and BMP pathway signaling, as well as the secretion of essential cytokines and chemoattractants.

An aspect of niche endothelial cells that requires further exploration concerns the IAHC microniche: IAHCs derived from the dorsal endothelium are complex structures containing a heterogeneous mix of cells along the developmental axis, from endothelial cells at the more basal layers, to early hematopoietic progenitor cells, to more mature progenitor cells (Yokomizo and Dzierzak, 2010; Boisset et al., 2015). The interplay of these cells and their contribution to dHSC formation is still unknown and may be of considerable interest for future research.

13.4.2 Signals From Ventral Versus Dorsal Microenvironments

Location is a crucial variable throughout blood development. From early hematopoiesis in the YS to the maturation of the first HSCs in the fetal liver, each location offers a unique microenvironment capable of supporting each step of early hematopoiesis. The relevance of location is not limited to the various developing organs but is even a determining factor at different points of the same tissue.

One notable example of this is the varying microenvironment surrounding the dorsal aorta (Ao). In the mouse AGM region, IAHCs can be found attached to both the dorsal side (AoD) and ventral floor (AoV) of the dorsal aorta, but to a large extent only the clusters of the AoV go on to produce dHSCs, despite cells of clusters from both sides having very similar transcriptomes (Baron et al., 2018; Taoudi and Medvinsky, 2007). Using an *ex vivo* culture system, it was shown that the difference in hematopoietic potential does not come from the AoV- or AoD-derived progenitors themselves, but from the surrounding ventral or dorsal tissue, which comprises gut progenitor cells on the ventral side and the neural tube on the dorsal side. More specifically, tissues of the AoV microenvironment secrete inductive signals in the early E10 embryo, including members of the hedgehog signaling pathway that are indispensable for embryonic patterning, while the AoD does not receive these same signals (Peeters et al., 2009). Through further studies it has become apparent that there are several rapidly changing signals that are spatially polarized between the AoV and AoD, with signals promoting HSC generation in the AoV, while inhibiting emergence at the AoD. Inductive signals are mediated by Stem cell factor (Scf), which is highly expressed in endothelial cells of the AoV and in the urogenital ridges of the AGM, along with Sonic hedgehog (Shh) and Indian hedgehog (Ihh) which stimulate HSC development at E10.5 but not at E11.5. Hh signaling molecules were found to be expressed throughout the AGM at E10, including high expression of the Hh targets Gli1, Gli2, and Gli3 in mesenchymal cells and low expression in hematopoietic and endothelial cells. These findings are evidence of a role of Hh signaling through mesenchymal cells, rather than directly through the hemogenic endothelium (Peeters et al., 2009). Shh secreted by the mesenchyme surrounding the Ao is thought to induce HSC production in the AoV, where an increased number of Hh targets are located. The role of Hh proteins in mouse embryos is furthermore in line with what has been detected in human cells. Blood cells isolated from human embryos show expression of Hh proteins, which increased the repopulation ability of the transplanted cells through modulation of Bone morphogenic protein 4 (Bmp4) (Bhardwaj et al., 2001). Shh in the ventral mesenchyme is in an antagonistic relationship with Bmp4, which acts to inhibit Shh but is in turn blocked by the ventrally polarized Noggin (Souilhol et al., 2016a).

Bmp4 is one of the best-defined hematopoietic factors secreted by the ventral mesenchyme in many different organisms, including zebrafish, frog, mouse, and human. In Xenopus, Bmp4 notably induces the expression of hematopoietic transcription factors such as Gata2 (Maeno et al., 1996). Bmp4 expression is furthermore downregulated within intra-aortic clusters, where there is a higher concentration of the Bmp4 inhibitor Noggin. This downregulation is thought to be necessary for cluster formation, as was shown through an endothelial knockout of Smad4, an important molecule in the Bmp4 signal pathway. Loss of Smad4 leads to an increase in Bmp4 expression, with a subsequently higher number of intra-aortic clusters (Lan et al., 2014).

13.4.3 Subaortic Mesenchyme

Mesenchymal cells surrounding the dorsal aorta have long been acknowledged for their contribution to the hematopoietic supportive niche. Early studies demonstrated the ability of quail embryonic dorsal aortae transplanted with surrounding mesenchyme into the midgut of chick embryos to generate para-aortic foci—early hematopoietic differentiation sites. When the dorsal aortae were transplanted after being cleared of any surrounding tissue, this effect was not replicated (Le Douarin et al., 1984).

In the mouse, mesenchymal precursors with multilineage differentiation potential have been identified in major embryonic hematopoietic sites including the AGM and the fetal liver. At E11, the precursors localize within the AGM to the urogenital ridges and to the tissue surrounding the dorsal aorta, where they continue to expand over the course of HSC emergence (Mendes et al., 2005). Factors secreted by the mesenchyme can affect HSC generation directly or indirectly and have been the focus of many research groups studying early hematopoiesis over the past 20 years.

Using transcriptional studies of AGM-derived stromal cell lines, which can be cultured and analyzed *in vitro*, signals that increase HSC repopulation activity and maintain the cells in the lab have been determined, including Bmp4 and the nerve growth factor β-NGF. When Bmp4 signaling is blocked by addition of an inhibitor, HSC activity is abolished (Durand et al., 2007). While many AGM regulators secreted by the mesenchyme have been found to positively contribute to HSC emergence, negative regulation of emerging HSC numbers also appears to be vital in the AGM. Delta-like homolog 1 (Dlk1) was identified as a potential AGM regulator through microarray expression analysis and was later shown to control the extent of HSC expansion through its expression in the subaortic mesenchyme. This negative regulation is undoubtedly an important facet of normal hematopoiesis, as aberrant Dlk1 has been associated with malignancy (Ma et al., 2012; Mascarenhas et al., 2009; Mirshekar-Syahkal et al., 2013). Dlk1, like many other hematopoiesis-associated genes, appears to be under the control of the master regulator Runx1. While Runx1 activity in the AGM is mainly associated with the hematopoietic program initiated in the endothelium of the dorsal aorta, the transcription factor is still being investigated in the supportive niche, including the mesenchyme. Here, it was shown that Gata3 directly targets the transcription factor Runx1. It is thought that Gata3 and Runx1 in the mesenchyme therefore act together with other hematopoietic factors to accurately time HSC emergence (Fitch et al., 2020).

Recently, RNA-seq studies (described above) have gained traction as a method to explore the AGM microenvironment and to identify previously undetected supportive factors. By sequencing dorsal and ventral AGM subsections and comparing their gene expression profiles to the transcriptome of the bone marrow-derived stromal line OP9, a cell line frequently used to support hematopoietic progenitor formation *in vitro*, it was possible to determine common transcriptional signatures that may play a role in HSC support. Indeed, functional validation confirmed Bmper, a modulator of the BMP pathway, as an HSC regulator secreted by the subaortic mesenchyme (McGarvey et al., 2017).

Genome-wide RNA tomography, which involves performing RNA-seq on individual tissue sections, is another powerful method recently used to profile the hematopoietic niche. A 2020 study by Yvernogeau and colleagues compared the spatial transcriptomics of zebrafish, chicken, mouse, and human embryonic aortic microenvironments to identify ligand-receptor cross-talks between niche and HSPCs. Interactions shared across species included ADM-RAMP2, a previously unidentified regulator of HSC emergence (Yvernogeau et al., 2020).

13.4.4 The Sympathetic Nervous System

Mesenchymal and endothelial cell involvement in definitive hematopoiesis has been the focus of numerous studies pertaining AGM niche signaling, and their importance in hematopoietic development has been widely accepted. However, in the last decade an initially unexpected role has also emerged for cells of the sympathetic nervous system (SNS), which are surrounded by mesenchyme and form clusters inside the AGM not far from the dorsal aorta (Figure 13.3).

The SNS became implicated in embryonic hematopoiesis through the transcription factor Gata3, which was found to be significantly upregulated in a microarray gene expression analysis in the AGM at the time of HSC emergence (Mascarenhas et al., 2009). Gata3 is part of the GATA family of transcription factors, which also includes the hematopoietic regulator Gata2, and is abundantly expressed in a variety of different tissues, including kidneys, skin, hair, the hematopoietic system with emphasis on T-cell development, and the SNS (reviewed in (Zaidan and Ottersbach, 2018)). Gata3-deficient mouse embryos notably die by E12 due to severe abnormalities in the nervous system and at hematopoietic sites including the fetal liver (Pandolfi et al., 1995). Despite Gata3 expression in adult HSCs and a potential role in regulating their cell cycle (Buza-Vidas et al., 2011; Frelin et al., 2013; Ku et al., 2012), Gata3 is absent from HSCs emerging in the AGM. Nevertheless, Fitch et al. found fewer IAHCs after Gata3 knockout at E11.5 and reduced numbers of HSCs (Fitch et al., 2012). Remarkably, they discovered that this phenotype was due to a role of Gata3 in SNS cells within the niche rather than due to a Gata3-mediated defect in the hematopoietic cells themselves. The observed decrease in HSCs appeared to be linked to Gata3 regulation of catecholamine signaling through its transcriptional control of tyrosine hydroxylase, which is responsible for catecholamine synthesis in SNS cells. The impaired hematopoietic stem and progenitor phenotype could notably be rescued through *in vitro* and *in vivo* addition of catecholamines (Fitch et al., 2012).

13.4.5 Immune-derived Signals

Signaling pathways involved in inflammation normally become activated in response to an insult or pathogen, which leads to a subsequent release of inflammatory cytokines, hormones, and other pro-inflammatory factors to combat the infection. Some of these inflammatory pathways markedly overlap with those pathways activated during hematopoietic development in the absence of disease, therefore named "sterile immune response," which is in accordance with the finding that cytokines play an active role during HSC generation. Cytokines are small proteins primarily classified as modulators of the immune response. They are largely known for their essential role in inflammation in response to infection or injury, but more recently have been recognized for their wide-ranging involvement during development. The diverse effects and pleiotropic nature of cytokines, from growth factors to chemoattractants, have made them an interesting point of research in the field of embryonic hematopoiesis.

The family of cytokines called interleukins (ILs) has been strongly implicated in HSC regulation. The Il1 receptor (Il1r), for example, was found to be expressed in the ventral mesenchyme and dorsal aorta of the AGM, and its knockout resulted in decreased HSC activity (Orelio et al., 2008). Furthermore, Il3 appears to be one of the most critical cytokines involved in embryonic hematopoiesis. This signaling protein was first linked to HSC production in the AGM at E11 when researchers showed that adding Il3 to Runx1 haploinsufficient mice strikingly led to HSC deficiency rescue, implicating Il3 as a potential target of the master hematopoietic transcription factor Runx1. CD34+ckit+ cells of the AGM, which include hematopoietic cells within IAHCs expressed the Il3r, and blocking Il3 led to a significant decrease in HSC activity (Robin et al., 2006). Further studies using chemical inhibitors revealed that Il3 is part of the JAK/STAT signaling cascade, which enhances cell survival (Mascarenhas et al., 2016).

Due to its critical role in HSC emergence, Il3 has been studied for its potential effects on co-culture or co-aggregate studies for hematopoietic differentiation along with FMS-like tyrosine kinase 3 ligand (Flt3l) and Scf (Robin et al., 2006; Taoudi et al., 2008). Studies by the Medvinsky group, whose work has been central to the dissection of EHT stages (see above), revealed that the earliest E9.5 HSC precursors, pro-HSCs, express the Scf receptor cKit at low levels, and addition of Scf only was enough to produce engraftable dHSCs in culture. Il3 largely did not affect dHSC output from precursors until E11.5, while Scf continued to be a potent promoter of hematopoietic fate even at this time point (Rybtsov et al., 2014). This finding helped to reinforce the position of Scf as an essential cytokine in AGM hematopoiesis.

One of the microarray gene expression screens that showed Il3 pathway upregulation also identified enhanced thrombopoietin (Thpo) expression around the E11 dorsal aorta (Mascarenhas et al., 2009). Thpo is a cytokine and hormone originally associated with platelet production in adults.

Mascarenhas et al. showed that like Il3, Thpo activated the JAK/STAT pathway and enhanced hematopoietic progenitor and HSC emergence in the AGM (Mascarenhas et al., 2016; Mascarenhas et al., 2009). More recently, Thpo involvement in hematopoiesis has been attributed to the formation of hematopoietic clusters. By transducing E10.5 AGM cells expressing low levels of the hematopoietic marker CD45 with Sox17, a transcription factor and hematopoietic regulator, Harada and colleagues were able to maintain hematopoietic cell clusters similar to hemogenic endothelium. Central to this cluster formation was the addition of Thpo to the transduced cells cultured with OP9 stromal cells. As might be anticipated given the strong response to Thpo stimulation, the Thpo receptor Mpl was found to be highly expressed on Sox17-transduced undifferentiated cells in culture. Furthermore, this discovery was in line with the observation that Mpl is expressed in IAHCs and a subset of endothelial cells in E10.5 dorsal aortae *in vivo* (Harada et al., 2017; Petit-Cocault et al., 2007).

Another cytokine involved in the JAK/STAT pathway that appears to participate in early HSC generation, is interferon-α (Ifna). Hematopoietic cells in the E11.5 AGM express the IFN-α receptor Ifnar1, and treatment of HSCs with Ifna results in increased engraftment and repopulation ability. The data indicate a role for Ifna in the maturation of AGM HSCs as they become closer transcriptionally and functionally to fetal liver HSCs (Kim et al., 2016). This stands in contrast to Ifng, which functions prior to Ifna during the hematopoietic process upon induction by Notch signaling. In zebrafish, Ifng partially controls HSC emergence itself through activation of the JAK/STAT signaling pathway member Stat3 (Sawamiphak et al., 2014).

Notch pathway signaling in the early AGM is also linked to TGFβ, a highly researched cytokine in the fields of immunology, infection, and cancer biology. Studies of TGFβ and its ligands in hematopoiesis have been challenging in mice, as knockout embryos die by E10.5. However, investigations using zebrafish embryos have been more successful. Monteiro et al. performed a series of knockouts using morpholinos and found that ligands TGFβ1a/1b and TGFβ3, which are expressed in the zebrafish endothelium of the dorsal aorta and in the notochord, respectively, are necessary for HSPC production. Using a combination of chemical inhibitors and gene expression analysis, the group showed that vascular endothelial growth factor (Vegf) was responsible for *tgfb1a/1b* transcription in the endothelium, which subsequently led to a hemogenic endothelial program through a cascade of Notch ligand Jagged-1a activation that in turn led to upregulation of essential hematopoietic genes such as *runx1* and *gata2*. Once again, timing may be a critical variable, as evidence points toward a later activation of *tgfb3* as it acts on the hemogenic endothelial cells induced by *tgfb1a/1b* expression (Monteiro et al., 2016).

Ly6a (Sca-1) is expressed in AGM HSCs and can therefore be used in combination with a GFP transgene to investigate early HSC emergence (de Bruijn et al., 2002). OP9 co-culture experiments using Ly6a-GFP+ endothelial and intra-aortic hematopoietic AGM cells from murine embryos, demonstrated that Ly6a-GFP+ hematopoietic cells derived from intra-aortic clusters were strongly enriched for HSCs compared to Ly6a-GFP+ endothelial cells or any Ly6a-GFP- cells. This HSC-enriched population of cells expressed genes strongly associated with the innate immune response and inflammatory response, such as toll-like receptors (TLR) and interferon (IFN) regulators. The group further described compelling evidence for interferon-mediated regulation of HSPCs (Li et al., 2014). Non-pathogenic inflammation as a mediator of HSC emergence was further demonstrated in zebrafish. Here, He and colleagues found that upon TLR stimulation, nuclear factor kappa-light-chain-enhancer of activated B (NF-κB) is activated by cytokines such as tumor necrosis factor-α (TNF-α), which in turn stimulates HSPC-enhancing Notch signaling. This was discovered by knocking out TLR-4 myeloid differentiation primary response 88 (MyD88) in zebrafish embryos, which robustly reduced HSPC numbers (He et al., 2015). The link between TNF-α, Notch signaling, and the NF-κB inflammatory pathway had previously also been described by Espín-Palazón et al. in zebrafish (Espin-Palazon et al., 2014). The exact extent of modulation and integration between the Notch and NF-κB pathways still needs to be fully investigated.

13.4.6 Macrophages

Macrophages are phagocytes capable of performing important innate immunity functions as well as activating the acquired immune system. Different types of macrophages can fulfil either pro- or anti-inflammatory roles that either stimulate or dampen the immune response. In the embryo, macrophages arise early in all three hematopoietic waves, starting with the primitive wave at E7.5 in the extra-embryonic YS. These YS-derived macrophages, which are derived from the first or second hematopoietic wave, become widespread in the early embryo, and their many functions, including development, homeostasis and wound healing, are still being explored (Rae et al., 2007). Over the past decade, however, the theory that YS-derived macrophages contribute to the regulation of dHSC generation in the third hematopoietic wave has been substantiated.

Travnickova and colleagues used zebrafish embryos with transgenically labeled HSPCs and macrophages to confirm the presence of primitive macrophages in the AGM at the time of dHSC emergence, which was further verified in human embryos using immunohistochemistry. Live imaging of the transgenic zebrafish showed a striking amount of direct interaction between macrophages and HSPCs, with almost 96% of HSPCs being in contact with one or more macrophages. Accordingly, when macrophages were chemically or genetically inhibited, this prevented definitive hematopoiesis. The group next proposed that the role of macrophages during HSC emergence was the release of matrix metalloproteases (MMPs), which are responsible for extracellular matrix (ECM) degradation. MMPs such as MMP-9 are secreted by macrophages and break down the ECM around HSPCs, therefore enabling HSPC mobilization and migration, which subsequently allows the HSPCs to colonize the early hematopoietic organs in the zebrafish embryos (Travnickova et al., 2015).

Recently, the study of macrophages in mouse AGMs has yielded intriguing findings that outline a dynamic role for macrophages in definitive hematopoiesis. Using a reporter mouse line with labeled macrophage stimulating factor receptor *Csf1r* expression, Mariani et al. found that YS-derived macrophages appear to make up a vast majority of hematopoietic cells in the AGM, where they accumulate before HSC emergence begins. Similarly to what had previously been observed in zebrafish, the macrophages were directly interacting with IAHCs, and HSC generation was reduced upon macrophage inhibition. Interestingly, the group also detected a CD206-expressing macrophage subpopulation, which promoted EHT and displayed a distinctly pro-inflammatory gene signature including expression of chemokine, MMP, Tnf, and Bmp genes. These data indicate that in addition to ECM degradation, macrophages are likely to influence the EHT through a pleiotropic inflammatory phenotype (Mariani et al., 2019).

13.4.7 Systemic Factors

13.4.7.1 Blood Flow

HSCs first emerge not far from the developing embryonic heart, which begins to beat approximately 2 days before their emergence in the mouse AGM, even though oxygen still reaches the tissue by diffusion at this point and does not require blood flow–mediated oxygenation (Ji et al., 2003). The significance of early blood flow was therefore a point of interest in the field of embryonic development. Studies investigating the effect of blood flow on HSC generation found that the sheer stress induced by circulation causes release of nitric oxide (NO) by endothelial cells. Using "silent heart" zebrafish embryos, which do not exhibit blood flow and therefore lack NO activation by AGM endothelium, North et al. reported altered EHT. NO is thought to regulate the movement of endothelial cells, which therefore may directly affect the formation of IAHCs. Furthermore, NO may be involved in the specification of the vascular niche, including establishment of vasculature around the major arteries of HSC development (North et al., 2009). These findings were corroborated by another study published in the same year using mouse embryonic stem cells subjected to sheer stress. The cultured cells displayed augmented hematopoietic potential upon administration of fluid sheer stress, which was again mediated by NO (Adamo et al., 2009). Most recently, the North group developed a dorsal aorta-on-a-chip approach which described Yes-activated protein (YAP) as a mechanotransducer stimulated by blood flow. This YAP activation led to accelerated HSPC generation from hemogenic endothelium, which may be a vital piece of information in the pursuit of *in vitro* HSPC/HSC generation (Lundin et al., 2020).

13.4.7.2 Hormone Signaling

So far, this chapter has dealt with the close-range interactions that enable precise regulation of HSC/HSPC emergence in the murine AGM. For example, hormone secretion by the SNS within the AGM is a form of paracrine (close range) signaling necessary for the HSC generation process. However, endocrine signaling, i.e. signaling from distant parts of the embryo has also been described as an HSPC regulator.

During development, various stressors such as inflammation, hypoxia or metabolic stress may occur in the embryo. The stressor induces activation of the well-known hypothalamic–pituitary–adrenal/interrenal (HPA/I) stress response axis, which is also an essential stress response pathway in adults. Through knockout studies in zebrafish, Kwan et al. found that central nervous system (CNS) signaling via serotonin as part of the stress response regulates HSPC numbers in the AGM. Here, serotonergic signaling causes the release of glucocorticoids (GC) into the blood, which once in the AGM act directly through *runx1*-mediated transcriptional control of GC receptors on the progenitor cells. The CNS may therefore be responsible for the embryo's rapid reaction to physiological or environmental stressors, which allows for the adjustment of HSPC production in response to a given situation (Kwan et al., 2016). A more direct role for serotonin in AGM HSPC production was described in another study published one year later. Lv and colleagues reported that serotonin is synthesized in endothelial cells of the murine AGM itself, where it promotes HSPC proliferation and inhibits apoptosis of newly formed HSPCs. Taken together, these findings characterize serotonin as a vital HSPC regulator, both in terms of paracrine signaling in the AGM and through endocrine signaling in peripheral tissue (Lv et al., 2017).

Vitamin D is another hormone known to be involved in numerous different physiological functions, including the absorption of vital elements such as calcium. The vitamin D biosynthetic pathway initiates with the generation of the vitamin D precursor cholecalciferol (D3) in response to UV light, which is inactive, in contrast to its downstream product vitamin D. Mutations in vitamin D receptor (VDR) signaling have previously been linked to aberrant Hh signaling, an important pathway in definitive hematopoiesis (see above), which has led to investigations of vitamin D in the developing AGM. Indeed, increased levels of the inactive D3 in zebrafish embryos led to impaired HSPC production and niche specification of hemogenic endothelium through Hh axis inhibition. This finding implicates D3 as a negative regulator of HSPC emergence via Hh signaling (Cortes et al., 2015). However, another study by the same lab found that active vitamin D in the AGM causes transcription of the chemokine CXCL8 through vitamin D binding to the VDR. Subsequent downstream signaling through the AKT pathway, which controls proliferation, results in an increase in HSPC production in the AGM (Cortes et al., 2016). These findings attribute contrasting roles to the vitamin D pathway in HSC development, with components of the pathway acting as both negative and positive regulators.

Additional systemic factors include prostaglandin E2 (PGE2), which belongs to a group of fatty acid–derived signaling molecules called eicosanoids. Prostaglandins are potent vasodilators responsible for increasing blood flow in many different parts of the body. PGE2 was initially identified in a zebrafish screen for definitive hematopoiesis modulators, and subsequent studies in murine embryos showed an increase in repopulating HSCs after PGE2 exposure. The PGE2 regulators cyclooxygenase 1 and 2 were furthermore found to be expressed on AGM endothelial cells and HSPCs, respectively. PGE2-mediated regulation of HSPCs is thought to occur at least in part through PGE2 interaction with the Wnt pathway (Goessling et al., 2009; North et al., 2007). Cannabinoids are another group of fatty acid–derived molecules that interact with eicosanoids and are also linked to HSC development. Treatment of zebrafish embryos with cannabinoid receptor 2 agonists enhances HSC output via the P-selectin pathway at the time of HSC generation, while later stimulation increases the number of HSPCs in various hematopoietic niches, including the thymus and the caudal hematopoietic tissue, where HSCs expand after emergence (Esain et al., 2015). Evidently, lipid modulators are powerful regulators of definitive hematopoiesis that may be medically useful once applied to patient samples in a lab setting.

13.4.7.3 Metabolism

Recently, a link has been made between cholesterol metabolism and HSC emergence in zebrafish embryos. Cholesterol is transported by high-density lipoprotein in the bloodstream, which is made up of multiple components including Apolipoprotein A-I binding protein 2 (Aibp2). Removal of Aibp2 in zebrafish embryos, which is associated with higher levels of cholesterol, inhibited the formation of HSCs from HE through downstream regulation of Notch signaling (Gu et al., 2019).

Another metabolic pathway previously investigated for its role in HSC development is glucose metabolism, an essential part of cellular respiration. In zebrafish, high levels of glucose led to enhanced HSC formation at the time of emergence, both in terms of accelerated generation and increased HSC numbers. The mechanism behind this expansion is thought to be an increase in reactive oxygen species by augmented oxidative phosphorylation (as a result of the high glucose levels), which in turn activates hypoxia-inducible factor 1α (Hif1a) signaling. Hif1a is thought to be largely induced by low oxygen levels (hypoxia) in order to protect the embryo from hypoxic stress, which may explain the powerful effects of Hif1a-induction at the site of definitive hematopoiesis (Harris et al., 2013).

13.4.8 Hypoxia

Indeed, Hif1a has been identified as an essential modulator of HSC development. Imanirad et al. investigated the influence of hypoxia on the EHT and blood formation and found hypoxic conditions surrounding the dorsal aorta in mouse embryos at the time of HSC emergence. Endothelial-specific loss of Hif1a furthermore negatively affected progenitor numbers and development in the AGM and the placenta, a finding which is in line with the effect of Hif1a loss on adult bone marrow HSCs (Imanirad et al., 2014). As hypoxia is widespread in the early embryo, it is therefore likely that adaptive mechanisms must be in place in order for the embryo to develop properly and for definitive hematopoiesis to proceed in hypoxic stress conditions.

13.5 Future Directions

The field of embryonic hematopoiesis remains an auspicious area of research due to the unified objective of generating transplantable HSCs *in vitro* for the treatment of patients with blood cancers or immunodeficiencies.

An essential part of this process is understanding fetal hematopoiesis in humans specifically. Embryonic studies in humans have proven to be uniquely challenging due to the scarcity and availability of human embryos, as well as ethical concerns. Nevertheless, over the past decades progress has been made in dissecting human fetal hematopoiesis and how it compares to murine, chick or zebrafish systems (reviewed in (Ivanovs et al., 2017; Kumar et al., 2019)). An intriguing development in the field of human embryogenesis has been the utilization of scRNA-seq technology to maximize the information that can be obtained from a single embryo, a highly advantageous tool due to the short supply of human tissue. A recent study by the Liu lab describes the transcriptome of the cells along the EHT trajectory in human AGMs, including a description of previously unidentified HSC-primed HEC subpopulations (Zeng et al., 2019).

Discoveries made using scRNA-seq can be used to inform experiments involving human pluripotent stem cells (hPSCs). This cell culture method is largely implemented to achieve the goal of personalized (patient-specific) treatments and to further our understanding of human hematopoiesis. hPSCs simultaneously are promising for direct use in medicine and as a platform for modeling disease and testing the effect of potential drugs on HSPCs. One of the first studies that reported successful transplant of induced PSCs (iPSCs)-derived HSCs described the injection of human iPSCs with OP9-stromal cells into immunodeficient mice, which generated functional blood stem cells during teratoma formation (Amabile et al., 2013). Naturally, this approach is not easily applicable or scalable as a treatment, so the search for lab-generated HSCs has actively continued. Since then, major progress has been made in the differentiation of hPSCs into functional HSCs. Sugimura et al. describe the generation of repopulating HSCs from hPSC-derived HE through transduction with seven transcription factors identified as sufficient to produce HSCs that give rise to all erythroid, myeloid, and lymphoid lineages upon transplantation into irradiated recipient mice (Sugimura et al., 2017). While this undoubtedly represents a significant breakthrough in the *in vitro* generation of HSCs, the ultimate target continues to be the generation of HSCs in culture without the use of exogenously added transgenes.

Lastly, there has recently been increased interest in the impact of physical forces on HSC emergence in the early embryo. Much like signaling molecules, biomechanical properties can lead to the activation of certain pathways within the cell. Understanding the forces acting on blood stem cells *in vivo* may be relevant to the production of HSCs *in vitro*. As previously described in this chapter, biomechanical forces including blood flow do indeed appear to impact definitive hematopoiesis inside the AGM (Adamo et al., 2009; North et al., 2009). Using live-imaging in zebrafish, Lancino and colleagues found that the physical constriction of endothelial cells undergoing the EHT is anisotropic (directionally dependent) and that the cells align along the anterior-posterior axis in accordance with blood flow. Along with this, actin-myosin complexes in the surrounding vascular niche participate in this constriction. Together, it is likely that these cytoskeletal forces control the direction of HSC emergence (Lancino et al., 2018). Poullet et al. further performed experiments using 4D confocal microscopy (3D microscopy over time) to track aortic remodeling during the EHT. In accordance with Lancino et al., they found that movement of HSPCs into the aortic space is controlled by an actin ring around the cells undergoing EHT. They go on to suggest that growth of the dorsal aorta spontaneously causes stresses that can aid cell extrusion during the transition process (Poullet et al., 2019). Due to its promising nature, this mostly unexplored facet of HSC emergence will certainly be subject to more detailed studies in the years to come.

In conclusion, while much progress has been made in the hematopoietic field over the past few decades, it is currently still not possible to recapitulate the HSC production process *in vitro*. This failure to faithfully reproduce *in vivo* hematopoiesis could be due to the multitude of external influences, the nuanced complexity of transcriptional dynamics and a necessity for accurate timing of each variable. Recent research, including the study conducted by Vink et al., have highlighted the sensitivity of each participating factor, as only small changes in gene expression may lead to the production of a slightly different cell type with varying potencies (Vink et al., 2020). A robust HSC production protocol, as well as a reliable strategy for the isolation of HSCs that does not rely on transplantation, will therefore continue to be the subject of many studies in the years to come.

13.6 Conclusions

- Hematopoietic stem cells (HSCs) reside in their local tissue environment, called "niche," which is made up of supportive cell

types that regulate stem cell generation, maturation, maintenance, and differentiation via signaling pathways.
- HSCs can be found in a variety of different niches along the developmental trajectory, including the extra-embryonic yolk sac, the aorta-gonad-mesonephros (AGM) region, the embryonic head, the placenta, and the fetal liver.
- The first definitive HSCs emerge in the dorsal aorta within the AGM via the endothelial-to-hematopoietic transition (EHT).
- The EHT is regulated by a network of transcription factors, with Runx1 and Gata2 being the most important drivers.
- Within the AGM niche, mesenchymal tissue surrounding the dorsal aorta sends numerous key signals for HSC emergence along the dorso-ventral axis along with signals secreted by clusters of sympathetic nervous system cells.
- Immune-derived signals, such as cytokines and chemokines secreted by macrophages and other immune cells, have been shown to greatly affect HSC output.
- Blood flow, hormone signaling and oxygen levels are just some of the many physical conditions and systemic variables that can affect blood formation in the AGM.
- In future, research will continue to focus on the ultimate goal of accurately recapitulating HSC emergence *in vitro*, in order to produce HSCs for medical purposes, such as drug modeling and disease treatment.

References

Adamo, L., Naveiras, O., Wenzel, P.L., McKinney-Freeman, S., Mack, P.J., Gracia-Sancho, J., Suchy-Dicey, A., Yoshimoto, M., Lensch, M.W., Yoder, M.C., Garcia-Cardena, G., Daley, G.Q., 2009. Biomechanical forces promote embryonic haematopoiesis. Nature 459, 1131–1135.

Amabile, G., Welner, R.S., Nombela-Arrieta, C., D'Alise, A.M., Di Ruscio, A., Ebralidze, A.K., Kraytsberg, Y., Ye, M., Kocher, O., Neuberg, D.S., Khrapko, K., Silberstein, L.E., Tenen, D.G., 2013. In vivo generation of transplantable human hematopoietic cells from induced pluripotent stem cells. Blood 121, 1255–1264.

Baccin, C., Al-Sabah, J., Velten, L., Helbling, P.M., Grunschlager, F., Hernandez-Malmierca, P., Nombela-Arrieta, C., Steinmetz, L.M., Trumpp, A., Haas, S., 2020. Combined single-cell and spatial transcriptomics reveal the molecular, cellular and spatial bone marrow niche organization. Nat Cell Biol 22, 38–48.

Baron, C.S., Kester, L., Klaus, A., Boisset, J.C., Thambyrajah, R., Yvernogeau, L., Kouskoff, V., Lacaud, G., van Oudenaarden, A., Robin, C., 2018. Single-cell transcriptomics reveal the dynamic of haematopoietic stem cell production in the aorta. Nat Commun 9, 2517.

Baryawno, N., Przybylski, D., Kowalczyk, M.S., Kfoury, Y., Severe, N., Gustafsson, K., Kokkaliaris, K.D., Mercier, F., Tabaka, M., Hofree, M., Dionne, D., Papazian, A., Lee, D., Ashenberg, O., Subramanian, A., Vaishnav, E.D., Rozenblatt-Rosen, O., Regev, A., Scadden, D.T., 2019. A Cellular Taxonomy of the Bone Marrow Stroma in Homeostasis and Leukemia. Cell 177, 1915–1932 e1916.

Bertrand, J.Y., Chi, N.C., Santoso, B., Teng, S., Stainier, D.Y., Traver, D., 2010. Haematopoietic stem cells derive directly from aortic endothelium during development. Nature 464, 108–111.

Boiers, C., Carrelha, J., Lutteropp, M., Luc, S., Green, J.C., Azzoni, E., Woll, P.S., Mead, A.J., Hultquist, A., Swiers, G., Perdiguero, E.G., Macaulay, I.C., Melchiori, L., Luis, T.C., Kharazi, S., Bouriez-Jones, T., Deng, Q., Ponten, A., Atkinson, D., Jensen, C.T., Sitnicka, E., Geissmann, F., Godin, I., Sandberg, R., de Bruijn, M.F., Jacobsen, S.E., 2013. Lymphomyeloid contribution of an immune-restricted progenitor emerging prior to definitive hematopoietic stem cells. Cell Stem Cell 13, 535–548.

Boisset, J.C., Clapes, T., Klaus, A., Papazian, N., Onderwater, J., Mommaas-Kienhuis, M., Cupedo, T., Robin, C., 2015. Progressive maturation toward hematopoietic stem cells in the mouse embryo aorta. Blood 125, 465–469.

Boisset, J.C., van Cappellen, W., Andrieu-Soler, C., Galjart, N., Dzierzak, E., Robin, C., 2010. In vivo imaging of haematopoietic cells emerging from the mouse aortic endothelium. Nature 464, 116–120.

Burns, C.E., Traver, D., Mayhall, E., Shepard, J.L., Zon, L.I., 2005. Hematopoietic stem cell fate is established by the Notch-Runx pathway. Genes Dev 19, 2331–2342.

Buza-Vidas, N., Duarte, S., Luc, S., Bouriez-Jones, T., Woll, P.S., Jacobsen, S.E., 2011. GATA3 is redundant for maintenance and self-renewal of hematopoietic stem cells. Blood 118, 1291–1293.

Cai, Z., de Bruijn, M., Ma, X., Dortland, B., Luteijn, T., Downing, R.J., Dzierzak, E., 2000. Haploinsufficiency of AML1 affects the temporal and spatial generation of hematopoietic stem cells in the mouse embryo. Immunity 13, 423–431.

Chen, M.J., Yokomizo, T., Zeigler, B.M., Dzierzak, E., Speck, N.A., 2009. Runx1 is required for the endothelial to haematopoietic cell transition but not thereafter. Nature 457, 887–891.

Chou, S., Lodish, H.F., 2010. Fetal liver hepatic progenitors are supportive stromal cells for hematopoietic stem cells. Proc Natl Acad Sci U S A 107, 7799–7804.

Clarke, R.L., Yzaguirre, A.D., Yashiro-Ohtani, Y., Bondue, A., Blanpain, C., Pear, W.S., Speck, N.A., Keller, G., 2013. The expression of Sox17 identifies and regulates haemogenic endothelium. Nat Cell Biol 15, 502–510.

Corada, M., Orsenigo, F., Morini, M.F., Pitulescu, M.E., Bhat, G., Nyqvist, D., Breviario, F., Conti, V., Briot, A., Iruela-Arispe, M.L., Adams, R.H., Dejana, E., 2013. Sox17 is indispensable for acquisition and maintenance of arterial identity. Nat Commun 4, 2609.

Cortes, M., Chen, M.J., Stachura, D.L., Liu, S.Y., Kwan, W., Wright, F., Vo, L.T., Theodore, L.N., Esain, V., Frost, I.M., Schlaeger, T.M., Goessling, W., Daley, G.Q., North, T.E., 2016. Developmental Vitamin D Availability Impacts Hematopoietic Stem Cell Production. Cell Rep 17, 458–468.

Cortes, M., Liu, S.Y., Kwan, W., Alexa, K., Goessling, W., North, T.E., 2015. Accumulation of the Vitamin D Precursor Cholecalciferol Antagonizes Hedgehog Signaling to Impair Hemogenic Endothelium Formation. Stem Cell Reports 5, 471–479.

de Bruijn, M.F., Ma, X., Robin, C., Ottersbach, K., Sanchez, M.J., Dzierzak, E., 2002. Hematopoietic stem cells localize to the endothelial cell layer in the midgestation mouse aorta. Immunity 16, 673–683.

de Pater, E., Kaimakis, P., Vink, C.S., Yokomizo, T., Yamada-Inagawa, T., van der Linden, R., Kartalaei, P.S., Camper, S.A., Speck, N., Dzierzak, E., 2013. Gata2 is required for HSC generation and survival. J Exp Med 210, 2843–2850.

Durand, C., Robin, C., Bollerot, K., Baron, M.H., Ottersbach, K., Dzierzak, E., 2007. Embryonic stromal clones reveal developmental regulators of definitive hematopoietic stem cells. Proc Natl Acad Sci U S A 104, 20838–20843.

Eich, C., Arlt, J., Vink, C.S., Solaimani Kartalaei, P., Kaimakis, P., Mariani, S.A., van der Linden, R., van Cappellen, W.A., Dzierzak, E., 2018. In vivo single cell analysis reveals Gata2 dynamics in cells transitioning to hematopoietic fate. J Exp Med 215, 233–248.

Esain, V., Kwan, W., Carroll, K.J., Cortes, M., Liu, S.Y., Frechette, G.M., Sheward, L.M., Nissim, S., Goessling, W., North, T.E., 2015. Cannabinoid

Receptor-2 Regulates Embryonic Hematopoietic Stem Cell Development via Prostaglandin E2 and P-Selectin Activity. Stem Cells 33, 2596–2612.

Espin-Palazon, R., Stachura, D.L., Campbell, C.A., Garcia-Moreno, D., Del Cid, N., Kim, A.D., Candel, S., Meseguer, J., Mulero, V., Traver, D., 2014. Proinflammatory signaling regulates hematopoietic stem cell emergence. Cell 159, 1070–1085.

Fitch, S.R., Kapeni, C., Tsitsopoulou, A., Wilson, N.K., Gottgens, B., de Bruijn, M.F., Ottersbach, K., 2020. Gata3 targets Runx1 in the embryonic haematopoietic stem cell niche. IUBMB Life 72, 45–52.

Fitch, S.R., Kimber, G.M., Wilson, N.K., Parker, A., Mirshekar-Syahkal, B., Gottgens, B., Medvinsky, A., Dzierzak, E., Ottersbach, K., 2012. Signaling from the sympathetic nervous system regulates hematopoietic stem cell emergence during embryogenesis. Cell Stem Cell 11, 554–566.

Frelin, C., Herrington, R., Janmohamed, S., Barbara, M., Tran, G., Paige, C.J., Benveniste, P., Zuniga-Pflucker, J.C., Souabni, A., Busslinger, M., Iscove, N.N., 2013. GATA-3 regulates the self-renewal of long-term hematopoietic stem cells. Nat Immunol 14, 1037–1044.

Gama-Norton, L., Ferrando, E., Ruiz-Herguido, C., Liu, Z., Guiu, J., Islam, A.B., Lee, S.U., Yan, M., Guidos, C.J., Lopez-Bigas, N., Maeda, T., Espinosa, L., Kopan, R., Bigas, A., 2015. Notch signal strength controls cell fate in the haemogenic endothelium. Nat Commun 6, 8510.

Gao, X., Xu, C., Asada, N., Frenette, P.S., 2018. The hematopoietic stem cell niche: from embryo to adult. Development 145, 139691.

Gekas, C., Dieterlen-Lievre, F., Orkin, S.H., Mikkola, H.K., 2005. The placenta is a niche for hematopoietic stem cells. Dev Cell 8, 365–375.

Godin, I., Garcia-Porrero, J.A., Dieterlen-Lievre, F., Cumano, A., 1999. Stem cell emergence and hemopoietic activity are incompatible in mouse intraembryonic sites. J Exp Med 190, 43–52.

Goessling, W., North, T.E., Loewer, S., Lord, A.M., Lee, S., Stoick-Cooper, C.L., Weidinger, G., Puder, M., Daley, G.Q., Moon, R.T., Zon, L.I., 2009. Genetic interaction of PGE2 and Wnt signaling regulates developmental specification of stem cells and regeneration. Cell 136, 1136–1147.

Gu, Q., Yang, X., Lv, J., Zhang, J., Xia, B., Kim, J.D., Wang, R., Xiong, F., Meng, S., Clements, T.P., Tandon, B., Wagner, D.S., Diaz, M.F., Wenzel, P.L., Miller, Y.I., Traver, D., Cooke, J.P., Li, W., Zon, L.I., Chen, K., Bai, Y., Fang, L., 2019. AIBP-mediated cholesterol efflux instructs hematopoietic stem and progenitor cell fate. Science 363, 1085–1088.

Guiu, J., Shimizu, R., D'Altri, T., Fraser, S.T., Hatakeyama, J., Bresnick, E.H., Kageyama, R., Dzierzak, E., Yamamoto, M., Espinosa, L., Bigas,

A., 2013. Hes repressors are essential regulators of hematopoietic stem cell development downstream of Notch signaling. J Exp Med 210, 71–84.

Harada, K., Nobuhisa, I., Anani, M., Saito, K., Taga, T., 2017. Thrombopoietin contributes to the formation and the maintenance of hematopoietic progenitor-containing cell clusters in the aorta-gonad-mesonephros region. Cytokine 95, 35–42.

Harris, J.M., Esain, V., Frechette, G.M., Harris, L.J., Cox, A.G., Cortes, M., Garnaas, M.K., Carroll, K.J., Cutting, C.C., Khan, T., Elks, P.M., Renshaw, S.A., Dickinson, B.C., Chang, C.J., Murphy, M.P., Paw, B.H., Vander Heiden, M.G., Goessling, W., North, T.E., 2013. Glucose metabolism impacts the spatiotemporal onset and magnitude of HSC induction in vivo. Blood 121, 2483–2493.

He, Q., Zhang, C., Wang, L., Zhang, P., Ma, D., Lv, J., Liu, F., 2015. Inflammatory signaling regulates hematopoietic stem and progenitor cell emergence in vertebrates. Blood 125, 1098–1106.

Imanirad, P., Solaimani Kartalaei, P., Crisan, M., Vink, C., Yamada-Inagawa, T., de Pater, E., Kurek, D., Kaimakis, P., van der Linden, R., Speck, N., Dzierzak, E., 2014. HIF1alpha is a regulator of hematopoietic progenitor and stem cell development in hypoxic sites of the mouse embryo. Stem Cell Res 12, 24–35.

Ivanovs, A., Rybtsov, S., Ng, E.S., Stanley, E.G., Elefanty, A.G., Medvinsky, A., 2017. Human haematopoietic stem cell development: from the embryo to the dish. Development 144, 2323–2337.

Jaffredo, T., Gautier, R., Eichmann, A., Dieterlen-Lievre, F., 1998. Intraaortic hemopoietic cells are derived from endothelial cells during ontogeny. Development 125, 4575–4583.

Ji, R.P., Phoon, C.K., Aristizabal, O., McGrath, K.E., Palis, J., Turnbull, D.H., 2003. Onset of cardiac function during early mouse embryogenesis coincides with entry of primitive erythroblasts into the embryo proper. Circ Res 92, 133–135.

Khan, J.A., Mendelson, A., Kunisaki, Y., Birbrair, A., Kou, Y., Arnal-Estape, A., Pinho, S., Ciero, P., Nakahara, F., Ma'ayan, A., Bergman, A., Merad, M., Frenette, P.S., 2016. Fetal liver hematopoietic stem cell niches associate with portal vessels. Science 351, 176–180.

Kim, P.G., Canver, M.C., Rhee, C., Ross, S.J., Harriss, J.V., Tu, H.C., Orkin, S.H., Tucker, H.O., Daley, G.Q., 2016. Interferon-alpha signaling promotes embryonic HSC maturation. Blood 128, 204–216.

Kissa, K., Herbomel, P., 2010. Blood stem cells emerge from aortic endothelium by a novel type of cell transition. Nature 464, 112–115.

Ku, C.J., Hosoya, T., Maillard, I., Engel, J.D., 2012. GATA-3 regulates hematopoietic stem cell maintenance and cell-cycle entry. Blood 119, 2242–2251.

Kumar, A., D'Souza, S.S., Thakur, A.S., 2019. Understanding the Journey of Human Hematopoietic Stem Cell Development. Stem Cells Int 2019, 2141475.

Kumaravelu, P., Hook, L., Morrison, A.M., Ure, J., Zhao, S., Zuyev, S., Ansell, J., Medvinsky, A., 2002. Quantitative developmental anatomy of definitive haematopoietic stem cells/long-term repopulating units (HSC/RUs): role of the aorta-gonad-mesonephros (AGM) region and the yolk sac in colonisation of the mouse embryonic liver. Development 129, 4891–4899.

Kwan, W., Cortes, M., Frost, I., Esain, V., Theodore, L.N., Liu, S.Y., Budrow, N., Goessling, W., North, T.E., 2016. The Central Nervous System Regulates Embryonic HSPC Production via Stress-Responsive Glucocorticoid Receptor Signaling. Cell Stem Cell 19, 370–382.

Lancino, M., Majello, S., Herbert, S., De Chaumont, F., Tinevez, J.Y., Olivo-Marin, J.C., Herbomel, P., Schmidt, A., 2018. Anisotropic organization of circumferential actomyosin characterizes hematopoietic stem cells emergence in the zebrafish. Elife 7:e37355

Lancrin, C., Mazan, M., Stefanska, M., Patel, R., Lichtinger, M., Costa, G., Vargel, O., Wilson, N.K., Moroy, T., Bonifer, C., Gottgens, B., Kouskoff, V., Lacaud, G., 2012. GFI1 and GFI1B control the loss of endothelial identity of hemogenic endothelium during hematopoietic commitment. Blood 120, 314–322.

Lancrin, C., Sroczynska, P., Stephenson, C., Allen, T., Kouskoff, V., Lacaud, G., 2009. The haemangioblast generates haematopoietic cells through a haemogenic endothelium stage. Nature 457, 892–895.

Le Douarin, N.M., Dieterlen-Lievre, F., Oliver, P.D., 1984. Ontogeny of primary lymphoid organs and lymphoid stem cells. Am J Anat 170, 261–299.

Li, Y., Esain, V., Teng, L., Xu, J., Kwan, W., Frost, I.M., Yzaguirre, A.D., Cai, X., Cortes, M., Maijenburg, M.W., Tober, J., Dzierzak, E., Orkin, S.H., Tan, K., North, T.E., Speck, N.A., 2014. Inflammatory signaling regulates embryonic hematopoietic stem and progenitor cell production. Genes Dev 28, 2597–2612.

Li, Z., Lan, Y., He, W., Chen, D., Wang, J., Zhou, F., Wang, Y., Sun, H., Chen, X., Xu, C., Li, S., Pang, Y., Zhang, G., Yang, L., Zhu, L., Fan, M., Shang, A., Ju, Z., Luo, L., Ding, Y., Guo, W., Yuan, W., Yang, X., Liu, B., 2012.

Mouse embryonic head as a site for hematopoietic stem cell development. Cell Stem Cell 11, 663–675.

Lichtinger, M., Ingram, R., Hannah, R., Muller, D., Clarke, D., Assi, S.A., Lie, A.L.M., Noailles, L., Vijayabaskar, M.S., Wu, M., Tenen, D.G., Westhead, D.R., Kouskoff, V., Lacaud, G., Gottgens, B., Bonifer, C., 2012. RUNX1 reshapes the epigenetic landscape at the onset of haematopoiesis. EMBO J 31, 4318–4333.

Ling, K.W., Ottersbach, K., van Hamburg, J.P., Oziemlak, A., Tsai, F.Y., Orkin, S.H., Ploemacher, R., Hendriks, R.W., Dzierzak, E., 2004. GATA-2 plays two functionally distinct roles during the ontogeny of hematopoietic stem cells. J Exp Med 200, 871–882.

Lis, R., Karrasch, C.C., Poulos, M.G., Kunar, B., Redmond, D., Duran, J.G.B., Badwe, C.R., Schachterle, W., Ginsberg, M., Xiang, J., Tabrizi, A.R., Shido, K., Rosenwaks, Z., Elemento, O., Speck, N.A., Butler, J.M., Scandura, J.M., Rafii, S., 2017. Conversion of adult endothelium to immunocompetent haematopoietic stem cells. Nature 545, 439–445.

Lizama, C.O., Hawkins, J.S., Schmitt, C.E., Bos, F.L., Zape, J.P., Cautivo, K.M., Borges Pinto, H., Rhyner, A.M., Yu, H., Donohoe, M.E., Wythe, J.D., Zovein, A.C., 2015. Repression of arterial genes in hemogenic endothelium is sufficient for haematopoietic fate acquisition. Nat Commun 6, 7739.

Lundin, V., Sugden, W.W., Theodore, L.N., Sousa, P.M., Han, A., Chou, S., Wrighton, P.J., Cox, A.G., Ingber, D.E., Goessling, W., Daley, G.Q., North, T.E., 2020. YAP Regulates Hematopoietic Stem Cell Formation in Response to the Biomechanical Forces of Blood Flow. Dev Cell 52, 446–460 e445.

Lv, J., Wang, L., Gao, Y., Ding, Y.Q., Liu, F., 2017. 5-hydroxytryptamine synthesized in the aorta-gonad-mesonephros regulates hematopoietic stem and progenitor cell survival. J Exp Med 214, 529–545.

Ma, X., Zhang, Y., Yang, L., Xu, Z., Xiao, Z., 2012. The effects of increased expression of DLK1 gene on the pathogenesis of myelodysplastic syndromes. Clin Lymphoma Myeloma Leuk 12, 261–268.

Mariani, S.A., Li, Z., Rice, S., Krieg, C., Fragkogianni, S., Robinson, M., Vink, C.S., Pollard, J.W., Dzierzak, E., 2019. Pro-inflammatory Aorta-Associated Macrophages Are Involved in Embryonic Development of Hematopoietic Stem Cells. Immunity 50, 1439–1452 e1435.

Mascarenhas, M.I., Bacon, W.A., Kapeni, C., Fitch, S.R., Kimber, G., Cheng, S.W., Li, J., Green, A.R., Ottersbach, K., 2016. Analysis of Jak2 signaling reveals resistance of mouse embryonic hematopoietic stem cells to myeloproliferative disease mutation. Blood 127, 2298–2309.

Mascarenhas, M.I., Parker, A., Dzierzak, E., Ottersbach, K., 2009. Identification of novel regulators of hematopoietic stem cell development through refinement of stem cell localization and expression profiling. Blood 114, 4645–4653.

McGarvey, A.C., Rybtsov, S., Souilhol, C., Tamagno, S., Rice, R., Hills, D., Godwin, D., Rice, D., Tomlinson, S.R., Medvinsky, A., 2017. A molecular roadmap of the AGM region reveals BMPER as a novel regulator of HSC maturation. J Exp Med 214, 3731–3751.

McGrath, K.E., Frame, J.M., Fegan, K.H., Bowen, J.R., Conway, S.J., Catherman, S.C., Kingsley, P.D., Koniski, A.D., Palis, J., 2015. Distinct Sources of Hematopoietic Progenitors Emerge before HSCs and Provide Functional Blood Cells in the Mammalian Embryo. Cell Rep 11, 1892–1904.

Medvinsky, A., Dzierzak, E., 1996. Definitive hematopoiesis is autonomously initiated by the AGM region. Cell 86, 897–906.

Mendes, S.C., Robin, C., Dzierzak, E., 2005. Mesenchymal progenitor cells localize within hematopoietic sites throughout ontogeny. Development 132, 1127–1136.

Mirshekar-Syahkal, B., Haak, E., Kimber, G.M., van Leusden, K., Harvey, K., O'Rourke, J., Laborda, J., Bauer, S.R., de Bruijn, M.F., Ferguson-Smith, A.C., Dzierzak, E., Ottersbach, K., 2013. Dlk1 is a negative regulator of emerging hematopoietic stem and progenitor cells. Haematologica 98, 163–171.

Monteiro, R., Pinheiro, P., Joseph, N., Peterkin, T., Koth, J., Repapi, E., Bonkhofer, F., Kirmizitas, A., Patient, R., 2016. Transforming Growth Factor beta Drives Hemogenic Endothelium Programming and the Transition to Hematopoietic Stem Cells. Dev Cell 38, 358–370.

Moore, K.A., Pytowski, B., Witte, L., Hicklin, D., Lemischka, I.R., 1997. Hematopoietic activity of a stromal cell transmembrane protein containing epidermal growth factor-like repeat motifs. Proc Natl Acad Sci U S A 94, 4011–4016.

Morrison, S.J., Scadden, D.T., 2014. The bone marrow niche for haematopoietic stem cells. Nature 505, 327–334.

Muller, A.M., Medvinsky, A., Strouboulis, J., Grosveld, F., Dzierzak, E., 1994. Development of hematopoietic stem cell activity in the mouse embryo. Immunity 1, 291–301.

North, T.E., Goessling, W., Peeters, M., Li, P., Ceol, C., Lord, A.M., Weber, G.J., Harris, J., Cutting, C.C., Huang, P., Dzierzak, E., Zon, L.I., 2009. Hematopoietic stem cell development is dependent on blood flow. Cell 137, 736–748.

North, T.E., Goessling, W., Walkley, C.R., Lengerke, C., Kopani, K.R., Lord, A.M., Weber, G.J., Bowman, T.V., Jang, I.H., Grosser, T., Fitzgerald, G.A., Daley, G.Q., Orkin, S.H., Zon, L.I., 2007. Prostaglandin E2 regulates vertebrate haematopoietic stem cell homeostasis. Nature 447, 1007–1011.

Oatley, M., Bolukbasi, O.V., Svensson, V., Shvartsman, M., Ganter, K., Zirngibl, K., Pavlovich, P.V., Milchevskaya, V., Foteva, V., Natarajan, K.N., Baying, B., Benes, V., Patil, K.R., Teichmann, S.A., Lancrin, C., 2020. Single-cell transcriptomics identifies CD44 as a marker and regulator of endothelial to haematopoietic transition. Nat Commun 11, 586.

Okuda, T., van Deursen, J., Hiebert, S.W., Grosveld, G., Downing, J.R., 1996. AML1, the target of multiple chromosomal translocations in human leukemia, is essential for normal fetal liver hematopoiesis. Cell 84, 321–330.

Orelio, C., Haak, E., Peeters, M., Dzierzak, E., 2008. Interleukin-1-mediated hematopoietic cell regulation in the aorta-gonad-mesonephros region of the mouse embryo. Blood 112, 4895–4904.

Ottersbach, K., Dzierzak, E., 2005. The murine placenta contains hematopoietic stem cells within the vascular labyrinth region. Dev Cell 8, 377–387.

Palis, J., Chan, R.J., Koniski, A., Patel, R., Starr, M., Yoder, M.C., 2001. Spatial and temporal emergence of high proliferative potential hematopoietic precursors during murine embryogenesis. Proc Natl Acad Sci U S A 98, 4528–4533.

Pandolfi, P.P., Roth, M.E., Karis, A., Leonard, M.W., Dzierzak, E., Grosveld, F.G., Engel, J.D., Lindenbaum, M.H., 1995. Targeted disruption of the GATA3 gene causes severe abnormalities in the nervous system and in fetal liver haematopoiesis. Nat Genet 11, 40–44.

Peeters, M., Ottersbach, K., Bollerot, K., Orelio, C., de Bruijn, M., Wijgerde, M., Dzierzak, E., 2009. Ventral embryonic tissues and Hedgehog proteins induce early AGM hematopoietic stem cell development. Development 136, 2613–2621.

Petit-Cocault, L., Volle-Challier, C., Fleury, M., Peault, B., Souyri, M., 2007. Dual role of Mpl receptor during the establishment of definitive hematopoiesis. Development 134, 3031–3040.

Poullet, N., Golushko, I., Lorman, V., Travnickova, J., Bureau, C., Chalin, D., Rochal, S., Parmeggiani, A., Kissa, K., 2019. Mechanical instabilities of aorta drive blood stem cell production: a live study. Cell Mol Life Sci 77, 3453–3464.

Rae, F., Woods, K., Sasmono, T., Campanale, N., Taylor, D., Ovchinnikov, D.A., Grimmond, S.M., Hume, D.A., Ricardo, S.D., Little, M.H., 2007. Characterisation and trophic functions of murine embryonic macrophages based upon the use of a Csf1r-EGFP transgene reporter. Dev Biol 308, 232–246.

Robert-Moreno, A., Guiu, J., Ruiz-Herguido, C., Lopez, M.E., Ingles-Esteve, J., Riera, L., Tipping, A., Enver, T., Dzierzak, E., Gridley, T., Espinosa, L., Bigas, A., 2008. Impaired embryonic haematopoiesis yet normal arterial development in the absence of the Notch ligand Jagged1. EMBO J 27, 1886–1895.

Robin, C., Ottersbach, K., Durand, C., Peeters, M., Vanes, L., Tybulewicz, V., Dzierzak, E., 2006. An unexpected role for IL-3 in the embryonic development of hematopoietic stem cells. Dev Cell 11, 171–180.

Rybtsov, S., Batsivari, A., Bilotkach, K., Paruzina, D., Senserrich, J., Nerushev, O., Medvinsky, A., 2014. Tracing the origin of the HSC hierarchy reveals an SCF-dependent, IL-3-independent CD43(-) embryonic precursor. Stem Cell Reports 3, 489–501.

Rybtsov, S., Sobiesiak, M., Taoudi, S., Souilhol, C., Senserrich, J., Liakhovitskaia, A., Ivanovs, A., Frampton, J., Zhao, S., Medvinsky, A., 2011. Hierarchical organization and early hematopoietic specification of the developing HSC lineage in the AGM region. J Exp Med 208, 1305–1315.

Sawamiphak, S., Kontarakis, Z., Stainier, D.Y., 2014. Interferon gamma signaling positively regulates hematopoietic stem cell emergence. Dev Cell 31, 640–653.

Souilhol, C., Gonneau, C., Lendinez, J.G., Batsivari, A., Rybtsov, S., Wilson, H., Morgado-Palacin, L., Hills, D., Taoudi, S., Antonchuk, J., Zhao, S., Medvinsky, A., 2016a. Inductive interactions mediated by interplay of asymmetric signalling underlie development of adult haematopoietic stem cells. Nat Commun 7, 10784.

Souilhol, C., Lendinez, J.G., Rybtsov, S., Murphy, F., Wilson, H., Hills, D., Batsivari, A., Binagui-Casas, A., McGarvey, A.C., MacDonald, H.R., Kageyama, R., Siebel, C., Zhao, S., Medvinsky, A., 2016b. Developing

HSCs become Notch independent by the end of maturation in the AGM region. Blood 128, 1567–1577.

Sugimura, R., Jha, D.K., Han, A., Soria-Valles, C., da Rocha, E.L., Lu, Y.F., Goettel, J.A., Serrao, E., Rowe, R.G., Malleshaiah, M., Wong, I., Sousa, P., Zhu, T.N., Ditadi, A., Keller, G., Engelman, A.N., Snapper, S.B., Doulatov, S., Daley, G.Q., 2017. Haematopoietic stem and progenitor cells from human pluripotent stem cells. Nature 545, 432–438.

Swiers, G., Baumann, C., O'Rourke, J., Giannoulatou, E., Taylor, S., Joshi, A., Moignard, V., Pina, C., Bee, T., Kokkaliaris, K.D., Yoshimoto, M., Yoder, M.C., Frampton, J., Schroeder, T., Enver, T., Gottgens, B., de Bruijn, M., 2013. Early dynamic fate changes in haemogenic endothelium characterized at the single-cell level. Nat Commun 4, 2924.

Tamplin, O.J., Durand, E.M., Carr, L.A., Childs, S.J., Hagedorn, E.J., Li, P., Yzaguirre, A.D., Speck, N.A., Zon, L.I., 2015. Hematopoietic stem cell arrival triggers dynamic remodeling of the perivascular niche. Cell 160, 241–252.

Taoudi, S., Gonneau, C., Moore, K., Sheridan, J.M., Blackburn, C.C., Taylor, E., Medvinsky, A., 2008. Extensive hematopoietic stem cell generation in the AGM region via maturation of VE-cadherin+CD45+ pre-definitive HSCs. Cell Stem Cell 3, 99–108.

Taoudi, S., Medvinsky, A., 2007. Functional identification of the hematopoietic stem cell niche in the ventral domain of the embryonic dorsal aorta. Proc Natl Acad Sci U S A 104, 9399–9403.

Tikhonova, A.N., Dolgalev, I., Hu, H., Sivaraj, K.K., Hoxha, E., Cuesta-Dominguez, A., Pinho, S., Akhmetzyanova, I., Gao, J., Witkowski, M., Guillamot, M., Gutkin, M.C., Zhang, Y., Marier, C., Diefenbach, C., Kousteni, S., Heguy, A., Zhong, H., Fooksman, D.R., Butler, J.M., Economides, A., Frenette, P.S., Adams, R.H., Satija, R., Tsirigos, A., Aifantis, I., 2019. The bone marrow microenvironment at single-cell resolution. Nature 569, 222–228.

Tober, J., Yzaguirre, A.D., Piwarzyk, E., Speck, N.A., 2013. Distinct temporal requirements for Runx1 in hematopoietic progenitors and stem cells. Development 140, 3765–3776.

Travnickova, J., Tran Chau, V., Julien, E., Mateos-Langerak, J., Gonzalez, C., Lelievre, E., Lutfalla, G., Tavian, M., Kissa, K., 2015. Primitive macrophages control HSPC mobilization and definitive haematopoiesis. Nat Commun 6, 6227.

Tsai, F.Y., Keller, G., Kuo, F.C., Weiss, M., Chen, J., Rosenblatt, M., Alt, F.W., Orkin, S.H., 1994. An early haematopoietic defect in mice lacking the transcription factor GATA-2. Nature 371, 221–226.

Vink, C.S., Calero-Nieto, F.J., Wang, X., Maglitto, A., Mariani, S.A., Jawaid, W., Gottgens, B., Dzierzak, E., 2020. Iterative Single-Cell Analyses Define the Transcriptome of the First Functional Hematopoietic Stem Cells. Cell Rep 31, 107627.

Wang, Q., Stacy, T., Binder, M., Marin-Padilla, M., Sharpe, A.H., Speck, N.A., 1996. Disruption of the Cbfa2 gene causes necrosis and hemorrhaging in the central nervous system and blocks definitive hematopoiesis. Proc Natl Acad Sci U S A 93, 3444–3449.

Wilkinson, A.C., Ishida, R., Kikuchi, M., Sudo, K., Morita, M., Crisostomo, R.V., Yamamoto, R., Loh, K.M., Nakamura, Y., Watanabe, M., Nakauchi, H., Yamazaki, S., 2019. Long-term ex vivo haematopoietic-stem-cell expansion allows nonconditioned transplantation. Nature 571, 117–121.

Wolock, S.L., Krishnan, I., Tenen, D.E., Matkins, V., Camacho, V., Patel, S., Agarwal, P., Bhatia, R., Tenen, D.G., Klein, A.M., Welner, R.S., 2019. Mapping Distinct Bone Marrow Niche Populations and Their Differentiation Paths. Cell Rep 28, 302–311 e305.

Yokomizo, T., Dzierzak, E., 2010. Three-dimensional cartography of hematopoietic clusters in the vasculature of whole mouse embryos. Development 137, 3651–3661.

Yvernogeau L, Klaus A, Maas J, Morin-Poulard I, Weijts B, Schulte-Merker S, Berezikov E, Junker JP, Robin C., 2020. Multispecies RNA tomography reveals regulators of hematopoietic stem cell birth in the embryonic aorta. Blood. 136, 831–844.

Zaidan, N., Ottersbach, K., 2018. The multi-faceted role of Gata3 in developmental haematopoiesis. Open Biol 8, 180152.

Zeng, Y., He, J., Bai, Z., Li, Z., Gong, Y., Liu, C., Ni, Y., Du, J., Ma, C., Bian, L., Lan, Y., Liu, B., 2019. Tracing the first hematopoietic stem cell generation in human embryo by single-cell RNA sequencing. Cell Res 29, 881–894.

Zhou, F., Li, X., Wang, W., Zhu, P., Zhou, J., He, W., Ding, M., Xiong, F., Zheng, X., Li, Z., Ni, Y., Mu, X., Wen, L., Cheng, T., Lan, Y., Yuan, W., Tang, F., Liu, B., 2016. Tracing haematopoietic stem cell formation at single-cell resolution. Nature 533, 487–492.

Zhu, Q., Gao, P., Tober, J., Bennett, L., Chen, C., Uzun, Y., Li, Y., Howell, E.D., Mumau, M., Yu, W., He, B., Speck, N.A., Tan, K., 2020. Developmental trajectory of pre-hematopoietic stem cell formation from endothelium. Blood 136, 845–856.

14

Biology of Hematopoietic Stem Cells in the Adult

Rima Haddad*, **Francoise Pflumio and Marie-Laure Arcangeli***

Team Niche and Cancer in Hematopoiesis, U1274, INSERM, 18 route du Panorama, 92260 Fontenay-aux-Roses, France
Laboratory of Hematopoietic Stem Cells and Leukemia/Service Stem Cells and Radiation /iRCM/JACOB/DRF, CEA, 18 route du Panorama, 92260 Fontenay-aux-Roses, France
Université de Paris and Université Paris-Saclay, Inserm, iRCM/IBFJ CEA, UMR Stabilité Génétique Cellules Souches et Radiations, F-92265, Fontenay-aux-Roses, France
E-mail: rima.haddad@universite-paris-saclay.fr; marie-Laure.arcangeli@inserm.fr
*Corresponding Authors

14.1 Definition, Concepts, History

Blood cells contain more than 10 types of different cells (red cells, platelets, B cells, T cells, NK cells, macrophages, neutrophils, etc.). Some of them are very short living such as granulocytes whose lifespan are a few hours, a few days for platelets, and around three months for red blood cells. Such cells thus need continuous regeneration during the entire adult life span or upon stress-induced failure (for instance following myeloablation treatments). This hematopoietic cell replenishment is possible due to the proliferation and differentiation of several compartments of cells located in the bone marrow (BM); the most immature ones among them are the hematopoietic stem cells (HSCs). At steady state, HSCs are quiescent but after BM injury, HSCs enter the cell cycle. HSCs have unique properties that allow to reconstitute the entire hematopoietic system, a process named hematopoiesis. HSCs are multipotent: one HSC has the ability to generate all lineage (lymphoid and myeloid) restricted mature cells. HSCs can self-renew and this property relies on their ability to perform symmetric or asymmetric divisions, avoiding their disappearance when differentiation is necessary. HSCs can home and migrate into the BM, the tissue where they are mainly located during adulthood in "medullar niches" where crucial interactions will support the maintenance and hematopoietic development capabilities of HSC[1]. Research on HSCs provided founding principles, basic thinking, and practical tools, for the discovery and comprehension of stem cells from many other organs, such as for instance, intestine stem cells[2]. It also laid foundation for the idea of a hierarchical cell composition of leukemia, which are blood-derived cancers, with at the apex the so-called leukemia-initiating/stem cells that are reminiscent of HSCs in these abnormal tissues (reviewed by[3]).

Of note, whereas most studies on mouse HSCs are done with BM cells, adult human HSC sources are of three origins: BM recovered from adult hip surgeries (mostly elderly patients), mobilized blood (MB) recovered after treatment of patients with mobilized agents, and umbilical cord blood (UCB). These three types of human HSCs have different proliferation and differentiation abilities that are important to consider when comparisons are made with mouse adult HSC.

In this chapter, we will focus on several basic aspects of the biology of HSC and provide examples of works, which were in our opinion fundamental for the comprehension of their biology. We will underline some of the still unsolved or debated controversies. We shall also explain some of the specific contributions of our team in the field of HSC.

14.2 Characterization of HSC

14.2.1 Using Phenotype Analysis

All the cells of the body express surface/membrane proteins and many of those are specific of the cells of interest (type/lineage/maturation stage) as they mediate particular functions (such as receptor, ligand, adhesion, migration, etc.). Development of monoclonal antibodies mAbs directed specifically against these proteins and of fluorescent activated cell sorting were breakthroughs for the identification and follow up of hematopoietic cells. HSC have greatly benefited from this technology.

14.2.1.1 Mouse HSC (see also[4,5])

In the BM, two cell populations can be distinguished: (1) cells expressing markers of mature lineages (CD3, CD4, CD8, CD19, NK1.1, Gr1, CD11b, and Ter119; Lin$^+$) and (2) cells negative for these markers (Lin$^-$) containing HSC, progenitor and precursor cells. Such definition of immature cells only leads to enrich immature cell populations and detection of markers expressed on cells allows to specifically fractionate HSC and progenitors. In mouse, HSC and multipotent progenitors (MPPs) are further enriched in Lin$^-$ cells with high levels of the Stem cell antigen (Sca)-1 and c-Kit receptor (Lin$^-$Sca-1^{hi}c-Kit^{hi}, LSK). The LSK subset represents 0.5% of total BM cellularity but less than 10% LSK cells are capable of reconstituting a lethally irradiated mouse indicating high heterogeneity[6]. The first enriched HSC population was defined in C57BL/6 mice expressing low levels of Thy1.1 allele (also named CD90) thus being Thy1.1.^{lo} HSC frequency, obtained after transplantation of cells at limiting dilution in lethally irradiated mice, represents up to 18% of this cell fraction.[7] Using CD34 and FLT3 cell surface expression, HSC with long-term reconstitution capacity (LT-HSC, LSK CD34$^-$FLT3$^-$), can be distinguished from short-term (ST)-HSC (LSK CD34$^+$FLT3$^-$) with a limited (up to 8 weeks) repopulation activity in lethally irradiated mice and from MPP (LSK CD34$^+$FLT3$^+$). Morrison's group using SLAM cell surface glycoproteins from the immunoglobulin super family as an alternative approach showed that HSCs are defined as LSK CD150$^+$CD48$^-$ of which 50% are LT-HSC,[8] 30% of them do not express CD34 suggesting that the LSK CD150$^+$CD48$^-$CD34$^-$ cell population might be even more enriched in HSC.[9] Combining SLAM markers with absence of CD34 expression is the most commonly used strategy to isolate mouse HSC in worldwide laboratories.[4,10] Based on the cycling capacity of HSC (LSK CD150$^+$CD48$^-$CD34$^-$), there are two types of mouse HSCs: "dormant" HSCs divide every 145 days and "activated" HSCs divide once a month[9,11]. Recently using RNA sequencing coupled to reporter mice, the dormant HSCs were found to express specifically Gprc5c (G protein–coupled receptor, class C, group 5, member of the vitamin A/retinoic acid signaling pathway)[10], therefore allowing dormant HSCs to be distinguished from activated HSCs using antibody-labeling strategy and not cell cycle–specific dyes.

14.2.1.2 Human HSC

Human HSCs, like mouse HSCs, are devoid of lineage specific markers and a mix of mAbs directed against a panel of markers (up to 18[12]) typically found on mature lymphoid and myeloid cells provided tools for the enrichment of immature Lin$^-$ cells. The most commonly used surface marker to study human HSC is CD34, an adhesion-mediated glycoprotein[13].

Findings that CD34$^+$ cells—about 1% of BM and UCB cells—are enriched in progenitors and HSCs, provided a major tool for HSC research and importantly for clinical applications. Indeed this observation greatly helped enrichment of hematopoiesis reconstituting cells and depletion of tumor cells or of lymphoid Graft versus Host Disease-mediated cells in transplantation settings[14–16]. CD34 expression is not exclusive to hematopoietic stem and progenitor cells as other BM resident cells, including endothelial cells, also bear surface CD34 expression[17]. Of note, usage of autologous CD34$^+$ cells as a product for transplantation requires to be done with caution because malignant hematopoietic stem cells express CD34 in many blood disorders[3].

Several works[18,19], including recent publications[12,20] have shown that HSCs can be found also among CD34$^-$Lin$^-$ cells.[20] However, and as mentioned recently, *"a simple calculation using reported frequencies indicates that >99% of human HSC must be CD34.$^+$"*[21] Interestingly, frequencies of *in vivo* reconstituting cells remain very low in these human UCB CD34$^-$CD38$^-$CD45$^+$CD93$^+$ cells, maybe due to the necessity of secondary transplants to reveal the true potential of these cells. Thus, future work should focus on understanding CD34 expression in human HSC, keeping in mind that CD34$^-$ cells may be the quiescent ancestors of CD34$^+$ cells.[20]

CD34 marks a heterogenous cell population containing HSC and committed progenitors. Consequently, as for mouse cells, additional markers are to be combined to further enrich for more cells that are immature. The mainly used ones are CD38[22], Thy1/CD90[23,24], CD45RA, and CD133/prominin1[25]. CD34$^+$CD38$^-$ cells represent 0.1% of mononucleated BM cells of which 1/617 can reconstitute human hematopoiesis in adult immune deficient (ID) mice[26] whereas around 1/10 and 1/100 CD34$^+$CD38$^-$CD90$^+$Lin$^-$CD45RA$^-$ cells from UCB do so in newborn and adult recipients respectively ([24] and our own results). The integrina6 (CD49f) as well as CLEC9 are also described in CD34$^+$CD90$^{+/-}$CD38$^-$Lin$^-$ UCB[27,28]. The HSC frequency is 1/9.5 in CD34$^+$CD90$^+$CD49f$^+$ cells[28] and 1/13.5 in CD34$^+$ CLEC9$^+$ CD49f$^+$ cells[27].

In conclusion, many surface markers are available for human HSC enrichment and the current studies do not achieve purification levels equivalent to mouse HSC. These limitations greatly preclude precise studies on the determination/self-renewal of human HSCs as it is described for mouse HSC[29,30]. Further refinements in the human HSC phenotype are still to be done.

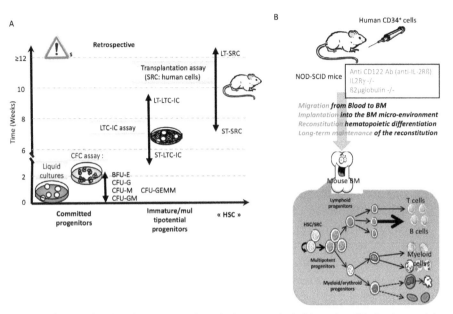

Figure 14.1 A functional way to get into the hematopoietic hierarchy. (**A**) *In vitro* and *in vivo* assays used to explore mouse and human hematopoietic progenitor/HSC. (**B**) SRC assays allow studying human HSC compartment. Immune-deficient mice, such as NOD-SCID (NS) are used as recipients for CD34^{+} cells isolated from cord blood, BM, or mobilized blood. Immune-deficiency of NS mice can be improved with treatment with anti-IL2Rb antibodies or combining IL2Rg or b$_2$-microglobulin gene deficiencies. Cells are usually transplanted by intravenous injection. HSC/SRC home to the mouse BM and recapitulate the hierachy of human hematopoiesis, including mature lymphoid and myeloid cells.

14.2.2 Using Functional Assays (see also [31])

HSCs and progenitor cells are best characterized by their functional properties: self-renewal and/or multilineage differentiation. The community commonly agrees that the more immature is a cell the highest its proliferation and the more diversified its cell progeny are, even though HSCs appear more and more to bear heterogeneous differentiation abilities (see below). Most of the experimental assays used to characterize immature cells are retrospective as they rely on cell proliferation and differentiation (Figure 14.1).

14.2.2.1 Colony-forming cells (CFCs) and long-term culture-initiating cells (LTC-ICs)

The CFC assay uses the property of committed progenitors to form colonies in semi-solid medium, such as methylcellulose, complemented with defined growth factors. It is a clonal assay: every colony derives from a single progenitor (Figure 14.1A). In mice and humans, it is mainly oriented toward myeloid and erythroid differentiation[32]. Myelo-erythroid progenitors are detected 1 (mouse) to 2 (human) weeks following culture initiation. Counting the number of colonies provides the frequency and the total number of committed progenitors of a given cell suspension. CFC assays also provide the quality (lineage, multi *vs.* mono-potent clones and size: small/mature or large/immature of colonies) of committed progenitors. Historically it is one of the first experimental assays developed to measure immature cells[33].

LTC-IC assays uncover even more immature cells, i.e., cells that require =3 weeks to proliferate and differentiate (Figure 14.1A). In humans, LTC-ICs are typically defined as cells that are able to give rise to CFCs after being co-cultured for 5 weeks with a feeder cell layer[34]. Originally, this assay was developed as a surrogate assay of the BM niche. While culturing the total BM cell suspension, a stromal layer developed together with hematopoietic cells and these mesenchymal cells supported immature and mature cells[35]. Afterwards stromal cell lines were isolated and the LTC-IC assay now lays on co-cultures of enriched immature cells with pre-established BM-derived adherent cells, such as MS-5 mouse stromal cells[36]. LTC-IC assay can provide quantitative data, using limiting dilution of cells and data analyzed with the Poisson statistics[31]. The frequency of LTC-ICs within a given cell population can be obtained by this way. LTC-ICs are comparable to the immature progenitors named Cobblestone Area-Forming Cells (CAFCs) that make paved colonies in close contact with stromal supporting cells[37].

Both CFC/LTC-IC assays can be extended in time, using for instance serial replatings of primary CFC or extended LTC culture period over 10 weeks. In both cases, it is used as a way to test the self-renewal capacity of immature cells and/or to reveal even more immature cells (thus getting closer to the HSC compartment)[38−41].

14.2.2.2 Using liquid cultures

Most liquid cultures of progenitors induce HSCs and progenitors to differentiate. These assays usually last 2 to 3 weeks and are done in the presence of lineage-oriented cytokines (reviewed by [42]) (Figure 14.1A). This is especially true for myelo/erythroid lineages. Lymphoid lineages are trickier to manipulate, as T- and B-cell development require interactions with a permissive microenvironment. During numerous years, the thymic 3D structure was described as crucial, and organotypic cultures of fetal thymic lobes with progenitors were used to study T-cell generation in mouse and man[43,44].

The description of the importance of Notch/ligands interactions during T-cell development[45] and the generation of stroma cell lines expressing Delta-like 1/4 (DLL1/4) allowed deciphering T-cell differentiation *in vitro* from mouse and human progenitors[46−49]. Mouse B cells can be efficiently produced *in vitro* from immature cells as mouse B-cell development relies on a stromal support that produce Interleukin-(IL-)7.[50] Human B-cell development also requires a stromal support as well as low concentration of serum in the culture medium[51]. However, human B-cell development in culture remains very inefficient and recent findings in understanding non-hematopoietic BM niche cell components, using single cell RNA-seq investigations, should help design novel *in vitro* conditions[52−54].

14.2.2.3 Using in vivo transplantation models

Measuring HSC content of a given cell population relies on *in vivo* transplantation into myelo-ablated mice (Figure 14.1). It is a mandatory assay to study HSC functions. Most studies use irradiation, as myelo-ablation regimen, but chemicals such as 5-fluorouracil or busulfan are alternatives. Such experimental approach lays on historical experiments by Till and McCulloch in the 1960's in which BM cells capable of clonal generation of multiple hematopoietic lineages and endowed with self-renewing properties were first uncovered (reviewed in[3]). Mouse HSC transplantation delineates different HSC populations: the ST- and the LT-HSC[55]. Studies of human HSCs are trickier as experimentation of human HSC properties *in vivo* in people is ethically impossible. Breakthroughs lay in studies in which immune-deficient (ID) mice were used as recipients of either human BM cells or fetal hematopoietic tissues with the idea of avoiding rejection by the host immune system[56−58]. Since then huge progresses have been made including great improvement in ID recipient mice as well as enrichment of tested cell fractions (for details [21]). As this assay is a surrogate assay for human HSC, the uncovered immature cell was named "SRC" for SCID-mouse-Repopulating-Cell, after the original ID mouse recipient[59]. As for CFC and LTC-IC, the SRC assay is quantitative (using limiting dilution experiments) and qualitative (using phenotype analysis of the progeny of SRC). The reconstitution of a human hierarchy in the BM of ID mice is real as human immature $CD34^+CD38^{lo}$ and/or $CD90^+$ cells recovered from mouse BM comprise all the progenitors (CFC, LTC-IC, lympho-myeloid cells, and multipotent T/B/NK/myeloid) described in normal BM cells[60,61]. Hematopoietic development following secondary transplantation into ID mice confirmed the engraftment of human HSCs in the mouse BM. Nevertheless, human hematopoietic reconstitution of ID mice does not totally mimic the full human BM mature cell diversity,[21] thus recent improvements have been made in terms of humanization of ID mice, that now can produce human growth factors such as the MISTRG mice (see for instance [62]) or recreate a human BM niche *in vivo* (for example[63], and reviewed in [64]).

14.2.3 Physiology of HSC

14.2.3.1 Self-renewal and quiescence properties

In mouse, BM HSC numbers are constant. Their maintenance is a balance between quiescence and self-renewal. Seventy percent of LSK/CD150$^+$CD48$^-$CD34$^-$Flt3$^-$ HSCs are in the G0 phase of the cell cycle, being Hoescht^{lo} Ki-67$^-$ (Ki-67 is a nuclear marker of mitotic cells). In contrast, only 10% of LSKCD150$^-$CD48$^+$CD34$^+$FLT3$^+$ MPP cells are quiescent.[9] When cell division rate of HSCs was measured using Bromodeoxyuridine (BrdU), an analog of thymidine that marks DNA of dividing cells, rarely dividing cells, such as HSCs, remain long labeled[9,11]. Several regulators of the cell cycle are involved in maintaining HSC quiescence. Negative regulators of the cell cycle are part of these major players (reviewed in [65]). For example, p57 is highly expressed by HSCs and its expression decreases with hematopoietic differentiation. Deletion of *p57* in LSK cells induces a loss of quiescence of HSCs leading to apoptosis, differentiation, and defect in HSCs' self-renewal potential.[66]

As described above, 95% of Gprc5c-EGFP$^+$ HSCs are quiescent, demonstrating a role of the vitamin A/retinoic acid signaling pathway in the control of HSC dormancy[10]. The retinoic acid pathway controls dormancy by regulating the levels of reactive oxygen species (ROS) in HSC. In Gprc5o-EGFP$^+$ HSCs, ROS levels are very low compared to Gprc5o-EGFP$^-$ HSCs in accordance with the fact that ROS levels must remain very low in HSCs[67]. Indeed, increased ROS levels act as a signal for activation and differentiation, and abnormal ROS levels lead to HSC exhaustion, HSC self-renewal defect, and eventually cell death. Upon activation, ROS levels increase in HSCs inducing cell cycle activation via the activation of p16ink4 and p19arf[68]. Recently, we showed that HSC exposure to low doses of irradiation (LDIR) induces HSC self-renewal defects through abnormal increase of ROS levels (as shown by increased 8oxodG lesions on 20 mGy-irradiated HSC DNA in Figure 14.2A) and abnormal p38MAPK activation[39] (Figure 14.2). After exposure to LDIR, 20 mGy-irradiated HSCs fail to serially replate in serial CFU-C assay (Figure 14.2B). The same consequences of LDIR have been

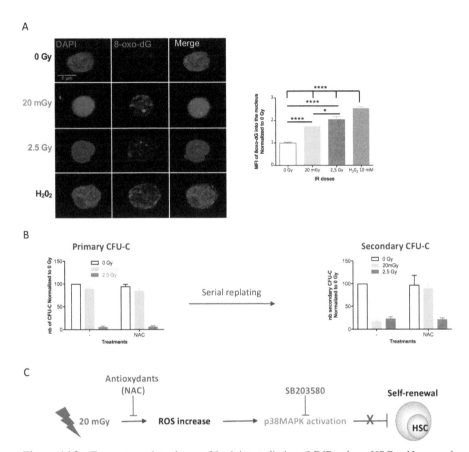

Figure 14.2 Exposure to low doses of ionizing radiation (LDIR) alters HSC self-renewal potential (adapted from Henry et al 2020). (**A**) After exposure to LDIR, a ROS increase, detected here by 8oxo-dG lesion on HSC DNA, are detected[39] (**B**) The capacity of HSC to serially generate colonies *in vitro* in CFU-C assay, after exposure to LDIR is diminished after the second plating compared to non-irradiated HSC (n>5). The defect can be rescued when HSC are treated with antioxidants (N-acetyl cytein, NAC) prior exposure to irradiation[39] (n = 3). (**C**) Our model: exposure to LDIR causes an increase in ROS that will activate p38MAPK signaling pathway, ultimately leading to a defect in HSC self-renewal potential. The consequences of LDIR exposure can be counteracted by treatment with antioxidants or p38MAPK inhibitor (SB203580).

reported in murine HSC[69]. In both models, the deleterious defects of LDIR can be overcome with antioxidants treatments of HSC, prior exposure to LDIR (Figure 14.2C,D).

14.2.3.2 HSC potential versus HSC fate

Transplantations of enriched HSC populations suggest that HSCs are able to generate the same proportion of myeloid and lymphoid cells (HSC potential). However, the reconstitution potential of single HSC appears biased toward myeloid or lymphoid commitment, supporting the idea that HSCs are functionally heterogeneous[70,71]. HSC functional heterogeneity is also revealed by differential self-renewal capacities of single HSC. Lineage preference and proliferation differences do not seem strictly dependent on extrinsic signals and can be rather cell autonomously predetermined. A very elegant study demonstrates that the monocytic transcription factor MafB, specifically controls HSCs' commitment toward myeloid lineage through modulation of Macrophage-Colony Stimulating factor (M-CSF) signaling[72]. Under M-CSF stimulation, MafB deficiency directs single HSC asymmetric division and favors the production of a PU1-expressing daughter HSC biased toward myeloid lineage.

Recently, new technologies such as single cell analyses, have questioned the strict delimitation between stem and progenitor cells but also the fate (what stem/progenitor cells do in vivo in steady state and stress conditions) rather than the potential of single HSC (all what such cells can do in presence of the right signals). They provided comprehensive cues to appreciate the actual global landscape of mouse and human early hematopoiesis (for reviews [73,74]). A revised model of hematopoiesis is proposed based on *in vitro* single cell assays. This model describes that lymphoid and myeloid lineage potentials should no longer be considered strictly dichotomic/exclusive as postulated by the CMP-CLP because lymphoid and myeloid potentials remain associated in the lymphoid-primed multipotent progenitor (LMPP) and multi-lymphoid progenitor (MLP).[75-80] Although multipotential differentiation is a fundamental hallmark of HSC and MPP, specific bias into differentiation potentials within these compartments has been revealed[4,10,81-83]. Indeed, using limiting dilution and single cell transplantation assays, several groups described murine HSCs that differ in their relative myeloid and lymphoid cell output or adopt lineage-restricted fate despite remaining multipotent[4,10,81-83]. More recently, single cell RNA sequencing assay demonstrated, in mouse as well as in human hematopoiesis, a continuum of differentiation from HSC, MPP cells, and more mature progenitors throughout the kinetic of lineage[84-87]. These studies have allowed, over the past decade, reconsidering the early stages of human and mouse hematopoiesis establishing hierarchical relationships between novel progenitor subsets and providing contribution of stem and multipotent progenitor cells to hematopoiesis ontogenesis and steady-state blood maintenance (for reviews [73,74]).

14.3 Regulation of HSC Functions

14.3.1 Extrinsic Regulators

Regulation of HSC maintenance occurs via the regulation of HSC quiescence, self-renewal, and retention into BM niches. To note, several paired receptors/ligands are involved in this process (for review [5]). Here, we describe briefly three major players of extrinsic regulation of HSC maintenance.

14.3.1.1 CXCL12/CXCR4

The CXCL12/CXCR4 axis is one of the major regulators of HSC maintenance in BM niches (for review [88]). CXCL12 (previously called SDF1) is secreted by endothelial, perivascular, reticular, and osteoblastic cells. HSCs express low levels of its receptor CXCR4 as do also some cells of the BM niche. This ligand/receptor is in charge of HSC attraction and retention into the BM[89]. Antagonizing CXCR4/CXCL12 interactions using CXCR4 inhibitor AMD3100/plerixafor or anti-CXR4 blocking antibody induces HSC mobilization in the peripheral blood. CXCR4/CXCL12 is also involved in HSC quiescence[90]. Mice deleted for CXCL12 regenerate faster and more efficiently hematopoiesis after 5-FU–induced myeloablation compared to controls[91]. A possible mechanism of CXCL12/CXCR4 regulation of HSC quiescence involves the induction of p57 gene expression[66].

14.3.1.2 Stem cell factor (SCF) and its receptor KIT

Another major player in HSC maintenance is SCF (also known as Mast Cell Growth Factor—MGDF) and its receptor KIT. Osteoblast, adipocyte, perivascular and endothelial cells, i.e., most non-hematopoietic components of HSC BM niches, produce SCF, whereas KIT is highly expressed by HSC. Analysis of SCF and KIT mutant mice has shown that this ligand/receptor pair is involved in HSC migration, quiescence, and self-renewal. SCF is also a major regulator of HSC adhesion since it activates VLA-4 and VLA-5 integrin expression at the surface of HSC.[92] SCF and CXCL12 functions in HSC maintenance have been recently enlightened in two elegant papers, using several genetic engineered mouse models suggesting that HSC maintenance is mostly controlled by CXCL12 and SCF secreted by endothelial cells[93,94].

14.3.1.3 Integrins and Adhesion molecules

Adhesion is also an important mechanism involved in the retention of HSC in the BM and thus indirectly in their maintenance. HSC express several different integrins but the most important members are VLA-4 (a_4b_1) and VLA-5 (a_5b_1). BM stromal cells express integrin ligands. VLA-4 mediates homing and adhesion of HSC via VCAM-1, fibronectin, and osteopontin expression by stromal cells[95]. Blocking VLA-4 interaction by the mean of antibodies or biomolecules induce HSC mobilization and prevents HSC homing in the BM[96,97]. Development of conditional gene deletion in mice uncovered the contribution of VLA-4 in HSC maintenance[97]. Similarly, VLA-5 has been involved in BM homing[98]. Other adhesion molecules take part in HSC maintenance. Recently connexin43 and 45 gap junction proteins were involved in CXCL12 secretion and thus in the BM retention of HSC.[99] Similarly, tight junction–associated proteins JAM-B and JAM-C are involved in the BM retention and maintenance of HSC[100]. JAM-C is highly expressed on HSC whereas BM niche cells express its receptor JAM-B. JAM-B–deficient mice have a defect in BM homeostasis, with increased number of cycling HSCs, in correlation with an increased CXCL12 secretion in the BM at steady state. Conversely, JAM-B–deficient mice fail to regenerate efficiently BM homeostasis after 5-FU–induced myeloablation[100–102]. These data are in agreement with the fact that conditional CXCL12-deficient mice exhibit a better hematopoietic regeneration after 5-FU–induced myeloablation[91].

Of note, the bone marrow microenvironment has been recently thoroughly explored, in particular, at single-cell resolution. These works outlined the diversity of cells and factors produced by these non-hematopoietic cells, and how remodeling happens upon stress signals, induced by leukemia development or chemotherapy treatment, inducing HSC to reprogram their differentiation[53,54,103,104].

14.3.2 Intrinsic Regulators (see also[105])

HSCs, like many cells, are regulated through cell surface receptor/ligand interactions inducing activation of transduction pathways that eventually results in gene regulation by nuclear transcription factors (TF). Numerous TFs are involved in such complex molecular network[105]. We have been interested in the Stem Cell Leukemia (SCL), also called T-cell Acute Leukemia-1 (TAL1) factor, a basic helix-loop-helix (bHLH) protein originally cloned from a patient with T-cell acute leukemia[106]. SCL/TAL1 involvement in mouse and human hematopoiesis and T-ALL development is deeply reviewed in [107,108]. Briefly, in adult murine hematopoiesis, SCL/TAL1 is expressed in HSC and its expression decreases with HSC differentiation to be maintained only in erythroid and megakaryocytic lineage[109]. Using hematopoietic-targeted conditional deletion of *scl* gene in mice, SCL/TAL1 was reported first to be dispensable for steady state hematopoiesis with no reported defect in HSC self-renewal[110]. *Scl/tal1* gene dosage expression seems however important in HSC self-renewal properties as HSC dysfunctions were described using $scl^{+/-}$ mice.[109] This had been suspected before since, using knock-down (KD)/shRNA strategies, we showed that *Scl*/Tal1-KD Sca1$^+$ mouse BM cells failed to support efficient long-term hematopoietic reconstitution after transplantation into lethally irradiated recipient mice[111]. Redundancy due to another bHLH factor named LYL1 also expressed in adult HSC is proposed to explain the discrepant results between *scl/tal1* KO and KD experiments[112]. HSCs from mice deficient for *lyl1* and *scl/tal1* failed to sustain hematopoiesis, whereas single-mutant HSC exhibited only a partial defect in hematopoietic reconstitution and self-renewal properties demonstrating that both *lyl1* and *scl/tal1* regulate HSC potentials[112]. In humans, *scl/tal1* levels are high in HSPCs, drastically decreased along lymphoid differentiation, highly expressed in erythroid/megakaryocytic progenitors, and decreased in granulo-monocytic progenitors. Our laboratory showed that enforced SCL could maintain/amplify SRCs as well as SRC-derived LTC-ICs and CFCs (after transplantation in ID mice)[113]. Inversely, and as for mouse, SCL/TAL1 KD dramatically impaired human SRC-derived–hematopoiesis[109,111]. In human HSCs, a target of SCL/TAL1 was identified asDDIT4/REDD1, a stress response factor, also implicated in the mTORC1 pathway[114]. SCL/TAL1, that represses REDD1/DDiT4 expression, thus probably plays a role in regulating the HSC stress response.

14.4 Ex Vivo Expansion of HSPC

Hematopoietic stem progenitor cells (HSPCs) derived from BM, MB, or UCB represent relevant graft options for allogeneic hematopoietic cell transplantation. However, the low number of HSPCs available from these tissues constitutes a major limitation to efficient hematopoietic cell transplantation. To overcome this limitation and improve the quality of the graft, several strategies based on the *ex vivo* manipulation of HSPC were set up and subjected to preclinical or clinical examination. The general idea was to increase their numbers while preserving their stem cell quality. This notion constitutes a fundamental prerequisite for cell therapy and gene therapy strategies based on the combination of HSPC proliferation/activation, while maintaining their stemness, during expansion and gene transfer. We will review here, early attempts and emerging approaches regarding *ex vivo* expansion strategies of HSPC and focus particularly on studies investigating the role of HOXB4 and low O_2 levels (also called hypoxia).

14.4.1 Extrinsic Factors for Ex Vivo HSPC Expansion

14.4.1.1 Usage of Cytokines and growth factors' combination

The pioneer work by D. Metcalf established the effects of colony-stimulating factors (CSF) using CFC assays (reviewed by [42]). The proliferative responses of HSPC are dependent on the simultaneous action of multiple cytokines. Since over 20 years, the types of "cocktail" of stimulatory cytokines/growth factors used in *ex vivo* HSPC expansion protocols greatly varied from one study to another but commonly include SCF, FMS-like tyrosine kinase 3 (FLT3)-ligand, Thrombopoietin (TPO), granulocyte (G)-CSF, granulocyte macrophage (GM)-CSF, IL-1, -3 and -6.[115-117] A phase I clinical trial based on the use of an efficient 3-cytokines combination that comprises SCF, TPO, and G-CSF, showed the safety and feasibility of UCB progenitor's expansion but a modest four-fold progenitor cell expansion[118].

14.4.1.2 Expansion in presence of stromal cells

As mesenchymal stem cells (MSCs) are essential components of HSPC niches, it was postulated that their presence might promote HSC self-renewal while preventing their differentiation. Such idea was also based on the fact that BM-derived stromal cells protect but also unveil immature cell potential[35,42]. Co-cultures of total UCB with BM-derived MSC in presence of SCF, G-SCF, and TPO markedly improved total nucleated cells, CD133[+] and CD34[+] cells, and HSPC output, based on CFC and CAFC assays, when compared to cultures without MSC support but with the same cytokines[119]. This preclinical study resulted in a successful phase I clinical trial where patients received double UCB transplant including one unit expanded *ex vivo* in presence of MSC obtained from either haploidentical BM or from an unrelated donor[120]. In this trial, the presence of MSC enhanced median numbers of HSPC by 40-fold without any apparent health complication.

The studies indicated here illustrate the fact that cytokine- or stroma-mediated expansion strategies are safe and generate substantial increase of HSPC numbers. However, these technologies did not provide definitive demonstration of increased functional ST- and LT-HSC engraftment or enhanced immune reconstitution[121,122]. This ascertainment reinforces the idea that *in vivo* studies such as transplantations in ID mice are fundamental for preclinical assessments that could potentially lead to clinical translation.

14.4.2 Developmental and Intrinsic Factors for Ex Vivo Human HSPC Expansion and or/Maintenance

In recent years, several factors were described to be involved in the ontogeny of the hematopoietic system and in the emergence and maintenance of HSPC. We will focus our attention on HOXB4, Notch, wingless-type (Wnt), and hypoxia/HIF pathways that have been shown to play a role in the regulation of HSPC expansion and maintenance.

14.4.2.1 Approaches Using Delivery of Intrinsic Factors
14.4.2.2 A. HOXB4-mediated expansion

The homeoprotein HOXB4 is a TF known to regulate HSPC self-renewal and expansion. In mice, retrovirus-driven transduction of the *hoxb4* coding sequence leads to a dramatic expansion of adult HSPC without modifying their differentiation potentials[123,124]. Active production of HOXB4 protein by engineered mouse BM stromal cells at the vicinity of human-enriched HSPC induced dramatic increase in total $CD34^+$ cells, expansion of CFC and immature LTC-IC progenitors and SRC/HSC. This expansion was associated with maintenance of lympho-myeloid cell production *in vivo*, an indicator of multipotent cells[125]. Whether these results could be extended to early lymphoid progenitors was tested, as these cells are critical for aplasia exit after allogenic BM transplantation. As for HSC, HOXB4 delivery also enhanced expansion of B-, T-, and NK-cell progenitors derived from $CD34^+CD38^{lo/-}$HPC and T-cell progenitors derived from $CD34^+CD45RA^{hi}CD7^+$ T/NK progenitor cells[78,126−128]. Altogether, these results elucidated how HOXB4 homeoprotein regulates the early stages of human hematopoiesis and provided basis for the development of new therapeutic strategies that include HOXB4-mediated expansion of human HSPC.

14.4.2.3 B. Other potential molecular targets (Notch and Wnt pathways)

Notch participates in the regulation of embryonic, fetal, and adult hematopoiesis. In adults, Notch is expressed in undifferentiated progenitors and is a crucial regulator of hematopoietic differentiation (for [129−131]).

Notch1 and Notch2 are expressed in HSC. The involvement of the Notch pathway in HSC maintenance and functions remains controversial but it was proposed that Notch ligands could contribute to HSC expansion in hematopoietic niches[132–135]. Similarly, to Notch signaling pathway, whether Wnt is involved in HSC self-renewal and quiescence has long been investigated (for review [136]). The function of b-catenin, a major component of Wnt pathway activation, is debated since loss of function experiments described either a defect in HSC self-renewal[137] or no defect in hematopoiesis[138]. Targeting the Wnt and Notch pathways were then proposed to be potentially efficient strategies for *ex vivo* expansion of human HSPC. HSPC co-cultures on stromal cells transduced with Wnt-5A or treated with Wnt-5A conditioned medium did not expand them more. However, treatment of mice engrafted with human HSPC with Wnt-5A increased multilineage reconstitution by at least three-fold compared with controls[139]. Glycogen synthase kinase-3 (GSK-3) negatively regulates several signaling pathways, including the Wnt and Notch pathways[140]. Administration of selective GSK-3 inhibitors, that are ATP-competitors, into recipient mice transplanted with human HSPC resulted in their enhanced sustained long-term repopulation ability[141]. Expansion of human HSPC during culture with a plastic-immobilized form of DLL1 (iDLL1) indicated that iDLL1 could efficiently activate endogenous Notch receptors and induce *ex vivo* expansion of murine and human HSPC. Culture of UCB-derived $CD34^+$ HSPC in presence of DLL1 and cytokines resulted in increased numbers of SRC/HSC with secondary *in vivo* hematopoietic reconstitution[142,143]. A phase I trial showed the feasibility and safety of Notch-mediated UCB expansion and a phase II and III clinical trials will be necessary to achieve clinically relevant effects[142,144].

14.4.2.4 Approaches using modification of human HSPC microenvironment: Focus on the Hypoxia/HIF pathway

Hypoxia plays major roles in the physiology of hematopoietic and immune niches (Nobel Prize in physiology/medicine for 2019 awarded to W. Kaelin, Jr., Sir P. Ratcliffe, and G. Semenza). Hypoxia stabilizes the hypoxia-inducible factor-1/2α (HIF-1/2α) proteins, which are master regulators of hypoxia transcriptional response at the cellular and systemic levels[145–149]. In humans, hypoxia promotes the formation of myelo-erythroid colonies and enhances maintenance, expansion, and proliferative capacities of HSPC tested *in vitro* and *in vivo* in ID mouse models[147,150–157]. Low O_2 levels also significantly impact lymphoid development from UCB-derived human hematopoietic cell progenitors. Indeed, hypoxia enhances human lymphoid

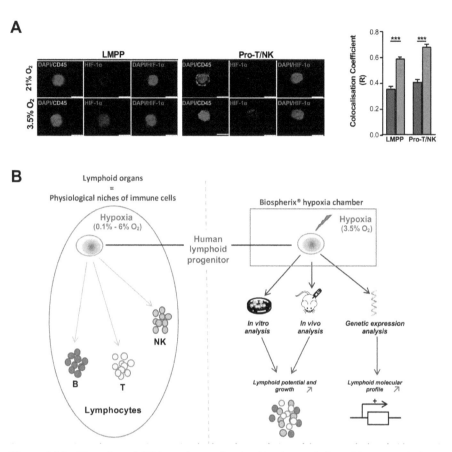

Figure 14.3 Hypoxia and HIF proteins are involved in the regulation of human early lymphoid progenitor functions (adapted from Chabi et al 2019). (**A**) HIF-1α protein expression. Cells derived from LMPP and Pro-TNK cells co-cultured under 21% O_2 and 3.5% O_2 for 7 days were stained with anti-human CD45 (**green**) and anti-human HIF-1α (**red**) and counterstained with DAPI. Cells were then analyzed by Leica TCS SP8 MP Confocal Microscope (Leica, Germany) (original magnification 63x, scale bar represents 7.5μM). Presence of HIF-1α protein in the nucleus is indicated by the colocalisation coefficient (R) (n = 50 cells).[158] (**B**) Landscape of hypoxia actions in human early lymphoid progenitors.

cell development from LMPP and MLP progenitors. Interestingly HIF-1α and HIF-2α did not overlap in their regulatory role in CB-derived lymphoid progenitors as HIF-1α rather enhanced LMPP-derived lymphoid cell production, whereas HIF-2α targeted Pro-T/NK-cell–derived lymphoid cell development[158] (Figure 14.3).

14.4.3 Chemical Compounds in the Era of Advanced Technologies for Ex Vivo Human HSPC Expansion

Evidence has accumulated suggesting that several small molecules may have substantial effects on HSPC expansion, thus representing promising tools in the field of HSC transplantation and gene therapy. 16-Dimethyl prostaglandin E_2 (dmPGE$_2$), the Aryl hydrocarbon receptor antagonist Stem-Regenin 1 (SR1), and the Pyrimidoindole derivative UM171 were identified by high-through screening of molecule libraries in CD34$^+$ cells. dmPGE$_2$ was shown to promote UCB HSPC function *in vitro* and *in vivo* in ID mouse models and dmPGE2-treated rhesus mobilized peripheral blood stem cells (mPBSC) induced stable long-term multilineage hematopoietic reconstitution in a nonhuman primate transplantation study, leading to a phase I clinical trial (NCT00890500)[159,160]. Cultures of UCB CD34$^+$ cells with SR1 resulted in a dramatic expansion in CD34$^+$-derived CFC (+65-fold), and in SRC (+17-fold)[161] supporting the clinical potential of SR1 in a phase I/II trial.[162] Moreover, UM171 improved *ex vivo* expansion of human LT-HSC (+100-fold) with functionally long-term reconstitution capability *in vivo* (SRC: +35-fold).[163] Phase I/II safety and feasibility studies (NCT02668315) have been achieved for this exiting molecule[163,164].

Epigenetic signatures are important hallmarks of HSPC fates and decisions and their regulation may account for the improvement of their intrinsic features. Histone deacetylase inhibitor (HDACi) such as valproic acid, increased the number of UCB CD34$^+$ cell–derived HSPCs and SRCs[165]. DNA Methyltransferase inhibitor (UNC0638) was shown to act synergistically with HDACi on *ex vivo* expansion of human CD34$^+$ cells by promoting maintenance of HSPCs *in vitro*[166]. It is to notice that the degree of expansion was substantially raised when both small molecules and cytokines were combined. The use of molecular compounds such as Nicotinamide, the Copper chelator tetraethylenepentamine, and C3a complement component showed enhancement of migration, homing, and engraftment capability of CD34$^+$ cells, a promotion of their expansion capacity and a delay of their differentiation. These macromolecules were examined for safety and feasibility in a phase I clinical trial or subjected to phase II/III clinical trial development protocols[120].

Very recently, a work demonstrated that incubation of UCB- and BM-derived CD34$^+$ cells with 3D zwitterionic hydrogels significantly enhanced clinically transferable *ex vivo* expansion of early HSPC (+73-fold) with *in vivo* long-term repopulating ability[167].

14.5 Conclusive Remarks and Perspectives

Knowledge of adult HSC biology is an "on-growing" field as very precise studies are currently using sophisticated mouse models. During the past 7 years, the extensive use of single cell assays such as single cell transplantations, barcoding experiments, or single cell RNA sequencing has provided a revolutionary understanding of the hematopoietic hierarchy, in mice and in humans. The combination of these innovative technological strategies was very recently published to elucidate clonal fate of MPP and HSC respectively *in vitro* and *in vivo* after transplantation, in regenerative hematopoiesis[168,169]. In the future, the adaptation of this technique to study native hematopoiesis will permit elucidating HSC contribution to daily blood cell production. In humans, relying on mostly *in vitro* functional tests, blood cells seem to be produced from multipotent and unipotent progenitors[79,170]. Therefore, there is a need to improve tools sustaining long-term human hematopoiesis. The humanized BM niche technology[64] will certainly have a huge impact in the field of human hematopoiesis and allow understanding human hematopoiesis at steady state as well as under stressful conditions.

To conclude, we can summarize with:

- HSC can be phenotypically characterized using a combination of markers among which some of them were very recently identified. However, to answer questions related to cognitive and clinical aspects of the scientific investigation, in *vitro* and *in vivo* functional assays as well as molecular analysis at the single cell level constitute important prerequisite to rigorously evaluate the fundamental properties of HSC i.e., multipotency, self-renewal capacity, and quiescence.
- Extrinsic factors provided by HSC microenvironment as well as intrinsic elements regulated in response to HSC interaction with its niche, give important cues to understand and predict HSC fate and function and can be used to optimize HSC manipulation for therapeutic purposes.

Acknowledgments

Our work is supported by INSERM, CEA, Université de Paris and Université Paris Saclay, Cancéropôle Ile de France, the Association Laurette Fugain, The European network RISK-IR, Electricité de France (EDF), CEA-Segment radiobiologie and the Association pour la Recherche contre le Cancer (ARC, équipe labellisée).

We apologize to all colleagues whose work could not be cited in this review due to word/citation limitations.

References

[1] Ehninger A, Trumpp A. The bone marrow stem cell niche grows up: mesenchymal stem cells and macrophages move in. *J Exp Med.* 2011;208(3):421–428.

[2] Barker N, Ridgway RA, van Es JH, et al. Crypt stem cells as the cells-of-origin of intestinal cancer. *Nature.* 2009;457(7229): 608–611.

[3] Dick JE. Stem cell concepts renew cancer research. *Blood.* 2008;112(13): 4793–4807.

[4] Cabezas-Wallscheid N, Klimmeck D, Hansson J, et al. Identification of regulatory networks in HSCs and their immediate progeny via integrated proteome, transcriptome, and DNA methylome analysis. *Cell Stem Cell.* 2014;15(4):507–522.

[5] Wilson A, Trumpp A. Bone-marrow haematopoietic-stem-cell niches. *Nat Rev Immunol.* 2006;6(2):93–106.

[6] Okada S, Nakauchi H, Nagayoshi K, Nishikawa S, Miura Y, Suda T. In vivo and in vitro stem cell function of c-kit- and Sca-1-positive murine hematopoietic cells. *Blood.* 1992;80(12):3044–3050.

[7] Uchida N, Weissman IL. Searching for hematopoietic stem cells: evidence that Thy-1.1lo Lin- Sca-1+ cells are the only stem cells in C57BL/Ka-Thy-1.1 bone marrow. *J Exp Med.* 1992;175(1):175–184.

[8] Kiel MJ, Yilmaz OH, Iwashita T, Yilmaz OH, Terhorst C, Morrison SJ. SLAM family receptors distinguish hematopoietic stem and progenitor cells and reveal endothelial niches for stem cells. *Cell.* 2005;121(7):1109–1121.

[9] Wilson A, Laurenti E, Oser G, et al. Hematopoietic stem cells reversibly switch from dormancy to self-renewal during homeostasis and repair. *Cell.* 2008;135(6):1118–1129.

[10] Cabezas-Wallscheid N, Buettner F, Sommerkamp P, et al. Vitamin A-Retinoic Acid Signaling Regulates Hematopoietic Stem Cell Dormancy. *Cell.* 2017;169(5):807–823 e819.

[11] Zhang J, Niu C, Ye L, et al. Identification of the haematopoietic stem cell niche and control of the niche size. *Nature.* 2003;425(6960):836–841.

[12] Takahashi M, Matsuoka Y, Sumide K, et al. CD133 is a positive marker for a distinct class of primitive human cord blood-derived CD34-negative hematopoietic stem cells. *Leukemia.* 2014;28(6):1308–1315.

[13] Civin CI, Strauss LC, Brovall C, Fackler MJ, Schwartz JF, Shaper JH. Antigenic analysis of hematopoiesis. III. A hematopoietic progenitor cell surface antigen defined by a monoclonal antibody raised against KG-1a cells. *J Immunol.* 1984;133(1):157–165.

[14] Civin CI, Trischmann T, Kadan NS, et al. Highly purified CD34-positive cells reconstitute hematopoiesis. *J Clin Oncol.* 1996;14(8):2224–2233.

[15] Silvestri F, Banavali S, Baccarani M, Preisler HD. The CD34 hemopoietic progenitor cell associated antigen: biology and clinical applications. *Haematologica.* 1992;77(3):265–273.

[16] Strauss LC, Trischmann TM, Rowley SD, Wiley JM, Civin CI. Selection of normal human hematopoietic stem cells for bone marrow transplantation using immunomagnetic microspheres and CD34 antibody. *Am J Pediatr Hematol Oncol.* 1991;13(2):217–221.

[17] Fina L, Molgaard HV, Robertson D, et al. Expression of the CD34 gene in vascular endothelial cells. *Blood.* 1990;75(12):2417–2426.

[18] Bhatia M, Bonnet D, Murdoch B, Gan OI, Dick JE. A newly discovered class of human hematopoietic cells with SCID-repopulating activity. *Nat Med.* 1998;4(9):1038–1045.

[19] Wang J, Kimura T, Asada R, et al. SCID-repopulating cell activity of human cord blood-derived CD34- cells assured by intra-bone marrow injection. *Blood.* 2003;101(8):2924–2931.

[20] Anjos-Afonso F, Currie E, Palmer HG, Foster KE, Taussig DC, Bonnet D. CD34(-) cells at the apex of the human hematopoietic stem cell hierarchy have distinctive cellular and molecular signatures. *Cell Stem Cell.* 2013;13(2):161–174.

[21] Doulatov S, Notta F, Laurenti E, Dick JE. Hematopoiesis: a human perspective. *Cell Stem Cell.* 2012;10(2):120–136.

[22] Terstappen LW, Huang S, Safford M, Lansdorp PM, Loken MR. Sequential generations of hematopoietic colonies derived from single nonlineage-committed CD34+CD38- progenitor cells. *Blood.* 1991;77(6):1218–1227.

[23] Baum CM, Weissman IL, Tsukamoto AS, Buckle AM, Peault B. Isolation of a candidate human hematopoietic stem-cell population. *Proc Natl Acad Sci U S A.* 1992;89(7):2804–2808.

[24] Majeti R, Park CY, Weissman IL. Identification of a hierarchy of multipotent hematopoietic progenitors in human cord blood. *Cell Stem Cell.* 2007;1(6):635–645.

[25] Gallacher L, Murdoch B, Wu DM, Karanu FN, Keeney M, Bhatia M. Isolation and characterization of human CD34(-)Lin(-) and CD34(+)Lin(-) hematopoietic stem cells using cell surface markers AC133 and CD7. *Blood*. 2000;95(9):2813–2820.

[26] Bhatia M, Wang JC, Kapp U, Bonnet D, Dick JE. Purification of primitive human hematopoietic cells capable of repopulating immune-deficient mice. *Proc Natl Acad Sci U S A*. 1997;94(10):5320–5325.

[27] Belluschi S, Calderbank EF, Ciaurro V, et al. Myelo-lymphoid lineage restriction occurs in the human haematopoietic stem cell compartment before lymphoid-primed multipotent progenitors. *Nat Commun*. 2018;9(1):4100.

[28] Notta F, Doulatov S, Laurenti E, Poeppl A, Jurisica I, Dick JE. Isolation of single human hematopoietic stem cells capable of long-term multilineage engraftment. *Science*. 2011;333(6039):218–221.

[29] Copley MR, Beer PA, Eaves CJ. Hematopoietic stem cell heterogeneity takes center stage. *Cell Stem Cell*. 2012;10(6):690–697.

[30] Yamamoto R, Morita Y, Ooehara J, et al. Clonal analysis unveils self-renewing lineage-restricted progenitors generated directly from hematopoietic stem cells. *Cell*. 2013;154(5):1112–1126.

[31] Coulombel L. Identification of hematopoietic stem/progenitor cells: strength and drawbacks of functional assays. *Oncogene*. 2004;23(43): 7210–7222.

[32] Paige CJ, Gisler RH, McKearn JP, Iscove NN. Differentiation of murine B cell precursors in agar culture. Frequency, surface marker analysis and requirements for growth of clonable pre-B cells. *Eur J Immunol*. 1984;14(11):979–987.

[33] Pluznik DH, Sachs L. The cloning of normal "mast" cells in tissue culture. *J Cell Physiol*. 1965;66(3):319–324.

[34] Coulombel L, Kalousek DK, Eaves CJ, Gupta CM, Eaves AC. Long-term marrow culture reveals chromosomally normal hematopoietic progenitor cells in patients with Philadelphia chromosome-positive chronic myelogenous leukemia. *N Engl J Med*. 1983;308(25):1493–1498.

[35] Dexter TM, Moore MA, Sheridan AP. Maintenance of hemopoietic stem cells and production of differentiated progeny in allogeneic and semiallogeneic bone marrow chimeras in vitro. *J Exp Med*. 1977;145(6):1612–1616.

[36] Issaad C, Croisille L, Katz A, Vainchenker W, Coulombel L. A murine stromal cell line allows the proliferation of very primitive human

CD34++/CD38- progenitor cells in long-term cultures and semisolid assays. *Blood*. 1993;81(11):2916–2924.

[37] Robinson SN, Ng J, Niu T, et al. Superior ex vivo cord blood expansion following co-culture with bone marrow-derived mesenchymal stem cells. . *Bone Marrow Transplant*. 2006;37(4):359–366.

[38] Hao QL, Thiemann FT, Petersen D, Smogorzewska EM, Crooks GM. Extended long-term culture reveals a highly quiescent and primitive human hematopoietic progenitor population. *Blood*. 1996;88(9):3306–3313.

[39] Henry E, Souissi-Sahraoui I, Deynoux M, et al. Human hematopoietic stem/progenitor cells display ROS-dependent long-term hematopoietic defects after exposure to low dose of ionizing radiations. *Haematologica*. 2019.

[40] Petzer AL, Hogge DE, Landsdorp PM, Reid DS, Eaves CJ. Self-renewal of primitive human hematopoietic cells (long-term-culture-initiating cells) in vitro and their expansion in defined medium. *Proc Natl Acad Sci U S A*. 1996;93(4):1470–1474.

[41] Rizo A, Dontje B, Vellenga E, de Haan G, Schuringa JJ. Long-term maintenance of human hematopoietic stem/progenitor cells by expression of BMI1. *Blood*. 2008;111(5):2621–2630.

[42] Metcalf D. Hematopoietic cytokines. *Blood*. 2008;111(2):485–491.

[43] Plum J, De Smedt M, Verhasselt B, et al. Human T lymphopoiesis. In vitro and in vivo study models. *Ann N Y Acad Sci*. 2000;917:724–731.

[44] Watanabe Y, Gyotoku J, Katsura Y. Analysis of the development of T cells by transferring precursors into cultured fetal thymus with a microinjector. *Thymus*. 1989;13(1-2):57–71.

[45] Radtke F, Wilson A, Stark G, et al. Deficient T cell fate specification in mice with an induced inactivation of Notch1. *Immunity*. 1999;10(5):547–558.

[46] Calvo J, BenYoucef A, Baijer J, Rouyez MC, Pflumio F. Assessment of human multi-potent hematopoietic stem/progenitor cell potential using a single in vitro screening system. *PLoS One*. 2012;7(11):e50495.

[47] La Motte-Mohs RN, Herer E, Zuniga-Pflucker JC. Induction of T-cell development from human cord blood hematopoietic stem cells by Delta-like 1 in vitro. *Blood*. 2005;105(4):1431–1439.

[48] Schmitt TM, Zuniga-Pflucker JC. Induction of T cell development from hematopoietic progenitor cells by delta-like-1 in vitro. *Immunity*. 2002;17(6):749–756.

[49] Van de Walle I, Waegemans E, De Medts J, et al. Specific Notch receptor-ligand interactions control human TCR-alphabeta/gammadelta development by inducing differential Notch signal strength. *J Exp Med.* 2013;210(4):683–697.

[50] Sudo T, Ito M, Ogawa Y, et al. Interleukin 7 production and function in stromal cell-dependent B cell development. *J Exp Med.* 1989;170(1):333–338.

[51] Berardi AC, Meffre E, Pflumio F, et al. Individual CD34+CD38lowCD 19-CD10- progenitor cells from human cord blood generate B lymphocytes and granulocytes. *Blood.* 1997;89(10):3554–3564.

[52] Balzano M, De Grandis M, Vu Manh TP, et al. Nidogen-1 Contributes to the Interaction Network Involved in Pro-B Cell Retention in the Peri-sinusoidal Hematopoietic Stem Cell Niche. *Cell Rep.* 2019;26(12):3257–3271 e3258.

[53] Baryawno N, Przybylski D, Kowalczyk MS, et al. A Cellular Taxonomy of the Bone Marrow Stroma in Homeostasis and Leukemia. *Cell.* 2019;177(7):1915–1932 e1916.

[54] Tikhonova AN, Dolgalev I, Hu H, et al. The bone marrow microenvironment at single-cell resolution. *Nature.* 2019;569(7755):222–228.

[55] Purton LE, Scadden DT. Limiting factors in murine hematopoietic stem cell assays. *Cell Stem Cell.* 2007;1(3):263–270.

[56] Kamel-Reid S, Dick JE. Engraftment of immune-deficient mice with human hematopoietic stem cells. *Science.* 1988;242(4886):1706–1709.

[57] Lapidot T, Pflumio F, Doedens M, Murdoch B, Williams DE, Dick JE. Cytokine stimulation of multilineage hematopoiesis from immature human cells engrafted in SCID mice. *Science.* 1992;255(5048): 1137–1141.

[58] Spangrude GJ, Heimfeld S, Weissman IL. Purification and characterization of mouse hematopoietic stem cells. *Science.* 1988;241(4861): 58–62.

[59] Larochelle A, Vormoor J, Hanenberg H, et al. Identification of primitive human hematopoietic cells capable of repopulating NOD/SCID mouse bone marrow: implications for gene therapy. *Nat Med.* 1996;2(12):1329–1337.

[60] Pflumio F, Izac B, Katz A, Shultz LD, Vainchenker W, Coulombel L. Phenotype and function of human hematopoietic cells engrafting immune-deficient CB17-severe combined immunodeficiency mice and nonobese diabetic-severe combined immunodeficiency mice after

transplantation of human cord blood mononuclear cells. *Blood.* 1996;88(10):3731–3740.

[61] Robin C, Pflumio F, Vainchenker W, Coulombel L. Identification of lymphomyeloid primitive progenitor cells in fresh human cord blood and in the marrow of nonobese diabetic-severe combined immunodeficient (NOD-SCID) mice transplanted with human CD34(+) cord blood cells. *J Exp Med.* 1999;189(10):1601–1610.

[62] Rongvaux A, Willinger T, Martinek J, et al. Development and function of human innate immune cells in a humanized mouse model. *Nat Biotechnol.* 2014;32(4):364–372.

[63] Reinisch A, Thomas D, Corces MR, et al. A humanized bone marrow ossicle xenotransplantation model enables improved engraftment of healthy and leukemic human hematopoietic cells. *Nat Med.* 2016;22(7):812–821.

[64] Dupard SJ, Grigoryan A, Farhat S, Coutu DL, Bourgine PE. Development of Humanized Ossicles: Bridging the Hematopoietic Gap. *Trends Mol Med.* 2020;26(6):552–569.

[65] Tesio M, Trumpp A. Breaking the cell cycle of HSCs by p57 and friends. *Cell Stem Cell.* 2011;9(3):187–192.

[66] Matsumoto A, Takeishi S, Kanie T, et al. p57 is required for quiescence and maintenance of adult hematopoietic stem cells. *Cell Stem Cell.* 2011;9(3):262–271.

[67] Suda T, Takubo K, Semenza GL. Metabolic regulation of hematopoietic stem cells in the hypoxic niche. *Cell Stem Cell.* 2011;9(4):298–310.

[68] Ito K, Hirao A, Arai F, et al. Regulation of oxidative stress by ATM is required for self-renewal of haematopoietic stem cells. *Nature.* 2004;431(7011):997–1002.

[69] Rodrigues-Moreira S, Moreno SG, Ghinatti G, et al. Low-Dose Irradiation Promotes Persistent Oxidative Stress and Decreases Self-Renewal in Hematopoietic Stem Cells. *Cell Rep.* 2017;20(13):3199–3211.

[70] Muller-Sieburg CE, Cho RH, Karlsson L, Huang JF, Sieburg HB. Myeloid-biased hematopoietic stem cells have extensive self-renewal capacity but generate diminished lymphoid progeny with impaired IL-7 responsiveness. *Blood.* 2004;103(11):4111–4118.

[71] Muller-Sieburg CE, Cho RH, Thoman M, Adkins B, Sieburg HB. Deterministic regulation of hematopoietic stem cell self-renewal and differentiation. *Blood.* 2002;100(4):1302–1309.

[72] Sarrazin S, Mossadegh-Keller N, Fukao T, et al. MafB restricts M-CSF-dependent myeloid commitment divisions of hematopoietic stem cells. *Cell.* 2009;138(2):300–313.

[73] Jacobsen SEW, Nerlov C. Haematopoiesis in the era of advanced single-cell technologies. *Nat Cell Biol.* 2019;21(1):2–8.

[74] Laurenti E, Gottgens B. From haematopoietic stem cells to complex differentiation landscapes. *Nature.* 2018;553(7689):418–426.

[75] Adolfsson J, Mansson R, Buza-Vidas N, et al. Identification of Flt3+ lympho-myeloid stem cells lacking erythro-megakaryocytic potential a revised road map for adult blood lineage commitment. *Cell.* 2005;121(2):295–306.

[76] Doulatov S, Notta F, Eppert K, Nguyen LT, Ohashi PS, Dick JE. Revised map of the human progenitor hierarchy shows the origin of macrophages and dendritic cells in early lymphoid development. *Nat Immunol.* 2010;11(7):585–593.

[77] Goardon N, Marchi E, Atzberger A, et al. Coexistence of LMPP-like and GMP-like leukemia stem cells in acute myeloid leukemia. *Cancer Cell.* 2011;19(1):138–152.

[78] Haddad R, Guardiola P, Izac B, et al. Molecular characterization of early human T/NK and B-lymphoid progenitor cells in umbilical cord blood. *Blood.* 2004;104(13):3918–3926.

[79] Karamitros D, Stoilova B, Aboukhalil Z, et al. Single-cell analysis reveals the continuum of human lympho-myeloid progenitor cells. *Nat Immunol.* 2018;19(1):85–97.

[80] Kohn LA, Hao QL, Sasidharan R, et al. Lymphoid priming in human bone marrow begins before expression of CD10 with upregulation of L-selectin. *Nat Immunol.* 2012;13(10):963–971.

[81] Benz C, Copley MR, Kent DG, et al. Hematopoietic stem cell subtypes expand differentially during development and display distinct lymphopoietic programs. *Cell Stem Cell.* 2012;10(3):273–283.

[82] Dykstra B, Kent D, Bowie M, et al. Long-term propagation of distinct hematopoietic differentiation programs in vivo. *Cell Stem Cell.* 2007;1(2):218–229.

[83] Pietras EM, Reynaud D, Kang YA, et al. Functionally Distinct Subsets of Lineage-Biased Multipotent Progenitors Control Blood Production in Normal and Regenerative Conditions. *Cell Stem Cell.* 2015;17(1):35–46.

[84] Buenrostro JD, Corces MR, Lareau CA, et al. Integrated Single-Cell Analysis Maps the Continuous Regulatory Landscape of Human Hematopoietic Differentiation. *Cell.* 2018;173(6):1535–1548 e1516.

[85] Giladi A, Paul F, Herzog Y, et al. Single-cell characterization of haematopoietic progenitors and their trajectories in homeostasis and perturbed haematopoiesis. *Nat Cell Biol.* 2018;20(7):836–846.

[86] Nestorowa S, Hamey FK, Pijuan Sala B, et al. A single-cell resolution map of mouse hematopoietic stem and progenitor cell differentiation. *Blood.* 2016;128(8):e20–31.

[87] Velten L, Haas SF, Raffel S, et al. Human haematopoietic stem cell lineage commitment is a continuous process. *Nat Cell Biol.* 2017;19(4):271–281.

[88] Lapidot T, Petit I. Current understanding of stem cell mobilization: the roles of chemokines, proteolytic enzymes, adhesion molecules, cytokines, and stromal cells. *Exp Hematol.* 2002;30(9):973–981.

[89] Peled A, Petit I, Kollet O, et al. Dependence of human stem cell engraftment and repopulation of NOD/SCID mice on CXCR4. *Science.* 1999;283(5403):845–848.

[90] Sugiyama T, Kohara H, Noda M, Nagasawa T. Maintenance of the hematopoietic stem cell pool by CXCL12-CXCR4 chemokine signaling in bone marrow stromal cell niches. *Immunity.* 2006;25(6):977–988.

[91] Tzeng YS, Li H, Kang YL, Chen WC, Cheng WC, Lai DM. Loss of Cxcl12/Sdf-1 in adult mice decreases the quiescent state of hematopoietic stem/progenitor cells and alters the pattern of hematopoietic regeneration after myelosuppression. *Blood.* 2011;117(2):429–439.

[92] Kovach NL, Lin N, Yednock T, Harlan JM, Broudy VC. Stem cell factor modulates avidity of alpha 4 beta 1 and alpha 5 beta 1 integrins expressed on hematopoietic cell lines. *Blood.* 1995;85(1):159–167.

[93] Ding L, Morrison SJ. Haematopoietic stem cells and early lymphoid progenitors occupy distinct bone marrow niches. *Nature.* 2013;495(7440):231–235.

[94] Ding L, Saunders TL, Enikolopov G, Morrison SJ. Endothelial and perivascular cells maintain haematopoietic stem cells. *Nature.* 2012;481(7382):457–462.

[95] Papayannopoulou T. Bone marrow homing: the players, the playfield, and their evolving roles. *Curr Opin Hematol.* 2003;10(3):214–219.

[96] Ramirez P, Rettig MP, Uy GL, et al. BIO5192, a small molecule inhibitor of VLA-4, mobilizes hematopoietic stem and progenitor cells. *Blood.* 2009;114(7):1340–1343.

[97] Scott LM, Priestley GV, Papayannopoulou T. Deletion of alpha4 integrins from adult hematopoietic cells reveals roles in homeostasis, regeneration, and homing. *Mol Cell Biol.* 2003;23(24):9349–9360.

[98] Peled A, Kollet O, Ponomaryov T, et al. The chemokine SDF-1 activates the integrins LFA-1, VLA-4, and VLA-5 on immature human

CD34(+) cells: role in transendothelial/stromal migration and engraftment of NOD/SCID mice. *Blood.* 2000;95(11):3289–3296.

[99] Schajnovitz A, Itkin T, D'Uva G, et al. CXCL12 secretion by bone marrow stromal cells is dependent on cell contact and mediated by connexin-43 and connexin-45 gap junctions. *Nat Immunol.* 2011;12(5):391–398.

[100] Arcangeli ML, Frontera V, Bardin F, et al. JAM-B regulates maintenance of hematopoietic stem cells in the bone marrow. *Blood.* 2011;118(17):4609–4619.

[101] Arcangeli ML, Bardin F, Frontera V, et al. Function of Jam-B/Jam-C interaction in homing and mobilization of human and mouse hematopoietic stem and progenitor cells. *Stem Cells.* 2014;32(4):1043–1054.

[102] Arcangeli ML, Frontera V, Aurrand-Lions M. Function of junctional adhesion molecules (JAMs) in leukocyte migration and homeostasis. *Arch Immunol Ther Exp (Warsz).* 2013;61(1):15–23.

[103] Ho YH, Del Toro R, Rivera-Torres J, et al. Remodeling of Bone Marrow Hematopoietic Stem Cell Niches Promotes Myeloid Cell Expansion during Premature or Physiological Aging. *Cell Stem Cell.* 2019;25(3):
407–418 e406.

[104] Mende N, Jolly A, Percin GI, et al. Prospective isolation of non-hematopoietic cells of the niche and their differential molecular interactions with HSCs. *Blood.* 2019;134(15):1214–1226.

[105] Orkin SH, Zon LI. Hematopoiesis: an evolving paradigm for stem cell biology. *Cell.* 2008;132(4):631–644.

[106] Begley CG, Aplan PD, Denning SM, Haynes BF, Waldmann TA, Kirsch IR. The gene SCL is expressed during early hematopoiesis and encodes a differentiation-related DNA-binding motif. *Proc Natl Acad Sci U S A.* 1989;86(24):10128–10132.

[107] Calvo J, Fahy L, Uzan B, Pflumio F. Desperately seeking a home marrow niche for T-cell acute lymphoblastic leukaemia. *Adv Biol Regul.* 2019;74:100640.

[108] Correia NC, Arcangeli ML, Pflumio F, Barata JT. Stem Cell Leukemia: how a TALented actor can go awry on the hematopoietic stage. *Leukemia.* 2016;30(10):1968–1978.

[109] Lacombe J, Herblot S, Rojas-Sutterlin S, et al. Scl regulates the quiescence and the long-term competence of hematopoietic stem cells. *Blood.* 2010;115(4):792–803.

[110] Mikkola HK, Klintman J, Yang H, et al. Haematopoietic stem cells retain long-term repopulating activity and multipotency in the absence of stem-cell leukaemia SCL/tal-1 gene. *Nature*. 2003;421(6922):547–551.

[111] Brunet de la Grange P, Armstrong F, Duval V, et al. Low SCL/TAL1 expression reveals its major role in adult hematopoietic myeloid progenitors and stem cells. *Blood*. 2006;108(9):2998–3004.

[112] Souroullas GP, Salmon JM, Sablitzky F, Curtis DJ, Goodell MA. Adult hematopoietic stem and progenitor cells require either Lyl1 or Scl for survival. *Cell Stem Cell*. 2009;4(2):180–186.

[113] Reynaud D, Ravet E, Titeux M, et al. SCL/TAL1 expression level regulates human hematopoietic stem cell self-renewal and engraftment. *Blood*. 2005;106(7):2318–2328.

[114] Benyoucef A, Calvo J, Renou L, et al. The SCL/TAL1 Transcription Factor Represses the Stress Protein DDiT4/REDD1 in Human Hematopoietic Stem/Progenitor Cells. *Stem Cells*. 2015;33(7):2268–2279.

[115] Kelly SS, Sola CB, de Lima M, Shpall E. Ex vivo expansion of cord blood. *Bone Marrow Transplant*. 2009;44(10):673–681.

[116] Piacibello W, Sanavio F, Garetto L, et al. Extensive amplification and self-renewal of human primitive hematopoietic stem cells from cord blood. *Blood*. 1997;89(8):2644–2653.

[117] Zandstra PW, Conneally E, Petzer AL, Piret JM, Eaves CJ. Cytokine manipulation of primitive human hematopoietic cell self-renewal. *Proc Natl Acad Sci U S A*. 1997;94(9):4698–4703.

[118] Shpall EJ, Quinones R, Giller R, et al. Transplantation of ex vivo expanded cord blood. *Biol Blood Marrow Transplant*. 2002;8(7):368–376.

[119] Robinson SN, Ng J, Niu T, et al. Superior ex vivo cord blood expansion following co-culture with bone marrow-derived mesenchymal stem cells. *Bone Marrow Transplant*. 2006;37(4):359–366.

[120] Norkin M, Lazarus HM, Wingard JR. Umbilical cord blood graft enhancement strategies: has the time come to move these into the clinic? *Bone Marrow Transplant*. 2013;48(7):884–889.

[121] Hofmeister CC, Zhang J, Knight KL, Le P, Stiff PJ. Ex vivo expansion of umbilical cord blood stem cells for transplantation: growing knowledge from the hematopoietic niche. *Bone Marrow Transplant*. 2007;39(1):11–23.

[122] Tung SS, Parmar S, Robinson SN, De Lima M, Shpall EJ. Ex vivo expansion of umbilical cord blood for transplantation. *Best Pract Res Clin Haematol.* 2010;23(2):245–257.

[123] Antonchuk J, Sauvageau G, Humphries RK. HOXB4-induced expansion of adult hematopoietic stem cells ex vivo. *Cell.* 2002;109(1): 39–45.

[124] Sauvageau G, Thorsteinsdottir U, Eaves CJ, et al. Overexpression of HOXB4 in hematopoietic cells causes the selective expansion of more primitive populations in vitro and in vivo. *Genes Dev.* 1995;9(14):1753–1765.

[125] Amsellem S, Pflumio F, Bardinet D, et al. Ex vivo expansion of human hematopoietic stem cells by direct delivery of the HOXB4 homeoprotein. *Nat Med.* 2003;9(11):1423–1427.

[126] Haddad R, Caignard A, Visentin G, Vigon I, Fichelson S, Amsellem S. The HOXB4 homeoprotein improves ex vivo generation of functional human NK-cell progenitors. *Leukemia.* 2007;21(8):1836–1839.

[127] Haddad R, Guimiot F, Six E, et al. Dynamics of thymus-colonizing cells during human development. *Immunity.* 2006;24(2):217–230.

[128] Haddad R, Pflumio F, Vigon I, et al. The HOXB4 homeoprotein differentially promotes ex vivo expansion of early human lymphoid progenitors. *Stem Cells.* 2008;26(2):312–322.

[129] Bigas A, Espinosa L. Hematopoietic stem cells: to be or Notch to be. *Blood.* 2012;119(14):3226–3235.

[130] Ntziachristos P, Lim JS, Sage J, Aifantis I. From fly wings to targeted cancer therapies: a centennial for notch signaling. *Cancer Cell.* 2014;25(3):318–334.

[131] Pajcini KV, Speck NA, Pear WS. Notch signaling in mammalian hematopoietic stem cells. *Leukemia.* 2011;25(10):1525–1532.

[132] Calvi LM, Adams GB, Weibrecht KW, et al. Osteoblastic cells regulate the haematopoietic stem cell niche. *Nature.* 2003;425(6960):841–846.

[133] Duncan AW, Rattis FM, DiMascio LN, et al. Integration of Notch and Wnt signaling in hematopoietic stem cell maintenance. *Nat Immunol.* 2005;6(3):314–322.

[134] Maillard I, Koch U, Dumortier A, et al. Canonical notch signaling is dispensable for the maintenance of adult hematopoietic stem cells. *Cell Stem Cell.* 2008;2(4):356–366.

[135] Mancini SJ, Mantei N, Dumortier A, Suter U, MacDonald HR, Radtke F. Jagged1-dependent Notch signaling is dispensable for hematopoietic

stem cell self-renewal and differentiation. *Blood.* 2005;105(6): 2340–2342.

[136] Reya T, Clevers H. Wnt signalling in stem cells and cancer. *Nature.* 2005;434(7035):843–850.

[137] Zhao C, Blum J, Chen A, et al. Loss of beta-catenin impairs the renewal of normal and CML stem cells in vivo. *Cancer Cell.* 2007;12(6): 528–541.

[138] Cobas M, Wilson A, Ernst B, et al. Beta-catenin is dispensable for hematopoiesis and lymphopoiesis. *J Exp Med.* 2004;199(2):221–229.

[139] Murdoch B, Chadwick K, Martin M, et al. Wnt-5A augments repopulating capacity and primitive hematopoietic development of human blood stem cells in vivo. *Proc Natl Acad Sci U S A.* 2003;100(6):3422–3427.

[140] Zhao C, Chen A, Jamieson CH, et al. Hedgehog signalling is essential for maintenance of cancer stem cells in myeloid leukaemia. *Nature.* 2009;458(7239):776–779.

[141] Trowbridge JJ, Xenocostas A, Moon RT, Bhatia M. Glycogen synthase kinase-3 is an in vivo regulator of hematopoietic stem cell repopulation. *Nat Med.* 2006;12(1):89–98.

[142] Delaney C, Heimfeld S, Brashem-Stein C, Voorhies H, Manger RL, Bernstein ID. Notch-mediated expansion of human cord blood progenitor cells capable of rapid myeloid reconstitution. *Nat Med.* 2010;16(2):232–236.

[143] Varnum-Finney B, Brashem-Stein C, Bernstein ID. Combined effects of Notch signaling and cytokines induce a multiple log increase in precursors with lymphoid and myeloid reconstituting ability. *Blood.* 2003;101(5):1784–1789.

[144] Dahlberg A, Delaney C, Bernstein ID. Ex vivo expansion of human hematopoietic stem and progenitor cells. *Blood.* 2011;117(23):6083–6090.

[145] Imanirad P, Solaimani Kartalaei P, Crisan M, et al. HIF1alpha is a regulator of hematopoietic progenitor and stem cell development in hypoxic sites of the mouse embryo. *Stem Cell Res.* 2014;12(1):24–35.

[146] Majmundar AJ, Wong WJ, Simon MC. Hypoxia-inducible factors and the response to hypoxic stress. *Mol Cell.* 2010;40(2):294–309.

[147] Rouault-Pierre K, Lopez-Onieva L, Foster K, et al. HIF-2alpha protects human hematopoietic stem/progenitors and acute myeloid leukemic cells from apoptosis induced by endoplasmic reticulum stress. *Cell Stem Cell.* 2013;13(5):549–563.

[148] Semenza GL. Hypoxia-inducible factors in physiology and medicine. *Cell.* 2012;148(3):399–408.

[149] Takubo K, Goda N, Yamada W, et al. Regulation of the HIF-1alpha level is essential for hematopoietic stem cells. *Cell Stem Cell.* 2010;7(3): 391–402.

[150] Cipolleschi MG, D'Ippolito G, Bernabei PA, et al. Severe hypoxia enhances the formation of erythroid bursts from human cord blood cells and the maintenance of BFU-E in vitro. *Exp Hematol.* 1997;25(11):1187–1194.

[151] Danet GH, Pan Y, Luongo JL, Bonnet DA, Simon MC. Expansion of human SCID-repopulating cells under hypoxic conditions. *J Clin Invest.* 2003;112(1):126–135.

[152] Guitart AV, Hammoud M, Dello Sbarba P, Ivanovic Z, Praloran V. Slow-cycling/quiescence balance of hematopoietic stem cells is related to physiological gradient of oxygen. *Exp Hematol.* 2010;38(10):847–851.

[153] Hammoud M, Vlaski M, Duchez P, et al. Combination of low O(2) concentration and mesenchymal stromal cells during culture of cord blood CD34(+) cells improves the maintenance and proliferative capacity of hematopoietic stem cells. *J Cell Physiol.* 2012;227(6):2750–2758.

[154] Ivanovic Z, Dello Sbarba P, Trimoreau F, Faucher JL, Praloran V. Primitive human HPCs are better maintained and expanded in vitro at 1 percent oxygen than at 20 percent. *Transfusion.* 2000;40(12): 1482–1488.

[155] Ivanovic Z, Hermitte F, Brunet de la Grange P, et al. Simultaneous maintenance of human cord blood SCID-repopulating cells and expansion of committed progenitors at low O2 concentration (3%). *Stem Cells.* 2004;22(5):716–724.

[156] Koller MR, Bender JG, Miller WM, Papoutsakis ET. Reduced oxygen tension increases hematopoiesis in long-term culture of human stem and progenitor cells from cord blood and bone marrow. *Exp Hematol.* 1992;20(2):264–270.

[157] Shima H, Takubo K, Iwasaki H, et al. Reconstitution activity of hypoxic cultured human cord blood CD34-positive cells in NOG mice. *Biochem Biophys Res Commun.* 2009;378(3):467–472.

[158] Chabi S, Uzan B, Naguibneva I, et al. Hypoxia Regulates Lymphoid Development of Human Hematopoietic Progenitors. *Cell Rep.* 2019;29(8):2307–2320 e2306.

[159] Cutler C, Multani P, Robbins D, et al. Prostaglandin-modulated umbilical cord blood hematopoietic stem cell transplantation. *Blood*. 2013;122(17):3074–3081.

[160] Goessling W, Allen RS, Guan X, et al. Prostaglandin E2 enhances human cord blood stem cell xenotransplants and shows long-term safety in preclinical nonhuman primate transplant models. *Cell Stem Cell*. 2011;8(4):445–458.

[161] Boitano AE, Wang J, Romeo R, et al. Aryl hydrocarbon receptor antagonists promote the expansion of human hematopoietic stem cells. *Science*. 2010;329(5997):1345–1348.

[162] Wagner JE, Jr., Brunstein CG, Boitano AE, et al. Phase I/II Trial of StemRegenin-1 Expanded Umbilical Cord Blood Hematopoietic Stem Cells Supports Testing as a Stand-Alone Graft. *Cell Stem Cell*. 2016;18(1):144–155.

[163] Fares I, Chagraoui J, Gareau Y, et al. Cord blood expansion. Pyrimidoindole derivatives are agonists of human hematopoietic stem cell self-renewal. *Science*. 2014;345(6203):1509–1512.

[164] Cohen S, Roy J, Lachance S, et al. Hematopoietic stem cell transplantation using single UM171-expanded cord blood: a single-arm, phase 1–2 safety and feasibility study. *Lancet Haematol*. 2020;7(2):e134-e145.

[165] Chaurasia P, Gajzer DC, Schaniel C, D'Souza S, Hoffman R. Epigenetic reprogramming induces the expansion of cord blood stem cells. *J Clin Invest*. 2014;124(6):2378–2395.

[166] Chen X, Skutt-Kakaria K, Davison J, et al. G9a/GLP-dependent histone H3K9me2 patterning during human hematopoietic stem cell lineage commitment. *Genes Dev*. 2012;26(22):2499–2511.

[167] Bai T, Li J, Sinclair A, et al. Expansion of primitive human hematopoietic stem cells by culture in a zwitterionic hydrogel. *Nat Med*. 2019;25(10):1566–1575.

[168] Rodriguez-Fraticelli AE, Weinreb C, Wang SW, et al. Single-cell lineage tracing unveils a role for TCF15 in haematopoiesis. *Nature*. 2020.

[169] Weinreb C, Rodriguez-Fraticelli A, Camargo FD, Klein AM. Lineage tracing on transcriptional landscapes links state to fate during differentiation. *Science*. 2020;367(6479).

[170] Notta F, Zandi S, Takayama N, et al. Distinct routes of lineage development reshape the human blood hierarchy across ontogeny. *Science*. 2016;351(6269):aab2116.

15

Epithelial Stem Cells in the Skin

**Romain Fontaine[1,2,*], Bénédicte Oulès[1,2,3], Mathieu Castela[1]
and Sélim Aractingi[1,2,3]**

[1]Institut Cochin, Inserm U1016-CNRS UMR8104, Paris, France
[2]Université de Paris, France
[3]Service de Dermatologie, Hôpital Cochin, Paris, France
E-mail: romain.fontaine@inserm.fr
*Corresponding Author

15.1 Introduction

Forming the outermost protection of the body, skin is organized from bottom to top in three different layers: the hypodermis and dermis of mesoderm origin, and the epidermis of ectoderm origin. The latter is a stratified squamous epithelium that serves as a natural barrier to various threats. With approximately 20 different resident cell types, it protects the body against external environmental insults such as microbial pathogens, chemical compounds, traumas, and oxidant UV light stress. It also provides mechanical resistance and protects against the loss of internal body fluids (Elias, 2007). In mammals, the formation of epidermis begins early in development, right after gastrulation, starting from a single layer of neurectoderm progenitor cells that remain at the embryo surface and will quickly undergo a stratification process (Gilbert, 2000). This mechanism will allow the formation of the so-called interfollicular epidermis (IFE), completed with various appendages such as the hair follicle (HF), the sebaceous glands (SbG), and the apocrine sweat gland (SwG), all of these composing the epidermis (Figure 15.1). Basically, the HF produces hair, the SbG lubricates the skin with sebum, and the SwG transfers fluid to temperate the body surface.

The IFE, which forms the main skin barrier, is composed of four cellular layers named in descending order: the stratum corneum (StC), the granular layer (Gr), the spinous layer (Sp), and the basal/germinal layer (BL);

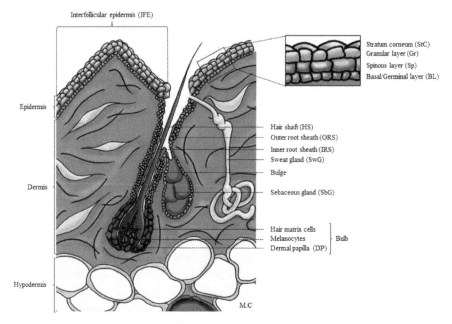

Figure 15.1   Structure of the epidermis and stem cell compartments. The epidermis is composed of the interfollicular epidermis (IFE), the hair follicles (HF), the sebaceous glands (SbG), and the sweat glands (SwG) (of note, an apocrine SwG is displayed here). The IFE is composed of four cellular layers: the stratum corneum (StC), the granular layer (Gr), the spinous layer (Sp), and the basal/germinal layer (BL); the latter is separating the epidermis from the dermis and is containing stem cells. The hair bulb consists of two structures called the dermal papilla (DP) and the hair matrix. Two other structures called the outer and the inner root sheaths (ORS and IRS) protect the hair shaft (HS). HF stem cells reside in the bulge (Copyright Mathieu Castela).

the latter is separating the epidermis from the dermis (Figure 15.1; Fuchs and Horsley, 2008). The dermis and the epidermis cooperate to enable the development of all epidermal appendages (Chuong et al., 1998). Mesoderm-derived progenitor cells contribute to the formation of the dermal fibroblasts, vasculature, and muscles attached to HF and fat cells (Blanpain and Fuchs, 2006). Melanocytes, the melanin-producing cells located in the BL of the epidermis are responsible for skin color and are derived from neural crest progenitor cells (Sommer, 2011; Thomas and Erickson, 2008). Neural crest-derived cells also contribute to the formation of sensory nerve endings of the skin (Dupin and Sommer, 2012).

In the adult, the epidermis is continuously replaced throughout life by a process called tissue homeostasis (Beck and Blanpain, 2012). Adult IFE, HF, SbG, and SwG possess distinct epidermal stem cells (SCs) capable of self-renewing for extended periods of time, differentiating into multiple lineages and maintaining tissue homeostasis (Beck and Blanpain, 2012). The coordinated action of these SCs together with other cell types, including dermal fibroblasts and immune cells, also enables efficient and harmonious repair after skin injury (Mascré et al., 2012; Dekoninck and Blanpain, 2019). Epidermal SC transplantation was proven to be efficient in human pathologies such as severe burns and genetic diseases (O'Connor et al., 1981; Mavilio et al., 2006), and the immediate benefits of the transplantation largely overcame the long-term risk of tumorigenesis due to mutations in SC regulation genes (Fuchs and Horsley, 2008; Nguyen et al., 2012; Blanpain, 2013). Recently, studies have pointed out the fact that the precise contribution of each SC niches to epidermal structure and signaling pathways regulating epidermal SC functions has not yet been totally elucidated. Indeed, various SCs or progenitors contribute to the maintenance of epidermal structures such as the isthmus (middle segment of the HF between the bulge and the SbG), the junctional zone (between the HF, SbG, and infundibulum), and the infundibulum (part of the HF that lies above the SbG and that is in continuity with the IFE) (Nijhof et al., 2006; Jensen et al., 2008; Raymond et al., 2010; Snippert et al., 2010; Beck and Blanpain, 2012). In the last decade, tremendous progresses have been made in understanding skin biology but outstanding issues need still to be addressed. Besides, one major point is to try to translate this knowledge into clinical applications (Fuchs, 2016).

15.2 The Interfollicular Epidermis

Established *in utero*, the IFE, is composed of multiple ordered layers of keratinocytes overlying a basement membrane. It is maintained throughout the entire postnatal life. It renews its surface by the continuous replacement of cells that are shed from the skin surface. It also undergoes renewal after wound injuries, i.e., the number of new cells generated in the BL must match the number lost every day.

15.2.1 Development of the IFE

During skin development (between E8.5 and E18.5 in mice), the single-layered surface ectoderm covering the mouse embryo initiates stratification and terminal differentiation to develop a functional multilayered epidermal barrier. Cells of the surface ectoderm, expressing keratins K8 and K18 (Moll et al., 1982), commit to an epidermal fate. This process depends on the expression of p63 since this epithelium remains single-layered in the absence of p63 expression (Koster and Roop, 2004). Then, keratinocytes in the newly established embryonic BL give rise to a second layer of cells called the periderm. This periderm initially protects against regions with a morphogenetically "active" basal epithelium (M'Boneko and Merker, 1988). Next, at around E9.5, embryonic basal keratinocytes express keratins K5/K14 and asymmetric cell division initiates the formation of an intermediate cell layer between the BL and the periderm from E11.5 to E14.5 (Byrne et al., 1994; Smart, 1970). Basal epidermal cells use their polarity to divide asymmetrically, generating a committed suprabasal cell and a proliferative basal cell (Lechler and Fuchs, 2005). Prior to E12.5, mitoses are predominantly planar, generating two symmetric basal cell daughters. After E12.5, asymmetric perpendicular divisions producing one basal cell and one suprabasal cell predominate until birth, with a peak when intermediate layer forms (Roshan and Jones, 2012; Smart, 1970; Lechler and Fuchs, 2005). Intermediate layer keratinocytes express K1, continue to divide during embryogenesis and mature to form the Sp when they exit the cell cycle (Smart, 1970). Sp cells expressing K1/K10 subsequently undergo further maturation into the Gr expressing filaggrin and loricrin, and cornified cells of the StC. At birth, the epidermis is a multilayered structure of dividing basal K5/K14 keratinocytes, overlaid by non-dividing suprabasal keratinocytes expressing K1 and K10 (Fuchs and Green, 1980). The final step in epidermal stratification involves the formation of the barrier acquisition with the formation of a 15-nm thick layer of proteins cross-linked by isopeptide and disulfide bonds, called the cornified cell envelope that serves as a scaffold for extruded lipid bilayers (Steven and Steinert, 1994).

15.2.2 IFE during Postnatal Growth

During postnatal growth, a concomitant expansion of the skin SCs and/or progenitors need to adjust the balance between renewal and differentiation to expand the numbers of epidermal cells. Recently, mainly using single-cell transcriptomics, it has been proposed that the mechanisms mediating skin expansion and controlling the imbalance of self-renewal over differentiation was mediated by a single population of equipotent developmental progenitors. They present a fixed fate imbalance of renewing divisions coupled with an ever-decreasing proliferation rate in order to ensure expansion of the basal progenitor cell pool while maintaining constant suprabasal thickness (Dekoninck et al., 2020).

15.2.3 IFE in Adulthood

During adult life, the classical paradigm suggested that IFE homeostasis was mediated by the presence of multiple clonal, small epidermal proliferation units (EPUs) distributed all along the IFE. Histological analysis of mouse IFE suprabasal layers suggested the presence of regular clusters of differentiating keratinocytes that formed a distinct spatial unit (Mackenzie, 1969). These EPUs were thought to contain a central long-lived, slow cycling, self-renewing basal keratinocyte SC, which divided asymmetrically to generate a short-lived population of around 10 surrounding transit amplifying (TA) cells. These TA cells then underwent fixed number of cell divisions (3–4 rounds), after which all of their progeny terminally differentiated to maintain the overlying differentiated cell layers. This hypothesis was supported by the clonal marking of IFE cells by retroviruses or mutagens, revealing long-lived columns of labeled IFE cells from the BL to the top of the cornified layer (Mackenzie, 1997; Ghazizadeh and Taichman, 2005; Ro and Rannala, 2004; 2005). However, numerous lineage tracing studies in transgenic mice do not support anymore the EPU model. In particular, quantitative analysis of lineage tracing data in IFE using a ubiquitous promoter reveals that the epidermis is maintained by a single population of cells cycling with stochastic fate and called committed progenitors (CPs). The outcome of each progenitor cell division is random, but the probabilities of generating either differentiated and/or progenitor daughter cells are equal, so that homeostasis is maintained (Clayton et al., 2007; Doupé and Jones, 2012). Nevertheless, it does not clearly explain the ability of the tissue to respond rapidly to wound healing and does not rule out the presence of a quiescent or slow-cycling SC population in the IFE (Jones and Simons, 2008). Recently, Mascré et al. have nicely demonstrated the existence of two distinct populations of epidermal progenitors that form a SC/CP cell hierarchy, differentially contributing to the homeostasis and repair of the tail epidermis. The authors made a quantitative analysis of lineage tracing using two different Cre recombinase-estrogen receptor transgenic mice marking the IFE progenitors included in the basal cell layer: aCre-ER under the control of the keratin 14 promoter (K14-Cre-ER/RosaYFP) enabled to study long-lived SCs, while a Cre-ER under the control of the involucrin promoter (Inv-Cre-ER/RosaYFP) enabled to study CPs. Then, to quantify the precise proliferation history *in vivo*, the proliferation dynamics of the IFE was measured using a pulse-chase strategy on the basis of a K5tTA/tet(O)-H2B-GFP transgenic mouse in which an epithelial K5 promoter drives repressible histone H2B-GFP expression. Upon feeding mice with doxycycline to shut off H2B-GFP expression, the protein is diluted two-fold following each division (Tumbar et al., 2004; Zhang et al., 2010). This technique allows the experiment to run for many more rounds of division than possible in a standard BrdU assay. In normal homeostatic conditions, SCs and CPs undergo asymmetric self-renewal with SCs dividing more slowly than CP cells. However, during wound healing, SC proliferation, migration, and differentiation could be uncoupled. SCs become active and proliferate, giving rise to new progenitors that expand and repair the wound, contributing to long-term repair, while CP cells conserve their homoeostatic mode of division, leading to their rapid depletion and making a minimal and transient contribution (Mascré et al., 2012; Aragona et al., 2017). These results have important implications for tissue regeneration, acute and chronic wound disorders. For now, these findings reconcile the contradictory theories of IFE maintenance, at least in the tail IFE, and are reminiscent of the situation encountered in others tissues. However, it has been pointed out that this study relied on clonal analysis of whole mounts of tail epidermis, but overlooked the fact that there are two programs of terminal differentiation (orthokeratotic and parakeratotic) within the tail IFE (Gomez et al., 2013).

Concerning wound healing, another process is of significance, namely the contribution of distant progenitors. Indeed, Asahara et al. were the first to show that after skin wounding, progenitors from the bone marrow were able to migrate to the dermis and participate in angiogenesis (Asahara et al., 1999). This interesting phenomenon has been nicely confirmed by several teams afterwards. Marrow endothelial progenitor cells (EPCs) are mobilized through the Stromal Derived Factor-1 (SDF1)/ C-C Chemokine Receptor 7 (CCR7) axis. This cell population persists for weeks after healing before disappearing from the skin. Mesenchymal stem cells (MSCs) are also mobilized from the marrow and differentiate into dermal fibroblasts through the Secondary Lymphoid-tissue Chemokine (SLC)/Chemokine C-C motif ligand (CCL21) pathway (Sasaki et al., 2008), contributing as well to the dermal granulation tissue. Tamai et al. were able to find other unexpected data. Indeed, when a skin wound was covered with an autologous graft, a peculiar population from the marrow, expressing the Platelet-Derived Growth Factor α (PDGFRα) was able to migrate and differentiate in the skin into keratinocytes (Tamai et al., 2010). In cell culture, using adequate protocols, another team was able to show that the bone marrow contains a population that grows into large cells, similar to embryonic stem (ES) cells. These cells express Stage Specific Embryonic Antigen SSEA3 protein and may differentiate into keratinocytes *in vivo* (Itoh et al., 2011). All these results show that distant progenitors may participate to skin healing and remodeling. During pregnancy, in women as well as in mice, fetal cells are transferred to the maternal circulation. These cells will persist for decades in maternal bone marrow and are well tolerated by the maternal immune system. In our lab, we were able to show that fetal progenitors are mobilized after skin wounding in gestating mice. These progenitors differentiate into endothelial cells forming full blood dermal vessels in the skin normal or delayed healing (Nassar et al., 2012). The fetal EPCs express different phenotypes as compared to the $CD11b^-$ Vascular Endothelial Growth Factor Receptor 2 $(VEGFR2)^+$ adult EPC phenotype. More recently, we found that C-C chemokine receptor 2 (Ccr2) was overexpressed on these cells upon maternal wounding in mice. Chemokine ligand 2 (Ccl2), its main ligand, was secreted by endothelial cells and macrophages. It enhanced the recruitment of the fetal progenitors to maternal wounds and Ccl2 administration improved delayed maternal wound healing in pregnant and postpartum mice, opening new strategies for tissue repair through natural stem cell therapy, a concept that could be further applied to other types of maternal diseases (Castela et al., 2017).

15.3 The Hair Follicle

The human HF is a unique appendage, composed of different concentric layers of cells, which results from epithelial–mesenchymal interactions initiated around the 3rd month of development in humans. It is a stocking-like structure, and the living part of the hair, called the hair bulb, is at the bottom part of this stocking. The HF possesses a dermal and an epithelial compartment to produce the hair. The dermal compartment comprises the connective tissue sheath and a large structure protruding in the bulb called the dermal papilla (DP), both of which are irrigated by micro-vessels. Around the DP is another bulb structure named the hair matrix, a collection of epithelial cells often interspersed with the melanocytic pigmentation unit responsible for hair color. The hair matrix epithelium is one of the fastest growing cell populations in the human body. Two other structures called the outer and the inner root sheaths (ORS and IRS, respectively) protect and mold the growing hair shaft (HS). The IRS is a rigid tube, follows the HS and acts as a channel that guides the HS to the skin surface. The ORS is contiguous with the epidermis and the

SbG (Figure 15.1). The HS is made up of keratins and is divided into three layers. The inner layer is called the medulla; the cortex makes up the majority of the HS, and the outer layer is called the cuticle. A muscle called erector pili muscle is attached to the HF. When this muscle contracts, it causes the hair to stand up. HF SCs reside in a niche that is located at the bottom of the non-cycling portion of the outer root sheath, called the bulge. These SCs give rise to all HF lineages and maintain homeostasis. They are responsible for hair production that is characterized by three phases including growth (anagen), cessation (catagen), and rest (telogen) phases (Bernard, 2006; Fuchs and Horsley, 2008).

15.3.1 HF Development

Many aspects of the HF morphogenesis are driven by epithelial–mesenchymal interactions and can be divided into three distinguishable stages: induction, organogenesis, and cyto differentiation. In mice, HF morphogenesis occurs in waves from E14.5 to E18.5 and begins with the downward invagination of the BL of the epithelium to form a microscopically recognizable precursor of the HF called the hair placode, then growing into the hair germ. As early as E12.5–E14.5, canonical Wnt signalization in the epithelial layer starts specification of this dermal placode. Soon after, a dermal condensate of mesenchymal cells (residing below) forms beneath this placode. Signaling between these two structures allows a proliferation of the overlying epithelium and downward extension of a new follicle into the dermis (also called "hair peg"). Then the epithelial cells envelope the dermal condensate that constitutes the mature DP. The DP will drive the HF formation as it extends into the dermis by inducing the highly proliferative surrounding epithelial matrix cells that will differentiate into the HS and its IRS. HF keratinocytes forming the IRS are the first epithelial cells in the follicle to terminally differentiate. In the center of this tube, terminally differentiated trichocytes will form the HS. The IRS itself is surrounded by a distinct outer layer of cylindric cells called the ORS. The ORS is contiguous with the epidermis and contains the bulge region (Schmidt-Ullrich and Paus,2005; Fuchs and Horsley, 2008; Driskell et al., 2011). IRS specification is regulated by Bone Morphogenetic Protein (BMP) signaling, Notch signaling, and transcription factors CDP (CCAAT-displacement protein) and GATA-3 (Blanpain and Fuchs, 2006). HS induction and specification is regulated by Wnt signaling and Lymphoid Enhancer-binding Factor-1 LEF1 (Ito et al, 2007). The SbG is also generated from the epithelial HF cells (Frances and Niemann, 2012). Melanocytes, originating from the neural crest, reside in the bulge and the matrix, and produce melanin pigments to color the HS (Nishimura, 2011). Of note, the HF compartmentalization and differentiation last up to 2.5 weeks after mouse birth (Ito et al., 2007).

15.3.2 HF Cycling

In mice, the hair cycle starts at around post-natal day 17 after the completion of hair morphogenesis. It begins with a catagen stage featured by a HF regression process. This first catagen lasts two to three days, and the lower two-thirds cycling portion of the HF are degraded by apoptosis. The DP remains intact and lifts upwards toward the permanent HF bulge containing epithelial and melanocyte SC (Paus and Foitzik, 2004; Nishimura, 2011). The molecular interaction between the HF SCs of the bulge and the DP is essential to form a new HF. Catagen phase is followed by the first phase of relative quiescence called the telogen phase lasting several days in the first telogen but usually more than 3 weeks in the second telogen. During this phase, HF consists of bulge and hair germ, a small epithelial structure generated at least in part by cells at the bulge base. The proximity between bulge and DP is maintained throughout this phase and the bulge anchors the old hair (called the club hair). Only when a critical concentration of hair growth–activating signals is reached, the next anagen phase is induced and a new hair is regrown. DP emits signals that induce SC activation and proliferation of hair germ cells that grow down together with the DP to generate a new complete follicle. SCs exit the bulge and proliferate downward, creating the long linear trail of cells of the ORS. Enveloping the DP at the HF base, matrix cells cycle rapidly but transiently before differentiating upward to specify a central hair shaft surrounded by its IRS, and companion layer that guide the shaft to its orifice at the skin surface (Ito et al., 2005; Zhang et al., 2009; Hsu et al., 2011; Sennett and Rendl, 2012). It has also been indicated that when quiescent bulge SCs become activated at the start of each new hair cycle, they exit the bulge niche and proliferate, moving downward to produce the matrix cells (Oshima et al., 2001). Bulge cells can be identified in the skin by their slow cycling property and specific surface expression of CD34 and α6-integrin (Cotsarelis et al., 1990; Beck and Blanpain, 2012). During each HF cycle, signals from the DP such as Transforming Growth factor beta 2 (TGFβ2), Fibroblast Growth Factors FGF7 or FGF10, and Wnt have been shown to promote anagen entry while others such as the Bone Morphogenetic Protein (BMP) 2 and BMP4 presumably produced by the subcutaneous fat layer or the DP have an inhibitory role on the cycle, maintaining the HF in a refractory phase (Mou et al., 2006; Plikus et al., 2009; Oshimori and Fuchs, 2012). The two initial hair cycles in mice are synchronized: the majority of HF are in a similar stage of the hair cycle at a given time. But as the mouse ages older, synchronicity is lost and hair domains are formed (Plikus et al., 2008). Indeed, hair cycling is reduced during aging but recently, it has been proposed using single-cell RNA sequencing that transcriptional changes in extracellular matrix genes and non-epithelial cell types, leading to structural perturbations in the SC niche, were a possible cause of hair loss. Although the number of HF SC declined, they were present, maintained their identity, and showed no overt signs of shifting to an epidermal fate. Aged skin was also defective at mobilizing SCs to regenerate HF after injury, highlighting the importance of the SC niche during homeostasis and tissue repair (Ge et al., 2020).

Another factor, IGF1 (Insulin-like Growth Factor 1), could play an important role in the regulation of HF cycling. IGF1 is expressed in the DP and dermal fibroblasts. Its receptor (IGF1R) is expressed in both the hair epidermis but also the DP making the IGF1/IGF1R pathway susceptible to play a crucial role in HF mesenchymal–epidermal interactions (Hodak et al., 1996; Rudman et al., 1997). Mice invalidated for IGF1R in K14 expressing cells were generated. A reduction in label-retaining SCs was reported as a consequence of altered IGF1R signaling (Stachelscheid et al., 2008). These data underline the importance ofKO mice for IGF1R in the K14 but also the K15 expressing cells. These results clearly indicate that IGF1R signaling is essential for SC maintenance as well as anagen-to-catagen transition in the hair cycleby promoting the expression of peculiar inhibitory signals (Roy et al., 2011; Castela et al., 2017).

15.4 Contribution of Bulge Stem Cells to the Epidermis

While the evidence that bulge SCs give rise to the regenerating follicle is unequivocal, it has also been postulated that these cells give rise to other epidermal lineages as well. Evidence that the bulge SCs, or their progeny, could be mobilized *in vivo* to regenerate an epidermis by the physical removal of the existing epidermis has accumulated20 years ago (Taylor et al., 2000; Tumbar et al., 2004). For example, bulge SCs were able to differentiate into all epidermal lineages upon transplantation into immunodeficient mice. *In vitro* holoclone descendants derived from a single bulge SC grafted in nude mice gave rise to epidermis, but also to multiple HF and SbG (Blanpain et al., 2004).

In 2004, Morris et al. designed a transgenic mouse with a K15 promoter to target mouse bulge cells with an inducible Cre recombinase construct or with the gene encoding enhanced green fluorescent protein (EGFP). They showed that bulge epithelial SCs gave rise to all the epithelial layers of the newly generated follicle during normal HF cycling, and that they reconstituted all epithelial cell types within the skin in a reconstitution assay (Morris et al., 2004). However, all of these observations and others have proven difficult to demonstrate that the SCs of the follicular bulge contribute to the epidermis in the absence of any traumatism. Soon after, the existence of a resident long-lived SC population in the epidermis has been demonstrated (Ghazizadeh and Taichman, 2001). Levy et al. took advantage of the properties of the Sonic hedgehog (Shh) gene to perform lineage analysis in undisturbed skin. At all stages of development analyzed, Shh (Sonic hedgehog) expression in the skin was restricted to the HF. A mouse expressing a GFPCre fusion under the control of Shh regulatory sequences was used to determine which parts of the follicle were derived from Shh-expressing cells. When crossed to the R26R reporter mice, cells that expressed GFPCre and Shh excised a transcriptional termination cassette and activated the expression of the beta-galactosidase gene, so that Shh-expressing cells and their descendants will express the beta-galactosidase. The authors showed that the SC resident in the follicular bulge regenerate the follicle but not the epidermis in the absence of trauma (Levy et al., 2005). In contrast, after skin traumatism such as wound healing, bulge-derived cells are able to transiently take part into interfollicular neo-epidermis repair. Most of this progeny is eliminated from the epidermis over several weeks, indicating that bulge SCs respond rapidly to epidermal wounding by generating short-lived "transient amplifying" cells responsible for acute wound repair (Ito et al., 2005; Nowak et al., 2008). To demonstrate this, Ito et al. have elegantly deleted bulge cells by targeting them with a suicide gene encoding herpes simplex virus thymidine kinase (HSV-TK) using a K15 promoter. Ablation leads to complete loss of all HF but survival of the epidermis in normal conditions. After a full-thickness 4-mm diameter wound or a 1-cm incisional wound on the backs of the mice, cells from the bulge are recruited into the epidermis and migrate in a centripetal manner toward the center of the wound, ultimately forming a marked radial pattern. Even a superficial wound using tape (removing the upper epidermal layers and leaving the basement membrane and dermis intact) stimulates bulge cell migration to the epidermis. The bulge-derived cells were generally located in the superficial epidermis after incisional wounding or tape-stripping, but localized to the basal layer after excisional wounding (Ito et al., 2005).

15.5 Contribution of Bulge Stem Cells to the Sebaceous Gland Lineage

Of the three lineages of the skin epithelium (the IFE, the HF, and the SbG), the SbG remains the least understood. The SbG are multi lobulated glands that secrete through holocrine differentiation of the sebocytes, an oily/waxy matter called sebum, to lubricate and waterproof the skin and hair of mammals. They are distributed throughout all skin sites except the palms and soles in humans. In hair-covered areas, they are connected to HF and bring the sebum to the skin surface along the HS. The structure consisting of hair, HF, arrector pili muscle, and SbG is known as a pilosebaceous unit.

Two models, not necessarily exclusive, have been proposed to explain how SbG cells might arise. Bulge SCs could serve as the residence of the multipotent progenitors which then migrate upward and differentiate to generate the SbG (Blanpain et al., 2004; Morris et al., 2004; Oshima et al., 2001; Taylor et al., 2000). Alternatively, a population of self-renewing progenitor cells resides within the SbG itself or close by the SbG, in order to maintain and generate the sebocytes (Ghazizadeh and Taichman, 2001; Beck and Blanpain, 2012).

The development of the pilosebaceous unit starts with the generation of the follicular compartment and remains permanently associated with the upper part of the HF. In mouse skin, the first sebocytes can be detected shortly after birth as compared to humans where SbG glands develop in the 13th–14th week of fetal life (Niemann and Horsley, 2012). During morphogenesis of the pilosebaceous unit in mouse epidermis, the spatial and temporal organization of distinct stem and progenitor compartments was analyzed with genetic lineage tracing experiments. SbG originates from bipotent embryonic progenitors common for both HF and SbG lineages that expressed Shh and Sox9 at one time of their specification (Levy et al., 2005; Nowak et al., 2008). Studies have also revealed a dynamic expression pattern for Sox9 and Lrig1 (Leucine-rich repeats and immunoglobulin-like domains protein 1). Sox9 and Lrig1 are initially co-expressed by epidermal progenitor cells and are restricted to different regions within the pilosebaceous unit when the specification of the sebocyte cell lineage takes place. The Sox9-expressing cells separate from the Lrig1$^+$ SC pool and are restricted to the bulge region of the HF. In contrast, cells remain at the upper part of the HF where mature sebocytes are about to emerge i.e., Lrig1$^+$ cells generate sebocytes (Nowak et al., 2008; Frances and Niemann, 2012; Niemann and Horsley, 2012).

In the adult HF, a unique population of transcriptional repressor Blimp1-expressing cells (B lymphocyte–induced maturation protein 1) that resides within the SbG was discovered. Its expression persists and it was suggested that the SbG could be maintained by unipotent progenitors during adult homeostasis. Using conditional gene targeting, loss of Blimp1 resulted in larger SbG, with enhanced pools of both slow-cycling progenitors and pro-liferative cells, accompanied by increased c-Myc expression (Horsley et al., 2006). However, this observation was later challenged by numerous reports demonstrating that Blimp1 was actually labelling differentiated cells in sev-eral skin compartments. Moreover, it has been suggested that cells located in the upper isthmus, the upper part of the HF between the SbG and the bulge, might contain a subset of SC that sustain SbG homeostasis. Cells reactive for the thymic epithelial progenitor cell marker MTS24 co-localized with expression of α6-integrin and K14, but not with the bulge-specific SC mark-ers CD34 or K15, thus identifying a new reservoir of HF keratinocytes with a proliferative capacity and gene expression profile suggestive of progenitor or SC (Nijhof et al., 2006). Likewise, a distinct population of murine HF keratinocytes residing in this upper isthmus was distinguished by low α6-integrin levels and negative expression for CD34. Purification of these cells gave rise to long-term, stable epidermal, follicular, and sebaceous lineages and could self-renew *in vivo* (Jensen et al., 2008). Finally, early central isthmus Lgr6$^+$ (Leucine-rich repeat–containing G-protein coupled receptor) cells were shown to established SbG and IFE, whereas contribution to hair lineages gradually decreased with age (Snippert et al, 2010). Of note, it was also demonstrated that another population of Lrig1$^+$ cells and enriched for Blimp1 transcript defined the HF junctional zone adjacent to the SbG and infundibulum. During homeostasis and under retinoic acid stimulation, they are bipotent, contributing to the SbG and IFE (Jensen et al., 2009).

In the adult, specific lineage tracing of bulge SCs during telogen phase, using K15CrePR, Lgr5CreER, and K19CreER demonstrated the ability of SCs characterized by these markers to contribute to the homeostasis of the hair follicle but not the SG or the IFE (Morris et al., 2004; Ito et al.,2005; Levy et al., 2005; Youssef et al., 2010). Thus, it was suggested that the SbG is maintained independently of bulge SCs in normal adult skin. However, in a most recent study, Petersson et al. used new bulge SC lineage tracing with a new K15CreERT2 transgenic mice crossed with the Rosa26–YFP or Rosa–LacZ reporter mice. In this study, the bulge was identified as a bipotent SC compartment that drives both cyclic regeneration of HF but also the continuous renewal of SbG. The aberrant signaling by transcription factors crucial for bulge SC activation and hair differentiation TCF/Lef1 resulted in this model in the development of ectopic SbG originating from bulge cells (Petersson et al., 2011; Beck and Blanpain, 2012). In conclusion, it seems that a subpopulation of bulge SCs may also contribute to the homeostasis of SbG even if the functional relationship between the diverse SC pools is very complex and not fully understood.

15.6 The Sweat Gland

Sweat Glands (SwG) are small tubular structures of the skin that produce sweat and are essential for thermoregulation. There are two main types of SwG: eccrine SwG, distributed almost all over the body, excrete directly onto the surface of the skin as a primary form of cooling; and apocrine SwG that are larger but distributed in specific body areas, and secrete sweat into the canal of the HF. In mice, they are restricted to foot pads.

Like mammary glands, SwG originate from epidermal progenitors. In mice, sweat buds emerge at E17.5, while mammary buds appear at E13.5. Both bud types are initiated by Ectodysplasin-A receptor–mediated signaling, and begin as morphologically similar, undifferentiated invaginations from epidermis (Cui and Schlessinger, 2006). The secretory coil of the SwG consists of an outer basal layer of $K5/K14^+$, smooth muscle actin$^+$ myoepithelial cells, and an inner suprabasal layer of $K8/K18/K19^+$ luminal cells (Langbein et al.,2005; Moll and Moll, 1992; Schön et al., 1999).

Lu et al. revealed that SwG glands develop from an epidermal bud of multipotent $K14^+$ progenitors that stratify to generate a transient but proliferative $K14l^{ow}/K18^+$ suprabasal ductal progenitors. Much of the morphogenetic potential to form SwG appears to come from these $K14^+$ basal and $K14^{low}/K18^+$ suprabasal ductal progenitors, which give rise to myoepithelial and luminal cells, respectively. In the adult, distinct types of unipotent progenitors within mature sweat ducts and glands emerge. They derive from a common multipotent $K14^+$ sweat bud progenitor. This progenitor then generates a nascent duct replete with distinct $K14^+$ basal and K18/K14-double positive suprabasal progenitors, both of which expand within their respective layers during sweat duct maturation. They also form the SwG. Moreover, SwG are specified by mesenchymal-derived BMP and FGF that signal to epithelial buds and suppress epithelial-derived SHH production. Finally, the authors also showed that adult sweat duct but not gland progenitors contribute to epidermal wound repair (Lu et al., 2012, 2016).

15.7 Concluding Remarks

- Multiple pools of SCs and progenitors reside in the skin.
- Adult IFE, HF, SbG, and SwG possess distinct epidermal SCs capable of self-renewing for extended periods of time, differentiating into multiple lineages and maintaining tissue homeostasis and repair.
- Proliferation kinetics, high-resolution lineage-tracing experiments on transgenic animal and transcriptomic analysis have provided useful quantitative insights on how these SCs work.
- The molecular mechanisms and signaling pathways controlling the stochastic fate choice of these cells still remain to be clarified.
- Skin is a privileged organ as its SCs can be efficiently "captured" for the development of engineered skin constructs or organoids as well as disease modeling, drug discovery, and regenerative medicine.

References

Aragona, M., Dekoninck, S., Rulands, S., Lenglez, S., Mascré, G., Simons, B. D., and Blanpain, C. (2017). Defining stem cell dynamics and migration during wound healing in mouse skin epidermis. Nat. Commun. 8, 14684.

Asahara, T., Masuda, H., Takahashi, T., Kalka, C., Pastore, C., Silver, M., Kearne, M., Magner, M., and Isner, J.M. (1999). Bone marrow origin of endothelial progenitor cells responsible for postnatal vasculogenesis in physiological and pathological neovascularization. Circ. Res. 85, 221–8.

Beck, B., and Blanpain, C. (2012). Mechanisms regulating epidermal stem cells. EMBO J. 31, 2067–75.

Bernard, B. A. (2006). The life of human hair follicle revealed. Med Sci (Paris) 22, 138–43.

Blanpain, C. (2013). Tracing the cellular origin of cancer. Nat. Cell Biol. 15, 126–34.

Blanpain, C., and Fuchs, E. (2006). Epidermal stem cells of the skin.Annu. Rev. Cell Dev. Biol. 22, 339–73.

Blanpain, C., Lowry, W. E., Geoghegan, A., Polak, L., and Fuchs, E. (2004). Self-renewal, multipotency, and the existence of two cell populations within an epithelial stem cell niche. Cell 118, 635–48.

Byrne, C., Tainsky, M., and Fuchs E. (1994). Programming gene expression in developing epidermis. Development 120, 2369–83.

Castela, M., Nassar, D., Sbeih, M., Jachiet, M., Wang, Z., and Aractingi S. (2017). Ccl2/Ccr2 signalling recruits a distinct fetal microchimeric population that rescues delayed maternal wound healing. Nat. Commun. 8, 15463.

Castela, M., Linay, F., Roy, E., Moguelet, P., Xu, J., Holzenberger, M., Khosrotehrani, K., and Aractingi S. (2017). Igf1r signalling acts on the anagen-to-catagen transition in the hair cycle. Exp. Dermatol. 9, 785-791.

Chuong, C. M., Jung, H. S., Noden, D., and Widelitz, R. B. (1998). Lineage and pluripotentiality of epithelial precursor cells in developing chicken skin. Biochem. Cell Biol. 76, 1069–77.

Clayton, E., Doupé, D. P., Klein, A. M., Winton, D. J., Simons, B. D., and Jones, P. H. (2007). A single type of progenitor cell maintains normal epidermis. Nature 446, 185–9.

Cotsarelis, G., Sun, T. T., and Lavker, R. M. (1990). Label-retaining cells reside in the bulge area of pilosebaceous unit: implications for follicular stem cells, hair cycle, and skin carcinogenesis. Cell 61, 1329–37.

Cui, C. Y., and Schlessinger, D. (2006). EDA signaling and skin appendage development. Cell Cycle 5, 2477–83.

Dekoninck, S., and Blanpain, C. (2019). Stem cell dynamics, migration and plasticity during wound healing. Nat. Cell Biol. 21, 18–24.

Dekoninck, S., Hannezo, E., Sifrim, A., Miroshnikova, Y.A., Aragona, M., Malfait, M., Gargouri, S., de Neunheuser, C., Dubois, C., Voet, T. et al. (2020). Defining the Design Principles of Skin Epidermis Postnatal Growth. Cell 181, 604–620.

Doupé, D. P., and Jones, P. H. (2012). Interfollicular epidermal homeostasis: dicing with differentiation. Exp. Dermatol. 21, 249–53.

Driskell, R. R., Clavel, C., Rendl, M., and Watt, F. M. (2011). Hair follicle dermal papilla cells at a glance. J. Cell Sci. 124, 1179–82.

Dupin, E., and Sommer, L. (2012). Neural crest progenitors and stem cells: from early development to adulthood. Dev. Biol. 366, 83–95.

Elias, P. M. (2007). The skin barrier as an innate immune element. Semin. Immunopathol. 29, 3–14.

Frances, D., and Niemann, C. (2012). Stem cell dynamics in sebaceous gland morphogenesis in mouse skin. Dev. Biol. 363, 138–46.

Fuchs, E., and Green, H. (1980). Changes in keratin gene expression during terminal differentiation of the keratinocyte. Cell 19, 1033–42.

Fuchs, E., and Horsley, V. (2008). More than one way to skin... Genes Dev. 22, 976–85.

Fuchs, E. (2016) Epithelial Skin Biology: Three Decades of Developmental Biology, a Hundred Questions Answered and a Thousand New Ones to Address. Curr. Top Dev. Biol. 116, 357–74.

Ghazizadeh, S., and Taichman, L. B. (2001). Multiple classes of stem cells in cutaneous epithelium: a lineage analysis of adult mouse skin. EMBO 20, 1215–22.

Ghazizadeh, S., and Taichman, L. B. (2005). Organization of stem cells and their progeny in human epidermis. J. Invest. Dermatol. 124, 367–72.

Gilbert, S. F. (2000). Paradigm shifts in neural induction. Rev. Hist. Sci. Paris 53, 555–79.

Gomez, C., Chua, W., Miremadi, A., Quist, S., Headon, D. J., and Watt, F. M. (2013). The Interfollicular Epidermis of Adult Mouse Tail Comprises Two Distinct Cell Lineages that Are Differentially Regulated by Wnt, Edaradd, and Lrig1. Stem Cell Reports 1, 19–27.

Haniffa, M. A., Wang, X. N., Holtick, U., Rae, M., Isaacs, J. D., Dickinson, A. M., Hilkens, C. M., and Collin, M. P. (2007). Adult human fibroblasts are potent immunoregulatory cells and functionally equivalent to mesenchymal stem cells. J. Immunol. 179, 1595–604.

Hodak, E., Gottlieb, A. B., Anzilotti, M., and Krueger, J. G. (1996). The insulin-like growth factor 1 receptor is expressed by epithelial cells with proliferative potential in human epidermis and skin appendages: correlation of increased expression with epidermal hyperplasia. J. Invest. Dermatol. 106, 564–70.

Horsley, V., O'Carroll, D., Tooze, R., Ohinata, Y., Saitou, M., Obukhanych, T., Nussenzweig, M., Tarakhovsky, A., and Fuchs, E. (2006). Blimp1 defines a progenitor population that governs cellular input to the sebaceous gland. Cell 126, 597–609.

Hsu, Y. C., Pasolli, H. A., and Fuchs, E. (2011). Dynamics between stem cells, niche, and progeny in the hair follicle. Cell 144, 92–105.

Ito M., Liu, Y., Yang, Z., Nguyen, J., Liang, F., Morris, R. J., and Cotsarelis, G. (2005). Stem cells in the hair follicle bulge contribute to wound repair but not to homeostasis of the epidermis. Nat Med. 11,1351–4.

Ito, M., Yang, Z., Andl, T., Cui, C., Kim, N., and Millar, S. E., and Cotsarelis, G. (2007). Wnt-dependent de novo hair follicle regeneration in adult mouse skin after wounding. Nature 447, 316–20.

Itoh, M., Kiuru, M., Cairo, M. S., and Christiano, A. M. (2011). Generation of keratinocytes from normal and recessive dystrophic epidermolysis bullosa-induced pluripotent stem cells. Proc. Natl. Acad. Sci. USA 108, 8797–802.

Jensen, K. B, Collins, C. A., Nascimento, E., Tan, D. W., Frye, M., Itami, S., and Watt F. M. (2009). Lrig1 expression defines a distinct multipotent stem cell population in mammalian epidermis. Cell Stem Cell 4, 427–39.

Jensen, U. B., Yan, X., Triel, C., Woo, S. H., Christensen, R., and Owens, D. M. (2008). A distinct population of clonogenic and multipotent murine follicular keratinocytes residing in the upper isthmus. J. Cell Sci. 121, 609–17.

Jones, P., and Simons, B. D. (2008). Epidermal homeostasis: do committed progenitors work while stem cells sleep? Nat. Rev. Mol. Cell Biol. 9, 82–8.

Koster, M. I. and Roop, D. R. (2004). Transgenic mouse models provide new insights into the role of p63 in epidermal development. Cell Cycle4, 411–3.

Langbein, L., Rogers, M. A., Praetzel, S., Cribier, B., Peltre, B., Gassler, N., and Schweizer, J. (2005). Characterization of a novel human type II epithelial keratin K1b, specifically expressed in eccrine sweat glands. J. Invest. Dermatol. 125, 428–44.

Lechler, T.and Fuchs, E. (2005). Asymmetric cell divisions promote stratification and differentiation of mammalian skin. Nature 437, 275–80.

Levy, V., Lindon, C., Harfe, B. D., and Morgan, B. A. (2005). Distinct stem cell populations regenerate the follicle and interfollicular epidermis. Dev. Cell 9, 855–61.

Lu, C. P., Polak, L., Rocha, A. S., Pasolli, H. A., Chen, S. C., Sharma, N., Blanpain, C., and Fuchs, E. (2012). Identification of stem cell populations in sweat glands and ducts reveals roles in homeostasis and wound repair. Cell 150, 136–50.

Lu, C.P., Polak, L., Keyes, B.E., and Fuchs, E. (2016). Spatiotemporal antagonism in mesenchymal-epithelial signaling in sweat versus hair fate decision. Science 6319, aah6102.

Mackenzie, I. C. (1997). Retroviral transduction of murine epidermal stem cells demonstrates clonal units of epidermal structure. J. Invest. Dermatol. 109, 377–83.

Mckenzie, J. C. (1969). Ordered structure of the stratum corneum of mammalian skin. Nature 222, 881–2.

Mascré, G., Dekoninck, S., Drogat, B., Youssef, K. K., Broheé, S., Sotiropoulou, P. A., Simons, B. D., and Blanpain, C. (2012). Distinct contribution of stem and progenitor cells to epidermal maintenance. Nature 489, 257–62.

Mavilio, F., Pellegrini, G., Ferrari, S., Di Nunzio, F., Di Iorio, E., Recchia, A., Maruggi, G., Ferrari, G., Provasi, E., and Bonini, C., et al. (2006). Correction of junctional epidermolysis bullosa by trans- plantation of genetically modified epidermal stem cells. Nat. Med. 12, 1397–402.

M'Boneko, V., and Merker, H. J. (1988). Development and morphology of the periderm of mouse embryos (days 9–12 of gestation). Acta Anat (Basel) 133, 325–36.

Moll, I., and Moll, R. (1992). Changes of expression of intermediate filament proteins during ontogenesis of eccrine sweat glands. J. Invest. Dermatol. 98, 777–85.

Moll, R., Franke, W. W., Schiller, D. L., Geiger, B and Krepler, R. (1982). The catalog of human cytokeratins: patterns of expression in normal epithelia, tumors and cultured cells. Cell 1, 11–24.

Morris, R. J., Liu, Y., Marles, L., Yang, Z., Trempus, C., Li, S., Lin, J. S., Sawicki, J. A., and Cotsarelis, G. (2004). Capturing and profiling adult hair follicle stem cells. Nat. Biotechnol. 22, 411–7.

Mou, C., Jackson, B., Schneider, P., Overbeek, P. A., and Headon, D. J. (2006). Generation of the primary hair follicle pattern. Proc. Natl. Acad. Sci. U S A 103, 9075–80.

Nassar, D., Droitcourt, C., Mathieu-d'Argent, E., Kim, M. J., Khosrotehrani, K., and Aractingi S. (2012). Fetal progenitor cells naturally transferred through pregnancy participate in inflammation and angiogenesis during wound healing. FASEB J. 26, 149–57.

Nguyen, L. V., Vanner, R., Dirks, P., and Eaves, C. J. (2012). Cancer stem cells: an evolving concept. Nat. Rev. Cancer 12, 133–43.

Niemann, C., and Horsley, V. (2012). Development and homeostasis of the sebaceous gland. Semin. Cell Dev. Biol. 23, 928–36.

Nijhof, J. G., Braun, K. M., Giangreco, A., van Pelt, C., Kawamoto, H., Boyd, R. L., Willemze, R., Mullenders, L. H., Watt, F. M., and de Gruijl, F. R., et al. (2006). The cell-surface marker MTS24 identifies a novel population of follicular keratinocytes with characteristics of progenitor cells. Development 133, 3027–37.

Nishimura, E. K. (2011). Melanocyte stem cells: a melanocyte reservoir in hair follicles for hair and skin pigmentation. Pigment Cell Melanoma Res. 24, 401–10.

Nowak, J. A., Polak, L., Pasolli, H. A., and Fuchs, E. (2008). Hair follicle stem cells are specified and function in early skin morphogenesis. Cell Stem Cell 3, 33–43.

O'Connor, N. E., Mulliken, J. B., Banks-Schlegel, S., Kehinde, O., and Green, H. (1981). Grafting of burns with cultured epithelium prepared from autologous epidermal cells. Lancet 1, 75–78.

Oshima, H., Rochat, A., Kedzia, C., Kobayashi, K., and Barrandon, Y. (2001). Morphogenesis and renewal of hair follicles from adult multipotent stem cells. Cell 104, 233–45.

Oshimori, N., and Fuchs, E. (2012). Paracrine TGF-β signaling counter-balances BMP-mediated repression in hair follicle stem cell activation. Cell Stem Cell 10, 63–75.

Paus, R., and Foitzik, K. (2004). In search of the "hair cycle clock": a guided tour. Differentiation 72, 489–511.

Petersson, M., Brylka, H., Kraus, A., John, S., Rappl, G., Schettina, P., and Niemann, C. (2011). TCF/Lef1 activity controls establishment of diverse stem and progenitor cell compartments in mouse epidermis. EMBO J. 30, 3004–18.

Plikus, M. V., Mayer, J. A., de la Cruz, D., Baker, R. E., Maini, P. K., Maxson, R., and Chuong, C. M. (2008). Cyclic dermal BMP signalling regulates stem cell activation during hair regeneration. Nature 451, 340–4.

Plikus, M. V., Widelitz, R. B., Maxson, R., and Chuong, C. M. (2009). Analyses of regenerative wave patterns in adult hair follicle populations reveal macro-environmental regulation of stem cell activity. Int. J. Dev. Biol. 53, 857–68.

Raymond, K., Richter, A., Kreft, M., Frijns, E., Janssen, H., Slijper, M., Praetzel-Wunder, S., Langbein, L., and Sonnenberg, A. (2010). Expression of the orphan protein Plet-1 during trichilemmal differentiation of anagen hair follicles. J. Invest. Dermatol. 130, 1500–13.

Ro, S., and Rannala, B. (2004). A stop-EGFP transgenic mouse to detect clonal cell lineages generated by mutation. EMBO Rep. 5, 914–20.

Ro, S., and Rannala, B. (2005). Evidence from the stop-EGFP mouse supports a niche-sharing model of epidermal proliferative units. Exp. Dermatol. 14, 838–43.

Roshan, A., and Jones, P. H. (2012). Act your age: tuning cell behavior to tissue requirements in interfollicular epidermis. Semin. Cell Dev. Biol. 23, 884–9.

Roy, E., Linay, F., Holzenberger, M., Oster, M., Aractingi, M., and Khos-rotehrani, K. (2011). Epidermal IGF-1 receptor is important in the maintenance of bulge stem cells, hair follicle cycle and response to skin wounds. 41st Annual Meeting of the European Society for Dermatological Research.

Rudman, S. M., Philpott, M. P., Thomas, G. A., and Kealey, T. (1997). The role of IGF-I in human skin and its appendages: morphogen as well as mitogen? J. Invest. Dermatol. 109, 770–7.

Sasaki, M., Abe, R., Fujita, Y., Ando, S., Inokuma, D., and Shimizu, H. (2008). Mesenchymal stem cells are recruited into wounded skin and contribute to wound repair by transdifferentiation into multiple skin cell type. J. Immunol. 180, 2581–7.

Schmidt-Ullrich, R., and Paus, R. (2005). Molecular principles of hair follicle induction and morphogenesis. Bioessays 27, 247–61.

Schön, M., Benwood, J., O'Connell-Willstaedt, T., and Rheinwald, J. G. (1999). Human sweat gland myoepithelial cells express a unique set of cytokeratins and reveal the potential for alternative epithelial and mesenchymal differentiation states in culture. J. Cell Sci. 112, 1925–36.

Sennett, R., and Rendl, M. (2012). Mesenchymal-epithelial interactions during hair follicle morphogenesis and cycling. Semin. Cell Dev. Biol. 23, 917–27.

Sipos, P. I., Rens, W., Schlecht, H., Fan, X., Wareing, M., Hayward, C., Hubel, C. A., Bourque, S., Baker, P. N., Davidge, S. T., et al. (2013). Uterine vasculature remodeling in human pregnancy involves functional macrochimerism by endothelial colony forming cells of fetal origin. Stem Cells 31, 1363–70.

Smart, I. H. (1970). Variation in the plane of cell cleavage during the process of stratification in the mouse epidermis. Br. J. Dermatol. 82, 276–82.

Snippert, H. J., Haegebarth, A., Kasper, M., Jaks, V., van Es, J. H., Barker, N., van de Wetering, M., van den Born, M., Begthel, H., and Vries, R. G., et al. (2010). Lgr6 marks stem cells in the hair follicle that generate all cell lineages of the skin. Science 327, 1385–9.

Sommer, L. (2011). Generation of melanocytes from neural crest cells. Pigment Cell Melanoma Res. 24, 411–21.

Stachelscheid, H., Ibrahim, H., Koch, L., Schmitz, A., Tscharntke, M., Wunderlich, F. T., Scott, J., Michels, C., Wickenhauser, C., Haase, I., et al. (2008). Epidermal insulin/IGF-1 signalling control interfollicular morphogenesis and proliferative potential through Rac activation. EMBO J. 27, 2091–101.

Steven, A. C., and Steinert, P. M. (1994). Protein composition of cornified cell envelopes of epidermal keratinocytes. J. Cell Sci. 107, 693–700.

Tamai, K., Yamazaki, T., Chino, T., Ishii, M., Otsuru, S., Kikuchi, Y., Iinuma, S., Saga, K., Nimura, K., Shimbo, T., et al. (2011). PDGFRalpha-positive cells in bone marrow are mobilized by high mobility group box 1 (HMGB1) to regenerate injured epithelia. Proc. Natl. Acad. Sci. USA 108, 6609–14.

Taylor, G., Lehrer, M. S., Jensen, P. J., Sun, T. T., and Lavker, R. M. (2000). Involvement of follicular stem cells in forming not only the follicle but also the epidermis. Cell 102, 451–61.

Thomas, A. J., and Erickson, C. A. (2008). The making of a melanocyte: the specification of melanoblasts from the neural crest. Pigment Cell Melanoma Res. 21, 598–610.

Tumbar, T., Guasch, G., Greco, V., Blanpain, C., Lowry, W. E., Rendl, M., and Fuchs, E. (2004). Defining the epithelial stem cell niche in skin. Science 303, 359–63.

Youssef, K. K., Van Keymeulen, A., Lapouge, G., Beck, B., Michaux, C., Achouri, Y., Sotiropoulou, P. A., and Blanpain, C. (2010). Identification of the cell lineage at the origin of basal cell carcinoma. Nat. Cell Biol. 12, 299–305.

Zhang, Y. V., Cheong, J., Ciapurin, N., McDermitt, D. J. and Tumbar, T. (2009). Distinct self-renewal and differentiation phases in the niche of infrequently dividing hair follicle stem cells. Cell Stem Cell 5, 267–78.

Zhang, Y. V., White, B. S., Shalloway, D. I., and Tumbar, T. (2010). Stem cell dynamics in mouse hair follicles: a story from cell division counting and single cell lineage tracing. Cell Cycle 9, 1504–10.

16

Mammary Stem Cells

Silvia Fre[1,*] and Ulysse Cherqui[2]

[1]Developmental Biology and Genetics Unit, CNRS UMR3215/
Inserm U934, PSL Research University, Institut Curie Centre de Recherche,
26 rue d'Ulm, 75248 Paris, Cedex 05, France
[2]Department of Molecular Cell Biology, Weizmann Institute of Science,
234 Herzl Street, 7610001 Rehovot, Israel
E-mail: silvia.fre@curie.fr
*Corresponding Author

16.1 Introduction

A fundamental question in developmental biology is how the distinct cell types of an organ are specified. Most epithelia renew through a process called tissue homeostasis, in which the number of cell divisions compensates for the number of cell losses. Tissue homeostasis, remodeling, and repair are ensured by stem cells (SCs) that by definition have the capacity to generate all cell types of the tissue in which they reside and are able to indefinitely self-renew. The building of an organism from a single cell to a multicellular, three-dimensional structure of characteristic shape and size is the result of coordinated gene action that directs the developmental fate of individual cells. The acquisition of different cell fates orchestrates an intricate interplay of cell proliferation, migration, growth, differentiation, and death, elaborating and bringing together cellular ensembles in a precise manner. During development, various cell types are generated and specified in a tightly regulated, temporal sequence to drive the formation of highly organised and complex tissues. Yet the precise mechanisms underlying stem cell self-renewal and differentiation, and how these relate to cellular organization during tissue morphogenesis, remain unclear. The growth of tumors involves the emancipation of cells from regular growth and differentiation constraints, resulting in the acquisition of a novel, malignant cell fate and consequently uncontrolled cell growth. Deregulation of homeostasis that keeps a balance between proliferation, apoptosis, and differentiation can lead to dysplastic pathologies. Conversely, the pathogenic state of a dysplastic condition may be ameliorated or cured by changing the dysplastic cell fate into a less pathogenic phenotype. Thus, the ability to change the fate of a given cell into a different one (via de-differentiation or lineage conversion) represents a fruitful therapeutic avenue.

16.2 The Mammary Epithelium and Its Stem Cells

The mammary gland represents an ideal system to study stem cell dynamics and lineage specification, as well as their contribution to tissue morphogenesis, maintenance, and remodeling. The mammary epithelium is comprised of a branched, bilayered ductal tree with an inner layer of polarised luminal cells (LCs), surrounding a central lumen, and an outer basal layer of cells (BCs) adjacent to the basement membrane. The luminal compartment is composed of two functionally distinct lineages defined by the expression of the steroid hormone receptors Estrogen alpha (ERα) and Progesterone (PR) (Visvader and Stingl, 2014): ERα-positive (hormone sensing) and ERα-negative (hormone responding) cells (Van Keymeulen et al., 2017; Wang et al., 2017), while the basal layer consists of elongated myoepithelial cells with contractile properties to facilitate milk flow through the ducts at lactation (Figure 16.1).

The early processes of mammary placode formation in embryogenesis and morphogenesis of the primordial mammary ductal tree before birth are still poorly defined. At puberty, hormones promote the elongation and extensive branching of the ductal tree that fills the mammary fat pad. This branched epithelium undergoes further proliferation and major tissue remodeling at pregnancy to generate secretory lobulo-alveolar structures that secrete milk. Alveoli will then be rapidly removed by programmed cell death when lactation ceases, returning the gland into a pre-pregnant state, a process repeated during recurrent reproductive cycles. This remarkable regeneration potential is attributed to the presence of mammary stem cells (MaSCs). As presumptive targets for transformation in breast cancer, the identity and cellular dynamics of adult MaSCs in the mammary gland have been the subject of intense investigation (Watson and Khaled, 2008, 2020).

The existence of a population of MaSCs in the adult mammary gland was initially shown by the ability of any fragment of the gland to reconstitute an entire gland upon transplantation. These experiments implicated the existence

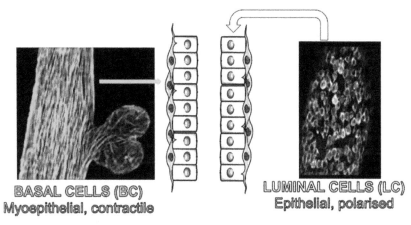

BASAL CELLS (BC)
Myoepithelial, contractile

LUMINAL CELLS (LC)
Epithelial, polarised

Figure 16.1 The mammary gland epithelium is composed of an outer layer of basal cells (in green) and an inner layer of luminal cells (in yellow). Images are whole mount staining of a mammary duct with the basal marker α-SMA (in green) or the luminal marker Keratin 8 (in yellow).

of multipotent stem cells able to give rise to all mammary lineages. In 1959, a seminal paper described the use of cleared mammary fat pads of host mice as a site for transplantation of normal or malignant mammary tissue (Deome et al., 1959). The transplantation technique relies on de-epithelialization or clearing of the inguinal fat pad, achieved via resection of the proximal portion of the fat pad in pre-pubertal female mice. Thereafter, transplantation is performed within the remaining fat pad, devoid of endogenous mammary epithelium. Remarkably, when fragments of the normal mammary epithelium of donor mice were transplanted into the cleared fat pad of recipient mice, they all presented the capacity of regenerating an entire and functional ductal epithelial tree.

Subsequent studies demonstrated that successful engraftment could be obtained with any portion of the mammary duct (Daniel et al., 1968; Hoshino, 1962; Smith and Medina, 1988), there by indicating that the repopulating MaSCs were widely distributed within the mammary tissue without a preferential niche. Later pivotal studies that explored the capacity of FACS-sorted mammary epithelial cells to reconstitute a whole functional gland when orthotopically transplanted in the cleared fat pad of host mice revealed that a small subset of basal cells is extremely efficient in mammary reconstitution experiments. Downstream analyses based on limiting dilutions eventually determined that a single cell could generate a complete mammary gland (Kordon and Smith, 1998; Shackleton et al., 2006; Stingl et al., 2006).

These studies revealed that the basal mammary compartment harbors multipotent MaSCs, initially believed to be responsible for homeostatic tissue maintenance throughout adult life. However, it has later been demonstrated that, notwithstanding the much higher transplantation efficiency of basal cells, specific subsets of luminal cells also can generate outgrowths in mammary reconstitution experiments (Chang et al., 2014; Rodilla et al., 2015; Van Keymeulen et al., 2011), revealing that this assay measures cell plasticity and not physiological *in vivo* stem cell potency. In other words, mammary transplantation can tell us what a specific cell can do, but not necessarily what it actually does in homeostatic conditions.

16.3 Lineage Tracing Analysis to Study Stem Cells In Vivo and In Situ

Lineage tracing is now considered the gold standard approach to study cellular hierarchies and cell fate *in vivo* (Kretzschmar and Watt, 2012; McKenna and Gagnon, 2019). This type of clonal analysis consists of using a genetic (hence heritable) label to target specific cells and follow their destiny and progeny *in vivo* and *in situ*, without removing the stem cells from their physiological niche (Kretzschmar and Watt, 2012). The most commonly used lineage tracing approach uses an inducible form of the Cre recombinase to trigger the permanent expression of a reporter gene in defined cells and at the desired time. Lineage tracing permits to identify stem cells and track their progeny, by genetically labelling cells in their physiological niche, so that their derived daughter cells retain marker expression. This strategy has been used to study cell fate and evaluate stem cell potency *in vivo* in most tissues. Such an approach has often been employed to define novel stem cell markers, such as Lgr5 in the small intestine (Barker et al., 2007), as well as to discover the existence of unipotent stem cells in the mammary gland, prostate, or lung (Hong et al., 2004; Lafkas et al., 2013; Lilja et al., 2018; Ousset et al., 2012; Rodilla et al., 2015; Tika et al., 2019; Van Keymeulen et al., 2011; Yang et al., 2018).

Initial genetic fate mapping studies using lineage-specific promoters, expressed in one of the two cellular compartments of the mammary epithelium, such as luminal (K8) or basal (K5, K14) cytokeratin promoters, indicated that in the adult homeostatic mouse mammary gland, BCs and LCs could only generate BCs and LCs, respectively, suggesting that tissue homeostasis is maintained by unipotent, lineage-restricted adult stem cells that give rise to only one type of progeny in the post-natal gland

(Prater et al., 2014; Van Keymeulen et al., 2011). However, another study evoked the existence of rare BCs that were found capable of producing both a basal and luminal progeny in adult mice, suggesting multipotency (Rios et al., 2014). Part of these discrepancies is explained by the use of different regions of the K5 promoter (BC marker), changing its cell specificity. These conflicting results have been eventually resolved through the use of clonal analysis at saturation, allowing the assessment of the fate of all cells of a given compartment (BCs with $K14^{rtTA}$-Cre^{TetO} and LCs with $K8^{rtTA}$-Cre^{TetO}), resulting in the definitive demonstration of the complete lack of multipotent stem cells in the postnatal mouse mammary gland (Wuidart et al., 2016). It should be noted that, as lineage tracing approaches are not applicable to humans, some differences in the cellular hierarchy might exist between the mouse mammary gland and the human breast.

Instead of using cytokeratin promoters, targeting in a rather generic way all cells in a given epithelial compartment, other groups have tested the potency of MaSCs by genetically marking specific cells with different promoters: Axin2-Cre,^{ERT2} marking Wnt/β-catenin–responsive cells throughout mammary gland development (van Amerongen et al., 2012); α-SMA with Acta2-Cre^{ERT2} mice (Prater et al., 2014) targeting postnatal myoepithelial cells, similarly to K5 or K14. Clonal analysis using the Dll1-Cre,^{ERT2} Lgr5-Cre^{ERT2} or Lgr6-Cre^{ERT2} lines could not reach a definitive consensus on the existence of unipotent or multipotent MaSCs, as these genes are predominantly expressed in BCs, but also in some LCs (Blaas et al., 2016; Chakrabarti et al., 2018; de Visser et al., 2012; Van Keymeulen et al., 2011). Rosa26-Cre^{ERT2} mice, using a ubiquitous promoter, have been used to achieve unbiased stochastic labeling of single proliferating cells (Davis et al., 2016; Scheele et al., 2017) (Figure 16.2).

Moreover, the promoters of different Notch receptors, as well as Sox9, Prominin-1, and ERα have been employed to gain insights into the cellular hierarchy within the luminal cell compartment. These studies highlighted the heterogeneity of LCs; for example, unlike *Notch3*, labeling both ERα-positive and ERα-negative LCs, the *Esr1* and *Prom1* genes mark exclusively ERα-positive LCs, whereas *Notch1* and *Sox9* target uniquely ERα-negative LCs in the postnatal gland (Lafkas et al., 2013; Rodilla et al., 2015; Van Keymeulen et al., 2017; Wang et al., 2017). Collectively, all these studies provided definitive evidence that in adult mice, BCs and LCs are entirely self-sustained by unipotent progenitors, and this holds true for ERα-positive and negative luminal subsets, representing two independent lineages (Figure 16.3).

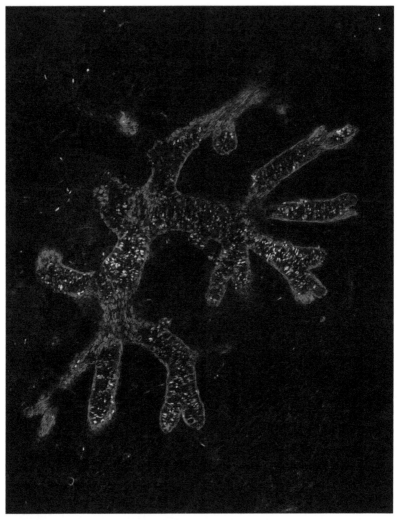

Figure 16.2   Multicolor lineage tracing of a lactating mammary gland.

Each of these cell populations maintains their respective lineage even after serial pregnancies, demonstrating long-term self-renewal capacity. To evaluate when the switch from multipotency to unipotency occurs during embryonic mammary development, our lab used a combination of *in vivo* multicolor clonal analysis with whole mount immuno-fluorescence and mathematical modeling at different embryonic and perinatal times

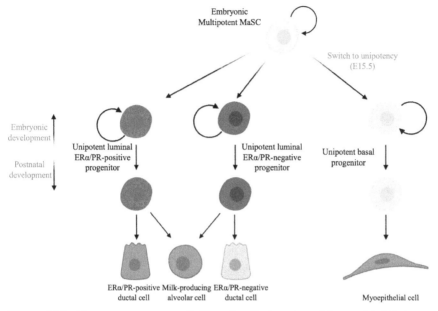

Figure 16.3 The mammary stem cell hierarchy. Embryonic multipotent MaSCs become lineage-restricted around embryonic day E15.5 and unipotent progenitors drive mammary gland development and homeostatic maintenance thereafter. Unipotent basal progenitors differentiate into basal cells, while unipotent luminal cells give rise to ERα-positive or ERα-negative luminal cells. Unipotent luminal progenitors differentiate into alveolar cells at pregnancy.

(Lilja et al., 2018). We found that multipotent MaSCs already coexist with unipotent progenitors by embryonic day E12.5, and that the switch from multipotency to unipotency seems to coincide with the initiation of branching morphogenesis at E15.5, suggesting a link between these two events (Lilja et al 2018). However, the mechanisms controlling the observed restriction in lineage potential during development are still largely unexplored.

16.4 Mammary Cell Plasticity

When adult unipotent stem cells or lineage-committed progenitors are dissociated from the tissue of origin and tested for their plasticity in transplantation assays, they display the capacity of producing different kinds of progeny, suggesting that the fate and multilineage differentiation potential of stem cells can change, depending on whether a stem cell resides within its niche and responds to normal homeostatic cues, or whether it is removed from its physiological niche and challenged to *de novo* tissue morphogenesis after transplantation. Indeed, it has long been observed that when SCs are taken out of context and transplanted, they exhibit multipotency and reveal extensive plasticity.

The remarkable capacity of lineage-restricted populations of committed progenitors to display such plasticity and to reacquire long-term self-renewing capacity and multipotency during regenerative conditions has now been shown for several stem cells. These experiments illustrate how the fate and differentiation potential of adult stem cells can change during tissue regeneration upon wounding. In some cases, unipotent progenitors acquire multipotency, whereas in others, committed cells revert back to a SC-like state to ensure tissue regeneration (see for example (Lehoczky et al., 2011; Park et al., 2012; van Es et al., 2012)). This extraordinary cellular plasticity may represent adaptive mechanisms for the self-preservation of epithelia following injuries and underlies the regenerative ability of epithelia in response to damage, which can also contribute to tumor development. Indeed, it is believed that adult MaSCs regain multilineage differentiation capacity during tumorigenesis, supporting the concept that cancer may arise from reactivation of embryonic developmental programs in postnatal tissues (Howard and Veltmaat, 2013; Spike et al., 2012; Zvelebil et al., 2013). Given their profound implications for regenerative medicine and cancer, the poorly understood molecular mechanisms underlying stem cell plasticity are the subject of intense investigation.

Notwithstanding their early lineage-restricted commitment, unipotent MaSCs, derived from multipotent embryonic mammary stem cells, retain a remarkably high degree of plasticity throughout adulthood, as demonstrated by their multilineage differentiation capacity when removed from their physiological niche. This has been shown under different stress conditions, such as transplantation, but also enzymatic dissociation or oncogene activation, when cells undergo lineage conversion, changing their physiological potency and commitment (Bouras et al., 2008; Centonze et al., 2020; Lilja et al., 2018; Rios et al., 2016; Van Keymeulen et al., 2011; Wuidart et al., 2018).

The mouse mammary gland represents thus a powerful paradigm to study stem cell plasticity and lineage specification during development, as it is composed of an epithelium maintained exclusively by unipotent lineage-restricted progenitors in homeostatic conditions (Davis et al., 2016; Lilja et al., 2018; Scheele et al., 2017; Seldin et al., 2017; Van Keymeulen et al., 2011; Wuidart et al., 2016; Wuidart et al., 2018), which derive from multipotent MaSCs and exhibit remarkable remodeling potential at puberty and pregnancy. This appears to be a common mechanism since several studies have recently shown that different glandular epithelia (mammary gland, prostate, sweat glands, and salivary glands) initially develop from multipotent SCs, which are progressively replaced by unipotent progenitors during development (Centonze et al., 2020; Choi et al., 2012; Lu et al., 2012; Rodilla et al., 2015; Van Keymeulen et al., 2011).

16.5 Molecular Signals Governing MaSCs and Gland Morphogenesis

16.5.1 The Notch Pathway

Aberrant Notch activation in the mammary gland has been shown to impair proper mammary gland development and to induce mammary tumors. Notch has first been identified as an oncogene in Mouse Mammary Tumor Virus (MMTV)-infected mice (Gallahan, 1987). Since this discovery, activated Notch signaling and upregulation of tumor-promoting Notch target genes have been observed in human breast cancer. In addition, high expression of Notch ligands and receptors has been shown to correlate with aggressive stage and treatment resistance in this malignancy (Reedijk et al., 2005). Several reports correlate human breast cancer with aberrant Notch activation and, importantly, Notch activity has been associated with breast tumors with poor clinical outcome (Reedijk, 2012). Perturbation of stem cell maintenance, cell fate specification, differentiation, proliferation, motility, and survival has been associated to aberrant Notch activity in breast cancer progression (Reedijk, 2012). In recent years, there has been an increasing interest in targeting Notch signaling for the treatment of breast cancer and, as a consequence, some Notch pathway inhibitors are now entering the initial stages of clinical trials (Brennan and Clarke, 2013). Notwithstanding the clear involvement of Notch signaling in mammary tumorigenesis, the specific expression of Notch pathway components in this tissue and the molecular mechanisms underlying its role are still unclear. Initial studies based on RNA expression indicated that the Notch ligand Delta-like1 is expressed in mammary basal cells, whereas the receptors are found in the luminal compartment (Bouras et al., 2008). The Notch pathway plays a major role in stem cell maintenance and lineage commitment in various tissues and it has been shown to promote a luminal fate in the mammary gland. Bouras and colleagues used gain and loss of function Notch mutant mice to show that Notch activation can shift the BC/LC ratio toward luminal differentiation and promote LCs proliferation. Conversely, when the Notch effector RBPJ

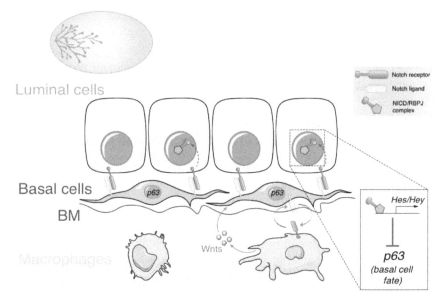

Figure 16.4   Notch signaling promotes a luminal fate. The mammary epithelium comprises an inner layer of Notch-expressing luminal cells and an outer layer of Delta-expressing basal cells adjacent to the basement membrane (BM). Notch activation in LCs leads to the transcriptional repression of the basal fate determinant p63 (modified from Lloyd-Lewis et al., 2019).

was knocked-down, a significant increase in the number of basal cells expressing p63 was observed and it correlated with reduced luminal cells, indicating that Notch activation primes mammary cells toward a luminal fate (Figure 16.4). More recently, our lab analyzed the expression of different Notch receptors by lineage tracing using NotchCreERT2 mouse lines and found that the Notch1, Notch2, and Notch3 paralogues are expressed in distinct clonogenic luminal progenitor cells, capable of self-renewal, but with different proliferative capacity and *in vivo* behavior (Lafkas et al., 2013; Rodilla et al., 2015). Indeed, these luminal cells present different molecular features, morphology, localization, and behavior, unraveling the complexity and diversity of the luminal mammary epithelium. Specifically, while Notch3-expressing luminal cells represent a rare and highly clonogenic subset of quiescent luminal progenitors (Lafkas et al., 2013), Notch1-marked cells are actively cycling and are a self-sustained lineage of unipotent progenitors, with self-renewal capacity and strong hormonal dependency (Rodilla et al., 2015). Intriguingly, Notch1-expressing luminal progenitors invariably lack expression of the steroid hormone receptors Estrogen (ER) and

Progesterone (PR), while Notch3-marked cells can be both ERα-positive and ERα-negative. Sale and colleagues have also shown that the Notch2 receptor marks alveolar progenitors (Sale et al., 2013). Taken together, these studies exemplify how the expression of different Notch receptors represents an ideal tool to dissect the hierarchical relationships of distinct luminal cells.

Mechanistically, we have recently shown that Notch signaling is directly implicated in regulating cell fate in the mammary gland, by promoting the commitment of embryonic MaSCs to a luminal fate, thus defining the identity of mammary progenitors. We found that constitutive Notch1 activation imposes a premature differentiation to multipotent embryonic MaSCs and, remarkably, ectopic Notch1 expression in BCs can also lineage convert committed unipotent adult cells (Lilja et al., 2018), implying that similar mechanisms control embryonic cell fate determination and adult cell plasticity.

The Notch pathway therefore plays a key role in MaSCs, where it is essential for lineage commitment toward the luminal fate (Figure 16.4). However, the additional molecular and cellular changes required for cell "reprogramming" are still undefined.

In fact, the restriction of multipotent stem cells to unipotent progenitors has not been explained with respect to the transcriptional regulators that control these events at the single cell level. Moreover, how "reprogramming" is orchestrated dynamically, both spatially and temporally, remains unknown. A lack of robust markers, combined with high cellular heterogeneity, has substantially hampered the analysis into the mechanisms driving the hierarchical organization of mammary epithelial cell lineages during development. In addition, *in situ* visualization of early, oncogene-induced changes in cellular behavior in living animals are significantly impeded by the rare nature of these events and the transient emergence of newly transformed clones prior to tumor establishment. Moreover, as previous studies have been limited to the static analysis of fixed tissues, the dynamic processes of MaSCs' plasticity and cell fate remain very poorly understood.

16.5.2 The Wnt Pathway

Wnt/β-catenin signaling is crucial for embryonic mammary placode development and for postnatal ductal and lobulo-alveolar growth, promoting the self-renewal of MaSCs and maintenance of the basal lineage (Badders et al., 2009). It has been shown that Wnt induction pattern correlates with mammary line and placode anatomical sites (DasGupta and Fuchs, 1999). In addition,

LEF1, a transcription factor inducing Wnt target gene expression, is key for mammary bud formation. Indeed, at E13.5, *Lef1-/-* embryos display a reduced number of buds (van Genderen et al., 1994). The importance of the Wnt pathway was further confirmed with *K14-Dkk1* mice, where overexpression of the Wnt inhibitor Dickkopf1 (Dkk1) precluded mammary bud formation (Andl et al., 2002). Ectopic expression of Wnt1 was also shown to accelerate mammary development and branching during embryogenesis (Ahn et al., 2013). Thus, multiple lines of evidence indicate that Wnt signals are necessary for early development of the mammary gland. In addition, Wnt signaling has been linked to the ability of MaSCs to self-renew (Zeng and Nusse, 2010) and promotes mammary branching (Imbert et al., 2001).

16.5.3 The FGF pathway

FGF signaling also provides the embryonic mammary gland with essential inputs. By mediating reciprocal signaling between the mammary epithelium and its surrounding stroma, an essential component of mammary gland morphogenesis, the FGF pathway controls early placode formation. Indeed,knock-out mice for the FGF receptor Fgfr2b or its ligand Fgf10 fail to develop mammary placodes, with the exception of the fourth inguinal bud, which develops normally until E18.5 but does not properly branch afterwards (Mailleux et al., 2002; Veltmaat et al., 2006). Early in development, the source of Fgf10 is the ventral dermomyotome of the somites, suggesting that a morphogen gradient dictates mammary placode development. Mosaic inactivation of Fgfr2 showed that Fgfr2 null mammary cells were out competed during ductal proliferation (Lu et al., 2008). A reduction in mammary outgrowths and frequency of Terminal End Buds (TEBs) during puberty was also observed following the deletion of both Fgfr1 and Fgfr2, due to a decline in basal progenitors (Pond et al., 2013). This study presented evidence demonstrating that the FGF pathway is essential for proliferation of basal unipotent progenitors. More recently, luminal cells were shown to respond to Fgf2 and Fgf9 by increasing mammary branching through paracrine signaling (Sumbal and Koledova, 2019). Thus, FGF signaling is necessary for embryonic mammary bud formation and plays an important role in branching morphogenesis at puberty by stimulating cell proliferation in TEBs.

16.5.4 The TGF β Pathway

TGFβ signaling plays essential roles in the mammary gland across all developmental and remodeling stages, as it has been associated with the ER and PR pathways during puberty, as well as cell death during involution (Moses and Barcellos-Hoff, 2011). During puberty, it acts as a negative regulator of mammary morphogenesis, as shown by mammary growth arrest upon over expression of TGFβ (Daniel et al., 1996). Molecular mechanisms restricting the epithelial growth out side the borders of the fat pad, as well as those limiting fat pad invasion, are crucial since epithelium-free space is indispensable for subsequent pregnancy-induced alveologenesis and further fat pad occupation. In this process, transforming growth factor beta 1 (TGFβ1) has been identified as a major negative regulator of both ductal elongation and side branching, by inhibiting epithelial proliferation (Ingman and Robertson, 2008).

TGFβ signaling can be inhibited by estrogen and progesterone. Upon decreased levels of hormonal signaling, TGFβ can regulate gland development by inducing apoptosis, extracellular matrix remodeling, and inhibiting proliferation(Moses and Barcellos-Hoff, 2011). Indeed, TGFβ signaling increases during involution, when its activity is required to mediate apoptosis. The role of TGFβ in preventing proliferation has been linked to growth inhibition in luminal cells. TGFβ production by epithelial cells was proposed to be hormone-dependent. Indeed, at puberty and pregnancy, the expression of TGFβ1 in ERα-positive LCs can prevent these cells from responding to hormonal signaling and stimulate proliferation of ERα-negative LCs (Ewan et al., 2002). TGFβ expression was also shown to inhibit branching morphogenesis during mammary development by altering the local microenvironment(Nelson et al., 2006). The TGFβ pathway plays thus an important role during development to regulate the proliferation of specific epithelial cell populations.

16.5.5 The Hedgehog Pathway

The Hh pathway functions through the binding of one of three secreted ligands, Sonic hedgehog (*Shh*), Indian hedgehog (*Ihh*), and Desert hedgehog (*Dhh*), to the Patch 1 or 2 (*Ptch1* or *Ptch2*) receptors. Although key effectors of the Hh pathway, such as *Shh, Ihh, the Ptch1* receptor or the *Gli2* transcription factor show expression in the embryonic gland, their loss of function does not disrupt mammary bud development (Hatsell and Cowin, 2006; Lewis et al., 1999; Michno et al., 2003). Hatsell and Cowin (2006) followed Hh activity during mammary gland morphogenesis using Gli1-lacZ reporter mice. They found that Gli1-lacZ expression was absent in embryonic mammary buds. Additionally, *Gli1* or *Gli2* null embryos presented normal mammogenesis. In contrast, loss of *Gli3* induces aberrant placode development, lack of mammary bud pairs and disrupted Wnt activity. In particular, formation of placode pairs 3 and 5 requires the repressor function of the Hedgehog signaling pathway regulator Gli3 (Chandramouli et al., 2013). Interestingly, it has been suggested that repression of Hh target genes is a hallmark of embryonic buds, as compared with other skin appendages, such as hair follicles, which display robust Hh signaling activity. Thus, repression of Hedgehog signaling activity is important for mammary embryonic development (Hatsell and Cowin, 2006).

16.6 Concluding Remarks

The mammary epithelium is an ideal paradigm to study stem cell potential and lineage specification during development, as it is maintained exclusively by unipotent lineage-restricted progenitors in homeostatic conditions, which derive from multipotent embryonic MaSCs, capable of generating all mammary lineages.

The last 10 years have seen extraordinary progress in understanding mammary gland development. Much of this is a consequence of technological advances, such as high-throughput and single cell sequencing and sophisticated microscopy. It should also be noted that much of the work in the past has used approaches, such as FACS and fat pad transplantations, which resulted in conclusions about MaSCs markers and characteristics that do not necessarily reflect their capacity *in situ* and *in vivo*.

Importantly, despite extensive cell heterogeneity in breast cancer, tumor cells retain many features of the normal mammary tissue. MaSCs are responsible for tissue morphogenesis and remodeling during reproductive life, but they have also been shown to be at the origin of breast cancer. Therefore, a comprehensive analysis of the identity, potency, and plasticity of normal MaSCs is essential for discovering the signals underlying cellular transformation in breast cancer.

While far from being exhaustive, this chapter summarizes the latest results on mammary stem cells, providing:

- a brief description of the mammary epithelium and its cell types;
- a comparison of the methods used to study mammary stem cells, in particular transplantation assays and lineage tracing;
- an explanation of stem cell plasticity and its implications in regenerative medicine and cancer using the example of the highly plastic MaSCs; and

- an overview of the major molecular pathways involved in mammary gland development and stem cell specification, with a special emphasis on Notch signaling.

A generality that can be drawn regarding the specific action of Notch signals in development is that modulation of the Notch pathway in cells that are not terminally differentiated provides them with the plasticity to change developmental fate. Based on this rationale, we can imagine that modulation of Notch signals in tumor cells may force cell fate changes that could have a significant therapeutic value.

References

Ahn, Y., Sims, C., Logue, J.M., Weatherbee, S.D., and Krumlauf, R. (2013). Lrp4 and Wise interplay controls the formation and patterning of mammary and other skin appendage placodes by modulating Wnt signaling. Development *140*, 58 3–593.

Andl, T., Reddy, S.T., Gaddapara, T., and Millar, S.E. (2002). WNT signals are required for the initiation of hair follicle development. Dev Cell *2*, 643–653.

Badders, N.M., Goel, S., Clark, R.J., Klos, K.S., Kim, S., Bafico, A., Lindvall, C., Williams, B.O., and Alexander, C.M. (2009). The Wnt receptor, Lrp5, is expressed by mouse mammary stem cells and is required to maintain the basal lineage. PLoS One *4*, e6594.

Barker, N., van Es, J.H., Kuipers, J., Kujala, P., van den Born, M., Cozijnsen, M., Haegebarth, A., Korving, J., Begthel, H., Peters, P.J., et al. (2007). Identification of stem cells in small intestine and colon by marker gene Lgr5. Nature *449*, 1003–1007.

Blaas, L., Pucci, F., Messal, H.A., Andersson, A.B., Josue Ruiz, E., Gerling, M., Douagi, I., Spencer-Dene, B., Musch, A., Mitter, R., et al. (2016). Lgr6 labels a rare population of mammary gland progenitor cells that are able to originate luminal mammary tumours. Nat Cell Biol *18*, 1346–1356.

Bouras, T., Pal, B., Vaillant, F., Harburg, G., Asselin-Labat, M.L., Oakes, S.R., Lindeman, G.J., and Visvader, J.E. (2008). Notch signaling regulates mammary stem cell function and luminal cell-fate commitment. Cell Stem Cell *3*, 429–441.

Brennan, K., and Clarke, R.B. (2013). Combining Notch inhibition with current therapies for breast cancer treatment. Therapeutic advances in medical oncology *5*, 17–24.

Centonze, A., Lin, S., Tika, E., Sifrim, A., Fioramonti, M., Malfait, M., Song, Y., Wuidart, A., Van Herck, J., Dannau, A., et al. (2020). Heterotypic cell-cell communication regulates glandular stem cell multipotency. Nature *584*, 608–613.

Chakrabarti, R., Celia-Terrassa, T., Kumar, S., Hang, X., Wei, Y., Choudhury, A., Hwang, J., Peng, J., Nixon, B., Grady, J.J., et al. (2018). Notch ligand Dll1 mediates cross-talk between mammary stem cells and the macrophageal niche. Science *360*.

Chandramouli, A., Hatsell, S.J., Pinderhughes, A., Koetz, L., and Cowin, P. (2013). Gli activity is critical at multiple stages of embryonic mammary and nipple development. PLoS One *8*, e79845.

Chang, T.H., Kunasegaran, K., Tarulli, G.A., De Silva, D., Voorhoeve, P.M., and Pietersen, A.M. (2014). New insights into lineage restriction of mammary gland epithelium using parity-identified mammary epithelial cells. Breast Cancer Res *16*, R1.

Choi, N., Zhang, B., Zhang, L., Ittmann, M., and Xin, L. (2012). Adult murine prostate basal and luminal cells are self-sustained lineages that can both serve as targets for prostate cancer initiation. Cancer Cell *21*, 253–265.

Daniel, C.W., De Ome, K.B., Young, J.T., Blair, P.B., and Faulkin, L.J., Jr. (1968). The in vivo life span of normal and preneoplastic mouse mammary glands: a serial transplantation study. Proc Natl Acad Sci U S A *61*, 53–60.

Daniel, C.W., Robinson, S., and Silberstein, G.B. (1996). The role of TGF-beta in patterning and growth of the mammary ductal tree. J Mammary Gland Biol Neoplasia *1*, 331–341.

DasGupta, R., and Fuchs, E. (1999). Multiple roles for activated LEF/TCF transcription complexes during hair follicle development and differentiation. Development *126*, 4557–4568.

Davis, F.M., Lloyd-Lewis, B., Harris, O.B., Kozar, S., Winton, D.J., Muresan, L., and Watson, C.J. (2016). Single-cell lineage tracing in the mammary gland reveals stochastic clonal dispersion of stem/progenitor cell progeny. Nat Commun *7*, 13053.

de Visser, K.E., Ciampricotti, M., Michalak, E.M., Tan, D.W., Speksnijder, E.N., Hau, C.S., Clevers, H., Barker, N., and Jonkers, J. (2012). Developmental stage-specific contribution of LGR5(+) cells to basal and luminal epithelial lineages in the postnatal mammary gland. J Pathol *228*, 300–309.

Deome, K.B., Faulkin, L.J., Jr., Bern, H.A., and Blair, P.B. (1959). Development of mammary tumors from hyperplastic alveolar nodules transplanted into gland-free mammary fat pads of female C3H mice. Cancer research *19*, 515–520.

Ewan, K.B., Shyamala, G., Ravani, S.A., Tang, Y., Akhurst, R., Wakefield, L., and Barcellos-Hoff, M.H. (2002). Latent transforming growth factor-beta activation in mammary gland: regulation by ovarian hormones affects ductal and alveolar proliferation. Am J Pathol *160*, 2081–2093.

Gallahan, D., C. Kozak, and R. Callahan (1987). A new common integration region (int-3) for mouse mammary tumor virus on mouse chromosome 17. J Virol *61*, 218–220.

Hatsell, S.J., and Cowin, P. (2006). Gli3-mediated repression of Hedgehog targets is required for normal mammary development. Development *133*, 3661–3670.

Hong, K.U., Reynolds, S.D., Watkins, S., Fuchs, E., and Stripp, B.R. (2004). In vivo differentiation potential of tracheal basal cells: evidence for multipotent and unipotent subpopulations. Am J Physiol Lung Cell Mol Physiol *286*, L643–649.

Hoshino, K. (1962). Morphogenesis and growth potentiality of mammary glands in mice. I. Transplantability and growth potentiality of mammary tissue of virgin mice. J Natl Cancer Inst *29*, 835–851.

Howard, B.A., and Veltmaat, J.M. (2013). Embryonic mammary gland development; a domain of fundamental research with high relevance for breast cancer research. Preface. J Mammary Gland Biol Neoplasia *18*, 89–91.

Imbert, A., Eelkema, R., Jordan, S., Feiner, H., and Cowin, P. (2001). Delta N89 beta-catenin induces precocious development, differentiation, and neoplasia in mammary gland. J Cell Biol *153*, 555–568.

Ingman, W.V., and Robertson, S.A. (2008). Mammary gland development in transforming growth factor beta1 null mutant mice: systemic and epithelial effects. Biol Reprod *79*, 711–717.

Kordon, E.C., and Smith, G.H. (1998). An entire functional mammary gland may comprise the progeny from a single cell. Development *125*, 1921–1930.

Kretzschmar, K., and Watt, F.M. (2012). Lineage tracing. Cell *148*, 33–45.

Lafkas, D., Rodilla, V., Huyghe, M., Mourao, L., Kiaris, H., and Fre, S. (2013). Notch3 marks clonogenic mammary luminal progenitor cells in vivo. The Journal of cell biology *203*, 47–56.

Lehoczky, J.A., Robert, B., and Tabin, C.J. (2011). Mouse digit tip regeneration is mediated by fate-restricted progenitor cells. Proc Natl Acad Sci U S A *108*, 20609–20614.

Lewis, M.T., Ross, S., Strickland, P.A., Sugnet, C.W., Jimenez, E., Scott, M.P., and Daniel, C.W. (1999). Defects in mouse mammary gland development caused by conditional haploinsufficiency of Patched-1. Development *126*, 5181–5193.

Lilja, A.M., Rodilla, V., Huyghe, M., Hannezo, E., Landragin, C., Renaud, O., Leroy, O., Rulands, S., Simons, B.D., and Fre, S. (2018). Clonal analysis of Notch1-expressing cells reveals the existence of unipotent stem cells that retain long-term plasticity in the embryonic mammary gland. Nat Cell Biol *20*, 677–687.

Lu, C.P., Polak, L., Rocha, A.S., Pasolli, H.A., Chen, S.C., Sharma, N., Blanpain, C., and Fuchs, E. (2012). Identification of stem cell populations in sweat glands and ducts reveals roles in homeostasis and wound repair. Cell *150*, 136–150.

Lu, P., Ewald, A.J., Martin, G.R., and Werb, Z. (2008). Genetic mosaic analysis reveals FGF receptor 2 function in terminal end buds during mammary gland branching morphogenesis. Dev Biol *321*, 77–87.

Mailleux, A.A., Spencer-Dene, B., Dillon, C., Ndiaye, D., Savona-Baron, C., Itoh, N., Kato, S., Dickson, C., Thiery, J.P., and Bellusci, S. (2002). Role of FGF10/FGFR2b signaling during mammary gland development in the mouse embryo. Development *129*, 53–60.

McKenna, A., and Gagnon, J.A. (2019). Recording development with single cell dynamic lineage tracing. Development *146*.

Michno, K., Boras-Granic, K., Mill, P., Hui, C.C., and Hamel, P.A. (2003). Shh expression is required for embryonic hair follicle but not mammary gland development. Dev Biol *264*, 153–165.

Moses, H., and Barcellos-Hoff, M.H. (2011). TGF-beta biology in mammary development and breast cancer. Cold Spring Harb Perspect Biol *3*, a003277.

Nelson, C.M., Vanduijn, M.M., Inman, J.L., Fletcher, D.A., and Bissell, M.J. (2006). Tissue geometry determines sites of mammary branching morphogenesis in organotypic cultures. Science *314*, 298–300.

Ousset, M., Van Keymeulen, A., Bouvencourt, G., Sharma, N., Achouri, Y., Simons, B.D., and Blanpain, C. (2012). Multipotent and unipotent progenitors contribute to prostate postnatal development. Nature cell biology *14*, 1131–1138.

Park, D., Spencer, J.A., Koh, B.I., Kobayashi, T., Fujisaki, J., Clemens, T.L., Lin, C.P., Kronenberg, H.M., and Scadden, D.T. (2012). Endogenous bone marrow MSCs are dynamic, fate-restricted participants in bone maintenance and regeneration. Cell Stem Cell *10*, 259–272.

Pond, A.C., Bin, X., Batts, T., Roarty, K., Hilsenbeck, S., and Rosen, J.M. (2013). Fibroblast growth factor receptor signaling is essential for normal mammary gland development and stem cell function. Stem Cells *31*, 178–189.

Prater, M.D., Petit, V., Alasdair Russell, I., Giraddi, R.R., Shehata, M., Menon, S., Schulte, R., Kalajzic, I., Rath, N., Olson, M.F., et al. (2014). Mammary stem cells have myoepithelial cell properties. Nat Cell Biol *16*, 942–950, 941–947.

Reedijk, M. (2012). Notch signaling and breast cancer. Advances in experimental medicine and biology *727*, 241–257.

Reedijk, M., Odorcic, S., Chang, L., Zhang, H., Miller, N., McCready, D.R., Lockwood, G., and Egan, S.E. (2005). High-level coexpression of JAG1 and NOTCH1 is observed in human breast cancer and is associated with poor overall survival. Cancer Res *65*, 8530–8537.

Rios, A.C., Fu, N.Y., Cursons, J., Lindeman, G.J., and Visvader, J.E. (2016). The complexities and caveats of lineage tracing in the mammary gland. Breast Cancer Res *18*, 116.

Rios, A.C., Fu, N.Y., Lindeman, G.J., and Visvader, J.E. (2014). In situ identification of bipotent stem cells in the mammary gland. Nature *506*, 322–327.

Rodilla, V., Dasti, A., Huyghe, M., Lafkas, D., Laurent, C., Reyal, F., and Fre, S. (2015). Luminal Progenitors Restrict Their Lineage Potential during Mammary Gland Development. PLoS Biol *13*, e1002069.

Sale, S., Lafkas, D., and Artavanis-Tsakonas, S. (2013). Notch2 genetic fate mapping reveals two previously unrecognized mammary epithelial lineages. Nature cell biology *15*, 451–460.

Scheele, C.L., Hannezo, E., Muraro, M.J., Zomer, A., Langedijk, N.S., van Oudenaarden, A., Simons, B.D., and van Rheenen, J. (2017). Identity and dynamics of mammary stem cells during branching morphogenesis. Nature *542*, 313–317.

Seldin, L., Le Guelte, A., and Macara, I.G. (2017). Epithelial plasticity in the mammary gland. Curr Opin Cell Biol *49*, 59–63.

Shackleton, M., Vaillant, F., Simpson, K.J., Stingl, J., Smyth, G.K., Asselin-Labat, M.L., Wu, L., Lindeman, G.J., and Visvader, J.E. (2006). Generation of a functional mammary gland from a single stem cell. Nature *439*, 84–88.

Smith, G.H., and Medina, D. (1988). A morphologically distinct candidate for an epithelial stem cell in mouse mammary gland. J Cell Sci *90 (Pt 1)*, 173–183.

Spike, B.T., Engle, D.D., Lin, J.C., Cheung, S.K., La, J., and Wahl, G.M. (2012). A mammary stem cell population identified and characterized in late embryogenesis reveals similarities to human breast cancer. Cell Stem Cell *10*, 183–197.

Stingl, J., Eirew, P., Ricketson, I., Shackleton, M., Vaillant, F., Choi, D., Li, H.I., and Eaves, C.J. (2006). Purification and unique properties of mammary epithelial stem cells. Nature *439*, 993–997.

Sumbal, J., and Koledova, Z. (2019). FGF signaling in mammary gland fibroblasts regulates multiple fibroblast functions and mammary epithelial morphogenesis. Development *146*.

Tika, E., Ousset, M., Dannau, A., and Blanpain, C. (2019). Spatiotemporal regulation of multipotency during prostate development. Development *146*.

van Amerongen, R., Bowman, A.N., and Nusse, R. (2012). Developmental stage and time dictate the fate of Wnt/beta-catenin-responsive stem cells in the mammary gland. Cell Stem Cell *11*, 387–400.

van Es, J.H., Sato, T., van de Wetering, M., Lyubimova, A., Nee, A.N., Gregorieff, A., Sasaki, N., Zeinstra, L., van den Born, M., Korving, J., et al. (2012). Dll1+ secretory progenitor cells revert to stem cells upon crypt damage. Nature cell biology *14*, 1099–1104.

van Genderen, C., Okamura, R.M., Farinas, I., Quo, R.G., Parslow, T.G., Bruhn, L., and Grosschedl, R. (1994). Development of several organs that require inductive epithelial-mesenchymal interactions is impaired in LEF-1-deficient mice. Genes Dev *8*, 2691–2703.

Van Keymeulen, A., Fioramonti, M., Centonze, A., Bouvencourt, G., Achouri, Y., and Blanpain, C. (2017). Lineage-Restricted Mammary Stem Cells Sustain the Development, Homeostasis, and Regeneration of the Estrogen Receptor Positive Lineage. Cell Rep *20*, 1525–1532.

Van Keymeulen, A., Rocha, A.S., Ousset, M., Beck, B., Bouvencourt, G., Rock, J., Sharma, N., Dekoninck, S., and Blanpain, C. (2011). Distinct stem cells contribute to mammary gland development and maintenance. Nature *479*, 189–193.

Veltmaat, J.M., Relaix, F., Le, L.T., Kratochwil, K., Sala, F.G., van Veelen, W., Rice, R., Spencer-Dene, B., Mailleux, A.A., Rice, D.P., et al. (2006). Gli3-mediated somitic Fgf10 expression gradients are required for the induction and patterning of mammary epithelium along the embryonic axes. Development *133*, 2325–2335.

Visvader, J.E., and Stingl, J. (2014). Mammary stem cells and the differentiation hierarchy: current status and perspectives. Genes Dev *28*, 1143–1158.

Wang, C., Christin, J.R., Oktay, M.H., and Guo, W. (2017). Lineage-Biased Stem Cells Maintain Estrogen-Receptor-Positive and -Negative Mouse Mammary Luminal Lineages. Cell Rep *18*, 2825–2835.

Watson, C.J., and Khaled, W.T. (2008). Mammary development in the embryo and adult: a journey of morphogenesis and commitment. Development *135*, 995–1003.

Watson, C.J., and Khaled, W.T. (2020). Mammary development in the embryo and adult: new insights into the journey of morphogenesis and commitment. Development *147*.

Wuidart, A., Ousset, M., Rulands, S., Simons, B.D., Van Keymeulen, A., and Blanpain, C. (2016). Quantitative lineage tracing strategies to resolve multipotency in tissue-specific stem cells. Genes Dev *30*, 1261–1277.

Wuidart, A., Sifrim, A., Fioramonti, M., Matsumura, S., Brisebarre, A., Brown, D., Centonze, A., Dannau, A., Dubois, C., Van Keymeulen, A., et al. (2018). Early lineage segregation of multipotent embryonic mammary gland progenitors. Nat Cell Biol *20*, 666–676.

Yang, Y., Riccio, P., Schotsaert, M., Mori, M., Lu, J., Lee, D.K., Garcia-Sastre, A., Xu, J., and Cardoso, W.V. (2018). Spatial-Temporal Lineage Restrictions of Embryonic p63(+) Progenitors Establish Distinct Stem Cell Pools in Adult Airways. Dev Cell *44*, 752–761 e754.

Zeng, Y.A., and Nusse, R. (2010). Wnt proteins are self-renewal factors for mammary stem cells and promote their long-term expansion in culture. Cell Stem Cell *6*, 568–577.

Zvelebil, M., Oliemuller, E., Gao, Q., Wansbury, O., Mackay, A., Kendrick, H., Smalley, M.J., Reis-Filho, J.S., and Howard, B.A. (2013). Embryonic mammary signature subsets are activated in Brca1-/- and basal-like breast cancers. Breast Cancer Res *15*, R25.

17

The Intestinal Stem Cells in Homeostasis and Repair

Aline Stedman

Sorbonne Université, UPMC Université Paris 06, IBPS, CNRS UMR7622, Inserm U 1156, Laboratoire de Biologie du Développement; Paris 75005, France
E-mail: aline.stedman@sorbonne-universite.fr

17.1 Introduction

The intestine performs two major functions: one is to defend the organism against the external world and, therefore to constitute an impermeable barrier against potentially toxic xenobiotics or pathogen microorganisms present in its lumen. The other is to digest and uptake the essential nutrients from the transiting diet, and host the symbiotic microorganisms that participate in this digestive function. The intestinal epithelium lies on the front position to achieve both functions. Epithelial cells of the intestine organize into a single layer with their apical surface lining the intestinal lumen. Basally, they face the vascular, muscular, nervous, and immune components of the intestinal mucosa. The absorptive lineage of the intestine comprises the enterocytes in the small intestine, and the colonocytes in the colon, recognizable by their microvilli brushes at the apical membrane. The secretory lineage comprises the goblet cells, which produce the mucus layer covering the epithelium, the Paneth cells which secrete antimicrobial peptides, hormone-producing enteroendocrine cells, and the less abundant chemosensory tuft cells. Other rare cellular subtypes, such as novel enteroendocrine cells (Grün et al., 2015) or pH-sensing colonocytes (Parikh et al., 2019) were also recently uncovered by single cell profiling studies.

In the small intestine, the epithelial surface extends into long digit–like protrusions called villi, which evaginate into the intestinal lumen. In addition, all along its length, the epithelium folds into small invaginations buried into the mucosa, called the crypts of Lieberkühn. This specific architecture reflects the functional organization of the intestine. Indeed, new epithelial cells are continuously being produced within crypts, where they amplify through several rounds of cell division, before differentiating and migrating upward to reach the surface. After approximately 3 days of migration, epithelial cells die and are shed into the intestinal lumen. As a result, proliferative cells are confined within crypts, surrounded by a specific mucosal environment and distant from the luminal content, while functionally mature cells are mainly found at the surface, or in villi. The only exceptions are Paneth cells, which migrate downward to the crypt base where they persist for around 6–8 weeks (Figure 17.1) (Ireland et al., 2005).

17.2 Different Overlapping Populations of Stem Cells Ensure Homeostasis of the Adult Intestine

17.2.1 Stem Cells Lie Within Crypts

The intestinal epithelium is the fastest cell-renewing tissue in mammals, as its complete turnover takes only 3–5 days. It is also characterized by a remarkable regeneration potential. As an example, like most proliferative tissues, the intestine is sensitive to ionizing radiation, but it can recover from doses of radiation that cause complete hematopoietic failure (10–12 Gy in mice), so that after only 5–7 days it has regained its normal histology. The existence of adult stem cells at the origin of the two main intestinal lineages, and providing the tissue with such remarkable regenerative capacities was postulated as early as in the 70s, however their identification was hampered by the lack of specific reliable markers and tools to functionally assess their stemness. In 1974, using electron microscopy, Cheng and Leblond identified cells with very specific columnar morphology at the base of small intestinal crypts, that they called Columnar Base Cells (CBCs) (Cheng, Leblond 1974). Indirect tracing approaches based on autoradiography and random mutagenesis showed that CBCs had clonogenic capacities and could give rise to the four main epithelial cell types of the intestine (Cheng, Leblond 1974; Bjerknes, Cheng 1999). The authors postulated that CBCs were the stem cells of the intestine. Another approach designed to track intestinal stem cells was label retention. This strategy was based on the assumption that like

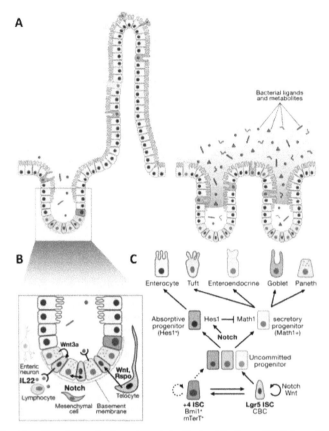

Figure 17.1 Functional organization of the intestinal epithelium. (**A**) Schematic view of the intestinal epithelium. In the small intestine (**left**), the epithelium organizes into villi, where differentiated cells reside, and crypts containing the proliferative stem and progenitor cells. The colon (**right**) is devoid of villi and contains large crypts. The mucus layer, depicted in blue, and the bacterial products, in red, densify along the intestinal length. (**B**) Intestinal stem cells (*Lgr5* expressing columnar base cells (CBCs) in green, and *Bmi1* or *mTert* expressing +4 stem cells (SCs) in dark orange) reside at the bottom of epithelial crypts, surrounded by a specific niche environment providing ISCs with important paracrine factors, like Wnt or Notch ligands, which regulate their fate. (**C**) Hierarchy of the intestinal lineage. Two main populations of intestinal stem cells co-exist, Lgr5$^+$ CBCs, which account for most of intestinal renewal at steady state, and a rarer population of +4 SCs, marked by the expression of *Bmi1* and *mTert*, which can compensate for the loss of Lgr5 SCs. Both populations can interconvert, self-renew and give rise to uncommitted progenitors which following several rounds of cell division adopt either an absorptive or a secretory fate. The Wnt and Notch pathways are essential regulators of ISCs fate. Along their differentiation, cells move upwards to finally exit the crypt compartment, with the exception of Paneth cells which stay in the crypt bottom intermingled with ISCs.

in the hematopoietic tissue, stem cells of the intestine should be quiescent. The experiment, conducted by the team of Chris Potten, consisted in injuring the intestinal epithelium by irradiation to trigger cell division, and exposing the cells to the radioactive DNA analog tritiated thymidine. All cycling cells should incorporate the label before diluting it after several rounds of cell division. In contrast, upon exiting the cell cycle, quiescent cells should remain radioactive. Potten's team could show that cells preferentially located above Paneth cells, a position also referred to as the +4 position, were label-retaining cells and proposed that these cells were the bona fide stem cells (Potten et al., 2002). From these experiments, it was concluded that intestinal stem cells (ISCs) reside in the crypts and two schools of thought persisted: one claiming that CBCs were ISCs, and the other that the bona fide ISCs corresponded to the +4 label–retaining cells.

17.2.2 Columnar Base Cells are Lgr5$^+$ Intestinal Stem Cells

A real breakthrough in the field of ISC biology came from the finding of the first reliable marker for ISCs, the Lgr5 gene. As the Wnt pathway is the main driver of cell proliferation in intestinal crypts, the lab of Hans Clevers made the guess that ISCs should express high levels of Wnt targets. A microarray-based screen was performed to identify Wnt-responsive genes with expression restricted to crypts. It led to the identification of Lgr5, a gene encoding an orphan G-coupled transmembrane receptor. β-galactosidase staining on Lgr5–LacZ transgenic mice revealed a very specific expression pattern in CBCs bothin small intestine and colonic crypts (Barker et al., 2007). Lineage tracing of these cells using mice carrying an inducible Lgr5-Cre knocked-in allele in conjunction with the Rosa26-LacZ Cre reporter, showed that they were long living and multipotent, as they gave rise to all intestinal lineages. Another demonstration of their stemness came from the work from Sato and co-authors, who showed that when put into 3D-culture, surrounded by an extracellular matrix and appropriate growth factors, single-sorted Lgr5$^+$ cells were able to proliferate, and to self-organize into multicellular budding structures. Strikingly, these organoids retain the topographical organization of the intestine, as they harbor crypt-like budding domains, where proliferating cells reside, separated by flat domains composed of mature differentiated cells. Like in the intestine, Lgr5$^+$ cells locate at a stereotyped position in the bottom of crypt-like domains. Tracing of these cells show that in this ex vivo context, they retain their self-renewal capacities (organoids can be maintained for over a year in culture), and remain multipotent (Figure 17.2) (Sato et al., 2009).

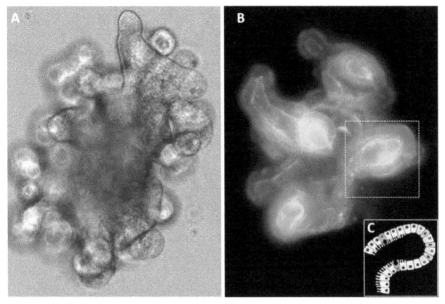

Figure 17.2 Intestinal Organoids. (A) Example of a 10-day-old mouse intestinal organoid. Buddings are crypt-like domains, which contain the stem and progenitor cells. Differentiated cells are found outside the buddings in flat villus-like domains. (**B**) Mouse intestinal organoid stained with Dapi (nuclei in blue), phalloidin (actin cytoskeleton in green). Similar to the *in vivo* intestine, organoids are composed of a single sheet of polarized epithelial cells which organize in three dimensions to form budding hollow structures with the apical surface (phalloidin) facing a central lumen, and the basal surface of the cells in regard with the external environment.

Combined with in vivo studies, this new model enabled to better characterize Lgr5$^+$ stem cells (Lgr5 ISCs). In striking contrast to adult stem cells of the bone marrow or the muscle, these ISCs are abundant (10–15 per crypt), cycling (divide once every 24 hours), and rely on a very active mitochondrial metabolism. Moreover, long-term clonal analysis of Lgr5 ISCs revealed that their self-renewal mainly relies on symmetric cell division, and that cell competition between symmetrically dividing neighboring stem cells leads to a progressive drift of crypts toward clonality (Snippert et al., 2010).

17.2.3 +4 Reserve Stem Cells

The finding of Lgr5 ISCs and the development of lineage tracing approaches in mouse accelerated the search for other markers of stem cells. In particular, the track for a stem cell population that would be quiescent and sit on top of the intestinal lineage hierarchy, by analogy with other tissues like the hematopoietic lineage, has been the subject of intense research. Sangiorgi and Capecchi found that *Bmi1*, a transcriptional repressor of the Polycomb family expressed in hematopoietic and neural quiescent stem cells, marked a rare population of cells, mainly found in crypts of the duodenum at the +4 position. Unexpectedly, they found that these cells were not strictly quiescent, but rather slow cycling. Moreover, lineage tracing showed that under steady-state conditions, their contribution to intestinal renewal was around 5 times less compared with Lgr5 ISCs (Sangiorgi & Capecchi 2008). Since the identification of *Bmi1*, other genes were found such as *Hopx*, *mTERT*, and *Lrig1* (Takeda et al., 2011; Montgomery et al., 2011; Powell et al., 2012), which mark slow-cycling stem cells at the +4 position. Expression of these genes only partially overlaps, suggesting that +4 ISCs have varying molecular signatures along the intestine, which might underly subtle functional differences between them. Similarly to Lgr5 ISCs, +4 ISCs are able to give rise to organoids when cultured *ex vivo*.

Hierarchy between these different stem cell populations is still a matter of debate, in particular because several studies have shown that Lgr5 ISCs express +4 genes such as *mTert* and vice versa (Munoz et al., 2012), questioning the interpretation of lineage-tracing experiments. However, it was shown that Lgr5 and +4 ISCs can inter convert, and that at steady state, +4 ISCs can compensate for the loss of Lgr5 ISCs following their genetic ablation (Tian et al., 2011). Altogether, these observations suggest that Lgr5 and +4 ISCs constitute different stem cell populations with no strict hierarchy, with +4 ISCs constituting a heterogenous reserve stem cell population.

17.3 The Multiple Strategies to Protect and Repair the Intestinal Barrier

17.3.1 Intestinal Stem Cells Under Stress

Intestinal repair was shown to rely on both Lgr5$^+$ and +4 ISCs. However, these cells differentially respond to stress, and therefore, depending on the type of damage inflicted, regeneration relies mostly on one or the other population. For example, genetically ablating Lgr5 ISCs is detrimental for intestinal repair in the context of acute ionizing radiation (over 10 Gy), but not following chemically-induced intestinal damage using administration of dextran sodium sulfate, where +4 ISCs can take over (Metcalfe et al., 2013). Coexistence of several stem cell populations might therefore provide the intestine with a very efficient and robust regeneration potential. Maintenance of ISC integrity throughout their lifespan is essential for tissue homeostasis and to avoid the propagation of detrimental mutations at the tissue level. Recent studies have shed light on the cytoprotective mechanisms at play in ISCs. Their location deep within crypts, far away from the intestinal lumen constitutes a first mechanism of protection. Interestingly, in the crypt-less zebrafish intestine, stem cells are more susceptible to the cytostatic effect of butyrate, a metabolite present in the intestinal lumen, than stem cells in mouse where butyrate is uptaken by enterocytes before being able to reach the crypt compartment (Kaiko et al., 2016). Intrinsic cytoprotective programs also play a major role in ISC protection. For instance, DNA repair in Lgr5 ISCs is driven by high-fidelity homologous recombination, as opposed to the error-prone non-homologous end-joining pathway (Hua et al., 2012). Autophagy, which is involved in the lysosomal degradation of harmful altered intracellular components, is upregulated in ISCs and was recently shown to be essential for their maintenance both at steady-state and upon acute stress (Trentesaux 2020). In comparison with more differentiated cell types, adult stem cells are also equipped with specific stress sensing and resolving systems. Proteostasis is for example tightly regulated in ISCs: defects in ribosome production or proteins misfolding activate ribosomal and ER stress pathways which in turn leads to the elimination of damaged ISCs to their exhaustion through apoptosis and premature differentiation (Heijmans, van Lidth de Jeude, Koo et al., 2013; Stedman et al., 2015).

17.3.2 A Highly Plastic Lineage

Recent lineage tracing experiments also showed that upon tissue damage, committed progenitors and even differentiated cells could replenish the stem cell pool, thereby contributing to intestinal repair. The Notch ligand Dll1 marks the common progenitors for intestinal secretory cells. Using lineage tracing strategies, Van Es et al., showed that in case of irradiation-induced tissue damage (6.0 Gy irradiation), Dll1 progenitors were able to revert to a stem-like state and give rise to the whole intestinal lineage (Van Es et al., 2012). High Wnt activation participates in this dedifferentiation process, and is sufficient, *ex vivo*, to trigger the formation of organoids from Dll1 single sorted cells. Intestinal plasticity is not limited to the secretory lineage, as progenitors of both enterocytes and enteroendocrine cells were also found to regain stemness following genetic ablation of Lgr5 stem cells (Tetteh et al., 2016, Yan et al., 2017). Further, more terminally differentiated cells can also participate in intestinal repair as long as they locate near the crypt base. Indeed, genetic labeling of Paneth cells and their progeny using Lyz1CreER/+; R26R-tdT mice showed that upon high dose of ionizing radiation (12 Gy), a subset of irradiated Paneth cells were able to acquire stem cell properties (Yu et al., 2018).

17.3.3 The Extracellular Matrix Pulls the Trigger

To gain better insight into the molecular mechanisms involved in intestinal regeneration, a transcriptomic analysis was performed on colonic epithelial cells under repair after chemically induced mucosal ulceration. This strategy brought into light that a specific expression program is transiently activated in the epithelium, characterized by the upregulation of fetal intestinal markers like Sca1 and Trop2, the concurrent suppression of adult stem cell markers, and the upregulation of genes encoding for components of the extracellular matrix, such as collagen type I. This study, further shows that remodeling of the extracellular matrix at the level of the mucosal wound, is sufficient to trigger YAP1 activation within epithelial cells (Yui et al., 2018). The Hippo–YAP/TAZ pathway regulates cell proliferation and survival in response to mechanical stimuli, including cellular matrix stiffness and cell density. YAP1 (Yes-associated protein 1), a transcriptional co-activator of the pathway, had previously been shown to be dispensable for intestinal homeostasis but detrimental for intestinal repair (Cai et al., 2010). Yui et al., demonstrated that YAP1 is essential for there programming of the wounded intestinal epithelium in response to extracellular matrix remodeling.

17.3.4 Mechano-sensitive Facultative Stem Cells

In order to gain a cellular resolution of the wounded intestine profile, single cell RNA sequencing was performed on the intestine following acute stress induced by 12-Gy irradiation (Ayyaz et al., 2019). This strategy enabled the identification of a new cell type marked by the expression of Clusterin (Clu), a chaperone protein, and key for crypts regeneration. Clu+ cells are rare (less than 0.1% of intestinal epithelial cells) quiescent Lgr5 ISC daughter cells, present in both small intestine and colonic crypts. Upon damage, these cells upregulate YAP1, and consequently activate to reconstitute new crypts. The dialogue established between epithelial cells and the underlying basal membrane is therefore essential to ensure proper intestinal repair. Interestingly, similar transcriptional dynamics implying the contact of ISCs with the extracellular matrix, and consequent YAP1 activation, were described as key events of organoid formation (Serra et al., 2019).

17.4 The Fine-tuning of Intestinal Stem Cells Fate

17.4.1 Extrinsic Modes of Intestinal Stem Cell Regulation

17.4.1.1 Intestinal stem cells neighborhood

The 3D-environment in which ISCs reside greatly impacts their fate. As described above, composition of the extracellular matrix can directly affect cell behavior through mechano-sensitive pathways. Cells inhabiting the ISC niche are crucial too, as they provide ISCs with important growth factors and metabolites. In the small intestine, Paneth cells physically interact with Lgr5 ISCs through their lateral membrane. Because of this intimate contact, they have long been assumed to be the main niche component of ISCs. Co-culture of single sorted ISCs with Paneth cells greatly improves their organoids forming efficiency. Moreover in absence of Paneth cells, ISCs cannot give rise to organoids. Paneth cells are therefore necessary and sufficient for ISC self-renewal and differentiation *ex vivo* (Sato et al., 2011; Durand et al., 2012). However, *in vivo*, lack of Paneth cells, for example in *Math1* deficient mice, does not alter ISC properties, suggesting that other, non-epithelial cells of the niche can compensate for their loss (Durand et al., 2012; Kim et al., 2012). In fact, myofibroblasts, subepithelial telocytes, as well as enteric neurons and immune cells surrounding the crypts were all shown to contribute to the regulation of ISC fate (Shoshkes-Carmel et al., 2018; Puzan et al., 2018; Lindemans et al., 2015; Gronke et al., 2019). How these different inputs are coordinated to instruct ISC behavior, how they get modulated through space and time, or in various disease contexts, but also how the different intestinal stem/progenitor cells could be induced by different cellular niches are still fundamental questions to address.

17.4.1.2 The Wnt pathway

The canonical Wnt/bcatenin pathway plays a major function in intestinal development and homeostasis, in particular in driving stem and progenitor cells proliferation. Extracellular WNT ligands bind to the Frizzled (FZD) receptors, leading to the nuclear translocation of the transcriptional co-activator β-catenin. In the absence of WNT, β-catenin is constantly targeted to degradation by the destruction complex that includes the tumor suppressor adenomatous polyposis coli (APC). Both epithelial and mesenchymal cells inside and surrounding the crypts produce Wnt ligands, such that Wnt activity follows a gradient with its highest activity in the stem cell zone (Gregorieff et al., 2005; Farin et al., 2012; Farin et al., 2016) Consistently, Lgr5 ISCs contain high levels of nuclear β-catenin and express numerous Wnt targets such as *Cyclin D* or *c-Myc* which promote their proliferation and growth. In mice lacking critical effectors of the pathway, crypts size is drastically reduced, and proliferation of ISCs and progenitors is abolished (Korinek et al., 1998; Van Es et al., 2012; Fevr et al., 2007). Inversely, overactivation of the pathway leads to abnormal crypts expansion and reduced differentiation. Mutations in the *APC* gene are found in a vast majority of human colonic tumors and are considered critical driving events for cancer initiation (Fodde et al., 2001). In mouse, conditionally inactivating *APC* in ISCs is sufficient to generate large adenomas (Andreu et al., 2005). Another important aspect of Wnt signaling is its major role in intestinal regeneration. Indeed, upregulation of Wnt in the damaged epithelium is essential to mobilize ISCs and thereby induce the robust expansion of crypts necessary for epithelial restitution (Ashton et al., 2010; Miyoshi et al., 2012; Zou et al., 2018). New evidence suggests that it might also participate in ISC cytoprotection and resistance to stress (Raveux et al., 2020).

17.4.1.3 The Notch pathway

Studies of mice deficient for essential Notch effectors demonstrated that this pathway plays a crucial role in regulating the balance between self-renewal and differentiation in ISCs. Multiple agonists of the pathway have been localized to the stem and progenitor zone of crypts. The Notch1 and Notch2 receptors are specifically expressed in crypt stem cells (Fre et al., 2011), while its ligands Jagged 1, Dll1, and Dll4 are expressed in Paneth cells and progenitors. Blockade of Notch signaling by conditional inactivation of Notch 1 (Kim et al., 2014), Dll1, and Dll4 (Pellegrinet et al., 2011), or by treatment with γ-secretase inhibitors (Milano et al., 2004; Van Es et al., 2005; VanDussen et al., 2012) induces Lgr5$^+$ CBC loss, abolishes proliferation in crypts and leads to an excessive production of secretory cells at the expense of absorptive enterocytes. Conversely, over activation of Notch signaling, by ectopically expressing a constitutively active form of Notch in the NICD-GFP mice, represses the secretory fate and expands the crypts (Fre et al., 2007). Hence, Notch not only regulates self-renewal and proliferation of ISCs, but it also impacts on cell fate decisions. The role of Notch on proliferation requires Wnt signals, while effects on differentiation are Wnt independent (Fre et al., 2009), and involves the active repression by the Notch target Hes1, of the master regulator of secretory fate Math1/Atoh1.

17.4.2 ISC Energy Metabolism: How to Boost Stemness

A lot of additional signaling pathways regulate ISC fate. Among them, diet and energy metabolism have emerged as new crucial regulators. As an example, calorie restriction induces crypts expansion and increases ISC number. This effect was shown to result from the non-autonomous influence of Paneth cells on their neighboring ISCs. The mammalian target of rapamycin (mTOR) senses nutrients and growth factors to coordinate cell growth and metabolism. Calorie restriction in mouse was shown to reduce mTOR signaling within Paneth cells, causing them to release the paracrine factor ADP ribose. ADP ribose in turn activates mTOR activity within ISCs, triggering their proliferation. (Yilmaz et al., nature 2012; Igarashi & Guarente2016). High-fat diet was also found to enhance ISC proliferation, this time through the cell-autonomous activation of the nuclear receptor peroxisome (PPAR)-delta, which among other targets regulates the expression of genes involved in the control of cell proliferation (Beyaz et al., 2018). Metabolic by-products secreted by niche cells can as well influence ISC function. For example, Paneth cells' energetic metabolism relies mostly on glycolysis. Interestingly, they secrete the glycolysis end-product lactate, which is then uptaken by neighboring Lgr5 ISCs asa substrate for oxidative phosphorylation. Furthermore, inhibition of mitochondrial activity in Lgr5ISCs or inhibition of glycolysis in Paneth cells, similarly affect ISC self-renewal, showing that the metabolic interplay between ISCs and their niche is essential to support epithelial homeostasis (Rodríguez-Colman et al., 2017).

17.5 Bugs in the Niche

The intestinal lumen is populated by a very dense and complex microbial ecosystem, which comprises more than trillion of bacteria, viruses and fungi, referred to as the intestinal microbiota. The intestinal microbiota has emerged as an essential component of our body homeostasis, as it is implicated in vital functions such as body growth, supply of essential vitamins and neurotransmitters, immune system maturation, and digestion of indigestible food (Sekirov et al., 2010). In this paragraph, we focus more specifically on the bacterial flora which constitutes the majority of the microorganisms inhabiting the intestinal lumen. Progresses made in high-throughput sequencing have enabled to better characterize the composition of the bacterial flora along the intestinal tract, and showed that it diversifies and densifies along the anteroposterior axis to reach its peak in the colon, with approximately 10^{12} bacteria per g (dry weight) of colonic contents, where Bacteroidetes and

Firmicutes represent the dominant phyla.In human, changes in its composition also called dysbiosis, are correlated with a multitude of diseases, including diabetes, obesity, and cancer (Karlsson et al., 2013; Vonaesch et al., 2018). Much of the current knowledge on bacterial functions stems from the use of mouse models with different microbial status, including germ-free, antibiotic-treated, or mono-colonized mice (gavaged with a specific bacteria). These models showed that in addition to important systemic roles, intestinal bacteria are essential locally, for intestinal homeostasis and regeneration. Even more fascinating, early life exposition to commensal flora seems to be instrumental for refining the methylation pattern (Yu et al., 2015) and transcriptome (Abo et al., 2020) in ISCs. Among the genes that were upregulated by the primo-colonization of commensals, Abo and collaborators identified Erdr1, a soluble factor that enhances Wnt signaling, and boosts ISC regeneration.

17.5.1 Bacterial Metabolites: A Complex Language Still to be Deciphered

Most of the commensal bacteria are anaerobic and reside in the lumen of the gut, far away from the epithelial lining. However, they can influence ISC fate at distance, in particular through the release of metabolites in the intestinal lumen. Such metabolites result from the transformation by bacteria of host-derived substrates like bile acids, or dietary components. For example, fermentative bacteria in the colon metabolize carbohydrates or dietary fibers, and as a result, produce a large amount of short chain fatty acids (acetate, butyrate, and propionate). Butyrate, a potent HDAC inhibitor, is abundant in the colonic lumen. It can be utilized as a respiratory substrate by colonocytes (Roediger et al., 1982). When it reaches the crypts bottom, for example in cases of mucosal injury, it was found to promote cell-cycle arrest and differentiation of colonic stem and progenitor cells (Kaiko et al., 2016). Another product of bacterial fermentation, lactate, was shown to enhance Lgr5 ISC proliferation and self-renewal at steady state, but also following damage. Lactate binding to its specific receptor Gpr81, expressed by Paneth cells and by a subset of stromal cells, was shown to increase Wnt3a production, thereby supporting ISC proliferation (Lee et al., 2018).

17.5.2 Innate Immunity and ISCs-bacteria Crosstalk

Both immune and epithelial cells of the intestine express a repertoire of innate immune receptors of the PRR family (Pattern Recognition Receptors), involved in the sensing of conserved bacterial or viral motifs outside

(Toll Like Receptors, TLRs) and inside (TLRs and NOD receptors) the cells. Classically, upon binding to their microbial ligands, PRRs activate a pro-inflammatory signaling cascade involved in getting rid of invading pathogens (Abreu, Fukata, & Arditi 2005; Kawai & Akira 2011). Miceknocked-out for *Myd88*, a signaling adaptor used by all TLRs, are not able to regenerate their intestine upon damage (Rakoff-Nahoum et al., 2004; Araki et al., 2005), suggesting that PRR activation by beneficial symbionts might also trigger pathways responsible for intestinal homeostasis. In line with this hypothesis, the IL22 cytokine can be released from intestinal dendritic cells and resident lymphocytes upon bacterial stimuli, and promote ISC proliferation and tissue repair, in part through activation of STAT3 within ISCs (Pickert et al., 2009; Hou et al., 2018). A direct ISCs-microbiota dialogue through PRRs might also be at play. Indeed, we found that among other PRRs, Lgr5 ISCs express high levels of Nod2, an intracellular receptor specific for MDP (Muramyl Dipeptide, the main component of bacterial cell wall). Using mice *in vivo* models as well as organoids, we showed that Nod2 activation protects ISCs from oxidative stress-induced apoptosis following exposure to the genotoxic compound doxorubicin, or irradiation (Nigro et al., 2014). In this context, Nod2 activation does not trigger the canonical pro-inflammatory cascade, but rather induces the autophagy machinery, leading to the rapid clearance of damaged mitochondria, and excessive levels of toxic reactive oxygen species within ISCs (Levy et al., 2020). Whether such cytoprotection operates at a distance or involves the close contact of specific bacteria with the epithelium, remains unanswered. Interestingly, a small cluster of bacteria were identified within colonic murine crypts, defined as the crypt-specific core microbiota (CSCM), suggesting that a close mutualistic interaction between a subset of bacteria and ISCs could take place at homeostasis (Pedron et al., 2012). Moreover, a shift of bacteria toward the epithelium is observed in a lot of instances of mucosal breaches, for example, upon intestinal mechanical injury by drug ingestion or pathogens invasion, or during cancer, suggesting the interaction between intestinal crypts and microbes is a dynamic process throughout an individual life, and is key to the understanding of certain pathologies like chronic inflammation, cancer, or metabolic diseases.

17.6 Perspectives on ISCs-based Regenerative Medicine

The recent advances made in intestinal stem cells research have undoubtedly influenced the field of adult stem cell biology, challenging old concepts and bringing into light new ideas in the field. As discussed in this chapter, study on the intestinal lineage has helped uncover the high degree of cell plasticity that can reach an adult tissue. It also highlighted the important regulatory functions exerted by biomechanics and metabolism on stem cell fate. A lot of promising research avenues remain, that could not be described in this chapter, including the role of epigenetic modifications on intestinal homeostasis, the ontogeny of ISCs, their modifications upon aging, or intestinal cancer stem cells. Research in the field greatly beneficiated from the breakthrough of intestinal organoids. Organoids are compatible with mechanical, chemical,or genetic perturbations and as such, they also provide a new platform for clinical applications like drug screening, personalized medicine, or cell-therapy (Fatehullah et al., 2016; Clevers 2016). Efficient protocols have been developed to generate human intestinal organoids from patient biopsies, or patient-derived induced pluripotent stem cells (Sato et al., 2011; Spence et al., 2011). Being able to heal a patient intestinal mucosa by producing genetically corrected organoids and engrafting them in order to reconstitute a normal mucosa is the goal of organoids-based cell therapy. The feasibility of organoids-based transplantation has first been addressed by showing that Lgr5 ISCs from mouse colonic organoids injected into the damaged colon of a recipient mousecould participate in epithelial repair and form new long-living colonic crypts (Yui et al., 2012; Fordham et al., 2013). More recently, human-derived colonic organoids were also shown capable of engraftment into a host mouse colon (Sugimoto et al., 2018). In this context, Human Lgr5 ISCs generated new long-living crypts, with characteristics of human crypts in terms of size and cell cycle length. The use of organoids for stem cell therapy in human requires further improvements among which the development of safe protocols for efficient organoids graft, new clinical-grade 3D matrices to replace Matrigel, which is a broadly used animal-derived matrix and therefore not suitable for clinical applications but is also very costly to produce. Synthetic polymers based on poly-ethylene glycol (PEG) hydrogel were developed that enable the culture and intestinal organoids and their tranplantability (Gjorevskiet al., 2016; Cruz-Acunaet al., 2017). Importantly, by playing with the density of the hydrogel, or by supplementing it with different adhesion molecules, the authors were able to modulate ISCs choice between self-renewal and differentiation. Finally, organoids derived from colonic tumor biopsies share a lot of similarities with primary tumor cells and recapitulate tumors drug responses (Van de Wetering et al., 2015; Abbasi et al., 2018), suggesting that cancer research could also benefit greatly from this technology. Consequently, biobanks of patient-derived organoids are now being established worldwide (Van de Wetering et al., 2015).

17.7 Concluding Remarks

- Different overlapping stem cell populations with no strict hierarchy sustain the adult mammalian intestinal homeostasis.
- The intestinal lineage is extremely plastic as both cellular dedifferentiation and activation of reserve stem cell populations can repair the tissue upon damage.
- A "ménage à trois" between intestinal stem cells, immune cells, and the microbiota regulates intestinal homeostasis.
- Intestinal organoids development contributed to bring new insights into the biology of adult stem cells, and bring new hopes in regenerative medicine.

A lot of exciting questions, some that are now starting to be addressed, remain. These include the ontogeny of intestinal stem cells, the role of epigenetics in intestinal plasticity, or the impact of the intestinal microbiota on ISCs in pathological contexts.

References

Abbasi J. (2018). Patient-derived organoids predict cancer treatment response. JAMA. 319:1427.

Abo H, Chassaing B, Harusato A, Quiros M, Brazil JC, Ngo VL, Viennois E, Merlin D, Gewirtz AT, Nusrat A, Denning TL. (2020). Erythroid differentiation regulator-1 induced by microbiota in early life drives intestinal stem cell proliferation and regeneration. Nat Commun. 11(1):513.

Abreu MT, Fukata M, Arditi M. (2005). TLR signaling in the gut in health and disease. (2005). J Immunol. 174(8):4453–4460.

Andreu P, Colnot S, Godard C, Gad S, Chafey P, Niwa-Kawakita M, Laurent-Puig P, Kahn A, Robine S, Perret C, Romagnolo B. (2005). Crypt-restricted proliferation and commitment to the Paneth cell lineage following Apc loss in the mouse intestine. Development. 132(6): 1443–51.

Araki A, Kanai T, Ishikura T, Makita S, Uraushihara K, Iiyama R, Totsuka T, Takeda K, Akira S, Watanabe M. (2005). MyD88-deficient mice develop severe intestinal inflammation in dextran sodium sulfate colitis. J Gastroenterol. (1):16–23.

Ashton GH, Morton JP, Myant K, Phesse TJ, Ridgway RA, Marsh V, Wilkins JA, Athineos D, Muncan V, Kemp R, Neufeld K, Clevers H, Brunton V, Winton DJ, Wang X, Sears RC, Clarke AR, Frame MC,

Sansom OJ. (2010). Focal adhesion kinase is required for intestinal regeneration and tumorigenesis downstream of Wnt/c-Myc signaling. Dev Cell. 19(2):259–69.

Ayyaz A, Kumar S, Sangiorgi B, Ghoshal B, Gosio J, Ouladan S, Fink M, Barutcu S, Trcka D, Shen J, Chan K, Wrana JL, Gregorieff A. (2019). Single-cell transcriptomes of the regenerating intestine reveal a revival stem cell. Nature. 569(7754):121–125.

Barker N, van Es JH, Kuipers J, Kujala P, van den Born M, Cozijnsen M, Haegebarth A, Korving J, Begthel H, Peters PJ, Clevers H. (2007). Identification of stem cells in small intestine and colon by marker gene Lgr5. Nature. 449(7165): 1003–7.

Beyaz S, Mana MD, Roper J, Kedrin D, Saadatpour A, Hong SJ, Bauer-Rowe KE, Xifaras ME, Akkad A, Arias E, Pinello L, Katz Y, Shinagare S, Abu-Remaileh M, Mihaylova MM, Lamming DW, Dogum R, Guo G, Bell GW, Selig M, Nielsen GP, Gupta N, Ferrone CR, Deshpande V, Yuan GC, Orkin SH, Sabatini DM, Yilmaz ÖH. (2016). High-fat diet enhances stemness and tumorigenicity of intestinal progenitors. Nature. 531(7592):53–8.

Bjerknes M, Cheng H. (1999). Clonal analysis of mouse intestinal epithelial progenitors. Gastroenterology. 116 : 7–14.

Cai J, Zhang N, Zheng Y, de Wilde RF, Maitra A, Pan D. (2010). The Hippo signaling pathway restricts the oncogenic potential of an intestinal regeneration program. Genes Dev.; 24(21):2383–2388.

Cheng, H. & Leblond, C. P. (1974). Origin, differentiation and renewal of the four main epithelial cell types in the mouse small intestine. V. Unitarian gheory of the origin of the four epithelial cell types. Am. J. Anat. 141, 537–561.

Clevers H. (2016). Modeling Development and Disease with Organoids. Cell. 165:1586–1597.

Cruz-Acuña R, Quirós M, Farkas AE, Dedhia PH, Huang S, Siuda D, García-Hernández V, Miller AJ, Spence JR, Nusrat A, García AJ. (2017). Synthetic hydrogels for human intestinal organoid generation and colonic wound repair. Nat Cell Biol. (11):1326–1335.

Durand A, Donahue B, Peignon G, Letourneur F, Cagnard N, Slomianny C, Perret C, Shroyer NF, Romagnolo B. (2012). Functional intestinal stem cells after Paneth cell ablation induced by the loss of transcription factor Math1 (Atoh1). Proc Natl Acad Sci U S A. 109(23):8965–70

Farin HF, Jordens I, Mosa MH, Basak O, Korving J, Tauriello DV, de Punder K, Angers S, Peters PJ, Maurice MM, Clevers H. (2016). Visualization of a short-range Wnt gradient in the intestinal stem-cell niche. Nature. 530(7590):340–3.

Farin HF, Van Es JH, Clevers H. (2012). Redundant sources of Wnt regulate intestinal stem cells and promote formation of Paneth cells. Gastroenterology. 143(6):1518–1529.e7.

Fatehullah A, Tan SH, Barker N. (2016). Organoids as an in vitro model of human development and disease. Nature cell biology. 18:246–254.

Fevr T, Robine S, Louvard D, Huelsken J. (2007). Wnt/beta-catenin is essential for intestinal homeostasis and maintenance of intestinal stem cells. Mol Cell Biol. 27(21):7551–7559.

Fodde R, Smits R, Clevers H. (2001). APC, signal transduction and genetic instability in colorectal cancer. Nat Rev Cancer. 1(1):55–67.

Fordham RP, Yui S, Hannan NR, Soendergaard C, Madgwick A, Schweiger PJ, Nielsen OH, Vallier L, Pedersen RA, Nakamura T, Watanabe M, Jensen KB. (2013). Transplantation of expanded fetal intestinal progenitors contributes to colon regeneration after injury. Cell Stem Cell.13(6):734–44.

Fre S, Hannezo E, Sale S, Huyghe M, Lafkas D, Kissel H, Louvi A, Greve J, Louvard D, Artavanis-Tsakonas S. (2011). Notch lineages and activity in intestinal stem cells determined by a new set of knock-in mice. PLoS One 6: e25785

Fre S, Hannezo E, Sale S, Huyghe M, Lafkas D, Kissel H, Louvi A, Greve J, Louvard D, Artavanis-Tsakonas S. (2011). Notch lineages and activity in intestinal stem cells determined by a new set of knock-in mice. PLoS One. 6(10):e25785.

Fre S, Pallavi SK, Huyghe M, Laé M, Janssen KP, Robine S, Artavanis-Tsakonas S, Louvard D. (2009). Notch and Wnt signals cooperatively control cell proliferation and tumorigenesis in the intestine. Proc Natl Acad Sci U S A. 106(15):6309–14.

Gjorevski N, Sachs N, Manfrin A, Giger S, Bragina ME, Ordóñez-Morán P, Clevers H, Lutolf MP. (2016). Designer matrices for intestinal stem cell and organoid culture. Nature. 539(7630):560–564.

Gregorieff A, Pinto D, Begthel H, Destrée O, Kielman M, Clevers H. (2005). Expression pattern of Wnt signaling components in the adult intestine. Gastroenterology. 129(2):626–638.

Gronke K, Hernández PP, Zimmermann J, Klose CSN, Kofoed-Branzk M, Guendel F, Witkowski M, Tizian C, Amann L, Schumacher F, Glatt H, Triantafyllopoulou A, Diefenbach A. (2019). Interleukin-22 protects intestinal stem cells against genotoxic stress. Nature. 566(7743): 249–253.

Grün D, Lyubimova A, Kester L, Wiebrands K, Basak O, Sasaki N, Clevers H, van Oudenaarden A. (2015). Single-cell messenger RNA sequencing reveals rare intestinal cell types. Nature. 525(7568):251–5.

H.J. Snippert, L.G. van der Flier, T. Sato, J.H. van Es, M. van den Born, C. Kroon-Veenboer, N. Barker, A.M. Klein, J.v. Rheenen, B.D. Simons, H. Clevers. (2010). Cell. 143, 134–144

Heijmans J, van Lidth de Jeude JF, Koo BK, Rosekrans SL, Wielenga MC, van de Wetering M, Ferrante M, Lee AS, Onderwater JJ, Paton JC, Paton AW, Mommaas AM, Kodach LL, Hardwick JC, Hommes DW, Clevers H, Muncan V, van den Brink GR. (2013). ER stress causes rapid loss of intestinal epithelial stemness through activation of the unfolded protein response. Cell Rep. 3(4):1128–39.

Hou Q, Ye L, Liu H, Huang L, Yang Q, Turner JR, Yu Q. (2018). Lactobacillus accelerates ISCs regeneration to protect the integrity of intestinal mucosa through activation of STAT3 signaling pathway induced by LPLs secretion of IL-22. Cell Death Differ. (9):1657–1670.

Hua G, Thin TH, Feldman R, Haimovitz-Friedman A, Clevers H, Fuks Z, Kolesnick R. (2012). Crypt base columnar stem cells in small intestines of mice are radioresistant. Gastroenterology. 143(5):1266–1276.

Igarashi M, Guarente L. (2016). mTORC1 and SIRT1 Cooperate to Foster Expansion of Gut Adult Stem Cells during Calorie Restriction. Cell. 166(2):436–450.

Ireland H, Houghton C, Howard L, Winton DJ. (2005). Cellular inheritance of a Cre-activated reporter gene to determine Paneth cell longevity in the murine small intestine. Dev Dyn. 233(4):1332–1336.

Kaiko GE, Ryu SH, Koues OI, Collins PL, Solnica-Krezel L, Pearce EJ, Pearce EL, Oltz EM, Stappenbeck TS. (2016). The Colonic Crypt Protects Stem Cells from Microbiota-Derived Metabolites. Cell. 165(7):1708–1720.

Karlsson F, Tremaroli V, Nielsen J, Bäckhed F. (2013). Assessing the human gut microbiota in metabolic diseases. Diabetes. 62(10):3341–9.

Kawai T, Akira S. (2011). Toll-like receptors and their crosstalk with other innate receptors in infection and immunity. Immunity. 34(5):637–650.

Kim TH, Escudero S, Shivdasani RA. (2012). Intact function of Lgr5 receptor-expressing intestinal stem cells in the absence of Paneth cells. Proc Natl Acad Sci U S A. 109(10):3932–7.

Kim TH, Li F, Ferreiro-Neira I, Ho LL, Luyten A, Nalapareddy K, Long H, Verzi M, Shivdasani RA. (2014). Broadly permissive intestinal chromatin underlies lateral inhibition and cell plasticity. Nature 506: 511–515.

Korinek V., Barker N., Moerer P., van Donselaar E., Huls G., Peters P.J., Clevers H. (1998). Depletion of epithelial stem-cell compartments in the small intestine of mice lacking Tcf-4. Nat. Genet. 19:379–383.

Lee YS, Kim TY, Kim Y, Lee SH, Kim S, Kang SW, Yang JY, Baek IJ, Sung YH, Park YY, Hwang SW, O E, Kim KS, Liu S, Kamada N, Gao N, Kweon MN. (2018). Microbiota-Derived Lactate Accelerates Intestinal Stem-Cell-Mediated Epithelial Development. Cell Host Microbe. 24(6):833–846.e6.

Levy A, Stedman A, Deutsch E, Donnadieu F, Virgin HW, Sansonetti PJ, Nigro G. (2020). Innate immune receptor NOD2 mediates LGR5+ intestinal stem cell protection against ROS cytotoxicity via mitophagy stimulation. Proc Natl Acad Sci U S A. 117(4):1994–2003.

Lindemans CA, Calafiore M, Mertelsmann AM, O'Connor MH, Dudakov JA, Jenq RR, Velardi E, Young LF, Smith OM, Lawrence G, Ivanov JA, Fu YY, Takashima S, Hua G, Martin ML, O'Rourke KP, Lo YH, Mokry M, Romera-Hernandez M, Cupedo T, Dow L, Nieuwenhuis EE, Shroyer NF, Liu C, Kolesnick R, van den Brink MRM, Hanash AM. (2015). Interleukin-22 promotes intestinal-stem-cell-mediated epithelial regeneration. Nature. 528(7583):560–564.

Metcalfe C, Kljavin NM, Ybarra R, de Sauvage FJ. (2014). Lgr5+ stem cells are indispensable for radiation-induced intestinal regeneration. Cell Stem Cell. 14(2):149–59.

Milano J, McKay J, Dagenais C, Foster-Brown L, Pognan F, Gadient R, Jacobs RT, Zacco A, Greenberg B, Ciaccio PJ. (2004). Modulation of notch processing by gamma-secretase inhibitors causes intestinal goblet cell metaplasia and induction of genes known to specify gut secretory lineage differentiation. Toxicol Sci. 82(1):341–58.

Miyoshi H., Ajima R., Luo C.T., Yamaguchi T.P., Stappenbeck T.S. (2012). Wnt5a potentiates TGF-β signaling to promote colonic crypt regeneration after tissue injury. Science. 338:108–113.

Montgomery RK, Carlone DL, Richmond CA, Farilla L, Kranendonk ME, Henderson DE, Baffour-Awuah NY, Ambruzs DM, Fogli LK, Algra S, Breault DT. (2011). Mouse telomerase reverse transcriptase (mTert) expression marks slowly cycling intestinal stem cells. Proc Natl Acad Sci U S A. 108(1):179–84

Muñoz J, Stange DE, Schepers AG, van de Wetering M, Koo BK, Itzkovitz S, Volckmann R, Kung KS, Koster J, Radulescu S, Myant K, Versteeg R, Sansom OJ, van Es JH, Barker N, van Oudenaarden A, Mohammed S, Heck AJ, Clevers H. (2012). The Lgr5 intestinal stem cell signature:

robust expression of proposed quiescent '+4' cell markers. EMBO J. 31(14):3079–91.

Nigro G, Rossi R, Commere PH, Jay P, Sansonetti PJ. (2014). The cytosolic bacterial peptidoglycan sensor Nod2 affords stem cell protection and links microbes to gut epithelial regeneration. Cell Host Microbe. 15(6):792–798.

Parikh K, Antanaviciute A, Fawkner-Corbett D, Jagielowicz M, Aulicino A, Lagerholm C, Davis S, Kinchen J, Chen HH, Alham NK, Ashley N, Johnson E, Hublitz P, Bao L, Lukomska J, Andev RS, Björklund E, Kessler BM, Fischer R, Goldin R, Koohy H, Simmons A. (2019). Colonic epithelial cell diversity in health and inflammatory bowel disease. Nature. 567(7746):49–55.

Pédron, T., Mulet, C., Dauga, C., Frangeul, L., Chervaux, C., Grompone, G., & Sansonetti, P. J. (2012). A crypt-specific core microbiota resides in the mouse colon. MBio, 3, e00116–e00112.

Pellegrinet L, Rodilla V, Liu Z, Chen S, Koch U, Espinosa L, Kaestner KH, Kopan R, Lewis J, Radtke F. (2011). Dll1- and dll4-mediated notch signaling are required for homeostasis of intestinal stem cells. Gastroenterology 140: 1230–1240.e7.

Pickert G, Neufert C, Leppkes M, Zheng Y, Wittkopf N, Warntjen M, Lehr HA, Hirth S, Weigmann B, Wirtz S, Ouyang W, Neurath MF, Becker C. (2009). STAT3 links IL-22 signaling in intestinal epithelial cells to mucosal wound healing. J Exp Med. 206(7):1465–72.

Potten, C. S., Owen, G. & Booth, D. (2002). Intestinal stem cells protect their genome by selective segregation of template DNA strands. J. Cell Sci. 115, 2381–2388.

Powell AE, Wang Y, Li Y, Poulin EJ, Means AL, Washington MK, Higginbotham JN, Juchheim A, Prasad N, Levy SE, Guo Y, Shyr Y, Aronow BJ, Haigis KM, Franklin JL, Coffey RJ. (2012). The pan-ErbB negative regulator Lrig1 is an intestinal stem cell marker that functions as a tumor suppressor. Cell. 149(1):146–58.

Puzan M, Hosic S, Ghio C, Koppes A. (2018). Enteric Nervous System Regulation of Intestinal Stem Cell Differentiation and Epithelial Monolayer Function. Sci Rep. 8(1):6313

Rakoff-Nahoum S, Paglino J, Eslami-Varzaneh F, Edberg S, Medzhitov R. (2004). Recognition of commensal microflora by toll-like receptors is required for intestinal homeostasis. Cell. 118(2):229–241.

Raveux A, Stedman A, Coqueran S, Vandormael-Pournin S, Owens N, Romagnolo B, Cohen-Tannoudji M. (2020). Compensation between

Wnt-driven tumorigenesis and cellular responses to ribosome biogenesis inhibition in the murine intestinal epithelium. Cell Death Differ. 10.1038/s41418-020-0548-6.

Rodríguez-Colman MJ, Schewe M, Meerlo M, Stigter E, Gerrits J, Pras-Raves M, Sacchetti A, Hornsveld M, Oost KC, Snippert HJ, Verhoeven-Duif N, Fodde R, Burgering BM. (2017). Interplay between metabolic identities in the intestinal crypt supports stem cell function. Nature. 543(7645):424–427.

Roediger WE. (1982). Utilization of nutrients by isolated epithelial cells of the rat colon. Gastroenterology. 83(2):424–429.

Sangiorgi, E. & Capecchi, M. R. (2008). Bmi1 is expressed in vivo in intestinal stem cells. Nat. Genet. 40, 915–920.

Sato T, Stange DE, Ferrante M, Vries RG, Van Es JH, Van den Brink S, Van Houdt WJ, Pronk A, Van Gorp J, Siersema PD, Clevers H. Long-term expansion of epithelial organoids from human colon, adenoma, adenocarcinoma, and Barrett's epithelium. (2011). Gastroenterology. 141(5):1762–72.

Sato T, van Es JH, Snippert HJ, Stange DE, Vries RG, van den Born M, Barker N, Shroyer NF, van de Wetering M, Clevers H. (2011). Paneth cells constitute the niche for Lgr5 stem cells in intestinal crypts. Nature. 469(7330):415–8.

Sato T, Vries RG, Snippert HJ, van de Wetering M, Barker N, Stange DE, van Es JH, Abo A, Kujala P, Peters PJ, Clevers H. (2009). Single Lgr5 stem cells build crypt-villus structures in vitro without a mesenchymal niche. Nature. 459(7244):262–5.

Sekirov I, Russell SL, Antunes LC, Finlay BB. (2010). Gut microbiota in health and disease. Physiol Rev. 90(3):859–904.

Serra D, Mayr U, Boni A, Lukonin I, Rempfler M, Challet Meylan L, Stadler MB, Strnad P, Papasaikas P, Vischi D, Waldt A, Roma G, Liberali P. (2019). Self-organization and symmetry breaking in intestinal organoid development. Nature. 569(7754):66–72.

Shoshkes-Carmel M, Wang YJ, Wangensteen KJ, Tóth B, Kondo A, Massasa EE, Itzkovitz S, Kaestner KH. (2018). Subepithelial telocytes are an important source of Wnts that supports intestinal crypts. Nature. 557(7704):242–246.

Spence JR, Mayhew CN, Rankin SA, Kuhar MF, Vallance JE, Tolle K, Hoskins EE, Kalinichenko VV, Wells SI, Zorn AM, Shroyer NF, Wells JM. (2011). Directed differentiation of human pluripotent stem cells into intestinal tissue in vitro. Nature. 470(7332):105–9.

Stedman A, Beck-Cormier S, Le Bouteiller M, Raveux A, Vandormael-Pournin S, Coqueran S, Lejour V, Jarzebowski L, Toledo F, Robine S, Cohen-Tannoudji M. (2015). Ribosome biogenesis dysfunction leads to p53-mediated apoptosis and goblet cell differentiation of mouse intestinal stem/progenitor cells. Cell Death Differ. (11):1865–76.

Sugimoto S, Ohta Y, Fujii M, Matano M, Shimokawa M, Nanki K, Date S, Nishikori S, Nakazato Y, Nakamura T, Kanai T, Sato T. (2018). Reconstruction of the Human Colon Epithelium In Vivo. Cell Stem Cell. 22(2):171–176.e5.

Takeda N, Jain R, LeBoeuf MR, Wang Q, Lu MM, Epstein JA. (2011). Interconversion between intestinal stem cell populations in distinct niches. Science. 334(6061):1420–1424.

Tetteh PW, Basak O, Farin HF, Wiebrands K, Kretzschmar K, Begthel H, van den Born M, Korving J, de Sauvage F, van Es JH, van Oudenaarden A, Clevers H. (2016). Replacement of Lost Lgr5-Positive Stem Cells through Plasticity of Their Enterocyte-Lineage Daughters. Cell Stem Cell. 18(2):203–13.

Tian H, Biehs B, Warming S, Leong KG, Rangell L, Klein OD, de Sauvage FJ. (2011). A reserve stem cell population in small intestine renders Lgr5-positive cells dispensable. Nature. 478(7368):255–9.

Trentesaux C, Fraudeau M, Pitasi CL, Lemarchand J, Jacques S, Duche A, Letourneur F, Naser E, Bailly K, Schmitt A, Perret C, Romagnolo B. (2020). Essential role for autophagy protein ATG7 in the maintenance of intestinal stem cell integrity. Proc Natl Acad Sci U S A. 117(20):11136–11146.

Van de Wetering M, Francies HE, Francis JM, Bounova G, Iorio F, Pronk A, van Houdt W, van Gorp J, Taylor-Weiner A, Kester L, McLaren-Douglas A, Blokker J, Jaksani S, Bartfeld S, Volckman R, van Sluis P, Li VS, Seepo S, Sekhar Pedamallu C, Cibulskis K, Carter SL, McKenna A, Lawrence MS, Lichtenstein L, Stewart C, Koster J, Versteeg R, van Oudenaarden A, Saez-Rodriguez J, Vries RG, Getz G, Wessels L, Stratton MR, McDermott U, Meyerson M, Garnett MJ, Clevers H. (2015). Prospective derivation of a living organoid biobank of colorectal cancer patients. Cell. 161(4):933–45.

Van Es JH, Haegebarth A, Kujala P, Itzkovitz S, Koo BK, Boj SF, Korving J, van den Born M, van Oudenaarden A, Robine S, Clevers H. (2012). A critical role for the Wnt effector Tcf4 in adult intestinal homeostatic self-renewal. Mol Cell Biol. (10):1918–27.

Van Es JH, Sato T, van de Wetering M, Lyubimova A, Yee Nee AN, Gregorieff A, Sasaki N, Zeinstra L, van den Born M, Korving J, Martens ACM, Barker N, van Oudenaarden A, Clevers H. (2012). Dll1+ secretory progenitor cells revert to stem cells upon crypt damage. Nat Cell Biol. (10):1099–1104.

Van Es JH, van Gijn ME, Riccio O, van den Born M, Vooijs M, Begthel H, Cozijnsen M, Robine S, Winton DJ, Radtke F, Clevers H. (2005). Notch/gamma-secretase inhibition turns proliferative cells in intestinal crypts and adenomas into goblet cells. Nature. 435(7044):959–63.

VanDussen KL, Carulli AJ, Keeley TM, Patel SR, Puthoff BJ, Magness ST, Tran IT, Maillard I, Siebel C, Kolterud Å, Grosse AS, Gumucio DL, Ernst SA, Tsai YH, Dempsey PJ, Samuelson LC. (2012). Notch signaling modulates proliferation and differentiation of intestinal crypt base columnar stem cells. Development. 139(3):488–97.

Vonaesch P, Anderson M, Sansonetti PJ. Pathogens, microbiome and the host: emergence of the ecological Koch's postulates. (2018). FEMS Microbiol Rev. 42(3):273–292.

Yan KS, Gevaert O, Zheng GXY, Anchang B, Probert CS, Larkin KA, Davies PS, Cheng ZF, Kaddis JS, Han A, Roelf K, Calderon RI, Cynn E, Hu X, Mandleywala K, Wilhelmy J, Grimes SM, Corney DC, Boutet SC, Terry JM, Belgrader P, Ziraldo SB, Mikkelsen TS, Wang F, von Furstenberg RJ, Smith NR, Chandrakesan P, May R, Chrissy MAS, Jain R, Cartwright CA, Niland JC, Hong YK, Carrington J, Breault DT, Epstein J, Houchen CW, Lynch JP, Martin MG, Plevritis SK, Curtis C, Ji HP, Li L, Henning SJ, Wong MH, Kuo CJ. (2017). Intestinal Enteroendocrine Lineage Cells Possess Homeostatic and Injury-Inducible Stem Cell Activity. Cell Stem Cell. 21(1):78–90.e6.

Yilmaz ÖH, Katajisto P, Lamming DW, Gültekin Y, Bauer-Rowe KE, Sengupta S, Birsoy K, Dursun A, Yilmaz VO, Selig M, Nielsen GP, Mino-Kenudson M, Zukerberg LR, Bhan AK, Deshpande V, Sabatini DM. (2012). mTORC1 in the Paneth cell niche couples intestinal stem-cell function to calorie intake. Nature. 486(7404):490–5.

Yu DH, Gadkari M, Zhou Q, Yu S, Gao N, Guan Y, Schady D, Roshan TN, Chen MH, Laritsky E, Ge Z, Wang H, Chen R, Westwater C, Bry L, Waterland RA, Moriarty C, Hwang C, Swennes AG, Moore SR, Shen L. (2015). Postnatal epigenetic regulation of intestinal stem cells requires DNA methylation and is guided by the microbiome. Genome Biol. 16:211.

Yu S, Tong K, Zhao Y, Balasubramanian I, Yap GS, Ferraris RP, Bonder EM, Verzi MP, Gao N. (2018). Paneth Cell Multipotency Induced by Notch Activation following Injury. Cell Stem Cell. 23(1):46–59.e5.

Yui S, Azzolin L, Maimets M, Pedersen MT, Fordham RP, Hansen SL, Larsen HL, Guiu J, Alves MRP, Rundsten CF, Johansen JV, Li Y, Madsen CD, Nakamura T, Watanabe M, Nielsen OH, Schweiger PJ, Piccolo S, Jensen KB. (2018). YAP/TAZ-Dependent Reprogramming of Colonic Epithelium Links ECM Remodeling to Tissue Regeneration. Cell Stem Cell. 22(1):35–49.e7.

Yui S, Nakamura T, Sato T, Nemoto Y, Mizutani T, Zheng X, Ichinose S, Nagaishi T, Okamoto R, Tsuchiya K, Clevers H, Watanabe M. (2012). Functional engraftment of colon epithelium expanded in vitro from a single adult Lgr5? stem cell. Nat Med. 18(4):618–23.

Zou WY, Blutt SE, Zeng XL, Chen MS, Lo YH, Castillo-Azofeifa D, Klein OD, Shroyer NF, Donowitz M, Estes MK. (2018). Epithelial WNT Ligands Are Essential Drivers of Intestinal Stem Cell Activation. Cell Rep. 22(4):1003–1015.

18

Neural Stem Cells

Nathalie Kubis[1] and Martin Catala[2,*]

[1]Université de Paris, INSERM U1148, Laboratory for Vascular Translational Science, F- 75018 Paris, France and Service de Physiologie Clinique – Explorations Fonctionnelles, DMU DREAM, APHP, Hôpital Lariboisière, F-75010 Paris, France
[2]Sorbonne Université, CNRS, Inserm U1156, Institut de Biologie Paris Seine, Laboratoire de Biologie du Développement/UMR7622, 9 Quai St-Bernard, 75005 Paris, France
E-mail: martin.catala@sorbonne-universite.fr
*Corresponding Author

18.1 Introduction

The central nervous system (CNS) consists of two distinct compartments, the gray and white matters. Neuronal cell bodies lie only in the gray matter whereas myelinated processes of neurons, namely the axons, travel mainly within the white matter. Glial cells (astrocytes, oligodendrocytes, and microglial cells) are distributed into these two compartments. Eventually, the ventricular cavities containing the cerebrospinal fluid (CSF) are limited by a highly specialized epithelium, the ependyma. Mitoses are observed predominantly before birth. However, even in the adult, few mitotic figures can be evidenced (Allen, 1912), although the fate of these dividing cells has remained obscure for a long time. At the end of the 19th century, tissues were divided into three groups according to Giulio Bizzozero (1846–1901). During a lecture presented in the XIth Medical Congress in Rome (Bizzozero 1894), this famous Italian professor of general pathology distinguished (i) tissues for which cells divide all along the life of the organisms, (ii) tissues for which cells are post-mitotic after birth but can re-enter cell division after injury, and (iii) tissues for which no cell division can be observed. As examples

461

of the third category, Bizzozero mentioned the striated muscles and nervous tissues. However, his conclusions were not so firmly established since he wrote (page 729): "After this period [embryonic life] I do not believe—until the contrary is proved—that in mammals either the number of nerve cells or of striped muscular fibres can increase." Moreover, Santiago Ramón y Cajal added: "Once development was ended, the fonts of growth and regeneration of the axons and dendrites dried up irrevocably. In adult centers, the nerve paths are something fixed and immutable: everything may die, nothing may be regenerated. It is for the science of the future to change, if possible, this harsh decree." (See Colucci-D'amato et al., 2006 for a review). These features are known as the central dogma of neurobiology. This dogma was first challenged by Joseph Altman in 1962, who claimed that although mitosis could not be evidenced in neurons, postnatal neurogenesis could not be excluded since: "...new neurons might arise from nondifferentiated precursors, such as ependymal cells." The dogma was definitively shattered in the 1990s with the discovery of stem cells capable of producing new neurons in the CNS of vertebrates including humans. We chose to address the following points:

– How to identify postnatal neurogenesis in mammals?
– Which are the main regions that host these stem cells and what is their histological organization?
– What is the ecology of these neural stem cells?

18.2 Evidencing Postnatal Neurogenesis in Mammals

18.2.1 The S Phase

A neuron is a post-mitotic cell, which differentiates as the result of an ultimate mitosis. Cell division or mitosis is the last phase of the cell cycle. The latter consists of four distinct phases: the first interphasic period (G1), the synthesis phase (S) in which DNA is duplicated, the second interphasic period (G2), and the M phase in which the cell divides into two daughter cells. It is important to note that a cell can divide only if the mother cell has duplicated its DNA. If a substance that integrates into the DNA and able to be detected can be injected, cells that have passed through an S phase can be tagged and identified. If the animal is killed early after injection of the tracer, all the cells that have passed through an S phase will be labeled. But, if the time of sacrifice is very long after injection, the cells that continue to divide will dilute their marker in their DNA because of repeated S phases. If the delay is long enough, it will not be possible to distinguish these cells from the background. In contrast, if a cell divides after receiving the injection of a tracer that has incorporated into its DNA and leaves immediately the cell cycle, the concentration of the marker will remain stable and be evidenced. This method allows marking specifically cells that divide and differentiate into post-mitotic neurons. This technique has revealed that neurogenesis occurred in the CNS and consequently demonstrated retrospectively the presence of precursor cells that can generate neurons.

This method (and examples will be given in the next paragraph) assumes that the cells that will be labeled undergo an S phase and therefore they are included in the cell cycle. All stem cells exist in a quiescent state (outside the cell cycle) and once they get activated, they enter the cell cycle either leading to self-renewal or are engaged in cell differentiation. This "quiescence" property has also been demonstrated for neural stem cells (Codega et al., 2014; Mich et al., 2014). Quiescent cells are by definition outside the cell cycle and do not undergo DNA synthesis. Consequently, they will not be marked by these techniques. It is therefore advisable to keep in mind this technical limit and to realize that some true stem cells will escape individualization after using these techniques.

18.2.2 Markers of Neurogenesis

Thymidine labeled with tritium, a radioactive isotope, is integrated into DNA during S phase like nonradioactive thymidine. After killing the animal, the histological sections are revealed by autoradiography. Joseph Altman uses this technique for the first time to demonstrate neurogenesis in rats. In 1962, he showed a local production of neurons after stereotactic electrical lesions of the lateral geniculate nucleus (Altman, 1962). He will then evidence physiological neurogenesis in the rat hippocampus (Altman, 1965) and olfactory bulb (Altman, 1969).

In the early 1980s, Fernando Nottebohm and his group studied neurogenesis in some birds. Since female canaries start to sing like males after injection of male hormones, Goldman and Nottebohm (1984) analysed the vocal control nucleus in treated female canaries and showed production of neurons that incorporate into neural circuits and that are functional (Paton and Nottebohm, 1985). In the canary, such a neurogenesis takes place spontaneously with seasonal cycles in parallel with behavioral changes in singing well known in this species (Alvarez-Buylla et al., 1990). These results gained in birds led Nottebohm to speculate about potential functional neurogenesis in humans (1985).

After Nottebohm's suggestion, Pasko Rakic (1985) conducted an experiment in rhesus monkeys aged from 6 months to 11 years using tritiated thymidine. He concluded that there is no postnatal neurogenesis in primates (1985). Thus, from Rakic's data, a seasonal neurogenesis was considered specific of birds, and absence of neurogenesis was suggested for mammals. However, the autoradiographic technique used by Rakic had some limitations inasmuch it could not confirm the exact nature of the cells that were produced, since it was not possible to couple autoradiography and immunohistochemistry.

This technical problem is removed by the use of another base analogue of DNA, bromodeoxyuridine (BrdU). BrdU is incorporated in DNA during S phase and can be detected thanks to anti BrdU antibodies. This leads to the possibility to perform double or triple marking on the same section and to adequately characterize the cell that has been produced by mitosis.

The use of BrdU allowed to demonstrate postnatal neurogenesis in the rat (Kuhn et al., 1996) or mouse hippocampus (Kempermann et al., 1997). The first demonstration of postnatal neurogenesis in a primate was performed in humans! Eriksson et al. (1998) took advantage of BrdU injection to patients with cancer. At that time, the injection of the tracer allowed the study of the kinetics of the tumoral cells. The authors used an injection performed for both diagnostic and prognostic purposes.Autopsy was performed in patients who died 58–72 days after injection, and their brain was scrutinized for BrdU-positive cell, evidencing that labeled cells were born after the injection. By combining immunostaining against BrdU with various markers of cell differentiation, they were able to assign an identity to the observed structures.Double positive cells with neuronal markers such as NeuN, calbindin (for specific types of neurons), or specific neuron enolase, were detected in the subventricular zone and in the dentate gyrus of the hippocampus.

Such an outcome in humans has led to reconsider conventional data obtained in monkeys by Rakic. Using BrdU Rakic himself (Kornack and Rakic, 1999) showed neurogenesis in the dentate gyrus of the rhesus monkey, and postnatal neurogenesis which was until then considered as absent in both primates and humans,was then established. Moreover, this neurogenesis runs continuously throughout the seasons of life contrary to the data obtained in birds by Nottebohm and his team.Later on, neurogenesis was described in the neocortex of primates (Gould et al., 1999; 2001). These controversies regarding the existence of neurogenesis in different brain regions are probably explained by differences in technical methods used in those studies.

The possibility of determining the birth date of neurons in humans has allowed answering these controversies as we will develop below.

18.2.3 The Bomb and the Brain

The concentration of isotope 14 of carbon (^{14}C) in the Earth's atmosphere has remained stable for very long periods of time until the use of nuclear weapons during World War II and atmospheric nuclear tests that were banned from 1963 onwards. Thus, there was an increase in the rate of ^{14}C in the atmosphere from mid-1950s to mid-1960s followed by a gradual fall. Currently, the atmospheric concentration of ^{14}C has not yet returned to baseline. Because carbon is an essential component of all biological molecules including DNA, measuring ^{14}C in the cell nucleus can be used to determine the age of neurons in cadavers born after 1955. Indeed, if no postnatal neurogenesis occurs, all neurons are born during the nine months of gestation and their ^{14}C content will correspond to that of the atmospheric concentration of the birth year. A difference in concentration would indicate that postnatal neurogenesis has taken place. This was used by Jonas Frisén's group of the Karolinska Institute in Stockholm (Spalding et al., 2005) to confirm or rule out the presence of stem cells in different brain regions. They studied the brain of subjects deceased from non-neurological causes. Different brain regions were dissected, the nervous tissue was dissociated and neurons were sorted out in order to measure the content of ^{14}C in their nuclei. Cortical neurons of the occipital lobe bear the same age as the studied subject demonstrating the absence of postnatal neurogenesis (Spalding et al., 2005), as for all the other cortical areas (Bhardwaj et al., 2006; Ernst et al., 2014). However, for technical reasons, the authors could not exclude a transitional neurogenesis consisting of neurons whose life would be less than 4.2 months (Bhardwaj et al., 2006). Anyway, a functional cortical neurogenesis in humans was excluded. Very surprisingly, Bergmann et al. (2012) by measuring the date of birth of neurons in the human olfactory bulb showed that it coincided with the age of the subject. Thus, in humans, there is no postnatal olfactory neurogenesis, which is a major difference with the other species studied so far. By contrast, hippocampal neurogenesis does occur in humans (Spalding et al., 2013) as in rodents and primates. The dynamics of this cell production is estimated at 700 new neurons per day (Spalding et al., 2013). These results confirm that the hippocampal neurogenesis is not a trivial phenomenon even in humans. Finally, neurogenesis is also present in the lateral wall of the lateral human ventricles (Ernst et al., 2014) as in rodents. But these newly formed cells migrate to generate interneurons (mainly expressing calretinin) of the striatum (caudate nucleus and putamen) in human controls (Ernst et al., 2014).

18.2.4 *In Vitro* Culture for Neural Stem Cells

One of the most widespread technic to evidence stem cells in the CNS is *in vitro* culture. Since this technique has played a major role in the concept of neural stem cells, we will describe the first work published by Brent A. Reynolds and Samuel Weis (1992) (Neuroscience Research Group, Calgary, Canada). These authors dissected striata from adult mice aged 3 to 18 months. These brain regions were dissociated using an enzyme and the isolated cells were cultured on a non-adhesive substrate in serum-free culture medium containing 20 ng/mL of EGF. After two days of culture, only 15 out of 1000 cells survive and proliferate. After 6 to 8 days of culture, the surviving and dividing cells detach and form spheres, which almost exclusively contain cells expressing nestin. These cells express neither neuronal nor glial markers. If a neurosphere is dissociated and the isolated cells are returned to culture under the same conditions, new neurospheres can be generated. On the other hand, if a neurosphere is placed on poly-L-ornithine–coated glass cover slips (i.e. adhesive substrate), the cells differentiate gradually. Twenty-one days later, some cells express neuronal (neuron-specific enolase, gamma aminobutyric acid, substance P) or glial markers (GFAP). These two conditions indicate that the original cells undergo (i) self-renewal and (ii) multipotent differentiation.

This technique has become very popular to evidence neural stem cells (Chojnacki et al., 2009). A Medline query found around 3000 articles recognized by the keyword "neurosphere" (1995–early 2021). However, if we analyze the methodology of these different studies, we see the lack of standardization of protocols. Thus, the culture medium may contain only EGF, bFGF (Fibroblast Growth Factor), or combine different factors among which EGF, bFGF, FGF2, LIF (Leukemia Inhibitory Factor), PDGF (Platelet Derived Growth Factor), and NSF (Neural Survival Factor) (Chaichana et al., 2006). Likewise, hormones (insulin, progesterone) can be added to the culture medium (Chaichana et al., 2006). The difference in the composition of the culture medium can obviously lead to disparate or even conflicting results (Chaichana et al., 2006; Jensen and Parmar, 2006; Chojnacki et al., 2009). For example, by combining the action of EGF with that of bFGF, Weiss et al. (1996) observed neurospheres produced from the entire rostro-caudal axis of the central nervous system of adult mice. They concluded that neural stem cells are scattered along this axis. We now know that some of the regions involved are devoid of stem cells rendering some of the previous conclusions obsolete.

Several other problems weaken conclusions after using this technique of neurosphere assay:

– EGF can stimulate proliferating precursors that are not quiescent stem cells to generate neurospheres (Doetsch et al., 2002). Thus, the observation of the latter is not sufficient to assess the existence of stem cells.

– A neurosphere is supposed to reflect the development of a single stem cell. Thus, such a sphere should be generated by a single cell. However, when the density of neurospheres is high, they tend to aggregate to form a single cell mass (Mori et al., 2007; Singec et al., 2006). Moreover, this aggregation is favored by the manipulations of the experimenter when one examines the cultures under a microscope (Coles-Takabe et al., 2008). It is obvious that if a cell mass is not clonal, the conclusions obtained by the experiment cannot be used to reflect the existence of a single stem cell.

– Another problem associated with these *in vitro* techniques is that quiescent stem cells do not generate neurospheres (Codega et al., 2014; Mich et al., 2014). So, the absence of neurospheres does not allow to conclude to the absence of stem cells.

These intrinsic problems of *in vitro* techniques have fueled controversies that are still on going.

18.3 Neural Stem Cells Lie in Specific CNS Regions

The second aspect that we wish to address in this chapter is the histological organization of the CNS regions that contain stem cells. Four regions have been studied so far: the subventricular zone of the hemispheres, the subcallosal zone, the subgranular zone of the hippocampus, and the spinal cord.

18.3.1 Stem Cells in the Subventricular Zone

The ventricles are central cavities of the CNS containing cerebrospinal fluid. One can distinguish four ventricles in the brain that continue through the central canal of the spinal cord. The cerebral hemispheres contain the two lateral ventricles, the latter communicating with the diencephalic third ventricle through Monro's foramina. The fourth ventricle lies at the level of the hindbrain. Its communication with the third ventricle is achieved by the mesencephalic aqueduct. The cerebrospinal fluid leaves the ventricular system through the roof of the fourth ventricle to diffuse in the meningeal spaces.

18.3.1.1 Mouse brain

The lateral wall of the hemispherical ventricles located at the level of the striatum contains a rich stem cell zone, the subependymal layer or subventricular zone (Figure 18.1A,B). It is worth noting that the medial wall of the lateral ventricles is devoid of such a potential. The histology of the subventricular zone of the lateral ventricle wall was particularly studied in the mouse. The most apical layer (namely the one bordering the ventricles) is composed of ependymal cells (Doetsch et al., 1997) and cytoplasmic processes from cells expressing GFAP and called B1 cells (Mirzadeh et al., 2008) (Figure 18.2A,B). Ependymal cells belong to two types, E1 cells that are multiciliated while E2 cells, which are smaller and fewer, have only two cilia (Mirzadeh et al., 2008). All the cilia of ependymal cells are motile. Their axis is composed of nine peripheral microtubule doublets and a central pair (the so-called 9+2 cilium). Ventricular surface of the B1cells is very narrow, 11 times more than that of E1 cells (Mirzdaeh et al., 2008). B1 cells are isolated or grouped in clusters of up to 40 cells (Mirzadeh et al., 2008). Ependymal cells located around such a cluster look like a rosette making a figure reminiscent of a pinwheel according to Mirzadeh et al. (2008) (Figure 18.2B). The ventricular expansion of B1 cells bears a primary cilium in which the axis is composed of 9 peripheral microtubule doublets and no central ones (the so-called 9+0 cilium) (Figure 18.2A,B). Beneath the ependymal layer, multiple cell types are found. GFAP expressing cells are called B cells and show two histological features (Doetsch et al., 1997). First, B1 cells, whose extensions penetrate between ependymal cells as we have described, possess a long basal extension that makes them resembling embryonic radial glia cells and that terminates at the surface of blood vessels (Mirzadeh et al., 2008). B2 cells are smaller than B1 cells, display a multipolar appearance, are devoid of apical ventricular process and possess along basal extension forming an astrocytic foot on the vessel wall (Doetsch et al., 1997; Mirzadeh et al., 2008). C cells have an indented nucleus, are larger and have higher electron-density than B cells (Doetsch et al., 1997). Another cell type is represented by groups of cells forming electron-dense chains or A cells (Doestch et al., 1997). They express PSA-NCAM and Tuj1 and are actually considered as immature neurons (Doestch et al., 1997).

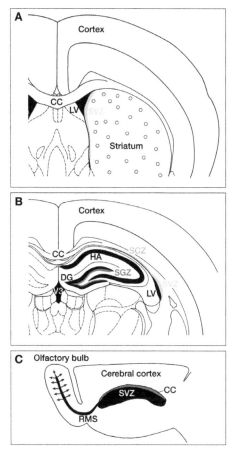

Figure 18.1 Anatomical localization of the cerebral zones containing neural stem cells. (**A**) Coronal section through the corpus callosum (CC), the cerebral cortex, and the striatum. The subventricular zone (SVZ) is located at the lateral margin of the lateral ventricle (LV). (**B**) Coronal section through the CC, cerebral cortex, and hippocampus. The subgranular zone (SGZ) is located in the internal part of the dentate gyrus (DG) of the hippocampus. The SVZ lies laterally to the LV. The subcallosal zone (SCZ) occupies the region between Ammon's horn (AH) and CC. V3 = third ventricle. (**C**) Para-sagittal section through the cerebral hemisphere and the olfactory bulb. The SVZ lies underneath the CC and extends to the olfactory bulb via the Rostral Migratory Stream (RMS).

The general organization of the subventricular region shows chains of A cells occupying the central region. This chain is surrounded by the extensions of B cells. C cells are less numerous and partially covered by the extensions of B cells (Doestch et al., 1997). Chains of A cells connect the Rostral Migratory Stream (RMS), a cellular structure that joins the ventricular

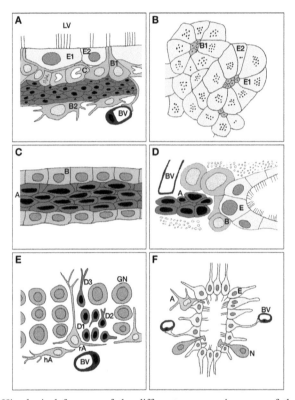

Figure 18.2   Histological features of the different neurogenic zones of the mouse brain.
(**A**) The subventricular zone corresponds to the lateral margin of the LV. This zone is limited
by ependymal cells (E1 and E2). E1 are multiciliated cells whereas E2 contains only two cilia.
Three types of cell compose the neurogenic zone. B cells can be subdivided into two major
groups: B1 cells are in close connection with ependymal cells and send a long ciliated process
that contacts the ventricular lumen. B2 cells are located at the interface with the striatum. C
cells are large cells separated from the ependymal by B1 cells processes. A cells (**red**) are
arranged as cellular chains in the center of the zone. BV = blood vessel. (**B**) Ventricular view
of the ependymal layer. E1 and E2 cells form a sheet and they organize as pinwheel around
the ventricular processes of B1 cells. (**C**) The RMS is only composed by B cells that surround
a central mass of migrating A cells. (**D**) In the subcallosal zone, the organization is similar to
that of the SVZ. However, C cells are very infrequent and B cells do not enwrap completely
A cells. (**E**) The subgranular zone is located beneath the granular zone composed by granular
neurons (GN). This zone contains astrocytes that behave either as horizontal processes (hA)
or radial processes (rA). Astrocytes give rise to D1 cells that differentiate into D2, D3, and
then GN. (**F**) The central canal of the mouse spinal cord. Ependymal cells (E) contain 1 to 4
cilia and their basal process allows to distinguish some cells that connect blood vessels (BV)
and cells of the ventral and dorsal poles in which the process is very long. A few astrocytes
(A) possess a cytoplasmic-ciliated process that contacts the lumen of the canal. A few neurons
display also processes that project into the lumen.

pole to the olfactory bulb (Lois and Alvarez- Buylla, 1994; Lois et al., 1996) (Figure 18.1C). This stream consists of an axis of A cells surrounded by a sleeve of B cells (Lois et al., 1996) (Figure 18.2C). The RMS is not only an area devoted to migration but also contains neural stem cells (Gritti et al., 2002).

The cells of the subventricular zone give rise to glial cells and neurons that populate the olfactory bulb (Lois and Alvarez-Buylla, 1993; 1994). When mice are treated with an anti-mitotic substance, namely cystosine-beta-D-arabinofuranoside, A and C cells die whereas B cells resist (Doetsch et al., 1999 a and b). If these treated mice survive a few days after stopping this antimitotic treatment, C cells are the first to be restored before A cells (Doestch et al., 1999a and b). These results suggest that B cells are able to regenerate the set of cells in the subventricular zone in the temporal sequence B > C > A. B1 cells divide by symmetrical division; they can self-renew or give rise to identical daughter cells (B2 and/or C) (Obernier et al., 2018). C cells are therefore considered as transit amplifying cells. In order to be certain that B cells are stem cells in this area, Doetsch et al. (1999b) used a transgenic mouse expressing the receptor for avian leukemia under the control of the GFAP promoter. Thus, in this mouse, only B cells express this receptor in the subventricular zone and are prone to infection by this type of virus. Then Doestch et al. (1999b) constructed a defective avian virus (i.e. a virus able to infect a cell but unable to diffuse from cell to cell) containing a reporter gene. By injecting this virus into the ventricles, Doestch et al. (1999b) infected only B1 cells of the subventricular zone. After infection, the cells in which the virus is integrated divide and generate cells that stay in the subventricular zone and also cells that become olfactory neurons (Doetsch et al., 1999). This result indicates that B1 cells are the stem cells of the subventricular zone.

B1 cells can be studied *in vitro* after isolating them using different molecular markers. This technique makes it possible to demonstrate heterogeneity within this cell population which therefore co-exists in two distinct states: quiescent and activated (Codega et al., 2014; Mich et al., 2014). Quiescent cells could not form neurospheres and are mitotically inactive in contrast to activated cells that display the exact contrary (Codega et al., 2014; Mich et al., 2014). Quiescent cells do not express EGF-receptor whereas activated cells express it (Codega et al., 2014; Mich et al., 2014; Llorens-Bobadilla et al., 2015). Single-cell transcriptomic analysis provides valuable information on the genes expressed by a single cell and helps to point differences between quiescent and activated stem cells. It is of course outside the scope of this chapter to list the differentially expressed genes in these two populations. However, we must mention that a recent study has shown more heterogeneity within quiescent cells, which therefore do not constitute one but at least four subgroups (Xie et al., 2020).

The subventricular zone in mouse is not a homogeneous structure but it is regionalized. Indeed, Merkle et al. (2007) identified specific regions in this area using a stereotaxic approach. For that purpose, they injected adenoviruses expressing Cre enzyme in Z/EG mice at birth. This mouse line is genetically modified in such way that, in presence of Cre, recombination takes place allowing production of the reporter protein, GFP (Green Fluorescent Protein). This recombination is stable and inherited by all the cells that are derived from the initial recombinated cells. Thus, it is possible to follow the progeny of these initial cells. The olfactory bulb is formed by different types of neurons. If we follow the progeny of stem cells present in different subregions of the subventricular zone, we find that each region gives rise to a specific type of olfactory neuron (Merkle et al., 2007). In addition, if stem cells from one region are transplanted into another region of the subventricular zone, the transplanted cells retain their own differentiation potentials and do not change their fate according to their new environment (Merkle et al., 2007). Deciphering the code of this heterogeneity is a very active field of research (Fiorelli et al., 2015; Poiana et al., 2020).

18.3.1.2 Human brain

The study of the subventricular region of the human being has led to histological surprises. From birth to 6 months, its appearance is similar to the one that is described in mice (Sanai et al., 2011). Then, from 6 to 18 months, there is a gradual decrease of the number of the cells of this layer (Sanai et al., 2011); so thereafter, the subventricular zone of human takes a radically different appearance from that of rodent adults. In human adults, indeed, the ependymal layer is based on a pauci-cellular layer (the so-called gap zone) (Sanai et al., 2004; Quiñones-Hinojosa et al., 2006; Sanai et al., 2011). A dense cell ribbon consisting of cell bodies of astrocytes is present under this gap zone (Sanai et al., 2004; Quiñones-Hinojosa et al., 2006; Sanai et al., 2011). Ki67, a cellular marker for cell proliferation, labels some astrocytes suggesting that they are integrated into the cell cycle (Sanai et al., 2004). When placed in culture, about 3% of astrocytes produce multipotent neurospheres (Sanai et al., 2004). Some neuroblasts are identified in the subventricular zone of the human adult but they never form chains of migrating cells contrary to what is commonly observed in the mouse (Sanai et al., 2004). There are regional variations in the structure of the subventricular zone according to the ventrodorsal and anteroposterior axes in humans (Quiñones-Hinojosa et al., 2006). We will not describe them any further in details. The use of iododeoxyuridine (which is incorporated in DNA during the S phase like BrdU) in patients demonstrated the presence of labeled neurons lying in the caudate nucleus and putamen (i.e. striatum) (Ernst et al., 2014). In this CNS structure, the study of carbon 14 content of neuronal nuclei shows that striatal interneurons are renewed during life, whereas projection medium spiny neurons constitute a population that is set up before birth (Ernst et al., 2014). In conclusion, cells located in the subventricular zone in humans produce striatal neurons unlike murine cells which generate olfactory neurons. This important difference between species should invite our readers to remain cautious and to avoid any generalization of murine results to humans.

The RMS in the human fetus is particularly developed (Wang et al., 2011). A cellular structure reminiscent of RMS, can be demonstrated in adults by Nissl staining or by the use of an anti-GFAP antibody (Sanai et al., 2004; Wang et al., 2011). However, few or no neuroblasts in the RMS are shown in contrast to what is observed in mice (Sanai et al., 2011; Wang et al., 2011). Neuroblasts never form chains of migration but appear isolated or paired (Wang et al., 2011). A supplementary tract of migrating cells was further demonstrated in the human being before the age of 18 months. This tract called Medial Migratory Stream (MMS) bifurcates from the proximal region of the RMS and goes to the ventromedial prefrontal cortex (Sanai et al., 2011). Such a structure has never been evidenced in other species but humans. Its functional significance is still unknown.

18.3.2 Stem Cells in the Subcallosal Zone

A second region containing stem cells was demonstrated in the adult rodent brain. It lies between the hippocampus and corpus callosum and is located around the caudal and dorsomedial extension of the lateral ventricles. This zone is called the subcallosal area (Seri et al., 2006) (Figure 18.1B). Histologically, this region consists of epithelial cavities separated from each other and bounded by ependymal multiciliated cells (Seri et al., 2006). The same cell components as in the ventricular zone are found (Figure 18.2D), namely B stem cells, transit amplifying C cells, and A cells (Seri et al., 2006). However, several features should be noted that contrast with the subventricular zone: (i) C cells are less numerous, (ii) B cells do not completely surround A cell clusters that may come into contact with axons or ependymal cells

(Seri et al., 2006), (iii) cells in this area give rise to oligodendrocytes that colonize the corpus callosum, and last but not least (iv) never give rise to neurons *in vivo* (Seri et al., 2006). In contrast, these cells when cultured *in vitro* can differentiate into neurons (Seri et al., 2006).

18.3.3 Stem Cells in the Dentate Gyrus of the Hippocampus

The hippocampus is a structure that belongs to the CNS limbic system (Felten et al., 2016). It consists of two radically different regions: the dentate gyrus and Ammon's horn, forming a three-layer cortex. The dentate gyrus contains granule neurons that project to the third field of Ammon's horn (CA3). Ammon's horn is constituted by projection pyramidal neurons whose axonal extensions contact targets that are remote from the hippocampus. Hippocampus plays a major role in memory processes. As we wrote already, the dentate gyrus is the site of continuing neurogenesis in mammals including humans (Figure 18.1B). This structure is formed of several layers of cells: (i) the hilus, the innermost region that contains few cells, (ii) the subgranular zone that is the site of stem or progenitor cells, previously included in the (iii) the granular layer, formed of functional neurons and finally (iv) the most superficial layer that is called the molecular cells. The following histological description of dentate gyrus is that of Alvarez-Buylla. There are three types of cells: astrocytes, D cells, and granular neurons (Figure 18.2E). Subgranular astrocytes express GFAP (Seri et al., 2001; 2004), divide (Seri et al., 2001), give rise to mature neurons of the granular layer (Seri et al., 2001), are partially resistant to antimitotic treatment (Seri et al., 2001), and restore the subgranular zone after such treatment is stopped (Seri et al., 2001). Thus, astrocytes play probably the role of stem cells of the hippocampus. Two histological forms of astrocytes can be described: the radial astrocytes whose extensions mingle with the granular layer and horizontal astrocytes whose processes orient parallel to the granular layer (Seri et al., 2004). It is currently unknown whether these morphological differences reflect a functional one. D cells are produced by astrocytes and form clusters of progenitors whose maturity progressively increases (Seri et al., 2004). According to their shape, one can distinguish D1 cells that are round or ovoid, D2 cells bearing a short process,and D3 cells that are immature granule cells (Seri et al., 2004) (Figure 18.2E). At the end of the ripening process, the D3 cells are indistinguishable from granule neurons. All these cells are intimately linked to vascular structures of the hippocampus (Palmer et al., 2000).

A recent study of hippocampal neurogenesis carried out on human brains has led to some very surprising results. The authors (Sorrells et al., 2018), using different markers by immunofluorescence, concluded that no neurogenesis takes place in hippocampus after the age of 18 in humans. With a different methodology, Boldrini et al., (2018) come to a diametrically opposite conclusion. For them, neurogenesis is continuous in the human hippocampus during the whole life, but is less abundant with ageing. This decrease is linked with a depletion in the pool of quiescent stem cells. This scientific controversy has led to many reactions. For some authors (Kempermann et al., 2018; Lucassen et al., 2020), the absence of neurogenesis observed by Sorrells et al. (2018) is an artefact linked to technical conditions. In order to avoid the problems inherent in post-mortem studies, Seki et al. (2019) studied hippocampus removed surgically either from patients with refractory epilepsy or from subjects operated upon for a brain tumor or a vascular malformation in this region. They showed that a limited neurogenesis does exist in all subjects whether or not they had epilepsy. Epileptic patients display an abnormal appearance of their hippocampus limiting the conclusions gained by their study. But, in the other subjects who have been studied, the presence of numerous immature neurons suggests that their half-life is long and that their maturation is protracted. It is interesting to note that a decrease in hippocampal neurogenesis is observed in all mammalian species studied (mice, non-human primates, and humans) (Seki, 2020). This decline occurs always at the same age (2–3 years of life) whatever the species and whatever its lifespan (Seki, 2020). This result suggests the existence of a clock that gauges ageing. This controversy on hippocampal neurogenesis in humans shows the importance of technical procedures in studies. The recent results invite to continue fundamental researches in order to try understanding the mechanisms leading to the maturational delay of these neuroblasts.

18.3.4 Neural Stem Cells in the Spinal Cord

The problem of stem cells in the spinal cord is more complex. Weiss *et al.* (1996) have demonstrated the presence of spinal cord cells able to generate neurospheres with both neuronal and glial differentiations *in vitro*. These cells self-renew (Weiss et al., 1996) and therefore display the characteristics of multipotent stem cells. Neurospheres from spinal cord can be generated in mouse (Weis et al., 1996), rat (Shihabuddin et al., 1997), and human (Dromart et al., 2008). However, we have described in the part dedicated to the neurosphere technique, that this criterion is not sufficient to confirm the existence of stem cells acting *in vivo*. Indeed, there is no post-natal neurogenesis in the rat spinal cord (Horner et al., 2000), but present proliferating cells can give rise to both astrocytes and oligodendrocytes (Horner et al., 2000). Furthermore, these proliferating cells can generate neurons after being transplanted into the dentate gyrus (Shihabuddin et al., 2000). This indicates that stem cells in the adult spinal cord have the potential to differentiate into neurons.

These findings led to active research on the histological description of the central canal of the spinal cord. One of the main problems concerning the ependymal region of the spinal cord is the lack of consensus to its histological description (Alfaro-Cervello et al., 2012). Furthermore, histological aspect of this region is different according to the species that is studied. Mouse spinal ependymal cells are not multiciliated as in the rest of the central nervous system but they display 1 to 4 cilia with a majority of biciliated cells (Alfaro-Cervello et al., 2012) (Figure 18.2F). The cilia of these cells contain 9 peripheral microtubule doublets and 2 central microtubules (9+2) (Alfaro-Cervello et al., 2012). Their basal body has a very special aspect, which is different from the one observed in other stem cells of the CNS (Alfaro-Cervello et al., 2012) and characterized by a very dense pericentriolar material. Some ependymal cells have a basal cytoplasmic process in connection with blood vessels (Meletis et al., 2008; Alfaro-Cervello et al., 2012) (Figure 18.2F). This peculiar histological aspect is described as tanycytes by some authors. The most ventral and dorsal ependymal cells have a very long cytoplasmic process directed toward the pial surface (Meletis et al., 2008; Alfaro-Cervello et al., 2012) (Figure 18.2F). A few astrocytes are present in contact with the ependymal layer. These cells develop a cytoplasmic process that creeps between ependymal cells and that contacts the liquid of the central canal. This extension carries a primary cilium (Alfaro-Cervello et al., 2012) (Figure 18.2F). Finally, immature neurons (DCX+ and NeuN–) are also present with a globular extension that projects into the cavity of the central canal (Alfaro-Cervello et al., 2012) (Figure 18.2F).

The central canal of the spinal cord has also been studied by the same group in *Macaca fascicularis* (crab-eating macaque) (Alfaro-Cervello et al., 2014). The histological appearance differs from that observed and described in mice. In this species, three types of ependymal cells are present: multi-ciliated, uniciliated, and biciliated. Multiciliated cells (absent in mice) are predominantly found on the lateral walls of the canal. They display 20 to 30 apical cilia (9+2 type) (Alfaro-Cervello et al., 2014). As their name suggests, uni- and biciliated ependymal cells contain one or two cilia (9+2 type). They predominate on the ventral and dorsal regions of the canal. Their basal pole extends with a very long process like that described in mice. Astrocytes with inter-ependymal process and neuron contacting the central canal are also present in this monkey as in mice. However, Alfaro-Cervello et al. (2014) could not assess the presence of primary cilium on astrocytes due to the rarity of this cell population.

Histological aspect of the central canal in humans is quite difficult to establish. Indeed, this canal is thought to be obliterated (Alfaro-Cervello et al., 2014) or partially obliterated (Dromart et al., 2008). However, this obliteration could be a post-mortem artefact (Dromart et al., 2008).

In the intact spinal cord of the adult rodent, very few cells proliferate within the ependyma (Meletis et al., 2008; Barnabé-Heider et al., 2010; Alfaro-Cervello et al., 2012) and their progeny remains locally (Meletis et al., 2008; Alfaro-Cervello et al., 2012). In contrast after a spinal injury, ependymal cells proliferate actively (Johansson et al., 1999; Mothe and Tator, 2005; Meletis et al., 2008; Barnabé-Heider et al., 2010). However, ependymal cells represent only a minor proportion of proliferating cells in the adult spinal cord. Indeed, the majority of proliferating cells in this case are oligodendrocyte progenitors that generate mature oligodendrocytes (Barnabé-Heider et al., 2010). However, only ependymal cells are able to generate neurospheres that self-renew (Meletis et al., 2008). These findings lead to the conclusion that neural stem cells in the spinal cord are likely to be ependymal cells. But this region is devoid of any neuronal potential *in vivo*.

18.4 Ecology of Neural Stem Cells

18.4.1 Neural Stem Cells Live in a Niche

Studying neural stem cells in the different brain regions shows that some of these cells do not produce locally neurons *in vivo* in physiological conditions but they have this ability when cultured *in vitro*. This is the case for the mouse subcallosal layer (Seri et al., 2006) or the subventricular zone of the human adult (Sanai et al., 2004). Moreover, mouse C cells (transient amplifying progenitors) of the subventricular layer, when cultured *in vitro* in the presence of EGF, are converted into B cells (stem cells) (Doestch et al., 2002). At last, stem cells differentiation highly depends on their environment: (i) when stem cells from the dentate gyrus of adult rats are transplanted into another dentate gyrus they give rise to granular neurons as expected (Suhonen et al., 1996); (ii) when transplanted in the cerebellum, a region that is devoid of neurogenesis in the adult, these cells never produce neurons

(Suhonen et al., 1996); (iii) when transplanted in the olfactory bulb, they generate dopaminergic neurons, a normal component of this region (Suhonen et al., 1996). It is important to note that dopaminergic neurons are never present in the dentate gyrus but in the olfactory bulb. All these results indicate that the fate of neural stem cells is not fixed and depends heavily on their environment. It is commonly accepted that neural stem cells occupy a privileged environment, namely the niche that allows them to persist as stem cells. During stimulation, external factors acting on the stem cells, promote both proliferation and differentiation.

The niche is therefore essential to maintain the characteristic cellular environment. Only the niches of the subventricular zone and of the subgranular layer begin to be described and analysed. Ependymal cells, neighboring cells and blood vessels constitute the niche of stem cells in the subventricular layer. Instead, stem cells of the subgranular zone occupy a niche formed by blood vessels, granular neurons, and neighboring cells. Thus, one can find that the composition of the two niches is different and that cellular interactions could also be different in the two locations.

18.4.2 Interaction With the Cerebrospinal Fluid (CSF)

CSF is present in the ventricular cavities bordered by ependymal cells and around the central nervous system in the subarachnoid spaces.CSF is produced in part by the choroid plexuses (Telano and Baker 2020), which are highly vascularized structures that protrude into the cavities of the four ventricles. More recently, an extra-choroidal origin of CSF (i.e. ependyma, glia limitans, and lepto-meningeal vessels) has also been demonstrated (Trillo-Contreras et al., 2019; Li et al., 2020). The respective percentage of these different components in the production of CSF still remains to be determined in both rodents and humans. Circulation of the CSF is still a subject of intense debates. The classical model has been described by the American neurosurgeon Walter Dandy (1886–1946). He postulates that the flow is continuous from the ventricle to the subarachnoid spaces. It is now established that this model is obsolete and must now be relegated to the history of concepts and ideas. *In vivo* studies by cine MRI have indeed shown that CSF circulation occurs in a pulsatile manner depending on the phases of the cardiac cycle (Kapsalaki et al., 2012). Nevertheless, the notion of this directional and pulsatile circulation is criticized and some authors propose that CSF does not circulate even if it undergoes pulsatile movements during the cardiac cycle (Orešković and Klarica, 2014).

The problem of CSF movements is even more complex. Indeed, the different models we have presented assume that the intraventricular CSF is located in a sealed compartment, meaning that this fluid cannot cross the so-called ependymal "barrier" that lines the ventricles. However, it has long been known that CSF water diffuses through ependyma and participates in the formation of interstitial fluid in the brain (Bering 1954; Feldberg and Fleischhauer 1960, Milhorat et al., 1970).

Finally, more recently, a new CSF circulation has been evidenced even though many questions still remain unresolved (Mestre et al., 2020). The CSF located in the subarachnoid spaces enters the brain parenchyma through the arterial perivascular spaces. Then it crosses the astrocytic feet thanks to the aquaporins that are expressed at this level and reaches the interstitial compartment of the central nervous system. Finally, interstitial fluid moves to the venous perivascular space that communicates with the subarachnoid spaces. This circulation has been called glymphatic because it resembles a lymphatic-like circulation. However, the involved "vessels" are not true vascular vessel but flow is limited by glial cells. This term seems thus unfortunate and is very criticized because it leads to confusion.

In mice, stem cells located in the wall of the lateral ventricles are in direct contact with the CSF. On the contrary, stem cells of the hippocampus are more distant from this fluid. However, the different ways of circulation (or diffusion) of CSF that we have above described can account for an action of this fluid on hippocampal stem cells. There are still many works to be done in this field of research.

CSF is a very complex fluid whose composition begins to be unveiled. New techniques allowing the analysis of CSF proteome have demonstrated the presence of around 2600 proteins (Pan et al., 2007, Schutzer et al., 2010), 56% of which are found in this liquid but not in plasma (Schutzer et al., 2010). Some proteins are transported by blood and cross blood-brain or blood–CSF barrier (Zhao et al., 2007), others are secreted by choroid plexus (Silva-Vargas et al., 2016) or subcommissural organ (Guerra et al., 2015). Some of these proteins play a trophic or a signaling role and consequently CSF should not be considered as a simple liquid ensuring the drainage of wasting products but as a nutritive active fluid (Illes 2017).

Embryonic CSF (extracted from E12.5 neural tube) is a potent activator of neurogenic activity on ventricular stem cells of adult mice, whereas adult CSF is not (Alonso et al., 2017). It is important to note that protein composition of CSF changes during aging (Zhang et al., 2005; Baird et al., 2012; Silva-Vargas et al., 2016). The concentration of some proteins decreases while it increases for others. It should also be noted that ventricular CSF of an adult human promotes *in vitro* proliferation of neural stem cells while lumbar CSF from the same subject inhibits this proliferation (de Sonnaville et al., 2020). Lumbar CSF promotes astroglial differentiation and inhibits neuronal one (Buddensiek et al., 2010). A better analysis of the differences observed could lead to evidence molecules involved in the regulation of neural stem cells.

18.4.3 Interactions Within the Niche

BMPs (Bone Morphogenetic Proteins) are a family of secreted molecules belonging to the TGF (Transforming Growth Factor) beta superfamily. They act on membrane receptors containing cytoplasmic serine/threonine kinase domain. Noggin is a secreted protein able to bind to BMPs. Noggin acts thus as an extracellular inhibitor of BMPs since the molecular complex generated cannot bind to the BMP receptor. BMPs are produced by subventricular stem cells and C cells (Transient Amplifying Progenitors) but not by ependymal cells and neuroblasts (A cells) (Lim et al., 2000). A, B, and C cells of the subventricular zone express certain types of BMP receptors (Lim et al., 2000). BMPs inhibitboth neurogenesis and oligodendrogenesis and promote terminal astrocytic differentiation (Lim et al., 2000; Bonaguidi et al., 2005). Neighboring ependymal cells secrete Noggin that antagonizes BMP signaling and promotes the maintenance of stem cell in their subventricular niche (Lim et al., 2000).

In contrast, BMP is produced by cells of the meninges in the region of the dentate gyrus (Chloe et al., 2013) and BMP receptors by cells of the dentate gyrus (Chloe et al., 2013). Noggin is secreted by some cells of the granular layer, the cells of the subgranular layer, and of the hilus (Bonaguidi et al., 2008). Inhibition of BMP signal leads to a loss of neurogenesis in the adult mouse (Colak et al., 2008). This loss is due to a massive stimulation of stem cells causing their proliferation leading to exhaustion of the stock explaining a subsequent impact on neurogenesis (Bonaguidi et al., 2008; Mira et al., 2010).

Thus, the BMP signal shows that the observed effects are diametrically opposite in the two stem cell niches: BMP signal is essential to maintain stem cells in the dentate gyrus while inhibition of BMP signal is mandatory for this function in the subventricular zone.

The Notch pathway is also important for neural stem cells. This pathway acts both in the subventricular zone (Aguirre et al., 2010) and in the subgranular zone (Lavado and Oliver, 2014). Notch activation is necessary to maintain neural stem cells in a quiescent state.

18.4.4 When Endothelial Cells Meet Neural Stem Cells

The presence of endothelial cells in a culture of adult stem cells from mouse subventricular zone induces an increase in precursor proliferation and represses their differentiation (Shen et al., 2004). This result shows the interactions played by endothelial cells toward neural stem cells. Before describing such interactions, we should provide a brief histological description of the relationships between these two groups of cells. Hippocampal cells that divide after incorporation of BrdU form clusters of cells associated with a blood vessel (Palmer et al., 2000). In the subventricular zone, the blood vessels are numerous, large, and form a dense plexus (Shen et al., 2008; Tavazoie et al., 2008). Blood vessels of the subventricular zone are enwrapped in a basal lamina like all the vessels of the body. However, this perivascular sheet extends in the subventricular zone forming a specific extra-vascular basal lamina of this area (Mercier et al., 2002). Proteins of the basal lamina such as laminin extend from the vessel to constitute a filiform extension called a stem, which presents in thickened zones called bulbs (Mercier et al., 2002). This extravascular basal lamina has a shape reminiscent of fractals which explains the name fractone proposed by Mercier et al. (2002) for its description. Extra-vascular lamina surrounds the cells in ventricular zone and ends in contact with ependymal cells (Mercier et al., 2002). Proliferating B cells are located in contact with the blood vessels (Shen et al., 2008; Tavazoie et al., 2008) by cytoplasmic extensions whose morphology are different from the astrocytic feet (Tavazoie et al., 2008). Furthermore, the vessels of the subventricular zone have no pericytes and are characterized by a leaky blood–brain barrier as evidenced by the passage of sodium fluorescein (376 Da) (Tavazoie et al., 2008). Vessels in the subventricular zone are not all equivalent. Some are associated with increased cell proliferation, while others are not (Tavazoie et al., 2008). The regulation of such a phenomenon is still unknown.

Schematically, we can summarize and classify the activity of vessels on stem cells according to three major mechanisms.

The first one is the contribution of molecules coming from the blood and diffusing to reach stem cells, such as diffusion, which would be facilitated since blood–brain barrier is leaky at this level. We will present a few examples of this type of regulation. During gestation in mice, neurogenesis in the subventricular zone is increased leading to a greater migration of neurons in the olfactory bulb (Shingo et al., 2003). This action is mediated by circulating prolactin (Shingo et al., 2003). Transfusing blood from a young mouse to an elderly one induces an increase in neurogenesis and an improvement of cognitive performances, an effect described as rejuvenation (Katsimpardi et al., 2014). This effect is due to circulating factors; Glial Derived Factor (GDF) 11 seems to play a crucial role (Katsimpardi et al., 2014). Transfusion of blood from an elderly mouse into a young one has the opposite effect leading to a reduction in neurogenesis (Villeda et al., 2011). Among the circulating factors implicated, the chemokine CCL11 (C-C motif chemokine ligand 11) plays an important role (Villeda et al., 2011).

The second mechanism implicated at this level is based upon cell–cell and cell–matrix interactions. The vascular process of neural stem cell expresses the laminin receptor alpha 6 beta 1 integrin (Shen et al., 2008). This adhesion molecule allows the cell to adhere to the perivascular extracellular matrix. Treating brain with antibodies against this type of integrin leads to a disruption of vascular adhesion, retraction of the process, and migration away from the vessel (Shen et al., 2008). This modification is associated with an increase of proliferation of stem cells suggesting that vessels play a role in generating quiescence (Shen et al., 2008). Cell–cell contact between endothelial cells and stem cells are also very important for stem cell physiology. These interactions are assured by Ephrin B2 and Jagged 1 expressed by endothelial cells (Ottone et al., 2014). In their absence, neural stem cells will be activated resulting in depletion of these cells (Ottone et al., 2014).

The last mechanisms that could account for this interplay is represented by the factors secreted by endothelial cells and acting on neural stem cells. Pigment Epithelium Derived Factor (PEDF) is secreted by both endothelial and ependymal cells and increases renewal of neural stem cells (Ramírez-Castillejo et al., 2006). Similarly, betacellulin is produced by both endothelial and choroid plexus cells and increases the proliferation of stem cells in the two main neurogenic regions (Gómez-Gaviro et al., 2012). Furthermore, mice lacking the gene encoding betacellulin are defective for neuroblastic regeneration (Gómez-Gaviro et al., 2012). In contrast, other factors are involved in the maintenance of quiescence: such is the case for NT3 (neurotrophin 3) (Delgado et al., 2014).

Interactions between endothelial cells and neural stem cells are multiple and sometimes play an antagonistic role. This suggests the importance of specialized vascularization for the balance between quiescence, activation, proliferation, and differentiation of stem cells in the CNS. Much work is still needed in order to draw up a picture of these interactions.

18.5 Concluding Remarks

The central nervous system contains true stem cells that act physiologically to allow limited replacement of neurons. These cells are arranged in four major brain regions whose histological appearance varies. Such variability is also observed for the regulation of these cells. These phenomena lead to develop specific studies of each region and should invite caution about abusive generalizations. In humans, the subventricular zone does not produce olfactory neurons in adults in physiological conditions contrary to what is observed in other mammals. Many fundamental works remain to be done in order to try to understand the physiological function of these systems. However, the existence of such cells encourages trying to establish tissue repair techniques in order to correct neurological disorders.

Key Points

- Stem cells that can differentiate in neurons are present in the adult CNS of mammals.
- Two regions are neurogenic in adult mammals: the subventricular zone of the telencephalon and the subgranular zone of the dentate gyrus.
- In mice, stem cells from the subventricular zone migrate and differentiate into olfactory neurons.
- In humans, stem cells from the subventricular zone give rise to striatal interneurons.
- In mice as in humans, stem cells from the subgranular zone generate granule cells of the dentate gyrus.
- Molecular regulations of neural stem cells begin to be unravelled.
- Numerous studies are still needed to try understanding a complete scenario accounting for neural stem cells regulation.

References

Allen, E. (1912). The cessation of mitosis in the central nervous system of the albino rat. J. Comp. Neurol. *22*, 547–568.

Bizzozero G. (1894). An address on the growth and regeneration of the organism. Br. Med. J. *i*, 728–732.

Colucci-D'Amato, L., Bonavita, V., di Porzio. U. (2006). The end of the central dogma of neurobiology: stem cells and neurogenesis in adult CNS. Neurol. Sci. *27*, 266–270.

Codega, P., Silva-Vargas, V., Paul, A., Maldonado-Soto, A.R., DeLeo, A.M., Pastrana, E., Doetsch, F. (2014). Prospective identification and purification of quiescent adult neural stem cells from their *in vivo* niche. Neuron. *82*, 545–559.

Mich, J.K., Signer, R.A.J., Nakada, D., Pineda, A., Burgess, R.J., Vue, T.Y., Johnson, J.E., Morrison, S.J. (2014). Prospective identification of functionally distinct stem cells and neurosphere-initiating cells in adult mouse forebrain. Elife. May 7; 3:e02669. doi: 10.6554/eLife.02669.

Altman, J. (1962). Are new neurons formed in the brains of adult mammals? Science. *135*, 1127–1128.

Altman, J., Das, G.D. (1965). Autoradiographic and histological evidence of postnatal hippocampal neurogenesis in rats. J. Comp. Neurol. *124*, 319–335.

Altman, J. (1969). Autoradiographic and histological studies of postnatal neurogenesis. IV. Cell proliferation and migration in the anterior forebrain, with special reference to persisting neurogenesis in the olfactory bulb. J. Comp. Neurol. *137*, 433–457.

Goldman, S.A., Nottebohm, F. (1983). Neuronal production, migration, and differentiation in a vocal control nucleus of the adult female canary brain. Proc. Natl. Acad. Sci. USA. *80*, 2390–2394.

Paton, J.A., Nottebohm, F.N. (1984). Neurons generated in the adult brain are recruited into functional circuits. Science. *225*, 1046–1048.

Alvarez-Buylla, A., Kim, J.R., Nottebohm, F. (1990). Birth of projection neurons in adult avian brain may be related to perceptual or motor learning. Science. *249*, 1444–1446.

Nottebohm, F. (1985). Neuronal replacement in adulthood. Ann. NY. Acad. Sci. *457*, 143–161.

Rakic, P. (1985). Limits of neurogenesis in Primates. Science. *227*, 1054–1056.

Kuhn, H.G., Dickinson-Anson, H., Gage, F.H. (1996). Neurogenesis in the dentate gyrus of the adult rat: age-related decrease of neuronal progenitor proliferation. J. Neurosci. *16*, 2027–2033.

Kempermann, G., Kuhn, H.G., Gage, F.H. (1997). More hippocampal neurons in adult mice living in an enriched environment. Nature *386*, 493–495.

Eriksson, P.S., Perfilieva, E., Björk-Eriksson, T., Alborn, A.-M., Nordborg, C., Peterson, D.A., Gage, F.H. (1998). Neurogenesis in the adult human hippocampus. Nature Medicine. *4*, 1313–1317.

Kornack, D.R., Rakic, P. (1999). Continuation of neurogenesis in the hippocampus of the adult macaque monkey. Proc. Natl. Acad. Sci. USA. *96*, 5768–5773.

Gould, E., Reeves, A.J., Graziano, M.J.A., Gross, C.G. (1999). Neurogenesis in the neocortex of adult primates. Science.*286*, 548–552.

Gould, E., Vail, N., Wagers, M., Gross, C.G. (2001). Adult-generated hippocampal and neocortical neurones in macaques have a transient existence. Proc. Natl. Acad. Sci. USA. *98*, 10910–10917.

Spalding, K.L., Bhardwaj, R.D., Buchholz, B.A., Druid, H., Frisén, J. (2005). Retrospective birth dating of cells in humans. Cell. *122*, 133–143.

Bhardwaj, R.D., Curtis, M.A., Spalding, K.L., Buchholz, B.A., Fink, D., Björk-Eriksson, T., Nordborg, C., Gage, F.H., Druid, H., Eriksson, P.S., Frisén, J. (2006). Neocortical neurogenesis in humans is restricted to development. Proc. Natl. Acad. Sci. USA. *103*, 12564–12568.

Ernst, A., Alkass, K., Bernard, S., Salehpour, M., Perl, S., Tisdale, J., Possnert, G., Druid, H., Frisén, J. (2014). Neurogenesis in the striatum of the adult human brain. Cell. *156*, 1072–1083.

Bergmann, O., Liebl, J., Bernard, S., Alkass, K., Yeung, M.S., Steier, P., Kutschera, W., Johnson, L., Landén, M., Druid, H., Spalding, K.L., Frisén, J. (2012). The age of olfactory bulb neurons in humans. Neuron *74*, 634–639.

Spalding, K.L., Bergmann, O., Alkass, K., Bernard, S., Salehpour, M., Huttner, H.B., Boström, E., Westerlund, I., Vial, C., Buchholz, B.A., Possnert, G., Mash, D.C., Druid, H., Frisén, J. (2013). Dynamics of hippocampal neurogenesis in adult humans. Cell *153*, 1219–1227.

Reynolds, B.A., Weiss, S. (1992). Generation of neurons and astrocytes from isolated cells of the adult mammalian central nervous system. Science. *255*, 1707–1710.

Chojnacki, A.K., Mak, G.K., Weiss, S. (2009). Identity crisis for adult periventricular neural stem cells: subventricular zone astrocytes, ependymal cells or both? Nature Reviews Neuroscience. *10*, 153–163.

Chaichana, K., Zamora-Berridi, G., Camara-Quintana, J., Quiñones-Hinojosa, A. (2006). Neurosphere assays: growth factors and hormone differences in tumor and nontumor studies. Stem Cells. *24*, 2851–2857.

Jensen, J.B., Parmar, M. (2006). Strengths and limitations of the neurosphere culture system. Mol. Neurobiol. *34*, 153–161.

Weiss, S., Dunne, C., Hewson, J., Wohl, C., Wheatley, M., Peterson, A.C., Reynolds, B.A. (1996). Multipotent CNS stem cells are present in the adult mammalian spinal cord and ventricular neuroaxis. J. Neurosci. *16*, 7599–7609.

Doetsch, F., Petreanu, L., Caillé, I., Garcia-Verdugo, J.M., Alvarez-Buylla, A. (2002). EGF converts transit-amplifying neurogenic precursors in the adult brain into multipotent stem cells. Neuron. *36*, 1021–1034.

Mori, H., Fujitani, T., Kanemura, Y., Kino-Oka, M., Taya, M. Observational examination of aggregation and migration during early phase of neurosphere culture of mouse neural stem cells. (2007). J. Biosci. Bioeng. *104*, 231–234.

Singec, I., Knoth, R., Meyer, R.P., Maciaczyk, J., Volk, B., Nikkhah, G., Frotscher, M., Snyder, E.Y. (2006). Defining the actual sensitivity and specificity of the neuropshere assay in stem cell biology. Nature Methods. *3*, 801–806.

Coles-Takabe, B.L.K., Brain, I., Purpura, K.A., Karpowicz, P., Zandstra, P.W., Morshead, C.M., van der Kooy, D. (2008). Don't look: growing clonal versus nonclonal neural stem cell colonies. Stem Cells. *26*, 2938–2944.

Doetsch, F., García-Verdugo, J.M., Alvarez-Buylla, A. (1997). Cellular composition and three-dimensional organization of the subventricular germinal zone in the adult mammalian brain. J. Neurosci. *17*, 5046–5061.

Mirzadeh, Z., Merkle, F.T., Soriano-Navarro, M., García-Verdugo, J.M., Alvarez-Buylla, A. (2008). Neural stem cells confer unique pinwheel architecture to the ventricular surface in neurogenic regions of the adult brain. Cell Stem Cell. *3*, 265–278.

Codega, P., Silva-Vargas, V., Paul, A., Maldonado-Soto, A.R., Deleo, A.M., Pastrana, E., Doetsch, F. (2014). Prospective identification and purification of quiescent adult neural stem cells from their in vivo niche. Neuron. *82*, 545–549.

Mich, J.K., Signer, R.A.J., Nakada, D., Pineda, A., Burgess, R.J., Vue, T.Y., Johnson, J.E., Morrison, S.J. (2014). Prospective identification of functionally distinct stem cells and neurosphere – initiating cells in adult mouse forebrain. Elife. May 7; 3:e02669. doi: 10.7554/eLife.02669.

Gritti, A., Bonfanti, L., Doetsch, F., Caillé, I., Alvarez-Buylla, A., Lim, D.A., Galli, R., Garcia Verdugo, J.M., Herrera, D.G., Vescovi, A.L. (2002). Multipotent neural stem cells reside into the rostral extension and olfactory bulb of adult rodents. J. Neurosci. *22*, 437–445.

Lois, C., Alvarez-Buylla, A. (1994). Long-distance neuronal migration in the adult mammalian brain. Science. *264*, 1145–1148.

Lois, C., García-Verdugo, J.-M., Alvarez-Buylla, A. (1996). Chain migration of neuronal precursors. Science. *271*, 978–981.

Lois, C., Alvarez-Buylla, A. (1993). Proliferating subventricular zone cells in the adult mammalian forebrain can differentiate into neurons and glia. Proc. Natl. Acad. Sci. USA. *90*, 2074–2077.

Obernier, K., Cebrian-Silla, A., Thomson, M., Parraguez J.I., Anderson, R., Guinto, C., Rodriguez, J.R., Garcia-Verdugo, J.-M., Alvarez-Buylla, A. (2018). Adult neurogenesis is sustained by symmetric self-renewal and differentiation. Cell Stem Cell *22*, 221–234.

Doetsch, F., García-Verdugo, J.M., Alvarez-Buylla, A. (1999a). Regeneration of a germinal layer in the adult mammalian brain. Proc. Natl. Acad. Sci. USA. *96*, 11619–11624.

Doestch, F., Caillé, I., Lim, D.A., García-Verdugo, J.-M., Alvarez-Buylla, A. (1999b). Subventricualr zone astrocytes are neural stem cells in the adult mammalian brain. Cell. *97*, 703–716.

Llorens-Bobadilla, E., Zhao, S., Baser, A., Saiz-Castro, G., Zwaldo, K., Martin-Villalba, A. (2015). Single-cell transcriptomics reveals a population of dormant neural stem cells that become activated upon brain injury. Cell Stem Cell *17*, 329–340.

Xie, X.P., Laks, D.R., Sun, D., Poran, A., Laughney, A.M., Wang, Z., Sam, J., Belenguer, G., Fariñas, I., Elemento, O., Zhou, X., Parada, L.F. (2020). High-resolution mouse subventricular zone stem-cell niche transcriptome reveals features of lineage, anatomy, and aging. Proc. Natl. Acad. Sci. USA. *117*, 31448–31458.

Merkle, F.T., Mirzadeh, Z., Alvarez-Buylla, A. (2007). Mosaic organization of neural stem cells in the adult brain. Science *317*, 381–384.

Fiorelli, R., Azim, K., Fischer, B., Raineteau, O. (2015). Adding a spatial dimension to postnatal ventricular-subventricular zone neurogenesis. Development *142*, 2109–2120.

Poiana, G., Gioia, R., Sineri, S., Cardarelli, S., Lupo, G., Cacci, E. (2020). Transcriptional regulation of adult neural stem-progenitor cells: tales from the subventricular zone. Neural Regen. Res. *15*, 1773–1783.

Sanai, N., Nguyen, T., Ihrie, R.A., Mirzadeh, Z., Tsai, H.-H., Wong, M., Gupta, N., Berger, M.S., Huang, E., Garcia-Verdugo, J.M., Rowitch, D.H., Alvarez-Buylla, A. (2011). Corridors of migrating neurons in the human brain and their decline during infancy. Nature. *478*, 382–386.

Quiñones-Hinojosa, A., Sanai, N., Soriano-Navarro, M., Gonzalez-Perez, O., Mirzadeh, Z., Gil-Perotin, S., Romero-Rodriguez, R., Berger, M.S., Garcia-Verdugo, J.M., Alvarez-Buylla, A. (2006). Cellular composition and cytoarchitecture of the adult human subventricular zone: a niche of neural stem cells. J. Comp. Neurol. *494*, 415–434.

Sanai, N., Tramontin, A.D., Quiñones-Hinojosa, A., Barbaro, N.M., Gupta, N., Kunwar, S., Lawton, M.T., McDermott, M.W., Parsa, A.T., García-Verdugo, J.M., Berger, M.S., Alvarez-Buylla, A. (2004). Unique astrocyte ribbon in adult human brain contains neural stem cells but lacks chain migration. Nature. *427*, 740–744.

Wang, C., Liu, F., Liu, Y.Y., Zhao, C.H., You, Y., Wang, L., Zhang, J., Wei, B., Ma, T., Zjang, Q., Zhang, Y., Chen, R., Song, H., Yang, Z. (2011). Identification and characterization of neuroblasts in the subventricular zone and rostral migratory stream of the adult human brain. Cell Res. *21*, 1534–1550.

Seri, B., Herrera, D.G., Gritti, A., Ferron, S., Collado, L., Vescovi, A., Garcia-Verdugo, J.M., Alvarez-Buyla, A. (2006). Composition and organization of the SCZ: a large germinal layer containing neural stem cells in the adult mammalian brain. Cereb. Cortex. *16*, 1103–1111.

Seri, B., García-Verdugo, J.M., McEwen, B.S., Alvarez-Buylla, A. (2001). Astrocytes give rise to new neurons in the adult mammalian hippocampus. J. Neurosci. *21*, 7153–7160.

Seri, B., García-Verdugo, J.M., Collado-Porente, L., McEwen, B.S., Alvarez-Buylla, A. (2004). Cell types, lineage, and architecture of the germinal zone in the adult dentate gyrus. J. Comp. Neurol. *478*, 359–378.

Felten, D.L., O'Bannion, M.K., Maida, M.E. (2016). Netter's atlas of neuroscience. 3rd edition. Elsevier, Philadelphia.

Palmer, T.D., Willhoite, A.R., Gage, F.H. (2000). Vascular niche for adult hippocampal neurogenesis. J. Comp. Neurol. *425*, 479–494.

Sorrells, S.F., Paredes, M.F., Cebrina-Silla, A., Sandoval, K., Qi, D., Kelley, K.W., James, D., Mayer, S., Chang, J., Auguste, K.I., Chang, E.F., Guttierrez, A.J., Kriegstein, A.R., Mathern, G.W., Oldham, E.F., Huang, E.J., Garcia-Verdugo, J.M., Yang, Z., Alvarez-Buylla, A. (2018). Human hippocampal neurogenesis drops sharply in children to undetectable levels in adults. Nature *555*, 377–381.

Boldrini, M., Fulmore, C.A., Tartt, A.N., Simeon, L.R., Pavlova, I., Poposka, V., Rosoklija, G.B., Stankov, A., Arango, V., Dwork, A.J., Hen, R., Mann, J.J. (2018). Human hippocampal neurogenesis persists throughout aging. Cell Stem Cell. *22*, 589–599.

Kempermann, G., Gage, F.H., Aigner, L., Song, H., Curtis, M.A., Thuret, S., Kuhn, H.G., Jessberger, S., Frankland, P.W., Toni, N., Schinder, A.F.,

Zhao, X., Lucassen, P.J., Frisén, J. (2018). Human adult neurogenesis: evidence and remaining questions. Cell Stem Cell *23*, 25–30.

Lucassen, P.J., Toni, N., Kempermann, G., Frisén, J., Gage, F.H., Swaab, D.F. (2020). Limits to human neurogenesis – really? Mol. Psychiatry *25*, 2207–2209.

Seki, T., Hori, T., Miyata, H., Maehara, M., Namba, T. (2019). Analysis of proliferating neuronal progenitors and immature neurons in the human hippocampus surgically removed from control and epileptic patients. Sci. Rep. Dec 3; 9(1):18194. doi: 10.1038/s41598–019-54684-z.

Seki, T. (2020). Understanding the real state of human adult hippocampal neurogenesis from studies of rodents and non-human primates. Front. Neurosci. Aug 11;14:839. doi: 10.3389/fnins.2020.00839.

Shihabuddin, M.S., Ray, J., Gage, F.H. (1997). FGF-2 is suffcient to isolate progenitors found in the adult mamamlian spinal cord. Exp. Neurol. *148*, 577–586.

Dromard, C., Guillon, H., Rigau, V., Ripoll, C., Sabourin, J.C., Perrin, F.E., Scamps, F., Bozza, S., Sabatier, P., Lonjon, N., Duffau, H., Vachiery-Lahaye, F., Prieto, M., Tran Van Ba, C., Deleyrolle, L., Boularan, A., Langley, K. Gaviria, M., Privat, A., Hugnot, J.P., Bauchet, L. (2008). Adult human spinal cord harbors neural precursors cells thta generate neurons and glian cells in vitro. J. Neurosci. Res. *86*, 1916–1926.

Horner, P.J., Power, A.E., Kempermann, G., Kuhn, H.G., Palmer, T.D., Winkler, J., Thal, L.J., Gage, F.H. (2000). Proliferation and differntiation of progenitor cells thorugout the intect adult spinal cord. J. Neurosci. *20*, 2218–2228.

Shihabuddin, M.S., Horner, P.J., Ray, J., Gage, F.H. (2000). Adult spinal cord stem cells generate neurons after transplantation in the adult dentate gyrus. J. Neurosci. *20*, 8727–8735.

Alfaro-Cervello, C., Soriano-Navarro, M., Mirzadeh, Z., Alvarez-Buylla, A., Garcia-Verdugo, J.M. (2012). Biciliated ependymal cell proliferation contributes to spinal cord growth. J. Comp. Neurol. *520*, 3528–3552.

Meletis, K., Barnabé-Heider, F., Carlén, M., Evergren, E., Tomilin, N., Shuliapov, O., Frisén, J. (2008). Spinal cord injury reveals multilineage differentiation of ependymal cells. PLoS Biol. 2008; 6(7):e182. doi:10.1371/ journal.pbio.0060182.

Alfaro-Cervello, C., Cebrian-Silla, A., Soriano-Navarro, M., Garcia-Tarraga, P., Matías-Guiu, J., Gonez-Pinedo, U., Molina Aguilar, P., Alvarez-Nuylla, A., Luqin, M.-R., Garcia-Verdugo, J.M. (2014). The adult macaque spinal cord central canal zone contains proliferative cells and closely resembles the human. J. Comp. Neurol. *522*, 1800–1817.

Barnabé-Heider, F., Görit, C., Sabelström, H., Takebayashi, H., Pfriegher, W., Meletis, K., Frisén, J. (2010). Origin of new glial cells in intact and injured adult spinal cord. Cell Stem Cell. *7*, 470–482.

Johansson, C.B., Momma, S., Clarke, D.L., Risling, M., Lendahl, U., Frisén, J. (1999). Identification of a neural stem cell in the adult mammalian central nervous system. Cell. *96*, 25–34.

Mothe, A.J., Tator, C.H. (2005). Proliferation, migration, and differentiation of endogenous ependymal region stem/progenitor cells following minimal spinal cord injury in the adult rat. Neuroscience. *131*, 177–187.

Suhonen, J.O., Peterson, D.A., Ray, J., Gage, F.H. (1996). Differentiation of adult hippocampus-derived progenitors into olfactory neurons in vivo. Nature. *383*, 624–627.

Telano, L.N., Baker, S. (2020). Physiology, cerebral spinal fluid. StatPearls. Jul2020.

Trillo-Contreras, J.L., Toledo-Aral, J.J., Echevarría, M., Villadiego, J. (2019). AQP1 and AQP4 contribution to cerebospinal fluid homeostasis. Cells. Feb 24; *8(2)*, 197. doi: 10.3390/cells8020197.

Li, Q., Aalling, N.N., Förstera, B., Ertürk, A., Nedergaard, M., Møllgård, K., Xavier, A.L.R. (2020). Aquaporin 1 and the Na+/K+/2Cl- cotransporter 1 are present in the leptomeningeal vasculature of the adult rodent central nervous system. Fluids Barriers CNS. Feb 11; *17(1)*, 15. doi: 10.1186/s12987-020-0176-z.

Proulx, S.T. (2021). Cerebrospinal fluid outflow: a review of the historical and contemporary evidence for arachnoid villi, perineural routes, and udral lymphatics. Cell Mol. Life Sci. Jan 11. doi: 10.1007/s00018-020-03706-5. Online ahead of print.

Kapsalaki, E., Svolos, P., Tsougos, I., Theodorou, K., Fezoulidis, I., Fountas, K.N. (2012). Quantification of normal CSF flow through the aqueduct using PC-cine MRI at 3T. Acta Neurochir. Suppl. *113*, 39–42.

Ores?kovicì, D., Klarica, M. (2014). A new look at cerebrospinal fluid movement. Fluids Barriers CNS. Jul 27; *11*:16. doi: 10.1186/2045-8118-11-16.

Bering, E.A.Jr. (1954). Water exchange in the brain and cerebrospinal fluid. Studies on the intraventricular instillation of Deuterium (heavy water). J. Neurosurg. *11*, 234–242.

Feldberg, W., Fleischhauer, K. (1960). Penetration of Bromphenol blue from the perfused cerebral ventricles into the brain tissues. J. Physiol. *150*, 451–462.

Milhorat, T.H., Clark, R.G., Hammock, M.K., McGrath, P.P. (1970). Structural, ultrastructural, and permeability changes in the ependyma and surrounding brain favouring equilibration in progressive hydrocephalus. Arch. Neurol. *22*, 397–407.

Mestre, H., Mori, Y., Nedergaard, M. (2020). The brain's glymphatic system: current controversies. Trends Neurosci. *43*, 458–466.

Pan, S., Zhu, D., Quinn, J.F., Peskind, E.R., Montine, T.J., Lin, B., Goodlett, D.R., Taylor, G., Eng, J., Zhang, J. (2007). A combined dataset of human cerebrospinal fluid proteins identified by multi-dimensional chromatography and tandem mass spectrometry. Proteomics. *7*, 469–473.

Schutzer, S.E., Liu, T., Natelson, B.H., Angal, T.E., Schepmoes, A.A., Purvine, S.O., Hixson, M.S., Camp, D.G., Coyle, P.K., Smith, R.D., Bergquist, J. (2010). Establishing the proteome of normal human cerebrospinal fluid. PLoS One Jun 11; 5(6):e10980. doi:10.1371/journal.pone.0010980.

Zhao L-R, Navalitloha Y, Singhal S, Mehta J, Piao C-S, Guo W-P, Kessler JA, Groothuis DR. Hematopoietic growth factors pass through the blood-brain barrier in intact rats. Exp Neurol 2007; 204:569–573.

Silva-Vargas V, Maldonado-Soto AR, Mizrak D, Codega P, Doetsch F. Age-dependent niche signals from choroid plexus regulate adult neural stem cells. Cell Stem Cell 2016; 19:643–652.

Guerra MM, González C, Caprile T, Jara M, Vío K, Muñoz RI, Rodríguez S, Rodríguez EM. Understanding how the subcommissural organ and other periventricular secretory structures contribute via the cerebrospinal fluid to neurogenesis. Front Cell Neurosci 2015 Dec 23; 9:480. doi:10.3389/fncel.2015.00480.

Illes, S. (2017). More than a drainage fluid: the role of CSF in signaling in the brain and other effects on brain tissue. Handb. Clin. Neurol. *146*, 33–46.

Alonso MI, Lamus F, Carnicero E, Moro JA, de la Mano A, Fernández JMF, Desmond ME, Gato A. (2017). Embryonic cerebrospinal fluid increases neurogenic activity in the brain ventricular-subventricular zone of adult mice. Front Neuroanat Dec 19; 11:124. doi: 10.3389/fnana.2017.00124.

Zhang, J., Goodlett, D.R., Peskind, E.R., Quinn, J.F., Zhou, Y., Wang, Q., Pan, C., Yi, E., Eng, J., Aebersold, R.H., Montine, T.J. (2005). Quantitative proteomic analysis of age-related changes in human cerebrospinal fluid. Neurobiol. Aging *26*, 207–227.

Baird, G.S., Nelson, S.K., Keeney, T.R., Stewart, A., Williams, S., Kraemer, S., Peskind, E.R., Montine. (2012). Age-dependent changes in the cerebrospinal fluid proteome by slow off rate modified aptamer array. Am. J. Pathol. *180*, 446–456.

Silva-Vargas, V., Maldonado-Soto, A.R., Mizrak, D., Codega, P., Doetsch, F. (2016). Age-dependent niche signals from the choroid plexus regulate adult neural stem cells. Cell Stem Cell *19*, 643–652.

de Sonnaville, S.F.A.M., van Strien, M.E., Middeldorp, J., Sluijs, J.A., van den Berge, S.A., Moeton, M., Donega, V., van Berkel, A., Deering, T., De Filippis, L., Vescovi, A.L., Aronica, E., Glass, R., van den Berg, W.D.J., Swaab, D.F., Robe, P.A., Hol, E.M. (2020). The adult human subventricular zone: partial ependymal coverage and proliferative capacity of cerebrospinal fluid. Brain Commun. Oct 13; 2(2):fcaa150. doi: 10.1093/braincomms/fcaa150.

Buddensiek J, Dressel A, Kowalski M, Runge U, Schroeder H, Hermann A, Kirsch M, Storch A, Sabolek M. Cerebrospinal fluid promotes survival and astroglial differentiation of adult human neural progenitor cells but inhibits proliferation and neuronal differentiation. BMC Neurosci 2010 Apr 8; 11:48. doi: 10.1186/1471-2202-11-48.

Lim, D.A., Tramontin, A.D., Trevejo, J.M., Herrera, D.G., García-Verdugo, J.M., Alvarez-Buylla, A. (2000). Noggin antagonizes BMP signaling to create a niche for adult neurogenesis. Neuron. *28*, 713–726.

Bonaguidi, M.A., McGuire, T., Hu, M., Kan, L., Samanta, J., Kessler, J.A. (2005). LIF and BMP signaling generate separate and discrete types of GFAP-expressing cells. Development. *132*, 5503–5514.

Chloe, Y., Kozlova, A., Graf, D., Pleasure, S.J. (2013). Bone morphogenetic protein signaling is a major determinant of dentate development. J. Neurosci. *33*, 6766–6775.

Bonaguidi, M.A., Pen,g C.-Y., McGuire, T., Falciglia, G., Gobeske, K.T., Czeisler, C., Kessler, J.A. (2008). Noggin expands neural stem cells in the adult hippocampus. J. Neurosci. *28*, 9194–9204.

Colak, D., Mori, T., Brill, M.S., Pfeifer, A., Falk, S., Deng, C., Monteiro, R., Mummery, C., Sommer, L., Götz, M. (2008). Adult neurogenesis requires Smad4-mediated bone morphogenetic protein signaling in stem cells. J. Neurosci. *28*, 434–446.

Mira, H., Andreu, Z., Suh, H., Lie, D.C., Jessberger, S., Consiglio, A., San Emeterio, J., Hortigüella, R., Marqués-Torrejón, M.A., Nakashima, K., Colak, D., Götz, M., Fariñas, I., Gage, F.H. (2010). Signaling through BMPR-IA regulates quiescence and long term activity of neural stem cells in the adult hippocampus. Cell Stem Cell. *7*, 78–89.

Aguirre, A., Rubio, M.E., Gallo, V. (2010). Notch and EGFR pathway inter-action regulates neural stem cell number and self-renewal. Nature 467, 323–327.

Lavado, A., Oliver, G. (2014). Jagged1 is necessary for postnatal and adult neurogenesis in the dentate gyrus. Dev. Biol. *388*, 11–21.

Shen, Q., Goderie, S.K., Jin, L., Karanth, N., Sun, Y., Abramova, N., Vincent, P., Pumiglia, K., Temple, S. (2004). Endothelial cells stimualte and expand neurogenesis of neural stem cells. Science. *304*, 1338–1340.

Shen, Q., Wang, Y., Kokovay, E., Lin, G., Chuang, S.-M., Goderie, S.K., Roysam, S., Temple, S. (2008). Adult SVZ stem cells lie in a vascular niche: a quantitative analysis of niche cell-cell interactions. Cell Stem Cell. *3*, 289–300.

Tavazoie, M., Van der Veken, L., Silva-Vargas, V., Louissaint, M., Colonna, L., Zaidi, B., Garcia-Verdugo, J.M., Doestch, F. (2008). A vascular niche for adult neural stem cells. Cell Stem Cell. *3*, 279–288.

Mercier, F., Kitasako, J.T., Hatton, G.I. (2002). Anatomy of the brain neuro-genic zones revisited: fractones and the fibroblast/macrophage network. J. Comp. Neurol. 451, 170–188.

Shingo, T., Gregg, C., Enwere, E., Fujiwara, H., Hassam, R., Geary, C., Cross, J.C., Weiss, S. (2003). Pregnancy-stimulated neurogenesis in the adult forebrain mediated by prolactin. Science *299*, 117–120.

Katsimpardi, L., Litterman, N.K., Schein, P.A., Miller, C.M., Loffredo, F.S., Wojtkiewicz, G.R., Chen, J.W., Lee, R.T., Wagers, A.J., Rubin, L.L. (2014). Vascular and neurogenic rejuvenation of the aging mouse brain by young systemic factor. Science *344*, 630–634.

Villeda, S.A., Luo, J., Mosher, K.I., Zou, B., Britschgi, M., Bieri, G., Stan, T.M., Fainberg, N., Ding, Z., Eggel, A., Lucin, K.M., Czirr, E., Park, J.-S., Coullard-Després, S., Aigner, L., Li, G., Peskind, E.R., Kaye, J.A., Quinn, J.F., Galasko, D.R., Xie, X.S., Rando, T.A., Wyss-Coray, T. (2011). The ageing systemic milieu negatively regulates neurogenesis and cognitive function. Nature *477*, 90–94.

Ottone, C., Krusche, B., Whitby, A., Clements, M., Quadrato, G., Pitulesco, M.E., Adams, R.H., Parrinello, S. (2014). Direct cell-cell contact with the vascular niche maintains quiescent neural stem cells. Nat. Cell. Biol. *16*, 1045–1056.

Ramírez-Castillejo, C., Sánchez-Sánchez, F., Andreu-Agulló, C., Ferrón, S.R., Aroca-Aguilar, J.D., Sánchez, P., Mira, H., Escribano, J., Fariñas, I. (2006). Pigment epithelium-derived factor is a niche signal for neural stem cell renewal. Nature Neurosci. *9*, 331–339.

Gómez-Gaviro, M.V., Scott, C.E., Sesay, A.K., Matheu, A., Booth, S., Galichet, C., Lovell-Bage, R. Betacellulin promotes cell proliferation in the neural stem cell niche and stimulates neurogenesis. Proc. Natl. Acad. Sci. USA. *109*, 1317–1322.

Delgado, A.C., Ferrón, S.R., Vicente, D., Porlan, E., Perez-Villalba, A., Trujillo, C.M., D'Ocón, P., Fariñas, I. (2014). Endothelial NT-3 delivered by vasculature and CSF promotes quiescence of subependymal neural stem cells through nitric oxide induction. Neuron *83*, 572–585.

19

Non-hematopoietic Stem Cells of Bone and Bone Marrow

Pierre Charbord

Sorbonne Université, CNRS, Inserm U1156, Institut de Biologie Paris Seine, Laboratoire de Biologie du Développement/UMR7622, 9 Quai St-Bernard, 75005 Paris, France
E-mail: pierre.charbord@sorbonne-universite.fr

19.1 Historical Background

Alexander Friedenstein, has opened the field in the 1960s and 1970s (review in (Friedenstein et al., 1970)). He found that bone marrow (BM) cells (from different species including humans) seeded in liquid cultures containing serum generated after one to two weeks discrete colonies consisting in plastic-adherent, non-phagocytic, and elongated cells of fibroblastic appearance. The clonogenic cells were not labeled following a single administration of H^3-thymidine to adult animals. A few weeks after transplantation, each colony seeded under the renal capsule gave rise to either fibrous tissue, bone, or to bone containing marrow. Using chimeric animals, Friedenstein further showed that marrow hematopoietic cells within the bony spaces were of recipient origin, contrarily to stromal cells that were from the donor. These data indicated that the BM of various animal species harbored mesenchymal colony–forming cells and that some of these developed a bone-forming tissue, where circulating HSCs could home and subsequently develop hematopoiesis in the BM logettes. These results did not receive the attention they deserved probably because they were published piecemeal over several years in relatively low-impact journals by a Russian researcher (at this time Soviet Union citizen) outside the mainstream of research.

In 1991, Arnold Caplan, drawing mainly from his experience on cartilage and bone development, posited the existence of a "Mesenchymal Stem cell" (MSC) that would give rise to different mesodermal lineages including not only osteoblasts (O), chondrocytes (C), adipocytes (A), and BM stromal cells, but also muscle and tendon cells and cells from the dermis (Caplan, 1991). Furthermore, he proposed to purify the human mesenchymal cells by their adherence to plastic and subsequent culture through many passages. Finally, he suggested that the MSCs would be useful as a cell therapy for the treatment of skeletal defects. Over the years, specific culture conditions have been established enabling cell proliferation or cell differentiation into A or O or C. Proliferation is optimal using screened serum. A and O differentiation requires supplementation with dexamethasone while C is obtained in micropellets in the presence of Transforming Growth Factor β (TGFβ). The reader will find full details in (Delorme and Charbord, 2007).

That mesenchymal cells forming the passaged adherent layers derived from plastic-adherent cells comprised cells at different stages of differentiation from a potential small population of stem cells was soon apparent, leading to an active research of such stem cells. One study identified a subpopulation of rapidly expanding and self-renewing cells within primary layers of human BM (Colter et al., 2000; Colter et al., 2001). These cells were obtained from week 2 adherent layers sub-cultured at very low cell density, and generated A, O, and C colonies when cultured in appropriate conditions. In another study, mesenchymal cells from human primary layers were collected at the end of the first passage (Lee et al., 2010). Cells were transduced using a lentiviral vector expressing the enhanced green fluorescent protein (EGFP). Single fluorescent cells cultured on a confluent irradiated feeder layer gave rise to high-proliferative–potential colonies that could be differentiated into the A, O, and C lineages. After pooling colonies and re-sorting, some of the single cells gave rise again to high-proliferative–potential colonies. In another study, cells from week 2 human primary layers were labeled with a membrane fluorescent marker (Russell et al., 2010). Fluorescent clones were obtained by limiting dilution. Proliferative clones were further differentiated into A, O, and C lineages, which revealed large heterogeneity with a majority of trilineal clones, but also bilineal and even unilineal ones. The highly proliferative clones appeared to be those with the greater differentiation potential. Taken together, these data confirmed that within an apparently morphologically homogeneous cell layer resided, even after some weeks, stem-like cells, tripotential and with limited self-renewal capacity. However, such cells were difficult to unravel and to quantify.

In parallel studies, in 1977, Michael Dexter et al. described a long-term culture system where murine Hematopoietic Stem Cells (HSCs) were generated over weeks providing the development of a feeder layer of stromal cells (Dexter et al., 1977). This system was adapted to humans a few years later (Gartner and Kaplan, 1980). These studies were made in the field of hematopoiesis by researchers whose interests were far removed from those of investigators in orthopedic research. Were human BM stromal cells generated in the long-term culture "Dexter" system, in a complex medium containing horse and fetal calf sera and hydrocortisone, similar to human BM mesenchymal cells, generated from plastic-adherent cells cultured in a simpler medium containing fetal beef serum ("Caplan" system)? One difference appeared to be the vascular smooth muscle (VSM) differentiation of mesenchymal cells observed with the "Dexter" system, but not with the "Caplan" one. One study showed that stromal cells generated in "Dexter" medium were mesenchymal cells differentiating over time into VSM cells expressing Smooth Muscle-actin (αSM-actin), calponin, metavin-culin, heavy chain caldesmon, and SM-myosin, making them similar to VSM cells found in the aortic intima in atherosclerotic lesions (Galmiche et al., 1993). Another study showed that stromal cells from week 3–6 adherent layers of human cultures when seeded in semi-solid medium supplemented with inflammatory cytokines gave rise to colonies, in turn able to generate in "Dexter" medium in the presence of Fibroblast Growth Factor 2 (FGF2) cell lines with hematopoietic-supportive capacity (Sensebe et al., 1995). These cells expressed mesenchymal markers such as interstitial collagens, fibronectin, laminin, and tenascin, and over time, VSM markers (Li et al., 1995). It was also shown that stromal cells expressing the Stro-1 membrane antigen and isolated from day 7 "Dexter" cultures, as well as a few immortalized murine stromal cell lines, were able to differentiate under adequate culture conditions into A, O, and C and VSM cells and to support hematopoiesis (Chateauvieux et al., 2007; Dennis et al., 2002; Gao et al., 2010). Finally, clones generated from sorted human BM cells could be differentiated into A, O, and C cells and into myocardin[+] VSM cells when cultured in "Dexter" like conditions (Delorme et al., 2009). Taken together, these studies suggested the presence in the "Dexter" culture of a few stem-like cells, quadripotential and with hematopoiesis-supportive capacity. The difference between "Dexter" and "Caplan" culture medium is probably a higher concentration of cytokines in the former, in particular TGFβ (Sensebe et al., 1997).

In the mid-1990s, the team of Darwin Prockop showed that murine plastic-adherent culture-expanded mesenchymal cells were transplantable, being detected after several months in irradiated recipients in cartilage, bone, BM, and spleen (Pereira et al., 1995). However, the percentage of donor cells found in tissues was low, estimated by DNA polymerase chain reaction assays from 2% to 12% of the cellular content. Moreover, transplantation in mice of mesenchymal cells issued from mice with a phenotype resembling brittle bone disease (due to the expression of a human minigene for collagen I) resulted in a small but significant increase in collagen and mineral content of bone one month after injection (Pereira et al., 1998). These data suggested that some of the plastic-adherent mesenchymal cells were able to home to bone, BM, and cartilage tissues, and, to some extent, improve the functionality of impaired bone. Such studies gave great impetus to the use of mesenchymal cells in pre-clinical and clinical studies.

In 1999, Mark Pittenger et al. showed that a few single cells directly isolated from human BM gave rise to colonies made of highly proliferating cells (19 to 21 population doublings from the initial cell) that could be differentiated into A, O, and C in adequate culture conditions (Pittenger et al., 1999). Membrane antigen analysis indicated the lack of expression of the pan-hematopoietic marker CD45 and of the endothelial marker CD31. Another study indicated the heterogeneity of such highly proliferative clones, one third of which were tripotent (Muraglia et al., 2000).

In 2007, the team of Paolo Bianco demonstrated that human BM CD45^{neg} but CD146$^+$ Colony-Forming Unit-fibroblasts (CFU-F), culture-expanded, attached to hydroxy-apatite tricalcium phosphate particles and embedded in a fibrin gel, when transplanted under the skin of immunocompromised mice formed a tissue consisting of bone and fibrosis at week 4, then of bone and sinusoids developed within the fibrous tissue at week 7, and finally by week 8 of bone, sinusoids, and BM logettes filled with hematopoietic cells in place of the fibrous tissue (Sacchetti et al., 2007). The endothelium of the sinusoids was murine, contrasting with abluminal reticular stromal cells that were human. Some of the transplants were recovered, collagenase-digested, and the resulting cells were culture-expanded, and then sorted according to CD146 expression. The CD146$^+$ cells, but not the CD146^{neg} cells, generated CFU-Fs at a frequency similar to, or greater than the CFU-Fs used to generate the explants. These data expanded the observations made by Friedenstein, demonstrating the self-renewal of some human BM mesenchymal cells by showing that one colony of cells with a definite membrane phenotype was able after implantation under the skin of immune deficient mouse to give at least one colony of cells presenting the same phenotype.

In 2009, Morikawa et al. identified murine cells that were CD45^{neg} but expressed the platelet-derived growth factor receptor α (Pdgfra) and the membrane antigen Sca-1 (consequently called PαS cells) (Morikawa et al., 2009a). Cells were extracted from femur and tibia bone fragments crushed with a pestle and enzymatically digested. Culture-expanded clones could be differentiated, in adequate culture conditions, into O and C, and some of them also into A and endothelial-like cells expressing CD31 and VE-cadherin. PαS cells from transgenic EGFP mice were transplanted into lethally irradiated mice (that also received an HSC transplant). Mice were sacrificed after 12 weeks; perilipin^{+} adipocytes, osteocalcin^{+} endosteal osteoblasts, and perivascular cells expressing the hematopoietic chemokine CXCL12 and associated to, but distinct from, αSM-actin^{+} VSM were found on femoral sections. Finally, GFP^{+} PαS were transplanted into wild-type mice and after 16 weeks the bone and BM were recovered and collagenase-digested. A few sorted GFP^{+} cells were clonogenic and able to differentiate into the three mesenchymal lineages. These data confirmed in the mouse the results obtained by Bianco et al. in humans, i.e., the presence in bone and BM of clonogenic cells able to self-renew and to generate *in vitro* and *in vivo* A, O, C, and stromal cells.

In sum, by the late 2000s, it was established that BM mesenchymal cells could be cultured with relative ease and that some of the cells were bona fide stem cells. In view of the widespread inaccurate use in pre-clinical and clinical publications of "mesenchymal *stem* cells" for plastic-adherent cells cultured in "Caplan" conditions, the International Society for Cellular Therapy (ISCT) suggested to use the term "mesenchymal *stromal* cells" unless the stemness of the cells had been clearly evidenced (Horwitz et al., 2005). Moreover, ISCT suggested a minimal set of standards to define these cells, adherence to plastic, specific set of membrane antigens (expression of CD105, CD73, CD90 and lack of expression of CD45, CD34, HLA-DR, and "lineage" markers of granulocytes and monocytes and B- and T-lymphocytes), and multipotent differentiation potential *in vitro* (Dominici et al., 2006). We will see that over the next years, new nomenclature will be proposed, fitting with the explosive demonstration of bone and BM mesenchymal stem cells using murine systems.

Over the years, it has been shown that stromal layers were capable of inhibiting the growth, or modifying the behavior of different categories of cells implicated in adaptive and innate immunity. This immunomodulatory property is of great clinical relevance and is the subject of a myriad of studies. However, it is a property unrelated to stemness. Different reports have shown that dermal fibroblasts were able to suppress the mixed lymphocyte reaction similarly to BM mesenchymal cells, leading to the hypothesis that immunosuppression is a property of the mesenchyme irrespective of its location (Haniffa et al., 2009; Haniffa et al., 2007; Jones et al., 2007). Recent data indicate that murine fibroblasts together with endothelial and epithelial cells are key regulators of organ-specific immune responses (Krausgruber et al., 2020). Gene expression profiles of these "structural" cells within the same organ were more similar to each other than cells of the same type across different organs. A network of structural cell to immune cell interactions was inferred based on known receptor–ligand pairs. Identification of the transcription factors responsible for the gene expression revealed a set of genes poised for rapid response in case of infection or cytokine administration. In short, the immune function of mesenchymal cells appears to be not only unrelated to stemness, but also shared with other cell types, altogether modulating the immune response at the organ level.

19.2 Bona fide Murine Stem Cells

In 2010, Méndez-Ferrer et al. showed that $BMCD45^{neg}/nestin^+$ cells from nestin-GFP transgenic mice gave rise to CFU-Fs contrarily to $CD45^{neg}/nestin^{neg}$ cells that were devoid of clonogenic activity (Méndez-Ferrer et al., 2010). When cultured in adequate media CFU-Fs differentiated into A, O, and C. In liquid culture containing chicken embryo extract, and supplemented with various non-hematopoietic growth factors, the $CD45^{neg}/nestin^+$ cells gave rise to non-adherent spheres (called mesenspheres) that, after two weeks, spontaneously differentiated into the A, O, and C lineages. Moreover, individual mesenspheres implanted subcutaneously into wild-type mice gave rise 2 months after transplantation to ossicles often associated with hematopoietic activity. Ossicles recovered from the mice and enzymatically digested yielded new mesenspheres, which after implantation into secondary mice gave rise to secondary ossicles. Lineage-tracing studies using Nes-Cre^{ERT2} mice in which the Cre recombinase was under the control of nestin promoter elements revealed the presence 8 months after tamoxifen treatment of bone-lining osteoblasts, osteocytes, and chondrocytes. Finally, the nestin compartment was depleted by intercrossing a Cre recombinase inducible diphteria toxin receptor line with Nes-Cre^{ERT2} mice and treating the mice with tamoxifen and diphteria toxin. In the treated mice, the most immature hematopoietic precursors, including HSCs, were reduced by 50%. These data clearly indicated that BM harbored mesenchymal self-renewing and multipotential cells, which contributed physiologically to osteochondral lineages and to the *in vivo* maintenance of HSCs.

In subsequent years,the presence of non-hematopoietic stem cells in bone, BM, and cartilage of mice was demonstrated according to several criteria listed below. Applying those criteria is a difficult task implying a range of techniques and approaches that were progressively implemented by a number of teams and are available only in the last decade. Table 19.1 is a non exhaustive list of papers reporting stem cells fulfilling most of these criteria:

(a) Rarity: the estimated stem cell frequency is variable from one report to another, due to cell preparation (flushed BM $\pm$ bone enzymatically digested), mode of expression (value related to total nucleated or mononuclear cells), and degree of immaturity. Most reported values are within the range 0.001–0.01% of the total Bone + BM cell content.

(b) Clonogenicity has been evidenced as reported previously (CFU-F, mensphere).

(c) Quiescence *in vivo* has been shown by showing that cell cycle transcripts were downregulated in the stem cell population as compared with non-stem cells (Méndez-Ferrer et al., 2010), by measuring the incorporation of thymidine nucleoside analogs (Mizuhashi et al., 2018; Zhou et al., 2014),or by analyzing the expression of Ki67 and DNA content (Qian et al., 2013).

(d) Transplantability after isolation by phenotype was assessed by injecting intravenously or intra-femorally the stem cell population in lethally (then together with HSC) or sublethally irradiated mice, and detecting donor cells on BM sections. The detection time was variable from 4 weeks (Zhou et al., 2014) to 4 months (Morikawa et al., 2009a; Park et al., 2012). Of note, in the study of Park et al. the homing capacity to bone of the Mx1$^+$ cells was 15–20% that of HSCs (double transplantation in irradiated animals).

(e) Self-renewal capacity: it was assessed in most reports either *in vitro* by serial cultures of colonies, and/or *in vivo* by recovering subcutaneous ossicles of grafts under the kidney capsules and assaying the clonogenic cells from the recovered specimens. In some instances, self-renewing cells were recovered from tissues after injury such as bone fracture sites (Park et al., 2012; Worthley et al., 2015).

(f) Commitment to differentiation: it was assessed in all reports, either *in vitro* by culturing culture-expanded clonogenic cells in A, O, and C conditions, and/or *in vivo* by following the progeny of stem cells implanted under the kidney capsule or in subcutaneous grafts.

Table 19.1 Non-hematopoietic Bone and BM Murine Stem Cells

Report	Phenotype	Location	Stemness Attribute
Morikawa et al., 2009	CD45–/TER119–/CD140a+/Sca-1+	Bone; BM: endosteal, perivascular	a; b; d; e; f (A, O, C vtvv)
Mendez-Ferrer et al., 2010	CD45–/Nestin+	BM: perivascular	a; b; c; e; f (A, O, C vt; OC vv)
Park et al., 2012	CD45–/TER119–/CD31–/CD105+/CD140a+/Mx1+	BM:endosteal	a; b; d; e; f (A, O, C vt; O vv); g (O, calvaria)
Pinho et al., 2013	CD45–/TER119–/CD31–/CD51+/CD140a+	BM:perivascular	a; b; e; f (A, O, C vt)
Qian et al., 2013	CD45–/TER119–/Ebf2+	BM: endosteal	a; b; c; e; f (A, O, C vt; O vv)
Zhou et al., 2014	CD45–/TER119–/CD140a+/CD51+/Lepr+	BM: perivascular	a; b; c; d; e; f (A, O, C vt); g (O: fracture; C: cartilage defect)
Worthley et al., 2015	CD45–/TER119–/CD140a+/Grem+	Bone (adjacent to growth plate and trabecular)	a; b; e; f (O, C vt; O, C, Stroma vv), g (O,C: fracture); h
Chan et al., 2015	CD45–/TER119–/Tie2–/CD90–/CD249–/CD105–/CD51+/CD200+	Bone; Growth plate	a; b; e; f (O, C, Stroma vtvv); g (O: fracture)
Kramann et al., 2015	Gli+	BM: endosteal, perivascular	a; b; f (A, O, C vt)
Mizuhashi et al., 2018	Pthrp+	Growth plate	b; c; e; f (A, O, C vt; OC v); h
Debnath et al., 2018	CD45–/TER119–/CD105–/CD90–/CD51low/CD200+/Ctsk+	Periosteum	b; e; f (A, O, C vt; O vv); g (O: fracture); h
Breitbach et al., 2018	CD45–/CD73+	Bone; BM: endothelium and few perivascular	a; b; f (A, O, C vt)
Zhao et al., 2019	N-Cadh+	BM: endosteal, perivascular	b; f (A, O, C vt; A, O vv adult; C vvdvpt), g (C: cartilage defect)
Newton et al., 2019	CD49e+/CD73+	Growth plate	b; f (A, O, C vt; C vv)

Abbreviations: a, b, c, d, e, f, g, h, A, O, C, BM: see text; vt: in vitro; vv: in vivo; vtvv: in vitro and in vivo; dvpt: development.

In transgenic mice with inducible promoter lineage-tracing studies allowed determining the nature and timing of differentiation following tamoxifen administration. One should note that trilineal differentiation was observed *in vitro* in all reports except two (Chan et al., 2015; Worthley et al., 2015), often contrasting with more restricted differentiation potential in studies *in vivo*.

(g) Regenerative capacity: it was assessed by injuring the tissue of interest (such bone fracture or perforation of the cartilage) and following the repair in transgenic animals after tamoxifen induction (Debnath et al., 2018; Park et al., 2012; Worthley et al., 2015; Zhao et al., 2019; Zhou et al., 2014).

(h) Depletion *in vivo*: this was assessed by inducing loss of the gene characteristic of the stem cell population thereby reducing the size and/or the functional capacity of its progeny (Debnath et al., 2018; Mizuhashi et al., 2018; Worthley et al., 2015). The most usual strategy was to use a diphteria toxin receptor as explained above. This strategy has allowed altering the periosteum and decreasing the total bone volume when depleting the Cstk gene (Mizuhashi et al., 2018) and the Grem1 gene (Worthley et al., 2015), respectively.

(i) Lack of expression of hematopoietic markers ($CD45^{neg}$ and hematopoietic $Lineage^{neg}$ cells)and of the endothelial marker CD31 in all cases except one (Breitbach et al., 2018).

The stem cells described above represent very different cell identities. One major difference is their location that already informs on their potential. BM cells may be perivascular, endosteal, or both. Stem cells located in the endosteal region appear to be osteogenic, even though they may give rise to the three mesenchymal lineages *in vitro* (Park et al., 2012; Qian et al., 2013). Perivascular stem cells located in the peri-sinusoidal space are tripotential and HSC-supportive (Méndez-Ferrer et al., 2010; Pinho et al., 2013; Zhou et al., 2014). These cells can be sorted according to the expression of CD140a (Pdgfra), CD51 (Itgav), and CD295 (Lepr). Although located mainly in the endosteal region, $Ncadh^+$ ($Ncadherin^+$) cells are also tripotential but may differ from perivascular stem cells in their stromal function since they are supportive not only of chemotherapy-sensitive HSCs but also of "reserve" HSCs resistant to chemotherapy (Zhao et al., 2019). $Gli1^+$ cells are located both in the endosteal and perivascular region and are tripotential; in fact these cells are present around the vasculature of many tissues and appear responsible by differentiating into $\alpha SM\text{-}actin^+$ myofibroblasts for the development of fibrosis after injury of the organ of their location (Kramann et al., 2015). Stem cells extracted from bone represent distinct cell types. The $CD140a^+/Sca\text{-}1^+$ cells are tripotential, but also, as already indicated, able to differentiate into endothelial-like cells (Morikawa et al., 2009a). The $CD140a^+$ and $Grem^+$ (gremlin 1-positive) cells are osteo and chondrogenic, but also able to generate BM reticular stromal cells whose HSC-supportive ability has not been studied (Worthley et al., 2015); moreover, such reticular cells are also observed within the intestinal subepithelial mesenchymal sheath. The $CD200^+/CD51^+$ Skeletal Stem Cells (SSC) generate $CD200^{neg}/CD51^+/CD105^+$ Bone-Cartilage-Stroma Progenitors (BCSP) and $CD200^+/CD51^{neg}/CD105^+/CD90^+$ ProChondrogenic Progenitors (PCP) (Chan et al., 2015). BCSP will in turn generate different types of stromal cells that will acquire or lose some membrane antigens (CD51, CD105, and CD249) and would play different roles as hematopoiesis-supportive cells. In sum, bone stem cells may appear to represent cells more immature than the BM ones since endowed with larger differentiation potential. $CD200^+/CD51^{low}/Ctsk^+$ (cathepsin K-positive) periosteal stem cells are osteogenic and do not express markers of perivascular or endosteal stem cells, in agreement with the restriction of hematopoiesis to the inner bone cavity (Debnath et al., 2018). $Pthrp^+$ (Parathyroid hormone–related protein-positive) and $CD49e^+/CD73^+$ stem cells located in the growth plate are chondrogenic (Mizuhashi et al., 2018; Newton et al., 2019), although the $Pthrp^+$ cells in the resting zone may also become osteogenic with the formation of secondary ossification centers. The $CD73^+$ cells described by Breitbach et al. appear to constitute a special population present not only in bone where it is able to generate clonogenic tripotential cells *in vitro*, but also in a subpopulation of BM sinusoidal $CD31^+$ endothelial cells and very rarely in $CD31^{neg}$ perivascular cells (Breitbach et al., 2018). A schematic representation of bone during childhood is shown on Figure 19.1A. Central BM with large sinusoids filled with red cells delimiting the marrow logettes filled with hematopoietic cells is shown on Figure 19.1B; the network of Leptin receptor–positive perivascular cells delineates the sinusoids from the logettes; of course, stem cells constitute only a minute subset of this population of Lepr+ cells.

All the afore described stem cells are obtained in the adult, or at least post-natally. During development, other stem cells have been described. It has been shown that in the limb bud the majority of $Lineage^{neg}$ cells were $CD140a^+/CD51^+$ (Nusspaumer et al., 2017). During fetal long bone development, another population of $CD140a^+/Sca1^+$ (PaS) is detected and this population is predominant around birth. In turn, this population is replaced

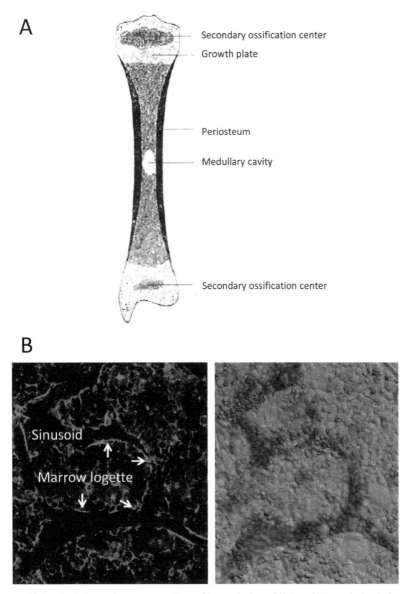

Figure 19.1 **A**: Schematic representation of bone during childhood. Bone is in dark gray, cartilage in light gray. Figure modified from Rouvière H: Anatomie Humaine, 2nd Edition, 1927 (Masson editor, Paris). **B**: Central BM with large sinusoids filled with red cells delimiting the marrow logettes filled with hematopoietic cells (section of mouse femur). **Left panel**: the network of Leptin receptor–positive perivascular cells labeled red (*arrows*) delineates the sinusoids from the logettes; these cells are located on the abluminal side of endothelial cells limiting the sinus lumen (not shown). **Right panel**: corresponding bright field view.

by that of CD200$^+$/CD51$^+$ SSCs that become predominant in the juvenile. In another report Osx$^+$ (Osterix$^+$) cells were detected in the perichondrium before formation of the marrow cavity (Mizoguchi et al., 2014). At E13.5 and early time after birth, Osx$^+$ cells were detected not only in bone tissues, but also in perivascular stromal cells. By contrast, in the adult, these cells were scarce, the labeling being restricted to bone lining osteoblasts. The Osx$^+$ cells detected around birth possessed stem cell properties (tripotential, clonogenic, and self-renewing cells).

Murine strains, resulting from the cross of a driver mouse in which a critical gene promoter drives the expression of an eventually inducible Cre recombinase with a reporter mouse expressing a fluorescent or luminescent protein, have proven essential to identify many of the aforementioned stem cells. Such genetic models were required when the critical gene coded for an intracellular or secreted protein such as the intermediate filament Nestin, the Interferon-induced GTP-binding protein Mx1, the transcription factors Ebf2 or Gli1, the Bone morphogenic protein antagonist Gremlin-1 (Grem), the Parathyroid hormone–related protein (Pthrp), or the protease Cathepsin K (Ctsk). However, genetic models were also used in cases in which the critical gene coded for a cellular membrane protein such as Leptin receptor (Lepr), N-cadherin (Ncadh), or CD73, due to the lack of adequate antibodies and the ease of detection of the reporters. However, the use of such strains is not without bias, such as off-target effects, side effects of the inducer molecule (tamoxifen), or limited recombination frequency. The reader will find extensive discussions of these problems in (Méndez-Ferrer et al., 2015) and (Chen et al., 2017). A recent report has studied the membrane profile of some of the genetically identified cells, confirming little content of Nestin-GFP$^+$ and Lepr-Cre$^+$ in osteoblastic progenitors (Mende et al., 2019).

19.3 Stem Cells in Human Adult Bone and BM

The first membrane marker used for the generation of clonogenic cells in the human adult BM was that recognized by the monoclonal antibody Stro1 (Simmons and Torok-Storb, 1991). It was shown that Stro-1$^+$ cells could generate adherent layers including adipocytes, fibroblastoid cells, and cells expressing αSM and γSM actin. These cells grown under osteogenic condition generated fully functional osteoblasts (Gronthos et al., 2003). As indicated above Stro-1$^+$ cells recovered from "Dexter" type cultures supplemented with FGF2 showed trilineal differentiation potential and hematopoietic-supportive capacity (Dennis et al., 2002); ultrastructural studies indicated that some of the cells were polarized, one pole showing myofilaments and fibronexus and the opposite one displaying lipid-laden vesicles. Taken together, these data indicated that the Stro-1$^+$ cells were indeed bona fide human BM mesenchymal stem cells. However, the antigen has not been fully characterized, which was one of the reasons that led to search for other markers allowing the isolation of a population with high cloning efficiency in CFU-F assays. The membrane antigens recognized were diverse: low-affinity neurotrophin receptor (CD271), ecto-enzyme CD73, cytokine and growth factor receptors (CD105, CD130), cell adhesion molecules (CD49a, CD49b, CD56, CD63, CD90, CD106, CD146, CD166, CD200, and integrin $\alpha v \beta 5$), neural ganglioside GD2, carbohydrate SSEA-4, a tetherin known for blocking viral infection (CD317), and some biochemically ill-defined plasma membrane components (D7-FIB, MSCA-1) (Battula et al., 2009; Delorme et al., 2008; Deschaseaux and Charbord, 2000; Filshie et al., 1998; Gang et al., 2007; Gronthos et al., 2003; James et al., 2015; Jones et al., 2002; Martinez et al., 2007; Quirici et al., 2002; Stewart et al., 2003). In most cases, the population tested was devoid of CD45$^+$ hematopoietic cells.

None of the markers indicated above was specific for BM mesenchymal stem cells, with the possible exception of Stro-1. As for other stem cells such as HSCs, combination of markers was investigated to find out which could yield highly enriched clonogenic mesenchymal stem cells. An early attempt was to select for Stro-1^{bright}/CD106$^+$ (VCAM1$^+$), which resulted in a 950-fold enrichment of CFU-F relative to their incidence in the unseparated mononuclear cell fraction (Gronthos et al., 2003). These cells lacked detectable expression of the Ki-67 antigen and were able to differentiate into osteoblasts. Other marker combinations included membrane antigens used for the isolation of murine stem cells. In one report, cells sorted according to the lack of expression of hematopoietic (CD45), endothelial (CD31), and erythroid (CD71) markers and the expression of CD146 and CD105 gave rise to mensespheres that when cultured in a cytokine-enriched medium containing chicken embryo extract remained unattached to the plastic and could be propagated over several passages while still containing CD146$^+$/CD105$^+$ cells (Isern et al., 2013); after digestion and replating, the mesenspheres gave rise to A, O, and C. In another study the staining pattern of novel antibodies was compared to that of CD271 on Lineage^{neg} BM mononuclear cells (Li et al., 2014). CD295 (Leptin receptor) or CD106 expression paralleled that of CD271, suggesting that no further enrichment should be expected if associating these markers to CD271. On the contrary, CD140a expression showed a more orthogonal expression and two populations were observed when plotting CD140a vs. CD271 (CD140a^{high}/CD271$^+$ and CD140a^{low}/CD271$^+$ cells). The CD140a^{low}/CD271$^+$ fraction contained clonogenic cells with trilineal differentiation potential *in vitro* and *in vivo* and with hematopoietic-supportive capacity. In another study, CD271 was used in association with CD146 after depletion of the CD45$^+$ cells to isolate human clonogenic cells with trilineal differentiation potential *in vitro* and *in vivo* (Tormin et al., 2011). The clonogenic cells were selected both from the CD271$^+$/CD146^{high} and the CD271$^+$/CD146^{low} fractions. In the BM, perivascular cells expressed both antigens, while endosteal cells expressed only CD271, and αSM-actin$^+$ smooth muscle cells in the media of small arteries expressed only CD146. The lack of expression of CD146 by endosteal cells was suggested to be related to the lower oxygen level observed near the bone surface (as compared to more central marrow) since CD146 expression decreased in cells cultured in hypoxic condition. In sum, different combinations of membrane antigens (Stro-1, CD105, CD106, CD140a, CD146, and CD271) appear adequate to sort very small (approximately 0.01–0.1% of the mononuclear cell fraction) BM clonogenic populations able to differentiate into the mesenchymal lineages and with hematopoiesis-supportive capacity. As in the mouse system, the diversity of membrane phenotypes probably reflects the heterogeneity of the human BM stem cell compartment according to location (endosteal, perivascular, or both) and to some of the stem cell attributes.

Contrarily to cells from BM collected from aspirates, cells from bone are obtained by enzymatic treatment of crushed bone fragments. One study has shown that the CFU-F frequency was far greater in bone than in BM (Mabuchi et al., 2013), confirming reports in mice (Mende et al., 2019; Morikawa et al., 2009a). Sorted CD271$^+$/CD90$^+$/CD106$^+$ cells yielded a majority of rapidly expanding clones (Mabuchi et al., 2013); the highly enriched (one of three isolated cells) clonogenic cells were self-renewing, multipotential and showed high cellular motility. The team of Irving Weissman identified the human counterparts of the murine SSCs (Chan et al., 2018). Two cell populations (PDPN$^+$/CD73$^+$/CD164$^+$/CD146^{neg} and PDPN$^+$/CD146^{low}) were multipotent and capable of endochondral differentiation when implanted under the kidney capsule. These cell subsets generated osteogenic (PDPN^{neg}/CD146$^+$/CD90$^{hi/low}$) or chondrogenic (PDPN$^+$/CD146^{neg}/CD164$^{+/neg}$/CD73$^{+/neg}$) progenitors. The osteoblast progenitors showed the largest HSC-supportive capacity.

19.4 A Stem Cell for Mesenchymal Stem Cells?

One pending question in the field is whether non-hematopoietic bone and BM stem cells would derive from more primitive stem cells found during development and, eventually, in adults. Three such hypothetical cells have been brought forth, that mesenchymal stem cells would derive from a common HSC/stromal precursor, from neural crest stem cells, or from stem cells with mesenchymal and endothelial properties.

In the 1990s some data have suggested the existence of a common HSC/stromal precursor in the human adult BM. Some of the supportive data have been retracted. Details can be found in another review (Charbord, 2010), and the hypothesis of a common precursor does not stand according to present day criteria of evidence.

Some of the markers used to identify clonogenic mesenchymal stromal cells in human or mouse BM are expressed widely in the nervous system: nestin, CD56, CD271, GD2. In addition, some arguments suggest the neuroectodermal origin of mesenchymal stem cells. Cell tracking studies have shown a long while ago that the connective tissue above the aortic arch derives from the neuroectoderm. More recently, it has been shown that neural crest stem cells generated from embryonic stem cells can give rise under appropriate conditions to A, O, C, and VSM cells (Lee et al., 2007). Another report indicated that the neuroepithelium supplies the earliest wave of MSC differentiation in the mouse embryo (Takashima et al., 2007). In addition, few mouse BM MSCs were proven to be neuroectodermal in origin (Morikawa et al., 2009b; Nagoshi et al., 2008), and comparison of neonatal mouse BM Nestin[pos]/Pdgfra[pos] to Nestin[pos]/Pdgfra[neg] mesenchymal stromal cells revealed enrichment in the Schwann precursor phenotype in the Nestin[pos]/Pdgfra[neg] population (Isern et al., 2014). However, in view of the wealth of reports on human or mouse adult mesenchymal cells, the articles indicating a neuroectodermal origin are scarce, and a mesodermal origin may be safely asserted in most cases (Dennis and Charbord, 2002).

The team of Igor Slukvin has provided data supporting the hypothesis that mesodermal cells with mesenchymal and endothelial potential constitute stem cells for mesenchymal stem cells. Human pluripotent stem cells (embryonic stem cells or induced pluripotent cells) were grown on the fetal calvaria OP9 cell line known as an efficient differentiation system for cells of mesodermal lineages (Vodyanik et al., 2010). Single cells plated in semisolid medium in the presence of FGF2 gave rise to compact colonies from which could be derived cell lines. Cells from colonies and lines expressed mesenchymal markers (CD146, CD271, CD140a, CD90, CD106, CD56, CD166), but a minority of cells also expressed endothelial markers such as TEK and EMCN, but no hematopoietic membrane antigens such as CD45. Of note, CD73 was not expressed in cells from clones, but was upregulated in cells from lines. The lines could be differentiated in A, O, and C *in vitro* and were able, when plated on Matrigel matrix, to organize into a network of typical vascular tubes. It was shown that APLNR (Apelin receptor) was upregulated in cultured pluripotent cells concomitantly with mesodermal commitment, and selection of APLNR^{neg} cells resulted in almost entire deletion of compact colony formation. The clonogenic cells generating the spheroid compact colonies were called mesenchymoangioblasts due to their dual mesenchymal and endothelial potential. A second paper reported the generation of pericytes and smooth muscle cells from mesenchymoangioblasts (Kumar et al., 2017). The progenies were obtained by culturing the cells in the presence of different growth factors such as FGF2 and platelet-derived growth factor β for immature pericytes, or TGFβ1 and sphingosyl-phosphorylcholine for immature smooth muscle cells. Pericytes were either of the capillary/proinflammatory type (NG2^{+}, desmin^{neg}) or of the arteriolar/contractile type (NG2^{high}, aSM actin^{+}, desmin^{+}). Smooth muscle cells were either of the proliferative/synthetic type (NG2^{low}, calponin^{+}, aSM actin^{+}, desmin^{low}, SM myosin heavy chain^{neg}) or of the contractile type (NG2^{neg}, calponin^{high}, aSM actin^{high}, desmin^{high}, SM myosin heavy chain^{+}). When cultured with human umbilical vascular endothelial cells pericytes increased the length and longevity of the vascular tubes. In sum, these data suggest that mesenchymoangioblasts are the precursors not only of mesenchymal stem cells but also of mural vascular cells. These data also suggest that BM mesenchymal clonogenic human cells cultured under "Dexter" conditions may be akin to mesenchymoangioblasts, which would explain, as already indicated, their generation of VSM cells (Charbord et al., 2002; Chateauvieux et al., 2007; Delorme et al., 2009; Galmiche et al., 1993; Li et al., 1995). Such cells may also be the founders of some stromal murine lines generated from BM or other developmental sites of hematopoiesis (Charbord et al., 2002; Chateauvieux et al., 2007).

19.5 Models of Stem Cell Differentiation: Hierarchical or Plastic?

Generation of CFU-Fs from human BM or from primary layers has suggested a hierarchy of differentiation potentials, the osteoblastic being the more frequent "default" pathway (Muraglia et al., 2000), although all lineage combinations could be found suggesting a complex lineage tree from tripotent to bipotent, and unipotent cells (Russell et al., 2010). Murine SSCs were shown to give rise to Bone-Chondro-Stromal Progenitors (BCSP), in turn differentiating in unipotent progenitors for chondrocytes or different types of stromal cells implicated at varying levels in hematopoiesis support (Chan et al., 2015). The differentiation pathways were less complex for human SSCs, but still a tree was suggested with stem cells at the apex and multiple downstream bifurcation points, reminiscent of what has been described for HSCs.

However, multiple observations have challenged this simple hierarchical model, indicating that mesenchymal cells can shift from one differentiation pathway to another under modified external conditions. It is long known that chondrocytes in the growth plate undergo hypertrophy before turning into osteoblasts. Rocky Tuan's team has shown that cloned osteoblasts can turn into chondrocytes or adipocytes (Song and Tuan, 2004; Song et al., 2006), and we have shown that clones of human BM cells differentiated into VSM cells can turn into A, O, or C in adequate culture conditions (Delorme et al., 2009). The team of Dov Zipori has shown that murine BM mesenchymal cells with a given differentiation potential can change their potential following cell isolation: for example, a cell isolated from a cell population with only osteo-chondrogenic potential could generate a tripotent clone that had acquired the adipogenic potential (Shoshani et al., 2014). The gene expression program was profoundly altered in cells cultured at low density; in particular the genes whose promoter bore the repressive histone mark H3K27me3 were upregulated in sparse-cultured cells. Another report showed that Pro Chondrogenic Progenitors (PCP) derived from murine SSC implanted under the kidney capsule together with Bone-Cartilage-Stroma Progenitors (BCSP) gave rise to osteoblasts instead of chondrocytes (Chan et al., 2015); conversely, BCSP implanted in mice with systemic inhibition of VEGF signaling generated cartilage instead of bone and BM. Finally, it has been shown that growth plate chondrogenic stem cells become osteogenic with the formation of secondary ossification centers (Mizuhashi et al., 2018). Taken together, these data indicate that lineage determination in the mesenchymal system is not irreversible but plastic, the switch from one lineage to the other possibly occurring at the differentiated, progenitor or stem cell level.

Lineage priming is a property of certain stem cells that already express a number of transcripts characteristic of committed cells in their progeny (review in (Orkin and Zon, 2008)). We have shown that clones of human

BM cells not induced to differentiate were primed to the lineages into which the cells could differentiate when submitted to appropriate culture conditions (Delorme et al., 2009). Specifically, these cells expressed at the transcript level most of the genes affiliated to the osteoblastic and vascular smooth muscle pathways, many of the chondrogenic genes and a few of the adipogenic genes. Lineage priming has been confirmed in murine BM mesenchymal cell clones (Hamidouche et al., 2016), and in individual cells from a population of Ebf2$^+$ murine BM mesenchymal stem cells (Qian et al., 2013), or from BM mesenchymal lines (Freeman et al., 2015). Plasticity may be related to lineage priming, since differentiation in the primed lineages would not entail the set-up of whole molecular program but the modulation of only a few of the program components. Moreover, lineage priming indicates that promoters of the key transcription factors are in an open configuration, and therefore accessible to certain chromatin transcriptional coactivators such as TAZ that may modify the balance in favor of adipogenic or osteoblastic pathway (Hong et al., 2005). A recent study using HSCs as model stem cells and a very sensitive assay for detection and quantification of transcription factors has shown that single HSC can express transcription factors that are antagonistic in terms of differentiation lineage induction (Wheat et al., 2020). Analysis of the transcription factors along differentiation trajectories led to a model of reversible transition states whereby the transcriptional state of a given cell is not fully predictive of the past or future states of parent or daughter cell. Such a model may be applied to mesenchymal stem cells by virtue of the observed lineage priming and changes in cell fates upon serial cell isolation (Shoshani et al., 2014).

A last argument in favor of the plasticity of mesenchymal stem cells comes from the study of the phenotype of murine BM tripotent and clonogenic cells (Hamidouche et al., 2016). Clones generated from single cells displayed inter-clonal heterogeneity. In particular Sca-1 expression was clone-specific and stable during expansion over several weeks. Sorted subpopulations with high, medium or low Sca-1 expression were able to regenerate the full expression profile characteristic of each clone after a few days in culture. Study of the promoter of Ly6a (gene coding for Sca-1) showed that the expression level of Sca-1 was related to the promoter occupancy by the activating histone mark H3K4me3. A computational model suggested that the oscillatory nature of Sca-1 expression might be explained by a positive feedback between promoter H3K4me3 modification and gene activation by upstream transcription factor(s). The bistable state resulting from this feedback is another facet of the mesenchymal cell plasticity.

19.6 Concluding Remarks

Half a century of research has enabled discovering a new category of bona fide non-hematopoietic stem cells located in the bone and BM. These stem cells display specific features:

- they contribute to the generation of bone and cartilage during development, including childhood, and to the repair of bone and cartilage in cases of fracture or lesion;
- they generate BM stromal cells with hematopoiesis-supportive capacity; this property has been directly demonstrated in certain cases, but can be inferred in almost all cases according to functional studies;
- they can be maintained in culture, although being difficult to detect and quantify in the *in vitro* conditions; and
- they are mesodermal in origin, and probably the progeny of stem cells with mesenchymal and endothelial features, themselves the progeny of pluripotent stem cells.

Lineage determination in the mesenchymal system is not irreversible but plastic; the switch from one lineage to the other possibly occurring at the differentiated, progenitor or stem cell level. Stochastic and reversible models would best account for the shift.

Acknowledgments

The author is indebted to Nathalie Chevallier for sharing the illustration on bone marrow biopsy, and to Thierry Jaffredo for careful review of the manuscript.

References

Battula, V.L., Treml, S., Bareiss, P.M., Gieseke, F., Roelofs, H., de Zwart, P., Müller, I., Schewe, B., Skutella, T., Fibbe, W.E., et al. (2009). Isolation of functionally distinct mesenchymal stem cell subsets using antibodies against CD56, CD271, and mesenchymal stem cell antigen-1. Haematologica *94*, 173–184.

Breitbach, M., Kimura, K., Luis, T.C., Fuegemann, C.J., Woll, P.S., Hesse, M., Facchini, R., Rieck, S., Jobin, K., Reinhardt, J., et al. (2018). In Vivo Labeling by CD73 Marks Multipotent Stromal Cells and Highlights Endothelial Heterogeneity in the Bone Marrow Niche. Cell Stem Cell *22*, 262–276.e267.

Caplan, A.I. (1991). Mesenchymal stem cells. J Orthop Res *9*, 641–650.

Chan, C.K., Seo, E.Y., Chen, J.Y., Lo, D., McArdle, A., Sinha, R., Tevlin, R., Seita, J., Vincent-Tompkins, J., Wearda, T., et al. (2015). Identification and specification of the mouse skeletal stem cell. Cell *160*, 285–298.

Chan, C.K.F., Gulati, G.S., Sinha, R., Tompkins, J.V., Lopez, M., Carter, A.C., Ransom, R.C., Reinisch, A., Wearda, T., Murphy, M., et al. (2018). Identification of the Human Skeletal Stem Cell. Cell *175*, 43–56 e21.

Charbord, P. (2010). Bone marrow mesenchymal stem cells: historical overview and concepts. Human gene therapy *21*, 1045–1056.

Charbord, P., Oostendorp, R., Pang, W., Herault, O., Noel, F., Tsuji, T., Dzierzak, E., and Peault, B. (2002). Comparative study of stromal cell lines derived from embryonic, fetal, and postnatal mouse blood-forming tissues. Experimental hematology *30*, 1202–1210.

Chateauvieux, S., Ichante, J.L., Delorme, B., Frouin, V., Pietu, G., Langonne, A., Gallay, N., Sensebe, L., Martin, M.T., Moore, K.A., et al. (2007). Molecular profile of mouse stromal mesenchymal stem cells. Physiol Genomics *29*, 128–138.

Chen, K.G., Johnson, K.R., and Robey, P.G. (2017). Mouse Genetic Analysis of Bone Marrow Stem Cell Niches: Technological Pitfalls, Challenges, and Translational Considerations. Stem cell reports *9*, 1343–1358.

Colter, D.C., Class, R., DiGirolamo, C.M., and Prockop, D.J. (2000). Rapid expansion of recycling stem cells in cultures of plastic-adherent cells from human bone marrow. Proc Natl Acad Sci U S A *97*, 3213–3218.

Colter, D.C., Sekiya, I., and Prockop, D.J. (2001). Identification of a sub-population of rapidly self-renewing and multipotential adult stem cells in colonies of human marrow stromal cells. Proc Natl Acad Sci U S A *98*, 7841–7845.

Debnath, S., Yallowitz, A.R., McCormick, J., Lalani, S., Zhang, T., Xu, R., Li, N., Liu, Y., Yang, Y.S., Eiseman, M., et al. (2018). Discovery of a periosteal stem cell mediating intramembranous bone formation. Nature *562*, 133–139.

Delorme, B., and Charbord, P. (2007). Culture and characterization of human bone marrow mesenchymal stem cells. Methods Mol Med *140*, 67–81.

Delorme, B., Ringe, J., Gallay, N., Le Vern, Y., Kerboeuf, D., Jorgensen, C., Rosset, P., Sensebe, L., Layrolle, P., Haupl, T., et al. (2008). Specific plasma membrane protein phenotype of culture-amplified and native human bone marrow mesenchymal stem cells. Blood *111*, 2631–2635.

Delorme, B., Ringe, J., Pontikoglou, C., Gaillard, J., Langonne, A., Sensebe, L., Noel, D., Jorgensen, C., Haupl, T., and Charbord, P. (2009).

Specific Lineage-Priming of Bone Marrow Mesenchymal Stem Cells Provides the Molecular Framework for Their Plasticity. Stem Cells *27*, 1142–1151.

Dennis, J.E., Carbillet, J.P., Caplan, A.I., and Charbord, P. (2002). The STRO-1+ marrow cell population is multipotential. Cells Tissues Organs *170*, 73–82.

Dennis, J.E., and Charbord, P. (2002). Origin and differentiation of human and murine stroma. Stem Cells *20*, 205–214.

Deschaseaux, F., and Charbord, P. (2000). Human marrow stromal precursors are alpha 1 integrin subunit-positive. J Cell Physiol *184*, 319–325.

Dexter, T.M., Allen, T.D., and Lajtha, L.G. (1977). Conditions controlling the proliferation of haemopoietic stem cells in vitro. J Cell Physiol *91*, 335–344.

Dominici, M., Le Blanc, K., Mueller, I., Slaper-Cortenbach, I., Marini, F., Krause, D., Deans, R., Keating, A., Prockop, D., and Horwitz, E. (2006). Minimal criteria for defining multipotent mesenchymal stromal cells. The International Society for Cellular Therapy position statement. Cytotherapy *8*, 315–317.

Filshie, R.J., Zannettino, A.C., Makrynikola, V., Gronthos, S., Henniker, A.J., Bendall, L.J., Gottlieb, D.J., Simmons, P.J., and Bradstock, K.F. (1998). MUC18, a member of the immunoglobulin superfamily, is expressed on bone marrow fibroblasts and a subset of hematological malignancies. Leukemia *12*, 414–421.

Freeman, B.T., Jung, J.P., and Ogle, B.M. (2015). Single-Cell RNA-Seq of Bone Marrow-Derived Mesenchymal Stem Cells Reveals Unique Profiles of Lineage Priming. PLoS One *10*, e0136199.

Friedenstein, A.J., Chailakhjan, R.K., and Lalykina, K.S. (1970). The development of fibroblast colonies in monolayer cultures of guinea-pig bone marrow and spleen cells. Cell Tissue Kinet *3*, 393–403.

Galmiche, M.C., Koteliansky, V.E., Briere, J., Herve, P., and Charbord, P. (1993). Stromal cells from human long-term marrow cultures are mesenchymal cells that differentiate following a vascular smooth muscle differentiation pathway. Blood *82*, 66–76.

Gang, E.J., Bosnakovski, D., Figueiredo, C.A., Visser, J.W., and Perlingeiro, R.C. (2007). SSEA-4 identifies mesenchymal stem cells from bone marrow. Blood *109*, 1743–1751.

Gao, J., Yan, X.L., Li, R., Liu, Y., He, W., Sun, S., Zhang, Y., Liu, B., Xiong, J., and Mao, N. (2010). Characterization of OP9 as authentic mesenchymal stem cell line. J Genet Genomics *37*, 475–482.

Gartner, S., and Kaplan, H.S. (1980). Long-term culture of human bone marrow cells. Proc Natl Acad Sci U S A *77*, 4756–4759.

Gronthos, S., Zannettino, A.C., Hay, S.J., Shi, S., Graves, S.E., Kortesidis, A., and Simmons, P.J. (2003). Molecular and cellular characterisation of highly purified stromal stem cells derived from human bone marrow. J Cell Sci *116*, 1827–1835.

Hamidouche, Z., Rother, K., Przybilla, J., Krinner, A., Clay, D., Hopp, L., Fabian, C., Stolzing, A., Binder, H., Charbord, P., et al. (2016). Bistable Epigenetic States Explain Age-Dependent Decline in Mesenchymal Stem Cell Heterogeneity. Stem Cells.

Haniffa, M.A., Collin, M.P., Buckley, C.D., and Dazzi, F. (2009). Mesenchymal stem cells: the fibroblasts' new clothes? Haematologica *94*, 258–263.

Haniffa, M.A., Wang, X.N., Holtick, U., Rae, M., Isaacs, J.D., Dickinson, A.M., Hilkens, C.M., and Collin, M.P. (2007). Adult human fibroblasts are potent immunoregulatory cells and functionally equivalent to mesenchymal stem cells. J Immunol *179*, 1595–1604.

Hong, J.H., Hwang, E.S., McManus, M.T., Amsterdam, A., Tian, Y., Kalmukova, R., Mueller, E., Benjamin, T., Spiegelman, B.M., Sharp, P.A., et al. (2005). TAZ, a transcriptional modulator of mesenchymal stem cell differentiation. Science *309*, 1074–1078.

Horwitz, E.M., Le Blanc, K., Dominici, M., Mueller, I., Slaper-Cortenbach, I., Marini, F.C., Deans, R.J., Krause, D.S., and Keating, A. (2005). Clarification of the nomenclature for MSC: The International Society for Cellular Therapy position statement. Cytotherapy *7*, 393–395.

Isern, J., García-García, A., Martín, A.M., Arranz, L., Martín-Pérez, D., Torroja, C., Sánchez-Cabo, F., and Méndez-Ferrer, S. (2014). The neural crest is a source of mesenchymal stem cells with specialized hematopoietic stem cell niche function. eLife *3*, e03696.

Isern, J., Martin-Antonio, B., Ghazanfari, R., Martin, A.M., Lopez, J.A., del Toro, R., Sanchez-Aguilera, A., Arranz, L., Martin-Perez, D., Suarez-Lledo, M., et al. (2013). Self-renewing human bone marrow mesenspheres promote hematopoietic stem cell expansion. Cell reports *3*, 1714–1724.

James, S., Fox, J., Afsari, F., Lee, J., Clough, S., Knight, C., Ashmore, J., Ashton, P., Preham, O., Hoogduijn, M., et al. (2015). Multiparameter Analysis of Human Bone Marrow Stromal Cells Identifies Distinct Immunomodulatory and Differentiation-Competent Subtypes. Stem cell reports *4*, 1004–1015.

Jones, E.A., Kinsey, S.E., English, A., Jones, R.A., Straszynski, L., Meredith, D.M., Markham, A.F., Jack, A., Emery, P., and McGonagle, D. (2002). Isolation and characterization of bone marrow multipotential mesenchymal progenitor cells. Arthritis Rheum *46*, 3349–3360.

Jones, S., Horwood, N., Cope, A., and Dazzi, F. (2007). The antiproliferative effect of mesenchymal stem cells is a fundamental property shared by all stromal cells. J Immunol *179*, 2824–2831.

Kramann, R., Schneider, R.K., DiRocco, D.P., Machado, F., Fleig, S., Bondzie, P.A., Henderson, J.M., Ebert, B.L., and Humphreys, B.D. (2015). Perivascular Gli1+ progenitors are key contributors to injury-induced organ fibrosis. Cell Stem Cell *16*, 51–66.

Krausgruber, T., Fortelny, N., Fife-Gernedl, V., Senekowitsch, M., Schuster, L.C., Lercher, A., Nemc, A., Schmidl, C., Rendeiro, A.F., Bergthaler, A., et al. (2020). Structural cells are key regulators of organ-specific immune responses. Nature *583*, 296–302.

Kumar, A., D'Souza, S.S., Moskvin, O.V., Toh, H., Wang, B., Zhang, J., Swanson, S., Guo, L.W., Thomson, J.A., and Slukvin, II (2017). Specification and Diversification of Pericytes and Smooth Muscle Cells from Mesenchymoangioblasts. Cell reports *19*, 1902–1916.

Lee, C.C., Christensen, J.E., Yoder, M.C., and Tarantal, A.F. (2010). Clonal analysis and hierarchy of human bone marrow mesenchymal stem and progenitor cells. Experimental hematology *38*, 46–54.

Lee, G., Kim, H., Elkabetz, Y., Al Shamy, G., Panagiotakos, G., Barberi, T., Tabar, V., and Studer, L. (2007). Isolation and directed differentiation of neural crest stem cells derived from human embryonic stem cells. Nat Biotechnol *25*, 1468–1475.

Li, H., Ghazanfari, R., Zacharaki, D., Ditzel, N., Isern, J., Ekblom, M., Méndez-Ferrer, S., Kassem, M., and Scheding, S. (2014). Low/negative expression of PDGFR-a identifies the candidate primary mesenchymal stromal cells in adult human bone marrow. Stem cell reports *3*, 965–974.

Li, J., Sensebe, L., Herve, P., and Charbord, P. (1995). Nontransformed colony-derived stromal cell lines from normal human marrows. II. Phenotypic characterization and differentiation pathway. Exp Hematol *23*, 133–141.

Mabuchi, Y., Morikawa, S., Harada, S., Niibe, K., Suzuki, S., Renault-Mihara, F., Houlihan, D.D., Akazawa, C., Okano, H., and Matsuzaki, Y. (2013). LNGFR(+)THY-1(+)VCAM-1(hi+) cells reveal functionally distinct subpopulations in mesenchymal stem cells. Stem cell reports *1*, 152–165.

Martinez, C., Hofmann, T.J., Marino, R., Dominici, M., and Horwitz, E.M. (2007). Human bone marrow mesenchymal stromal cells express the neural ganglioside GD2: a novel surface marker for the identification of MSCs. Blood *109*, 4245–4248.

Mende, N., Jolly, A., Percin, G.I., Günther, M., Rostovskaya, M., Krishnan, S.M., Oostendorp, R.A.J., Dahl, A., Anastassiadis, K., Höfer, T., et al. (2019). Prospective isolation of nonhematopoietic cells of the niche and their differential molecular interactions with HSCs. Blood *134*, 1214–1226.

Méndez-Ferrer, S., Michurina, T.V., Ferraro, F., Mazloom, A.R., Macarthur, B.D., Lira, S.A., Scadden, D.T., Ma'ayan, A., Enikolopov, G.N., and Frenette, P.S. (2010). Mesenchymal and haematopoietic stem cells form a unique bone marrow niche. Nature *466*, 829–834.

Méndez-Ferrer, S., Scadden, D.T., and Sánchez-Aguilera, A. (2015). Bone marrow stem cells: current and emerging concepts. Ann N Y Acad Sci *1335*, 32–44.

Mizoguchi, T., Pinho, S., Ahmed, J., Kunisaki, Y., Hanoun, M., Mendelson, A., Ono, N., Kronenberg, H.M., and Frenette, P.S. (2014). Osterix marks distinct waves of primitive and definitive stromal progenitors during bone marrow development. Dev Cell *29*, 340–349.

Mizuhashi, K., Ono, W., Matsushita, Y., Sakagami, N., Takahashi, A., Saunders, T.L., Nagasawa, T., Kronenberg, H.M., and Ono, N. (2018). Resting zone of the growth plate houses a unique class of skeletal stem cells. Nature *563*, 254–258.

Morikawa, S., Mabuchi, Y., Kubota, Y., Nagai, Y., Niibe, K., Hiratsu, E., Suzuki, S., Miyauchi-Hara, C., Nagoshi, N., Sunabori, T., et al. (2009a). Prospective identification, isolation, and systemic transplantation of multipotent mesenchymal stem cells in murine bone marrow. The Journal of experimental medicine *206*, 2483–2496.

Morikawa, S., Mabuchi, Y., Niibe, K., Suzuki, S., Nagoshi, N., Sunabori, T., Shimmura, S., Nagai, Y., Nakagawa, T., Okano, H., et al. (2009b). Development of mesenchymal stem cells partially originate from the neural crest. Biochem Biophys Res Commun *379*, 1114–1119.

Muraglia, A., Cancedda, R., and Quarto, R. (2000). Clonal mesenchymal progenitors from human bone marrow differentiate in vitro according to a hierarchical model. J Cell Sci *113 (Pt 7)*, 1161–1166.

Nagoshi, N., Shibata, S., Kubota, Y., Nakamura, M., Nagai, Y., Satoh, E., Morikawa, S., Okada, Y., Mabuchi, Y., Katoh, H., et al. (2008). Ontogeny and multipotency of neural crest-derived stem cells in mouse bone marrow, dorsal root ganglia, and whisker pad. Cell Stem Cell *2*, 392–403.

Newton, P.T., Li, L., Zhou, B., Schweingruber, C., Hovorakova, M., Xie, M., Sun, X., Sandhow, L., Artemov, A.V., Ivashkin, E., et al. (2019). A radical switch in clonality reveals a stem cell niche in the epiphyseal growth plate. Nature *567*, 234–238.

Nusspaumer, G., Jaiswal, S., Barbero, A., Reinhardt, R., Ishay Ronen, D., Haumer, A., Lufkin, T., Martin, I., and Zeller, R. (2017). Ontogenic Identification and Analysis of Mesenchymal Stromal Cell Populations during Mouse Limb and Long Bone Development. Stem cell reports *9*, 1124–1138.

Orkin, S.H., and Zon, L.I. (2008). Hematopoiesis: an evolving paradigm for stem cell biology. Cell *132*, 631–644.

Park, D., Spencer, J.A., Koh, B.I., Kobayashi, T., Fujisaki, J., Clemens, T.L., Lin, C.P., Kronenberg, H.M., and Scadden, D.T. (2012). Endogenous bone marrow MSCs are dynamic, fate-restricted participants in bone maintenance and regeneration. Cell Stem Cell *10*, 259–272.

Pereira, R.F., Halford, K.W., O'Hara, M.D., Leeper, D.B., Sokolov, B.P., Pollard, M.D., Bagasra, O., and Prockop, D.J. (1995). Cultured adherent cells from marrow can serve as long-lasting precursor cells for bone, cartilage, and lung in irradiated mice. Proc Natl Acad Sci U S A *92*, 4857–4861.

Pereira, R.F., O'Hara, M.D., Laptev, A.V., Halford, K.W., Pollard, M.D., Class, R., Simon, D., Livezey, K., and Prockop, D.J. (1998). Marrow stromal cells as a source of progenitor cells for nonhematopoietic tissues in transgenic mice with a phenotype of osteogenesis imperfecta. Proc Natl Acad Sci U S A *95*, 1142–1147.

Pinho, S., Lacombe, J., Hanoun, M., Mizoguchi, T., Bruns, I., Kunisaki, Y., and Frenette, P.S. (2013). PDGFRalpha and CD51 mark human Nestin+ sphere-forming mesenchymal stem cells capable of hematopoietic progenitor cell expansion. The Journal of experimental medicine *210*, 1351–1367.

Pittenger, M.F., Mackay, A.M., Beck, S.C., Jaiswal, R.K., Douglas, R., Mosca, J.D., Moorman, M.A., Simonetti, D.W., Craig, S., and Marshak, D.R. (1999). Multilineage potential of adult human mesenchymal stem cells. Science *284*, 143–147.

Qian, H., Badaloni, A., Chiara, F., Stjernberg, J., Polisetti, N., Nihlberg, K., Consalez, G.G., and Sigvardsson, M. (2013). Molecular characterization of prospectively isolated multipotent mesenchymal progenitors provides new insight into the cellular identity of mesenchymal stem cells in mouse bone marrow. Mol Cell Biol *33*, 661–677.

Quirici, N., Soligo, D., Bossolasco, P., Servida, F., Lumini, C., and Deliliers, G.L. (2002). Isolation of bone marrow mesenchymal stem cells by anti-nerve growth factor receptor antibodies. Exp Hematol *30*, 783–791.

Russell, K.C., Phinney, D.G., Lacey, M.R., Barrilleaux, B.L., Meyertholen, K.E., and O'Connor, K.C. (2010). In vitro high-capacity assay to quantify the clonal heterogeneity in trilineage potential of mesenchymal stem cells reveals a complex hierarchy of lineage commitment. Stem Cells *28*, 788–798.

Sacchetti, B., Funari, A., Michienzi, S., Di Cesare, S., Piersanti, S., Saggio, I., Tagliafico, E., Ferrari, S., Robey, P.G., Riminucci, M., et al.(2007). Self-renewing osteoprogenitors in bone marrow sinusoids can organize a hematopoietic microenvironment. Cell *131*, 324–336.

Sensebe, L., Li, J., Lilly, M., Crittenden, C., Herve, P., Charbord, P., and Singer, J.W. (1995). Nontransformed colony-derived stromal cell lines from normal human marrows. I. Growth requirement and myelopoiesis supportive ability. Exp Hematol *23*, 507–513.

Sensebe, L., Mortensen, B.T., Fixe, P., Herve, P., and Charbord, P. (1997). Cytokines active on granulomonopoiesis: release and consumption by human marrow myoid [corrected] stromal cells. British journal of haematology *98*, 274–282.

Shoshani, O., Ravid, O., Massalha, H., Aharonov, A., Ovadya, Y., Pevsner-Fischer, M., Leshkowitz, D., and Zipori, D. (2014). Cell isolation induces fate changes of bone marrow mesenchymal cells leading to loss or alternatively to acquisition of new differentiation potentials. Stem Cells *32*, 2008–2020.

Simmons, P.J., and Torok-Storb, B. (1991). CD34 expression by stromal precursors in normal human adult bone marrow. Blood *78*, 2848–2853.

Song, L., and Tuan, R.S. (2004). Transdifferentiation potential of human mesenchymal stem cells derived from bone marrow. Faseb J *18*, 980–982.

Song, L., Webb, N.E., Song, Y., and Tuan, R.S. (2006). Identification and functional analysis of candidate genes regulating mesenchymal stem cell self-renewal and multipotency. Stem Cells *24*, 1707–1718.

Stewart, K., Monk, P., Walsh, S., Jefferiss, C.M., Letchford, J., and Beresford, J.N. (2003). STRO-1, HOP-26 (CD63), CD49a and SB-10 (CD166) as markers of primitive human marrow stromal cells and their more differentiated progeny: a comparative investigation in vitro. Cell Tissue Res *313*, 281–290.

Takashima, Y., Era, T., Nakao, K., Kondo, S., Kasuga, M., Smith, A.G., and Nishikawa, S. (2007). Neuroepithelial cells supply an initial transient wave of MSC differentiation. Cell *129*, 1377–1388.

Tormin, A., Li, O., Brune, J.C., Walsh, S., Schütz, B., Ehinger, M., Ditzel, N., Kassem, M., and Scheding, S. (2011). CD146 expression on primary nonhematopoietic bone marrow stem cells is correlated with in situ localization. Blood *117*, 5067–5077.

Vodyanik, M.A., Yu, J., Zhang, X., Tian, S., Stewart, R., Thomson, J.A., and Slukvin, II (2010). A mesoderm-derived precursor for mesenchymal stem and endothelial cells. Cell Stem Cell *7*, 718–729.

Wheat, J.C., Sella, Y., Willcockson, M., Skoultchi, A.I., Bergman, A., Singer, R.H., and Steidl, U. (2020). Single-molecule imaging of transcription dynamics in somatic stem cells. Nature *583*, 431–436.

Worthley, D.L., Churchill, M., Compton, J.T., Tailor, Y., Rao, M., Si, Y., Levin, D., Schwartz, M.G., Uygur, A., Hayakawa, Y., et al. (2015). Gremlin 1 identifies a skeletal stem cell with bone, cartilage, and reticular stromal potential. Cell *160*, 269–284.

Zhao, M., Tao, F., Venkatraman, A., Li, Z., Smith, S.E., Unruh, J., Chen, S., Ward, C., Qian, P., Perry, J.M., et al. (2019). N-Cadherin-Expressing Bone and Marrow Stromal Progenitor Cells Maintain Reserve Hematopoietic Stem Cells. Cell reports *26*, 652–669.e656.

Zhou, B.O., Yue, R., Murphy, M.M., Peyer, J.G., and Morrison, S.J. (2014). Leptin-receptor-expressing mesenchymal stromal cells represent the main source of bone formed by adult bone marrow. Cell Stem Cell *15*, 154–168.

20

Dental Stem Cells

Anne-Margaux Collignon[1,2,3], Caroline Gorin[1,3],
Catherine Chaussain[1] and Anne Poliard[1,*]

[1]Laboratoire Pathologies, Imagerie et Biothérapies Orofaciales, UR 2496, UFR d'Odontologie Montrouge, Université de Paris, 1 rue Maurice Arnoux, 92120 Montrouge, France
[2]AP-HP, Dental Department, Charles Foix, Louis Mourier and Bretonneau Hospitals, France
[3]Both these authors contributed equally
E-mail: anne.poliard@parisdescartes.fr
*Corresponding Author

20.1 Introduction

For two decades now, the stem cell field has tremendously evolved with the highlighting of the presence of what has been called "stem cells" in most adult organs such as for example, Mesenchymal Stem/Stromal Cells (MSCs) in all vascularized organs and the possibility of generating pluripotent cells from adult somatic cells: the induced pluripotent stem cells (iPSCs). These discoveries have paved the way for setting up novel repair/regenerative therapies in all medical fields that have raised enormous hopes. Numerous studies have been undertaken and are ongoing for determining the most efficient ways of using these different "stem cell" populations in pathological contexts. There are over 950 registered MSC trials listed by the FDA with about 200 phases 1 and 2 and 10 phases 3 (https://celltrials.org/public-cells-data/msc-trials-2011-2018/65). If the first phase 2 clinical trials using iPSCs has been launched in Japan for Parkinson's disease and in California for age-related macular degeneration, very few clinical applications have however been approved so far. MSCs and other stem cells do offer remarkable potentials but our understanding of their biology and medical applications are not yet ready for unregulated, widespread use.

Odontology has been at the forefront in this field with the identification of mesenchymal multipotent MSC-like populations in several tissues of the dental organ (Mayo et al. 2014; Zhao et Chai 2015) and the possibility of efficiently deriving iPSCs from these cells (Yan et al. 2010; El Ayachi et al. 2018). These findings are stimulating research on the molecular mechanisms involved in the recruitment and differentiation of these MSC-like populations and on the development of innovative therapies involving them for treating lesions of the tooth (Iohara et al. 2013; Nakashima et Iohara 2017; Xuan et al. 2018), but also of the craniofacial bone (Giuliani et al. 2013), for generating complete teeth by tissue engineering (for a recent review (Yelick et Sharpe 2019)) or treating neurodegenerative defects (Gervois et al. 2016; Ueda et al. 2020).

20.2 State of the Art

Tooth formation is regulated by sequential and reciprocal inductive interactions between ectomesenchymal cells, derived from the neural crest (NC) and the dental epithelium (Balic et Thesleff 2015). The ectomesenchymal cells participate in dental follicle and papilla formation while the dental epithelium leads to the formation of the enamel organ (Jernvall et Thesleff 2000). In the mature tooth (Figure 20.1), the dental papilla will further give rise to odontoblasts, which produce dentin and most of the cells of the pulp tissue while the dental follicle will participate in the periodontal ligament and cementoblast formation. The enamel organ gives rise to the enamel-secreting ameloblasts.

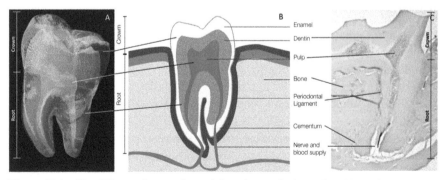

Figure 20.1   The different tissues of an adult molar tooth shown in: (A) a micro- scanner view and (B) a schematic representation of a human molar tooth (C) a Masson trichrome staining of a mice molar section.

Although it has long been known that the dental pulp (DP) contained cells capable of tissue repair upon lesions, the nature of the cells involved in this task had remained a matter of debate in the literature. Unlike bone, dentin does not undergo remodeling; dentinal damages induce the formation of a dentinal protective barrier to the dental pulp. The cells involved in this process depend on the degree of the damage. Mild injury to the dentin, which does not reach the odontoblast layer, stimulates deposition of a reactionary dentin, produced by pre-existing odontoblasts and structurally similar to physiological dentin. Severe pulpo-dentinal lesions, which destroys dentin and the underlying odontoblasts, requires generation of new odontoblast-like cells from undifferentiated MSCs located in the dental pulp. These cells, after recruitment from the pulpal tissue, will produce a poorly organized reparative dentin displaying a mainly atubular structure, in contrast with the physiological dentin, with cells trapped within the matrix (Sloan et Waddington 2009), and therefore often referred to osteodentin, which provides tooth protection from bacterial damage.

The stem cell story of the tooth really began two decades ago when Gronthos and his collaborators demonstrated the existence of an MSC-like stem cell population within the DP tissue of mature and also decidual teeth. These cells were designated Dental Pulp Stem Cells (DPSCs), when isolated from an adult tooth (Gronthos et al. 2000) and SHEDs (for Stem cells from Human Exfoliated Deciduous tooth), when isolated from deciduous teeth, (Miura et al. 2003) or iDPSCs (for immature Dental Pulp Stem Cells), when obtained from primary teeth of non-exfoliated deciduous teeth (Kerkis et al. 2006) or from natal tooth (Pisal et al. 2018). Soon after, MSC-like cells were also identified in other mesenchymal tissues of the tooth organ as the periodontal ligament (PDLSC), apical papilla (SCAP) or dental follicle (DFSC) (Huang et al. 2009). Although these dental tissues are initially all derived from the NC (Chai et al. 2000), the characteristics and *in vivo* behavior of these various cell populations differ (Taraslia et al. 2018). Amongst these different MSC-like cell populations, those derived from the pulp have been the most studied, thanks to their relatively easiness of access together with the ability to harvest a significant number of cells through minimally invasive processes (natural tooth exfoliation in the childhood or extraction for orthodontic reasons in teenagers or young adults).

The research data mentioned in this chapter were obtained with dental pulp stem cells derived from adult permanent teeth (DPSCs) and from deciduous teeth (SHEDs). Although DPSCs and SHEDs have not been systematically compared, they appear to share similar features and common differentiation properties. Indeed, SHEDs present a higher proliferation and differentiation potential, i.e., are more plastic, than their counterpart from adult tooth (Nakamura et al. 2009; Wang et al. 2012). For the sake of clarity, the term DPSC will be used in the rest of this manuscript unless otherwise specified.

DPSCs follow the criteria that characterize multipotent MSCs, as initially defined and recently revised by the International Society for Cellular Therapy (Dominici et al. 2006; Viswanathan et al. 2019) and therefore they also express the common MSC-associated surface markers CD146, CD90, CD105 while lacking the expression of hematopoietic markers CD14, CD34, CD45, CD19, and HLA-DR. They can be isolated either by enzymatic digestion (Gronthos et al. 2000) or by outgrowth of the pulp tissue (Kerkis et al. 2006) and were showed to be endowed with properties similar to the MSCs described in the bone marrow (BMSCs) (Gronthos et al. 2002; Nakajima et al. 2018). As other MSC-like cells, they display a spindle-shaped morphology and correspond to a heterogeneous population of cells endowed with a multilineage differentiation potential. As such they are capable of *in vitro* differentiation toward various fates in response to specific stimuli, such as osteoblasts or odontoblasts (Laino et al. 2006; Lacerda-Pinheiro et al. 2012), adipocytes (Gronthos et al. 2002), chondrocytes (Iohara et al. 2006), neurons, (Kiraly et al. 2011) , endotheliocytes (d'Aquino et al. 2007), corneal cells (Monteiro et al. 2009), melanocytes (Stevens et al. 2008), or hepatocytes (Ishkitiev et al. 2015). These potentials differ from those of the golden MSC standard, the BMSC, likely related to the different embryonic origin of those two cell types (neural crest vs mesoderm) (Yianni et Sharpe 2019).

In vivo, DPSCs are capable of forming a pulpo-dentinal complex with pulp and dentin-forming cells after ectopic implantation (Gronthos et al. 2000), a property expected for DPSCs. These observations paved the way for developing models of pulp regeneration first in ectopic sites such as a tooth slice (Cordeiro et al. 2008) or emptied root canal (Huang et al. 2010), and then in models closer to the clinical situation, in response to a DP aggression in rat, canine, and porcine pulp injury models (Iohara et al. 2011; Mangione et al. 2017; Nakashima et Iohara 2017). These studies have led recently to the first clinical assays in humans that confirmed the potential of DPSCs for pulpal lesion therapies as an alternative to endodontic treatments (Nakashima et al. 2017; Xuan et al. 2018). In relation to their ability to form a mineralized tissue, DPSCs are also capable of promoting bone repair when seeded in an appropriate osteogenic microenvironment (Seo et al. 2008). In a pilot clinical study, implantation of DPSCs in a mandibular bone defect has been shown to promote a faster and more extensive bone healing than in the untreated control side but with generation of a compact rather than alveolar bone (Giuliani et al. 2013). However, despite their name, multipotent behavior and to some extent, capacity of self-renewal in culture, DPSCs as all MSC-like cells found in vascularized tissues, do not meet the full criteria to qualify as bona fide "stem cells" such as those found in the embryo, the embryonic stem cells (ES cells). In addition, numerous studies have also demonstrated that the MSC multipotent behavior *in vitro* does not reflect the actual *in vivo* potential of the cells in a particular microenvironment. Therefore, readers should keep these restrictions in mind when referring to the term "stem cells" (Sacchetti et al. 2016).

An important property of DPSCs, which rapidly arose from *in vivo* studies, is their strong pro-angiogenic action. This is particularly interesting in the context of tissue engineering applications since, if blood supply cannot be established rapidly, insufficient oxygen and nutrient transport will prevail and necrosis/apoptosis of any grafted cells or tissue will occur. DPSCs display, a pro-angiogenic activity both *in vitro* and *in vivo* (Bronckaers et al. 2013; Gorin et al. 2016), which appears superior to that of other types of MSC-like cells (Hilkens et al. 2014). This pro-angiogenic property seems essentially mediated through a paracrine action, promoting both angiogenesis and recruitment of host progenitors. Indeed, MSC-like cells secrete large quantities of various bioactive molecules, among which Vascular Endothelial Growth Factor A (VEGF-A) is a pro-angiogenic and pro-vasculogenic molecule (Caplan et Correa 2011). However, DPSCs can also display a pericyte-like phenotype *in vitro* under appropriate culture conditions or in 3D vessel formation models, stabilizing newly formed vessels (Janebodin et al. 2013; Delle Monache et al. 2019; Atlas et al., 2021). In addition, DPSCs can functionally integrate within the brain vasculature of athymic nude mice, after neurogenic *in vitro* priming, expressing markers and morphological vasculature features with characteristics of both pericytes and endothelial cells (Luzuriaga et al. 2019).

DPSCs also display a neuronal potential *in vitro* as *in vivo* (Martens et al. 2014; Luo et al. 2018). For instance, they are capable of responding to a neuronal environment, by undergoing functional neuronal differentiation after implantation in chicken embryo heads (Arthur et al. 2008). In addition, neuronally pre-differentiated DPSCs injected into the brain cortex of an animal model of brain injury integrate the host brain and display neuronal properties, such as typical neuronal marker expression or voltage-dependent sodium and potassium channels (Kiraly et al. 2011). DPSCs have also been shown to promote regeneration of neural tissue in rat sciatic nerve defects (Sanen et al. 2017). Preclinical data in support of using DPSCs in acute ischemic stroke have been obtained recently (Gervois et al. 2016; Zhang et al. 2018). In all cases analyzed, a low survival rate of the implanted DPSCs was observed, suggesting the involvement of paracrine mechanisms in the repair (Ueda et al. 2020), and in particular that of exosomes (Stanko et al. 2018), although evidence for DPSC differentiation in neurons and astrocytes was also recorded (Song et al. 2017).

Another important feature of MSC-like cells and of DPSCs is their immunomodulatory properties (Zhou et al. 2020). This has attracted an enormous interest in the context of cell therapy and tissue repair engineering strategies but not only. MSCs target several subsets of innate and adaptive immune human cells such as for instance helper T lymphocytes, dendritic cells, macrophages, and regulatory T lymphocytes. These effects are mediated by MSC-secreted soluble factors (PGE2, TGFβ, Nitric oxide, etc.) or via a cell-to-cell-dependent manner inducing T-cell apoptosis through the Fas/FasL pathway (Zhao et al. 2012). *In vivo,* these immunomodulatory properties have been exploited for the experimental treatment of auto-immune disorders in animal models especially through anti-inflammatory cytokines production and regulatory T cells promotion (De la Rosa-Ruiz et al. 2019). BM-MSCs are now frequently used in the clinics for preventing graft-versus-host disease (Weiss et Dahlke 2019). MSCs maintain homeostasis in injured tissues by regulating immune response and promoting tissue regeneration. These functions have been explained by their interactions with immune cells and tissue-specific progenitor cells. Monocytes and macrophages are key immune cells which, in case of injury, are rapidly recruited and will undergo prompt phenotypic and functional changes to adopt first pro-inflammatory and then anti-inflammatory and pro-wound healing phenotypes in response to the microenvironment. The MSCs–macrophages crosstalk has emerged as an essential step in tissue repair (Pajarinen et al. 2019) but the cellular and molecular mechanisms underlying its action are still poorly understood.

20.3 Major Unsolved Problems and Ongoing Controversies in the Field

Over the last decade, data have accumulated, demonstrating the interest of DPSCs for developing cell therapies for a series of pathologies (Botelho et al. 2017). Moving these cells routinely toward therapeutics however still requires answers to major challenges. First of all, since "DPSCs" constitute a heterogeneous cell population composed of a small number of true "stem cells" mixed with more advanced progenitors, to reproducibly obtain well-defined cell populations efficient in a given therapeutic context, is mandatory. To this end, the nature of different cell subpopulations must be thoroughly characterized through possible identification of a set of cell surface markers, as developed for the hematopoietic stem cells (Eaves 2015), with the goal of defining the most efficient subpopulation for a given therapeutic application (Ducret et al. 2019). Also, since most tissue engineering applications are likely to require an important number of cells that cannot be obtained without previous *in vitro* amplification, well-defined Good Manufacturing Practice (GMP) culture conditions must be standardized and international standards for the production of human DPSCs, must be set up for allowing comparing the different studies (Spina et al. 2016). Recent data suggested that *in vitro* priming of the cells upon expansion culture can affect their fate during repair (Collignon et al. 2019; Delle Monache et al. 2019). This further implies the necessity of standardizing defined culture conditions for each specific *in vivo* application.

Up to now, few studies have been focused on the structure and fate of repaired tissues over the long term. Questions, such as: "what are the cellular and molecular mechanisms underlying the repair processes" or "does the newly formed tissue behaves as the physiological one in response to new aggression and ageing," will have to be addressed before routine clinical transfer of these new therapeutic approaches can be set up.

20.3.1 Heterogeneity in Pulpal Cell Populations

DP is formed during the course of development by a mixture of cells derived from the cranial NC, cells originating from the paraxial mesenchyme and mesenchymal cells from the first branchial arch (Chai et al. 2000). The adult DP has long been known to be composed of odontoblasts, neural, vascular and immune cells, fibroblasts and a mixture of "immature/undifferentiated" cells, those now known as "stem cells" (Gronthos et al. 2000) involved in tissue homeostasis and repair. Recent data support the hypothesis of a neural crest origin of the more immature DPSCs (Janebodin et al. 2011). Over the years, different MSC populations have been described in the human DP suggesting the existence, as in the bone marrow, of a hierarchy of progenitors endowed with various differentiation potentials (Huang et al. 2009; Ducret et al. 2019).

DPSCs are usually obtained from the pulp by enzymatic digestion (Gronthos et al. 2000) or by explant outgrowth (Kerkis et al. 2006). Both procedures allow expansion of similar mesenchymal progenitors (Hilkens et al. 2013), but explant outgrowth cultures seems to give rise to a more immature cell subpopulation (Kerkis et Caplan 2012).

In the absence of specific marker allowing their purification, many laboratories still isolate DPSCs on the classical MSC capacity for plastic adherence and colony formation, as in the initial publication (Gronthos et al. 2000). Other laboratories have enriched on cell surface molecule expression (STRO-1, CD146, CD34, CD105), on the capacity of a DNA binding dye exclusion or on the size (Gronthos et al. 2002; Laino et al. 2006; Iohara et al. 2009), respectively. Not surprisingly, the resulting subpopulations display different features, degrees of "stemness" and *in vitro* or *in vivo* potentials. For instance, DPSCs isolated on their colony forming unit capacity contribute to the formation of a pulp-like vascularized tissue (Huang et al. 2010) or differentiate into neuronal cells after appropriate *in vitro* induction (Arthur et al. 2008), while a CD34+ DPSC fraction is prone to engage into an osteogenic or endothelial program (d'Aquino et al. 2007) and a CD31-/CD146- side population pulpal cells display a highly vasculogenic potential (Sugiyama et al. 2011). Each of these stem cell–enriched populations has, in general only, been used in one specific protocol and it is therefore impossible to apprehend their full potential outside this particular context. It is thus mandatory to further clarify the nature of these different subpopulations and their capacity for tissue repair, to define the most efficient DPSC population for each potential tissue engineering application.

Other factors contributing to the phenotypic heterogeneity of the cells, are the variations in DPSC culture conditions and media composition (percentage of fetal calf serum, supplementation or not with growth factors, inoculation density, number of passages, etc.) between the various laboratories studying DPSCs. Such variations impact the cell phenotype, likely favoring certain subpopulations over others. DPSCs clearly undergo phenotypic changes with subculture, losing markers while acquiring others and in the end, eventually, losing the stem cell phenotype (Kanafi et al. 2014; Bakopoulou et al. 2017). Trying both to improve and standardize the culture conditions of DPSCs, for their use in tissue engineering strategies, is therefore another major challenge for progressing toward therapeutic protocols.

Notably, in the ES cell field, maintenance of the stem cell potential and homogeneity of differentiation have greatly benefited from the development of synthetic culture media. The same might be true for DPSCs.

Using a typical ES cell culture medium, Atari et al. (Atari et al. 2012) have maintained a DPSC homogeneous and stable subpopulation displaying a pluripotent-like phenotype over the long term. Another interesting variable to be explored is the culture in hypoxic conditions (3–5%). Data have suggested that a moderate hypoxia positively impact DPSC proliferation (Sakdee et al. 2009; Ahmed et al. 2016) and the maintenance of an immature stem cell phenotype (Janebodin et al. 2011; Gorin et al. 2016; Collignon et al. 2019). However, the effect of such culture conditions on the *in vivo* engraftment, survival, and repair potential of DPSCs has not yet been assessed in the various preclinical models used.

20.4 True Nature of DPSCs

MSCs were first described almost 50 years ago in the bone marrow (Frieden-stein et al. 1974; Caplan 1991). Since then, cells with a similar immunophe-notype and functional properties, MSC-like cells, have been described in multiple organs and in particular in the teeth (Huang et al. 2009). The multipotent potential demonstrated by these cells have risen an enormous interest for their use in tissue engineering and repair strategies, as already mentioned herein but less emphasis was put on their *in vivo* origin. Freshly established cultures of DPSCs are known, since their discovery, to present a cell surface phenotype similar to BMSCs (Shi et Gronthos 2003) with STRO-1 expression together with a series of CD markers, such as CD90, CD105, or CD73. They were thus proposed by the International Society for Cell Therapy for the identification of human MSCs (Dominici et al. 2006; Viswanathan et al. 2019). These markers however do not always reflect the DPSC *in vivo* phenotype (Gomez-Salazar et al. 2020). In normal pulp sections, STRO-1 expressing cells are only found in the microvasculature (Shi et Gronthos 2003), an observation compatible with the hypothesis that DPSCs, as MSCs from other tissues, might correspond to pericytes, the cells that encircle the endothelial cells of capillaries and microvessels (Crisan et al. 2008). But the fact that DPSCs are heterogeneous also raised the question of the existence of other potential stem cell niches in the pulp (Harichane et al. 2011). And indeed, not all MSCs appear to originate from pericytes. In a model of incisor growth or damage, lineage-tracing experiments have demonstrated the con-tributions of NG2$^+$ pericytes originating from Gli1$^+$ cells and glia-derived MSCs (Kaukua et al. 2014; Zhao et al. 2014) to odontoblast-like cells during reparative dentinogenesis. In the molar tooth, such Gli1+ cells are not found in the native pulp, but some are localized in its apical region (Men et al. 2020).

The existence of non-pericytic–derived MSC progenitors has been evidenced in vascularized tissue (James et Peault 2019; Gomez-Salazar et al. 2020). The cells of the tunica adventitia (or adventitial cells) are also perivascular cells but, in contrast to pericytes, surround the largest vessels and are not associated with endothelial cells. Although they differ in immunophenotype from pericytes (CD146$^+$/CD34$^-$ versus respectively CD146$^-$/CD34$^+$), adventitial cells are similarly endowed with typical mesenchymal progenitors properties (Crisan et al. 2012). Do both these populations originally deriving from the cranial NC (Creuzet 2009) contribute to the complete DPSC population, and newly formed odontoblasts in the tooth is still an open question.

20.5 Fate of Implanted Dental Stem Cells in Tissue Repair

The development of safe therapies through tissue engineering requires an understanding of the cellular and molecular mechanisms underlying the action of the implanted cells in tissue repair or regeneration. In each potential therapeutic context, determining whether the stem cells are contributing to the formation of a neo-tissue by direct differentiation or by recruiting local host progenitors through their secretome or both, has to be established since both mechanisms have been described for DPSCs and MSC-like cells in general (Cordeiro et al. 2008; Kichenbrand et al. 2019). Although several studies have demonstrated the *in vivo* capacity for osteogenic differentiation of various types of MSCs (Seo et al. 2008; Lo et al. 2012), others have shown a massive apoptosis of implanted cells in the week following the graft (Moya et al. 2017). This poor survival and engraftment of MSC-like cells after transplantation, in bone as in other tissues, is likely due to the hostile microenvironment (ischemic tissue, oxidative stress, hypoxia, loss of nutrients and extracellular matrix) they encounter upon transplant. Consequently, it is now generally admitted that the main effect of stem cells in tissue repair proceeds through a paracrine action, a hypothesis supported by data showing that the repair promoted by implanted MSC-like cells could be mimicked by their conditioned medium (Sugimura-Wakayama et al. 2015; Fujio et al. 2017). Recent experiments have however suggested that "priming" i.e., pre-treating, DPSCs and MSC-like cells in general, by various types of factors, may be an efficient way to improve their survival and direct repair properties (Collignon et al. 2019; Salazar-Noratto et al. 2020). For instance, exposure to sub-lethal conditions allows cells to gradually adapt to changes in their environment, mounting an anti-stress response that activates pro-survival pathways (Baldari et al. 2017). Understanding and leveraging the molecular mechanisms underlying these actions may dramatically improve the overall therapeutic potential of MSCs.

20.6 Contribution of the Team to the Field: DP Stem Cells and Craniofacial Bone Repair

Our team has acquired over the years an expertise on mouse, rat, porcine, and human DP stem cells as a tool for tissue engineering of the cranio-facial defects (Lacerda-Pinheiro et al. 2012; Coyac et al. 2013; Souron et al. 2013; Chamieh et al. 2016; Mangione et al. 2017; Collignon et al. 2018, 2019; Novais et al. 2019). Our studies are focused on several topics: (1) optimizing scaffolds to support and facilitate the tissue repair process promoted by DPSCs, (2) characterizing their precise roles, direct or/and indirect, in the repair process as well as the fate of the DPSCs implanted in a wounded tissue and (3) optimizing the DPSC intrinsic properties to improve their *in vivo* reparative action. A summary of our latest observations is described in the following paragraphs.

20.6.1 Optimization of the Scaffold to Support the DPSC-promoted Tissue Repair

Current bone repair techniques use a combination of autologous, allogeneic, and prosthetic materials. Even if these approaches achieve a degree of functional restoration, they possess inherent limitations (donor site morbidity, immune rejection, pathogen transfer, etc.). The emergence over the years 2000 of the new medical branch "Regenerative medicine," with the development of tissue engineering approaches, involving a combination of various (stem) cells, scaffolds and signaling molecules, has raised new hopes for treating complex medical issues, such as large craniofacial bone defects (Hutmacher 2000). One element of the tissue engineering triad, the scaffold, is particularly important in the case of large bone defects. It must be a conveniently implanted, resorbable material and should display osteoinductive and/or osteoconductive properties, providing a mechanical support to promote bone repair/regeneration into a defect site (Bueno et Glowacki 2009). The scaffold is most frequently a suitable biomaterial associated with cells and ways of potentiation (growth factors...). A large variety of materials, including naturally originated ones, polymers, ceramics, and composites, as well as nanomaterials, are currently explored as biomaterials *in vivo* (Khademhosseini et Langer 2016).

Our team has chosen to use collagen hydrogels for their properties of biocompatibility, tissue integration, and mostly because fibrillar type I collagen is the most abundant protein in the human body. It is predominantly found in the extracellular matrix of connective tissues and has appeared as an ideal support for bone cells, since the barrier membranes firstly introduced in the mid-80s. *In vitro* reassembled acid-solubilized collagen type I hydrogels are highly hydrated and display a mechanical instability that has limited their use. To overcome this pitfall, a plastic compression of the collagen hydrogels can be achieved that allows their densification and mechanical stabilization (Abou Neel et al. 2006). This matrix manufacturing method maintains the fibrillar nature of collagen, similar to that of native bone matrix (Engler et al. 2006) and preserves the porous microstructures necessary for a biological function (Brown et al. 2005). This allows cell incorporation into a neutralized collagen solution prior to gel fibrillogenesis and densification to achieve a homogeneous three-dimensional embedding within the dense collagen scaffold (Coyac et al. 2013). These scaffolds then provide the structure for cell growth and differentiation without compromising their viability. They can also be associated with potentiating factors like entrapped molecules, which can then stimulate or recruit resident cells of the engineered tissue. In this context, ceramics, growth factors, and cytokines have been associated to the biomaterial to promote the migration and engraftment of embedded cells and contribute to tissue regeneration (Miri et al. 2016). Among these molecules, inorganic bioactive glass particles (Bioglass 45S5®) has been recently shown to enhance bone formation *in vivo* when incorporated in a collagen scaffold (Marelli et al. 2011). Along this line, our team has recently started to evaluate the efficacy of DPSC delivery in dense collagen gels functionalized by such bioactive glass. The data are still under investigation, but, in agreement with previous studies (Marelli et al. 2011; van Gestel et al. 2015), it already appears that the hybrid biomaterial DPSC/specific bioglass/dense collagen scaffolds accelerate DPSC *in vitro* differentiation and mineralization as well as bone formation, when implanted into a mouse critical calvaria defect compared to the "simple" dense collagen hydrogels (Park et al. 2021). The structure of the bone formed under these conditions together with its vascularization are currently under analysis to further characterize this biomaterial for a potential development in bone tissue engineering therapies.

20.6.2 Fate of DPSCs/SHEDs Implanted in a Critical Defect of the Calvaria

In view of determining the exact involvement of DPSCs in the process of calvaria critical defect repair, we have used molar DPSCs isolated from a Wnt1-CRE-Rosa^{Tomato} mouse, a NC lineage tracing reporter model (Collignon et al. 2019). The *Wnt1* gene is known to become activated in all NC cells, at the time of their emergence from the neural tube (Echelard et al. 1994). In the Wnt1-CRE-Rosa^{Tomato} system, once Wnt1 is expressed, the CRE recombinase will excise a STOP cassette which will allow overexpression of the Tomato fluorescent protein in all the NC cells and their progeny, and therefore also in DPSCs. These cells can then be followed *in vitro* or *in vivo* through their red fluorescence. We have compared the kinetics of bone repair process in the calvaria defect mediated by DPSCs cultured either in "osteo" conditions i.e., a classical culture medium, alpha-MEM, supplemented with BMP2 and FGF2 under a classical oxygen atmosphere of 19% (normoxia) to facilitate their differentiation in osteoblast/osteocyte (Chamieh et al. 2016; Collignon et al. 2018) with DPSCs cultured in "stem" conditions i.e., a same classical culture medium, alpha-MEM, supplemented with FGF2 and Leukemia Inhibitory Factor (LIF) under an oxygen atmosphere of 5% (hypoxia) (Collignon et al. 2019) to favor their stem cell phenotype. Indeed, culture in 3–5% hypoxia (also named physioxia) has been shown to more closely resemble *in vivo* stem cell niches than classical normoxia and to favor stemness, growth kinetics, genetic stability, and cell survival in traumatized microenvironments (Haque et al. 2013). In addition, hypoxic treatment of MSC-like cells has also been shown to favor rapid vascularization, an essential property for graft survival (Gorin et al. 2016; Antebi et al. 2018).

The use of Wnt1-CRE-Rosa^{Tomato} DPSCs has allowed us to show that both culture conditions i.e., "priming," led to efficient bone repair but with an apparent different action of DPSCs. Indeed, "Stem primed" DPSCs survived in the scaffold implanted in the defect for up to 3 months, and readily differentiate into chondrocytes and hypertrophic ones. These cells recruited which recruited host osteoprogenitors and finally underwent apoptosis, thus displaying involvement in a classical endochondral bone repair process (Collignon et al. 2019) (Figure 20.2). In contrast, "Osteo primed" DPSCs become rapidly undetectable within the scaffold and replaced by host cells differentiating into osteoblast-like cells and osteocytes, thus undergoing an intramembranous bone repair process. We are currently studying the mechanisms underlying

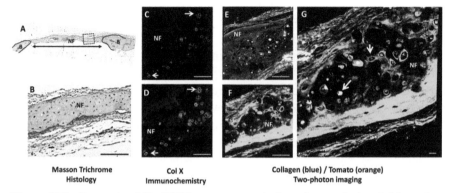

Figure 20.2 Stem-primed T-mDPSCs promote repair through an endochondral bone formation process. Histochemical, immunofluoresence and two-photon microscopic analyses of mDPSC cellularized scaffold sections at 12-weeks following implantation. (A, B): Masson trichrome histological staining shows a collagen matrix with cells embedded in a cartilage-like newly formed tissue (NF) and native bone (B). Double Arrow covers the bone critical defect (C): immunostaining with anti DsRed recognizing the Tomato protein specifically reveals the implanted cells in the newly formed tissue (NF). (D) immunostaining of type X collagen (Col X), a marker of hypertrophic chondrocyte, shows specific labelling (green) of the T-mDPSCs in the NF. (E, F, G) The two-photon microscopy analysis directly show Tomato mDPSC (red) surrounded by a collagen matrix (blue) in the NF. Scale bars: 50 μm.

this DPSC maintenance/disappearance as well as the nature and origin of the host cells that invade the scaffold within 2 weeks (unpublished data). MSC-progenitors capable of participating in bone repair have been described in the calvaria sutures and in the periosteum (Doro et al. 2017; Debnath et al. 2018), but the invading cells could also originate from the systemic circulation. These different possibilities are now explored through the use of transgenic mice in which these cell populations can be specifically followed.

Another important question to explore in these repair processes, is the molecular mechanisms underlining the dialogue established between the local and implanted cells. It is now well established that, in the context of tissue repair, MSC-like cells display an immunomodulatory action and are capable of promoting the polarization of macrophages from the M1 pro-inflammatory phenotype toward an anti-inflammatory and pro-repair M2 phenotype (Horwood 2016). This switch is known to play an active part in the tissue repair/regeneration process (Pajarinen et al. 2019). Most likely, different cell priming impact the DPSC intrinsic state (such as their epigenome and secretome) and that, in turn, may modify their response to the local microenvironment and the dialogue with host surrounding cells. This could also potentially differentially influence the inflammatory response. We are currently exploring these important points, both in the context of the critical defect of the calvaria, but also in a rat mandibular defect, a model where the bone is subjected to intense mechanical stresses as compared to the calvaria.

20.6.3 Pro-angiogenic Potential of DPSCs/SHEDs

Rapid vascularization and establishment of a functional vascular network is a cornerstone for the clinical outcome of MSC-based therapies and pre-vascularization of tissue constructs is considered as the most promising approach to address this issue (Rademakers et al. 2019; Song et al. 2019). For instance, during intramembranous bone regeneration, exposure to hypoxia in the initial inflammatory phase stimulates osteoblasts to release several factors, including VEGF, *via* the HIF-1α pathway, inducing endothelial migration and proliferation and vessel permeability (Wang et al. 2007). The new vessels increase the supply of nutrients, oxygen, and minerals necessary for osteogenesis and may recruit osteoprogenitors to the injury site. Furthermore, endothelial cells also produce osteogenic factors (e.g., BMP-2 and BMP-4) that promote osteoblast differentiation, while differentiating osteoblasts secrete angiogenic factors (e.g., PDGF-BB and VEGF) to further support angiogenesis by a positive feedback loop (Duan et al. 2016).

DPSCs have shown their capacity to induce capillary formation *in vitro* through the release of angiogenic factors (Bronckaers et al. 2013). Along this line, using a tooth slice model containing DPSCs seeded in a collagen hydrogel and implanted subcutaneously in immunodeficient mice, we have recently shown that hypoxia- or FGF-2-primed DPSCs allowed the formation of a functional vascularized pulp-like tissue. In this context, DPSCs released both more VEGF, which induces capillary formation, and HGF, which potentiates the effect of VEGF on the host endothelial cells (Gorin et al. 2016) (Figure 20.3). On another hand, DPSCs appear also able to participate more directly in the neoangiogenic process. Indeed, MSC-like cells and pericytes are known to share perivascular distribution and regeneration properties (Crisan et al. 2008; de Souza et al. 2016). DPSCs, in particular, have been localized within the vascular niche and reported to display pericyte-like properties and specific interactions with endothelial cells (ECs) as mentioned earlier in this chapter. In addition, they present an immunophenotypic profile similar to pericytes, which appears to depend on their isolation method and culture conditions (Janebodin et al. 2013; Gorin et al. 2016; Ducret et al. 2019).

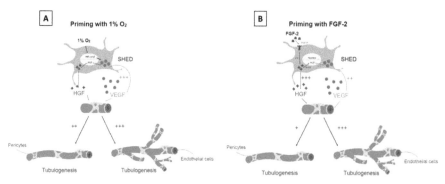

Figure 20.3   Diagrams showing the action mechanisms of (A) hypoxic (1%) or (B) FGF-2 priming on SHEDs, SHEDs synthesize HGF, a pro-angiogenic factor in normal culture conditions. In response to severe hypoxia (A), SHED start to secrete VEGF via the HIF-1α pathway, potentiating their initial pro-angiogenic properties, and enhancing endothelial cell tubulogenesis. When treated with FGF-2 (B) SHED increase both VEGF and HGF, synthesis and independently stimulate the two distinct pro-angiogenic pathways, HGF potentiating the formation of capillaries induced by VEGF.

We have recently started to characterize the interactions established between DPSCs and ECs seeded in collagen scaffolds in 3D *in vitro* culture. We have firstly highlighted that the presence of DPSCs close to ECs generate the formation of capillaries inside the constructs and that secondly, DPSCs were recruited to capillary surfaces through the PDGF-BB signaling pathway, extending protrusions to enwrap ECs to participate in the generation of a typical basement membrane, in a same way as pericytes (Armulik et al. 2011; Shen et McCloskey 2017). This particular localization close to the vessels of certain DPSCs suggests that these cells can play a major role in capillary stabilization. Our *in vivo* analyses have further shown that ectopic subcutaneous implantation of such *in vitro*-generated mature microvascular networks allow a better vascularization, survival/protection from apoptosis and innervation of the DPSCs implanted scaffold (Atlas et al, 2021).

20.7 Perspectives

DPSCs appear to also be capable of directly contributing to vessel stabilization by assuming a pericyte-like function during the angiogenic process (Atlas et al, 2021). How various priming can modulate this property has still to be evaluated. Tissue-specific stem cell priming appears as an important novel route for improving stem cell potential for tissue repair and regeneration (for a recent review on MSC priming see Noronha et al. 2019). A detailed analysis of the cellular and molecular mechanisms underlining the priming action is required before routine clinical transfer can be envisioned. In particular, a thorough evaluation of how the response to local inflammation, the stem cell epigenome and metabolome are impacted by priming is mandatory to allow a development of reproductive therapeutical protocols.

Different potential of cell candidates have currently been identified for craniofacial repair or regeneration, involving various types of oral MSC-like cells known to display bone repair potentials, but also those derived from pluripotent stem cells (Jungbluth et al. 2019). Which of these populations will be most suitable for a clinical transfer is not yet clear at present. The answer certainly depends on further research data, especially on the mechanisms underlining the repair and the dialogue with host cells, on the type of defects to repair, and last but not least on economical contingencies and public health choices. To safely, and predictably utilize these cells for medical applications, future research will have to also further focus on how to maintain the cells as stem cells, as well as on their behavior and fate on the long term. Indeed, nothing is still known on the precise functionality of a stem cell-repaired tissue over time. Despite, these challenges, the promising results of tissue engineering experiments involving stem cells in preclinical models pave the way for new treatment possibilities in a field where current therapies present significant limits and in this context, DPSC already have a place of choice.

20.8 Concluding Remarks

- DPSCs can be obtained safely and easily from non-essential teeth without significant morbidity or ethical concerns.
- DPSCs as many tissues of the craniofacial area, are neural crest derived progenitor cells which present similar properties to mesenchymal stromal/stem cells–like cell populations.
- DPSCs strong pro-angiogenic, pro-neurogenic, and immunomodulatory characters can be valorized in the dental context but also for craniofacial injuries such as bone, neuronal, or ischemic lesions.
- DPSCs are able to survive and to readily differentiate *in situ* to promote bone tissue repair, depending on the type of priming they have been submitted to.
- DPSC transplantation after pulpectomy has recently been shown to successfully rebuild a functionalized physiological pulp structure *in situ*.

Acknowledgments

The authors thank all the members of UR2496 for their help, and particularly J. Lesieur, S. Ribes, B. Barouck, J. Sadoine and Dr L.Slimani. The ANR Pulp Cell and the Fondation des Gueules Cassées financially supported our studies.

References

Abou Neel EA, Cheema U, Knowles JC, Brown RA, Nazhat SN. Use of multiple unconfined compression for control of collagen gel scaffold density and mechanical properties. Soft Matter. 2006;2(11):986.

Ahmed NE-MB, Murakami M, Kaneko S, Nakashima M. The effects of hypoxia on the stemness properties of human dental pulp stem cells (DPSCs). Sci Rep. 2016;6(1):35476.

Antebi B, Rodriguez LA, Walker KP, Asher AM, Kamucheka RM, Alvarado L, et al. Short-term physiological hypoxia potentiates the therapeutic function of mesenchymal stem cells. Stem Cell Res Ther. 2018;9(1):265.

d'Aquino R, De Rosa A, Lanza V, Tirino V, Laino L, Graziano A, et al. Human mandible bone defect repair by the grafting of dental pulp stem/progenitor cells and collagen sponge biocomplexes. Eur Cell Mater. 2009;18:75–83.

d'Aquino R, Graziano A, Sampaolesi M, Laino G, Pirozzi G, De Rosa A, et al. Human postnatal dental pulp cells co-differentiate into osteoblasts and endotheliocytes: a pivotal synergy leading to adult bone tissue formation. Cell Death Differ. 2007;14:1162–71.

Armulik A, Genové G, Betsholtz C. Pericytes: Developmental, Physiological, and Pathological Perspectives, Problems, and Promises. Dev Cell. 2011;21(2):193–215.

Arthur A, Rychkov G, Shi S, Koblar SA, Gronthos S. Adult human dental pulp stem cells differentiate toward functionally active neurons under appropriate environmental cues. Stem Cells. 2008;26:1787–95.

Arthur A, Shi S, Zannettino ACW, Fujii N, Gronthos S, Koblar SA. Implanted Adult Human Dental Pulp Stem Cells Induce Endogenous Axon Guidance: Human Dental Pulp Stem Cells Induce Neuroplasticity. STEM CELLS. 2009;27(9):2229–37.

Atari M, Gil-Recio C, Fabregat M, Garcia-Fernandez D, Barajas M, Carrasco MA, et al. Dental pulp of the third molar: a new source of pluripotent-like stem cells. J Cell Sci. 2012;125:3343–56.

Atlas Y, Gorin C, Novais A, Marchand MF, Chatzopoulou E, Lesieur J, et al. Microvascular maturation by mesenchymal stem cells in vitro improves blood perfusion in implanted tissue constructs. Biomaterials. 2021;268:120594.

Bakopoulou A, Apatzidou D, Aggelidou E, Gousopoulou E, Leyhausen G, Volk J, et al. Isolation and prolonged expansion of oral mesenchymal stem cells under clinical-grade, GMP-compliant conditions differentially affects "stemness" properties. Stem Cell Res Ther. 2017;8(1):247.

Baldari S, Di Rocco G, Piccoli M, Pozzobon M, Muraca M, Toietta G. Challenges and Strategies for Improving the Regenerative Effects of Mesenchymal Stromal Cell-Based Therapies. Int J Mol Sci. 2017;18(10):2087.

Balic A, Thesleff I. Tissue Interactions Regulating Tooth Development and Renewal. Curr Top Dev Biol. 2015;115:157–86.

Botelho J, Cavacas MA, Machado V, Mendes JJ. Dental stem cells: recent progresses in tissue engineering and regenerative medicine. Ann Med. 2017;49(8):644–51.

Bronckaers A, Hilkens P, Fanton Y, Struys T, Gervois P, Politis C, et al. Angiogenic properties of human dental pulp stem cells. PLoS One. 2013;8:e71104.

Brown RA, Wiseman M, Chuo C-B, Cheema U, Nazhat SN. Ultrarapid Engineering of Biomimetic Materials and Tissues: Fabrication of Nano- and Microstructures by Plastic Compression. Adv Funct Mater. 2005;15(11):1762–70.

Bueno EM, Glowacki J. Cell-free and cell-based approaches for bone regeneration. Nat Rev Rheumatol. 2009;5(12):685–97.

Caplan AI. Mesenchymal stem cells. J Orthop Res. 1991;9(5):641–50.

Caplan AI, Correa D. The MSC: an injury drugstore. Cell Stem Cell. 2011;9:11–5.

Chai Y, Jiang X, Ito Y, Bringas P, Han J, Rowitch DH, et al. Fate of the mammalian cranial neural crest during tooth and mandibular morphogenesis. Development. 2000;127:1671–9.

Chamieh F, Collignon A-M, Coyac BR, Lesieur J, Ribes S, Sadoine J, et al. Accelerated craniofacial bone regeneration through dense collagen gel scaffolds seeded with dental pulp stem cells. Sci Rep. 2016;6:38814.

Collignon A-M, Castillo-Dali G, Gomez E, Guilbert T, Lesieur J, Nicoletti A, et al. Mouse Wnt1-CRE-RosaTomato Dental Pulp Stem Cells Directly Contribute to the Calvarial Bone Regeneration Process. Stem Cells. 2019;37(5):701–11.

Collignon A-M, Lesieur J, Anizan N, Azzouna RB, Poliard A, Gorin C, et al. Early angiogenesis detected by PET imaging with 64Cu-NODAGA-RGD is predictive of bone critical defect repair. Acta Biomater. 2018;82: 111–21.

Cordeiro MM, Dong Z, Kaneko T, Zhang Z, Miyazawa M, Shi S, et al. Dental pulp tissue engineering with stem cells from exfoliated deciduous teeth. J Endod. 2008;34:962–9.

Coyac BR, Chicatun F, Hoac B, Nelea V, Chaussain C, Nazhat SN, et al. Mineralization of dense collagen hydrogel scaffolds by human pulp cells. J Dent Res. 2013;92(7):648–54.

Creuzet SE. Neural crest contribution to forebrain development. Semin Cell Dev Biol. 2009;20:751–9.

Crisan M, Corselli M, Chen WC, Peault B. Perivascular cells for regenerative medicine. J Cell Mol Med. 2012;16:2851–60.

Crisan M, Yap S, Casteilla L, Chen CW, Corselli M, Park TS, et al. A perivascular origin for mesenchymal stem cells in multiple human organs. Cell Stem Cell. 2008;3:301–13.

De la Rosa-Ruiz MDP, Álvarez-Pérez MA, Cortés-Morales VA, Monroy-García A, Mayani H, Fragoso-González G, et al. Mesenchymal Stem/Stromal Cells Derived from Dental Tissues: A Comparative In Vitro Evaluation of Their Immunoregulatory Properties Against T cells. Cells. 8 (12)/:491.

Debnath S, Yallowitz AR, McCormick J, Lalani S, Zhang T, Xu R, et al. Discovery of a periosteal stem cell mediating intramembranous bone formation. Nature. 2018;562(7725):133–9.

Delle Monache S, Martellucci S, Clementi L, Pulcini F, Santilli F, Mei C, et al. In Vitro Conditioning Determines the Capacity of Dental Pulp Stem Cells to Function as Pericyte-Like Cells. Stem Cells Dev. 2019;28(10):695–706.

Dennis SC, Berkland CJ, Bonewald LF, Detamore MS. Endochondral ossification for enhancing bone regeneration: converging native extracellular matrix biomaterials and developmental engineering in vivo. Tissue Eng Part B Rev. 2015;21(3):247–66.

Dissanayaka WL, Zhu L, Hargreaves KM, Jin L, Zhang C. Scaffold-free Prevascularized Microtissue Spheroids for Pulp Regeneration. J Dent Res. 2014;93(12):1296–303.

Dominici M, Le Blanc K, Mueller I, Slaper-Cortenbach I, Marini F, Krause D, et al. Minimal criteria for defining multipotent mesenchymal stromal cells. The International Society for Cellular Therapy position statement. Cytotherapy. 2006;8:315–7.

Doro DH, Grigoriadis AE, Liu KJ. Calvarial Suture-Derived Stem Cells and Their Contribution to Cranial Bone Repair. Front Physiol. 2017;8:956.

Duan X, Bradbury SR, Olsen BR, Berendsen AD. VEGF stimulates intramembranous bone formation during craniofacial skeletal development. Matrix Biol. 2016;52–54:127–40.

Ducret M, Farges J-C, Pasdeloup M, Perrier-Groult E, Mueller A, Mallein-Gerin F, et al. Phenotypic Identification of Dental Pulp Mesenchymal Stem/Stromal Cells Subpopulations with Multiparametric Flow Cytometry. Methods Mol Biol Clifton NJ. 2019;1922:77–90.

Eaves CJ. Hematopoietic stem cells: concepts, definitions, and the new reality. Blood. 2015;125(17):2605–13.

Echelard Y, Vassileva G, McMahon AP. Cis-acting regulatory sequences governing Wnt-1 expression in the developing mouse CNS. Dev Camb Engl. 1994;120(8):2213–24.

El Ayachi I, Zhang J, Zou X-Y, Li D, Yu Z, Wei W, et al. Human dental stem cell derived transgene-free iPSCs generate functional neurons via embryoid body-mediated and direct induction methods. J Tissue Eng Regen Med. 2018;12(4):e1836–51.

Engler AJ, Sen S, Sweeney HL, Discher DE. Matrix elasticity directs stem cell lineage specification. Cell. 25 2006;126(4):677–89.

Friedenstein AJ, Chailakhyan RK, Latsinik NV, Panasyuk AF, Keiliss-Borok IV. Stromal cells responsible for transferring the microenvironment of the hemopoietic tissues. Cloning in vitro and retransplantation in vivo. Transplantation. 1974;17(4):331–40.

Fujio M, Xing Z, Sharabi N, Xue Y, Yamamoto A, Hibi H, et al. Conditioned media from hypoxic-cultured human dental pulp cells promotes bone healing during distraction osteogenesis: Hypoxic culture condition increases the angiogenic factors from hDPCs. J Tissue Eng Regen Med. 2017;11(7):2116–26.

Gervois P, Wolfs E, Ratajczak J, Dillen Y, Vangansewinkel T, Hilkens P, et al. Stem Cell-Based Therapies for Ischemic Stroke: Preclinical Results and the Potential of Imaging-Assisted Evaluation of Donor Cell Fate and Mechanisms of Brain Regeneration: Med Res Rev. 2016;36(6):1080–126.

van Gestel NAP, Geurts J, Hulsen DJW, van Rietbergen B, Hofmann S, Arts JJ. Clinical Applications of S53P4 Bioactive Glass in Bone Healing and Osteomyelitic Treatment: A Literature Review. BioMed Res Int. 2015;2015:1–12.

Giuliani A, Manescu A, Langer M, Rustichelli F, Desiderio V, Paino F, et al. Three years after transplants in human mandibles, histological and in-line holotomography revealed that stem cells regenerated a compact rather than a spongy bone: biological and clinical implications. Stem Cells Transl Med. 2013;2:316–24.

Gomez-Salazar M, Gonzalez-Galofre ZN, Casamitjana J, Crisan M, James AW, Péault B. Five Decades Later, Are Mesenchymal Stem Cells Still Relevant? Front Bioeng Biotechnol. 2020;8:148.

Gorin C, Rochefort GY, Bascetin R, Ying H, Lesieur J, Sadoine J, et al. Priming Dental Pulp Stem Cells With Fibroblast Growth Factor-2 Increases Angiogenesis of Implanted Tissue-Engineered Constructs Through Hepatocyte Growth Factor and Vascular Endothelial Growth Factor Secretion. Stem Cells Transl Med. 2016;5(3):392–404.

Gronthos S, Brahim J, Li W, Fisher LW, Cherman N, Boyde A, et al. Stem cell properties of human dental pulp stem cells. J Dent Res. 2002;81:531–5.

Gronthos S, Mankani M, Brahim J, Robey PG, Shi S. Postnatal human dental pulp stem cells (DPSCs) in vitro and in vivo. Proc Natl Acad Sci U A. 2000;97:13625–30.

Haque N, Rahman MT, Abu Kasim NH, Alabsi AM. Hypoxic Culture Conditions as a Solution for Mesenchymal Stem Cell Based Regenerative Therapy. Sci World J. 2013;2013:1–12.

Harichane Y, Hirata A, Dimitrova-Nakov S, Granja I, Goldberg A, Kellermann O, et al. Pulpal progenitors and dentin repair. Adv Dent Res. 2011;23:307–12.

Hilkens P, Fanton Y, Martens W, Gervois P, Struys T, Politis C, et al. Pro-angiogenic impact of dental stem cells in vitro and in vivo. Stem Cell Res. mai 2014;12(3):778–90.

Hilkens P, Gervois P, Fanton Y, Vanormelingen J, Martens W, Struys T, et al. Effect of isolation methodology on stem cell properties and multilineage differentiation potential of human dental pulp stem cells. Cell Tissue Res. 2013;353:65–78.

Horwood NJ. Macrophage Polarization and Bone Formation: A review. Clin Rev Allergy Immunol. 2016;51(1):79–86.

Huang GT, Gronthos S, Shi S. Mesenchymal stem cells derived from dental tissues vs. those from other sources: their biology and role in regenerative medicine. J Dent Res. 2009;88:792–806.

Huang GT, Yamaza T, Shea LD, Djouad F, Kuhn NZ, Tuan RS, et al. Stem/progenitor cell-mediated de novo regeneration of dental pulp with newly deposited continuous layer of dentin in an in vivo model. Tissue Eng Part A. 2010;16:605–15.

Hutmacher DW. Scaffolds in tissue engineering bone and cartilage. Biomaterials. 2000;21(24):2529–43.

Iohara K, Imabayashi K, Ishizaka R, Watanabe A, Nabekura J, Ito M, et al. Complete pulp regeneration after pulpectomy by transplantation of CD105+ stem cells with stromal cell-derived factor-1. Tissue Eng Part A. 2011;17:1911–20.

Iohara K, Murakami M, Takeuchi N, Osako Y, Ito M, Ishizaka R, et al. A novel combinatorial therapy with pulp stem cells and granulocyte colony-stimulating factor for total pulp regeneration. Stem Cells Transl Med. 2013;2(7):521–33.

Iohara K, Zheng L, Ito M, Ishizaka R, Nakamura H, Into T, et al. Regeneration of dental pulp after pulpotomy by transplantation of CD31(-)/CD146(-) side population cells from a canine tooth. Regen Med. 2009;4:377–85.

Iohara K, Zheng L, Ito M, Tomokiyo A, Matsushita K, Nakashima M. Side population cells isolated from porcine dental pulp tissue with self-renewal and multipotency for dentinogenesis, chondrogenesis, adipogenesis, and neurogenesis. Stem Cells. 2006;24:2493–503.

Ishkitiev N, Yaegaki K, Imai T, Tanaka T, Fushimi N, Mitev V, et al. Novel management of acute or secondary biliary liver conditions using hepatically differentiated human dental pulp cells. Tissue Eng Part A. 2015;21(3–4):586–93.

James AW, Péault B. Perivascular Mesenchymal Progenitors for Bone Regeneration. J Orthop Res. 2019;37(6):1221–8.

Janebodin K, Horst OV, Ieronimakis N, Balasundaram G, Reesukumal K, Pratumvinit B, et al. Isolation and characterization of neural crest-derived stem cells from dental pulp of neonatal mice. PLoS One. 2011;6:e27526.

Janebodin K, Zeng Y, Buranaphatthana W, Ieronimakis N, Reyes M. VEGFR2-dependent Angiogenic Capacity of Pericyte-like Dental Pulp Stem Cells. J Dent Res. juin 2013;92(6):524–31.

Jernvall J, Thesleff I. Reiterative signaling and patterning during mammalian tooth morphogenesis. Mech Dev. 2000;92:19–29.

Jungbluth P, Spitzhorn L-S, Grassmann J, Tanner S, Latz D, Rahman MS, et al. Human iPSC-derived iMSCs improve bone regeneration in minipigs. Bone Res. 2019;7(1):32.

Kanafi M, Majumdar D, Bhonde R, Gupta P, Datta I. Midbrain Cues Dictate Differentiation of Human Dental Pulp Stem Cells Towards Functional Dopaminergic Neurons: DPSCs DIFFERENTIATION TO DOPAMINERGIC NEURONS. J Cell Physiol. 2014;229(10):1369–77.

Karbanova J, Soukup T, Suchanek J, Mokry J. Osteogenic differentiation of human dental pulp-derived stem cells under various ex-vivo culture conditions. Acta Medica (Hradec Kralove). 2010;53:79–84.

Kaukua N, Shahidi MK, Konstantinidou C, Dyachuk V, Kaucka M, Furlan A, et al. Glial origin of mesenchymal stem cells in a tooth model system. Nature. 2014;513(7519):551–4.

Kerkis I, Caplan AI. Stem cells in dental pulp of deciduous teeth. Tissue Eng Part B Rev. 2012;18:129–38.

Kerkis I, Kerkis A, Dozortsev D, Stukart-Parsons GC, Gomes Massironi SM, Pereira LV, et al. Isolation and characterization of a population of immature dental pulp stem cells expressing OCT-4 and other embryonic stem cell markers. Cells Tissues Organs. 2006;184:105–16.

Khademhosseini A, Langer R. A decade of progress in tissue engineering. Nat Protoc. 2016;11(10):1775–81.

Kichenbrand C, Velot E, Menu P, Moby V. Dental Pulp Stem Cell-Derived Conditioned Medium: An Attractive Alternative for Regenerative Therapy. Tissue Eng Part B Rev. 2019;25(1):78–88.

Kiraly M, Kadar K, Horvathy DB, Nardai P, Racz GZ, Lacza Z, et al. Integration of neuronally predifferentiated human dental pulp stem cells into rat brain in vivo. Neurochem Int. 2011;59:371–81.

Király M, Porcsalmy B, Pataki A, Kádár K, Jelitai M, Molnár B, et al. Simultaneous PKC and cAMP activation induces differentiation of human dental pulp stem cells into functionally active neurons. Neurochem Int. 2009;55(5):323–32.

Ko IK, Lee SJ, Atala A, Yoo JJ. In situ tissue regeneration through host stem cell recruitment. Exp Mol Med. 2013;45(11):e57-e57.

Komada Y, Yamane T, Kadota D, Isono K, Takakura N, Hayashi S, et al. Origins and properties of dental, thymic, and bone marrow mesenchymal cells and their stem cells. PLoS One. 2012;7:e46436.

Lacerda-Pinheiro S, Dimitrova-Nakov S, Harichane Y, Souyri M, Petit-Cocault L, Legres L, et al. Concomitant multipotent and unipotent dental pulp progenitors and their respective contribution to mineralised tissue formation. Eur Cell Mater. 2012;23:371–86.

Laino G, Carinci F, Graziano A, d'Aquino R, Lanza V, De Rosa A, et al. In vitro bone production using stem cells derived from human dental pulp. J Craniofac Surg. 2006;17:511–5.

Lo DD, Hyun JS, Chung MT, Montoro DT, Zimmermann A, Grova MM, et al. Repair of a Critical-sized Calvarial Defect Model Using Adipose-derived Stromal Cells Harvested from Lipoaspirate. J Vis Exp. 2012;(68):4221.

Luo L, He Y, Wang X, Key B, Lee BH, Li H, et al. Potential Roles of Dental Pulp Stem Cells in Neural Regeneration and Repair. Stem Cells Int. 2018;2018:1–15.

Luzuriaga J, Pastor-Alonso O, Encinas JM, Unda F, Ibarretxe G, Pineda JR. Human Dental Pulp Stem Cells Grown in Neurogenic Media Differentiate Into Endothelial Cells and Promote Neovasculogenesis in the Mouse Brain. Front Physiol. 2019;10:347.

Mangione F, EzEldeen M, Bardet C, Lesieur J, Bonneau M, Decup F, et al. Implanted Dental Pulp Cells Fail to Induce Regeneration in Partial Pulpotomies. J Dent Res. 2017;96(12):1406–13.

Marelli B, Ghezzi CE, Mohn D, Stark WJ, Barralet JE, Boccaccini AR, et al. Accelerated mineralization of dense collagen-nano bioactive glass hybrid gels increases scaffold stiffness and regulates osteoblastic function. Biomaterials. 2011;32(34):8915–26.

Martens W, Sanen K, Georgiou M, Struys T, Bronckaers A, Ameloot M, et al. Human dental pulp stem cells can differentiate into Schwann cells and promote and guide neurite outgrowth in an aligned tissue-engineered collagen construct in vitro. FASEB J Off Publ Fed Am Soc Exp Biol. 2014;28(4):1634–43.

Maruyama T, Jeong J, Sheu T-J, Hsu W. Stem cells of the suture mesenchyme in craniofacial bone development, repair and regeneration. Nat Commun. 2016;7(1):10526.

Mayo V, Sawatari Y, Huang C-YC, Garcia-Godoy F. Neural crest-derived dental stem cells—Where we are and where we are going. J Dent. 2014;42(9):1043–51.

Men Y, Wang Y, Yi Y, Jing D, Luo W, Shen B, et al. Gli1+ Periodontium Stem Cells Are Regulated by Osteocytes and Occlusal Force. Dev Cell. 2020;54(5):639–654.e6.

de Mendonca Costa A, Bueno DF, Martins MT, Kerkis I, Kerkis A, Fanganiello RD, et al. Reconstruction of large cranial defects in nonimmunosuppressed experimental design with human dental pulp stem cells. J Craniofac Surg. 2008;19:204–10.

Menicanin D, Bartold PM, Zannettino ACW, Gronthos S. Identification of a Common Gene Expression Signature Associated with Immature Clonal Mesenchymal Cell Populations Derived from Bone Marrow and Dental Tissues. Stem Cells Dev. 2010;19(10):1501–10.

Miri AK, Muja N, Kamranpour NO, Lepry WC, Boccaccini AR, Clarke SA, et al. Ectopic bone formation in rapidly fabricated acellular injectable dense collagen-Bioglass hybrid scaffolds via gel aspiration-ejection. Biomaterials. 2016;85:128–41.

Miura M, Gronthos S, Zhao M, Lu B, Fisher LW, Robey PG, et al. SHED: stem cells from human exfoliated deciduous teeth. Proc Natl Acad Sci U A. 2003;100:5807–12.

Monteiro BG, Serafim RC, Melo GB, Silva MC, Lizier NF, Maranduba CM, et al. Human immature dental pulp stem cells share key characteristic features with limbal stem cells. Cell Prolif. 2009;42:587–94.

Moya A, Larochette N, Paquet J, Deschepper M, Bensidhoum M, Izzo V, et al. Quiescence Preconditioned Human Multipotent Stromal Cells Adopt a Metabolic Profile Favorable for Enhanced Survival under Ischemia. Stem Cells. 2017;35(1):181–96.

Nakajima K, Kunimatsu R, Ando K, Ando T, Hayashi Y, Kihara T, et al. Comparison of the bone regeneration ability between stem cells from human exfoliated deciduous teeth, human dental pulp stem cells and human bone marrow mesenchymal stem cells. Biochem Biophys Res Commun. 2018;497(3):876–82.

Nakamura S, Yamada Y, Katagiri W, Sugito T, Ito K, Ueda M. Stem cell proliferation pathways comparison between human exfoliated deciduous teeth and dental pulp stem cells by gene expression profile from promising dental pulp. J Endod. 2009;35:1536–42.

Nakashima M, Iohara K. Recent Progress in Translation from Bench to a Pilot Clinical Study on Total Pulp Regeneration. J Endod. 2017;43(9S):S82–6.

Nakashima M, Iohara K, Murakami M, Nakamura H, Sato Y, Ariji Y, et al. Pulp regeneration by transplantation of dental pulp stem cells in pulpitis: a pilot clinical study. Stem Cell Res Ther. 2017;8(1):61.

Nito C, Sowa K, Nakajima M, Sakamoto Y, Suda S, Nishiyama Y, et al. Transplantation of human dental pulp stem cells ameliorates brain damage following acute cerebral ischemia. Biomed Pharmacother. 2018;108: 1005–14.

Noronha N de C, Mizukami A, Caliári-Oliveira C, Cominal JG, Rocha JLM, Covas DT, et al. Priming approaches to improve the efficacy of mesenchymal stromal cell-based therapies. Stem Cell Res Ther. 2019;10(1):131.

Novais A, Lesieur J, Sadoine J, Slimani L, Baroukh B, Saubaméa B, et al. Priming Dental Pulp Stem Cells from Human Exfoliated Deciduous Teeth with Fibroblast Growth Factor-2 Enhances Mineralization Within Tissue-Engineered Constructs Implanted in Craniofacial Bone Defects. Stem Cells Transl Med. 2019;8(8):844–57.

Pajarinen J, Lin T, Gibon E, Kohno Y, Maruyama M, Nathan K, et al. Mesenchymal stem cell-macrophage crosstalk and bone healing. Biomaterials. 2019;196:80–9.

Paquet J, Deschepper M, Moya A, Logeart-Avramoglou D, Boisson-Vidal C, Petite H. Oxygen Tension Regulates Human Mesenchymal Stem Cell Paracrine Functions: Oxygen Tension and Mesenchymal Stem Cell Secretome. STEM CELLS Transl Med. 2015;4(7):809–21.

Park H, Collignon A-M, Lepry WC, Ramirez-GarciaLuna JL, Rosenzweig DH, Chaussain C, et al. Acellular dense collagen-S53P4 bioactive glass hybrid gel scaffolds form more bone than stem cell delivered constructs. Mater Sci Eng C Mater Biol Appl. 2021;120:111743.

Pisal RV, Suchanek J, Siller R, Soukup T, Hrebikova H, Bezrouk A, et al. Directed reprogramming of comprehensively characterized dental pulp stem cells extracted from natal tooth. Sci Rep. 2018;8(1):6168.

Rademakers T, Horvath JM, Blitterswijk CA, LaPointe VLS. Oxygen and nutrient delivery in tissue engineering: Approaches to graft vascularization. J Tissue Eng Regen Med. 2019;13(10):1815–29.

Sacchetti B, Funari A, Remoli C, Giannicola G, Kogler G, Liedtke S, et al. No Identical "Mesenchymal Stem Cells" at Different Times and Sites: Human Committed Progenitors of Distinct Origin and Differentiation Potential Are Incorporated as Adventitial Cells in Microvessels. Stem Cell Rep. 2016;6(6):897–913.

Sakai VT, Zhang Z, Dong Z, Neiva KG, Machado MA, Shi S, et al. SHED differentiate into functional odontoblasts and endothelium. J Dent Res. 2010;89:791–6.

Sakdee JB, White RR, Pagonis TC, Hauschka PV. Hypoxia-amplified proliferation of human dental pulp cells. J Endod. 2009;35:818–23.

Salazar-Noratto GE, Luo G, Denoeud C, Padrona M, Moya A, Bensidhoum M, et al. Understanding and leveraging cell metabolism to enhance mesenchymal stem cell transplantation survival in tissue engineering and regenerative medicine applications. STEM CELLS. 2020;38(1): 22–33.

Sanen K, Martens W, Georgiou M, Ameloot M, Lambrichts I, Phillips J. Engineered neural tissue with Schwann cell differentiated human dental pulp stem cells: potential for peripheral nerve repair? J Tissue Eng Regen Med. 2017;11(12):3362–72.

Schlundt C, El Khassawna T, Serra A, Dienelt A, Wendler S, Schell H, et al. Macrophages in bone fracture healing: Their essential role in endochondral ossification. Bone. 2018;106:78–89.

Seo BM, Sonoyama W, Yamaza T, Coppe C, Kikuiri T, Akiyama K, et al. SHED repair critical-size calvarial defects in mice. Oral Dis. 2008;14:428–34.

Shen EM, McCloskey KE. Development of Mural Cells: From In Vivo Understanding to In Vitro Recapitulation. Stem Cells Dev. 2017;26(14):1020–41.

Shi S, Gronthos S. Perivascular niche of postnatal mesenchymal stem cells in human bone marrow and dental pulp. J Bone Min Res. 2003;18:696–704.

Sloan AJ, Waddington RJ. Dental pulp stem cells: what, where, how? Int J Paediatr Dent. 2009;19(1):61–70.

Song M, Lee J-H, Bae J, Bu Y, Kim E-C. Human Dental Pulp Stem Cells are more Effective than Human Bone Marrow-Derived Mesenchymal Stem Cells in Cerebral Ischemic Injury. Cell Transplant. 2017;26(6):1001–16.

Song W, Chiu A, Wang L-H, Schwartz RE, Li B, Bouklas N, et al. Engineering transferrable microvascular meshes for subcutaneous islet transplantation. Nat Commun. 2019;10(1):4602.

Souron JB, Petiet A, Decup F, Tran XV, Lesieur J, Poliard A, et al. Pulp Cell Tracking by Radionuclide Imaging for Dental Tissue Engineering. Tissue Eng Part C Methods. 2014;20(3):188–97

de Souza LEB, Malta TM, Kashima Haddad S, Covas DT. Mesenchymal Stem Cells and Pericytes: To What Extent Are They Related? Stem Cells Dev. 2016;25(24):1843–52.

Spina A, Montella R, Liccardo D, De Rosa A, Laino L, Mitsiadis TA, La Noce M. NZ-GMP Approved Serum Improve hDPSC Osteogenic Commitment and Increase Angiogenic Factor Expression. Front Physiol. 2016;7:354

Stanko P, Altanerova U, Jakubechova J, Repiska V, Altaner C. Dental Mesenchymal Stem/Stromal Cells and Their Exosomes. Stem Cells Int. 2018;2018:1–8.

Stevens A, Zuliani T, Olejnik C, LeRoy H, Obriot H, Kerr-Conte J, et al. Human dental pulp stem cells differentiate into neural crest-derived melanocytes and have label-retaining and sphere-forming abilities. Stem Cells Dev. 2008;17:1175–84.

Sugimura-Wakayama Y, Katagiri W, Osugi M, Kawai T, Ogata K, Sakaguchi K, et al. Peripheral Nerve Regeneration by Secretomes of Stem Cells from Human Exfoliated Deciduous Teeth. Stem Cells Dev. 2015;24(22): 2687–99.

Sugiyama M, Iohara K, Wakita H, Hattori H, Ueda M, Matsushita K, et al. Dental pulp-derived CD31(-)/CD146(-) side population stem/progenitor cells enhance recovery of focal cerebral ischemia in rats. Tissue Eng Part A. 2011;17:1303–11.

Sui B, Wu D, Xiang L, Fu Y, Kou X, Shi S. Dental Pulp Stem Cells: From Discovery to Clinical Application. J Endod. sept 2020;46(9S):S46–55.

Taraslia V, Lymperi S, Pantazopoulou V, Anagnostopoulos AK, Papassideri IS, Basdra EK, et al. A High-Resolution Proteomic Landscaping of Primary Human Dental Stem Cells: Identification of SHED- and PDLSC-Specific Biomarkers. Int J Mol Sci. 2018;19(1).

Ueda T, Inden M, Ito T, Kurita H, Hozumi I. Characteristics and Therapeutic Potential of Dental Pulp Stem Cells on Neurodegenerative Diseases. Front Neurosci. 2020;14:407.

Vannella KM, Wynn TA. Mechanisms of Organ Injury and Repair by Macrophages. Annu Rev Physiol. 2017;79(1):593–617.

Viswanathan S, Shi Y, Galipeau J, Krampera M, Leblanc K, Martin I, et al. Mesenchymal stem versus stromal cells: International Society for Cell & Gene Therapy (ISCT§) Mesenchymal Stromal Cell committee position statement on nomenclature. Cytotherapy. 2019;21(10):1019–24.

Wang X, Sha XJ, Li GH, Yang FS, Ji K, Wen LY, et al. Comparative characterization of stem cells from human exfoliated deciduous teeth and dental pulp stem cells. Arch Oral Biol. 2012;57:1231–40.

Wang Y, Wan C, Deng L, Liu X, Cao X, Gilbert SR, et al. The hypoxia-inducible factor α pathway couples angiogenesis to osteogenesis during skeletal development. J Clin Invest. 2007;117(6):1616–26.

Weiss ARR, Dahlke MH. Immunomodulation by Mesenchymal Stem Cells (MSCs): Mechanisms of Action of Living, Apoptotic, and Dead MSCs. Front Immunol. 2019;10:1191.

Xuan K, Li B, Guo H, Sun W, Kou X, He Xs, et al. Deciduous autologous tooth stem cells regenerate dental pulp after implantation into injured teeth. Sci Transl Med. 2018;10(455).

Yan X, Qin H, Qu C, Tuan RS, Shi S, Huang GT. iPS cells reprogrammed from human mesenchymal-like stem/progenitor cells of dental tissue origin. Stem Cells Dev. 2010;19:469–80.

Yelick PC, Sharpe PT. Tooth Bioengineering and Regenerative Dentistry. J Dent Res. 2019;98(11):1173–82.

Yianni V, Sharpe PT. Perivascular-Derived Mesenchymal Stem Cells. J Dent Res. 2019;98(10):1066–72.

Zhang X, Zhou Y, Li H, Wang R, Yang D, Li B, et al. Transplanted Dental Pulp Stem Cells Migrate to Injured Area and Express Neural Markers in a Rat Model of Cerebral Ischemia. Cell Physiol Biochem. 2018;45(1): 258–66.

Zhao H, Chai Y. Stem Cells in Teeth and Craniofacial Bones. J Dent Res. nov 2015;94(11):1495–501.

Zhao H, Feng J, Seidel K, Shi S, Klein O, Sharpe P, et al. Secretion of Shh by a Neurovascular Bundle Niche Supports Mesenchymal Stem Cell Homeostasis in the Adult Mouse Incisor. Cell Stem Cell. 2014;14(2): 160–73.

Zhao Y, Wang L, Jin Y, Shi S. Fas Ligand Regulates the Immunomodulatory Properties of Dental Pulp Stem Cells. J Dent Res. 2012;91(10): 948–54.

Zhou L, Liu W, Wu Y, Sun W, Dörfer CE, Fawzy El-Sayed KM. Oral Mesenchymal Stem/Progenitor Cells: The Immunomodulatory Masters. Stem Cells Int. 2020;2020:1–16.

21

Stem Cells and Retina: From Regeneration to Cell Therapy

Olivier Goureau, Giuliana Gagliardi and Gael Orieux

Institut de la Vision, Sorbonne Université, INSERM U968, CNRS
UMR_7210, Paris, France
Contact: Olivier Goureau
Institut de la Vision, 17, Rue Moreau 75012 Paris, France
Email: olivier.goureau@inserm.fr

21.1 Introduction

Vision begins at the retina, a light-sensitive tissue that lines the back
of the eye. The retina is organized in vertebrates as a cell-type-specific
laminated structure, comprised of six classes of specialized neurons (retinal ganglion cells, bipolar cells, amacrine cells, horizontal cells, cone
and rod photoreceptors) and one glial type (Müller cells). The neural retina consists of three major laminar structures, the ganglion cell
layer (GCL) (the inner-most layer), the inner nuclear layer (INL), and
the outer nuclear layer (ONL). The INL is separated from the GCL
and ONL by two plexiform layers, the inner plexiform layer (IPL) and
outer plexiform layer (OPL), which are packed with synaptic connections.
The retinal pigmented epithelium (RPE), which is positioned between the
neural retina and the choroid (i.e., the vascular meshwork at the outer
part of the neural retina), is an indispensable component of the eye,
as it ensures proper functioning of the photoreceptors within the neural
retina.

Like other parts of the central nervous system (CNS), the retina can
undergo degeneration due to genetic mutations or following damage. In
the context of the CNS, the retina stands out for its ability in some

non-mammalian vertebrates to fully or partially regenerate during development and sometimes in the adult. Various sources of cells have been reported to support regeneration and some of them have been described in mammals. This feature contributes to the idea that (i) the mammalian retina is not fully devoid of endogenous potential for repair and (ii) more extensively the retina could represent a model to develop and test new strategies for CNS regeneration. Although these approaches are still far from becoming clinically applicable, huge progress has been made concerning the ability to produce retinal cells from embryonic stem (ES) cells or induced pluripotent stem (iPS) cells. ES and iPS cells have the capacity for self-renewal, proliferation, and differentiation into all specialized cell types of the adult organism. Different protocols have been established allowing the generation of various retinal cell types, such as photoreceptors or retinal pigmented epithelial cells from ES or iPS cells. Integration and functional recovery have been reported after cell transplantation in different animal models of visual defects, thus holding great promise for restoring vision by means of cell therapy.

21.2 Eye Morphogenesis and Retinogenesis

The vertebrate eye is derived during embryogenesis from three types of tissue: the retina and RPE arise from the neural ectoderm, the cornea and the lens from the surface ectoderm, and the sclera from the mesoderm. Eye development initiates during gastrulation with specification of the eye field, a unique domain of retinal precursors located in the anterior neural plate, characterized on the molecular level by the expression of "eye-field transcription factors" (EFTFs). During midline formation, the single eye field separates giving rise to two optic areas. Following eye field formation in each of these two areas, the neuroepithelium of the ventral forebrain (diencephalon) evaginates, resulting in the formation of the optic vesicle. The specification of neural retina and RPE domains takes place during the evagination process. Surface ectoderm is induced to form the lens placode after contacting the distal portion of the optic vesicle. The lens placode invaginates into the optic vesicle resulting in the formation of the lens vesicle and the bilayered optic cup. The neural retina develops from the inner surface of the optic cup, while the RPE emerges from the outer surface of the optic cup. Several intrinsic and extracellular factors control the different steps of eye organogenesis from the delimitation of the eye field territory to the patterning of the optic vesicles and optic cups (Fuhrmann, 2010; Miesfeld and Brown, 2019).

Table 21.1 Major transcription factors involved in the specification/differentiation of each retinal cell type and subtype

Cell Type	bHLH-type factors	Homeobox-type factors	Others factors (Forkead, zinc finger, Nuclear receptors)
Ganglion cells	Atoh7, Neurod1	Pou4f1/2/3, Isl1, Pax6	–
Amacrine cells	Ptf1a, Neurod1/2/4/6, Bhlhb5, Barhl2	Six3,Pax6, Isl1	Foxn4, Nr4a2
Horizontal cells	Ptf1a, Neurod4	Prox1, Six3, Pax6,	Foxn4, Sall3
Cone photoreceptors	Ascl1, Neurod1	Otx2, Crx,	Sall3, RXRγ, RORβ, Nr2f1/2, Prdm1
Rod photoreceptors	Ascl1, Neurod1	Otx2, Crx, Nrl	Nr2e3, RORγ, Prdm1
Bipolar cells	Ascl1, Bhlbh4/5, Neurod4	Vsx2, Irx5	–
Müller glial cells	Hes1/5, Hesr2	Rax	–

The neural retina is generated from common multipotent retinal progenitor cells (RPCs), which give rise to all types of cells in an orderly manner that is generally conserved among many species. Retinal ganglion cells (RGCs), amacrine and horizontal cells and cone photoreceptors are generated at relatively early stages, while rod photoreceptors, bipolar, and Müller cells differentiate mainly at later stages (Bassett and Wallace, 2012; Brzezinski and Reh, 2015; Stenkamp, 2015). Several lines of evidence demonstrate that this process is regulated by a combination of extrinsic and intrinsic influences (Bassett and Wallace, 2012; Boije et al., 2014; Ohsawa and Kageyama, 2008). Some secreted factors are involved in guiding RPCs toward different cell fates and expression of a combination of different intrinsic transcription factors, such as bHLH-type and homeobox-type factors, are required for correct specification of RPCs (Table 21.1). Recently, the association between RNA-Seq and single-cell technologies (scRNA-Seq) allowed to get a better understanding of cell fate determination at a cellular resolution in mice (Clark et al., 2019) and also in human retina (Lu et al., 2020; Sridhar et al., 2020). For example, gene expression changes at different stages of the human fetal retinogenesis allowed to characterize the developmental trajectories of RPCs: first cluster of RPCs expressing high level of ATOH7 leading to generation of RGCs and amacrine/horizontal cells and two secondary clusters of RPCs originating from the first cluster giving rise either to amacrine cells or photoreceptors and bipolar cells (Sridhar et al., 2020).

21.3 Adult Retinal Stem Cells and Neurogenic Potential

Adult neurogenesis is widely accepted in vertebrates, although the process was initially ignored in the context of mammals. Regarding the retina, significant progress has been made in our understanding of the neural retina's regenerative potential. The incredible capability of newts to regenerate lost tissue has been known for at least three or four centuries and studies of eye regeneration date back several decades (Morgan, 1901; Stone, 1950). In the vertebrate eye, several cell types have been reported to display stem cell properties (Karl and Reh, 2010; Langhe and Pearson, 2020; Locker et al., 2009). Some of them have been described in virtually all vertebrates, such as the Müller glial cells, while others have been described in only some taxonomic groups such as RPE cells (in amphibians and, to some extent in birds and mammals). Besides these post-mitotic and mature cell types, a specific neurogenic niche is localized in the circumferential germinal zone (CGZ), also called the ciliary marginal zone (CMZ),[1] corresponding to the most peripheral part of the retina in non-mammalian vertebrates. This region does not exist formally in mammals but the presence of retinal stem cells in the ciliary body epithelium has been reported in both rodents and humans (Coles et al., 2004; Tropepe, 2000).

21.4 The CMZ: a Retinal Stem Cell Niche

21.4.1 The CMZ in Cold-blooded Vertebrates

In both fishes and amphibians, the CMZ is composed of different cell types, spatially organized within three or four major regions (Fernández-Nogales et al., 2019). The most undifferentiated and self-renewing retinal stem cells are located in the most peripheral tip of the CMZ and characterized by the expression of some EFTFs, such as Rx, Vsx2, Six6, Six3, or Pax6 (Fernández-Nogales et al., 2019). These retinal stem cells generate progenitors, which migrate more centrally, and exhibit faster but limited capacity of

---

[1]To simplify the annotation, the term CMZ will be used indifferently of CMZ or CGZ

proliferation. Finally, these cells develop centrally in a post-mitotic zone of immature retinal precursors (Fernández-Nogales et al., 2019).

Since fishes and amphibians display continuous growth throughout life, one essential feature of the retina is the continuous process of neurogenesis throughout adulthood. This persistent retinogenesis from the CMZ allows for retinal growth by the addition of concentric rings of newly generated neurons at the periphery. The maintenance of the retinal stem cell niche in the CMZ is controlled by a balance of two opposite gradients of Wnt, contributing to retinal stem cell maintenance in the most peripheral part of the CMZ, and sonic hedgehog (shh) signaling that promotes cell-cycle exit from the RPCs (Borday et al., 2012). Following damage, proliferation and neurogenesis are stimulated in the CMZ, giving rise to all types of retinal cells with the exception of the rod photoreceptors which appear to be derived from a specific "stem cell" type distributed throughout the entire retina (Ail and Perron, 2017; Fischer et al., 2013).

21.4.2 The CMZ in Birds and Mammals

By comparison with the previously cited organisms, in the avian and mammalian eye, retinogenesis is complete around hatching or birth. Although progenitors in the CMZ have been identified in birds and capable to proliferate modestly after retinal injury, CMZ-derived cells in the mature injured retina fail to efficiently contribute to tissue repair (Fernández-Nogales et al., 2019; Fischer et al., 2013). While the CMZ per se does not formally exist in the adult mammalian eye, the pigmented epithelium of the ciliary body (CPE) in both rodents and humans (Coles et al., 2004; Tropepe, 2000) has been shown to contain a very sparse population of cells expressing some non-specific RPC markers, like Pax6. The potential of these CPE-derived cells to express various markers of virtually all retinal cell types has been reported (Coles et al., 2004; Tropepe, 2000), but their ability to contribute significantly to adult retinal repair after injury has not been reported. The contribution of CMZ to retinogenesis in mammals has been described only during development (Bélanger et al., 2017; Marcucci et al., 2016). Conversely, expression of LRP2 in the CMZ has been shown to force quiescence state in adult CMZ (Christ et al., 2015). Considering the relatively poor capacity of CPE-derived cells to proliferate in vitro (Tropepe, 2000; Wohl et al., 2012) and the low efficiency of transplantation in the adult retina (Klassen et al., 2007), these cells do not represent currently the best candidates for clinical approaches.

21.5 The Neurogenic Potential of the RPE

21.5.1 RPE Cell Transdifferentiation in Amphibian and Chick Embryos

In some amphibians and chick embryos, but not in fish, the RPE has the capacity to transdifferentiate into neural retinal cells. This mechanism, defined as the stable switch of one mature cell type to another, is so effective that a functional retina can be generated after complete removal of the neural retina, especially in urodela (salamanders, newts) (Fuhrmann, 2014; Wohl et al., 2012). In anurians (frogs, xenopus), the RPE's ability for transdifferentiation is much less efficient and is retained as long as the vascular membrane remains (Yoshii et al., 2007). After retinectomy, a subset of RPE cells detaches, loses its pigments and proliferates to form a transitory germinative neuroepithelium that subsequently differentiates into all neuroretinal cell types. Extracellular matrix components like Matrix Metalloproteinase (MMP), inflammatory cytokines (IL-1β and TNF-α) and growth factors like FGF2 or IGF-1 represent essential cues involved in the process of converting RPE cells into neural progenitors, while other factors, like inhibitors of FGF pathway, have been shown to prevent RPE transdifferentiation (Mitsuda et al., 2005; Naitoh et al., 2017). In addition, expression of many intrinsic factors is modulated. For instance, Sox2, normally downregulated in mature RPE cells is upregulated during the transdifferentiation process, as in the case of Pax6 and Rx. Mitf, a key factor for RPE identity, and Otx2, are concomitantly downregulated (Fuhrmann, 2014; Ng et al., 2009). An unresolved issue lies in determining the timeframe for RPE transdifferentiation in certain species. Progressive irreversibility of RPE commitment or emergence of inhibitory cues, especially from the neural are some of the proposed hypothesis (Fuhrmann, 2014).

21.5.2 RPE Transdifferentiation in Mammals

The RPE's ability to transdifferentiate into neural retinal cells in mammals is reported essentially in vitro (Engelhardt et al., 2005). RPE cells retain the ability to enter the cell cycle and adopt neural cell morphology both in vitro (Engelhardt et al., 2005) and in vivo (Al-Hussaini et al., 2008). A subpopulation of adult human RPE can be activated into a source of self-renewing and multipotent cells expressing stem cell markers, like KLF4 and C-MYC (Salero et al., 2012). These RPE-derived stem cells can differentiate into various epithelial or mesenchymal lineages, like the neural or osteogenic lineages (Salero et al., 2012) and have been used as source of cells for RPE cell replacement in animal models (Davis et al., 2017; Stanzel et al., 2014).

21.6 The Müller Glial Cells: A Retinal Stem Cells Reservoir in Adulthood?

The retinal-specific Müller glial (RMG) cells are the principal glial cell type in the retina. The molecular features of these cells lead some authors to consider them as a gradual transition between neural progenitors and mature glial cells (Jadhav et al., 2009).

21.6.1 Fish

Once the developmental phase is complete, the fish retina grows by addition of concentric rings of new retinal cells from the CMZ, and from specific endogenous stem cells located in the INL (Fernández-Nogales et al., 2019; Lenkowski and Raymond, 2014). A former lineage-tracing study demonstrates that RMG cells were able to differentiate and give rise to photoreceptors (Bernardos et al., 2007), thus suggesting that these RMG cell-derived progenitors could be the so-called adult "endogenous stem cell" located in the INL. In addition to these properties of the RMG cells in the non-injured retina, many studies demonstrated the ability of RMG cell-derived progenitors to efficiently regenerate the retina, notably in zebrafish (Goldman, 2014; Lahne et al., 2020; Langhe and Pearson, 2020). However, regenerative potential can differ between species. The group of J. Wittbrodt reported that in medaka, RMG cells were able to proliferate after injury but failed to self-renew and ultimately only restored photoreceptors (Lust and Wittbrodt, 2018). Interestingly, the authors showed that, unlike zebrafish's RMG cells, proliferating RMG cells in medaka failed to maintain Sox2 expression after injury. Numerous signals released from the damaged retina, including Shh, Transforming Growth Factor-β (TGFβ), Platelet-Derived Growth Factor or inflammation and some matrix-metalloproteinases have been shown to stimulate the regeneration process, especially the generation of new photoreceptors (Lahne et al., 2020; Langhe and Pearson, 2020). The EGF-signaling and Dkk/β-catenin pathways, have even been demonstrated to be sufficient and necessary for this process (Ramachandran et al., 2011; Wan et al., 2012) and in most studies the criterion for successful initiation of regeneration is the re-entry of RMG cells in the cell cycle (Fimbel et al., 2007; Ramachandran et al., 2010). Interestingly, a recent study suggested that only some RMG cells have the capacity to contribute to retinal regeneration depending locally on specific FGF8 and Notch opposite signaling (Wan and Goldman, 2017). Injury-induced expression of Midkine-a by RMGs was shown recently to be required for progression of RMG cells through cell cycle and subsequent retinal regeneration (Nagashima et al., 2020). In contrast, TGFβ3 and miR-216a which are expressed in quiescent RMG cells, are downregulated in response to injury permitting, with Notch suppression, RMG cell dedifferentiation, and proliferation (Elsaeidi et al., 2018; Kara et al., 2019; Lee et al., 2020). Moreover, overexpression of TGFβ3 in RMG cells inhibited injury-induced RMG cell proliferation (Lee et al., 2020). In parallel, key stem cell factors, such as c-Myc, Oct4, and Lin28 or neural progenitor markers like Ascl1a, are upregulated during RMG cell dedifferentiation (Ramachandran et al., 2010). Non-specific markers of RPCs, such as Pax6, are also upregulated (Thummel et al., 2010). In addition, numerous microRNA, such as miR-124, miR-9-9, and mir-216a have been implicated in RMG cell-dependent retinal regeneration (Kara et al., 2019; Wohl and Reh, 2016). Downregulation of Let-7 microRNA, resulting from Ascl1a and Lin28 induction is associated with expression of many regeneration-associated genes (Ramachandran et al., 2010). Finally the multipotency of RMG cell-derived progenitors leads to the generation of all retinal cell types, allowing the successful repair of the injured retina (Lahne et al., 2020; Langhe and Pearson, 2020).

21.6.2 Amphibians and Birds

A series of works, particularly from the Fisher's group, have demonstrated that RMG cells are capable of dedifferentiating and proliferating in the chicken retina in response to acute retinal injury (Wilken and Reh, 2016). After injury, RMG cells enter the cell-cycle and upregulate the expression of some "EFTFs" like Pax6, Vsx2, or Six3. RMG cell response involves mTOR, Jak/Stat, MAPK, and Erk signaling and a variety of extracellular cues of diffusible factors such as the Notch pathway, Hh, FGF2, ciliary neurotrophic factor, and insulin-like growth factors (Lahne et al., 2020; Langhe and Pearson, 2020; Wilken and Reh, 2016). However, the majority of RMG cell-derived progenitors remains undifferentiated and the others differentiate preferentially into RMG cells, thus failing to generate new neurons (Fischer and Bongini, 2010).

21.6.3 Mammals

In the mammalian central nervous system (CNS), tissue damage is usually followed by a reactive gliosis. Similar to astrocytes in the rest of the CNS, RMG cells usually become activated after retinal injury and build a glial scar. However, different studies reported that RMG cells could generate retinal neurons after the proliferative phase, notably with the addition of growth factors, such as EGF, retinoic acid, or Shh (Lahne et al., 2020; Langhe and Pearson, 2020). However, the number of new generated neurons was remarkably low. One explanation could be the absence of Ascl1 expression in RMG cells after injury, a key factor of retinal regeneration in fish (Karl and Reh, 2010). Interestingly, as in fish, Let-7 microRNA inhibition using antagomiRs is able, in vitro, to induce Ascl1 expression in RMG cells leading to conversion into RPCs (Wohl et al., 2019). In combination with a histone deacetylase inhibitor, overexpression of Ascl1 in RMG cells, enabled the generation of new neurons from RMG cells after retinal injury in adult mice (Jorstad et al., 2017). This observation has been attributed to poor chromatin accessibility and limitation of Ascl1 chromatin binding by STAT (Jorstad et al., 2020). The Hippo pathway may be one of the factors explaining the low regenerative capacity of RMGs in mammals. Indeed, re-entry of RMG in the cell cycle in response to injury is repressed by the Hippo pathway and blocking of the Hippo signaling-induced loss of RMG cell identity and their proliferation (Rueda et al., 2019). The activation of the Wnt pathway using β-catenin gene transfer has been shown to efficiently stimulate RMG cell proliferation in vivo and transition into neurogenic fate (Yao et al., 2016). Recently, the same group demonstrated that combining Wnt pathway stimulation with Otx2, Crx, and Nrl gene overexpression efficiently lead to RMG-dependent retinal regeneration and visual restoration, via a Lin28 and Let-7 dependent pathways (Yao et al., 2018). All these data highlight the poor ability of mammals RMG to spontaneously contribute to retinal regeneration after injury. Even if massive genetic manipulation demonstrated the latent potential or RMG to do so, like in cold-blood vertebrates, clinical applications appear to be challenging to date. For this reason, a cell replacement strategy using extra-ocular cells as a source of retinal cells is likely to have the greatest therapeutic potential.

21.7 Cell Replacement for Retinal Repair

The impaired or complete loss of function of photoreceptor cells or supporting RPE cells is the main cause of irreversible blindness in industrialized countries caused by retinal diseases, such as inherited retinopathies and age-related macular degeneration (AMD). Since currently available pharmacological treatments are limited to delaying the onset or slowing down the progress of visual impairment, innovative therapeutic strategies to rescue the degenerated retina are under investigation such as gene therapy strategies (gene supplementation vs. endogenous gene suppression). For late stage diseases, when photoreceptor degeneration is advanced, strategies aiming at restoring light-sensitivity of the retina are currently investigated using prosthetic implants, optogenetic tools, and photosensitive switches (Roska and Sahel, 2018). Cell replacement after transplantation represents another complementary and more generic approach to replace lost cells (RPE cells and/or photoreceptors). Unlike other approaches, cell therapy attempts at recreating the pre-existing functional retinal system by replacing the degenerated cells with new cells. Over the past two decades, a wide range of cell types has been evaluated for its ability to restore vision after transplantation such as bone marrow-derived mesenchymal stem cells, neural stem cells, retinal stem cells, and RPCs, but their ability to generate new retinal neurons appears limited in vivo (Ramsden et al., 2013).

Recently pluripotent stem cells, which have the ability to be expanded indefinitely in culture while retaining their pluripotent status and can differentiate into cells from all the three germ layers (endoderm, mesoderm, and ectoderm), appeared as an attractive source of cells in regenerative medicine. Human embryonic stem (ES) cells have been the best studied since their first isolation from the inner cell mass of a blastocyst (Thomson, 1998). In 2006, the group of S. Yamanaka generated another type of pluripotent stem cells by reprogramming fibroblasts with four specific transcription factors (Takahashi and Yamanaka, 2006). Starting with a pool of 24 candidate genes which had already been reported to have a role in pluripotency, phenotype, or morphology of ES cells, they demonstrated that only four factors, i.e., Oct3/4, Sox2, c-Myc, and Klf4, were necessary and sufficient to obtain pluripotent stem cells from mouse fibroblasts. Following the same reprogramming strategy, Yamanaka's team generated human pluripotent stem (iPS) cells derived from human dermal fibroblasts (Takahashi et al., 2007). For this stunning discovery Shinya Yamanaka was awarded the 2012 Nobel Prize (Physiology/Medicine). Although ES cells are still the "gold standard" for pluripotency, the iPS cell technology has been a revolution in the field since it bypasses some ethical issues. Today a lot of protocols are available for differentiation of various cell types, including retinal cells, particularly photoreceptors and RPE cells (Aghaizu et al., 2017; Gagliardi et al., 2019; Jin et al., 2019).

21.7.1 RPE Cells Derived From Human ES and iPS Cells

Human ES and iPS cells have repeatedly been shown to be capable of spontaneous differentiation into RPE cells through removal of FGF2 (growth factor responsible for the maintenance of pluripotency) from the medium or by guided differentiation of embryonic bodies (Leach and Clegg, 2015). Different groups reported encouraging morphological and functional results in animal models of retinal degeneration after transplantation of RPE cells derived from both human ES and iPS cells (Ben M'Barek and Monville, 2019; Jones et al., 2017). RPE cell transplantation with healthy RPE cells derived from ES cells or iPS cells is not limited to animal models, since clinical trials for macular degeneration have already started in order to replace the lost or dysfunctional RPE in AMD patients (Zarbin et al., 2019). First phase I/II trials were launched to evaluate the safety and tolerability of a subretinal injection of human ES cell-derived RPE as cell suspension in patients with advanced dry AMD and Stargardt's disease (Mehat et al., 2018; Schwartz et al., 2015). Safety and tolerability prospective clinical trials are also currently underway to evaluate subretinal injection of human ES cell- or iPS cell-derived RPE cell sheets (da Cruz et al., 2018; Kashani et al., 2018; Mandai et al., 2017).

21.7.2 Photoreceptors Derived From Human ES and iPS Cells

Based on the identification of the specific molecular signals required for eye field specification and retinal differentiation, Lamba et al. provided the first protocol for the efficient generation of retinal cells from human ES cells. Successive culture steps of a serum-free floating culture of embryoid bodies (called SFEB system) followed by adherent cultures (plating on different specific coating) allowed the generation of RPCs and photoreceptors thanks to addition of a BMP inhibitor (NOGGIN or Lefty-A), a WNT inhibitor (DKK1), and the insulin-like growth factor (IGF-1) in a specific window of time. Another combination of signals, using small molecules (SB431542 and CKI-7) to replace morphogens (Lefty-A and DKK1), has also been demonstrated to trigger differentiation of human ES and iPS cells into retinal cells and toward the photoreceptor lineage (Osakada et al., 2008). In 2009, a method developed by Meyer et al., 2011, based on a passive strategy for the neural induction, avoids the use of recombinant proteins and takes advantage of the endogenous secretion of NOGGIN and DKK1 by culturing human ES cells in neural differentiation media.

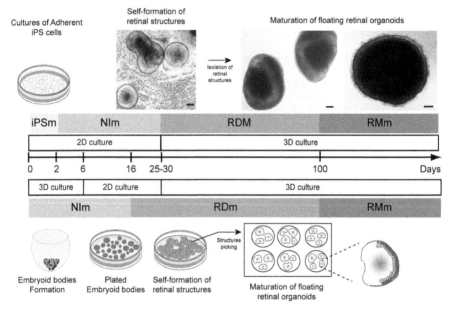

Figure 21.1   Representation of key steps for the generation of retinal organoids from human ES or iPS cells. Top panel illustrates the different steps to generate retinal organoids from adherent cell cultures (detailed in Reichman et al., 2017). Bottom panel shows the different 2D and 3D steps using ES or iPS cell-derived embryoid bodies (detailed in Zhong et al., 2014). iPSm: iPS media; NIm: Neural induction media; RDm: retinal differentiation media; RMm: retinal maturation media. Scale bars: 100 μm

However, even if adherent culture systems are more convenient they do not recapitulate entirely the aspect present in vivo, such as cell architecture and cell-cell interactions. During the last 10 years, innovative studies that mark the exciting advances in the field involve the generation of three-dimensional retinal structures, which partially or completely recapitulate retinal development in vitro. As previously demonstrated using mouse ES cells (Eiraku et al., 2011), the self-formation of optic cups have been obtained with human ES cells cultured as suspended aggregates (similar to EBs) in the presence of matrigel and after addition of specific molecules (an anti-apoptotic agent, a WNT agonist, a SHH agonist and serum) during a specific temporal window (Nakano et al., 2012). Over time, the RPCs present in the newly formed optic cups, divided and differentiated into the major classes of retinal cells (ganglion, amacrine and horizontal cells, and photoreceptors), and the cells appeared in their proper location within the laminated retina. In 2015, the same group developed a "step-wise induction-reversal method," with an early treatment with BMP4 that allowed the co-formation of neural retina with adjacent RPE from human ES cells. Interestingly, the interaction between the two tissues recreates a growth zone containing a stem cell niche (Kuwahara et al., 2015). In parallel, different groups have improved or developed distinct protocols to generate these 3D retinal structures, called retinal organoids (Gagliardi et al., 2019; Llonch et al., 2018) (Figure 21.1). For example, one strategy was to add an adherent intermediate culture stage (3D/2D/3D culture) (Meyer et al., 2011). Indeed, sequential culture of the ES or iPS cells in specific media, similar to the EBs in suspension, followed by plating into a laminin substrate and by novel floating cultures of the emerging neural clusters, led to the generation of optic vesicle-like structures, containing a homogenous population of RPCs. The progenitors in these structures were able to differentiate and to form laminated retinal organoids containing all retinal cell types. Recent advances in single-cell RNA-sequencing technologies have made it possible to fully characterize different retinal populations during organoid differentiation and demonstrate the similarity between retinogenesis in vitro (organoids) and in vivo (human fetal tissue) (Mao et al., 2019; Sridhar et al., 2020). Several studies have shown the ability of ganglion cells present in relatively young organoids to emit action potentials reflecting functional maturity (Rabesandratana et al., 2018). Some studies have reported a response to light by photoreceptors in organoids older than 25 weeks (Zhong et al., 2014), or altered calcium levels in photoreceptors after application of cyclic nucleotide cGMP, a characteristic of photo transduction, both of which reflect functional maturation of the photoreceptors (Gagliardi et al., 2018; Reichman et al., 2017).

All these protocols, however, depend on the addition of exogenous factors like serum or matrigel which are not compatible with further therapeutic applications, following current Good Manufacturing Practices (GMP) guidelines. In order to use a more defined cell culture system which is safer for clinical applications, Reichman et al., 2014 established a 2D/3D culture protocol bypassing the formation of EBs and the use of substrates or exogenous molecules. Of note, to be completely xeno-free (without any animal component in the culture medium) and compatible with the GMP-standards, the protocol was then adapted to feeder-free conditions (Reichman et al., 2017).

The presence of photoreceptor precursors that can be used for cell transplantation, within retinal organoids, makes these structures a very interesting source of therapeutical cells. Indeed, previous studies in animal models have shown that transplantation of mature (adult) photoreceptors is ineffective, as the photoreceptors must be at a specific ontogenetic stage (immature precursors) at the time of transplantation (MacLaren et al., 2006). Early studies of transplantation of these photoreceptor precursors isolated from the retina of newborn rodents in different blind rodent models led researchers to conclude that the transplanted cells could survive in the degenerating retina, differentiate into mature photosensitive photoreceptors, establish certain synaptic connections with appropriate partners, leading to partial recovery of visual function (Barber et al., 2013; MacLaren et al., 2006; Pearson et al., 2012; Santos-Ferreira et al., 2015). However, these spectacular results were recently reinterpreted using different strategies that unequivocally distinguish donor from recipient cells. It has been reported that cells identified in early works as transplanted cells were actually host photoreceptors that had integrated some of the contents of the transplanted cells by cytoplasmic material transfer rather than differentiated donor cells that were fully and perfectly integrated into the host tissue (Pearson et al., 2016; Santos-Ferreira et al., 2016; Singh et al., 2016). The elucidation of material exchange processes could lead to new therapeutic options for cell therapy applications, in which donor cells can be considered as a vector to deliver a missing protein or other molecules to the remaining photoreceptors rather than replacing lost photoreceptors.

Recent publications described the transplantation of photoreceptor precursors from human ES cell or iPS cell-derived retinal organoids into different animal models using different strategies (Gagliardi et al., 2019; Gasparini et al., 2019). Some approaches consist in transplanting a retina sheet derived from retinal organoids. Grafting of such a retinal sheet could lead to functional restoration (McLelland et al., 2018; Shirai et al., 2016; Tu et al., 2019); however, the presence of retinal cells other than photoreceptors, as well as the organization of these photoreceptors in the form of rosettes within these sheets hinders the ability of grafted photoreceptors to reconnect in the host tissue and restore visual function. On the contrary, the injection of a cell suspension of photoreceptor precursors into the subretinal space, after appropriate selection (cell sorting by cytometry or magnetic sorting), allows to transplant a homogeneous population of photoreceptors and facilitates contact with the neurons of the host tissue (Gagliardi et al., 2018; Lakowski et al., 2018). The morphological and functional maturation of the grafted immature photoreceptors, which requires the development of light-sensitive outer segments, remains an obstacle to be overcome. To date, although the expression of mature photoreceptor markers allowing photosensitivity is described in the literature (Elsaeidi et al., 2018), the morphological and functional maturation of transplanted cells is often disappointing or even absent. New avenues of research are beginning to emerge. Scaffolds which are already widely developed for RPE replacement strategies (Ben M'Barek et al., 2020; da Cruz et al., 2018; Kashani et al., 2018), are now being tested to promote the maturation and polarization of photoreceptor precursors in order to transplant photoreceptors with light-sensitive outer segments (Jung et al., 2018). An alternative strategy has been to combine cell therapy and optogenetics by artificially conferring photosensitivity to the donor cells through the expression of specific microbial opsins. The insertion of such light sensors has made it possible to make the grafted photoreceptor precursors photosensitive, independently of the formation of outer segments, the expression of photo-transduction actors, and the presence of RPE (Garita-Hernandez et al., 2019).

21.8 Concluding Remarks

This review highlights some of the significant progress in the field of stem cell research on the development of prospective cell therapy treatments for retinal diseases:

- Adult retinal stem cells and regenerative potential are reported only in cold-blood vertebrates.
- One future direction for endogenous cell replacement and retinal regeneration in mammals will be the combination of methods involving in vivo reprogramming (gene transfer) with modulation of environmental cues required for the regenerative process.
- In terms of cell transplantation, human pluripotent stem cells (ES and iPS cells) appear as the best source of retinal cells (RPE or photoreceptors) for cell replacement strategies.
- RPE cell replacement is already moving to a clinical research stage using ES or iPS cell-derived RPE cells or RPE sheets with both autologous and allogenic grafts.
- Current photoreceptor replacement strategies with ES or iPS cell-derivatives, transplanted either as retinal sheet or as a photoreceptor cell suspension, has shown some limitations. The achievement of fully successful photoreceptor replacement will rely on co-implementation of multiple strategies in all the steps leading to a cell therapy application, from cell product manufacturing, to surgical delivering and functional restoration.

While significant hurdles need to be addressed before achieving any therapeutic ends, most of these open challenges do not concern exclusively the development of retinal cell therapies, but they are common to all stem cell-based treatments. Collaborative efforts from not only distinct stem cell domains, but also different scientific disciplines (biologists, engineers, and clinicians) should accelerate the progress of regenerative medicine. While many challenges remain, these developments are likely to bring iPS cell applications to clinical trials for photoreceptor degenerative diseases and they offer a genuine hope for patients with currently untreatable retinal diseases.

References

Aghaizu, N. D., Kruczek, K., Gonzalez-Cordero, A., Ali, R. R. and Pearson, R. A. (2017). Pluripotent stem cells and their utility in treating photoreceptor degenerations. *Prog. Brain Res.*231, 191-223.

Ail, D. and Perron, M. (2017). Retinal Degeneration and Regeneration-Lessons From Fishes and Amphibians. *Curr. Pathobiol. Rep.* 5, 67-78.

Al-Hussaini, H., Kam, J. H., Vugler, A., Semo, M. and Jeffery, G. (2008). Mature retinal pigment epithelium cells are retained in the cell cycle and proliferate in vivo. *Mol. Vis.* 14, 1784-91.

Barber, A. C., Hippert, C., Duran, Y., West, E. L., Bainbridge, J. W. B., Warre-Cornish, K., Luhmann, U. F. O., Lakowski, J., Sowden, J. C., Ali, R. R., et al. (2013). Repair of the degenerate retina by photoreceptor transplantation. *Proc. Natl. Acad. Sci. U. S. A.* 110, 354-359.

Bassett, E. A. and Wallace, V. A. (2012). Cell fate determination in the vertebrate retina. *Trends Neurosci.* 35, 565-73.

Bélanger, M.-C., Robert, B. and Cayouette, M. (2017). Msx1-Positive Progenitors in the Retinal Ciliary Margin Give Rise to Both Neural and Non-neural Progenies in Mammals. *Dev. Cell* 40, 137-150.

Ben M'Barek, K. and Monville, C. (2019). Cell Therapy for Retinal Dystrophies: From Cell Suspension Formulation to Complex Retinal Tissue Bioengineering. *Stem Cells Int.* 2019, 4568979.

Ben M'Barek, K., Bertin, S., Brazhnikova, E., Jaillard, C., Habeler, W., Plancheron, A., Fovet, C.-M., Demilly, J., Jarraya, M., Bejanariu, A., et al. (2020). Clinical-grade production and safe delivery of human ESC derived RPE sheets in primates and rodents. *Biomaterials* 230, 119603.

Bernardos, R. L., Barthel, L. K., Meyers, J. R. and Raymond, P. A. (2007). Late-stage neuronal progenitors in the retina are radial Müller glia that function as retinal stem cells. *J. Neurosci.* 27, 7028-40.

Boije, H., MacDonald, R. B. and Harris, W. A. (2014). Reconciling competence and transcriptional hierarchies with stochasticity in retinal lineages. *Curr. Opin. Neurobiol.* 27, 68-74.

Borday, C., Cabochette, P., Parain, K., Mazurier, N., Janssens, S., Tran, H. T., Sekkali, B., Bronchain, O., Vleminckx, K., Locker, M., et al. (2012). Antagonistic cross-regulation between Wnt and Hedgehog signalling pathways controls post-embryonic retinal proliferation. *Development* 139, 3499-509.

Brzezinski, J. A. and Reh, T. A. (2015). Photoreceptor cell fate specification in vertebrates. *Development* 142, 3263-3273.

Christ, A., Christa, A., Klippert, J., Eule, J. C., Bachmann, S., Wallace, V. A., Hammes, A. and Willnow, T. E. (2015). LRP2 Acts as SHH Clearance Receptor to Protect the Retinal Margin from Mitogenic Stimuli. *Dev. Cell* 35, 36-48.

Clark, B. S., Stein-O'Brien, G. L., Shiau, F., Cannon, G. H., Davis-Marcisak, E., Sherman, T., Santiago, C. P., Hoang, T. V., Rajaii, F., James-Esposito, R. E., et al. (2019). Single-Cell RNA-Seq Analysis of Retinal Development Identifies NFI Factors as Regulating Mitotic Exit and Late-Born Cell Specification. *Neuron* 102, 1111-1126.e5.

Coles, B. L. K., Angénieux, B., Inoue, T., Del Rio-Tsonis, K., Spence, J. R., McInnes, R. R., Arsenijevic, Y. and van Der Kooy, D. (2004). Facile isolation and the characterization of human retinal stem cells. *Proc. Natl. Acad. Sci. U. S. A.* 101, 15772-7.

da Cruz, L., Fynes, K., Georgiadis, O., Kerby, J., Luo, Y. H., Ahmado, A., Vernon, A., Daniels, J. T., Nommiste, B., Hasan, S. M., et al. (2018). Phase 1 clinical study of an embryonic stem cell-derived retinal pigment epithelium patch in age-related macular degeneration. *Nat. Biotechnol.* 36, 328-337.

Davis, R. J., Alam, N. M., Zhao, C., Müller, C., Saini, J. S., Blenkinsop, T. A., Mazzoni, F., Campbell, M., Borden, S. M., Charniga, C. J., et al. (2017). The Developmental Stage of Adult Human Stem Cell-Derived Retinal Pigment Epithelium Cells Influences Transplant Efficacy for Vision Rescue. *Stem Cell Reports* 9, 42-49.

Eiraku, M., Takata, N., Ishibashi, H., Kawada, M., Sakakura, E., Okuda, S., Sekiguchi, K., Adachi, T. and Sasai, Y. (2011). Self-organizing optic-cup morphogenesis in three-dimensional culture. *Nature* 472, 51-56.

Elsaeidi, F., Macpherson, P., Mills, E. A., Jui, J., Flannery, J. G. and Goldman, D. (2018). Notch Suppression Collaborates with Ascl1 and Lin28 to Unleash a Regenerative Response in Fish Retina, But Not in Mice. *J. Neurosci.* 38, 2246-2261.

Engelhardt, M., Bogdahn, U. and Aigner, L. (2005). Adult retinal pigment epithelium cells express neural progenitor properties and the neuronal precursor protein doublecortin. *Brain Res.* 1040, 98-111.

Fernández-Nogales, M., Murcia-Belmonte, V., Chen, H. Y. and Herrera, E. (2019). The peripheral eye: A neurogenic area with potential to treat retinal pathologies? *Prog. Retin. Eye Res.* 68, 110-123.

Fimbel, S. M., Montgomery, J. E., Burket, C. T. and Hyde, D. R. (2007). Regeneration of inner retinal neurons after intravitreal injection of ouabain in zebrafish. *J. Neurosci.* 27, 1712-24.

Fischer, A. J. and Bongini, R. (2010). Turning Müller glia into neural progenitors in the retina. *Mol. Neurobiol.* 42, 199-209.

Fischer, A. J., Bosse, J. L. and El-Hodiri, H. M. (2013). The ciliary marginal zone (CMZ) in development and regeneration of the vertebrate eye. *Exp. Eye Res.* 116, 199-204.

Fuhrmann, S. (2010). Eye morphogenesis and patterning of the optic vesicle. *Curr. Top. Dev. Biol.* 93, 61-84.

Fuhrmann, S., Zou, C. and Levine, E. M. (2014). Retinal pigment epithelium development, plasticity, and tissue homeostasis. *Exp. Eye Res.* 123, 141-50.

Gagliardi, G., Ben M'Barek, K., Chaffiol, A., Slembrouck-Brec, A., Conart, J.-B., Nanteau, C., Rabesandratana, O., Sahel, J.-A., Duebel, J., Orieux, G., et al. (2018). Characterization and Transplantation of CD73-Positive Photoreceptors Isolated from Human iPSC-Derived Retinal Organoids. *Stem Cell Reports* 11, 665-680.

Gagliardi, G., Ben M'Barek, K. and Goureau, O. (2019). Photoreceptor cell replacement in macular degeneration and retinitis pigmentosa: A pluripotent stem cell-based approach. *Prog. Retin. Eye Res.* 71, 1-25.

Garita-Hernandez, M., Lampič, M., Chaffiol, A., Guibbal, L., Routet, F., Santos-Ferreira, T., Gasparini, S., Borsch, O., Gagliardi, G., Reichman, S., et al. (2019). Restoration of visual function by transplantation of optogenetically engineered photoreceptors. *Nat. Commun.* 10, 4524.

Gasparini, S. J., Llonch, S., Borsch, O. and Ader, M. (2019). Transplantation of photoreceptors into the degenerative retina: Current state and future perspectives. *Prog. Retin. Eye Res.* 69, 1-37.

Goldman, D. (2014). Müller glial cell reprogramming and retina regeneration. *Nat. Rev. Neurosci.* 15, 431-42.

Jadhav, A. P., Roesch, K. and Cepko, C. L. (2009). Development and neurogenic potential of Müller glial cells in the vertebrate retina. *Prog. Retin. Eye Res*. 28, 249-62.

Jin, Z.-B., Gao, M.-L., Deng, W.-L., Wu, K.-C., Sugita, S., Mandai, M. and Takahashi, M. (2019). Stemming retinal regeneration with pluripotent stem cells. *Prog. Retin. Eye Res*. 69, 38-56.

Jones, M. K., Lu, B., Girman, S. and Wang, S. (2017). Cell-based therapeutic strategies for replacement and preservation in retinal degenerative diseases. *Prog. Retin. Eye Res*. 58, 1-27.

Jorstad, N. L., Wilken, M. S., Grimes, W. N., Wohl, S. G., Vandenbosch, L. S., Yoshimatsu, T., Wong, R. O., Rieke, F. and Reh, T. A. (2017). Stimulation of functional neuronal regeneration from Müller glia in adult mice. *Nature* 548, 103-107.

Jorstad, N. L., Wilken, M. S., Todd, L., Finkbeiner, C., Nakamura, P., Radulovich, N., Hooper, M. J., Chitsazan, A., Wilkerson, B. A., Rieke, F., et al. (2020). STAT Signaling Modifies Ascl1 Chromatin Binding and Limits Neural Regeneration from Müller Glia in Adult Mouse Retina. *Cell Rep*. 30, 2195-2208.e5.

Jung, Y. H., Phillips, M. J., Lee, J., Xie, R., Ludwig, A. L., Chen, G., Zheng, Q., Kim, T. J., Zhang, H., Barney, P., et al. (2018). 3D Microstructured Scaffolds to Support Photoreceptor Polarization and Maturation. *Adv. Mater*. 30, e1803550.

Kara, N., Kent, M. R., Didiano, D., Rajaram, K., Zhao, A., Summerbell, E. R. and Patton, J. G. (2019). The miR-216a-Dot1l Regulatory Axis Is Necessary and Sufficient for Müller Glia Reprogramming during Retina Regeneration. *Cell Rep*. 28, 2037-2047.e4.

Karl, M. O. and Reh, T. a (2010). Regenerative medicine for retinal diseases: activating endogenous repair mechanisms. *Trends Mol. Med*. 16, 193-202.

Kashani, A. H., Lebkowski, J. S., Rahhal, F. M., Avery, R. L., Salehi-Had, H., Dang, W., Lin, C.-M., Mitra, D., Zhu, D., Thomas, B. B., et al. (2018). A bioengineered retinal pigment epithelial monolayer for advanced, dry age-related macular degeneration. *Sci. Transl. Med*. 10, eaao4097.

Klassen, H., Kiilgaard, J. F., Zahir, T., Ziaeian, B., Kirov, I., Scherfig, E., Warfvinge, K. and Young, M. J. (2007). Progenitor cells from the porcine neural retina express photoreceptor markers after transplantation to the subretinal space of allorecipients. *Stem Cells* 25, 1222-30.

Kuwahara, A., Ozone, C., Nakano, T., Saito, K., Eiraku, M. and Sasai, Y. (2015). Generation of a ciliary margin-like stem cell niche from self-organizing human retinal tissue. *Nat. Commun*. 6, 6286.

Lahne, M., Nagashima, M., Hyde, D. R. and Hitchcock, P. F. (2020). Reprogramming Müller Glia to Regenerate Retinal Neurons. *Annu. Rev. Vis. Sci.* 6:171-193.

Lakowski, J., Welby, E., Budinger, D., Di Marco, F., Di Foggia, V., Bainbridge, J. W. B., Wallace, K., Gamm, D. M., Ali, R. R. and Sowden, J. C. (2018). Isolation of Human Photoreceptor Precursors via a Cell Surface Marker Panel from Stem Cell-Derived Retinal Organoids and Fetal Retinae. *Stem Cells* 36, 709-722.

Lamba, D. A., Karl, M. O., Ware, C. B. and Reh, T. A. (2006). Efficient generation of retinal progenitor cells from human embryonic stem cells. *Proc. Natl. Acad. Sci.* 103, 12769-12774.

Langhe, R. and Pearson, R. A. (2020). Rebuilding the Retina: Prospects for Müller Glial-mediated Self-repair. *Curr. Eye Res.* 45, 349-360.

Leach, L. L. and Clegg, D. O. (2015). Concise Review: Making Stem Cells Retinal: Methods for Deriving Retinal Pigment Epithelium and Implications for Patients With Ocular Disease. *Stem Cells* 33, 2363-2373.

Lee, M.-S., Wan, J. and Goldman, D. (2020). Tgfb3 collaborates with PP2A and notch signaling pathways to inhibit retina regeneration. *Elife* 9:e55137.

Lenkowski, J. R. and Raymond, P. A. (2014). Müller glia: Stem cells for generation and regeneration of retinal neurons in teleost fish. *Prog. Retin. Eye Res.* 40, 94-123.

Llonch, S., Carido, M. and Ader, M. (2018). Organoid technology for retinal repair. *Dev. Biol.* 433, 132-143.

Locker, M., Borday, C. and Perron, M. (2009). Stemness or not stemness? Current status and perspectives of adult retinal stem cells. *Curr. Stem Cell Res. Ther.* 4, 118-30.

Lu, Y., Shiau, F., Yi, W., Lu, S., Wu, Q., Pearson, J. D., Kallman, A., Zhong, S., Hoang, T., Zuo, Z., et al. (2020). Single-Cell Analysis of Human Retina Identifies Evolutionarily Conserved and Species-Specific Mechanisms Controlling Development. *Dev. Cell* 53, 473-491.e9.

Lust, K. and Wittbrodt, J. (2018). Activating the regenerative potential of Müller glia cells in a regeneration-deficient retina. *Elife* 7:e32319.

MacLaren, R. E., Pearson, R. A., MacNeil, A., Douglas, R. H., Salt, T. E., Akimoto, M., Swaroop, A., Sowden, J. C. and Ali, R. R. (2006). Retinal repair by transplantation of photoreceptor precursors. *Nature* 444, 203-207.

Mandai, M., Watanabe, A., Kurimoto, Y., Hirami, Y., Morinaga, C., Daimon, T., Fujihara, M., Akimaru, H., Sakai, N., Shibata, Y., et al.

(2017). Autologous Induced Stem-Cell-Derived Retinal Cells for Macular Degeneration. *N. Engl. J. Med.* 376, 1038-1046.

Mao, X., An, Q., Xi, H., Yang, X.-J., Zhang, X., Yuan, S., Wang, J., Hu, Y., Liu, Q. and Fan, G. (2019). Single-Cell RNA Sequencing of hESC-Derived 3D Retinal Organoids Reveals Novel Genes Regulating RPC Commitment in Early Human Retinogenesis. *Stem Cell Reports* 13, 747-760.

Marcucci, F., Murcia-Belmonte, V., Wang, Q., Coca, Y., Ferreiro-Galve, S., Kuwajima, T., Khalid, S., Ross, M. E., Mason, C. and Herrera, E. (2016). The Ciliary Margin Zone of the Mammalian Retina Generates Retinal Ganglion Cells. *Cell Rep.* 17, 3153-3164.

McLelland, B. T., Lin, B., Mathur, A., Aramant, R. B., Thomas, B. B., Nistor, G., Keirstead, H. S. and Seiler, M. J. (2018). Transplanted hESC-derived retina organoid sheets differentiate, integrate, and improve visual function in retinal degenerate rats. Investig. *Ophthalmol. Vis. Sci.* 59, 2586-2603.

Mehat, M. S., Sundaram, V., Ripamonti, C., Robson, A. G., Smith, A. J., Borooah, S., Robinson, M., Rosenthal, A. N., Innes, W., Weleber, R. G., et al. (2018). Transplantation of Human Embryonic Stem Cell-Derived Retinal Pigment Epithelial Cells in Macular Degeneration. *Ophthalmology* 125, 1765-1775.

Meyer, J. S., Howden, S. E., Wallace, K. A., Verhoeven, A. D., Wright, L. S., Capowski, E. E., Pinilla, I., Martin, J. M., Tian, S., Stewart, R., et al. (2011). Optic vesicle-like structures derived from human pluripotent stem cells facilitate a customized approach to retinal disease treatment. *Stem Cells* 29, 1206-18.

Miesfeld, J. B. and Brown, N. L. (2019). Eye organogenesis: A hierarchical view of ocular development. *Curr. Top. Dev. Biol.* 132, 351-393.

Mitsuda, S., Yoshii, C., Ikegami, Y. and Araki, M. (2005). Tissue interaction between the retinal pigment epithelium and the choroid triggers retinal regeneration of the newt Cynops pyrrhogaster. *Dev. Biol.* 280, 122-32.

Morgan, T. H. (1901). Regeneration and liability to injury. *Science.* 14, 235-248.

Nagashima, M., D'Cruz, T. S., Danku, A. E., Hesse, D., Sifuentes, C., Raymond, P. A. and Hitchcock, P. F. (2020). Midkine-a Is Required for Cell Cycle Progression of Müller Glia during Neuronal Regeneration in the Vertebrate Retina. *J. Neurosci.* 40, 1232-1247.

Naitoh, H., Suganuma, Y., Ueda, Y., Sato, T., Hiramuki, Y., Fujisawa-Sehara, A., Taketani, S. and Araki, M. (2017). Upregulation of matrix metalloproteinase triggers transdifferentiation of retinal pigmented epithelial cells in

Xenopus laevis: A Link between inflammatory response and regeneration. *Dev. Neurobiol.* 77, 1086-1100.

Nakano, T., Ando, S., Takata, N., Kawada, M., Muguruma, K., Sekiguchi, K., Saito, K., Yonemura, S., Eiraku, M. and Sasai, Y. (2012). Self-Formation of Optic Cups and Storable Stratified Neural Retina from Human ESCs. *Cell Stem Cell* 10, 771-785.

Ng, L., Ma, M., Curran, T. and Forrest, D. (2009). Developmental expression of thyroid hormone receptor beta2 protein in cone photoreceptors in the mouse. *Neuroreport* 20, 627-31.

Ohsawa, R. and Kageyama, R. (2008). Regulation of retinal cell fate specification by multiple transcription factors. *Brain Res.* 1192, 90-8.

Osakada, F., Ikeda, H., Mandai, M., Wataya, T., Watanabe, K., Yoshimura, N., Akaike, A., Akaike, A., Sasai, Y. and Takahashi, M. (2008). Toward the generation of rod and cone photoreceptors from mouse, monkey and human embryonic stem cells. *Nat. Biotechnol.* 26, 215-24.

Pearson, R. A., Barber, A. C., Rizzi, M., Hippert, C., Xue, T., West, E. L., Duran, Y., Smith, A. J., Chuang, J. Z., Azam, S. A., et al. (2012). Restoration of vision after transplantation of photoreceptors. *Nature* 485, 99-103.

Pearson, R. A., Gonzalez-Cordero, A., West, E. L., Ribeiro, J. R., Aghaizu, N., Goh, D., Sampson, R. D., Georgiadis, A., Waldron, P. V, Duran, Y., et al. (2016). Donor and host photoreceptors engage in material transfer following transplantation of post-mitotic photoreceptor precursors. *Nat. Commun.* 7, 13029.

Rabesandratana, O., Goureau, O. and Orieux, G. (2018). Pluripotent Stem Cell-Based Approaches to Explore and Treat Optic Neuropathies. Front. *Neurosci.* 12, 1-22.

Ramachandran, R., Fausett, B. V and Goldman, D. (2010). Ascl1a regulates Müller glia dedifferentiation and retinal regeneration through a Lin-28-dependent, let-7 microRNA signalling pathway. *Nat. Cell Biol.* 12, 1101-1107.

Ramachandran, R., Zhao, X.-F. and Goldman, D. (2011). Ascl1a/Dkk/beta-catenin signaling pathway is necessary and glycogen synthase kinase-3beta inhibition is sufficient for zebrafish retina regeneration. *Proc. Natl. Acad. Sci. U. S. A.* 108, 15858-63.

Ramsden, C. M., Powner, M. B., Carr, A.-J. F., Smart, M. J. K., da Cruz, L. and Coffey, P. J. (2013). Stem cells in retinal regeneration: past, present and future. *Development* 140, 2576-85.

Reichman, S., Terray, A., Slembrouck, A., Nanteau, C., Orieux, G., Habeler, W., Nandrot, E. F., Sahel, J.-A., Monville, C. and Goureau, O. (2014). From confluent human iPS cells to self-forming neural retina and retinal pigmented epithelium. *Proc. Natl. Acad. Sci.* 111, 8518-8523.

Reichman, S., Slembrouck, A., Gagliardi, G., Chaffiol, A., Terray, A., Nanteau, C., Potey, A., Belle, M., Rabesandratana, O., Duebel, J., et al. (2017). Generation of Storable Retinal Organoids and Retinal Pigmented Epithelium from Adherent Human iPS Cells in Xeno-Free and Feeder-Free Conditions. *Stem Cells* 35, 1176-1188.

Roska, B. and Sahel, J. (2018). Restoring vision. *Nature* 557, 359-367.

Rueda, E. M., Hall, B. M., Hill, M. C., Swinton, P. G., Tong, X., Martin, J. F. and Poché, R. A. (2019). The Hippo Pathway Blocks Mammalian Retinal Müller Glial Cell Reprogramming. *Cell Rep.* 27, 1637-1649.e6.

Salero, E., Blenkinsop, T. A., Corneo, B., Harris, A., Rabin, D., Stern, J. H. and Temple, S. (2012). Adult human RPE can be activated into a multipotent stem cell that produces mesenchymal derivatives. *Cell Stem Cell* 10, 88-95.

Santos-Ferreira, T., Postel, K., Stutzki, H., Kurth, T., Zeck, G. and Ader, M. (2015). Daylight Vision Repair by Cell Transplantation. *Stem Cells* 33, 79-90.

Santos-Ferreira, T., Llonch, S., Borsch, O., Postel, K., Haas, J. and Ader, M. (2016). Retinal transplantation of photoreceptors results in donor-host cytoplasmic exchange. *Nat. Commun.* 7, 13028.

Schwartz, S. D., Regillo, C. D., Lam, B. L., Eliott, D., Rosenfeld, P. J., Gregori, N. Z., Hubschman, J.-P., Davis, J. L., Heilwell, G., Spirn, M., et al. (2015). Human embryonic stem cell-derived retinal pigment epithelium in patients with age-related macular degeneration and Stargardt's macular dystrophy: follow-up of two open-label phase 1/2 studies. *Lancet* 385, 509-516.

Shirai, H., Mandai, M., Matsushita, K., Kuwahara, A., Yonemura, S., Nakano, T., Assawachananont, J., Kimura, T., Saito, K., Terasaki, H., et al. (2016). Transplantation of human embryonic stem cell-derived retinal tissue in two primate models of retinal degeneration. *Proc. Natl. Acad. Sci. U. S. A.* 113, E81-E90.

Singh, M. S., Balmer, J., Barnard, A. R., Aslam, S. A., Moralli, D., Green, C. M., Barnea-Cramer, A., Duncan, I. and MacLaren, R. E. (2016). Transplanted photoreceptor precursors transfer proteins to host photoreceptors by a mechanism of cytoplasmic fusion. *Nat. Commun.* 7, 13537.

Sridhar, A., Hoshino, A., Finkbeiner, C. R., Chitsazan, A., Dai, L., Haugan, A. K., Eschenbacher, K. M., Jackson, D. L., Trapnell, C., Bermingham-McDonogh, O., et al. (2020). Single-Cell Transcriptomic Comparison of Human Fetal Retina, hPSC-Derived Retinal Organoids, and Long-Term Retinal Cultures. *Cell Rep.* 30, 1644-1659.e4.

Stanzel, B. V, Liu, Z., Somboonthanakij, S., Wongsawad, W., Brinken, R., Eter, N., Corneo, B., Holz, F. G., Temple, S., Stern, J. H., et al. (2014). Human RPE stem cells grown into polarized RPE monolayers on a polyester matrix are maintained after grafting into rabbit subretinal space. *Stem Cell Reports* 2, 64-77.

Stenkamp, D. L. (2015). Development of the Vertebrate Eye and Retina. *Prog Mol Biol Transl Sci.* 134, 397-414.

Stone, L. S. (1950). Neural retina degeneration followed by regeneration from surviving retinal pigment cells in grafted adult salamander eyes. *Anat. Rec.* 106, 89-109.

Takahashi, K. and Yamanaka, S. (2006). Induction of pluripotent stem cells from mouse embryonic and adult fibroblast cultures by defined factors. *Cell* 126, 663-76.

Takahashi, K., Tanabe, K., Ohnuki, M., Narita, M., Ichisaka, T., Tomoda, K. and Yamanaka, S. (2007). Induction of pluripotent stem cells from adult human fibroblasts by defined factors. *Cell* 131, 861-72.

Thomson, J. A. (1998). Embryonic Stem Cell Lines Derived from Human Blastocysts. *Science* 282, 1145-1147.

Thummel, R., Enright, J. M., Kassen, S. C., Montgomery, J. E., Bailey, T. J. and Hyde, D. R. (2010). Pax6a and Pax6b are required at different points in neuronal progenitor cell proliferation during zebrafish photoreceptor regeneration. *Exp. Eye Res.* 90, 572-82.

Tropepe, V. (2000). Retinal Stem Cells in the Adult Mammalian Eye. *Science* 287, 2032-2036.

Tu, H. Y., Watanabe, T., Shirai, H., Yamasaki, S., Kinoshita, M., Matsushita, K., Hashiguchi, T., Onoe, H., Matsuyama, T., Kuwahara, A., et al. (2019). Medium- to long-term survival and functional examination of human iPSC-derived retinas in rat and primate models of retinal degeneration. *EBioMedicine* 39, 562-574.

Wan, J. and Goldman, D. (2017). Opposing Actions of Fgf8a on Notch Signaling Distinguish Two Muller Glial Cell Populations that Contribute to Retina Growth and Regeneration. *Cell Rep.* 19, 849-862.

Wan, J., Ramachandran, R. and Goldman, D. (2012). HB-EGF is necessary and sufficient for Müller glia dedifferentiation and retina regeneration. *Dev. Cell* 22, 334-47.

Wilken, M. S. and Reh, T. A. (2016). Retinal regeneration in birds and mice. *Curr. Opin. Genet. Dev.* 40, 57-64.

Wohl, S. G. and Reh, T. A. (2016). miR-124-9-9* potentiates Ascl1-induced reprogramming of cultured Müller glia. *Glia* 64, 743-62.

Wohl, S. G., Schmeer, C. W. and Isenmann, S. (2012). Neurogenic potential of stem/progenitor-like cells in the adult mammalian eye. *Prog. Retin. Eye Res.* 31, 213-42.

Wohl, S. G., Hooper, M. J. and Reh, T. A. (2019). MicroRNAs miR-25, let-7 and miR-124 regulate the neurogenic potential of Müller glia in mice. *Development* 146, dev179556.

Yao, K., Qiu, S., Tian, L., Snider, W. D., Flannery, J. G., Schaffer, D. V. and Chen, B. (2016). Wnt Regulates Proliferation and Neurogenic Potential of Müller Glial Cells via a Lin28/let-7 miRNA-Dependent Pathway in Adult Mammalian Retinas. *Cell Rep.* 17, 165-178.

Yao, K., Qiu, S., Wang, Y. V., Park, S. J. H., Mohns, E. J., Mehta, B., Liu, X., Chang, B., Zenisek, D., Crair, M. C., et al. (2018). Restoration of vision after de novo genesis of rod photoreceptors in mammalian retinas. *Nature* 560, 484-488.

Yoshii, C., Ueda, Y., Okamoto, M. and Araki, M. (2007). Neural retinal regeneration in the anuran amphibian Xenopus laevis post-metamorphosis: transdifferentiation of retinal pigmented epithelium regenerates the neural retina. *Dev. Biol.* 303, 45-56.

Zarbin, M., Sugino, I. and Townes-Anderson, E. (2019). Concise Review: Update on Retinal Pigment Epithelium Transplantation for Age-Related Macular Degeneration. *Stem Cells Transl. Med.* 8, 466-477.

Zhong, X., Gutierrez, C., Xue, T., Hampton, C., Vergara, M. N., Cao, L.-H., Peters, A., Park, T. S., Zambidis, E. T., Meyer, J. S., et al. (2014). Generation of three-dimensional retinal tissue with functional photoreceptors from human iPSCs. *Nat. Commun.* 5, 4047.

22

Glioblastoma Stem Cells

Nathalie Magne[1,*]**, Sandra E Joppé**[1,*]**, Franck Bielle**[1,2] **and
Emmanuelle Huillard**[1,#]

[1]Sorbonne Université, Institut du Cerveau - Paris Brain Institute - ICM,
Inserm, CNRS, APHP, Paris, France
[2]AP-HP, Hôpitaux Universitaires La Pitié Salpêtrière - Charles Foix,
Département de Neuropathologie Escourolle, F-75013, Paris, France
E-mail: emmanuelle.huillard@icm-institute.org
*These authors contributed equally
#Corresponding author

22.1 Introduction

Diffuse gliomas are malignant and infiltrative tumors originating in the brain.
They represent a highly heterogeneous group of tumors that display morphological features of glial cells, such as astrocytes and oligodendrocytes. The
presence or absence of mutation of *IDH1/2* genes distinguishes two main
types. IDH-wildtype Glioblastomas (GBMs) are the most aggressive tumors
compared to IDH-mutated gliomas (astrocytomas and oligodendrogliomas).
There is currently no curative treatment for diffuse gliomas. The majority
of GBMs arise de novo (called primary GBMs), without a detectable pre-
existing lesion, and a minority evolves from a lower-grade glioma (secondary
GBMs). The standard treatment for GBMs includes maximal resection
(achievable in a minority of patients) followed by radiotherapy plus concomitant and adjuvant chemotherapy. Despite this treatment, the majority of
GBMs recur within a year, resulting in poor patient survival.

Multiple genetic and epigenetic alterations have been identified in
GBMs. Most GBMs harbor alterations in components of three core sig-
naling pathways: receptor tyrosine kinase/Ras/phosphoinositide 3-kinase
(RTK/RAS/PI3K), TP53, and RB pathways. The most frequent alterations

include mutation of *TERT* promoter, polysomy of chromosome 7 and monosomy of chromosome 10, mutations, amplification, or deletion in *PTEN, EGFR, CDKN2A/B, TP53, PIK3CA, PIK3R1,* and *NF1* genes (Aldape et al., 2015).

Over the past 15 years, the development and refinement of genetically engineered mouse models, as well as technological advances enabling analysis at the single cell level have dramatically contributed to our understanding of GBM development and heterogeneity. Importantly, the discovery of tumor cells with stem cell features in GBMs has revolutionized the way we view GBMs and has laid the ground for many studies that have provided fundamental knowledge on how these tumors develop and recur.

This chapter will review current knowledge on glioblastoma stem cells, focusing on adult primary GBMs, which have been the most extensively characterized. We will discuss GBM initiation and development. We will review the identification and characterization of glioma stem cells in GBMs, how they interact with cells of the microenvironment, and discuss consequences of glioma stem cell behavior on GBM heterogeneity and treatment.

22.2 GBM initiation

22.2.1 Tumor Initiation and Evolution

Tumors arise from cells that have acquired and accumulated genetic or epigenetic alterations resulting in selective growth advantage. Tumorigenesis is an evolutionary process that includes multiple steps of genomic alterations, clonal selection, and bidirectional communication between tumor cells and the microenvironment. Barthel and colleagues (2018) have proposed a model of gliomagenesis in five sequential phases (Barthel et al., 2018). First, an initial growth occurs from the cell of origin, presumably due, in most cases, to gain of chromosome 7 genes (i.e., *MET, EGFR* oncogenes) and loss of chromosome 10 genes (i.e., *PTEN* suppressor gene) (Korber et al., 2019). Sustained oncogenic signaling then leads to cellular senescence, a protective barrier that tumor cells overcome by deleting the *CDKN2A/B* locus or mutating the *TP53* gene. Subsequently, continued proliferation promotes telomere dysfunction that activates the DNA damage repair machinery and leads to genomic instability. From this point, most cells either die or undergo replicative senescence, but a few reactivate telomerase (*TERT*) expression to restore genome stability and become immortalized. These processes will establish the initial cancer stem cell population that will fuel tumor growth and evolve during the course of tumor development, treatment, and recurrence.

22.2.2 Cells of Origin for GBMs

What is the cell population from which GBM arise? The cell of origin is defined as the cell population that is first transformed by genetic alterations and initiates tumor development. As stated above, GBMs display characteristics of glial cells, such as oligodendrocytes and astrocytes. Oligodendrocytes and astrocytes are generated from neural stem cells (NSCs). NSCs are cells of the nervous system endowed with the potential to self-renew, proliferate, and give rise to neurons and glial cells (Adams and Morshead, 2018). During brain development, NSCs generate neural progenitor cells (NPCs) that will in turn produce neurons, oligodendrocytes, and astrocytes, in a sequential manner. Moreover, embryonic NSCs give rise to adult NSCs that remain throughout adulthood in two neurogenic niches (dentate gyrus and subventricular zone, SVZ). Like their embryonic counterparts, these adult NSCs give rise to neural progenitors that will generate neurons and glial cells (Kriegstein and Alvarez-Buylla, 2009).

Unlike neurons, astrocytes and oligodendrocyte precursors (OPCs) are capable of proliferation. Astrocytes are the largest glial population and play major functions in the regulation of synaptic transmission, metabolic support, maintenance of the blood-brain barrier. Cortical astrocytes have been shown to proliferate in the postnatal brain (Ge et al., 2012).

OPCs (also called NG2-glia) account for the majority of proliferating cells in the adult brain (Vigano and Dimou, 2016). Although their main function is to renew oligodendrocytes, the myelin-forming cells of the central nervous system, OPCs have a broader potential: (i) they can adopt stem cell features *in vitro*; (ii) they can generate astrocytes and neurons in some studies; (iii) in response to a brain lesion, they can generate other lineages, such as Schwann cells, the myelinating cells of the peripheral nervous system and sometimes subsets of astrocytes (Richardson et al., 2011). OPCs also play important roles in myelin maintenance and repair and have been shown to modulate inflammatory cells and synaptic activity.

As described below, NSC/NPCs and OPCs are likely to be the cells of origin of GBMs. In genetically engineered mouse models, deletion of key GBM tumor suppressor pathways (*TP53*, *NF1*, *PTEN*, and *RB*) in NSC/NPCs leads to the development of GBM-like tumors (Alcantara Llaguno et al., 2009; Chow et al., 2011; Friedmann-Morvinski et al., 2012; Liu et al., 2011; Persson et al., 2010). However, when the same alterations are introduced in lineage-restricted cells (neuroblasts, neurons, and mature astrocytes), tumors do not form (Alcantara Llaguno et al., 2009; Alcantara

Llaguno et al., 2019; Chow et al., 2011). It is interesting to note that in these experimental models, tumors contain high levels of proliferative progenitor cells resembling OPCs (PDGFRa+, OLIG2+, NG2+) with only a minor expansion of the other cell types (Alcantara Llaguno et al., 2009; Liu et al., 2011; Persson et al., 2010). Accordingly, specific inactivation of *Tp53* and *Nf1* in NG2+ cells generates tumors that are identical phenotypically and transcriptomically to those initiated in NSCs/NPCs (Liu et al., 2011). Importantly, NG2+ cells sorted from these tumors are able to reconstitute a tumor in a recipient mouse (Persson et al., 2010). Together, these findings indicate that cells along the NSC-NPC-OPC lineage are the cells susceptible to transformation leading to GBM. These conclusions are supported by studies in GBM patients. First, genomic analysis of tumor biopsies and matched tumor-free SVZ demonstrated that tumor-free SVZ NSCs harbor low-level GBM driver mutations such as *TERT* promoter, *EGFR, PTEN*, and *TP53* mutations, suggesting that these cells evolve to GBM (Lee et al., 2018). Second, the transplantation of NG2+ cells from GBM patient biopsies induced the development of tumors in recipient mice, while the NG2- cells did not, confirming mouse studies (Al-Mayhani et al., 2011). Altogether, these data suggest a model in which the initial cell to undergo mutation is a NSC, which will accumulate mutations and, once reaching the OPC stage, will expand to form a tumor (Figure 22.1A).

22.3 Identification of GSCs

Once the tumor is established, complex interactions between tumor cells and their microenvironment drives tumor development and evolution. Almost 20 years ago, cells with characteristics of NSCs were identified in GBMs. The application of NSC isolation techniques to glioma cells (using limiting dilution and sphere formation assays) revealed the existence of stem-like tumor cells that proliferate, self-renew, are multipotent, and able to reconstitute, upon transplantation, a tumor with characteristics of the patients tumor (Galli et al., 2004; Ignatova et al., 2002; Singh et al., 2004). Accordingly, analysis of transcriptomes of single cells from GBM biopsies highlighted the presence of cells with NSC gene expression signatures (Patel et al., 2014). The similarities between these cells and normal NSCs laid the foundation for the glioma stem cell hypothesis. According to this theory, only a fraction of tumor cells in GBMs is tumorigenic and endowed with stem cell properties, termed glioma stem cells (GSCs). This term should be used with caution: it refers to a functional, cellular state rather than a cell type. Furthermore,

A **INITIATION**   B **EVOLUTION**

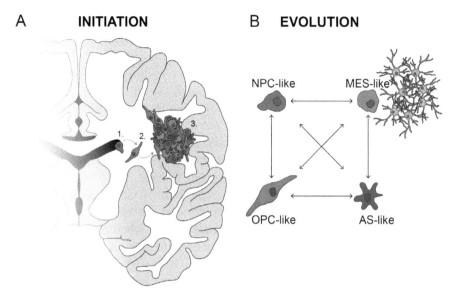

Figure 22.1 Model for the initiation and evolution of primary (IDH1-wildtype) GBMs. A. Sequence of cellular events leading to GBM. In this model, NSCs are the cells of mutation and OPC-like cells are the cells of tumor expansion. B. Once the tumor is formed, OPC-like cells make up most of the GSC pool, but as the tumor evolves and interacts with its microenvironment, GSC clones appear and show plasticity among them (Neftel et al., 2019). Accordingly, all GBMs contain cells with a proneural (OPC-like/NPC-like) transcriptional profile (regardless of the dominant subtype of the tumor) (Neftel et al., 2019; Patel et al., 2014). MES-like GSCs are enriched in hypoxic regions whereas OPC-like/NPC-like GSCs are more proliferative and enriched at the tumor infiltrative edge. Tumor-associated macrophages (brown cells) control transition to a MES phenotype. The temporal sequence of events is unknown but it is estimated that the founder cell arises about 2-7 years prior to diagnosis (Korber et al., 2019).

the presence of stem-like cells in tumors does not mean that NSCs are the obligate cell of origin for this tumor. Thus, like their normal counterparts, GSCs appear to be at the apex of a cellular hierarchy, generating glioma progenitors that will be responsible for tumor expansion.

Initially, CD133 (Prominin1) was successfully used to isolate a cell fraction from human biopsies that was able to reconstitute a tumor upon grafting into immunocompromised mice (Singh et al., 2004). Although this work was supported by subsequent studies (Bao et al., 2006), later studies demonstrated that: (i) some GBMs are devoid of CD133+ cells, (ii) the percentage of CD133+ cell is not correlated with poorer prognosis or stem cell *in vitro* properties, and (iii) CD133- cells can be tumorigenic

(Beier et al., 2007; Chen et al., 2010). Although other GSC cell surface markers were later identified (CD15, ITGA6, A2B5), none captures the full GSC population (Dirkse et al., 2019; Lathia et al., 2010; Son et al., 2009; Tchoghandjian et al., 2010). In fact, multiple GSC subpopulations, identified by distinct cell surface markers, are present in GSC cultures and display similar stem cell properties *in vitro* (Dirkse et al., 2019).

In patient-derived xenograft models, marker expression is highly variable between tumors and these markers are not specifically linked to a particular tumor phenotype (infiltrative or angiogenic) (Dirkse et al., 2019). The findings that multiple cell surface markers can identify GSCs point to the existence of multiple GSCs clones within tumors, as we will explain below (GSC heterogeneity and plasticity). More recently, by labeling GBM cells with DNA barcodes to investigate the behavior of GSC clones, Lan et al. were able to trace individual clones and their progeny in xenografts. By performing multiple rounds of transplantation, they observed the presence of multiple clones, of different sizes. This demonstrates that multiple, equipotent, tumor-initiating cells drive tumor growth, rather than a few tumor-initiating cells (Lan et al., 2017).

22.4 Regulation of GSC Activity

22.4.1 Cell-intrinsic Regulation

Cell-intrinsic factors and pathways involved in GSC self-renewal and proliferation have been extensively studied. Interestingly, many of them are key regulators of neural cell proliferation, migration, and fate during brain development, further highlighting the similarities between organogenesis and oncogenesis. For instance, developmental signaling pathways, such as NOTCH, WNT, BMP, or TGFβ, play key roles in GSC biology (Zhang and Lin, 2014). Examination of the transcriptional and epigenetic landscapes of GSCs revealed widespread activation of transcription factors in GSCs (Rheinbay et al., 2013). Strikingly, a module of only four developmental transcription factors (POU3F2, SALL2, SOX2, and OLIG2) was sufficient to reprogram differentiated GBM cells into sphere-forming and tumorigenic GSCs (Suva et al., 2014). In particular, factors involved in OPC specification and development are critical drivers of GSCs. This is the case of OLIG2, a master regulator of neural progenitor proliferation and oligodendrocyte development (Ligon et al., 2007; Lu et al., 2002), that has been shown to promote GBM development, partly by antagonizing p53 activity (Ligon et al.,

2007; Mehta et al., 2011). Likewise, the transcription factor ZFP36L1 was recently shown to promote oligodendroglial fate commitment in the mouse brain, abrogate tumor development in a mouse GBM model and induce OPC-associated genes in GSCs (Weng et al., 2019).

Epigenetic modifications, such as DNA methylation, histone modifications, and chromatin remodeling, are important regulatory mechanisms in development and cancer. Some are emerging as key regulators of GSC activity, such as long non-coding RNAs, micro-RNAs, and mRNA modifications. Many RNA-modifying proteins are altered in GBMs, impacting mRNA stability, translation, or splicing (Dong and Cui, 2020).

About half GBMs carry mutations in genes functionally linked to chromatin modifications (Brennan et al., 2013). For instance, chromatin regulators such as BMI-1, EZH2, and MLL1 have been shown to enhance the self-renewing potential and tumorigenicity of GSCs (Bruggeman et al., 2007; Suva et al., 2009). EZH2 directly regulates the expression of MYC (Suva et al., 2009), a major transcriptional regulator of stem cell biology that is required for the maintenance of GSC self-renewal and tumorigenic potential. Of note, MYC also participates to metabolic reprogramming in GSCs, by notably inducing the expression of genes involved in the mevalonate and purine biosynthetic pathways (Wang et al., 2017c; Wang et al., 2017d).

22.4.2 Interactions with the Tumor-associated Microenvironment

Similar to normal NSCs, GSCs lie in a complex microenvironment. This microenvironment is composed of non-neoplastic cell types and molecules, embedded in or adjacent to the tumor, among which are blood vessels, immune cells (microglia, the resident immune cells of the brain, peripheral macrophages, and infiltrating lymphocytes), extracellular matrix components, and neural cells (astrocytes, oligodendrocytes, and neurons). Within this microenvironment, GSCs are enriched in three different tumor regions called niches: near the abnormal angiogenic vasculature (perivascular niche), in necrotic (hypoxic) regions, and in the invasive tumor front, where GSCs use normal blood vessels for their dissemination into the parenchyma (Hambardzumyan and Bergers, 2015). In these niches, GSCs interact with multiple cell types to provide a supportive environment that will favor their stemness, self-renewal, and survival (Figure 22.2). For example, hypoxia signaling pathway participates to stemness and can be activated in the necrotic niche. Although perivascular GSCs have a direct access to oxygen, they can hijack hypoxia signaling to maintain stemness, a state described as pseudo-hypoxic (Pietras et al., 2014).

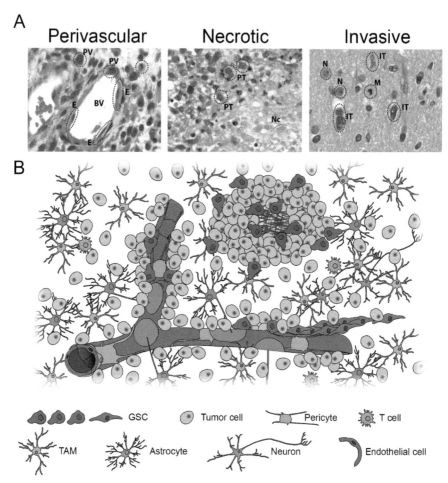

Figure 22.2 GSCs' interactions within the different tumor niches. A. Hematoxylin and eosin stainings showing the perivascular, necrotic, and infiltrative regions on GBM sections. B. Scheme of the three niches showing the interactions of GSCs with cells of the microenvironment. TAM: tumor-associated microglia/macrophage; BV: blood vessel; E: endothelial cells; PV: perivascular tumor cell; M: mitotic cell; PT: perinecrotic tumor cell; Nc: necrosis; IT: invasive tumor cell; N: neuron.

Blood vessels are an essential component of the tumor microenvironment. Blood vessels are composed of two cellular compartments: endothelial cells, which make up the wall of the vessel, and pericytes, which cover and stabilize the vessel to regulate blood flow. In GBMs, GSCs interact closely with endothelial cells in a bidirectional manner. Endothelial cells promote gliomasphere growth *in vitro* and brain tumor propagation *in vivo* (Calabrese et al., 2007). GSCs express high levels of angiogenic factors (VEGF) to support the formation of new vessels from endothelial cells (Bao et al., 2006). Interestingly, it was shown that endothelial cells recruit GSCs and induce their transdifferentiation to pericytes in a TGF-β dependent manner. Accordingly, most tumor pericytes carry the same genetic alterations as matched GSCs (Cheng et al., 2013). The blood-brain barrier (BBB) is a specialized vasculature structure comprising endothelial cells, pericytes, and astrocyte endfeet that regulates the transfer of molecules between the blood and the brain. Glioma cells can displace astrocyte endfeet to have a direct access to the vessels, leading to a loss of tight junctions, disruption of astrocyte-vascular coupling, thereby causing a local breach in the BBB (Watkins et al., 2014).

The immune system constitutes a tumor-suppressive barrier that GBMs have to bypass in order to grow. As the BBB is compromised, immune cells, including macrophages, myeloid-derived suppressor cells, T cells and NK cells (Gieryng et al., 2017), infiltrate the brain. The glioma environment is immuno-suppressive: it contains high levels of regulatory T cells and cytokines secreted by tumor cells that limit the antitumor activity of T cells. In addition, NK cells are nonfunctional in GBMs. Tumor-associated microglia/macrophages (TAMs), which regroup brain resident microglia and peripheral macrophages, represent up to 30% of all cells in human GBM. TAM infiltration positively correlates with tumor grade (Gieryng et al., 2017). Glioma cells secrete chemoattractants for TAMs (Gieryng et al., 2017; Gutmann and Kettenmann, 2019). In turn, TAMs can exert antitumoral or protumoral functions but whether these functions are related to different TAM populations is debated. TAMs release cytokines that will promote angiogenesis, extracellular matrix remodeling, survival, migration, glycolytic metabolism, and growth of tumor cells (Broekman et al., 2018; Chen et al., 2019; Markovic et al., 2009; Zhang et al., 2018).

The glioma microenvironment also contains neurons, astrocytes, and oligodendroglial cells. In the normal brain, neuronal stimulation has been shown to induce OPC proliferation and differentiation (Gibson et al., 2014). In the same manner, recent studies showed that electrical activity, via the release of Neuroligin-3, stimulates the growth of glioma cells (Venkatesh et al., 2015; Venkatesh et al., 2017). Intriguingly, glioma cells express synaptic genes and establish functional synapses with neurons (Venkatesh et al., 2019).

In pathological situations, astrocytes become hypertrophic, upregulate intermediate filaments and can enter in proliferation. These cells, called reactive astrocytes, can be found near or within GBMs. They can interact directly with glioma cells, establishing gap junctions and tubular extensions called tunneling nanotubes, which protects tumor cells from chemotherapy and radiotherapy (Osswald et al., 2015). Through the secretion of numerous cytokines or chemokines, reactive astrocytes promote GBM progression by promoting glioma cell proliferation and migration (Guan et al., 2018). Compared to normal astrocytes, tumor-associated reactive astrocytes display a distinct transcriptional program. In particular, they increase JAK/STAT signaling that maintains an anti-inflammatory environment (Henrik Heiland et al., 2019). Very few studies have investigated the relationships between OPCs and GSCs. A recent study showed that OPC conditioned medium induced stemness markers, increased sphere numbers and chemo-radioresistance of GBM cultures (Hide et al., 2018).

22.5 GSC Heterogeneity and Plasticity

GBMs show a high degree of inter-tumoral heterogeneity: proliferation index, growth or response to therapy can vary significantly from one patient to another. In addition, there is extensive cellular heterogeneity within each tumor. Over the past decade, there has been tremendous progress in understanding the basis of this intra-tumoral heterogeneity.

Three main GBM subtypes have been identified, based on their gene expression and DNA methylation profiles, as well as genetic alterations: Mesenchymal (MES), classical (CL), and proneural (PN) subgroups (Klughammer et al., 2018; Phillips et al., 2006; Verhaak et al., 2010). These subgroups are not mutually exclusive entities: multiple GBM subgroups can be detected within the same tumor (Sottoriva et al., 2013) and convert from one type to another (often PN to MES transition) during tumor progression and recurrence (Patel et al., 2014; Sottoriva et al., 2013). Along these lines, clones harboring distinct genetic alterations (*EGFR, PDGFRA*) were found to co-exist and be actively proliferating within the same tumors (Snuderl et al., 2011). Analysis of single cell transcriptomes further highlighted intratumoral heterogeneity in primary GBMs: individual tumors contain a spectrum of molecular subtypes and sometimes hybrid cellular states (Patel et al., 2014). In agreement with these observations, many studies reported transcriptional and functional heterogeneity between GSC cultures in vitro (Bhat et al., 2013; Chen et al., 2010; Cusulin et al., 2015). Together, these findings raised the hypothesis that the different tumor subtypes could arise from transcriptionally or genetically different GSCs.

Recent scRNAseq studies have identified four recurrent cellular states among glioma cells (NPC-like, OPC-like, astrocyte-like, and mesenchymal-like states) that recapitulate distinct neural cell types (Neftel et al., 2019; Pine et al., 2020). Each of the four states contains cycling cells, and is present, but in different proportions, in all GBMs analyzed (Neftel et al., 2019). Strikingly, isolated cell populations from each state could initiate a tumor in mice and give rise to all four states in a distribution similar to that found in the patient sample (Neftel et al., 2019). This shows that GSCs can shift from one state to another, demonstrating GSC plasticity, which is the ability to reversibly acquire different cellular phenotypes (Figure 22.1B).

What are the factors that underlie GSC plasticity? It was recently shown that the distribution of cellular states within the tumor is influenced by genetic alterations: *EGFR* and *CDK4* lead to predominance of astrocyte-like and NPC-like programs in mouse cells, respectively (Neftel et al., 2019). In addition, the microenvironment seems to greatly modulate the tumorigenic potential and cellular state of GSC. Indeed, the finding that cells from a given state are able to give rise to all four states after transplantation into host brains adds support to this notion (Neftel et al., 2019). As previously mentioned, GSCs cultures are phenotypically heterogeneous and can undergo dynamic and reversible transitions *in vitro* and *in vivo* (Dirkse et al., 2019). It was shown that phenotypic transitions are influenced by hypoxia, acidic stress, nutrient deprivation or macrophages/microglial cells (Bhat et al., 2013; Dirkse et al., 2019;Flavahan et al., 2013; Hjelmeland et al., 2011). GSC populations are thus highly capable to adapt to a new microenvironment, providing a mechanism for overcoming selective pressure and enhancing tumor growth. Hence, genetics and the microenvironment influence the fate decisions of multiple populations of GSCs, resulting in intra-tumoral heterogeneity.

22.6 GSC Resistance to Therapies

22.6.1 GSCs Are Resistant to Radiotherapy and Chemotherapy

Glioblastomas are difficult to treat. The first-line treatment consists in surgically removing the tumor mass and treating patients with concomitant radio and chemotherapies. Furthermore, the infiltrative properties of tumor cells make it difficult to completely remove tumor tissue from the brain parenchyma and the BBB can impede drug delivery. Regardless of the treatment, GBMs inevitably recur. There is now ample evidence showing that GSCs are responsible for tumor resistance and recurrence.

GSCs have developed a full arsenal of mechanisms to counter conventional therapies. They have a highly efficient DNA repair system, a slow division rate (making them less sensitive to standard therapies targeting proliferative cells), and express high levels of drug transporters such as ABC transporters (that pump drugs out of the cell) (Auffinger et al., 2015). Exposure of GBM cultures or xenografts to irradiation or temozolomide (TMZ), an alkylating agent used in first-line chemotherapy, increases the fraction of GSCs, whether defined by their functional properties, a cell surface marker, or a transcriptomic profile (Bao et al., 2006; Bleau et al., 2009). In a genetically engineered mouse model of GBM, Parada and colleagues demonstrated that quiescent stem-like cells can be activated and reconstitute a tumor after chemotherapy treatment, thus being responsible for tumor recurrence (Chen et al., 2012). The DNA repair enzyme O6-methylguanine DNA methyltransferase (MGMT) directly removes cytotoxic DNA lesions caused by TMZ. The expression of MGMT in GSCs and its regulation are key contributors to chemoresistance (Beier et al., 2011).

A consequence of the high heterogeneity and plasticity of GSC is that subsets of cells can escape or adapt to therapy and be responsible for increased tumor malignancy, resistance, and recurrence (Wang et al., 2016). However, TMZ treatment does not lead to major changes in the distribution of phenotypic GSC subpopulations, suggesting that subpopulations of GSCs identified by cell surface markers are not selectively enriched by treatment (Dirkse et al., 2019). In accordance with this finding, marker expression is not different between initial and recurrent tumors (Dirkse et al., 2019). In contrast, by lineage tracing experiments of barcoded GSCs, Lan and colleagues found that while TMZ treatment eradicated the majority of GSC clones, it promoted the expansion of pre-existing drug-resistant GSCs (Lan et al., 2017). Liau and colleagues provided evidence that, following treatment with inhibitors of RTKs, GSCs reversibly transition to a slow-dividing, drug-resistant state characterized by induction of high NOTCH signaling, stemness, and quiescence programs, enabling the cells to adapt to treatment (Liau et al., 2017). It was also shown that treatment induces transitions in GBM transcriptional subgroups. For instance, PN to MES shift can be induced by radiation and at recurrence (Minata et al., 2019; Phillips et al., 2006; Segerman et al., 2016). This can even lead to the recurrence of GBM into gliosarcoma, in which tumor cells have lost glial features to gain a fibroblastic phenotype (Smith et al., 2018). Subgroup conversions are associated with worse survival (Wang et al., 2016; Wang et al., 2017b) and are accompanied by enhanced DNA repair machinery (Bao et al., 2006), higher expression of ABC transporters or activation of oncogenic drivers (CD109, YAP/TAZ) that promote clonogenic growth and GSC properties (Lee et al., 2020; Minata et al., 2019). Furthermore, under clinically relevant doses of TMZ, differentiated tumor cells can convert to GSC-like cells that display invasive and tumorigenic properties similar to that of parental GSCs (Auffinger et al., 2014). In addition, TMZ treatment has been shown to induce markers of endothelial cells and the formation of tumor-derived vessels in orthotopic xenografts (Baisiwala et al., 2019). Thus, conventional radiotherapy and chemotherapy contribute to GSC plasticity and tumor heterogeneity.

22.6.2 Strategies to Target GSCs

Since GBMs are driven by cells with stem cell features, treatments could be developed that target these specific features of tumor cells. However, GSC heterogeneity and plasticity would make ineffective a strategy targeting a specific type of GSC: the treatment would deplete a GSC subpopulation but spare another, which will lead to tumor recurrence. How to target GSCs? One strategy may consist in targeting the self-renewal property of GSCs. Many transcription factors (i.e., SOX2, OLIG2, ASCL1) or chromatin modifiers (i.e., HDAC, BRD4, EZH2) regulate the self-renewal and identity of GSCs (Lathia et al., 2015). However, these factors are nuclear proteins, which are not easily accessible to drugs or inhibitors. Targeting developmental pathways active in GSCs (NOTCH, SHH, WNT, EGFR) can represent interesting approaches. Yet these pathways are ubiquitous and may result in severe side effects. Despite a well-tolerated treatment, clinical trials using gamma secretase inhibitors to inhibit the NOTCH pathway was not efficient at preventing tumor recurrence (Xu et al., 2016). Other approaches include targeting the DNA damage response (i.e., PARP, CHK2 inhibitors) and the DNA repair pathway (MGMT) responsible for specific GSC chemoresistance or radioresistance (Ferri et al., 2020). A differentiation therapy, whereby GSCs are forced to differentiate, represents a promising approach: for instance BMP4 treatment has been shown to induce GSC differentiation and abrogates tumor development in a mouse model of GBM (Piccirillo et al., 2006). It was recently shown that the transcription factor ASCL1 acts as a pioneer factor that opens chromatin to activate a neuronal differentiation program in GSCs and decrease their tumorigenicity (Park et al., 2017).

Modulation of the tumor microenvironment, in order to decrease niche support of GSCs, may represent a promising strategy. Anti-angiogenic therapies (using anti-VEGF antibodies or inhibitors of receptor tyrosine kinases) initially show radiographic responses but fail to show survival benefit in patients with GBM. Tumor cells seem to utilize alternative mechanisms (vessel co-option) to recruit blood vessels to escape these treatments (Wang et al., 2017a). Immunotherapies aim at overcoming the immune-suppressive environment of the tumor and restoring T-cell function. Use of checkpoint inhibitors (i.e., PD-1/PD-L1, CTLA-4) are effective for patients with melanoma and lung cancer, but have so far shown limited efficacy in brain tumors (Brahm et al., 2020). Vaccine approaches are also developed: targeting the neoepitope EGFRvIII in a subset of patients shows encouraging results in a phase 2 trial but need to be confirmed in larger cohorts (Reardon et al., 2020). More recently, cell-based immunotherapy (CAR-T cell) has demonstrated very interesting results in a patient suffering from a very advanced and aggressive glioblastoma (Brown et al., 2016), appearing therefore as a promising approach in glioblastoma patients. All these strategies are likely to remove a given GSC subpopulation, but as we have seen, because of the diversity and plasticity of GSC populations, a successful treatment for GBMs will require combinatory treatments.

22.7 Concluding Remarks

- Glioblastomas originate from the transformation of cells belonging to the oligodendrocyte lineage.
- Glioblastomas contain glioma stem cells (GSCs) that are responsible for tumor development and recurrence.
- GSCs reactivate developmental pathways and their activity is regulated by cells of the microenvironment.
- GSCs are phenotypically heterogeneous and can switch phenotypes, making them moving targets when considering therapies.
- Drivers of GSC plasticity need to be identified and targeted in combination with conventional therapies.

Acknowledgments

Research in the Huillard team is funded by the Ligue Nationale contre le Cancer, Fondation ARC, ARTC, and program "Investissements d'avenir" ANR-10-IAIHU-06. We acknowledge the contribution of SiRIC CURAMUS (INCA-DGOS-Inserm_12560) which is financially supported by the French National Cancer Institute, the French Ministry of Solidarity and Health and

Inserm. FB's research work is funded by a grant Emergence 2019 from Sorbonne Université.

References

Adams, K.V., and Morshead, C.M. (2018). Neural stem cell heterogeneity in the mammalian forebrain. Prog Neurobiol 170, 2-36.

Al-Mayhani, M.T., Grenfell, R., Narita, M., Piccirillo, S., Kenney-Herbert, E., Fawcett, J.W., Collins, V.P., Ichimura, K., and Watts, C. (2011). NG2 expression in glioblastoma identifies an actively proliferating population with an aggressive molecular signature. Neuro-oncology 13, 830-845.

Alcantara Llaguno, S., Chen, J., Kwon, C.H., Jackson, E.L., Li, Y., Burns, D.K., Alvarez-Buylla, A., and Parada, L.F. (2009). Malignant astrocytomas originate from neural stem/progenitor cells in a somatic tumor suppressor mouse model. Cancer cell 15, 45-56.

Alcantara Llaguno, S., Sun, D., Pedraza, A.M., Vera, E., Wang, Z., Burns, D.K., and Parada, L.F. (2019). Cell-of-origin susceptibility to glioblastoma formation declines with neural lineage restriction. Nature neuroscience 22, 545-555.

Aldape, K., Zadeh, G., Mansouri, S., Reifenberger, G., and von Deimling, A. (2015). Glioblastoma: pathology, molecular mechanisms and markers. Acta neuropathologica 129, 829-848.

Auffinger, B., Spencer, D., Pytel, P., Ahmed, A.U., and Lesniak, M.S. (2015). The role of glioma stem cells in chemotherapy resistance and glioblastoma multiforme recurrence. Expert review of neurotherapeutics 15, 741-752.

Auffinger, B., Tobias, A.L., Han, Y., Lee, G., Guo, D., Dey, M., Lesniak, M.S., and Ahmed, A.U. (2014). Conversion of differentiated cancer cells into cancer stem-like cells in a glioblastoma model after primary chemotherapy. Cell death and differentiation 21, 1119-1131.

Baisiwala, S., Auffinger, B., Caragher, S.P., Shireman, J.M., Ahsan, R., Lee, G., Hasan, T., Park, C., Saathoff, M.R., Christensen, A.C., et al. (2019). Chemotherapeutic Stress Induces Transdifferentiation of Glioblastoma Cells to Endothelial Cells and Promotes Vascular Mimicry. Stem Cells Int 2019, 6107456.

Bao, S., Wu, Q., McLendon, R.E., Hao, Y., Shi, Q., Hjelmeland, A.B., Dewhirst, M.W., Bigner, D.D., and Rich, J.N. (2006). Glioma stem cells promote radioresistance by preferential activation of the DNA damage response. Nature 444, 756-760.

Barthel, F.P., Wesseling, P., and Verhaak, R.G.W. (2018). Reconstructing the molecular life history of gliomas. Acta neuropathologica 135, 649-670.

Beier, D., Hau, P., Proescholdt, M., Lohmeier, A., Wischhusen, J., Oefner, P.J., Aigner, L., Brawanski, A., Bogdahn, U., and Beier, C.P. (2007). CD133(+) and CD133(-) glioblastoma-derived cancer stem cells show differential growth characteristics and molecular profiles. Cancer research 67, 4010-4015.

Beier, D., Schulz, J.B., and Beier, C.P. (2011). Chemoresistance of glioblastoma cancer stem cells–much more complex than expected. Molecular cancer 10, 128.

Bhat, K.P., Balasubramaniyan, V., Vaillant, B., Ezhilarasan, R., Hummelink, K., Hollingsworth, F., Wani, K., Heathcock, L., James, J.D., Goodman, L.D., et al. (2013). Mesenchymal differentiation mediated by NF-kappaB promotes radiation resistance in glioblastoma. Cancer cell 24, 331-346.

Bleau, A.M., Hambardzumyan, D., Ozawa, T., Fomchenko, E.I., Huse, J.T., Brennan, C.W., and Holland, E.C. (2009). PTEN/PI3K/Akt pathway regulates the side population phenotype and ABCG2 activity in glioma tumor stem-like cells. Cell stem cell 4, 226-235.

Brahm, C.G., van Linde, M.E., Enting, R.H., Schuur, M., Otten, R.H.J., Heymans, M.W., Verheul, H.M.W., and Walenkamp, A.M.E. (2020). The Current Status of Immune Checkpoint Inhibitors in Neuro-Oncology: A Systematic Review. Cancers 12.

Brennan, C.W., Verhaak, R.G., McKenna, A., Campos, B., Noushmehr, H., Salama, S.R., Zheng, S., Chakravarty, D., Sanborn, J.Z., Berman, S.H., et al. (2013). The somatic genomic landscape of glioblastoma. Cell 155, 462-477.

Broekman, M.L., Maas, S.L.N., Abels, E.R., Mempel, T.R., Krichevsky, A.M., and Breakefield, X.O. (2018). Multidimensional communication in the microenvirons of glioblastoma. Nature reviews Neurology 14, 482-495.

Brown, C.E., Alizadeh, D., Starr, R., Weng, L., Wagner, J.R., Naranjo, A., Ostberg, J.R., Blanchard, M.S., Kilpatrick, J., Simpson, J., et al. (2016). Regression of Glioblastoma after Chimeric Antigen Receptor T-Cell Therapy. The New England journal of medicine 375, 2561-2569.

Bruggeman, S.W., Hulsman, D., Tanger, E., Buckle, T., Blom, M., Zevenhoven, J., van Tellingen, O., and van Lohuizen, M. (2007). Bmi1 controls tumor development in an Ink4a/Arf-independent manner in a mouse model for glioma. Cancer cell 12, 328-341.

Calabrese, C., Poppleton, H., Kocak, M., Hogg, T.L., Fuller, C., Hamner, B., Oh, E.Y., Gaber, M.W., Finklestein, D., Allen, M., et al. (2007). A perivascular niche for brain tumor stem cells. Cancer cell 11, 69-82.

Chen, J., Li, Y., Yu, T.S., McKay, R.M., Burns, D.K., Kernie, S.G., and Parada, L.F. (2012). A restricted cell population propagates glioblastoma growth after chemotherapy. Nature 488, 522-526.

Chen, P., Zhao, D., Li, J., Liang, X., Li, J., Chang, A., Henry, V.K., Lan, Z., Spring, D.J., Rao, G., et al. (2019). Symbiotic Macrophage-Glioma Cell Interactions Reveal Synthetic Lethality in PTEN-Null Glioma. Cancer cell 35, 868-884 e866.

Chen, R., Nishimura, M.C., Bumbaca, S.M., Kharbanda, S., Forrest, W.F., Kasman, I.M., Greve, J.M., Soriano, R.H., Gilmour, L.L., Rivers, C.S., et al. (2010). A hierarchy of self-renewing tumor-initiating cell types in glioblastoma. Cancer cell 17, 362-375.

Cheng, L., Huang, Z., Zhou, W., Wu, Q., Donnola, S., Liu, J.K., Fang, X., Sloan, A.E., Mao, Y., Lathia, J.D., et al. (2013). Glioblastoma stem cells generate vascular pericytes to support vessel function and tumor growth. Cell 153, 139-152.

Chow, L.M., Endersby, R., Zhu, X., Rankin, S., Qu, C., Zhang, J., Broniscer, A., Ellison, D.W., and Baker, S.J. (2011). Cooperativity within and among Pten, p53, and Rb pathways induces high-grade astrocytoma in adult brain. Cancer cell 19, 305-316.

Cusulin, C., Chesnelong, C., Bose, P., Bilenky, M., Kopciuk, K., Chan, J.A., Cairncross, J.G., Jones, S.J., Marra, M.A., Luchman, H.A., et al. (2015). Precursor States of Brain Tumor Initiating Cell Lines Are Predictive of Survival in Xenografts and Associated with Glioblastoma Subtypes. Stem Cell Reports 5, 1-9.

Dirkse, A., Golebiewska, A., Buder, T., Nazarov, P.V., Muller, A., Poovathingal, S., Brons, N.H.C., Leite, S., Sauvageot, N., Sarkisjan, D., et al. (2019). Stem cell-associated heterogeneity in Glioblastoma results from intrinsic tumor plasticity shaped by the microenvironment. Nature communications 10, 1787.

Dong, Z., and Cui, H. (2020). The Emerging Roles of RNA Modifications in Glioblastoma. Cancers 12.

Ferri, A., Stagni, V., and Barila, D. (2020). Targeting the DNA Damage Response to Overcome Cancer Drug Resistance in Glioblastoma. International journal of molecular sciences 21.

Flavahan, W.A., Wu, Q., Hitomi, M., Rahim, N., Kim, Y., Sloan, A.E., Weil, R.J., Nakano, I., Sarkaria, J.N., Stringer, B.W., et al. (2013). Brain tumor initiating cells adapt to restricted nutrition through preferential glucose uptake. Nature neuroscience 16, 1373-1382.

Friedmann-Morvinski, D., Bushong, E.A., Ke, E., Soda, Y., Marumoto, T., Singer, O., Ellisman, M.H., and Verma, I.M. (2012). Dedifferentiation of neurons and astrocytes by oncogenes can induce gliomas in mice. Science 338, 1080-1084.

Galli, R., Binda, E., Orfanelli, U., Cipelletti, B., Gritti, A., De Vitis, S., Fiocco, R., Foroni, C., Dimeco, F., and Vescovi, A. (2004). Isolation and characterization of tumorigenic, stem-like neural precursors from human glioblastoma. Cancer research 64, 7011-7021.

Ge, W.P., Miyawaki, A., Gage, F.H., Jan, Y.N., and Jan, L.Y. (2012). Local generation of glia is a major astrocyte source in postnatal cortex. Nature 484, 376-380.

Gibson, E.M., Purger, D., Mount, C.W., Goldstein, A.K., Lin, G.L., Wood, L.S., Inema, I., Miller, S.E., Bieri, G., Zuchero, J.B., et al. (2014). Neuronal activity promotes oligodendrogenesis and adaptive myelination in the mammalian brain. Science 344, 1252304.

Gieryng, A., Pszczolkowska, D., Walentynowicz, K.A., Rajan, W.D., and Kaminska, B. (2017). Immune microenvironment of gliomas. Lab Invest 97, 498-518.

Guan, X., Hasan, M.N., Maniar, S., Jia, W., and Sun, D. (2018). Reactive Astrocytes in Glioblastoma Multiforme. Mol Neurobiol 55, 6927-6938.

Gutmann, D.H., and Kettenmann, H. (2019). Microglia/Brain Macrophages as Central Drivers of Brain Tumor Pathobiology. Neuron 104, 442-449.

Hambardzumyan, D., and Bergers, G. (2015). Glioblastoma: Defining Tumor Niches. Trends Cancer 1, 252-265.

Henrik Heiland, D., Ravi, V.M., Behringer, S.P., Frenking, J.H., Wurm, J., Joseph, K., Garrelfs, N.W.C., Strahle, J., Heynckes, S., Grauvogel, J., et al. (2019). Tumor-associated reactive astrocytes aid the evolution of immunosuppressive environment in glioblastoma. Nature communications 10, 2541.

Hide, T., Komohara, Y., Miyasato, Y., Nakamura, H., Makino, K., Takeya, M., Kuratsu, J.I., Mukasa, A., and Yano, S. (2018). Oligodendrocyte Progenitor Cells and Macrophages/Microglia Produce Glioma Stem Cell Niches at the Tumor Border. EBioMedicine 30, 94-104.

Hjelmeland, A.B., Wu, Q., Heddleston, J.M., Choudhary, G.S., MacSwords, J., Lathia, J.D., McLendon, R., Lindner, D., Sloan, A., and Rich, J.N. (2011). Acidic stress promotes a glioma stem cell phenotype. Cell death and differentiation 18, 829-840.

Ignatova, T.N., Kukekov, V.G., Laywell, E.D., Suslov, O.N., Vrionis, F.D., and Steindler, D.A. (2002). Human cortical glial tumors contain neural stem-like cells expressing astroglial and neuronal markers in vitro. Glia 39, 193-206.

Klughammer, J., Kiesel, B., Roetzer, T., Fortelny, N., Nemc, A., Nenning, K.H., Furtner, J., Sheffield, N.C., Datlinger, P., Peter, N., et al. (2018). The DNA methylation landscape of glioblastoma disease progression shows extensive heterogeneity in time and space. Nature medicine 24, 1611-1624.

Korber, V., Yang, J., Barah, P., Wu, Y., Stichel, D., Gu, Z., Fletcher, M.N.C., Jones, D., Hentschel, B., Lamszus, K., et al. (2019). Evolutionary Trajectories of IDH(WT) Glioblastomas Reveal a Common Path of Early Tumorigenesis Instigated Years ahead of Initial Diagnosis. Cancer cell 35, 692-704 e612.

Kriegstein, A., and Alvarez-Buylla, A. (2009). The glial nature of embryonic and adult neural stem cells. Annual review of neuroscience 32, 149-184.

Lan, X., Jorg, D.J., Cavalli, F.M.G., Richards, L.M., Nguyen, L.V., Vanner, R.J., Guilhamon, P., Lee, L., Kushida, M.M., Pellacani, D., et al. (2017). Fate mapping of human glioblastoma reveals an invariant stem cell hierarchy. Nature 549, 227-232.

Lathia, J.D., Gallagher, J., Heddleston, J.M., Wang, J., Eyler, C.E., Macswords, J., Wu, Q., Vasanji, A., McLendon, R.E., Hjelmeland, A.B., et al. (2010). Integrin alpha 6 regulates glioblastoma stem cells. Cell stem cell 6, 421-432.

Lathia, J.D., Mack, S.C., Mulkearns-Hubert, E.E., Valentim, C.L., and Rich, J.N. (2015). Cancer stem cells in glioblastoma. Genes & development 29, 1203-1217.

Lee, C.A.A., Banerjee, P., Wilson, B.J., Wu, S., Guo, Q., Berg, G., Karpova, S., Mishra, A., Lian, J.W., Tran, J., et al. (2020). Targeting the ABC transporter ABCB5 sensitizes glioblastoma to temozolomide-induced apoptosis through a cell-cycle checkpoint regulation mechanism. The Journal of biological chemistry 295, 7774-7788.

Lee, J.H., Lee, J.E., Kahng, J.Y., Kim, S.H., Park, J.S., Yoon, S.J., Um, J.Y., Kim, W.K., Lee, J.K., Park, J., et al. (2018). Human glioblastoma arises from subventricular zone cells with low-level driver mutations. Nature 560, 243-247.

Liau, B.B., Sievers, C., Donohue, L.K., Gillespie, S.M., Flavahan, W.A., Miller, T.E., Venteicher, A.S., Hebert, C.H., Carey, C.D., Rodig, S.J., et al. (2017). Adaptive Chromatin Remodeling Drives Glioblastoma Stem Cell Plasticity and Drug Tolerance. Cell stem cell 20, 233-246 e237.

Ligon, K.L., Huillard, E., Mehta, S., Kesari, S., Liu, H., Alberta, J.A., Bachoo, R.M., Kane, M., Louis, D.N., Depinho, R.A., et al. (2007). Olig2-regulated lineage-restricted pathway controls replication competence in neural stem cells and malignant glioma. Neuron 53, 503-517.

Liu, C., Sage, J.C., Miller, M.R., Verhaak, R.G., Hippenmeyer, S., Vogel, H., Foreman, O., Bronson, R.T., Nishiyama, A., Luo, L., et al. (2011). Mosaic analysis with double markers reveals tumor cell of origin in glioma. Cell 146, 209-221.

Lu, Q.R., Sun, T., Zhu, Z., Ma, N., Garcia, M., Stiles, C.D., and Rowitch, D.H. (2002). Common developmental requirement for Olig function indicates a motor neuron/oligodendrocyte connection. Cell 109, 75-86.

Markovic, D.S., Vinnakota, K., Chirasani, S., Synowitz, M., Raguet, H., Stock, K., Sliwa, M., Lehmann, S., Kalin, R., van Rooijen, N., et al. (2009). Gliomas induce and exploit microglial MT1-MMP expression for tumor expansion. Proceedings of the National Academy of Sciences of the United States of America 106, 12530-12535.

Mehta, S., Huillard, E., Kesari, S., Maire, C.L., Golebiowski, D., Harrington, E.P., Alberta, J.A., Kane, M.F., Theisen, M., Ligon, K.L., et al. (2011). The central nervous system-restricted transcription factor Olig2 opposes p53 responses to genotoxic damage in neural progenitors and malignant glioma. Cancer cell 19, 359-371.

Minata, M., Audia, A., Shi, J., Lu, S., Bernstock, J., Pavlyukov, M.S., Das, A., Kim, S.H., Shin, Y.J., Lee, Y., et al. (2019). Phenotypic Plasticity of Invasive Edge Glioma Stem-like Cells in Response to Ionizing Radiation. Cell reports 26, 1893-1905 e1897.

Neftel, C., Laffy, J., Filbin, M.G., Hara, T., Shore, M.E., Rahme, G.J., Richman, A.R., Silverbush, D., Shaw, M.L., Hebert, C.M., et al. (2019). An Integrative Model of Cellular States, Plasticity, and Genetics for Glioblastoma. Cell 178, 835-849 e821.

Osswald, M., Jung, E., Sahm, F., Solecki, G., Venkataramani, V., Blaes, J., Weil, S., Horstmann, H., Wiestler, B., Syed, M., et al. (2015). Brain tumour cells interconnect to a functional and resistant network. Nature 528, 93-98.

Park, N.I., Guilhamon, P., Desai, K., McAdam, R.F., Langille, E., O'Connor, M., Lan, X., Whetstone, H., Coutinho, F.J., Vanner, R.J., et al. (2017). ASCL1 Reorganizes Chromatin to Direct Neuronal Fate and Suppress Tumorigenicity of Glioblastoma Stem Cells. Cell stem cell 21, 209-224.e207.

Patel, A.P., Tirosh, I., Trombetta, J.J., Shalek, A.K., Gillespie, S.M., Wakimoto, H., Cahill, D.P., Nahed, B.V., Curry, W.T., Martuza, R.L., et al.

(2014). Single-cell RNA-seq highlights intratumoral heterogeneity in primary glioblastoma. Science.

Persson, A.I., Petritsch, C., Swartling, F.J., Itsara, M., Sim, F.J., Auvergne, R., Goldenberg, D.D., Vandenberg, S.R., Nguyen, K.N., Yakovenko, S., et al. (2010). Non-stem cell origin for oligodendroglioma. Cancer cell 18, 669-682.

Phillips, H.S., Kharbanda, S., Chen, R., Forrest, W.F., Soriano, R.H., Wu, T.D., Misra, A., Nigro, J.M., Colman, H., Soroceanu, L., et al. (2006). Molecular subclasses of high-grade glioma predict prognosis, delineate a pattern of disease progression, and resemble stages in neurogenesis. Cancer cell 9, 157-173.

Piccirillo, S.G., Reynolds, B.A., Zanetti, N., Lamorte, G., Binda, E., Broggi, G., Brem, H., Olivi, A., Dimeco, F., and Vescovi, A.L. (2006). Bone morphogenetic proteins inhibit the tumorigenic potential of human brain tumour-initiating cells. Nature 444, 761-765.

Pietras, A., Katz, A.M., Ekström, E.J., Wee, B., Halliday, J.J., Pitter, K.L., Werbeck, J.L., Amankulor, N.M., Huse, J.T., and Holland, E.C. (2014). Osteopontin-CD44 signaling in the glioma perivascular niche enhances cancer stem cell phenotypes and promotes aggressive tumor growth. Cell stem cell 14, 357-369.

Pine, A.R., Cirigliano, S.M., Nicholson, J.G., Hu, Y., Linkous, A., Miyaguchi, K., Edwards, L., Singhania, R., Schwartz, T.H., Ramakrishna, R., et al. (2020). Tumor Microenvironment Is Critical for the Maintenance of Cellular States Found in Primary Glioblastomas. Cancer discovery 10, 964-979.

Reardon, D.A., Desjardins, A., Vredenburgh, J.J., O'Rourke, D.M., Tran, D.D., Fink, K.L., Nabors, L.B., Li, G., Bota, D.A., Lukas, R.V., et al. (2020). Rindopepimut with Bevacizumab for Patients with Relapsed EGFRvIII-Expressing Glioblastoma (ReACT): Results of a Double-Blind Randomized Phase II Trial. Clinical cancer research : an official journal of the American Association for Cancer Research 26, 1586-1594.

Rheinbay, E., Suva, M.L., Gillespie, S.M., Wakimoto, H., Patel, A.P., Shahid, M., Oksuz, O., Rabkin, S.D., Martuza, R.L., Rivera, M.N., et al. (2013). An aberrant transcription factor network essential for Wnt signaling and stem cell maintenance in glioblastoma. Cell reports 3, 1567-1579.

Richardson, W.D., Young, K.M., Tripathi, R.B., and McKenzie, I. (2011). NG2-glia as multipotent neural stem cells: fact or fantasy? Neuron 70, 661-673.

Segerman, A., Niklasson, M., Haglund, C., Bergstrom, T., Jarvius, M., Xie, Y., Westermark, A., Sonmez, D., Hermansson, A., Kastemar, M., et al. (2016). Clonal Variation in Drug and Radiation Response among Glioma-Initiating Cells Is Linked to Proneural-Mesenchymal Transition. Cell reports 17, 2994-3009.

Singh, S.K., Hawkins, C., Clarke, I.D., Squire, J.A., Bayani, J., Hide, T., Henkelman, R.M., Cusimano, M.D., and Dirks, P.B. (2004). Identification of human brain tumour initiating cells. Nature 432, 396-401.

Smith, D.R., Wu, C.C., Saadatmand, H.J., Isaacson, S.R., Cheng, S.K., Sisti, M.B., Bruce, J.N., Sheth, S.A., Lassman, A.B., Iwamoto, F.M., et al. (2018). Clinical and molecular characteristics of gliosarcoma and modern prognostic significance relative to conventional glioblastoma. Journal of neuro-oncology 137, 303-311.

Snuderl, M., Fazlollahi, L., Le, L.P., Nitta, M., Zhelyazkova, B.H., Davidson, C.J., Akhavanfard, S., Cahill, D.P., Aldape, K.D., Betensky, R.A., et al. (2011). Mosaic amplification of multiple receptor tyrosine kinase genes in glioblastoma. Cancer cell 20, 810-817.

Son, M.J., Woolard, K., Nam, D.H., Lee, J., and Fine, H.A. (2009). SSEA-1 is an enrichment marker for tumor-initiating cells in human glioblastoma. Cell stem cell 4, 440-452.

Sottoriva, A., Spiteri, I., Piccirillo, S.G., Touloumis, A., Collins, V.P., Marioni, J.C., Curtis, C., Watts, C., and Tavare, S. (2013). Intratumor heterogeneity in human glioblastoma reflects cancer evolutionary dynamics. Proceedings of the National Academy of Sciences of the United States of America 110, 4009-4014.

Suva, M.L., Rheinbay, E., Gillespie, S.M., Patel, A.P., Wakimoto, H., Rabkin, S.D., Riggi, N., Chi, A.S., Cahill, D.P., Nahed, B.V., et al. (2014). Reconstructing and reprogramming the tumor-propagating potential of glioblastoma stem-like cells. Cell 157, 580-594.

Suva, M.L., Riggi, N., Janiszewska, M., Radovanovic, I., Provero, P., Stehle, J.C., Baumer, K., Le Bitoux, M.A., Marino, D., Cironi, L., et al. (2009). EZH2 is essential for glioblastoma cancer stem cell maintenance. Cancer research 69, 9211-9218.

Tchoghandjian, A., Baeza, N., Colin, C., Cayre, M., Metellus, P., Beclin, C., Ouafik, L., and Figarella-Branger, D. (2010). A2B5 cells from human glioblastoma have cancer stem cell properties. Brain Pathol 20, 211-221.

Venkatesh, H.S., Johung, T.B., Caretti, V., Noll, A., Tang, Y., Nagaraja, S., Gibson, E.M., Mount, C.W., Polepalli, J., Mitra, S.S., et al. (2015).

Neuronal Activity Promotes Glioma Growth through Neuroligin-3 Secretion. Cell 161, 803-816.

Venkatesh, H.S., Morishita, W., Geraghty, A.C., Silverbush, D., Gillespie, S.M., Arzt, M., Tam, L.T., Espenel, C., Ponnuswami, A., Ni, L., et al. (2019). Electrical and synaptic integration of glioma into neural circuits. Nature 573, 539-545.

Venkatesh, H.S., Tam, L.T., Woo, P.J., Lennon, J., Nagaraja, S., Gillespie, S.M., Ni, J., Duveau, D.Y., Morris, P.J., Zhao, J.J., et al. (2017). Targeting neuronal activity-regulated neuroligin-3 dependency in high-grade glioma. Nature 549, 533-537.

Verhaak, R.G., Hoadley, K.A., Purdom, E., Wang, V., Qi, Y., Wilkerson, M.D., Miller, C.R., Ding, L., Golub, T., Mesirov, J.P., et al. (2010). Integrated genomic analysis identifies clinically relevant subtypes of glioblastoma characterized by abnormalities in PDGFRA, IDH1, EGFR, and NF1. Cancer cell 17, 98-110.

Vigano, F., and Dimou, L. (2015). The heterogeneous nature of NG2-glia. Brain Res.

Wang, J., Cazzato, E., Ladewig, E., Frattini, V., Rosenbloom, D.I., Zairis, S., Abate, F., Liu, Z., Elliott, O., Shin, Y.J., et al. (2016). Clonal evolution of glioblastoma under therapy. Nature genetics 48, 768-776.

Wang, N., Jain, R.K., and Batchelor, T.T. (2017a). New Directions in Anti-Angiogenic Therapy for Glioblastoma. Neurotherapeutics : the journal of the American Society for Experimental NeuroTherapeutics 14, 321-332.

Wang, Q., Hu, B., Hu, X., Kim, H., Squatrito, M., Scarpace, L., deCarvalho, A.C., Lyu, S., Li, P., Li, Y., et al. (2017b). Tumor Evolution of Glioma-Intrinsic Gene Expression Subtypes Associates with Immunological Changes in the Microenvironment. Cancer cell 32, 42-56 e46.

Wang, X., Huang, Z., Wu, Q., Prager, B.C., Mack, S.C., Yang, K., Kim, L.J.Y., Gimple, R.C., Shi, Y., Lai, S., et al. (2017c). MYC-Regulated Mevalonate Metabolism Maintains Brain Tumor-Initiating Cells. Cancer research 77, 4947-4960.

Wang, X., Yang, K., Xie, Q., Wu, Q., Mack, S.C., Shi, Y., Kim, L.J.Y., Prager, B.C., Flavahan, W.A., Liu, X., et al. (2017d). Purine synthesis promotes maintenance of brain tumor initiating cells in glioma. Nature neuroscience 20, 661-673.

Watkins, S., Robel, S., Kimbrough, I.F., Robert, S.M., Ellis-Davies, G., and Sontheimer, H. (2014). Disruption of astrocyte-vascular coupling and the blood-brain barrier by invading glioma cells. Nature communications 5, 4196.

Weng, Q., Wang, J., Wang, J., He, D., Cheng, Z., Zhang, F., Verma, R., Xu, L., Dong, X., Liao, Y., et al. (2019). Single-Cell Transcriptomics Uncovers Glial Progenitor Diversity and Cell Fate Determinants during Development and Gliomagenesis. Cell stem cell 24, 707-723 e708.

Xu, R., Shimizu, F., Hovinga, K., Beal, K., Karimi, S., Droms, L., Peck, K.K., Gutin, P., Iorgulescu, J.B., Kaley, T., et al. (2016). Molecular and Clinical Effects of Notch Inhibition in Glioma Patients: A Phase 0/I Trial. Clinical cancer research : an official journal of the American Association for Cancer Research 22, 4786-4796.

Zhang, Y., Yu, G., Chu, H., Wang, X., Xiong, L., Cai, G., Liu, R., Gao, H., Tao, B., Li, W., et al. (2018). Macrophage-Associated PGK1 Phosphorylation Promotes Aerobic Glycolysis and Tumorigenesis. Molecular cell 71, 201-215 e207.

Zhang, Z., and Lin, C.C. (2014). Taking advantage of neural development to treat glioblastoma. The European journal of neuroscience 40, 2859-2866.

23

Cardiac Tissue Engineering for Repair and Regeneration of the Heart

---

Pierre Joanne and Onnik Agbulut*

Sorbonne Université, Institut de Biologie Paris-Seine (IBPS), CNRS UMR
8256, Inserm ERL U1164, Biological Adaptation and Ageing, 75005,
Paris-France
E-mail: onnik.agbulut@sorbonne-universite.fr
*Corresponding author(s)

23.1 Introduction

Repairing defective organs is an old dream of mankind. With the advent
of knowledge in developmental biology, physiology, molecular and cellular
biology, and bioengineering this dream appears much closer to reality than
ever. Indeed, for several decades, stem cell-based therapies have emerged
as a promising therapeutic option for treating many incurable degenerative
disorders. In the cardiovascular field, attempt to regenerate the chronically
failing heart by stem cells has raised a tremendous interest. Stem cell therapy aims to repair injured heart through the delivery of cells that have the
capacity to remuscularize the wounded heart and consequently to restore, at
least partially, the functionality of the damaged region. While direct intra-
cardiac injection of these cells holds great promises (Liu et al., 2018),
this technique also demonstrated some limitations as recently highlighted
(Mallapaty, 2020), and is not expected to be applicable in all cases. To
overcome these limitations, alternative strategies were developed and have
contributed to the fast and exciting emergence of the field of cardiac tissue
engineering.

Cardiac tissue engineering is dedicated to the fabrication of a functional
construct that recapitulates the biological and physiological features of the

603

human heart (Hirt et al., 2014; Nguyen et al., 2019). The principle of this method mainly relies upon the promotion of cell growth on a bio-inspired scaffold that mimics the extracellular matrix (ECM) of the tissue which has to be engineered. Cells, materials, scaffolds, and biological cues are generally combined with techniques from the physicochemical or biological fields to fabricate an engineered heart tissue (EHT) to replace damaged or missing myocardial tissues. Tissue engineering provides to cells, a natural and suitable environment for an adequate cellular growth, organotypic assembly, and tissue formation. Indeed, in the context of cell therapy, this technology enhances cell survival and retention of implanted cells in damaged hearts after implantation because it avoids the proteolytic dissociation of the cells which is normally required prior to cells injection. The ability to create tightly controlled EHTs has progressed considerably in recent years. Ultimately, this field will give rise to the perspective of recreating an entire heart in vitro. Until we achieve this distant dream, the challenges to reconstruct a simple and functional fragment of this complex syncytium are numerous (Huang et al., 2018) and are at the heart of the field of cardiac tissue engineering. The different future challenges of this constantly evolving field will be presented in the subsequent parts of this chapter even if creating small but functional cardiac tissue elementary units is already possible (Rogozhnikov et al., 2016). These EHTs are used in several experimental assays, mainly to model different cardiac diseases/conditions (Mohammadi et al., 2017; Tiburcy et al., 2017), to screen for new therapeutics drugs (Vunjak Novakovic et al., 2014) or to improve or restore damaged heart that is often caused by several pathologies (Dai and Foley, 2014). Whether it is to repair or to model the heart, cardiac tissue engineering is closely linked to experimental studies that try to decipher or treat heart diseases (Veldhuizen et al., 2019).

23.2 Heart Diseases and Stem Cell-based Therapies

Because the heart is a vital organ, the pathologies involving the heart function are rarely benign. In heart failure, the heart has not the capacity to pump blood efficiently throughout the body leading to the appearance of a large panel of symptoms such as chest pain, fainting, or severe weakness, tachycardia or bradycardia, cardiac arrhythmias, difficulty of breathing, or sudden cardiac arrest. Ultimately, heart failure leads to the death of patients affected and heart transplantation is the only way to definitely cure this severe condition. Different conditions can damage or weaken the cardiac function and may cause heart failure. Among these different conditions, some of them were

Table 23.1 Principal heart diseases targeted in experimental studies that use cardiac tissue engineering as treatment

Heart Disease	Description	Tissue Engineering-Related References
Ischemic heart disease	This condition commonly results from the deposition of an atheromatous plaque in the coronary arteries, reducing the blood flow perfusing the cardiac muscle (stenosis). The sudden rupture of the plaque often results in the closure of the coronary artery leading to myocardial infarction.	(Madonna et al., 2019; Richards et al., 2020)
Congenital heart disease	Congenital heart disease is characterized by malformations of the structures of the heart. It is generally inherited (genetic causes) or due to adverse exposure to toxic molecules such as drugs or alcohol during the prenatal development.	(Boyd et al., 2019; Mantakaki et al., 2018)
Cardiomyopathy	Cardiomyopathies are a group of heterogeneous diseases affecting the heart muscle. It is characterized by an abnormal structure or function of the heart in the absence other cardiac disease. The different types of cardiomyopathy include dilated, hypertrophic, restrictive, and arrhythmogenic cardiomyopathies.	(Tang and Alvarez, 2017; Turnbull et al., 2018)
Cardiac arrhythmias	Arrhythmias are characterized as irregularities of the heartbeat. Common types of arrythmias include atrial fibrillation, bradycardia, and supraventricular tachycardia. Different causes can induce arrhythmias, among which genetic causes are frequently found.	(Colatsky et al., 2016; Kawatou et al., 2017; Ma et al., 2014)
Valvular heart disease	Valves open and close in function of blood pressure and are essential to drive the blow flow into the physiological direction. A lot of several conditions can affect the function of the valves (congenital defects, genetic, toxicity of medications, infections, etc).	(Cheung et al., 2015; Oveissi et al., 2020)
Other causes and cardiotoxicity	Mainly different chronic diseases can contribute to heart failure (e.g., diabetes, hemochromatosis, amyloidosis). Some substances are cardiotoxic such as alcohol and some drugs (especially drugs used for chemotherapy) for example.	(Richards et al., 2020; Song et al., 2011; Takeda et al., 2018)

particularly targeted in applications of cardiac tissue engineering and were listed in Table 23.1.

Among these different heart diseases, ischemic heart disease was particularly targeted in studies using tissue engineering and cell therapy. Ischemic heart disease is generally due to the closure of coronary arteries leading to the death of cardiomyocytes irrigated by these coronaries. Then, the cardiac muscle in the infarct zone degenerates and, after a strong inflammation, is progressively replaced by a fibrotic tissue with very low contractile function. In this context, the replacement of the scar tissue due to an infarction with functional contractile cells appears as a very interesting way to cure this debilitating disease. Even if this idea appears simple, replacing cells into the functioning heart is a very challenging issue because the heart is a complex organ with numerous types of cells and many physical constraints.

23.3 The Heart: A Complex Organ

The human heart can be basically viewed as a pump allowing the blood circulation into the vessels that irrigate all the organs of the body. But it is important to note that the "pump" function of this organ needs to be permanently maintained and adjusted in real time in function of the oxygen and nutrient demand of the body. Human heart is composed of two parts (right and left), and each part is composed of two chambers (atrium and ventricle) separated by the atrioventricular valves. The veins carrying oxygen-depleted blood are connected to the heart at the level of the right atria. After passing the right atrioventricular valve, blood is ejected from the right ventricle toward the pulmonary artery after the opening of the pulmonary valve and will thus join the lungs where it is oxygenated. Through the pulmonary veins, oxygenated blood joins the left atrium and then the left ventricle from where it can be expelled toward the aorta (after the opening of the aortic valve), and thus irrigate the whole organism. Cardiomyocytes are a specialized type of cells which is responsible for the contraction of the heart. These are specific muscle cells that are very different from their skeletal or smooth counterparts. Moreover, cardiomyocytes from the different chambers of the heart exhibit specific features such as gene regulation, protein expression, and electrophysiological profiles. For an efficient contraction, the different chambers of the heart have to be tightly controlled and synchronized. Specialized cardiomyocytes belonging to the conduction system are dedicated to this task by generating (the pacemaker's cells of the sinoatrial node) and conducting electrical impulses (the Purkinje fibers). In addition to this different type of cardiomyocytes, heart is also composed of other cell types including cardiac fibroblasts, endothelial cells, pericytes, and smooth muscle cells. Resident macrophages are also found in the heart and play an important role in homeostasis, contraction, but also during heart disease and aging (Ma et al., 2018). In terms of numbers, the proportion of cardiomyocytes, fibroblasts, and endothelial cells is approximately 30:15:55 in the adult heart (Pinto et al., 2016). However, the mass of cardiomyocytes represents the majority of the cardiac muscle mass, indicating the central role of these cells for heart function.

Compared to skeletal muscles which have a huge regenerative capacity, mediated by a pool of dedicated stem cells (i.e., satellite cells or myoblasts), the cardiac muscles in complex organisms, like human, have a reduced ability to efficiently regenerate. Thus, contrary to the skeletal, the loss of an important number of cells in the heart following an injury cannot be compensated by the activation, multiplication, and differentiation of stem cells. However, cardiac stem cells have been reported in 2003 (Beltrami et al., 2003) and these cells were described to express c-Kit. Even if it was demonstrated that neonatal c-Kit+ cardiac stem cells are able to differentiate into cardiomyocytes, this capacity seems to be lost in adult, making these cells marginal contributors in cardiomyogenesis (Van Berlo et al., 2014). Moreover, the entire field of c-Kit+ cardiac stem cells remains largely controversial since most of the papers relative to their use in clinical perspective was withdrawn for suspected fraud (Reardon, 2018). To date, the role of c-Kit in cardiac biology is still under investigation and appears more complex than previously reported (Gude et al., 2018). Other types of cardiac progenitor cells were described based on the expression of Sca-1, Islet-1, or Tbx18, but possible overlapping between these cell populations makes interpretation difficult and recent lineage tracing studies indicate that these cells minimally contribute to cardiomyogenesis (Vagnozzi et al., 2018). It is also suggested that pre-existing cardiomyocytes may undergo dedifferentiation followed by duplication and redifferentiation (Senyo et al., 2013). Other cell candidates for cardiomyogenesis are fetal microchimeric cells. In this process, fetal cells pass through the placenta to the mother and participate in the repair of many tissues, including the heart (Kara et al., 2012). Even if the precise mechanism of cardiac cell turnover is not known, recent results based on ^{14}C birth dating indicates that approximately 1% of cardiomyocytes are renewed each year before the age of 20. This turnover ratio declines with age to 0.5% in elderly individuals (Lázár et al., 2017).

In addition to the different cell types and their relative organization in the heart, the ECM also plays an essential role (Valiente-Alandi et al., 2016). By surrounding and connecting cardiac cells, ECM components participate in the cell-to-cell interaction, force transmission as well as to the relaxation of this organ. ECM is also essential for the anisotropic organization of cardiomyocytes which promotes development of a functional cardiac syncytium. At the cellular level, ECM participates in cell geometry, cell mechanics as well as in chromatin organization and gene expression by interacting with mechanotransduction pathways. The ECM of the heart is mainly composed of collagen (collagen type I and III essentially), fibronectin, and laminin. A large number of growth factors, cytokines and proteases are also tightly associated with the ECM and participate in its remodeling and/or in cell signaling. It should also be noted that the composition of the ECM dramatically evolves during development, aging, or in response to a physiological modification or a disease (Lockhart et al., 2011). To conclude, it is well known that all the different cell populations, embedded in the ECM of the heart, form together a dense and highly organized syncytium that maintain the cardiac function under homeostatic and disease conditions.

23.4 Cell Types for Heart Repair: An Endless Story?

In the history of cardiac cell therapy (Samak and Hinkel, 2019), one of the very first challenge to achieve an efficient remuscularization of the heart was to define the best type of cells to be grafted in the diseased organ. Obviously, these cells are those endowed with a real cardiomyogenic differentiation potential, i.e., those able to generate new contractile cardiomyocytes to regenerate chronically failing hearts. About 20 years ago, it emerged that adult stem cells can be the best candidates to this aim because they are undifferentiated and have the capacity to self-renew and differentiate in at least one cell type. Among the different stem cells that can be found in adult, first, bone marrow-derived stem cells (BM-SCs, either hematopoietic or mesenchymal stem cells) and skeletal myoblasts attracted attention, because of their safety, autologous availability, and effectiveness. Moreover, they are already largely used in clinical or preclinical studies for the treatment of several diseases. Skeletal myoblasts, after their delivery in the damaged area of the myocardium, undergo an activation that triggers their migration, proliferation, fusion, differentiation and consequently, because of their remarkable capacity of resistance to ischemia, actively participate in remuscularization of the necrotic area. However, a potential limitation to the efficacy of skeletal myoblasts is their exclusive commitment into the skeletal muscle lineage without any differentiation into cardiac or endothelial cells. As a consequence, the benefits of skeletal muscle transplantation have been considered marginal and are most likely due to the paracrine effects of the transplanted cells rather than to a real regeneration of the scarred myocardium originating from the grafted cells. BM-SCs were also considered in cardiac cell therapy purpose based mainly on their multipotency. It is widely documented that BM-SCs are able to participate in the regeneration of many injured tissues. In cardiovascular field, since the 2000s, these stem cells were preclinically tested in various conditions, especially in myocardial infarction models such as rats (Agbulut et al., 2006) or sheep (Bel et al., 2003). They were also tested in heart failure such as doxorubicin-induced heart failure mice model (Agbulut et al., 2003) or in dilated cardiomyopathy (Pouly et al., 2004). Although bone-marrow derived cells were used in numerous clinical trials, results obtained by the last phase III clinical trial (Mathur et al., 2017) seems to not demonstrate a clear benefit over early revascularization procedure (Menasché, 2020), even if the final results of this clinical trial are not already available. Mesenchymal stem cells (MSCs) are another type of adult stem cells that were also scrutinized extensively and that continue to be regarded with interest for cardiac cell therapy. Thus, different clinical trials are presently conducted with these cells (as for example with DREAM-HF (Borow et al., 2019). This interest toward MSCs principally relies on their immunomodulatory, anti-inflammatory, and angiogenic properties. Indeed, even if MSCs are not susceptible to participate in the remuscularization of the cardiac muscle, their secretion seems to be interesting to stimulate the endogenous repair mechanisms as it was demonstrated in several studies (Hamdi et al., 2011, 2013; Kompa et al., 2020).

In line with this observation, a new paradigm recently emerged alongside with the remuscularization: the paracrine signaling paradigm (Menasché, 2020). Actually, most of the effects mediated by the different cells described above were attributed to "paracrine signals" secreted by the cells injected into the heart. Indeed, experimental studies revealed that most of the injected cells were not able to survive even if functional effects were frequently reported. Thus, it appears that the secretome of these cells could stimulate the intrinsic repair ability of the heart, essentially through the modulation of angiogenesis, fibrosis, inflammation, and apoptosis. Currently, in line with this new paradigm, several studies have been dedicated to assay "cell-free" cardiac therapy. It was, for example, shown that cardiovascular progenitors or their secreted extracellular vesicles both enhance the cardiac function in a mouse model of infarction (Kervadec et al., 2016). Contrarily to what was expected, the secretome of different cardiac cells do not have the same efficiency (El Harane et al., 2018) and the choice of cells therefore remains crucial for this application. In addition to the huge number of studies concerning adult stem cells (BM-SCs, skeletal myoblasts, or MSCs) and the emerging new paradigm

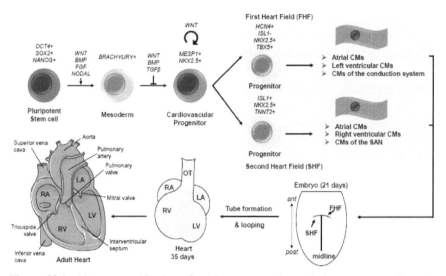

Figure 23.1   Lineage specification of cardiomyocytes: from pluripotent stem cells to the formation of the heart. Different signaling pathways (BMP, Nodal, Wnt, and FGF) induce the differentiation of pluripotent stem cells toward mesoderm. The downregulation of canonical Wnt pathway promotes the cardiogenic commitment of mesodermal precursors through the expression of MESP1. These cardiovascular progenitors then undergo further lineage specification and differentiate into progenitor of the first heart field (FHF) and second heart field (SHF) that form the cardiac crescent. During embryogenesis, naïve cardiac mesoderm progenitors migrate in the splanchnic mesoderm and form the cardiac crescent (second week of human gestation). Then, cardiac crescent fuses at the midline and gives rise to the linear heart tube. The cardiomyocytes (CMs) of the left ventricle (LV) mainly derives from the FHF. Additional cells from the SHF expand the linear heart tube at the arterial and venous poles and contribute to the formation of outflow tract (OT), right ventricle (RV), and a large part of the left and right atria (LA and RA). After fusion and looping of the heart tube, the chambers of the developing heart are fully septated and connected to the outflow tract (≈35 days of human gestation) and will further grow until adulthood. The figure includes material from SMART Servier medical Art (https://smart.servier.com/) under a Creative Commons (CC) license 3.0.

of the paracrine signaling (Menasché, 2020), it seems reasonable to envisage cells from cardiac origin as the best candidates for cardiac cell therapy. Besides cardiac stem cells and their controversial results (already presented and discussed above), the use of pluripotent stem cells differentiated into cardiomyocytes has opened a very stimulating and attractive way (Desgres and Menasché, 2019).

23.5 The Rise of Pluripotent Stem Cells-derived Cardiomyocytes

Human pluripotent stem cells (hPSCs), including human embryonic stem cells (hESCs (Thomson et al., 1998)) and human induced pluripotent stem cells (hiPSCs (Takahashi et al., 2007)) can be propagated indefinitely without entering senescence, and virtually differentiated in any cell types derived from three germ layers (endoderm, mesoderm, and ectoderm). Thus, with the development of *in vitro* culture systems, it is now relatively easy to produce these cells on large scale even in a basic research facility (Kempf et al., 2016). Based on developmental studies (see Figure 23.1), huge efforts to determine the molecular and environmental cues driving PSC differentiation toward cardiac lineages have been made since the 2000s (Mummery et al., 2012; Protze et al., 2019). By using agonists or antagonists of relevant signaling pathways involved in cardiogenesis, such as ActivinA/Nodal and Bone Morphogenetic Protein (BMP), cardiomyocytes derived from hPSCs can be efficiently obtained and used in the perspective of clinical applications (Laflamme et al., 2007). The first developed protocols based on the use of these biological molecules (specific growth factors and cytokines) were further optimized through the manipulation of the Wnt pathway (Willems et al., 2011). Few years after, a protocol entirely based on the use of small artificial molecules related to the Wnt pathway was developed and massively adopted in the field because of its robustness and its reduced cost (Lian et al, 2012). It was further improved to be fully compliant with clinical requirements that avoid the use of animal-origin components (Burridge et al., 2014). This protocol is based on the use of the GSK3 inhibitor CHIR99021 that induces an activation of the Wnt pathway during the first days of culture followed by an inhibition of the Wnt pathway (several molecules are available for this step such as IWP2 for example). To obtain an efficient differentiation a precise time window between the application of these two molecules have to be respected (Lian et al, 2012). In this protocol, the activation of the Wnt pathway followed by its inhibition regulates the BMP and Activin/Nodal signalings. Another critical factor of this Wnt-based protocol is the starting cell density which appears to be variable and has to be finely tuned on each hPSCs cell line (Laco et al., 2018). Even if these cardiac differentiation protocols were efficiently optimized and generally give rise to more than 80% of cardiomyocytes, they also generate different population of cells as it was demonstrated by a single cell transcriptomic analysis (Friedman et al., 2018). The identity of these residual cells has been described as related to cardiac outflow tract cells, a structure composed of several types of cells such as fibroblast, endothelial cells, and smooth muscle cells (Friedman et al., 2018).

But in a clinical and medical perspective, cell populations have to be fully characterized. Indeed, the potential tumorigenicity of undifferentiated hPSCs may represent an important safety issue. Thus, to increase the purity of differentiated hPSCs, several methods of purification were developed (Ban et al., 2017). Classically, standard methods to purify specific cell types are based on expression of specific membrane markers, such as VCAM1 (Schwach and Passier, 2016) or SIRPA (Dubois et al., 2011) in the case of cardiomyocytes derived from PSCs. After a specific immunostaining, the purification step can be done using flow cytometry or magnetic cell sorting. Both methods necessitate preparation of single cell suspensions, a delicate step that can induce cell death. Moreover, these methods of sorting induce also an important cell loss that reduces drastically the number of available cells. Cells can also be genetically engineered to express a selection gene whose expression is driven by a cardiac specific promoter (Ritner et al., 2011). However, this strategy modifies the genome of the recipient cells and appears to be weakly compatible with clinical applications. Finally, in 2013, a Japanese team developed a purification method based on the specific metabolism of cardiomyocytes derived from hPSCs (Tohyama et al., 2013). Indeed, these cells have the capacity to use lactate as a source of energy unlike other type of cells. Thus, after replacement of glucose with lactate in culture medium, cardiomyocytes can be easily enriched because of the death of noncardiomyocytes cells. This method has been massively adopted and has the advantage to be highly compatible with large-scale production of hPSC-derived cardiomyocytes. Another very interesting strategy consists in depleting nonmyocytes from differentiated cells. Thus, several approaches to eliminate residual undifferentiated cells by cytotoxic antibodies (Tan et al., 2009) or genotoxic drugs (Smith et al., 2012) have been designed but were not for the moment transferred to the clinic.

To conclude, since the development of large-scale production, differentiation, and purification of hPSC-derived cardiomyocytes, it seems that the quest to find the best cell sources for cardiac cell therapy has come to a halt. However, several challenges need to be overcome because there is, for example, no real consensus concerning standardized protocols, making difficult a pertinent comparison of the results. Moreover, the "classic" Wnt-based protocols did not control the subtypes and the maturation stage of the cardiomyocytes obtained, two major drawbacks in the fields of regenerative medicine and disease modeling.

23.6 Cardiomyocytes Subtypes of hPSC-derived Cardiomyocytes

Although the diversity of cardiomyocytes is still under investigation (Cui et al., 2019), one can consider that the heart is mainly composed by three archetypes of cardiomyocytes, i.e., ventricular, atrial, and pacemaker cardiomyocytes. During development, they arise from the mesoderm and more specifically from the first and the second heart field (FHF and SHF). Progenitors from the FHF will then contribute to the formation of left ventricle and atria while progenitors from the SHF will contribute to the formation of right ventricle, outflow tract, and also atria. As knowledge on cardiac development increases, a growing number of studies was conducted to control the differentiation of hPSCs toward a specific subtype of cardiomyocytes (Protze et al., 2019; Zhao et al., 2020). As a consequence, different factors were discovered and, in this complex and still not fully understood differentiation process, retinoic acid was found to be one of the key factors that control the subtype specification of cardiomyocytes. Retinoic acid is a metabolite of vitamin A and binds to retinoic acid receptors (RARs) which are nuclear proteins that can mediate transcription of genes involved in stemness and differentiation (Mezquita and Mezquita, 2019). The treatment of differentiating hPSCs by the Activin/Nodal-based protocol with retinoic acid induced the differentiation of cardiomyocytes with an atrial signature based on the measure of their gene and protein expression and their electrophysiological features (Lee et al., 2017). In addition, the treatment of differentiating hPSCs with an inhibitor of retinoic acid pathway induces an enrichment of ventricular cardiomyocytes in the obtained population (Zhang et al., 2011). To conclude this section, the protocols of cardiac differentiation are continuously improved to give rise to the different subpopulation encountered in the adult heart tissue.

23.7 Maturation of hPSC-derived Cardiomyocytes

Even if the progress to obtain a pure population of hPSC-derived cardiomyocytes of a certain subtype has been noticeable, the major hurdle in this field is related to the immaturity of these cells. At every level, hPSC-derived cardiomyocyte greatly differs from the adult cell. Thus, their morphology (size, shape, and intracellular organization), electrophysiology, calcium handling properties, contractility, and metabolism have significant differences (for example see Table 1 of (Ge et al., 2019)), and hPSC-derived cardiomyocytes share more similarities with prenatal cardiomyocytes. Different attempts have been made to push maturation further (Karbassi et al., 2020). In conventional 2D cell culture, these assays have been demonstrated to be limited but are easier to be implemented in routine. They are essentially based on the use of modified culture medium supplemented with growth factors such as triiodothyronine (T3), corticosteroid (such as dexamethasone), and IGF1 (Birket et al., 2015) and/or the use of fatty acids instead of glucose as energy sources (Correia et al., 2017). A prolonged time of culture have also been proposed to improved cardiomyocytes maturation (Kamakura et al., 2013). However, the best to date results was achieved using a combined 3D engineered heart tissue approach (Karbassi et al., 2020). In this study, a fibrin hydrogel containing hPSC-derived cardiomyocytes and fibroblasts was casted in a specific system that allows electromechanical stimulation. Authors reported an outstanding maturation of cardiomyocytes with adult-like morphology (especially with the presence of T-tubules and well-organized sarcomeres and mitochondria), gene expression profile and electrophysiological properties. More interestingly, this EHT had a positive force-frequency relationship, a property of the adult cardiac muscle in which an increase in the frequency of contraction results also in an increase of the force generated by the cardiac muscle. This relationship is negative at birth and is also negative in non-mature EHT further demonstrating that a step has been taken with this study. It thus clearly appears that the best way to achieve a final maturation of hPSC-derived cardiomyocytes resides in the combination of several approaches, including biological, biochemical, mechanical, electrical and physical environment of the cells in order to mimic, as close as possible, the real cardiac tissue. A better understanding of postnatal heart growth will necessary benefits, but the knowledge is, for the moment, limited in this particular field.

23.8 Biomaterials for Stem Cell-based Cardiac Therapies

As mentioned, stem cell-based cardiac therapies aim at restoring some functionality in the damaged myocardium by repopulating them with new contractile elements. So far, however, clinical outcomes have shown limited, if any, efficacy, and even beyond the choice of the specific cell type that has been used, a major reason for these suboptimal results is likely the low rate of engraftment and high mortality of the transplanted cells into diseased heart. These two phenomena are caused by a mechanical leakage of cells, subsequently worsened by an interplay of biologic factors that include inflammation, ischemia due to poor vascularization of the injected areas, and apoptosis related to the detachment of anchorage-dependent cells from their ECM, so-called anoïkis. The recognition of these contributing factors provides a rationale for embedding cells into biomaterials that may better preserve cell survival and enhance cell engraftment after transplantation, consequently improving cardiac cell therapy compared with direct intramyocardial injection of isolated cells. Once the principle of using a scaffold has been adopted, some key parameters need to be determined, which include material composition, surface characteristics, methods of manufacturing, and cell seeding conditions. Safety of the overall construct with regard to inflammatory and immune responses, clinical applicability, cost of production, and manufacturing conditions compatible with regulatory guidelines for human applications are additional factors that need to be taken into consideration.

Myocardial tissue has a hierarchical structure, characterized by aligned cells embedded in 3D fibrous structure (Engelmayr et al., 2008). A scaffold with such a structure is crucial for cell survival, organotypic organization, and physiological function of seeded cardiac cells (Kitsara et al., 2017). In summary, the features of a therapeutically exploitable scaffold are the following: (i) the fibrous structure needs to be aligned to follow the anisotropic organization of cardiac tissue and to help the development of a functional cardiac syncytium; (ii) it should reproduce the elastic mechanical properties of the cardiac tissue. Basically, materials should have a stiffness of tens of kPa and a strain at break between 20-40% (Domian et al., 2017); (iii) it has to actively interact with cells at molecular level, to incorporate ligands and growth factors, to promote cell survival and cell-cell contacts; (iv) it must be biocompatible and biodegradable, thereby supporting both the appropriate cellular activity and the generation of cardiac tissues; and (v) finally, it needs to be macroporous with a high pore interconnection to favor cell infiltration, vascularization, and organotypic assembly. Different polymers and techniques have been used to fabricate micro-structured scaffolds to mimic the morphology of heart tissue. Natural polymers such as collagen, fibrin, alginate, or blended ECM are highly biocompatible and favor cell attachment and differentiation. However, they suffer from several drawbacks such as poor mechanical properties and rapid degradability which limit their utilization. Therefore, most of natural polymers require an extra cross-linking step to strengthen their structure, which affects not only their 3D structure and porosity, but also their biocompatibility due to the introduction of toxic chemical agents. Synthetic polymers exhibit interesting mechanical properties and their degradability can be tuned. Their major drawback is a poor interaction with cells and sometimes their immunogenicity. Recently, composite biomaterials associating a natural and a synthetic biopolymer have been developed, presenting better bioactivity and good mechanical properties (Silvestri et al., 2013). New trends in this domain consist in integrating therapeutic molecules within the scaffolds to stimulate cell survival, growth, and differentiation. Whatever the polymer and manufacturing techniques used (i.e., electrospinning, freeze drying, particles leaching, knitting or gas foaming, 3D printing), the use of engineered biomaterials for cell delivery provides a real solution to cell injection challenges for cell therapy purposes. In addition, biomaterials can provide mechanical support for injured cardiac tissue to reduce adverse post myocardial infarction remodeling. In this perspective, in addition to cellularized biomaterials, acellular biomaterials have been developed from a variety of materials to provide structural support for the damaged ventricle, stimulate cellular recruitment into the material, and/or induce adaptive tissue remodeling and endogenous repair mechanisms. Indeed, epicardial implantation of acellular biomaterials has been extensively explored across a number of clinically relevant animal models of ischemic injury, which demonstrated overall that surgically implanted biomaterials can attenuate infarct expansion and left ventricular remodeling (Efraim et al., 2017; Serpooshan et al., 2013).

23.9 New Perspectives in Cardiac Tissue Engineering: Making Hearts in Tubes

Without animal models, the entire field of biology, especially translational medicine and many of the current medical advances would probably not exist. However, in successfully predicting the effects of various therapeutic drugs in humans, animal models have been disappointing because of some limitations and marked interspecies differences in cardiac physiology and anatomy. Thus, the use of rodent model is able to predict toxicity in only approximately about 40% of cases (Olson et al., 2000). In terms of efficacy of a drug, the failure is even more dramatic with approximately 90% of drugs developed in animal models that fail in humans (Wall and Shani, 2008). This results in massive economical and human losses during the clinical trials. Thus, even if the field of EHT has been incredibly developed by the need to repair damaged myocardium, several research teams focus their attention to draw innovative platforms for drug efficacy and toxicity testing to fill the weaknesses of animal drug testing (Fine and Vunjak-Novakovic, 2017). To date, the most exciting progress have been performed independently by the teams of Kevin Costa (Li et al., 2018) and Kevin Parker (Macqueen et al., 2018). They each publish in 2018 their results related to the fabrication of "mini-hearts" constituted of a unique functional fluid-ejecting chamber fabricated using a matrix and hPSC-derived cardiomyocytes that mimics a ventricular chamber and allows the measurements of physiological parameters such as volumes and pressures, but also electrophysiological properties such as action potential and conduction velocity. In the study of Kevin Costa, the hPSC-derived cardiomyocytes were embedded into a collagen-based hydrogel casted around the balloon of a 6-Fr silicone Foley catheter (Li et al., 2018). After several days, the balloon was deflated to make room for a cavity mimicking the ventricular chamber. These exciting works are now developed in a pharmacological company (Novoheart Limited). Concerning the work of Kevin Parker, a microfibers scaffold was first prepared with a ventricular shape and then seeded with hPSC-derived cardiomyocytes (Macqueen et al., 2018). The whole EHTs were then mounted in a dedicated bioreactor that allows to control pressure variations with valves to the mini-heart. This specific set-up was acquired at the end of 2019 by Novoheart to improve its own system. These exciting and innovative studies paved the way toward the fabrication of a full heart in a tube.

23.10 Concluding Remarks

Cardiac cell therapy holds a real promise for improving function of the chronically failing myocardium. It aims at restoring some functionality in damaged myocardium by incorporating new contractile elements. However, clinical outcomes of patients included in cell therapy trials, carried out over the past 20 years, have not met enough the expectations raised by the preceding experimental studies. Taking in consideration all possible scenarios to obtain these suboptimal results, i.e., source of the cells, limited capacities of proliferation and differentiation of the grafted cells, cell retention, and survival after graft; over the last decade, the scientific and medical communities switch from mere cell therapy to a more composite tissue engineering to raise all current obstacles of cell therapy. Indeed, as mentioned in this chapter, all recent cell therapy studies using engineered tissues confirm the relevance of this hypothesis and emphasize that the dream of repairing a damaged heart appears much closer to reality than ever.

Key Points

- Cardiovascular diseases are the leading cause of disability, limiting the activity and eroding the quality of life of millions of both middle age adults and elderly each year.
- Stem cell-based therapies aim to repair injured heart through the cardiac delivery of cells to restore its functionality.
- The efficacy of the cell transplant is largely dependent on the engraftment rate, which in turn, requires cells to receive an adequate blood supply to survive.
- Injection-based cell delivery is not satisfactory, primarily because it involves a proteolytic dissociation of the cells that sets the stage for their apoptotic death.
- Tissue engineering and biomaterials for delivery and support of transplanted cells have begun to provide potential solutions that recapitulate the biological and physiological features of the human heart.

References

Agbulut, O., Menot, M.-L., Li, Z., Marotte, F., Paulin, D., Hagège, A.A., Chomienne, C., Samuel, J.-L., and Menasché, P. (2003). Temporal patterns of bone marrow cell differentiation following transplantation in doxorubicin-induced cardiomyopathy. Cardiovasc. Res. 58, 451-459.

Agbulut, O., Mazo, M., Bressolle, C., Gutierrez, M., Azarnoush, K., Sabbah, L., Niederlander, N., Abizanda, G., Andreu, E.J., Pelacho, B., et al. (2006). Can bone marrow-derived multipotent adult progenitor cells regenerate infarcted myocardium? Cardiovasc. Res. 72, 175-183.

Ban, K., Bae, S., and Yoon, Y. sup (2017). Current strategies and challenges for purification of cardiomyocytes derived from human pluripotent stem cells. Theranostics 7, 2067-2077.

Bel, A., Messas, E., Agbulut, O., Richard, P., Samuel, J.L., Bruneval, P., Hagège, A.A., and Menasché, P. (2003). Transplantation of autologous fresh bone marrow into infarcted myocardium: A word of caution. Circulation 108.

Beltrami, A.P., Barlucchi, L., Torella, D., Baker, M., Limana, F., Chimenti, S., Kasahara, H., Rota, M., Musso, E., Urbanek, K., et al. (2003). Adult cardiac stem cells are multipotent and support myocardial regeneration. Cell 114, 763-776.

Van Berlo, J.H., Kanisicak, O., Maillet, M., Vagnozzi, R.J., Karch, J., Lin, S.C.J., Middleton, R.C., Marbán, E., and Molkentin, J.D. (2014). C-kit+ cells minimally contribute cardiomyocytes to the heart. Nature 509, 337-341.

Birket, M.J., Ribeiro, M.C., Kosmidis, G., Ward, D., Leitoguinho, A.R., van de Pol, V., Dambrot, C., Devalla, H.D., Davis, R.P., Mastroberardino, P.G., et al. (2015). Contractile Defect Caused by Mutation in MYBPC3 Revealed under Conditions Optimized for Human PSC-Cardiomyocyte Function. Cell Rep. 13, 733-745.

Borow, K.M., Yaroshinsky, A., Greenberg, B., and Perin, E.C. (2019). Phase 3 DREAM-HF Trial of Mesenchymal Precursor Cells in Chronic Heart Failure: A Review of Biological Plausibility and Implementation of Flexible Clinical Trial Design. Circ. Res. 125, 265-281.

Boyd, R., Parisi, F., and Kalfa, D. (2019). State of the Art: Tissue Engineering in Congenital Heart Surgery. Semin. Thorac. Cardiovasc. Surg. 31, 807-817.

Burridge, P.W., Matsa, E., Shukla, P., Lin, Z.C., Churko, J.M., Ebert, A.D., Lan, F., Diecke, S., Huber, B., Mordwinkin, N.M., et al. (2014). Chemically defined generation of human cardiomyocytes. Nat. Methods 11, 855-860.

Cheung, D.Y., Duan, B., and Butcher, J.T. (2015). Current progress in tissue engineering of heart valves: Multiscale problems, multiscale solutions. Expert Opin. Biol. Ther. 15, 1155-1172.

Colatsky, T., Fermini, B., Gintant, G., Pierson, J.B., Sager, P., Sekino, Y., Strauss, D.G., and Stockbridge, N. (2016). The Comprehensive in Vitro Proarrhythmia Assay (CiPA) initiative âĂŤ Update on progress. J. Pharmacol. Toxicol. Methods 81, 15-20.

Correia, C., Koshkin, A., Duarte, P., Hu, D., Teixeira, A., Domian, I., Serra, M., and Alves, P.M. (2017). Distinct carbon sources affect structural and functional maturation of cardiomyocytes derived from human pluripotent stem cells. Sci. Rep. 7, 8590.

Cui, Y., Zheng, Y., Liu, X., Yan, L., Fan, X., Yong, J., Hu, Y., Dong, J., Li, Q., Wu, X., et al. (2019). Single-Cell Transcriptome Analysis Maps the Developmental Track of the Human Heart. Cell Rep. 26, 1934-1950.e5.

Dai, Y., and Foley, A.C. (2014). Tissue Engineering Approaches to Heart Repair. Crit. Rev. Biomed. Eng. 42, 213-227.

Desgres, M., and Menasché, P. (2019). Clinical Translation of Pluripotent Stem Cell Therapies: Challenges and Considerations. Cell Stem Cell 25, 594-606.

Domian, I.J., Yu, H., and Mittal, N. (2017). On Materials for Cardiac Tissue Engineering. Adv. Healthc. Mater. 6.

Dubois, N.C., Craft, A.M., Sharma, P., Elliott, D.A., Stanley, E.G., Elefanty, A.G., Gramolini, A., and Keller, G. (2011). SIRPA is a specific cell-surface marker for isolating cardiomyocytes derived from human pluripotent stem cells. Nat. Biotechnol. 29, 1011-1018.

Efraim, Y., Sarig, H., Cohen Anavy, N., Sarig, U., de Berardinis, E., Chaw, S.Y., Krishnamoorthi, M., Kalifa, J., Bogireddi, H., Duc, T.V., et al. (2017). Biohybrid cardiac ECM-based hydrogels improve long term cardiac function post myocardial infarction. Acta Biomater. 50, 220-233.

Engelmayr, G.C., Cheng, M., Bettinger, C.J., Borenstein, J.T., Langer, R., and Freed, L.E. (2008). Accordion-like honeycombs for tissue engineering of cardiac anisotropy. Nat. Mater. 7, 1003-1010.

Fine, B., and Vunjak-Novakovic, G. (2017). Shortcomings of Animal Models and the Rise of Engineered Human Cardiac Tissue. ACS Biomater. Sci. Eng. 3, 1884-1897.

Friedman, C.E., Nguyen, Q., Lukowski, S.W., Helfer, A., Chiu, H.S., Miklas, J., Levy, S., Suo, S., Han, J.D.J., Osteil, P., et al. (2018). Single-Cell Transcriptomic Analysis of Cardiac Differentiation from Human PSCs Reveals HOPX-Dependent Cardiomyocyte Maturation. Cell Stem Cell 23, 586-598.e8.

Ge, F., Wang, Z., and Xi, J.J. (2019). Engineered Maturation Approaches of Human Pluripotent Stem Cell-Derived Ventricular Cardiomyocytes. Cells 9, 9.

Gude, N.A., Firouzi, F., Broughton, K.M., Ilves, K., Nguyen, K.P., Payne, C.R., Sacchi, V., Monsanto, M.M., Casillas, A.R., Khalafalla, F.G., et al. (2018). Cardiac c-Kit biology revealed by inducible transgenesis. Circ. Res. 123, 57-72.

Hamdi, H., Planat-Benard, V., Bel, A., Puymirat, E., Geha, R., Pidial, L., Nematalla, H., Bellamy, V., Bouaziz, P., Peyrard, S., et al. (2011). Epicardial adipose stem cell sheets results in greater post-infarction survival than intramyocardial injections. Cardiovasc. Res. 91, 483-491.

Hamdi, H., Boitard, S.E., Planat-Benard, V., Pouly, J., Neamatalla, H., Joanne, P., Perier, M.-C., Bellamy, V., Casteilla, L., Li, Z., et al. (2013).

Efficacy of epicardially delivered adipose stroma cell sheets in dilated cardiomyopathy. Cardiovasc. Res. 99, 640-647.

El Harane, N., Kervadec, A., Bellamy, V., Pidial, L., Neametalla, H.J., Perier, M.-C., Lima Correa, B., Thiébault, L., Cagnard, N., Duché, A., et al. (2018). Acellular therapeutic approach for heart failure: in vitro production of extracellular vesicles from human cardiovascular progenitors. Eur. Heart J. 39, 1835-1847.

Hirt, M.N., Hansen, A., and Eschenhagen, T. (2014). Cardiac tissue engineering: State of the art. Circ. Res. 114, 354-367.

Huang, N.F., Serpooshan, V., Morris, V.B., Sayed, N., Pardon, G., Abilez, O.J., Nakayama, K.H., Pruitt, B.L., Wu, S.M., Yoon, Y. sup, et al. (2018). Big bottlenecks in cardiovascular tissue engineering. Commun. Biol. 1.

Kamakura, T., Makiyama, T., Sasaki, K., Yoshida, Y., Wuriyanghai, Y., Chen, J., Hattori, T., Ohno, S., Kita, T., Horie, M., et al. (2013). Ultrastructural maturation of human-induced pluripotent stem cell-derived cardiomyocytes in a long-term culture. Circ. J. 77, 1307-1314.

Kara, R.J., Bolli, P., Karakikes, I., Matsunaga, I., Tripodi, J., Tanweer, O., Altman, P., Shachter, N.S., Nakano, A., Najfeld, V., et al. (2012). Fetal cells traffic to injured maternal myocardium and undergo cardiac differentiation. Circ. Res. 110, 82-93.

Karbassi, E., Fenix, A., Marchiano, S., Muraoka, N., Nakamura, K., Yang, X., and Murry, C.E. (2020). Cardiomyocyte maturation: advances in knowledge and implications for regenerative medicine. Nat. Rev. Cardiol. 17.

Kawatou, M., Masumoto, H., Fukushima, H., Morinaga, G., Sakata, R., Ashihara, T., and Yamashita, J.K. (2017). Modelling Torsade de Pointes arrhythmias in vitro in 3D human iPS cell-engineered heart tissue. Nat. Commun. 8.

Kempf, H., Andree, B., and Zweigerdt, R. (2016). Large-scale production of human pluripotent stem cell derived cardiomyocytes. Adv. Drug Deliv. Rev. 96, 18-30.

Kervadec, A., Bellamy, V., El Harane, N., Arakélian, L., Vanneaux, V., Cacciapuoti, I., Nemetalla, H., Périer, M.C., Toeg, H.D., Richart, A., et al. (2016). Cardiovascular progenitor-derived extracellular vesicles recapitulate the beneficial effects of their parent cells in the treatment of chronic heart failure. J. Hear. Lung Transplant. 35, 795-807.

Kitsara, M., Agbulut, O., Kontziampasis, D., Chen, Y., and Menasché, P. (2017). Fibers for hearts: A critical review on electrospinning for cardiac tissue engineering. Acta Biomater. 48, 20-40.

Kompa, A.R., Greening, D.W., Kong, A.M., McMillan, P.J., Fang, H., Saxena, R., Wong, R.C.B., Lees, J.G., Sivakumaran, P., Newcomb, A.E., et al. (2020). Sustained subcutaneous delivery of secretome of human cardiac stem cells promotes cardiac repair following myocardial infarction. Cardiovasc. Res.

Laco, F., Woo, T.L., Zhong, Q., Szmyd, R., Ting, S., Khan, F.J., Chai, C.L.L., Reuveny, S., Chen, A., and Oh, S. (2018). Unraveling the Inconsistencies of Cardiac Differentiation Efficiency Induced by the GSK3β Inhibitor CHIR99021 in Human Pluripotent Stem Cells. Stem Cell Reports 10, 1851-1866.

Laflamme, M.A., Chen, K.Y., Naumova, A. V., Muskheli, V., Fugate, J.A., Dupras, S.K., Reinecke, H., Xu, C., Hassanipour, M., Police, S., et al. (2007). Cardiomyocytes derived from human embryonic stem cells in pro-survival factors enhance function of infarcted rat hearts. Nat. Biotechnol. 25, 1015-1024.

Lázár, E., Sadek, H.A., and Bergmann, O. (2017). Cardiomyocyte renewal in the human heart: insights from the fall-out. Eur. Heart J. 38, 2333-2342.3

Lee, J.H., Protze, S.I., Laksman, Z., Backx, P.H., and Keller, G.M. (2017). Human Pluripotent Stem Cell-Derived Atrial and Ventricular Cardiomyocytes Develop from Distinct Mesoderm Populations. Cell Stem Cell 21, 179-194.e4.

Li, R.A., Keung, W., Cashman, T.J., Backeris, P.C., Johnson, B. V., Bardot, E.S., Wong, A.O.T., Chan, P.K.W., Chan, C.W.Y., and Costa, K.D. (2018). Bioengineering an electro-mechanically functional miniature ventricular heart chamber from human pluripotent stem cells. Biomaterials 163, 116-127.

Lian, X., Hsiao, C., Wilson, G., Zhu, K., Hazeltine, L.B., Azarin, S.M., Raval, K.K., Zhang, J., Kamp, T.J., and Palecek, S.P. (2012). Robust cardiomyocyte differentiation from human pluripotent stem cells via temporal modulation of canonical Wnt signaling. Proc. Natl. Acad. Sci. U. S. A. 109, E1848-1857.

Liu, Y.W., Chen, B., Yang, X., Fugate, J.A., Kalucki, F.A., Futakuchi-Tsuchida, A., Couture, L., Vogel, K.W., Astley, C.A., Baldessari, A., et al. (2018). Human embryonic stem cell-derived cardiomyocytes restore function in infarcted hearts of non-human primates. Nat. Biotechnol. 36, 597-605.

Lockhart, M., Wirrig, E., Phelps, A., and Wessels, A. (2011). Extracellular matrix and heart development. Birth Defects Res. Part A - Clin. Mol. Teratol. 91, 535-550.

Ma, Y., Mouton, A.J., and Lindsey, M.L. (2018). Cardiac macrophage biology in the steady-state heart, the aging heart, and following myocardial infarction. Transl. Res. 191, 15-28.

Ma, Z., Koo, S., Finnegan, M.A., Loskill, P., Huebsch, N., Marks, N.C., Conklin, B.R., Grigoropoulos, C.P., and Healy, K.E. (2014). Three-dimensional filamentous human diseased cardiac tissue model. Biomaterials 35, 1367-1377.

Macqueen, L.A., Sheehy, S.P., Chantre, C.O., Zimmerman, J.F., Pasqualini, F.S., Liu, X., Goss, J.A., Campbell, P.H., Gonzalez, G.M., Park, S.J., et al. (2018). A tissue-engineered scale model of the heart ventricle. Nat. Biomed. Eng. 2, 930-941.

Madonna, R., Van Laake, L.W., Botker, H.E., Davidson, S.M., De Caterina, R., Engel, F.B., Eschenhagen, T., Fernandez-Aviles, F., Hausenloy, D.J., Hulot, J.S., et al. (2019). ESC working group on cellular biology of the heart: Position paper for Cardiovascular Research: Tissue engineering strategies combined with cell therapies for cardiac repair in ischaemic heart disease and heart failure. Cardiovasc. Res. 115, 488-500.

Mallapaty, S. (2020). Revealed: two men in China were first to receive pioneering stem-cell treatment for heart-disease. Nature 581.

Mantakaki, A., Fakoya, A.O.J., and Sharifpanah, F. (2018). Recent advances and challenges on application of tissue engineering for treatment of congenital heart disease. PeerJ 2018.

Mathur, A., Arnold, R., Assmus, B., Bartunek, J., Belmans, A., Bönig, H., Crea, F., Dimmeler, S., Dowlut, S., Fernández-Avilés, F., et al. (2017). The effect of intracoronary infusion of bone marrow-derived mononuclear cells on all-cause mortality in acute myocardial infarction: rationale and design of the BAMI trial. Eur. J. Heart Fail. 19, 1545-1550.

Menasché, P. (2020). Cardiac cell therapy: Current status, challenges and perspectives. Arch. Cardiovasc. Dis. 113, 285-292.

Mezquita, B., and Mezquita, C. (2019). Two opposing faces of retinoic acid: Induction of stemness or induction of differentiation depending on cell-type. Biomolecules 9.

Mohammadi, M.H., Obregón, R., Ahadian, S., Ramón-Azcón, J., and Radisic, M. (2017). Engineered Muscle Tissues for Disease Modeling and Drug Screening Applications. Curr. Pharm. Des. 23.

Mummery, C.L., Zhang, J., Ng, E.S., Elliott, D.A., Elefanty, A.G., and Kamp, T.J. (2012). Differentiation of human embryonic stem cells and induced pluripotent stem cells to cardiomyocytes: A methods overview. Circ. Res. 111, 344-358.

Nguyen, A.H., Marsh, P., Schmiess-Heine, L., Burke, P.J., Lee, A., Lee, J., and Cao, H. (2019). Cardiac tissue engineering: State-of-the-art methods and outlook. J. Biol. Eng. 13.

Olson, H., Betton, G., Robinson, D., Thomas, K., Monro, A., Kolaja, G., Lilly, P., Sanders, J., Sipes, G., Bracken, W., et al. (2000). Concordance of the toxicity of pharmaceuticals in humans and in animals. Regul. Toxicol. Pharmacol. 32, 56-67.

Oveissi, F., Naficy, S., Lee, A., Winlaw, D.S., and Dehghani, F. (2020). Materials and manufacturing perspectives in engineering heart valves: a review. Mater. Today Bio 5.

Pinto, A.R., Ilinykh, A., Ivey, M.J., Kuwabara, J.T., DâĂŹantoni, M.L., Debuque, R., Chandran, A., Wang, L., Arora, K., Rosenthal, N.A., et al. (2016). Revisiting cardiac cellular composition. Circ. Res. 118, 400-409.

Pouly, J., Hagège, A.A., Vilquin, J.T., Bissery, A., Rouche, A., Bruneval, P., Duboc, D., Desnos, M., Fiszman, M., Fromes, Y., et al. (2004). Does the functional efficacy of skeletal myoblast transplantation extend to nonischemic cardiomyopathy? Circulation 110, 1626-1631.

Protze, S.I., Lee, J.H., and Keller, G.M. (2019). Human Pluripotent Stem Cell-Derived Cardiovascular Cells: From Developmental Biology to Therapeutic Applications. Cell Stem Cell 25, 311-327.

Reardon, S. (2018). US government halts heart stem-cell study. Nature.

Richards, D.J., Li, Y., Kerr, C.M., Yao, J., Beeson, G.C., Coyle, R.C., Chen, X., Jia, J., Damon, B., Wilson, R., et al. (2020). Human cardiac organoids for the modelling of myocardial infarction and drug cardiotoxicity. Nat. Biomed. Eng. 4, 446-462.

Ritner, C., Wong, S.S.Y., King, F.W., Mihardja, S.S., Liszewski, W., Erle, D.J., Lee, R.J., and Bernstein, H.S. (2011). An engineered cardiac reporter cell line identifies human embryonic stem cell-derived myocardial precursors carissa ritner. PLoS One 6.

Rogozhnikov, D., O'Brien, P.J., Elahipanah, S., and Yousaf, M.N. (2016). Scaffold Free Bio-orthogonal Assembly of 3-Dimensional Cardiac Tissue via Cell Surface Engineering. Sci. Rep. 6.

Samak, M., and Hinkel, R. (2019). Stem Cells in Cardiovascular Medicine: Historical Overview and Future Prospects. Cells 8, 1530.

Schwach, V., and Passier, R. (2016). Generation and purification of human stem cell-derived cardiomyocytes. Differentiation 91, 126-138.

Senyo, S.E., Steinhauser, M.L., Pizzimenti, C.L., Yang, V.K., Cai, L., Wang, M., Wu, T. Di, Guerquin-Kern, J.L., Lechene, C.P., and Lee, R.T. (2013).

Mammalian heart renewal by pre-existing cardiomyocytes. Nature 493, 433-436.

Serpooshan, V., Zhao, M., Metzler, S.A., Wei, K., Shah, P.B., Wang, A., Mahmoudi, M., Malkovskiy, A. V., Rajadas, J., Butte, M.J., et al. (2013). The effect of bioengineered acellular collagen patch on cardiac remodeling and ventricular function post myocardial infarction. Biomaterials 34, 9048-9055.

Silvestri, A., Boffito, M., Sartori, S., and Ciardelli, G. (2013). Biomimetic materials and scaffolds for myocardial tissue regeneration. Macromol. Biosci. 13, 984-1019.

Smith, A.J., Nelson, N.G., Oommen, S., Hartjes, K.A., Folmes, C.D., Terzic, A., and Nelson, T.J. (2012). Apoptotic Susceptibility to DNA Damage of Pluripotent Stem Cells Facilitates Pharmacologic Purging of Teratoma Risk. Stem Cells Transl. Med. 1, 709-718.

Song, H., Zandstra, P.W., and Radisic, M. (2011). Engineered heart tissue model of diabetic myocardium. Tissue Eng. - Part A 17, 1869-1878.

Takahashi, K., Tanabe, K., Ohnuki, M., Narita, M., Ichisaka, T., Tomoda, K., and Yamanaka, S. (2007). Induction of Pluripotent Stem Cells from Adult Human Fibroblasts by Defined Factors. Cell 131, 861-872.

Takeda, M., Miyagawa, S., Fukushima, S., Saito, A., Ito, E., Harada, A., Matsuura, R., Iseoka, H., Sougawa, N., Mochizuki-Oda, N., et al. (2018). Development of in vitro drug-induced cardiotoxicity assay by using three-dimensional cardiac tissues derived from human induced pluripotent stem cells. Tissue Eng. - Part C Methods 24, 56-67.

Tan, H.L., Fong, W.J., Lee, E.H., Yap, M., and Choo, A. (2009). mAb 84, a cytotoxic antibody that kills undifferentiated human embryonic stem cells via oncosis. Stem Cells 27, 1792-1801.

Tang, W.H.W., and Alvarez, P. (2017). Recent Advances in Understanding and Managing Cardiomyopathy. F1000Research 6.

Thomson, J.A., Itskovitz-Eldor, J., Shapiro, S.S., Waknitz, M.A., Swiergiel, J.J., Marshall, V.S., and Jones, J.M. (1998). Embryonic stem cell lines derived from human blastocysts. Science 282, 1145-1147.

Tiburcy, M., Hudson, J.E., Balfanz, P., Schlick, S., Meyer, T., Liao, M.L.C., Levent, E., Raad, F., Zeidler, S., Wingender, E., et al. (2017). Defined engineered human myocardium with advanced maturation for applications in heart failure modeling and repair. Circulation 135, 1832-1847.

Tohyama, S., Hattori, F., Sano, M., Hishiki, T., Nagahata, Y., Matsuura, T., Hashimoto, H., Suzuki, T., Yamashita, H., Satoh, Y., et al. (2013). Distinct metabolic flow enables large-scale purification of mouse and human pluripotent stem cell-derived cardiomyocytes. Cell Stem Cell 12, 127-137.

Turnbull, I.C., Mayourian, J., Murphy, J.F., Stillitano, F., Ceholski, D.K., and Costa, K.D. (2018). Cardiac tissue engineering models of inherited and acquired cardiomyopathies. In Methods in Molecular Biology, (Humana Press Inc.), pp. 145-159.

Vagnozzi, R.J., Sargent, M.A., Lin, S.C.J., Palpant, N.J., Murry, C.E., and Molkentin, J.D. (2018). Genetic lineage tracing of Sca-1+ cells reveals endothelial but not myogenic contribution to the murine heart. Circulation 138, 2931-2939.

Valiente-Alandi, I., Schafer, A.E., and Blaxall, B.C. (2016). Extracellular matrix-mediated cellular communication in the heart. J. Mol. Cell. Cardiol. 91, 228-237.

Veldhuizen, J., Migrino, R.Q., and Nikkhah, M. (2019). Three-dimensional microengineered models of human cardiac diseases. J. Biol. Eng. 13.

Vunjak Novakovic, G., Eschenhagen, T., and Mummery, C. (2014). Myocardial tissue engineering: In vitro models. Cold Spring Harb. Perspect. Med. 4.

Wall, R.J., and Shani, M. (2008). Are animal models as good as we think? Theriogenology 69, 2-9.

Willems, E., Spiering, S., Davidovics, H., Lanier, M., Xia, Z., Dawson, M., Cashman, J., and Mercola, M. (2011). Small-molecule inhibitors of the Wnt pathway potently promote cardiomyocytes from human embryonic stem cell-derived mesoderm. Circ. Res. 109, 360-364.

Zhang, Q., Jiang, J., Han, P., Yuan, Q., Zhang, J., Zhang, X., Xu, Y., Cao, H., Meng, Q., Chen, L., et al. (2011). Direct differentiation of atrial and ventricular myocytes from human embryonic stem cells by alternating retinoid signals. Cell Res. 21, 579-587.

Zhao, M.T., Shao, N.Y., and Garg, V. (2020). Subtype-specific cardiomyocytes for precision medicine: Where are we now? Stem Cells.

24

Stem Cells for Red Blood Cell Production

Laurence Guyonneau-Harmand[1,2] **and Hélène Lapillonne**[1,3]

[1]Sorbonne Université, Inserm, Centre de recherche Saint-Antoine, CRSA, F-75012, Paris, France
[2]Établissements Français du Sang Île-de-France, Unité d'ingénierie et de thérapie cellulaire, Créteil, F-75012, Paris, France
[3]APHP, Hôpital Trousseau, département d'hématologie, F-75012, Paris, France
E-mail: laurence.harmand@sorbonne-universite.fr

24.1 Introduction

Red blood cells (RBCs) represent about 84% of the cells in the human body. If they are the most common type of blood cells, they are particular because they lack nucleus that is eliminated during RBC maturation.

The RBC cell membrane structure, i.e., the flexible and oval biconcave disk, provides the deformability and stability properties which are essential for RBC physiological functions such as withstanding the blood flow and going through the capillary network. The RBC cytoplasm contains hemoglobin, an iron-containing oxygen-transport metalloprotein which can bind oxygen as well as carbon dioxide and other gases. One RBC contains approximately 270 millions of hemoglobin molecules (D'Alessandro et al., 2017) that collectively have an oxygen-binding capacity of 1.34 mL O_2 per gram (Dominguez de Villota et al., 1981).

RBCs take up oxygen in the lungs, carry it throughout the body and release it into the tissues in exchange for carbon dioxide, which is eliminated by the lungs. Each round of circulation takes about 1 minute. Newly-formed RBCs will last and circulate for approximately 120 days before being eliminated by the spleen. While circulating, RBCs are continuously subjected to sources of reactive oxygen species (ROS), endogenous and exogenous. High amount of ROS results into cell structure damages and leads to functional impairments (Mohanty et al., 2014). To counteract ROS damages, RBCs have a fully operational antioxidant system involving both enzymatic antioxidants like catalase (Gonzales et al., 1984) or peroxiredoxin-2 (Nagababu et al., 2013) and nonenzymatic low–molecular-weight antioxidants (like glutathione or ascorbic acid).

With their particular shape, membrane structure, and durability, RBCs are natural carriers and have been extensively studied as drug delivery systems, but their use is currently restricted mostly due to storage lesion. RBCs have been used as drug carriers for decades via two main paths. The first one consists in drugs encapsulation i.e., a transient hypotonic shock that allows drug internalization followed by membrane resealing (Ihler and Tsang, 1987). The second one uses the RBC surface binding properties i.e., surface conjugation chemical (covalent linkage), affinity binding (antibodies) (Muzykantov et al., 1993), direct binding (non-covalent linkage), or nanoparticle attachment (Sahoo et al., 2016).

24.2 Erythropoiesis

Blood cell formation is continuous throughout life and relies on a rare population of cells designated as the hematopoietic stem cells (HSC or CD34+) that are endowed with multipotent and self-renewable properties. Blood production is massive: a human body generates 2.5 millions of mature red blood cells every second (Palis, 2014).

HSCs are present in the bone marrow (BM) and are located in a specific microenvironment called the niche where they can stay in a quiescent state or respond to different stimuli by proliferating and differentiating into progenitors with more and more restricted potentialities.

During erythropoiesis, HSCs successively differentiate into a multipotent common myeloid progenitor (CMP), a bipotent erythro-megakaryocytic progenitor (MEP), and at last a unipotent erythroid progenitor (BFU-E). Erythroid progenitors undergo early erythropoiesis, with the expression of the erythroid transcriptional program. During this step, cell proliferation is high, allowing a massive amplification of the erythroid compartment in the bone marrow to produce 2.10^{11} red cells per day in physiological conditions. Erythroid cells are dependent on several cytokines such as interleukin 3 (IL-3), stem cell factor (SCF), and erythropoietin (EPO) for their proliferation and survival. The proliferative capacity of the cells decreases as they mature into erythroid colony forming units (CFU-E). After this step, they differentiate successively into pro-erythroblasts, basophilic erythroblasts, polychromatophilic erythroblasts, and orthochromatic erythroblasts thus undergoing late erythropoiesis which is characterized by a decreased proliferation and the induction of terminal maturation with hemoglobin synthesis and nucleus condensation. At the orthochromatic step, once the chromatin has been highly condensed, cells enucleate in specific niches called "erythroblast islands" made by macrophages (Chasis and Mohandas, 2008). The enucleation consists in the exocytosis of the nucleus containing pyrenocyte. The enucleated cell is called a reticulocyte. Reticulocytes are characterized by the persistence of cytoplasmic organelles (mitochondria, ribosomes, Golgi apparatus, vacuoles). They leave the bone marrow, join the bloodflow and undergo several maturation steps such as the loss of organelles, changes in membrane permeability (Mairbaurl et al., 2000), cellular mobility (Mel et al., 1977), reduction of the cellular surface area (Waugh et al., 1997), changes in the organization of membrane proteins (Liu et al., 2010), and removal of the membrane components (Rieu et al., 2000). All along RBC life in the blood stream, the spleen will play an important role for the continuous remodeling of the membrane (Minetti et al., 2018). Eventually, they mature within 48 hours into functional RBCs, acquiring the typical biconcave disc shape, which improves their ability to deform and to carry O_2.

One of the major growth factors regulating erythropoiesis is EPO. Its receptor EPOR is expressed on the surface membrane of erythroid cells and the number of molecules decreases as maturation occurs (Broudy et al., 1991). Epo or EpoR knockout mice die from severe anemia at embryonic day 13, even though residual BFU-E and CFU-E could be found in the liver. Therefore, EPO and EPOR are essential for the complete differentiation of erythroid progenitors and are not required for HSC commitment into the erythroid lineage (Wu et al., 1995). At the molecular level, activation of this pathway occurs when EPO binds to EPOR. Binding induces homodimerization of the receptor and activation of the Janus kinase 2 (JAK2) by phosphorylation (Witthuhn et al., 1993). The signal transducer and activator of transcription 5 (STAT5) is recruited, phosphorylated, and dissociates from EPOR to dimerize. Activated STAT5 moves to the nucleus and induces the transcription of different target genes such as BCL-XL (Socolovsky et al., 1999) and c-MYC (Lord et al., 2000). BCL-XL expression is known to stimulate survival and to complete erythroid maturation (Socolovsky et al., 2001) while c-MYC is a well-known oncogene which promotes cell proliferation over differentiation (Pelengaris et al., 2002).

Development of the red blood cell lineage also requires a precise and well-coordinated expression of transcription factors (TFs) working synergistically or antagonistically. TFs control the cell fate decisions by regulating gene expression. It includes the GATA transcription factors (GATA1 and GATA2); basic helix-loop-helix factors (TAL1 also named SCL); multiple adaptors (Friend of GATA1: FOG-1 and LDB1); Krüppel-containing factors (KLF1); LMO2; STAT5; the Hypoxia-inducible factor-1 (HIF1), and others (Dore and Crispino, 2011). Some of these TFs are master TFs involved in the core erythroid network (CEN). It includes both DNA binding TFs such as GATA1, TAL1, KLF1 and non-DNA binding TFs such as LDB1 and LMO2. Of interest, knockout mice lacking any of these TFs from the CEN display an impaired erythropoiesis and die after gestation (Nandakumar et al., 2016).

The GATA family of TFs is one of the main regulators of erythropoiesis. GATA2 regulates proliferation and maintenance of the early hematopoietic progenitors, while GATA1 is expressed in mature cells and responsible for terminal maturation of several lineages such as erythrocytes, megakaryocytes, eosinophils and mast cells (Dore and Crispino, 2011). GATA1 works with other TFs and co-factors of the CEN such as LMO2, LDB1, TAL1 to directly activate erythroid genes and to transcribe a cohort of components required to assemble the autophagy machinery and to remove organelles in the late stage of erythroid maturation (Barminko et al., 2016). Moreover, GATA1 and KLF1 directly controls β-globin gene expression, heme synthesis, and iron procurement pathway, leading to the establishment of functional erythrocytes (Dore and Crispino, 2011)..

Aside JAK2 pathway and the CEN, miRNAs are also involved in the genesis of RBCs by affecting the genes or transcription factors involved in erythropoiesis (Azzouzi et al., 2015). MiR-24 plays an important role in proliferation and differentiation of erythroid progenitors (Shiozaki et al., 1992, Moritz et al., 1997), MiR-451a regulates positively terminal erythroid differentiation and protects RBCs against oxidant stress (Pase et al., 2009, Patrick et al., 2010, Yu et al., 2010) and MIR-320 regulates the expression of the transferrin receptor (Chen et al., 2008).

Erythropoiesis is thus the result of a series of complex mechanisms tightly regulated by internal and external stimuli.

24.3 From RBCs to Transfusion and Back to Blood Substitutes

Transfusion of red blood cells has a major place in the treatment of anemia, in an acute context after hemorrhage, during hematopoietic malignancies with bone marrow involvement or during chemotherapy for hematopoietic and solid malignancies (Tzounakas et al., 2017). It is also used to manage pregnancy-related complications and severe childhood anaemia (Tzounakas et al., 2018). Transfusions depend on donations and there is a global shortage of RBC units. In 2017, the worldwide blood need was about 305+/-9 millions while the worldwide blood supply was 272+/- 4 millions of blood product units (Roberts et al., 2019). To overcome this shortage, different approaches have been explored to have access to complementary RBC sources.

The development of "blood alternates" or "blood substitutes" able to act as artificial oxygen, carrying out the important function of transporting oxygen and carbon dioxide throughout the body has been done since many years to overcome issues of blood transfusion and blood supply. These molecules have in theory big advantages; a faster and better oxygen delivery, an immediate availability, a universal compatibility. Moreover, they are free of infectious agents and because of their synthetic origin they display a significant reduction in ischaemic, inflammatory and reperfusion injuries (Modery-Pawlowski et al., 2013, Haldar et al., 2019).

The most promising oxygen carriers are Hemoglobin-Based Oxygen Carriers (HBOCs) and Perfluorocarbon-Based Oxygen Carriers (PFBOCs) (Kim and Greenburg, 2004, Modery-Pawlowski et al., 2013, Palmer and Intaglietta, 2014). HBOCs are engineered or purified hemoglobin, optimized for oxygen delivery. But these compounds exhibit two problems; first their high oxygen affinity does not allow offloading to the tissues and second, their intravascular circulation half time is too short to be useful ($T_{1/2}$ <1.5 h). PFBOCs are aqueous emulsions of perfluorocarbon derivatives with high gas solubility. Their oxygen delivery capacity is less than 30% of normal blood. Thus, oxygen enriched air breathing would be required to ensure normal oxygen delivery to patients but would cause adverse effects on the lungs. Some of these oxygen-therapeutics have been through clinical studies but numerous and severe adverse effects have emerged in phase II and III clinical trials. In sum, those blood substitutes are of high interest but clearly need to be improved to avoid safety concerns.

Another approach is to mimic *ex vivo* the *in vivo* production of RBCs. This is nowadays a widely followed approach, but the challenge is to produce enough functional RBCs to meet transfusion requirements. In the following text, we will discuss the different approaches currently followed. The main sources of cells are human HSCs, human embryonic stem cells (hESCs) and human induced pluripotent stem cells (hiPSCs) (Shah et al., 2014) which are used to differentiate into RBCs via culture protocols mainly consisting of three steps; commitment, expansion and maturation (Shah et al., 2014). Once produced, cultured RBCs must be evaluated *in vivo* for survival, functionality, and safety.

24.3.1 Ex Vivo Generated RBCs From HSCs

HSCs can be found in bone marrow, cord blood, and peripheral blood but in a lower number in this latter case, a granulocyte colony–stimulating factor–based mobilization can increase this number and a leukapheresis can dramatically increase this number. Neildez-Nguyen et al. were the first to report the generation of RBCs from cord blood HSCs. The culture method consisted of three steps with sequential addition of cytokines i.e., Flt3 ligand, thrombopoietin, stem cell factor, erythropoietin, and insulin-like growth factor I. The erythroblasts produced displayed the expression of fetal hemoglobin (HbF) and low enucleation (4%). When injected into immunodeficient mice, they were able to mature and to widely enucleate *in vivo* (Neildez-Nguyen et al., 2002). Three years later, Giarratana et al. (Giarratana et al., 2005) used a similar protocol that included a co-culture step with the murine stromal cell line MS-5 cell line to mimic the BM niche (Figure 24.1). Under these conditions, mature RBCs were produced by either cord blood or leukapheresis HSCs; they survived normally post infusion in immunodeficient NOD/SCID mice. In 2011, the same lab produced and injected 10^{10} cultured red blood cells into a human (NCT0929266). The RBCs generated by culture exhibited *in vivo* a half-life of 26 days, which is very close to the expected native RBCs half-life. It was the first proof of principle for *ex vivo* generated RBC transfusion (Giarratana et al., 2011). CB HSCs showed a tenfold higher amplification than HSCs obtained from leukapheresis. The latter seems to be a promising source for manufacturing autologous RBCs, especially in cases of alloimmunization or rare blood groups. However their poor expansion potential restricts their clinical application (Heshusius et al., 2019).

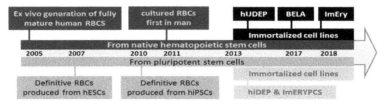

Figure 24.1   Chronology of cultured red blood cells generation.

24.3.2 Ex Vivo Generated RBCs From hESCs

hESCs are pluripotent stem cells derived from the inner cell mass of a blastocyst. They can be expanded *ex vivo* retaining their normal karyotype, pluripotency, and full-length telomeres (Amit et al., 2000) almost indefinitely (over 300 divisions), providing a potentially inexhaustible and donor-less source of cells for human therapy. In 2008, Lu et al. published a three-step protocol to differentiate and to mature hESCs into functional oxygen-carrying RBCs. The first step consisted in EB formation and hemangioblast precursor induction; the second step in hemangioblast expansion and the third step in erythroid differentiation and expansion. The protocol had two limitations i.e., the lack of enucleation and the expression of fetal hemoglobin (Lu et al., 2007)(Figure 24.1). Improved protocols (Lu et al., 2008, Qiu et al., 2008) were published but definitive erythropoiesis as well as robust expansion remained unattainable goals.

24.3.3 Ex Vivo Generated RBCs From hiPSCs

In 2006, Yamanaka developed a methodology to reprogram mature cells back to the inner cell mass pluripotent stem cells through forced expression of four transgenes (*Oct3/4*, *Sox2*, *c-Myc* and *Klf4*) (Takahashi et al., 2007). In principle, hiPSCs can be derived from any cell type. They are capable of self-renewal, large-scale expansion, and differentiation into all derivatives of the three germ layers *ex vivo* including germ cells. Their major advantage is to allow selection of donor's phenotype without posing the ethical dilemmas associated with the use of hESCs.

In 2010, Lapillonne et al. published a two-step cell culture protocol for the direct commitment of hiPSCs into definitive erythropoiesis (Figure 24.1). Unfortunately the amplification and enucleation rates were lower than those displayed by hESCs (10% vs. 66%). In addition, hemoglobin synthesis was blocked at the stage of HbF (Lapillonne et al., 2010). Erythroblasts from hiPSCs could achieve terminal maturation in terms of enucleation *in vitro* but complete maturation in terms of both enucleation and hemoglobin switch happened only *in vivo*, when cells were injected into immunodeficient mice. To improve *ex vivo* enucleation, Rouzbeh et al. proposed the inhibition of miR30-A (Rouzbeh et al., 2015).

Even if hESCs seem to be superior to hiPSCs in terms of expansion and enucleation rate, they do not reach the potential of cord blood HSCs, thus posing limitations for scale up production necessary for clinical applications.

24.4 The Cell Engineering Era

Available methods for *ex vivo* generation of RBCs from leukapheresis and cord blood HSCs do not yet provide a clinical and sustainable supply. The unlimited proliferation of pluripotent stem cells could have met transfusion purpose, but their low enucleation rate invalidates their use.

Alternative approaches have been therefore developed such as the generation of cell lines already committed into the erythroid lineage and capable of exponential amplification and terminal differentiation. The immortalization strategy should not be a bottleneck because RBCs are enucleated. For each of the stable lines of erythroid precursor presented below, the Hayflick limit has been over reached.

In 2013, Nakamura's laboratory developed the hiDEP and hUDEP cell lines (Figure 24.1). HiDEP are hematopoietic cells generated from hiPSC over-expressing the TAL1 gene and infected with a lentiviral vector containing a TET inducible expression system for HPV16-E6/E7. HuDEP are cord blood HSCs infected with a lentiviral vector containing a TET inducible expression system for HPV16-E6/E7 (Kurita et al., 2013). HiDEP and HUDEP cells can enucleate at a low level and produce hemoglobin with oxygen binding and dissociation abilities equivalent to red blood cells produced *ex vivo*. The karyotype of HUDEP-2 revealed a modal chromosome number of either 51 (range 49–53) (Moir-Meyer et al., 2018) or 50 (Vinjamur and Bauer, 2018), with a number of whole or partial trisomies and other chromosome anomalies, the slight variation between reports is likely due to the number of metaphases (clones) analyzed and the methods used.

The same year, Eto's laboratory developed ImERYPCs cell line (Figure 24.1). ImERYPCs was developed by transducing the c-MYC and BCL-XL genes into multipotent hematopoietic progenitor cells derived from hiPSC. The overexpression of c-MYC and BCL-XL enable sustained exponential self-replication of erythroblasts while turning off their differentiation.

ImERYPCs mature and recapitulate normal erythropoiesis. They displayed fetal-type hemoglobin and normal oxygen dissociation *in vitro* and showed high rates of enucleation following injection into immunodeficient mice (Hirose et al., 2013).

In 2017, Frayne's team (Trakarnsanga et al., 2017) created an immortalized adult erythroid line, named BEL-A (for Bristol Erythroid Line Adult), by introducing a Tet-inducible HPV16-E6/E7 expression system into bone marrow $CD34^+$ cells (Figure 24.1). The mean doubling time of the cells after day 100 in continuous culture was 20 hours. Morphological analysis of these cells showed that they are pro- to early basophilic erythroblasts. BEL-A RBCs had biochemical and structural features of normal erythropoiesis and developmental potential into functional, enucleated reticulocytes that survived *in vivo* expressing mainly hemoglobin A. Karyotype analysis of BEL-A by G-banding gave a modal chromosome number of 48, XX range 44–48 (50 metaphases) (Daniels et al., 2020). The most common trisomies were of chromosome 6 and 19 (46 and 45 metaphases respectively). Eleven metaphases had chromosome loss, most commonly monosomy of chromosome 8, 9, 12, 15, 17, 18 or 21. There is also a small number of partial chromosome losses.

BEL-A and HUDEP-2 were thoroughly compared and it appears that doubling time for BEL-A (18.7±2.0 hours) during the expansion phase was faster than HUDEP-2 (22.7±5.0 hours). Moreover, BEL-A recapitulated more closely adult erythropoiesis than HUDEP-2 (up to 40% of enucleation vs. less than 19% of enucleation), possibly due to a more optimal erythroid phenotype at the time of immortalization. These stable lines present outstanding results but as Daniels et al. pointed out, they are composed of several clones with abnormal karyotypes, which could lead to genetic and/or epigenetic changes (Daniels et al., 2020).

In 2018, Lee et al., generated the ImEry cell line, $CD71^+CD235a^+$ erythroblasts isolated from adult PB infected with lentiviruses containing c-MYC and BCL-XL. They produced the first proof of principle for the feasibility of scaling up immortalized erythroblast expansion in controlled bioreactors (i.e., stirred-bioreactor under controlled settings). The ImEry cell line displayed a doubling time of 17.9 hours. Of note, the team observed that serine was completely depleted by 71.5 hours in culture. Culturing in fed batch improved dramatically cell amplification, which implies that the culture could be limited by amino acid nutrients. ImEry exhibited identical O_2 binding capacity compared to adult erythroblasts (Lee et al., 2018). Amplification of ImEry cell line has already been well documented in stirred bioreactors. But at present, there is no published example of enucleation in bioreactor.

In sum, stable lines of erythroid precursors exhibit high heterogeneity of doubling times and enucleation potential. They also express greater amounts of HbF than native RBCs even if they mostly exhibited normal oxygen-binding capacity *in vitro*. Once karyotype abnormalities will be handled, immortalized erythroid precursors should allow scaled-up productions and transfusion purposes. Moreover, stable erythroblastic lines offer a much more manageable system for gene editing enabling the study of specific diseases (malaria) or to enhance RBC transfusion compatibility.

24.5 The Gene Editing Era

To generate a real representative model for malaria, Scully et al. knocked-out the DARC gene via CRISP/Cas 9 technology in an erythroblastic line generated with a Tet-inducible HPV16-E6/E7 expression system. This leads to the generation of erythroblastic lines from donors harboring unique genetic backgrounds or rare polymorphisms to better understand the genetic determinants of malaria susceptibility (Scully et al., 2019).

A common problem in transfused patients receiving lifelong and frequent transfusions such as thalassemia or sickle cell disease is alloimmunization which leads to difficulties in obtaining compatible RBC units and the development of delayed hemolytic. There are currently 30 identified blood group systems and over 300 RBC allo-antigens. Most alloimmunized patients develop antibodies belonging to the Rh, Kell, Duffy, Kidd, and MNS blood group systems (Peyrard et al., 2009). To hamper alloimmunization, Hawksworth et al. used CRISPR Cas 9 gene editing method to generate a single cell line deficient in multiple antigens responsible for the most common transfusion incompatibilities: ABO (Bombay phenotype), Rh (Rh null), Kell (K 0), Duffy (Fy null), GPB (S−s−U−) (Hawksworth et al., 2018). Unfortunately, they did not document their single cell line karyotype.

Gene editing and immortalized erythroblasts should allow, in the near future, the generation of *ex vivo* RBC disease models and the generation of universal safe RBCs.

24.6 Conclusion

During the last two decades, erythropoiesis *ex vivo* has been widely explored.

Stable erythroblast cell lines have been generated; they are capable of exponential amplification and to some extent enucleation but exhibit karyotype abnormalities. The never-ending scientific and technological evolution will provide the tools to improve safety and scale-up production to meet the transfusion purpose requirements.

In the near future, RBCs production *ex vivo* should definitely become the cornerstone of red cell substitution in the field of transfusion therapy. The first clinical applications should be in rare blood groups and/or chronic transfusion dependent patients; the second application should be as reliable drug carriers.

Concluding Remarks:

- Red blood cell transfusion is the oldest and most applied cellular therapy.
- The number of RBC transfusions is increasing year after year.
- An approach to produce RBC is the *in vitro* reproduction of normal erythropoiesis.
- Methods to produce RBCs *in vitro* use one of three major sources of stem cells i.e., hematopoietic stem/progenitor cells, human embryonic stem cells and induced pluripotent stem cells have been described, but allow low productions of mature RBCs.
- One transfusion unit = 2.10^{12} red blood cells
- Stable erythroblast cell lines have been generated. They are capable of exponential amplification and, to some extent, enucleation.
- Scaled-up productions are needed to meet transfusion purpose requirements.

References

Amit, M., Carpenter, M. K., Inokuma, M. S., Chiu, C. P., Harris, C. P., Waknitz, M. A., Itskovitz-Eldor, J. & Thomson, J. A. 2000. Clonally derived human embryonic stem cell lines maintain pluripotency and proliferative potential for prolonged periods of culture. *Dev Biol*, 227, 271-8.

Azzouzi, I., Moest, H., Wollscheid, B., Schmugge, M., Eekels, J. J. M. & Speer, O. 2015. Deep sequencing and proteomic analysis of the microRNA-induced silencing complex in human red blood cells. *Exp Hematol*, 43, 382-392.

Barminko, J., Reinholt, B. & Baron, M. H. 2016. Development and differentiation of the erythroid lineage in mammals. *Dev Comp Immunol*, 58, 18-29.

Broudy, V. C., Lin, N., Brice, M., Nakamoto, B. & Papayannopoulou, T. 1991. Erythropoietin receptor characteristics on primary human erythroid cells. *Blood*, 77, 2583-90.

Chasis, J. A. & Mohandas, N. 2008. Erythroblastic islands: niches for erythropoiesis. *Blood*, 112, 470-8.

Chen, S. Y., Wang, Y., Telen, M. J. & Chi, J. T. 2008. The genomic analysis of erythrocyte microRNA expression in sickle cell diseases. *PLoS One*, 3, e2360.

D'Alessandro, A., Dzieciatkowska, M., Nemkov, T. & Hansen, K. C. 2017. Red blood cell proteomics update: is there more to discover? *Blood Transfus*, 15, 182-187.

Daniels, D. E., Downes, D. J., Ferrer-Vicens, I., Ferguson, D. C. J., Singleton, B. K., Wilson, M. C., Trakarnsanga, K., Kurita, R., Nakamura, Y., Anstee, D. J. & Frayne, J. 2020. Comparing the two leading erythroid lines BEL-A and HUDEP-2. *Haematologica*, 105, e389-e394.

Dominguez de Villota, E. D., Ruiz Carmona, M. T., Rubio, J. J. & de Andres, S. 1981. Equality of the in vivo and in vitro oxygen-binding capacity of haemoglobin in patients with severe respiratory disease. *Br J Anaesth*, 53, 1325-8.

Dore, L. C. & Crispino, J. D. 2011. Transcription factor networks in erythroid cell and megakaryocyte development. *Blood*, 118, 231-9.

Giarratana, M. C., Kobari, L., Lapillonne, H., Chalmers, D., Kiger, L., Cynober, T., Marden, M. C., Wajcman, H. & Douay, L. 2005. Ex vivo generation of fully mature human red blood cells from hematopoietic stem cells. *Nat Biotechnol*, 23, 69-74.

Giarratana, M. C., Rouard, H., Dumont, A., Kiger, L., Safeukui, I., Le Pennec, P. Y., Francois, S., Trugnan, G., Peyrard, T., Marie, T., Jolly, S., Hebert, N., Mazurier, C., Mario, N., Harmand, L., Lapillonne, H., Devaux, J. Y. & Douay, L. 2011. Proof of principle for transfusion of in vitro-generated red blood cells. *Blood*, 118, 5071-9.

Gonzales, R., Auclair, C., Voisin, E., Gautero, H., Dhermy, D. & Boivin, P. 1984. Superoxide dismutase, catalase, and glutathione peroxidase in red blood cells from patients with malignant diseases. *Cancer Res*, 44, 4137-9.

Haldar, R., Gupta, D., Chitranshi, S., Singh, M. K. & Sachan, S. 2019. Artificial Blood: A Futuristic Dimension of Modern Day Transfusion Sciences. *Cardiovasc Hematol Agents Med Chem*, 17, 11-16.

Hawksworth, J., Satchwell, T. J., Meinders, M., Daniels, D. E., Regan, F., Thornton, N. M., Wilson, M. C., Dobbe, J. G., Streekstra, G. J., Trakarnsanga, K., Heesom, K. J., Anstee, D. J., Frayne, J. & Toye, A.

M. 2018. Enhancement of red blood cell transfusion compatibility using CRISPR-mediated erythroblast gene editing. *EMBO Mol Med*, 10.

Heshusius, S., Heideveld, E., Burger, P., Thiel-Valkhof, M., Sellink, E., Varga, E., Ovchynnikova, E., Visser, A., Martens, J. H. A., Von Lindern, M. & van den Akker, E. 2019. Large-scale in vitro production of red blood cells from human peripheral blood mononuclear cells. *Blood Adv*, 3, 3337-3350.

Hirose, S., Takayama, N., Nakamura, S., Nagasawa, K., Ochi, K., Hirata, S., Yamazaki, S., Yamaguchi, T., Otsu, M., Sano, S., Takahashi, N., Sawaguchi, A., Ito, M., Kato, T., Nakauchi, H. & Eto, K. 2013. Immortalization of Erythroblasts by c-MYC and BCL-XL Enables Large-Scale Erythrocyte Production from Human Pluripotent Stem Cells. *Stem Cell Reports*, 1, 499-508.

Ihler, G. M. & Tsang, H. C. 1987. Hypotonic hemolysis methods for entrapment of agents in resealed erythrocytes. *Methods Enzymol*, 149, 221-9.

Kim, H. W. & Greenburg, A. G. 2004. Artificial oxygen carriers as red blood cell substitutes: a selected review and current status. *Artif Organs*, 28, 813-28.

Kurita, R., Suda, N., Sudo, K., Miharada, K., Hiroyama, T., Miyoshi, H., Tani, K. & Nakamura, Y. 2013. Establishment of immortalized human erythroid progenitor cell lines able to produce enucleated red blood cells. *PLoS One*, 8, e59890.

Lapillonne, H., Kobari, L., Mazurier, C., Tropel, P., Giarratana, M. C., Zanella-Cleon, I., Kiger, L., Wattenhofer-Donze, M., Puccio, H., Hebert, N., Francina, A., Andreu, G., Viville, S. & Douay, L. 2010. Red blood cell generation from human induced pluripotent stem cells: perspectives for transfusion medicine. *Haematologica*, 95, 1651-9.

Lee, E., Lim, Z. R., Chen, H. Y., Yang, B. X., Lam, A. T., Chen, A. K., Sivalingam, J., Reuveny, S., Loh, Y. H. & Oh, S. K. 2018. Defined Serum-Free Medium for Bioreactor Culture of an Immortalized Human Erythroblast Cell Line. *Biotechnol J*, 13, e1700567.

Liu, J., Guo, X., Mohandas, N., Chasis, J. A. & An, X. 2010. Membrane remodeling during reticulocyte maturation. *Blood*, 115, 2021-7.

Lord, J. D., Mcintosh, B. C., Greenberg, P. D. & Nelson, B. H. 2000. The IL-2 receptor promotes lymphocyte proliferation and induction of the c-myc, bcl-2, and bcl-x genes through the trans-activation domain of Stat5. *J Immunol*, 164, 2533-41.

Lu, S. J., Feng, Q., Caballero, S., Chen, Y., Moore, M. A., Grant, M. B. & Lanza, R. 2007. Generation of functional hemangioblasts from human embryonic stem cells. *Nat Methods*, 4, 501-9.

Lu, S. J., Feng, Q., Park, J. S., Vida, L., Lee, B. S., Strausbauch, M., Wettstein, P. J., Honig, G. R. & Lanza, R. 2008. Biologic properties and enucleation of red blood cells from human embryonic stem cells. *Blood*, 112, 4475-84.

Mairbaurl, H., Schulz, S. & Hoffman, J. F. 2000. Cation transport and cell volume changes in maturing rat reticulocytes. *Am J Physiol Cell Physiol*, 279, C1621-30.

Mel, H. C., Prenant, M. & Mohandas, N. 1977. Reticulocyte motility and form: studies on maturation and classification. *Blood*, 49, 1001-9.

Minetti G., Achilli C., Perotti C., Ciana A. 2018 Continuous Change in Membrane and Membrane-Skeleton Organization During Development From Proerythroblast to Senescent Red Blood Cell. Front Physiol. 2018; 9: 286.

Modery-Pawlowski, C. L., Tian, L. L., Pan, V. & Sen Gupta, A. 2013. Synthetic approaches to RBC mimicry and oxygen carrier systems. *Biomacromolecules*, 14, 939-48.

Mohanty, J. G., Nagababu, E. & Rifkind, J. M. 2014. Red blood cell oxidative stress impairs oxygen delivery and induces red blood cell aging. *Front Physiol*, 5, 84.

Moir-Meyer, G., Cheong, P. L., Olijnik, A. A., Brown, J., Knight, S., King, A., Kurita, R., Nakamura, Y., Gibbons, R. J., Higgs, D. R., Buckle, V. J. & Babbs, C. 2018. Robust CRISPR/Cas9 Genome Editing of the HUDEP-2 Erythroid Precursor Line Using Plasmids and Single-Stranded Oligonucleotide Donors. *Methods Protoc*, 1.

Moritz, K. M., Lim, G. B. & Wintour, E. M. 1997. Developmental regulation of erythropoietin and erythropoiesis. *Am J Physiol*, 273, R1829-44.

Muzykantov, V. R., Smirnov, M. D. & Klibanov, A. L. 1993. Avidin attachment to red blood cells via a phospholipid derivative of biotin provides complement-resistant immunoerythrocytes. *J Immunol Methods*, 158, 183-90.

Nagababu, E., Mohanty, J. G., Friedman, J. S. & Rifkind, J. M. 2013. Role of peroxiredoxin-2 in protecting RBCs from hydrogen peroxide-induced oxidative stress. *Free Radic Res*, 47, 164-71.

Nandakumar, S. K., Ulirsch, J. C. & Sankaran, V. G. 2016. Advances in understanding erythropoiesis: evolving perspectives. *Br J Haematol*, 173, 206-18.

Neildez-Nguyen, T. M., Wajcman, H., Marden, M. C., Bensidhoum, M., Moncollin, V., Giarratana, M. C., Kobari, L., Thierry, D. & Douay, L. 2002. Human erythroid cells produced ex vivo at large scale differentiate into red blood cells in vivo. *Nat Biotechnol*, 20, 467-72.

Palis, J. 2014. Primitive and definitive erythropoiesis in mammals. *Front Physiol*, 5, 3.

Palmer, A. F. & Intaglietta, M. 2014. Blood substitutes. *Annu Rev Biomed Eng*, 16, 77-101.

Pase, L., Layton, J. E., Kloosterman, W. P., Carradice, D., Waterhouse, P. M. & Lieschke, G. J. 2009. miR-451 regulates zebrafish erythroid maturation in vivo via its target gata2. *Blood*, 113, 1794-804.

Patrick, D. M., Zhang, C. C., Tao, Y., Yao, H., Qi, X., Schwartz, R. J., Jun-Shen Huang, L. & Olson, E. N. 2010. Defective erythroid differentiation in miR-451 mutant mice mediated by 14-3-3zeta. *Genes Dev*, 24, 1614-9.

Pelengaris, S., Khan, M. & Evan, G. 2002. c-MYC: more than just a matter of life and death. *Nat Rev Cancer*, 2, 764-76.

Peyrard, T., Pham, B. N. & Rouger, P. 2009. [The red blood cell antigen terminologies]. *Transfus Clin Biol*, 16, 388-99.

Qiu, C., Olivier, E. N., Velho, M. & Bouhassira, E. E. 2008. Globin switches in yolk sac-like primitive and fetal-like definitive red blood cells produced from human embryonic stem cells. *Blood*, 111, 2400-8.

Rieu, S., Geminard, C., Rabesandratana, H., Sainte-Marie, J. & Vidal, M. 2000. Exosomes released during reticulocyte maturation bind to fibronectin via integrin alpha4beta1. *Eur J Biochem*, 267, 583-90.

Roberts, N., James, S., Delaney, M. & Fitzmaurice, C. 2019. The global need and availability of blood products: a modelling study. *Lancet Haematol*, 6, e606-e615.

Rouzbeh, S., Kobari, L., Cambot, M., Mazurier, C., Hebert, N., Faussat, A. M., Durand, C., Douay, L. & Lapillonne, H. 2015. Molecular signature of erythroblast enucleation in human embryonic stem cells. *Stem Cells*, 33, 2431-41.

Sahoo, K., Koralege, R. S., Flynn, N., Koteeswaran, S., Clark, P., Hartson, S., Liu, J., Ramsey, J. D., Pope, C. & Ranjan, A. 2016. Nanoparticle Attachment to Erythrocyte Via the Glycophorin A Targeted ERY1 Ligand Enhances Binding without Impacting Cellular Function. *Pharm Res*, 33, 1191-203.

Scully, E. J., Shabani, E., Rangel, G. W., Gruring, C., Kanjee, U., Clark, M. A., Chaand, M., Kurita, R., Nakamura, Y., Ferreira, M. U. & Duraisingh, M. T. 2019. Generation of an immortalized erythroid progenitor cell line from peripheral blood: A model system for the functional analysis of Plasmodium spp. invasion. *Am J Hematol*, 94, 963-974.

Shah, S., Huang, X. & Cheng, L. 2014. Concise review: stem cell-based approaches to red blood cell production for transfusion. *Stem Cells Transl Med*, 3, 346-55.

Shiozaki, M., Sakai, R., Tabuchi, M., Nakamura, T., Sugino, K., Sugino, H. & Eto, Y. 1992. Evidence for the participation of endogenous activin A/erythroid differentiation factor in the regulation of erythropoiesis. *Proc Natl Acad Sci U S A*, 89, 1553-6.

Socolovsky, M., Fallon, A. E., Wang, S., Brugnara, C. & Lodish, H. F. 1999. Fetal anemia and apoptosis of red cell progenitors in Stat5a-/-5b-/- mice: a direct role for Stat5 in Bcl-X(L) induction. *Cell*, 98, 181-91.

Socolovsky, M., Nam, H., Fleming, M. D., Haase, V. H., Brugnara, C. & Lodish, H. F. 2001. Ineffective erythropoiesis in Stat5a(-/-)5b(-/-) mice due to decreased survival of early erythroblasts. *Blood*, 98, 3261-73.

Takahashi, K., Tanabe, K., Ohnuki, M., Narita, M., Ichisaka, T., Tomoda, K. & Yamanaka, S. 2007. Induction of pluripotent stem cells from adult human fibroblasts by defined factors. *Cell*, 131, 861-72.

Trakarnsanga, K., Griffiths, R. E., Wilson, M. C., Blair, A., Satchwell, T. J., Meinders, M., Cogan, N., Kupzig, S., Kurita, R., Nakamura, Y., Toye, A. M., Anstee, D. J. & Frayne, J. 2017. An immortalized adult human erythroid line facilitates sustainable and scalable generation of functional red cells. *Nat Commun*, 8, 14750.

Tzounakas, V. L., Seghatchian, J., Grouzi, E., Kokoris, S. & Antonelou, M. H. 2017. Red blood cell transfusion in surgical cancer patients: Targets, risks, mechanistic understanding and further therapeutic opportunities. *Transfus Apher Sci*, 56, 291-304.

Tzounakas, V. L., Valsami, S. I., Kriebardis, A. G., Papassideri, I. S., Seghatchian, J. & Antonelou, M. H. 2018. Red cell transfusion in pae-diatric patients with thalassaemia and sickle cell disease: Current status, challenges and perspectives. *Transfus Apher Sci*, 57, 347-357.

Vinjamur, D. S. & Bauer, D. E. 2018. Growing and Genetically Manipulating Human Umbilical Cord Blood-Derived Erythroid Progenitor (HUDEP) Cell Lines. *Methods Mol Biol*, 1698, 275-284.

Waugh, R. E., Mckenney, J. B., Bauserman, R. G., Brooks, D. M., Valeri, C. R. & Snyder, L. M. 1997. Surface area and volume changes during maturation of reticulocytes in the circulation of the baboon. *J Lab Clin Med*, 129, 527-35.

Witthuhn, B. A., Quelle, F. W., Silvennoinen, O., Yi, T., Tang, B., Miura, O. & Ihle, J. N. 1993. JAK2 associates with the erythropoietin receptor and is tyrosine phosphorylated and activated following stimulation with erythropoietin. *Cell*, 74, 227-36.

Wu, H., Liu, X., Jaenisch, R. & Lodish, H. F. 1995. Generation of committed erythroid BFU-E and CFU-E progenitors does not require erythropoietin or the erythropoietin receptor. *Cell*, 83, 59-67.

Yu, D., Dos Santos, C. O., Zhao, G., Jiang, J., Amigo, J. D., Khandros, E., Dore, L. C., Yao, Y., D'Souza, J., Zhang, Z., Ghaffari, S., Choi, J., Friend, S., Tong, W., Orange, J. S., Paw, B. H. & Weiss, M. J. 2010. miR-451 protects against erythroid oxidant stress by repressing 14-3-3zeta. *Genes Dev*, 24, 1620-33.

25

Prospectives for Therapy With Stem Cells in Skeletal Muscular Diseases

Negroni Elisa, Butler-Browne Gillian, and Mouly Vincent

Sorbonne Université, Inserm, Institut de Myologie, Centre de Recherche en Myologie, F-75013 Paris, France
E-mail: elisa.negroni@sorbonne-universite.fr

25.1 Introduction

The principle for cell therapy in skeletal muscle was demonstrated in the 80s when normal myoblasts restored dystrophin in a mouse model for Duchenne Muscular Dystrophy (DMD). However, subsequent clinical trials in DMD patients using myoblasts failed to bring any clinical benefit to the patients, but they revealed many unexpected difficulties that hampered cell therapy for muscular dystrophies. Since then, stem cells with a myogenic potential have been described, and proposed as candidates for stem cell therapy in skeletal muscle. Up to now, the only clinical trials that showed some clinical benefit for the patients have concerned pathological situations, where a limited volume of muscles had to be targeted, such as in Oculo-Pharyngeal Muscular Dystrophy (OPMD). This chapter describes the recent advances in pre-clinical cell therapy approaches, with a special attention on human candidates potentially attractive for human clinical trials, and discusses future perspectives for cell therapy in skeletal muscle.

25.2 What Is Expected From a Good Cell Candidate?

Cell therapy in skeletal muscle is based on the delivery of precursor cells to muscle tissue that will contribute to muscle regeneration, either enhancing tissue repair and/or bringing a missing protein. Two strategies for obtaining therapeutic cells can be employed: (1) they can be obtained from a healthy donor (allogeneic graft); the cells will express the wild-type protein but will also induce an immune response when grafted in the patient, requiring an immune suppression treatment; (2) they can be obtained from the patient himself (autologous graft); the cells will be genetically modified to express the missing/mutated protein, avoiding immune suppression (unless the reintroduced protein becomes immunogenic). A third possible option (3) can be envisaged for some muscular dystrophies (MD) with limited muscle targets, such as Oculo-Pharyngeal Muscular Dystrophy (OPMD), where the use of non-modified cells in autologous but heterotopic graft can be envisaged, and this will be further herein detailed.

The choice of cell candidates should be based on precise characteristics to ensure sufficient efficiency:

1. The cell candidate must have been clearly identified and characterized in human, since the transfer from murine models to human is not always trivial, and markers are not always conserved between species.
2. If the cells are to be isolated from skeletal muscle, they must be numerous enough, since removing a large biopsy that will reduce mobility is obviously inappropriate, particularly if the cells are isolated from the patient (autologous graft). If the cells can be isolated from other human tissues, particularly in a non-invasive manner such as from blood, which may be preferable, those cells must present a demonstrated myogenic potential.
3. The cell candidate will have to be amplified at some stage, and the conditions for amplification should be suited to clinical conditions in order to obtain a sufficient amount of progenitors for regenerating the whole muscle target.
4. The cell candidate should be preferentially genetically modifiable in order to correct the consequences of a missing or modified gene, so that autologous cell therapy, the best-adapted strategy since it does not require long-term immune suppression, can be envisaged.
5. Although many pluripotent cells have been described, clinical requirements imply a stable myogenic fate. The final fate of delivered cells is essential in order to avoid differentiation into an inappropriate cell type. For instance, research on induced pluripotent stem (iPS) cells has been rapidly growing and personalized cell therapy using iPS cells has started to be considered as a realistic option (Schweitzer et al., 2020). However, the use of these cells in skeletal muscle cell therapy will only be possible when the potential differentiation in any other undesired cell type is fully mastered.

6. Finally, an ideal candidate should also be able to restore the muscle stem cell pool located between the basal lamina and the muscle fibers (aka satellite cells) in order to amplify the therapeutic effect in the next rounds of regeneration.

25.3 Candidate Cells

Choosing a candidate for stem cell therapy is not trivial: it must have a myogenic potential, but also be adapted to the target, and the mode of delivery will depend on the size of this target. Concerning the route of administration, two ways have been explored, intra-muscular or systemic delivery. Intra-muscular injection has been widely used for myoblasts, and seems appropriate if the muscle target to be treated is limited, due to the limited migratory potential of myoblasts. For a broader distribution to the whole body or a range of muscles, systemic or loco-regional delivery via the blood flow is better adapted. This requires that therapeutic cells can extravasate and be distributed within the target tissue.

25.3.1 Intramuscular Delivery

25.3.1.1 Myoblasts

Since myoblasts, derived from satellite cells, are the physiological precursors of muscle fibers, and they can be isolated from humans in clinical conditions using identified and validated surface markers such as CD56 among others, they are obvious candidates for stem cell therapy. The first article assessing myoblasts transfer therapy was published by the group of T. Partridge in 1989, and used cultured myoblasts in order to restore dystrophin expression in mdx mice (a model for DMD, in which dystrophin is absent) (Partridge et al., 1989). This first demonstration was followed by clinical trials in DMD patients. Few dystrophin-positive fibers were detected, even though some of them were revertant fibers expressing a truncated form of the protein (probably by natural exon-skipping, which is common in mdx model and can be observed in DMD patients as well (Fanin et al., 1993)). Clinical trials based on intra-muscular transplantation of allogeneic myoblasts (derived from an immunologically related donor, often a sibling) associated with immunosuppression by cyclosporine A (Gussoni et al., 1992) or cyclophosphamide (Karpati et al., 1993) did not result in clinical benefit for the patients (Partridge, 2000; Cossu and Sampaolesi, 2007), and dystrophin expression was very limited in treated muscles.

Following these failures, different limiting factors have been identified in myoblasts cell therapy in murine models. A precocious cell death occurs in the first few hours after myoblasts transplantation in mouse muscles, resulting in a loss of over 90% 24 hours post-injection using immortalized cells (Beauchamp et al., 1999). Neutrophils, natural killer cells and macrophages do not seem to be involved in this process (Sammels et al., 2004). Implanted myoblasts die by both apoptosis and necrosis (Skuk et al., 2003), suggesting more than one mechanism for this early cell death. This huge cell death is accompanied by a limited proliferation in situ. Human primary myoblasts seem to be more resistant, as observed in implantation into regenerating muscle of immunodeficient recipients, but the proliferation of myoblasts post transplantation is still limited and can only compensate for the precocious loss (Riederer et al., 2012). Moreover, grafted myoblasts migration is very limited within mouse (Lipton and Schultz, 1979; Rando et al., 1995) or monkey (Skuk et al., 1999; Quenneville et al., 2007) muscle. Extracellular matrix (ECM) remodeling is probably involved, since myoblasts dispersion is improved if the composition of ECM in laminins is increased (Silva-Barbosa et al., 2008) or if the secretion of matrix metalloproteases such as MMP-1, MMP-2 or MMP-9 is upregulated (El Fahime et al., 2002; Gargioli et al., 2008; Morgan et al., 2010; Pan et al., 2015). A systemic delivery could increase the dispersion of myoblasts within the body, but myoblasts are not able to cross the endothelial blood vessel barrier (Dellavalle et al., 2007).

Elegant studies have been carried out on quiescence-related grafting capacities of murine satellite cells. Freshly isolated satellite cells or isolated myofibers associated to few satellite cells participate very efficiently to the regeneration of host's muscle (Collins et al., 2005; Montarras et al., 2005; Kuang et al., 2007). Recently, these studies have been translated to humans (X. Xu et al., 2015; Garcia et al., 2018), confirming the ability of human satellite cells (freshly isolated or still associated to myofibers) to participate to mouse regeneration, even if less efficiently compared to mouse studies. The variable efficiency in grafting, still much higher than for myoblasts amplified in vitro, probably reflects the state of stem cells, from full quiescence to alert (Rodgers et al., 2014) or early activated. Even if these experiences cannot be easily adapted to a clinical context, they have opened up new avenues of research focusing on early activation and maintenance of quiescence that will be presented in the last section on this chapter.

Myoblasts are often exhausted in dystrophic conditions (such as DMD), due to repeated cycles of degeneration and regeneration which mobilize them repeatedly (Webster and Blau, 1990; Renault et al., 2002). This hampers their isolation, modification, and amplification for autologous strategies. Furthermore, myoblasts, irrespective of their origin, cannot be injected systemically, which is the ideal route to target large amounts of tissue (Galli et al., 2018). These reasons have encouraged the investigation and the identification of other types of progenitor/stem cells, distinct from satellite cells but exhibiting myogenic capacities, in a search for candidates with a better performance/efficacy to restore muscle function in MD.

25.3.2 New Candidates for a Systemic Delivery

25.3.2.1 BMSC and SP cells

Initially promising experiments, as well as the description of a DMD patient who received a bone-marrow transplant and presented a mild phenotype, have suggested that bone marrow stem cells (BMSCs) could participate to muscle regeneration in the mouse (Ferrari et al., 1998) by a nonconventional conversion to myogenesis (Xynos et al., 2010). However, the frequency of these fusion events is very low and equivalent to that of revertant fibers (Gussoni et al., 1999; Ferrari, Stornaiuolo and Mavilio, 2001).

Subpopulations of cells derived from BMSCs, and among them, the Side Population cells (SP cells), identified by their ability to exclude Hoechst 33342, have been assessed in vivo (Goodell et al., 1996). But again the percentage of donor-derived myonuclei reached only 1–2% (Torrente et al., 2001; Gavina et al., 2006), a number increased (up to 5–8%) after muscle damage or in dystrophic conditions following exercise (Bachrach et al., 2006). These numbers are still not compatible with a clinical benefit for the patient. However, these SP cells can favor the engraftment of myoblasts by secreting factors enhancing myoblasts proliferation and dispersion in vivo (Motohashi et al., 2008). More recently a subpopulation of SP cells expressing MCAM has been isolated from human foetal muscle (Lapan et al., 2012). These cells represent a very minor population, and further studies should demonstrate if these cells possess an advantage compared to human satellite cells/myoblasts, if they still exist in adult muscle and if their potential is maintained after amplification. Likewise, their potential systemic delivery should be also addressed.

25.3.2.2 CD133+ cells

CD133 is expressed in a minor subpopulation of hematopoietic stem cells, but are also present, although rare, within skeletal muscle. A comparison of the regenerative potential in vivo of muscle-derived CD133+ cells to that of satellite cells showed that CD133+ cells are more efficient than myoblasts to participate to mouse muscle regeneration and to restore the satellite cell pool (Negroni et al., 2009). Purifying these cells from blood would avoid a muscle biopsy in already devastated muscles. These blood-derived CD133+ cells can be delivered by a systemic route and give rise to multiple progeny, such as endothelial and skeletal myogenic lineages (Benchaouir et al., 2007; Shi et al., 2009). An autologous phase I clinical trial transplanting muscle-derived CD133+ cells intra-muscularly has been performed in DMD patients. Safety has been investigated and no side effects were observed (Torrente et al., 2007). However, this trial was not followed by successive attempts.

25.3.2.3 Mesoangioblasts/Pericytes

Mesoangioblasts are mesodermal progenitors initially isolated from dorsal aorta that have been identified in avian and mammalian species (De Angelis et al., 1999; Minasi et al., 2002; Cossu and Bianco, 2003). Human mesoangioblasts, suspected to correspond to cells previously defined as pericytes (Cossu and Sampaolesi, 2007), express markers such as nerve/glial antigen 2 (NG2) and alkaline phosphatase (ALP), and can proliferate and spontaneously differentiate *in vitro* into myotubes. More importantly, they can extravasate from blood vessels and migrate into recipient's muscle, and this is enhanced by TNF-α and SDF-1 (Galvez et al., 2006). In vivo intra-arterial injection of mouse mesoangioblasts in dystrophic mice (Sampaolesi et al., 2003) or dogs (Sampaolesi et al., 2006), resulted in a spectacular amelioration in muscle structure and function. A phase I/IIa clinical trial with intra-arterial transplantation of mesoangioblasts in DMD patients (obtained from HLA-matched siblings) has been recently conducted. This study was relatively safe, but a low engraftment and no functional improvement was observed, possibly due to the age of the patients and the already degraded state of their muscles by the time expanded mesoangioblasts could be infused (Cossu et al., 2015).

25.4 Therapeutic Application and Future Optimization

Muscle dystrophies are a very heterogeneous group of diseases ranging from dystrophies involving most of the musculature (but never all of it) such as DMD, to diseases with a very limited number of targets, such as OPMD. Cell candidates should be adapted to each type of dystrophy. For dystrophies such as DMD, progenitors should be delivered systemically, attracted to degenerative sites, crossing vessels and colonizing injured muscle tissue, and should be able then to differentiate. Mesoangioblasts respond to these requirements and are obvious candidates for DMD (Cossu et al., 2015). Candidates should also be amplifiable in clinical conditions, taking into account economic parameters. Autologous progenitors should be preferred to avoid immune suppression of the patients, since even immune privileged cells may not keep this property once they engage into the myogenic lineage and differentiation.

At the other end of the spectrum, amplified myoblasts may be adapted to focused treatments of localized forms of muscular dystrophy such as OPMD, a late onset dystrophy with eyelid and pharyngeal muscles as primary targets, and where muscles spared by the disease represent potential sources for autologous progenitors, as long as the effect of the mutation, present in all cells but with late onset consequences in skeletal muscle, is not rapidly occurring after transplantation. While myoblasts isolated from the affected cricopharyngeal muscle present defects in proliferation, those isolated from spared muscles of OPMD patients do not present this defect (Périé et al., 2006). A phase I clinical trial, using autologous myoblasts isolated from non-clinically affected muscles, expanded in vitro and transplanted into upper pharyngeal muscle of patients, has proven feasibility and safety. A dose-dependant functional improvement has been observed (Périé et al., 2014).

The optimization of cell therapy requires further upstream research, and recent data obtained using human myoblasts transplanted into cryo-injured muscle of immunodeficient mice (Figure 25.1) has given more insights concerning the behavior of human myoblasts in vivo: 72 hours post injection, almost all myoblasts have started to differentiate, limiting their migration; at 5 days post-transplantation, cell dispersion has ended (Riederer et al., 2012). These results suggest that a modulation of the environment favoring proliferation (without preventing at long term the differentiation) and migration can significantly increase the regenerative capacity.

Moreover, mastering the quiescence cues of muscle stem cells has been of fundamental importance since the superiority of fresh non-activated satellite cells has been shown in cell grafting experiments (Montarras et al., 2005) and many efforts have been made in that direction, to understand how the satellite cell niche contributes to maintain satellite cells in quiescence (Baghdadi et al., 2018a; Baghdadi et al., 2018b; Sampath et al., 2018; Verma et al., 2018). Regarding the microenvironment, one of the first evidence showing that the microenvironment is critical for stem cell specification was given few years ago, showing that human mesenchymal stem cells were able to adopt a specific phenotype after being cultured on matrices of

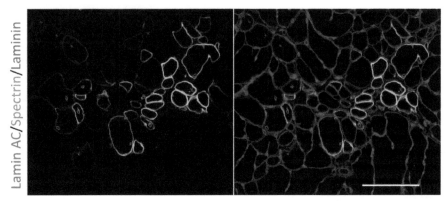

Lamin AC/Spectrin/Laminin

Figure 25.1 Grafting human myoblasts. The optimization of cell therapy requires upstream transplantation research. The behavior of human myoblasts *in vivo* can be monitored using immunodeficient mice. Human myoblasts injected into a muscle of an immunodeficient mouse fuse along with mouse fibers during a regeneration process triggered by a cryo damage. Human cells are identified with a Lamin A/C staining for human nuclei (**red**), human spectrin protein produced by human myoblasts is stained in green and laminin surrounding muscle fibers is stained in blue. Scale bar: 100 μm.

specific stiffness (Engler et al., 2006). 12kPa stiffer elasticity (that mimic muscle elasticity) drives human mesenchymal stem cells toward a myogenic commitment. Afterwards, Gilbert et al. showed that mouse satellite cells cultured on 12kPa soft hydrogel were able to self-renew in vitro, showing less differentiation, and contributed robustly to muscle regeneration when transplanted in vivo (Gilbert et al., 2010). More recently, soft biopolymers have been shown to be able to reduce the expression of differentiation markers in human myoblasts, sustaining stem cell marker expression (Monge et al., 2017). Further work investigating the importance of stemness in human cell grafting should be addressed to justify the use of these specific culture conditions for human clinical trials. In human cells, similar results could be obtained by modulating specific molecular pathways. The transient inhibition of the p38-signaling pathway prevents differentiation of human satellite cells, thus enabling their expansion (Charville et al., 2015). Moreover such treatment promotes a better engraftment and an improved colonisation of the host satellite cell niche. The p38-signaling pathway seems to be of particular relevance for muscle stem cells, since its transient inhibition along with culture on soft hydrogels (Cosgrove et al., 2014) or combined with the modulation of the FGF receptor signaling can ameliorate age-associated self-renewal defects in mouse satellite cells and could open up new strategies for localized autologous myoblast cell therapy for the elderly (Bernet et al., 2014).

Recent studies have also given insights in either how cells can be modulated at the epigenetic level to achieve a better amplification in vitro of myogenic cells (Judson et al., 2018) or how targeting metabolic changes occurring during the satellite cell's switch from quiescence towards differentiation could promote self-renewal (Theret et al., 2017), proliferation and prevent premature differentiation (L'honoré et al., 2018).

We cannot conclude without mentioning the research concerning human embryonic stem (hES) cells and human hiPS cells, whose interest in the field of the regenerative medicine has been increasing worldwide. hES cells retain the potential to differentiate into cells belonging to ectoderm, mesoderm, and endoderm lineages, and are poorly immunogenic (for a recent review on this topic, see (Haworth and Sharpe, 2020)). However, they may loose this immune privilege and be rejected once fully differentiated into the myogenic lineage, and may turn tumorigenic, leading to formation of teratoma (Bodnar et al., 2004). Multipotent mesenchymal precursors derived from hES cells have been transplanted and the formation of few myofibers was observed (Barberi et al., 2007).

hiPS cells more interestingly can be generated from adult dermal fibroblasts by transduction of four defined transcription factors: Oct3/4, Sox2, Klf4, and c-Myc (Takahashi et al., 2007). iPS can be driven to myogenesis, e.g., by a conditional expression of Pax7 (Darabi et al., 2012; Kim et al., 2017a; Kim et al., 2017b) or MyoD (Tedesco et al., 2012) and recently, myogenic differentiation was also induced in hiPS in the presence of defined factors (Xu et al., 2013, Mazaleyrat et al., 2020). Recently, the group of Giulio Cossu reprogrammed fibroblasts from dystrophic patients to generate hiPS cells and derive mesoangioblast-like cells. These cells, transduced with the myogenic regulator MyoD gene, showed a robust myogenic differentiation in vivo also after systemic delivery (Tedesco et al., 2012), and a very recent report suggests some immune tolerance after induction of hiPS cells by ectopic expression of MyoD (Benabdallah et al., 2020). hiPS cells represent a major improvement over hES cells since they can be isolated from the patient and thus be administered autologously. However, many issues, including a comprehensive understanding of the processes that regulate their differentiation towards the myogenic lineage (Xi et al., 2020) and overall safety, but also the cost of such personalized medicine as compared to other innovative therapeutic strategies, need to be addressed before they can be used in clinical trials.

25.5 Concluding Remarks

Muscle dystrophies are a very heterogeneous group of diseases: While some dystrophies involve most of the musculature such as DMD, others affect a very limited number of muscles, such as OPMD. Cell candidates should be adapted to each type of dystrophy. Myoblasts seem appropriate if the muscle target to be treated is limited and can be reached by local delivery (i.e., OPMD); for dystrophies such as DMD, progenitors should be delivered systemically, attracted to degenerative sites, crossing vessels and colonizing injured muscle tissue, and finally differentiate into new muscle fibers.

Autologous transfer is always simpler and less hazardous than allogeneic, which requires immune suppression. Autologous cells, whatever the mutation, must be ideally genetically corrected. However, the progresses made in genetic correction, including gene editing, will facilitate these therapeutic approaches.

Other therapeutic strategies are being developed, at the mRNA level such as exon skipping, or at the DNA level such as gene delivery or gene editing and all its derivatives. These approaches require vectors, which are not usually 100% efficient, and will not reach all muscles. A combination of these cellular and molecular approaches will probably be necessary to correct some of these complex diseases.

References

Bachrach, Estanislao, Antonio L. Perez, Yeong-Hoon Choi, Ben M. W. Illigens, Susan J. Jun, Pedro del Nido, Francis X. McGowan, et al. 2006. "Muscle Engraftment of Myogenic Progenitor Cells Following Intraarterial Transplantation." *Muscle & Nerve* 34 (1): 44–52. https://doi.org/10.1002/mus.20560.

Baghdadi, Meryem B., David Castel, Léo Machado, So-Ichiro Fukada, David E. Birk, Frederic Relaix, Shahragim Tajbakhsh, and Philippos Mourikis. 2018a. "Reciprocal Signalling by Notch-Collagen V-CALCR Retains Muscle Stem Cells in Their Niche." *Nature* 557 (7707): 714-18. https://doi.org/10.1038/s41586-018-0144-9.

Baghdadi, Meryem B., Joao Firmino, Kartik Soni, Brendan Evano, Daniela Di Girolamo, Philippos Mourikis, David Castel, and Shahragim Tajbakhsh. 2018b. "Notch-Induced MiR-708 Antagonizes Satellite Cell Migration and Maintains Quiescence." *Cell Stem Cell* 23 (6): 859-868.e5. https://doi.org/10.1016/j.stem.2018.09.017.

Barberi, Tiziano, Michelle Bradbury, Zehra Dincer, Georgia Panagiotakos, Nicholas D Socci, and Lorenz Studer. 2007. "Derivation of Engraftable Skeletal Myoblasts from Human Embryonic Stem Cells." *Nature Medicine* 13 (5): 642-48. https://doi.org/10.1038/nm1533.

Beauchamp, Jonathan R., Jennifer E. Morgan, Charles N. Pagel, and Terence A. Partridge. 1999. "Dynamics of Myoblast Transplantation Reveal a Discrete Minority of Precursors with Stem Cell-like Properties as the Myogenic Source." *The Journal of Cell Biology* 144 (6): 1113-22. https://doi.org/10.1083/jcb.144.6.1113.

Benabdallah, Basma, Cynthia Désaulniers-Langevin, Marie-Lyn Goyer, Chloé Colas, Chantale Maltais, Yuanyi Li, Jean V. Guimond, Jacques P. Tremblay, Elie Haddad, and Christian Beauséjour. 2020. "Myogenic Progenitor Cells Derived from Human Induced Pluripotent Stem Cell Are Immune-Tolerated in Humanized Mice." *Stem Cells Translational Medicine*, September. https://doi.org/10.1002/sctm.19-0452.

Benchaouir, Rachid, Mirella Meregalli, Andrea Farini, Giuseppe D'Antona, Marzia Belicchi, Aurélie Goyenvalle, Maurizio Battistelli, et al. 2007. "Restoration of Human Dystrophin Following Transplantation of Exon-Skipping-Engineered DMD Patient Stem Cells into Dystrophic Mice." *Cell Stem Cell* 1 (6): 646-57. https://doi.org/10.1016/j.stem.2007.09.016.

Bernet, Jennifer D., Jason D. Doles, John K. Hall, Kathleen Kelly Tanaka, Thomas A. Carter, and Bradley B. Olwin. 2014. "P38 MAPK Signaling Underlies a Cell-Autonomous Loss of Stem Cell Self-Renewal in Skeletal Muscle of Aged Mice." *Nature Medicine* 20 (3): 265-71. https://doi.org/10.1038/nm.3465.

Bodnar, Megan S., Juanito J. Meneses, Ryan T. Rodriguez, and Meri T. Firpo. 2004. "Propagation and Maintenance of Undifferentiated Human Embryonic Stem Cells." *Stem Cells and Development* 13 (3): 243-53. https://doi.org/10.1089/154732804323099172.

Charville, Gregory W., Tom H. Cheung, Bryan Yoo, Pauline J. Santos, Gordon K. Lee, Joseph B. Shrager, and Thomas A. Rando. 2015. "Ex Vivo Expansion and In Vivo Self-Renewal of Human Muscle Stem Cells." *Stem Cell Reports* 5 (4): 621-32. https://doi.org/10.1016/j.stemcr.2015.08.004.

Collins, Charlotte A., Irwin Olsen, Peter S. Zammit, Louise Heslop, Aviva Petrie, Terence A. Partridge, and Jennifer E. Morgan. 2005. "Stem Cell Function, Self-Renewal, and Behavioral Heterogeneity of Cells from the Adult Muscle Satellite Cell Niche." *Cell* 122 (2): 289–301. https://doi.org/10.1016/j.cell.2005.05.010.

Cosgrove, Benjamin D., Penney M. Gilbert, Ermelinda Porpiglia, Foteini Mourkioti, Steven P. Lee, Stephane Y. Corbel, Michael E. Llewellyn, Scott L. Delp, and Helen M. Blau. 2014. "Rejuvenation of the Muscle Stem Cell Population Restores Strength to Injured Aged Muscles." *Nature Medicine* 20 (3): 255-64. https://doi.org/10.1038/nm.3464.

Cossu, Giulio, and Paolo Bianco. 2003. "Mesoangioblasts–Vascular Progenitors for Extravascular Mesodermal Tissues." *Current Opinion in Genetics & Development* 13 (5): 537-42. https://doi.org/10.1016/j.gde.2003.08.001.

Cossu, Giulio, Stefano C. Previtali, Sara Napolitano, Maria Pia Cicalese, Francesco Saverio Tedesco, Francesca Nicastro, Maddalena Noviello, et al. 2015. "Intra-Arterial Transplantation of HLA-Matched Donor Mesoangioblasts in Duchenne Muscular Dystrophy." *EMBO Molecular Medicine* 7 (12): 1513-28. https://doi.org/10.15252/emmm.201505636.

Cossu, Giulio, and Maurilio Sampaolesi. 2007. "New Therapies for Duchenne Muscular Dystrophy: Challenges, Prospects and Clinical Trials." Trends in Molecular Medicine. Trends Mol Med. December 2007. https://doi.org/10.1016/j.molmed.2007.10.003.

Darabi, Radbod, Robert W. Arpke, Stefan Irion, John T. Dimos, Marica Grskovic, Michael Kyba, and Rita C. R. Perlingeiro. 2012. "Human ES- and IPS-Derived Myogenic Progenitors Restore DYSTROPHIN and Improve Contractility upon Transplantation in Dystrophic Mice." *Cell Stem Cell* 10 (5): 610-19. https://doi.org/10.1016/j.stem.2012.02.015.

De Angelis, L., L. Berghella, M. Coletta, L. Lattanzi, M. Zanchi, M. G. Cusella-De Angelis, C. Ponzetto, and G. Cossu. 1999. "Skeletal Myogenic Progenitors Originating from Embryonic Dorsal Aorta Coexpress Endothelial and Myogenic Markers and Contribute to Postnatal Muscle Growth and Regeneration." *The Journal of Cell Biology* 147 (4): 869-78. https://doi.org/10.1083/jcb.147.4.869.

Dellavalle, Arianna, Maurilio Sampaolesi, Rossana Tonlorenzi, Enrico Tagliafico, Benedetto Sacchetti, Laura Perani, Anna Innocenzi, et al. 2007. "Pericytes of Human Skeletal Muscle Are Myogenic Precursors Distinct from Satellite Cells." *Nature Cell Biology* 9 (3): 255-67. https://doi.org/10.1038/ncb1542.

El Fahime, E., P. Mills, J. F. Lafreniere, Y. Torrente, and J. P. Tremblay. 2002. "The Urokinase Plasminogen Activator: An Interesting Way to Improve Myoblast Migration Following Their Transplantation." *Experimental Cell Research* 280 (2): 169-78. https://doi.org/10.1006/excr.2002.5642.

Engler, Adam J., Shamik Sen, H. Lee Sweeney, and Dennis E. Discher. 2006. "Matrix Elasticity Directs Stem Cell Lineage Specification." *Cell* 126 (4): 677-89. https://doi.org/10.1016/j.cell.2006.06.044.

Fanin, M., E. P. Hoffman, F. A. Saad, A. Martinuzzi, G. A. Danieli, and C. Angelini. 1993. "Dystrophin-Positive Myotubes in Innervated Muscle Cultures from Duchenne and Becker Muscular Dystrophy Patients." *Neuromuscular Disorders*: NMD 3 (2): 119-27. https://doi.org/10.1016/0960-8966(93)90003-3.

Ferrari, G., G. Cusella-De Angelis, M. Coletta, E. Paolucci, A. Stornaiuolo, G. Cossu, and F. Mavilio. 1998. "Muscle Regeneration by Bone Marrow-Derived Myogenic Progenitors." *Science (New York, N.Y.)* 279 (5356): 1528-30. https://doi.org/10.1126/science.279.5356.1528.

Ferrari, G., A. Stornaiuolo, and F. Mavilio. 2001. "Failure to Correct Murine Muscular Dystrophy." *Nature* 411 (6841): 1014-15. https://doi.org/10.1038/35082631.

Galli, Francesco, Laricia Bragg, Linda Meggiolaro, Maira Rossi, Miriam Caffarini, Naila Naz, Sabrina Santoleri, and Giulio Cossu. 2018. "Gene and Cell Therapy for Muscular Dystrophies: Are We Getting There?" *Human Gene Therapy* 29 (10): 1098-1105. https://doi.org/10.1089/hum.2018.151.

Galvez, Beatriz G., Maurilio Sampaolesi, Silvia Brunelli, Diego Covarello, Manuela Gavina, Barbara Rossi, Gabriela Constantin, Gabriela Costantin, Yvan Torrente, and Giulio Cossu. 2006. "Complete Repair of Dystrophic Skeletal Muscle by Mesoangioblasts with Enhanced Migration Ability." *The Journal of Cell Biology* 174 (2): 231-43. https://doi.org/10.1083/jcb.200512085.

Garcia, Steven M., Stanley Tamaki, Solomon Lee, Alvin Wong, Anthony Jose, Joanna Dreux, Gayle Kouklis, et al. 2018. "High-Yield Purification, Preservation, and Serial Transplantation of Human Satellite Cells." *Stem Cell Reports* 10 (3): 1160-74. https://doi.org/10.1016/j.stemcr.2018.01.022.

Gargioli, Cesare, Marcello Coletta, Fabrizio De Grandis, Stefano M. Cannata, and Giulio Cossu. 2008. "PlGF-MMP-9-Expressing Cells Restore Microcirculation and Efficacy of Cell Therapy in Aged Dystrophic Muscle." *Nature Medicine* 14 (9): 973-78. https://doi.org/10.1038/nm.1852.

Gavina, Manuela, Marzia Belicchi, Barbara Rossi, Linda Ottoboni, Fabio Colombo, Mirella Meregalli, Maurizio Battistelli, et al. 2006. "VCAM-1 Expression on Dystrophic Muscle Vessels Has a Critical Role in the Recruitment of Human Blood-Derived CD133+ Stem Cells after Intra-Arterial Transplantation." *Blood* 108 (8): 2857-66. https://doi.org/10.1182/blood-2006-04-018564.

Gilbert, P. M., K. L. Havenstrite, K. E. G. Magnusson, A. Sacco, N. A. Leonardi, P. Kraft, N. K. Nguyen, S. Thrun, M. P. Lutolf, and H. M. Blau. 2010. "Substrate Elasticity Regulates Skeletal Muscle Stem Cell

Self-Renewal in Culture." *Science (New York, N.Y.)* 329 (5995): 1078-81. https://doi.org/10.1126/science.1191035.

Goodell, M. A., K. Brose, G. Paradis, A. S. Conner, and R. C. Mulligan. 1996. "Isolation and Functional Properties of Murine Hematopoietic Stem Cells That Are Replicating in Vivo." *The Journal of Experimental Medicine* 183 (4): 1797-1806. https://doi.org/10.1084/jem.183.4.1797.

Gussoni, E., G. K. Pavlath, A. M. Lanctot, K. R. Sharma, R. G. Miller, L. Steinman, and H. M. Blau. 1992. "Normal Dystrophin Transcripts Detected in Duchenne Muscular Dystrophy Patients after Myoblast Transplantation." *Nature* 356 (6368): 435-38. https://doi.org/10.1038/356435a0.

Gussoni, E., Y. Soneoka, C. D. Strickland, E. A. Buzney, M. K. Khan, A. F. Flint, L. M. Kunkel, and R. C. Mulligan. 1999. "Dystrophin Expression in the Mdx Mouse Restored by Stem Cell Transplantation." *Nature* 401 (6751): 390-94. https://doi.org/10.1038/43919.

Haworth, Richard, and Michaela Sharpe. 2020. "Accept or Reject: The Role of Immune Tolerance in the Development of Stem Cell Therapies and Possible Future Approaches." *Toxicologic Pathology*, April, 192623320918241. https://doi.org/10.1177/0192623320918241.

Judson, Robert N., Marco Quarta, Menno J. Oudhoff, Hesham Soliman, Lin Yi, Chih Kai Chang, Gloria Loi, et al. 2018. "Inhibition of Methyltransferase Setd7 Allows the In Vitro Expansion of Myogenic Stem Cells with Improved Therapeutic Potential." *Cell Stem Cell* 22 (2): 177-190.e7. https://doi.org/10.1016/j.stem.2017.12.010.

Karpati, G., D. Ajdukovic, D. Arnold, R. B. Gledhill, R. Guttmann, P. Holland, P. A. Koch, E. Shoubridge, D. Spence, and M. Vanasse. 1993. "Myoblast Transfer in Duchenne Muscular Dystrophy." *Annals of Neurology* 34 (1): 8-17. https://doi.org/10.1002/ana.410340105.

Kim, Jaemin, Alessandro Magli, Sunny S. K. Chan, Vanessa K. P. Oliveira, Jianbo Wu, Radbod Darabi, Michael Kyba, and Rita C. R. Perlingeiro. 2017a. "Expansion and Purification Are Critical for the Therapeutic Application of Pluripotent Stem Cell-Derived Myogenic Progenitors." *Stem Cell Reports* 9 (1): 12-22. https://doi.org/10.1016/j.stemcr.2017.04.022.

Kim, Jaemin, Vanessa K. P. Oliveira, Ami Yamamoto, and Rita C. R. Perlingeiro. 2017b. "Generation of Skeletal Myogenic Progenitors from Human Pluripotent Stem Cells Using Non-Viral Delivery of Minicircle DNA." *Stem Cell Research* 23: 87–94. https://doi.org/10.1016/j.scr.2017.07.013.

Kuang, Shihuan, Kazuki Kuroda, Fabien Le Grand, and Michael A. Rudnicki. 2007. "Asymmetric Self-Renewal and Commitment of Satellite Stem Cells in Muscle." Cell 129 (5): 999–1010. https://doi.org/10.1016/j.cell.2007.03.044.

Lapan, Ariya D., Anete Rozkalne, and Emanuela Gussoni. 2012. "Human Fetal Skeletal Muscle Contains a Myogenic Side Population That Expresses the Melanoma Cell-Adhesion Molecule." *Human Molecular Genetics* 21 (16): 3668-80. https://doi.org/10.1093/hmg/dds196.

L'honoré, Aurore, Pierre-Henri Commère, Elisa Negroni, Giorgia Pallafacchina, Bertrand Friguet, Jacques Drouin, Margaret Buckingham, and Didier Montarras. 2018. "The Role of Pitx2 and Pitx3 in Muscle Stem Cells Gives New Insights into P38α MAP Kinase and Redox Regulation of Muscle Regeneration." *ELife* 7. https://doi.org/10.7554/eLife.32991.

Lipton, B. H., and E. Schultz. 1979. "Developmental Fate of Skeletal Muscle Satellite Cells." *Science (New York, N.Y.)* 205 (4412): 1292-94. https://doi.org/10.1126/science.472747.

Mazaleyrat K, Badja C, Broucqsault N, Chevalier R, Laberthonnière C, Dion C, Baldasseroni L, El-Yazidi C, Thomas M, Bachelier R, Altié A, Nguyen K, Lévy N, Robin JD, Magdinier F. Multilineage Differentiation for Formation of Innervated Skeletal Muscle Fibers from Healthy and Diseased Human Pluripotent Stem Cells. Cells. 2020 Jun 23;9(6):1531. doi: 10.3390/cells9061531.

Minasi, Maria G., Mara Riminucci, Luciana De Angelis, Ugo Borello, Barbara Berarducci, Anna Innocenzi, Arianna Caprioli, et al. 2002. "The Meso-Angioblast: A Multipotent, Self-Renewing Cell That Originates from the Dorsal Aorta and Differentiates into Most Mesodermal Tissues." *Development (Cambridge, England)* 129 (11): 2773-83.

Monge, Claire, Nicholas DiStasio, Thomas Rossi, Muriel Sébastien, Hiroshi Sakai, Benoit Kalman, Thomas Boudou, et al. 2017. "Quiescence of Human Muscle Stem Cells Is Favored by Culture on Natural Biopolymeric Films." *Stem Cell Research & Therapy* 8 (1): 104. https://doi.org/10.1186/s13287-017-0556-8.

Montarras, Didier, Jennifer Morgan, Charlotte Collins, Frédéric Relaix, Stéphane Zaffran, Ana Cumano, Terence Partridge, and Margaret Buckingham. 2005. "Direct Isolation of Satellite Cells for Skeletal Muscle Regeneration." *Science (New York, N.Y.)* 309 (5743): 2064-67. https://doi.org/10.1126/science.1114758.

Morgan, J, Rouche A, Bausero P, Houssaïni A, Gross J, Fiszman My, and Alameddine Hs. 2010. "MMP-9 Overexpression Improves Myogenic Cell Migration and Engraftment." Muscle & Nerve. Muscle Nerve. October 2010. https://doi.org/10.1002/mus.21737.

Motohashi, Norio, Akiyoshi Uezumi, Erica Yada, So-ichiro Fukada, Kazuhiro Fukushima, Kazuhiko Imaizumi, Yuko Miyagoe-Suzuki, and Shin'ichi Takeda. 2008. "Muscle CD31(-) CD45(-) Side Population Cells Promote Muscle Regeneration by Stimulating Proliferation and Migration of Myoblasts." *The American Journal of Pathology* 173 (3): 781-91. https://doi.org/10.2353/ajpath.2008.070902.

Negroni, Elisa, Ingo Riederer, Soraya Chaouch, Marzia Belicchi, Paola Razini, James Di Santo, Yvan Torrente, Gillian S. Butler-Browne, and Vincent Mouly. 2009. "In Vivo Myogenic Potential of Human CD133+ Muscle-Derived Stem Cells: A Quantitative Study." *Molecular Therapy: The Journal of the American Society of Gene Therapy* 17 (10): 1771-78. https://doi.org/10.1038/mt.2009.167.

Pan, H, Vojnits K, Liu Tt, Meng F, Yang L, Wang Y, Huard J, Cox Cs, Lally Kp, and Li Y. 2015. "MMP1 Gene Expression Enhances Myoblast Migration and Engraftment Following Implanting into Mdx/SCID Mice." Cell Adhesion & Migration. Cell Adh Migr. 2015. https://doi.org/10.4161/19336918.2014.983799.

Partridge, T. A., J. E. Morgan, G. R. Coulton, E. P. Hoffman, and L. M. Kunkel. 1989. "Conversion of Mdx Myofibres from Dystrophin-Negative to -Positive by Injection of Normal Myoblasts." *Nature* 337 (6203): 176-79. https://doi.org/10.1038/337176a0.

Partridge, Terry. 2000. "The Current Status of Myoblast Transfer." *Neurological Sciences: Official Journal of the Italian Neurological Society and of the Italian Society of Clinical Neurophysiology* 21 (5 Suppl): S939-942. https://doi.org/10.1007/s100720070007.

Périé, Sophie, Kamel Mamchaoui, Vincent Mouly, Stéphane Blot, Belaïd Bouazza, Lars-Eric Thornell, Jean Lacau St Guily, and Gillian Butler-Browne. 2006. "Premature Proliferative Arrest of Cricopharyngeal Myoblasts in Oculo-Pharyngeal Muscular Dystrophy: Therapeutic Perspectives of Autologous Myoblast Transplantation." *Neuromuscular Disorders*: NMD 16 (11): 770-81. https://doi.org/10.1016/j.nmd.2006.07.022.

Périé, Sophie, Capucine Trollet, Vincent Mouly, Valérie Vanneaux, Kamel Mamchaoui, Belaïd Bouazza, Jean Pierre Marolleau, et al. 2014. "Autologous Myoblast Transplantation for Oculopharyngeal Muscular Dystrophy: A Phase I/IIa Clinical Study." *Molecular Therapy: The Journal of the American Society of Gene Therapy* 22 (1): 219-25. https://doi.org/10.1038/mt.2013.155.

Quenneville, Simon P., Pierre Chapdelaine, Daniel Skuk, Matin Paradis, Marlyne Goulet, Joël Rousseau, Xiao Xiao, Luis Garcia, and Jacques P.

Tremblay. 2007. "Autologous Transplantation of Muscle Precursor Cells Modified with a Lentivirus for Muscular Dystrophy: Human Cells and Primate Models." *Molecular Therapy: The Journal of the American Society of Gene Therapy* 15 (2): 431-38. https://doi.org/10.1038/sj.mt.6300047.

Rando, T. A., G. K. Pavlath, and H. M. Blau. 1995. "The Fate of Myoblasts Following Transplantation into Mature Muscle." *Experimental Cell Research* 220 (2): 383-89. https://doi.org/10.1006/excr.1995.1329.

Renault, Valérie, Lars-Eric Thornell, Per-Olof Eriksson, Gillian Butler-Browne, Vincent Mouly, and Lars-Eric Thorne. 2002. "Regenerative Potential of Human Skeletal Muscle during Aging." *Aging Cell* 1 (2): 132-39. https://doi.org/10.1046/j.1474-9728.2002.00017.x.

Riederer, Ingo, Elisa Negroni, Maximilien Bencze, Annie Wolff, Ahmed Aamiri, James P. Di Santo, Suse D. Silva-Barbosa, Gillian Butler-Browne, Wilson Savino, and Vincent Mouly. 2012. "Slowing down Differentiation of Engrafted Human Myoblasts into Immunodeficient Mice Correlates with Increased Proliferation and Migration." *Molecular Therapy: The Journal of the American Society of Gene Therapy* 20 (1): 146-54. https://doi.org/10.1038/mt.2011.193.

Rodgers, Joseph T., Katherine Y. King, Jamie O. Brett, Melinda J. Cromie, Gregory W. Charville, Katie K. Maguire, Christopher Brunson, et al. 2014. "MTORC1 Controls the Adaptive Transition of Quiescent Stem Cells from G0 to G(Alert)." *Nature* 510 (7505): 393-96. https://doi.org/10.1038/nature13255.

Sammels, Leanne M., Erika Bosio, Clayton T. Fragall, Miranda D. Grounds, Nico van Rooijen, and Manfred W. Beilharz. 2004. "Innate Inflammatory Cells Are Not Responsible for Early Death of Donor Myoblasts after Myoblast Transfer Therapy." *Transplantation* 77 (12): 1790-97. https://doi.org/10.1097/01.tp.0000131150.76841.75.

Sampaolesi, Maurilio, Stephane Blot, Giuseppe D'Antona, Nicolas Granger, Rossana Tonlorenzi, Anna Innocenzi, Paolo Mognol, et al. 2006. "Mesoangioblast Stem Cells Ameliorate Muscle Function in Dystrophic Dogs." *Nature* 444 (7119): 574-79. https://doi.org/10.1038/nature05282.

Sampaolesi, Maurilio, Yvan Torrente, Anna Innocenzi, Rossana Tonlorenzi, Giuseppe D'Antona, M. Antonietta Pellegrino, Rita Barresi, et al. 2003. "Cell Therapy of Alpha-Sarcoglycan Null Dystrophic Mice through Intra-Arterial Delivery of Mesoangioblasts." *Science (New York, N.Y.)* 301 (5632): 487-92. https://doi.org/10.1126/science.1082254.

Sampath, Srinath C., Srihari C. Sampath, Andrew T. V. Ho, Stéphane Y. Corbel, Joshua D. Millstone, John Lamb, John Walker, Bernd Kinzel, Christian Schmedt, and Helen M. Blau. 2018. "Induction of Muscle Stem Cell Quiescence by the Secreted Niche Factor Oncostatin M." Nature Communications 9 (1): 1531. https://doi.org/10.1038/s41467-018-03876-8.

Schweitzer, Jeffrey S., Bin Song, Todd M. Herrington, Tae-Yoon Park, Nayeon Lee, Sanghyeok Ko, Jeha Jeon, et al. 2020. "Personalized IPSC-Derived Dopamine Progenitor Cells for Parkinson's Disease." *The New England Journal of Medicine* 382 (20): 1926-32. https://doi.org/10.1056/NEJMoa1915872.

Shi, Ming, Masakazu Ishikawa, Naosuke Kamei, Tomoyuki Nakasa, Nobuo Adachi, Masataka Deie, Takayuki Asahara, and Mitsuo Ochi. 2009. "Acceleration of Skeletal Muscle Regeneration in a Rat Skeletal Muscle Injury Model by Local Injection of Human Peripheral Blood-Derived CD133-Positive Cells." *Stem Cells (Dayton, Ohio)* 27 (4): 949-60. https://doi.org/10.1002/stem.4.

Silva-Barbosa, Suse D., Gillian S. Butler-Browne, Wallace de Mello, Ingo Riederer, James P. Di Santo, Wilson Savino, and Vincent Mouly. 2008. "Human Myoblast Engraftment Is Improved in Laminin-Enriched Microenvironment." *Transplantation* 85 (4): 566-75. https://doi.org/10.1097/TP.0b013e31815fee50.

Skuk, D., B. Roy, M. Goulet, and J. P. Tremblay. 1999. "Successful Myoblast Transplantation in Primates Depends on Appropriate Cell Delivery and Induction of Regeneration in the Host Muscle." *Experimental Neurology* 155 (1): 22-30. https://doi.org/10.1006/exnr.1998.6973.

Skuk, Daniel, Nicolas J. Caron, Marlyne Goulet, Brigitte Roy, and Jacques P. Tremblay. 2003. "Resetting the Problem of Cell Death Following Muscle-Derived Cell Transplantation: Detection, Dynamics and Mechanisms." *Journal of Neuropathology and Experimental Neurology* 62 (9): 951-67. https://doi.org/10.1093/jnen/62.9.951.

Takahashi, Kazutoshi, Koji Tanabe, Mari Ohnuki, Megumi Narita, Tomoko Ichisaka, Kiichiro Tomoda, and Shinya Yamanaka. 2007. "Induction of Pluripotent Stem Cells from Adult Human Fibroblasts by Defined Factors." Cell 131 (5): 861-72. https://doi.org/10.1016/j.cell.2007.11.019.

Tedesco, Francesco Saverio, Mattia F. M. Gerli, Laura Perani, Sara Benedetti, Federica Ungaro, Marco Cassano, Stefania Antonini, et al. 2012. "Transplantation of Genetically Corrected Human IPSC-Derived Progenitors in Mice with Limb-Girdle Muscular Dystrophy." *Science Translational Medicine* 4 (140): 140ra89. https://doi.org/10.1126/scitranslmed.3003541.

Theret, Marine, Linda Gsaier, Bethany Schaffer, Gaëtan Juban, Sabrina Ben Larbi, Michèle Weiss-Gayet, Laurent Bultot, et al. 2017. "AMPKα1-LDH Pathway Regulates Muscle Stem Cell Self-Renewal by Controlling Metabolic Homeostasis." *The EMBO Journal* 36 (13): 1946-62. https://doi.org/10.15252/embj.201695273.

Torrente, Y., M. Belicchi, C. Marchesi, G. D'Antona, F. Cogiamanian, F. Pisati, M. Gavina, et al. 2007. "Autologous Transplantation of Muscle-Derived CD133+ Stem Cells in Duchenne Muscle Patients." *Cell Transplantation* 16 (6): 563-77. https://doi.org/10.3727/000000007783465064.

Torrente, Y., J. P. Tremblay, F. Pisati, M. Belicchi, B. Rossi, M. Sironi, F. Fortunato, et al. 2001. "Intraarterial Injection of Muscle-Derived CD34(+)Sca-1(+) Stem Cells Restores Dystrophin in Mdx Mice." *The Journal of Cell Biology* 152 (2): 335-48. https://doi.org/10.1083/jcb.152.2.335.

Verma, Mayank, Yoko Asakura, Bhavani Sai Rohit Murakonda, Thomas Pengo, Claire Latroche, Benedicte Chazaud, Linda K. McLoon, and Atsushi Asakura. 2018. "Muscle Satellite Cell Cross-Talk with a Vascular Niche Maintains Quiescence via VEGF and Notch Signaling." *Cell Stem Cell* 23 (4): 530-543.e9. https://doi.org/10.1016/j.stem.2018.09.007.

Webster, C., and H. M. Blau. 1990. "Accelerated Age-Related Decline in Replicative Life-Span of Duchenne Muscular Dystrophy Myoblasts: Implications for Cell and Gene Therapy." *Somatic Cell and Molecular Genetics* 16 (6): 557-65. https://doi.org/10.1007/BF01233096.

Xi, Haibin, Justin Langerman, Shan Sabri, Peggie Chien, Courtney S. Young, Shahab Younesi, Michael Hicks, et al. 2020. "A Human Skeletal Muscle Atlas Identifies the Trajectories of Stem and Progenitor Cells across Development and from Human Pluripotent Stem Cells." *Cell Stem Cell* 27 (1): 158-176.e10. https://doi.org/10.1016/j.stem.2020.04.017.

Xu, Cong, Mohammadsharif Tabebordbar, Salvatore Iovino, Christie Ciarlo, Jingxia Liu, Alessandra Castiglioni, Emily Price, et al. 2013. "A Zebrafish Embryo Culture System Defines Factors That Promote Vertebrate Myogenesis across Species." *Cell* 155 (4): 909-21. https://doi.org/10.1016/j.cell.2013.10.023.

Xu, Xiaoti, Karlijn J. Wilschut, Gayle Kouklis, Hua Tian, Robert Hesse, Catharine Garland, Hani Sbitany, et al. 2015. "Human Satellite Cell Transplantation and Regeneration from Diverse Skeletal Muscles." *Stem Cell Reports* 5 (3): 419-34. https://doi.org/10.1016/j.stemcr.2015.07.016.

Xynos, Alexandros, Paola Corbella, Nathalie Belmonte, Roberta Zini, Rossella Manfredini, and Giuliana Ferrari. 2010. "Bone Marrow-Derived Hematopoietic Cells Undergo Myogenic Differentiation Following a Pax-7 Independent Pathway." *Stem Cells (Dayton, Ohio)* 28 (5): 965-73. https://doi.org/10.1002/stem.418.

26

Legal Framework for Research on Human Embryonic Stem Cells in France and in Europe

Arnaud de Guerra, Samuel Arrabal*, and Emmanuelle Prada-Bordenave

Agence de la biomédecine, 1 avenue du Stade de France, SAINT-DENIS LA PLAINE Cedex 93212, France
E-mail: samuel.arrabal@biomedecine.fr
*Corresponding author

26.1 Introduction

The first Human Embryonic Stem Cells (hESC) lines have been obtained more than 20 years ago (Thomson et al., 1998) . These cells immediately raised great expectations regarding the possibilities they offered in terms of what is commonly called "regenerative medicine": being the only example, at this time (1998), of human pluripotent cells, they carried the ability to differentiate in any kind of tissue. The echo hESC research met in the public opinion and media have been huge, as the concept of regenerative medicine was simple to understand and hit the old myth of "fountain of youth." But scientists knew at this time that this would be a long way before reaching medical applications that could routinely be used at patient bedside, and this has been confirmed throughout the years. To date, in 2021, around 30 clinical trials using hESC-derived cells have been authorized worldwide and most of them are still recruiting patients. The main hurdles have been scientific, as it has proved very difficult to be able to obtain a pure population of differentiated cells, fitting with all the regulatory and safety requirements to be used for human therapeutics.

Researchers worldwide also had to deal with an unusual legal context regarding fundamental research, as it appeared that working on hESCs would require to comply with a precise legislation including the obligation to obtain authorizations from diverse institutions, either local or national. Why have the legislators decided to require such an exceptional provision as an authorization for performing research on hESCs?

hESCs are obtained from human embryos, which are considered to be ethically extremely sensible: for a range of persons with diverse sensibilities, the embryo, starting on at fertilization, has direct or indirect links that relates it with the status of a full human being. In a range of countries, including France, hESC research is then regulated because there has been no way until now to obtain hESC lines without destroying a human embryo (although there have been some reports, including a recent one, of obtaining lines at 8-cells embryo without destruction (Rodin et al., 2014) : one can nevertheless hardly see these embryos being than transferred in a woman uterus). Until recently, most legislation (not all, see for example United Kingdom below) considered that working with hESCs is similar to working with embryos, and must be submitted to the same special scrutiny. However, the most recent opinions of bioethics committees tend to change the way of considering the use of hESCs: if their embryonic origin is still taken into consideration in the way they are regulated, it is no longer the determining factor. It is their pluripotent character and the potentialities that this state confers, which today raise the most delicate ethical questions as to their possible applications. Therefore, the latest recommendations of the French National Committee on ethics strongly suggest that the legal regimes governing research on pre-implantation human embryo and that on hESCs should be separated.

Finally, hESC research would never have existed without Assisted Reproductive Technologies (ART) and *in vitro* fertilization (IVF) procedures that allowed to gain access to a source of pre-implantation embryos that is obviously not accessible otherwise. Hence the fact that the legislation upstream in the process, regarding ART aspects of embryos donation for example, has also to be considered when examining the legislative aspects of hESC research.

Here we propose an overview of some regulations concerning human embryo and hESC research throughout Europe. We will first consider in depth an example of a regulation that could be described as mid-way: the French legislation. We will then review three other and different ways of regulating the domain: Spain, UK, and Germany.

26.2 How to Regulate hESC Research? The French Example

France is the first country to have passed a general bioethics law in 1994, and the legislator has since been implicated several times in trying to find solutions to the numerous problems of this ever-changing domain. Human embryo and hESC research has been a particularly hard item to regulate.

The overall hESC research is regulated by the 2011 Bioethics Laws, currently under revision, which deal as well among others with organs, tissues and cells procurement and transplantation, and ART. Nevertheless, one central article of the law concerning hESCs has been replaced by a new article passed in July 2013 (art. L2151-5). This article addresses the general status of hESC research and has been thoroughly debated: it is now authorized, while in the 2011 equivalent article it was forbidden with possible dispensations. The bill under discussion provides for simplifying the administrative procedures for research on hESCs: it would be subject to declaration and no longer to authorization.

26.2.1 General Provisions Regarding hESC Research

Since the 2013 specific law (dated 6 August 2013), research on human embryos and ESCs is now authorized in France, provided that the submitted research protocol fulfills a range of requirements:

- It is scientifically relevant;
- Fundamental or not, it pursues a medical aim;
- It cannot be conducted with another kind of cells or organism than the hESCs or human embryos; and
- It must comply with the ethics principles for embryo and hESC research.

As previously mentioned, the bill under discussion should simplify the administrative procedure, by changing the authorization frame into a declaration. However, the requirements for implementing the protocol would remain unchanged, with the exception of the alternative condition, which would disappear.

One central—logic and shared with most countries—principle is that the authorization is specifically delivered for a specific research project, and not to a scientific team: obtaining the authorization entitles the applicant to conduct only the submitted project.

The institution in charge of delivering these authorizations is Agence de la biomédecine (ABM), a State agency under the supervision of the

Health Ministry (see next chapter). Research authorizations are issued for a maximum duration of 5 years. After almost 10 years of issuing such authorizations, it is now clear that in most of the cases this duration is not sufficient to complete a research project, and a particular kind of authorization has been set up for delivering an extension to the first authorization. Although the legal frame and the examination procedure are exactly the same, the file by itself is rather different as it asks for results obtained during the previous period of authorization, introducing an additional element for evaluating the application.

In addition to applying for a research authorization, and depending on their project, scientific teams might also apply for authorizations concerning:

- Importing/exporting the hESC lines on which the research will be conducted (this was vital in 2004, when research was for the first time authorized to be conducted by dispensation, and when no lines were available in France). It is especially checked whether the conditions in which embryos have been donated by couples before the hESC line derivation are in agreement with the prevailing conditions in France, especially at the level of the donation consent.
- Storage of these hESC lines

Of course, the application file includes the three authorizations, which are processed simultaneously.

26.2.2 Agence de la Biomédecine as the Central Piece of the National Regulation

Created by the 2004 Bioethics Laws, ABM is one of several administrative agencies shaping the healthcare landscape in France, depending directly from the Health Ministry. Its domains range from organs, tissues and cells transplantation to ART, through genetics. Regulating hESC and embryo research has been one of the sensible missions why it has originally been created.

When it comes to hESC research regulation, ABM plays a unique role as it is absolutely central in all the authorization (and control) processes. Apart from issuing authorizations, the Bioethics Laws gives ABM the mission to check whether these authorizations have been implemented following the content of the initial files and adequate rules considering the ethically delicate aspects of hESCs:

- The authorized research teams have to keep ABM aware of the implementation of their protocol: they must inform it as soon as they start, and they have to send an annual report describing where they are regarding the scientific project they initially submitted. Any change in the protocol must be communicated before implementation; if needed (i.e., if it is estimated that the change significantly modifies the protocol), ABM will then ask the team to apply for a change that will be examined in the same way as the initial demand. The change has then to be precisely justified. One reason for such a demand is, for example, a replacement of the project's lead researcher. Once the protocol is considered by the research team to be achieved, a final scientific report is to be sent to ABM that will consider the authorization no longer valid.

- Another entitlement for ABM in following the authorized researches is the possibility, quite unique again for fundamental research, to come on the actual research site and to inspect the conditions in which the protocol is implemented. Admittedly, the goal is not to give any opinion regarding the way the project is scientifically conducted: what is inspected is the way the hESC lines are being stored, including above all their safety (and occasionally manipulators' safety) and their traceability, as well as the way they are used (technical and material conditions in which the research takes place). Again behind this is the idea that hESC lines are something so ethically special that their use justifies a tight control from the State. ABM inspectors also check that the scientific project being conducted is indeed consistent with the initial application, and that all hESC lines used are destroyed once the project is over.

Finally, ABM is also in charge of maintaining a national registry of all embryos used for research, as well as a registry of all hESC lines created in France (available at http://www.agence-biomedecine.fr/Recherche-sur-l-embryon#4).

26.2.3 Authorization Delivery

Practically, the applications for authorization are examined following a procedure that has not changed since ABM is actually in charge of that task. Following the conditions defined by the law, two committees have a central role in the process, chronologically:

- A scientific committee, including around 12 high-level scientists, independent from ABM, specialists from different biological domains. They are nominated by ABM's General Director for 3 years, and chosen both for their excellence in their domains and for their capacity to adapt to the projects evaluation following the quite particular frame described above.

This "College" is thus in charge of evaluating all scientific aspects of the applications: scientific relevance, medical aims, no alternative.

- An ethics committee, called "conseil d'orientation" (CO), hosted by ABM, and nominated by a decree issued by the Health Ministry for 3 years. It is composed of Members of parliament, lawyers, representatives of patient associations and of the National Ethics Committee (CCNE), philosophers, and medical doctors. This Committee deals with the numerous ethical questions brought in by ABM domains (transplantation, ART, and genetics). It is also in charge of advising ABM's Director General in delivering various authorizations, including authorizations concerning hESC research, storage, and import/export. In this case, when deliberating, this committee considers the evaluations from the Scientific Committee and from the Inspection department (see above), and then evaluates the adequacy of the applications regarding ethics principles.

Although not standing at the same level (the Ethics Committee only is empowered by the law to deliver an advice to ABM's Director General), this double-committee system ensures a thorough evaluation of the applications, as potential ethical problems could also be raised by the Scientific Committee and as the Ethics Committee might ask for clarification considering some scientific aspects. Finally, it is up to ABM's Director General to take the decision to deliver the authorization, on the base of the Ethics Committee advice and, if appropriate, of other concerns that would not have been considered by the Committees.

The law sets a maximum duration of 4 months to deliver the decisions (starting from receipt of the application), otherwise the decision is considered to be a refusal (this is quite exceptional in French law, where usually a decision not delivered in a defined duration is considered to be positive). Fortunately, in almost 20 years, this has never happened. Practically, the process explained above hardly takes less than 3 months, mainly because of the time needed by evaluators from both Committees to work on the application files.

26.2.4 Donation of Human Embryos for Research

Although France has ratified the 1997 Oviedo convention only in 2011, its Bioethics laws have a lot of common principles. The stronger principle (but not unanimously shared) is the interdiction to create human embryos for the only purpose of research. Human embryos used for research then have to come from IVF procedures in the context of ART (which is widely assimilated as an healthcare) and have then to be donated following defined criteria (this is hence also the requirements for embryos used to derive hESC lines). According to the French law, three kinds of embryos could be eligible for donation:

- Supernumerary embryos from an IVF procedure: in France, one or two embryos per cycle are usually transferred to the women uterus. Remaining fitted embryos (eligible for a later transfer) are then kept frozen at the ART center to spare the woman undergoing another hyperstimulation procedure. Following the law, the couple is asked every year by the ART center whether they still intend to use their frozen embryos for their parenting project. If not, they can:

  o Ask for their destruction;
  o Donate them for another couple following an ART procedure; or
  o Donate them for research.

  In both last cases, two consents are required by the law, the second one being signed at least 3 months after the first one (giving the couple some more time to be sure about their decision).
  If used for derivation, these embryos will provide lines free of genetic anomalies (as opposed to the two following cases, see below), to conduct research on normal processes of differentiation. If derived in GMP (Good Manufacturing Practices) conditions, these hESC lines might then also possibly provide the safe material used in future cellular therapies.
- Embryos carrying genetic mutations put aside following a preimplantation genetic diagnosis (PGD) (Handyside et al., 1992). As the genetic mutation is then clearly identified and associated with the phenotype already present in the family, hESC lines derived from such embryos could therefore be used as valuable models in research of the specific genetic disease for which the PGD has been performed.
- Embryos derived from IVF but whose poor quality (based on criteria such as morphology, kinetic of development, etc.) rules out the transfer to woman's uterus.

In all three cases, the consent (or its confirmation) is sought after the couple received full information about the overall categories of research in which the embryo might be included. For poor quality embryos and embryos donated from PGD, since they will never be transferred, only a single consent is needed; the couple can then be informed precisely about the research to be done, as they will be used immediately in a defined project. For supernumerary frozen embryos, providing the information can be trickier, as they are usually donated at a moment where no research project is planned yet: all ART centres have the legal obligation to proceed with the donation, even if they have no link with a research team. Later, once the centre is contacted to provide embryos (sometime after several years), it might then prove difficult to come back to the couple to inform it about the project where their embryos have been used. Although this is asked by some guidelines (Williams, 2009), this is not the choice that has been made in France.

Despite these precisions, the law does not specify what has to be included in the consent forms. ABM, along with some ART professionals and researchers, has designed suggestions for these consent forms, using existing international guidelines that could be found on its website. Suggestions to ART professionals on procedures and ways of seeking the consent are also included in a memorandum to ART professionals (http://www.agence-biomedecine.fr/Recherche-sur-l-embryon).

Last but not least, the law specifically states that an embryo on which a research has been conducted in the conditions described in Art 2151-5 cannot be transferred in a women uterus. To date anyway, all the authorized projects included experiments implying the destruction of the embryo during the protocol.

26.2.5 Translating hESC Research Into Clinics

Since the first revision of the Bioethics laws in 2004 and the possibility to conduct research on hESCs (initially by dispensation), the delivery of the authorization has always depended (in various wordings) on the fact that the research protocol would bring medical progress. But strangely enough, no specific regulation has been envisioned by the law concerning a future transition to clinical research, using cells differentiated from hESCs into patients. The adopted frame has been the already-existing one concerning any kind of clinical trial for cellular therapy: ABM and ANSM (another State agency depending on the Health Ministry in charge, among other, to deliver authorizations for clinical trials for any health product but also marketing authorization for drugs) therefore jointly suggested to consider any request for authorization for protocols using cells differentiated from hESCs as a regular application for cellular therapy: authorization from ANSM issued after consultation of ABM.

26.3 Examples of Regulations in Europe: France's Neighbors and Significant Variations in Regulating hESC Research

When it comes to comparing and classifying the various legislations world-wide, most of the studies considered a limited range of criteria that would be authorized/forbidden:

1) creation of embryos for research,
2) somatic cell nuclear transfer, SCNT (also called therapeutic cloning),
3) research on supernumerary embryos from ART/possibility to derive hESC lines from ART embryos, and
4) research on hESCs only (embryos excluded).

Based on these criteria, it has been possible to break down countries legislation in a few categories:

- permissive: all four criteria above accepted
- permissive with restriction: only criteria 3 and 4
- restrictive: only criteria 4
- All kind of research prohibited

For Europe, this could be summed-up in the following legislative map.

Nevertheless, these kind of broad categories are only informative up to a certain degree. As described above for France, hESC regulations usually address a range of item that could hardly be examined simultaneously. More-over, this breakdown becomes less relevant with scientific and technological progress, the best example being the authorization for performing SCNT: since the discovery of induced pluripotent stem cells (iPSCs), the patient-specific hESC lines that could have been derived from a pseudo-embryo obtained by SCNT are less useful from a clinical perspective. Of course they still remain of fundamental value.

Here we propose to compare the way hESC research is regulated in three major European countries: Germany, Spain, and United Kingdom (UK). UK and Spain are considered as permissive countries, but have different overall organizations. Germany can be considered as a restrictive country regarding hESC regulation.

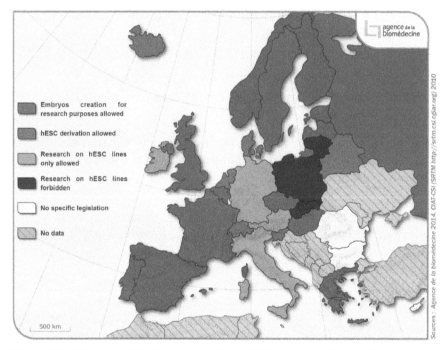

Figure 26.1 Diversity of legislations regarding hESC research in Europe (from European Science Fundation, Science Policy Briefing, May 2010).

26.3.1 Spain

26.3.1.1 Regulation Frame

The general frame for hESC research regulation is the Law 14/2007 (dated 3 July 2007) on Biomedical Research, mainly Title IV. It defines a coherent system describing what kind of research could be done, what kind of embryos could be used, the process to be followed to be authorized and the institutions and committees having a role in issuing the authorization. Some aspects especially regarding embryo donation, are included in the Law 16/2006 dealing with ART.

26.3.1.2 Criteria for Authorization

The criteria mentioned by the Law 14/2007 are clearly stated. They are relatively similar to the ones included in the French law, although emphasis and details are organized differently. The main requirements for a research to be authorized are the following:

- It must be relevant and feasible
- It has to abide with the ethical principles

These criteria, as in the French law, underline the fact that hESCs are ethically sensible and should be used in a restricted way.

Apart from that, the project must include the following elements:

- Description of possible (or lack of) links between the ART center providing the embryos and the research team.
- Commitment to provide the relevant public authority with necessary elements concerning the hESC lines used or created during the project's course as well as the number and origin of the pre-embryos possibly used, and the related consent forms.
- Commitment to provide the hESC lines possibly obtained in the course of the project for free to other research teams (it is not stated whether Spanish or from abroad).

26.3.1.3 Institutions

The central institution regarding hESC research is the Health Institute Carlos III (Instituto de Salud Carlos III, ISCIII), based in Madrid, being under the supervision of the Spanish Ministry for Science and Innovation. It has the following missions regarding embryos and hESCs:

- Guaranteeing access to cryopreserved embryos from ART donated to research
- Coordinating the National Cell Lines Bank (BNLC), including embryonic and adult stem cell lines, composed of three nodes (Granada, Valencia, and Barcelona)
- Maintaining a national registry of all research projects using human embryos and hESCs
- Promoting and coordinating research involving embryonic biological samples, including of course hESCs

Although central, the ISCIII has no role in the authorization delivery.

The Law 14/2007 creates a dedicated commission named "Guarantees Commission for the Donation and Use of Human cells and Tissues," hosted by ISCIII, which plays a central role in the authorization process. Its 12 members are nominated for 3 years by the Health Ministry, 6 of them being representatives from the autonomous regions, 6 being proposed by the national State administration (Health, Research, and Justice Ministries). The Commission has to grant its approval to any research project dealing with human pre-embryos (see below), hESCs (whether obtained from Spain or abroad), and SCNT for research purposes. Generally speaking, approval has to be sought for any techniques using human biological samples which can lead to obtain stem cells, which include iPSC lines.

The Commission is also in charge of producing various reports on the subject, especially about the projects using hESC lines imported from abroad. It must release an annual evaluation of the hESC research in Spain.

26.3.1.4 Process for Approval

According to the strong regional structure in Spain, any research using hESCs has to be approved at three levels:

- At the research center level, especially from its ethics committee. Overall, the research center or institute is accountable for the project to be in accordance with all legal and ethical requirements and for the provision of the related data to the national Guarantees Commission.
- At the regional level, in particular from the dedicated structure from the Autonomous Communities. These structures, according to the law, have to provide support and reference to the national Guarantees Commission.
- At the national level, from the Guarantees Commission for the Donation and Use of Human cells and Tissues.

This is quite a long process going through many steps. No maximal duration is defined for reviewing the application.

26.3.1.5 Embryos for Research

The Spanish law uses the notion of pre-embryo, which defines an embryo younger than 14 days having been obtained by IVF. This limit of 14 days post-fertilization (which is indeed a theoretical limit, as it is not possible yet to grow an embryo *in vitro* beyond around 7 days) corresponds to the development of the primitive streak and is also used in the English law (see below). Beyond 14 days post-fertilization, research is absolutely forbidden.

Despite what is often mentioned, the creation of pre-embryos for experimentation is strictly prohibited in Spain (the confusion comes from the fact that the law allows SCNT by putting the emphasis on the goal of collecting hESCs for therapeutic purposes, implicitly stating that it does not correspond to pre-embryo creation).

Research on the embryos is thus only permitted on supernumerary embryos (or pre-embryos) created for reproductive purposes. Only one consent (compared to France) is collected from the couple (or specifically from a woman alone), but the 2006 ART law is clear about the fact that the couple must be informed about the actual research project that will be conducted on its embryos (the name of the project has to be mentioned in the consent, (Aran et al., 2010).

This implies that the donation must take place when the embryo is actually recruited in the research project, as opposed to France where it can be donated well before, thus without knowing the project where the embryo will be included. If a written consent is collected and the embryo is not used, the consent has to be renewed every two years.

26.3.2 United Kingdom

26.3.2.1 Regulation Frame

Legislation in the UK is fundamentally different from Spain and France's regulation, as it sets a difference between the level of regulation for performing a research using human embryos and a research using "established" hESC lines.

Researches using human embryos, including derivation of hESC lines, are governed by the Human Fertilisation and Embryology (HFE) Act, first released in 1991, revised in 2001 and 2008. The HFE Act mainly regulates all items dealing with use of human gametes and embryos, either in healthcare (ART) or for research. While the 1991 version allowed embryo research for a limited range of purposes, the 2001 revision widened this range and opened the possibility to obtain hESCs from embryos. The 2008 amendment mainly brought precisions about the creation and use of "human admixed embryos"—interspecies embryos such as cytoplasmic hybrids.

Once the hESC lines are established, they are not considered as embryos anymore, they are then not concerned by the HFE Act and the level of regulation is totally different. The use of hESC lines is then determined by the "Code of Practice for the use of Human Stem Cell Lines," developed and regularly amended by the "Steering Committee for the UK Stem Cell Bank and for the use of Stem Cell lines" (last amended version dated April 2010).

Additionally, The Human Tissue (Quality and Safety for Human Application) Regulations 2007, fully implement the EU Tissues and Cells Directives (EUTCD), the primary aim of the Directive being to ensure the quality, safety and traceability of tissue and cells used for human application. The UK

2007 regulation clearly includes hESC lines, and states that licenses must be obtained from the Human Tissue Authority (HTA, see below) regarding testing, processing import/export and distribution. Again, this concerns only lines aimed at being used for human application.

26.3.2.2 Institutions

The initial 1991 HFE Act set up the Human Fertilisation and Embryo Authority (HFEA), which is an independent regulator (as opposed to ABM in France and ISCIII in Spain). HFEA is basically in charge of addressing the issues of IVF, donor insemination, and embryo research. The agency also regulates the storage of eggs, sperm, and embryos, and reviews all new developments in treatments and research, and advises ministers. Concerning embryo research, HFEA is in charge of delivering the licenses for any project using or creating human embryos (use of sperm, eggs in research is only licensed by the HFEA if they are used to create embryos). Of course the licenses include the projects aiming at hESC lines derivations, these licenses requiring to deposit a sample of each cell line generated at the UK Stem Cell Bank, which guarantees the quality and the centralization of any hESC lines produced in the United Kingdom.

As already mentioned, hESCs obtained from the embryos are then no longer under the scrutiny of HFEA, but falls under the competency of the "Steering Committee for the UK Stem Cell Bank and for the use of Stem Cell lines" ("steering committee"). This committee is a non-statutory body reporting annually to the Medical Research Council (MRC); its main missions are to support stem cell research, to ensure that it is conducted within a proper ethical framework transparent to the public, and also to oversee the UK Stem Cell Bank, where all hESC lines obtained in UK are stored (see HFEA). The steering committee approval has to be obtained to use any of the hESC lines stored in the UKSCB, or any hESCs that would be imported from abroad. This means that, practically, the steering committee is consulted for any research project using hESCs.

Things stop here as long as the cell line has no chance to be used for a human application. If it does, then the HTA is in charge of controlling their correct use, by releasing the licenses for any activity linked to the EU Tissues and Cells Directives: testing, processing, import/export and distribution. As explained before, the embryo procurement, theoretically covered by EUTCD as well, is guaranteed in UK by HFEA.

Finally, if the research ends up with a clinical trial, approval by the "Medicines and Healthcare products Regulatory Agency" (MHRA) Clinical

Trials Unit is required. As the trial involves a cell therapy derived from a stem cell line, a favorable opinion by the Gene Therapy Advisory Committee (GTAC) is required. Approval of the local National Health Service Research and Development (NHS R&D) office is also needed.

26.3.2.3 Process for Approval

a. Research Using Human Embryos: License From HFEA

After having obtained the approval from both the local Research Ethics Committee and the local National Health Service Research and Development (NHS R&D) office, the research team has to submit its application to HFEA. The HFEA grants research licenses for up to three years (can be renewed through the same process) for individual research projects, including derivation of new hESC lines. The process is split in several steps:

- Peer review: HFEA asks for evaluation of independent researchers to determine especially if the application:

    ○ requires human embryos to fulfil its aims and objectives
    ○ requires the numbers and types of embryos described in the application

- If the peer reviews are satisfactory, the HFEA initiates a visit to the research site, in order to:

    ○ review proposed project protocols,
    ○ inspect research laboratories, and
    ○ meet research teams.

- The inspectors' report, together with the application and the peer reviews are presented to the HFEA's Research License Committee, who decide whether to grant a research license.

This process shows some similarities with the French one, as evaluation by independent researchers and site inspections are performed.

b. Research Using Established hESC Lines: the "Steering Committee for the UK Stem Cell Bank and for the use of Stem Cell lines"

This "Steering Committee" is an independent national committee, not statutory, reporting annually to the Medical Research Council, including scientists, medical doctors, and specialists in ethics, theology, lay members, and representatives of public agencies. It oversees all research using hESCs in UK, through allowing access to hESC lines stored in the UK Stem Cell Bank or alternatively allowing import of hESC lines (it does also examine the application for established hESC exports).

Its role is to make sure that the cell lines have been ethically sourced, and are used for justified and valuable purposes: basically, the research project must increase knowledge about serious diseases and their treatment. This includes basic cell research which underpins these aims.

Again, a clear distinction is made between fundamental research and research aiming at human therapy. For example, a laboratory-based research does not have to request for a Research Ethics Committee approval to come before the Steering Committee. A fast track procedure has also been established for projects fulfilling a range of conditions, including the use of cell lines already approved by the Committee, the fact that the project has already been peer-reviewed, and that it is an exclusive *in vitro* project.

Basically, the process is lighter than the one aimed at obtaining licenses from HFEA. That reflects the will of the legislation establishing that research involving hESC lines does not need the same level of regulation to which embryo research is subject. Nevertheless, the Parliament made clear that hESCs should not be used for trivial purposes.

26.3.2.4 Embryos for Research

Having not signed the Oviedo Convention, UK is one of the few European countries where it is possible to create embryos for research (specifically forbidden in the Oviedo Convention). In addition, embryos that come from ART (donated for research by the couples undergoing an ART procedure) may be used for research purposes. These fall broadly in the same categories than the ones defined in the French law: poor quality embryos that will not be transferred, embryos discarded following a PGD, supernumerary embryos from an IVF procedure. Concerning the latter, a fundamental difference is that these supernumerary embryos could be used fresh, immediately after IVF, provided of course that enough embryos of good quality have been obtained to realize both the transfer (usually maximum two) and the donation for research. The drawback of this is that these donated embryos could have been frozen and used for a later transfer. The advantage is that the consent that has to be signed by the couple includes the precise research protocol in which the embryos will be used.

Whatever the embryo, the HFE Act sets a precise limitation concerning the period during when it could be used for research: as for Spain, this might not extend beyond 14 days after fertilization or beyond the appearance of the primitive streak, whichever occurs the first.

Being allowed to create embryos for research means that gametes used for the fertilization should also be donated. The donors have then to be precisely informed about the research project. As it has already been mentioned, such a possibility of embryo creation is almost unique in Europe and is linked to tough conditions: "embryos should not be created specifically for research purposes unless there is a demonstrable and exceptional need which cannot be met by the use of surplus embryos" (position taken by the House of Lords Committee on Stem Cell Research).

Finally, embryos could also be obtained by SNCT. HFE Act 2008 revision also considers the use of defined categories of "human admix embryos" (HAE), or interspecies embryos.

26.3.3 Germany

26.3.3.1 Regulation Frame

Research on hESCs is governed by one unique piece of law, the Stem Cell Act ("Stammzellgesetz"—"StZG") of June 2002, amended in August 2008. Literally, the importation and utilization of hESCs are banned, but by exception import and use of hESCs for research purposes can be authorized. The principle is very similar to what was included in the French Bioethics laws from 2004 to 2013 ("interdiction with dispensation") and that was has been thoroughly discussed in France.

26.3.3.2 Institutions

The institution in charge of managing the delivery of authorizations is the Robert Koch Institute (RKI), which is the German public health agency depending on the Federal Ministry of Health. In 2002, following the vote on the Stem Cell Act, it has been designated as the Competent Authority for the domain. The hESC research authorization office at RKI manages the whole process of authorization delivery, from reception to final decision. It performs it own evaluation apart from the one made by ZES (see below).

In order to provide support to the RKI decisions, the Stem Cell Act envisioned the creation of the "Central Ethics Committee for Stem Cell Research (ZES)," an independent committee including 9 members and 9 deputy members (in practice all members and deputy members participate to the meetings) specialized in medicine, biology, ethics, or theology. Members are appointed by the Ministry of Health for a 3-year period (extendable). ZES aim is to review and assess the applications for research projects involving hESCs.

26.3.3.3 Process for Approval

The conditions made for authorizing research projects are precise and their review is clearly dispatched between ZES and RKI.

RKI is the addressee of the application file. Upon reception, it sends it to ZES, which is in charge of checking whether it fulfills the conditions defined by section 5 of the Stem Cell Act, especially its section 5:

- Above all, the project must serve research goals of premium importance to generate scientific knowledge in basic research or to increase medical knowledge for the development of diagnosis.
- The relevant scientific questions must have been clarified as far as possible through in vitro models using animal cells or through animal experiments.
- A particular emphasis is also put on the fact that—according to the state of the art—no other material than hESCs can be used to answer the questions asked by the research protocol.

Taken together, these requirements strongly remind what is also included in the French Bioethics laws.

ZES has to assess whether these requirements are met by the research project and whether, accordingly, use of hESCs is ethically acceptable. ZES has 6 weeks to provide a written opinion to the RKI (and to the applicant). Nevertheless, this opinion is not binding for RKI. The RKI provides an own review that also deals, for example, with the scientific relevance of the proposed research, but also evaluates the presence of the prerequisites with regard to the hESC lines (e.g., source of embryos, lack of payment for embryo donation, compliance with legal regulations in the country of origin, etc.).

RKI has 8 more weeks to perform its review. It then releases its decision, which runs without time limitation.

Once the authorization is issued, there is no follow-up regarding the results of the research projects. RKI is neither entitled to inspect the research centers nor authorized to require research reports. Finally, RKI has to maintain a public register of authorized projects. This register contains a description of the research projects as well as the reasons that were essential for approving research. The registry also provides information on hESC lines used in the respective research projects.

26.3.3.4 Embryos for Research

As opposed to the three previous countries described, research on human embryos is strictly forbidden in Germany, as well as embryo destruction in the purpose of deriving hESC lines. As for the "Embryo Protection Act" (1991), the IVF of an ovum and the use of an embryo for any other purpose than inducing a pregnancy, including research, are prohibited under penalty.

The only researches allowed are researches using hESC lines derived abroad before 1 May 2007 (this novel cut-off date was one of the main subjects for the 2008 amendment, the initial date being 1 January 2002).

Logically but interestingly, import and use of hESC lines derived from PGD embryos are forbidden, which is reasoned by the intention to prevent any embryo selection. To our knowledge, few countries include this limitation in their hESC legislation.

26.4 Concluding Remarks

- As hESC lines derivation implied the destruction of a human embryo, most countries in Europe have developed specific legislation covering hESC research.
- Responses have been diverse and regulations, while belonging to a few broad categories, all have particularities due to national sensibilities.
- Whatever the regulation, researchers have been capable to adapt and all the countries described hold high-level scientific teams in the domain
- These regulations deal with fast-developing technologies, constantly evolving, while legislation cannot be changed as often. Regulatory agencies have then to manage an expanding gap between what is regulated and the new concepts brought in by science and technology developments.
- Future challenges in regulation should deal with pluripotency (either for iPSCs or hESCs), which is not deprived of ethical issues. For example, the—non existing yet—possibility to obtain gametes from hESCs or iPSCs (for review, Saitou 2021), the creation of inter-species chimeras (for review, De Los Angeles 2019), or the generation of embryoid-like structure *in vitro* (Moris 2020). These three topics are under discussion in the current revision of bioethics laws in France.

References

Aran, B., Rodriguez-Piza, I., Raya, A., Consiglio, A., Munoz, Y., Barri, P.N., Izpisua, J.C., and Veiga, A. (2010). Derivation of human embryonic stem cells at the Center of Regenerative Medicine in Barcelona. In Vitro. Cell Dev. Biol. Anim 46, 356-366.

Saitou M. Mammalian germ cell development: from mechanism to in vitro reconstitution. Stem Cell Reports. 2021 Feb 2:S2213-6711(21)00038-2.

De Los Angeles A. The Pluripotency Continuum and Interspecies Chimeras. Curr Protoc Stem Cell Biol. 2019 Sep; 50(1): e87.

Moris N, Anlas K, van den Brink SC, Alemany A, Schröder J, Ghimire S, Balayo T, van Oudenaarden A, Martinez Arias A. An in vitro model of early anteroposterior organization during human development. Nature. 2020 Jun; 582(7812): 410-415.

Handyside, A.H., Lesko, J.G., Tarin, J.J., Winston, R.M., and Hughes, M.R. (1992). Birth of a normal girl after in vitro fertilization and preimplantation diagnostic testing for cystic fibrosis. N. Engl. J. Med. 327, 905-909.

Rodin, S., Antonsson, L., Niaudet, C., Simonson, O.E., Salmela, E., Hansson, E.M., Domogatskaya, A., Xiao, Z., Damdimopoulou, P., Sheikhi, M., Inzunza, J., Nilsson, A.S., Baker, D., Kuiper, R., Sun, Y., Blennow, E., Nordenskjold, M., Grinnemo, K.H., Kere, J., Betsholtz, C., Hovatta, O., and Tryggvason, K. (2014). Clonal culturing of human embryonic stem cells on laminin-521/E-cadherin matrix in defined and xeno-free environment. Nat. Commun. 5, 3195.

Rugg-Gunn, P.J., Ogbogu, U., Rossant, J., and Caulfield, T. (2009). The challenge of regulating rapidly changing science: stem cell legislation in Canada. Cell Stem. Cell 4, 285-288.

Thomson, J.A., Itskovitz-Eldor, J., Shapiro, S.S., Waknitz, M.A., Swiergiel, J.J., Marshall, V.S., and Jones, J.M. (1998). Embryonic stem cell lines derived from human blastocysts. Science 282, 1145-1147.

Williams, D.A. (2009). National Institutes of Health releases new guidelines on human stem cell research. Mol. Ther. 17, 1485-1486.

27

Stem Cell Conceptual Clarifications

Lucie Laplane

CNRS, University Paris I Panthéon-Sorbonne (IHPST – UMR 8590),
Paris, France;
Gustave Roussy Cancer Center (UMR 1287), Villejuif, France
E-mail: lucie.laplane@gustaveroussy.fr

27.1 Introduction

Everybody knows what a stem cell is, and yet nobody knows what a stem cell is. Stem cells are traditionally characterized by two functional properties: self-renewal and differentiation. This characterization raises two questions. First, do these two properties adequately distinguish stem cells from non-stem cells? Second, what do we mean by self-renewal and differentiation? A recurrent problem with stem cells is that they encompass a great diversity of cells. Stem cells of different tissues, at different developmental stages, and in different species differ in their abilities to self-renew and to differentiate (Laplane, 2011). Moreover, these properties are not specific to stem cells. This raises a question about the category of stem cells: are all types of stem cells different exemplars of a same thing? Stated more philosophically, do they belong to a natural kind or is the concept of stem cell artificially regrouping cells that are biologically very different? Recent advances have undermined the traditional view of the stem cell as a fixed entity and led to an alternative view according to which stem cell refers to a cell state (Blau et al., 2001; Zipori, 2004; Mikkers and Frisen, 2005; Adler and Sánchez Alvarado, 2015; Clevers and Watt, 2018). I will contribute to these attempts to clarify what stem cells are by distinguishing four ways in which a cell can be a stem cell, and explore some practical consequences in oncology. I hope that this chapter also illustrates how philosophy can contribute to biology (Laplane et al., 2019b).

27.2 Stem Cell Definition

A definition, in its most traditional meaning, identifies a class of objects by some common characteristic properties, which take the form of a list of necessary and sufficient properties. For stem cells, whether the abilities to self-renew and differentiate are necessary and sufficient for a cell to be a stem cell is a topic of recurrent debates among biologists and philosophers (Seaberg and van der Kooy, 2003; Mikkers and Frisen, 2005; Shostak, 2006; Lander, 2009; Tajbakhsh, 2009; Fagan, 2013b; Lancaster, 2017).

A first difficulty comes from the non-specificity of self-renewal and differentiation abilities. Some non-stem cells can have a large differentiation potential like hematopoietic multipotent progenitors. Some non-stem cells can long-term self-renew, like embryonic macrophages and some lymphocytes. Moreover, self-renewal and differentiation potentials are quantitative properties rather than on-off properties that a cell would either have or not have. These potentials can be lost progressively during differentiation with no sharp gaps between stem and non-stem cells. Single-cell RNA sequencing analyses comfort this view of differentiation as a continuum rather than a series of discrete steps (e.g. Velten et al., 2017; Giladi et al., 2018; Karamitros et al., 2018).

A second source of difficulty comes from the fact that not all stem cells have extensive self-renewal and differentiation abilities. *In vivo* totipotent and pluripotent stem cells only exist transiently in the mammalian early embryo preventing long-term self-renewal, and some stem cells can only differentiate into one cell type (e.g., skeletal muscle stem cells). This conundrum gives rise to recurrent, sometimes virulent debates, between those who think that the definition should be revised and those who think that some cells, like mammalian totipotent and pluripotent stem cells *in vivo*, should not be called stem cells. For instance, Jonathan Slack, who has read several of the philosophical work that will be discussed in this Chapter, has argued that "real stem cells comprise two fundamentally different types: the pluripotent stem cells that exist only *in vitro*, and tissue-specific stem cells that exist *in vivo* in the postnatal organism" (Slack, 2018). He explicitly argues that "the cells of the early mammalian embryo that are the precursors of embryonic stem cells (ESCs), namely the cells of the inner cell mass and the epiblast, are not themselves stem cells [...] because their pluripotency is very short lived." For a large part, definitions are a matter of convention. Historically, the term stem cells was first coined to refer to the fertilized egg, from which all the cells of the organism stem (Maehle, 2011; Lancaster, 2017). Now several scientists argue that the concept would gain clarity if these early mammalian embryonic cells where not referred to as stem cells. In other species, though, these pluripotent cells can remain throughout the life of the organism and are thus called stem cells (Lai and Aboobaker, 2018), like neoblasts in planarians (Zeng et al., 2018). How different are these planarians pluripotent stem cells/ neoblasts from the pluripotent (non-stem) cells of mammalian embryo? Are the biological differences between them sufficient to support that the former are considered stem cells while the latter are not? In addition, assessing self-renewal ability might not always be easy. For example, there is a debate on how much hematopoietic stem cells (HSCs) actually self-renew during the life of an organism. Using mouse models that allow cell division tracking, some have argued that HSCs only divide four times before they enter permanent senescence (Bernitz et al., 2016; see also Wilson et al., 2008). Others have highlighted a great diversity among HSCs, suggesting that each "individual HSCs possess an almost unique capability to self-renew" (Haas et al., 2018), and among which some might never divide (and thus self-renew) during the adulthood (Morcos et al., 2020). How much should a cell self-renew to be considered a stem cell?

Stem cell definition thus faces three problems: (1) the name refers to cells that can vary greatly in their abilities to self-renew and to differentiate, (2) these two properties are unspecific and (3) the difference between stem cells and non-stem cells might be more quantitative than qualitative.

27.3 Self-Renewal and Differentiation Definition

As the concept of stem cell heavily relies on the notion of self-renewal and differentiation, ambiguities surrounding these notions should also be discussed. As highlighted by Melinda Fagan, 2013b, both notions are ambiguous in the way they use sameness (self-renewal) and difference (differentiation) to characterize stem cell reproduction abilities. As noticed by Shahragim Tajbakhsh, 2009 "self-renewal is a misnomer, because mutations accumulate in the genome with every round of DNA replication; thus, the daughter of a stem cell is marginally distinct, at least by this criterion, from its parent." No two cells are the same in every respect. Thus, strictly speaking, one could conclude that (a) stem cells never self-renew, and (b) stem cells always differentiate. Since both are wrong, one should conclude that self-renewal should not be understood as merely reproduction of the same and that differentiation should not be understood as merely a production of something different. Fagan thus proposed more stringent definitions:

> Sameness and difference of cells is relative to some set of charac-
> ters, such as size, shape, and concentration of a particular molecule.
> At a given time each cell has some value for a given character, and
> it is comparisons among these values (which I shall also refer to as
> traits) that determine sameness or difference in any particular case
> (Fagan, 2013b, p. 21).

Self-renewal should be understood as the production of a daughter cell or
two daughter cells that share the same values than the mother stem cell for
some particular characters (e.g., some cell surface proteins). Differentiation
should be understood as production of one or two daughter cells that have
different values than the mother stem cell for these same characters. Differentiation also implies a direction toward the differentiated state of the
cell type and a diversification (i.e., the production of multiple and heterogeneous cell types), leading Fagan to refine the definition of differentiation
as follows:

> Differentiation occurs within cell lineage L during interval t1-t2 if
> and only if some cells in L change their traits such that (i) cells of
> L at t2 vary more with respect to characters C than at t1 or (ii) cells
> of L at t2 have traits more similar to traits Cm1 … Cmk of mature
> cell types $\{1, \ldots k\}$ than at t1 (Fagan, 2013b, p. 24).

Fagan's formalizations of self-renewal and differentiation offer a clarification
of the use of these concepts in stem cell biology, and allow her to propose a
general definition of stem cells:

> A stem cell can be defined as the unique origin (stem) of lineage
> L for time interval n, characters C and mature cell characters M.
> Relative to these parameters, a stem cell has maximal self-renewal
> and differentiation potential (Fagan, 2013c, p. 1151)

This abstract definition has the advantage to accommodate the diversity
among stem cells, and the quantitative aspect of self-renewal and differentiation. Stem cells are not distinguished from non-stem cells because they
possess unique properties but because they have the greater ability to self-
renew and differentiate relatively to the lineage to which they belong. The
concept of stem cell is relative to a cell lineage, a period of time, and some
characters of interest.

27.4 Proving Stemness: The Uncertainty Principle

Defining stem cells by the abilities to self-renewal and differentiate generates an experimental conundrum. To prove that a cell is a stem cell, one needs to empirically validate the ability of the cell to self-renew and differentiate, which might be virtually impossible to do. (1) Different contexts (e.g., cell culture medium) might be needed to explore self-renewal as opposed to differentiation, and vice-versa. Fagan has argued that self-renewal and differentiation potential cannot both be measured for a single cell (Fagan, 2014); not even single-cell lineage tracing or single-cell transplantations makes such an attempt possible (Fagan, 2015). (2) One particular context can never show the whole potential of differentiation (Loeffler and Roeder, 2002). (3) It is logically impossible to prove self-renewal: it would need to prove that the stem cell can produce at least one new stem cell, but to prove that a new stem cell has been produced would require proving that it is able to self-renew. The proof is indefinitely deferred to the next generation of cell (Fagan, 2015). The consequence is a necessary uncertainty with regard to the stem cell identity of any given cell, which is referred to as "the uncertainty principle" in reference to Heisenberg's uncertainty principle (according to which mass and velocity of a particle cannot be simultaneously measured because it is impossible to measure one without interfering with the other) (Potten and Loeffler, 1990; Loeffler and Roeder, 2002; Zipori, 2004; Lawrence, 2004). Stem cell identification is necessarily retrospective: "Stem cell biologists literally don't know what they've got 'til it's gone" (Fagan, 2015, p. 190). Thus, a peculiarity of the stem cell is that one cannot know with certainty that a particular cell is a stem cell.

27.5 Natural Kind or Artificial Grouping?

The relative and functional nature of the definition of stem cells, the diversity between stem cells, the possible continuity in cell differentiation all raise a question surrounding the ontological status of stem cells: is the concept of stem cell referring to a biological grouping (what philosophers call "natural kinds"), or is it an artificial grouping (i.e., we see similarities between cells that are actually biologically very different) (Bird and Tobin, 2018)? Natural kinds rely on the sharing of some properties that make all the members of a kind belong to that kind. In biology, the existence of natural kinds is threatened by the inherent diversity of biological entities that results from evolution (Hull, 1965). A contribution of philosophers has been to show that diversity and uniqueness do not preclude that a grouping is natural. Biological entities of a natural kind might not all share a particular property or set of properties but instead a cluster of properties, referred to as a homeostatic property cluster (HPC) (Boyd, 1999). Each entities of the kind might possess a particular subset of these properties. The grouping is a natural kind if the cluster is non-accidental, i.e., the properties of the cluster are often co-instantiated for biological reasons. Wilson et al., 2007 defended the view that stem cells, despite their heterogeneity, are a natural kind because they share a HPC. They suggest the following HPC definition of stem cells:

- morphologically undifferentiated;
- ability of self-renewal (cell division with at least one daughter cell of the same type) over an extended period of time;
- ability to give rise to various differentiated cell types (pluripotency, or at least multipotency);
- developmentally derived from certain cells or tissues located in specific parts of tissues;
- particular complex profile of gene expression and presence of transcription factors;
- found in certain cellular-molecular microenvironment ("niche"), which influences the stem cell's behavior; and
- low rate of cell division (Wilson et al., 2007).

In accordance with the HPC view, none of these properties are necessary, and different subsets may be sufficient, to be defined as a stem cell. Wilson et al. offer an interesting proof of principle that stem cells can belong to a natural kind despite their heterogeneity and the non-specificity of self-renewal and differentiation. Stated another way, the great variability observed between stem cells might still be superficial and hide a common underlying essence. But they do not provide a demonstration that this is actually the case, leaving the debate wide open.

27.6 Does the Concept of Stem Cell Refer to an "Entity" or to a Cell "State"?

An ontological debate on stem cells has gained importance over the past decades: does the concept of stem cell refers to an "entity" or to a particular cell "state"? Traditionally stem cells are understood as a unique type of cell, with specific constitutive properties. A box that no other cells can enter:

either a cell is born a stem cell or not. Accumulating evidence have shown that differentiation might not be as irreversible as initially thought (for a historical perspective see: Maienschein, 2003; Maienschein, 2014; Kraft and Rubin, 2016). The possibility to reprogram the nucleus of a differentiated cell into a pluripotent stem cell through *in vitro* cloning (Gurdon, 1962), and to reprogram somatic cells into induced pluripotent stem cells (Takahashi and Yamanaka, 2006) offered in vitro demonstration that cell fate is reversible. Evidence of reprogramming is less obvious in vivo, and was thought, until recently, to be restricted to plants or animals with extensive regeneration ability (e.g., Salamander limb regeneration). Helen Blau's lab early advocated that there are no cell types, only cell states, and that cell identity is actively maintained by the environment rather than intrinsically wired (Blau et al., 1985; Blau and Baltimore, 1991; Kraft and Rubin, 2016). Recent evidence showed that dedifferentiation to a stem cell state can occur in mammals, notably, in epithelial tissues in context of repair (Blanpain and Fuchs, 2014;Donati and Watt, 2015). This has led several biologists to claim that the traditional entity view should be abandoned and replaced by the state view. According to this emerging view stemness is not an essence but a function that cells can express when they are in a certain state. Stemness is then perceived as a transient and reversible property that many different cells can acquire (Blau et al., 2001; Zipori, 2004; Mikkers and Frisen, 2005; Adler and Sánchez Alvarado, 2015; Hermann and Sainz, 2018; Clevers and Watt, 2018). Philosophers have since discussed this entity/state alternative on several occasions. Leychkis et al., 2009 and Fagan, 2013a claim that the debate cannot be empirically solved (because the current data are insufficient to determine if the concept of stem cell refers to entities or to a cell state), but that it can be theoretically or conceptually solved. Leychkis et al., 2009) highlight the shortcomings of the entity theory and the state theory, and argue that a theoretical clarification could resolve the debate between them. They suggest a refined genetic interpretation of both theories. They show that if the main claim of the entity theory is that the expression of some genes provides the best biological explanation of the potential for self-renewal and multilineage differentiation, that is, for stemness, then there is no disagreement with the state theory where stemness could be transient and acquired but would also be instantiated by the expression of these genes. Fagan, on her part, argues that shifting the focus from theories to models can better solve the stemness debate. She shows that the entity and the state views are two particular specifications of one model—her minimal stem cell model described in section 3. The state and entity views simply correspond to different interpretations or specifications of variables n (time interval), C (stem cell characters), M (mature cell characters). For instance, in the entity view, C would be the genetic signature, whereas in the state view, C could be something different, like localization in a niche. More recently Fagan has suggested that the concept of stem cell is not referring to a cell but to a place in a lineage: "stem cells are not cells," they "are instead the starting-points of lineages associated with particular experimental contexts" (Fagan, 2019). While Fagan and Leychkis et al. were interested in highlighting common grounds between the entity and state views, others have argued in favor of a transition from the entity view to a state view. Cheryle Lancaster has offered a historical reconstruction of the two views according to which the experimental work carried out since Till and McCulloch "has in fact demonstrated that stemness is a state, and that stem cells are not an entity" (Lancaster, 2017, p. 275). Anja Pichl, 2019 has criticized the maintenance of the entity view, showing that the reductionism on which it relies has a misleading influence on the understanding of stem cells, reinforcing problems in both science and society.[1]

27.7 Stemness Ontology: Not Two but Four

The debate between the entity and state views is mostly perceived and discussed as if there were only two possibilities: either the concept of stem cell refers to an entity (the old view) or it refers to a state (the new view), and only one of them can be biologically true (the new view). I see two issues with this interpretation.

First, in the case of stem cells, I see no reason that only one of these views would be correct. There are many stem cell types in the organisms. Jason Robert early defended that "no stem cell is a model for stem cells as such" (Robert, 2004, p. 1008). Different stem cells might correspond to different views. For example, in mammals, dedifferentiation can occur in context of repair of epithelial tissues such as the skin or intestine, while it does not occur in the hematopoietic system. Stem cells do not work the same ways in those tissues (Clevers and Watt, 2018; Post and Clevers, 2019). Recognizing this ontological diversity is of foremost empirical importance as it changes what one can expect from different stem cells, and how to identify or work with them.

---

[1]Ethical and societal issues are left aside from this Chapter. Hauskeller, Manzeschke, and Pichl have recently edited a book that gathers these different aspects (Hauskeller et al., 2019).

Second, the dichotomy between state and entity suggests there are only two possibilities, but a careful look at stem cell biology shows that this dichotomy conflates two questions that should be distinguished: (1) can stemness be acquired? (2) Does the microenvironment play a determinant role in stemness? When distinguished, these two questions lead to four alternative views of stem cells (recapitulated in Table 1) (Laplane, 2016). So far, I have not discussed the role of the microenvironment. Does it play a determinant role in stemness? Although there is now a widely established consensus that stem cells reside in structured environments that can play important roles, the traditional account of stem cells initially made little case of the environment. There are of course a few exceptions such as the work of Wolf and Trentin, 1968. Alexander Friedenstein was also pioneering experimental description of mesenchymal stromal cells at that time, building on the early idea of Alexander Maximow's view on the interaction between hematopoietic cells and their stroma at the very beginning of the twentieth century (Friedenstein, 1989). The notion of "stem cell niche" was introduced in the late 1970s by Ray Schofield, 1978, 1983, but raised little consideration at first. Despite these lines of work, in the early traditional view, stemness was conceived as a categorical property-a constitutive property that belongs to an object, so that the property can be explained without referring to anything else other than the object itself (Ellis, 2010). A classic example of categorical property is the atomic structure of chemical elements. It is unclear whether any stem cell type actually instantiates that possibility in healthy tissues, but some data suggest that stemness might become a categorical property in some malignant contexts. For example, in hematological malignancies, some mutations can lead to constitutive activation of signaling pathways that impede the regulation of the self-renewal/differentiation balance by the microenvironment (Staerk and Constantinescu, 2012; Deininger et al., 2017). In a mouse model of myeloproliferative neoplasm, it was shown that cancer cells damage the bone-marrow microenvironment, resulting in mutated cells showing an advantage over healthy cells (Arranz et al., 2014). Healthy HSCs-whose survival relies on their niche-are greatly impacted by the aforementioned damage while the mutated stem cells expand, suggesting that upon transformation these malignant stem cells gained independence from this niche. The central or dispensable role of the niche could change with the types of alterations acquired. Disease progression could thus lead to (or result from) a progressive loss of microenvironment-dependency (Méndez-Ferrer et al., 2020).

The classical view of the stem cell niche is that the niche regulates stemness activity, i.e., whether a stem cell stays quiescent or divides, whether it self-renews and/or differentiates (e.g.,Arai and Suda, 2008). In this view, stemness remains a constitutive properties of stem cells: the niche only regulates a property that is already there. Stemness, then, is what philosophy calls a dispositional property-a property that has a categorical basis (it is constitutive to the object) but whose expression depends on extrinsic stimuli (Mumford, 1998). Fragility, for example, is a dispositional property. Some objects are fragile, other are not. However, objects that are fragile only break if there is an impact, thus requiring extrinsic stimuli. With the current state of knowledge in stem cell biology, stemness appears as a dispositional property in non-malignant HSCs: non-stem cells of the hematopoietic system cannot return to a stem cell state, indicating categorical bases, but what they do is tightly regulated by the bone-marrow microenvironment. There is an alternative view of the stem cell niche, less popular but constantly gaining more traction, according to which the niche does more than just regulating the stemness activity, it can actually induce stemness in non-stem cells.[2] In this case stemness would be better described as a relational property-a property that is entirely determined by a relation between two entities. "Partner" is an example of relational property: you are the partner of your partner only as long as your relationship stands. In biology, an example comes from the study of the germline in the drosophila and mouse: when stem cells are removed, non-stem cells can return to the niche and, once there, return to a stem cell state (Brawley and Matunis, 2004; Kai and Spradling, 2004; Barroca et al., 2009).

The notions of niche and microenvironment can be ambiguous (Laplane et al., 2018, 2019a) but most often the stem cell niche refers to cells that are in close contact with stem cells, and the factors they secrete, such as the cap or hub cells in the drosophila germline. It is unclear that stemness reacquisition, when it can happen, necessarily requires the presence of such niche cells. For example, Gupta et al., 2011 sorted breast cancer cell lines into stem and non-stem cells populations separately cultured *in vitro*, and observed a return to equilibrium, with non-stem cells returning to a stem cell state. This experiment suggests that stemness can be reacquired in absence of specific niche cells. The authors interpret stemness reacquisition as an effect of stochastic gene expression. At all time, stochastic expression of stemness genes would allow a certain proportion of cells to be in a stem cell

---

[2]Interestingly, Schofield early view of the niche was very similar to this emerging view has he believed that stemness was not a property of a cell but a cell plus its environment and that cells could gain and loose stemness by entering and exiting the stem cell niche.

state. Another interpretation could be that cellular behavior is governed at the population level, like in ants where the level of pheromone in the colony can modulate the phenotypes of individual ants. In any case, when stemness can be reacquired without involving a relationship with a specific niche, it would be more correctly described as a systemic property—a property regulated at the system level that can be gain by any (or various) entities of the system. Agent Smith in Matrix movies is a systemic property as any human in the matrix can become an agent at any time.

27.8 Philosophy Matters

Does it matter whether stemness is a categorical, dispositional, relational, or systemic property? First, depending on what stemness is, different requirements might be needed to isolate and study them. If stemness is a dispositional or relational property, special attention should be payed to the relationship between stem cells and their microenvironment. Relational or systemic properties require attention to the factors that can induce stemness, which might be of very different nature in these two cases. Second, what can be expected from cells will also be different depending on the nature of stemness. As indicated by Pichl, 2019, the wrong idea of stemness can lead to illegitimate hopes of magical stem cell treatments. Third, these four stem cell ontologies have important consequence for cancer treatments, in particular in the context of the cancer stem cell (CSC) model (Laplane, 2016). The CSC model depicts cancer organization as similar to healthy tissues, with a subpopulation of CSCs that serves as a reservoir to fuel the production of cancer cells. This model leads to an important change in the way to conceive cancer and its treatment. In the traditional view of cancer, all cancer cells contribute to the tumor development and can deem therapy ineffective. In the CSC model, cancer non-stem cells are doomed to senescence and cell death. Their contribution to tumor development is thus limited over time. If CSCs are the only reservoir of tumorigenesis then their elimination should be "necessary and sufficient" to cure cancer (Reya et al., 2001). Necessary, because remaining CSCs could lead to relapse. Sufficient, because in absence of CSCs the population of cancer cells would no longer be able to maintain itself. The CSC model has led to the search of new cancer treatments oriented toward two strategies: targeting the CSCs or targeting their niche (Saygin et al., 2019). These two strategies heavily rely on presuppositions with regard to the ontology of stem cells.

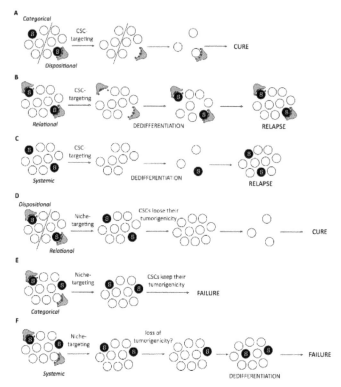

Figure 27.1   Diagrammatic representation of the success or failure of therapeutic strategies according to stemness ontology. Black dots with a S are CSCs, white dots are cancer non-stem cells.A-C illustrate the CSC targeting strategies. Targeting CSCs is necessary and sufficient to cure cancers if stemness is a categorical or a dispositional property (A), but insufficient if stemness is a relational property (B) or a systemic property (C). D-F illustrate the niche targeting strategy that relies on stemness being a disposition or a relational property (D), but might fail to be curative if stemness is a categorical property (E) or a systemic property (F).

CSC-targeting strategies rely on the assumption that stemness is a constitutive property of stem cells (either categorical or dispositional; see Figure 27.1A). If stemness is a relational property, then cancer non-stem cells, left untargeted, could reenter emptied niche, thereby recreating a new pool of CSCs (Figure 27.1B). A mouse model of colon cancer, in which it was possible to induce LGR5$^+$ CSCs cell death, has offered a recent illustration (Melo et al., 2017). After elimination of the CSCs, Melo et al. observed the reoccurrence of new CSCs in the primary tumors but not in the liver metastases, indicating that the niche in which primary tumors were located was necessary for stemness reacquisition. If stemness is a systemic property then one should expect the cancer population to react to the elimination of CSCs by recreating a new pool of CSCs (Figure 27.1C), as observed *in vitro* by Gupta et al., 2011. A recent study using a strategy similar to that of Melo et al. showed that eliminating POLR1A$^+$ cancer cells that have elevated rDNA transcription rates and protein synthesis capacity could overcome dedifferentiation. The capacity of non-stem cells to return to a stem cell state might become gradually limited due to the downregulation of the rDNA transcription machinery up to a point where dedifferentiation is no longer possible (Morral et al., 2020). A larger target, including actual and potential CSCs might thus provide alternative solutions in cases where stemness is a relational or systemic property. The niche-targeting strategies rely on a different assumption: that stemness relies on the interaction of stem cells with their niche, which is only the case if stemness is a dispositional or relational property (Figure 27.1D-F). As discussed above with the examples of malignant hemopathies, cancer transformation can come with a loss of regulation by the niche, which could compromise the efficiency of niche-targeting strategies. Alternatively, there might be ways for non-stem cells to return to a stem cell state in absence of specific structured niches, such as stochastic gene expression or intra-population regulations as discussed above.

The discovery of cell plasticity and of the capacity of cancer cells to return to a stem cell state in some cancers introduced significant skepticism among biologists toward the CSC model. Now it is frequent to hear researchers claiming that the model does not hold. I would like to make two comments with this regard. Firstly, that the CSC model does not hold in some cancers does not mean that it holds for no cancer at all. CSCs might be of different nature in different cancer. Secondly, the discovery that CSCs could be of a different nature than initially thought does not per se justify abandoning the model, but it does call for its revision. If stemness is a relational property, it can still be assumed that niche-targeting strategies might find therapeutic applications (Boesch et al., 2016). If stemness is a systemic property, the urgency is in understanding the mechanisms that allow the acquisition of stemness by cancer non-stem cells in order to inhibit the production of new CSCs.

27.9 Concluding Remarks

This chapter has highlighted the epistemological difficulties inherent to the question "what is a stem cell?" These difficulties concern the definition (or classification) of stem cells—the distinction between stem cells and non-stem cells—and the ontology of stem cells. Philosophers have contributed to the clarification of stem cell definition, and have refined the debate between the state and entity views. I have shown that stemness can be described as four different types of property. Stemness is most likely a different type of property in different types of stem cells (e.g., it is a dispositional property in HSCs and a relational property in the germline). This philosophical analysis bears considerable practical consequences, in particular in the field of oncology since the effectiveness of certain therapeutic strategies depends on the type of property stemness is. At the end of this chapter, a question remains open: if there are four different types of stem cells, then does it mean that stem cells do not belong to one but four natural kinds? Should the notion of stem cell be abandoned and replaced by four different concepts? The state of knowledge on stem cell biology does not allow any firm position on this regard. It is possible that all stem cells share a common essence but act differently because of the context in which they are. A better understanding of the evolution of stem cell types could provide important insights on whether stem cells share a common cell type ancestor or whether stemness occurred multiple times during evolution as various solutions to a similar problem (e.g., the maintenance of tissues in multicellular organisms).

Highlights

- The question "what is a stem cell?" is both biological and philosophical.
- Philosophical issues relate to the definition of stem cells and to their ontology (what they are).
- Self-renewal and differentiation are quantitative properties that are not specific to stem cells and that vary between stem cells. They do not easily discriminate stem cells from non-stem cells.
- There are four different ways in which cells are stem cells. In some stem cells stemness is a categorical property, in others a dispositional property, in others a relational property, and in yet others a systemic property.
- What type of property stemness is depends on whether non-stem cells can gain stemness/return to a stem cell state (if yes, stemness is a relational or systemic property; if not, stemness is a categorical or dispositional property), and whether the microenvironment plays a determinant role (if yes, stemness is a dispositional or relational property; if not, stemness is a categorical or systemic property) (see Table 27.1).

Table 27.1 What kind of property is stemness?

	Niche-independent	Niche-dependent
No dedifferentiation	Categorical property	Dispositional property
Dedifferentiation	Systemic property	Relational property

This table presents four conceptions of stemness according to the role of the niche and the possibility of dedifferentiation.

- Philosophy matters: depending on whether stemness is a categorical, dispositional, relational, or systemic property in a given cancer, different therapeutic strategies will be better suited for its treatment (see Figure 27.1).

Acknowledgments

I would like to thank the French stem cell biology community which provides me with a great amount of support and offers many opportunities to discuss my work. These discussions have prompted invaluable comments and questions. A great number of stem cell biologists, in France as abroad, show a real open mind and the will to question their science in every possible way. It is a pleasure to work with such an amazing scientific community. I take the opportunity to also thank the students in Stem Cell Biology who often raise great, sometime unexpected questions. Finally, I thank Claire McMurray for correcting English and providing feedbacks. This chapter have also benefited discussions with more philosophers than I can acknowledge here. I am very happy to see the philosophical literature on the biology of stem cells growing over the years. Lastly, I would like to thank the institutions that have provided me funding to develop this research over the years, in particular the MITI at CNRS, the Cancéropôle Île-de-France, the ANR and the McDonnell Foundation.

References

Adler CE, Sánchez Alvarado A (2015) Types or States? Cellular Dynamics and Regenerative Potential. Trends Cell Biol 25:687–696. doi: 10.1016/j.tcb.2015.07.008
Arai F, Suda T (2008) Quiescent stem cells in the niche. In: StemBook (ed) The Stem Cell Research Community. StemBook

Arranz L, Sánchez-Aguilera A, Martín-Pérez D, et al (2014) Neuropathy of haematopoietic stem cell niche is essential for myeloproliferative neoplasms. Nature 512:78–81. doi: 10.1038/nature13383

Barroca V, Lassalle B, Coureuil M, et al (2009) Mouse differentiating spermatogonia can generate germinal stem cells in vivo. Nat Cell Biol 11:190-6. doi: ncb1826 [pii] 10.1038/ncb1826

Bernitz JM, Kim HS, MacArthur B, et al (2016) Hematopoietic Stem Cells Count and Remember Self-Renewal Divisions. Cell 167:1296-1309.e10. doi: 10.1016/j.cell.2016.10.022

Bird A, Tobin E (2018) Natural Kinds. In: Zalta EN (ed) The Stanford Encyclopedia of Philosophy. Metaphysics Research Lab, Stanford University

Blanpain C, Fuchs E (2014) Stem cell plasticity. Plasticity of epithelial stem cells in tissue regeneration. Science (80-) 344:1242281. doi: 344/6189/1242281 [pii] 10.1126/science.1242281

Blau HM, Baltimore D (1991) Differentiation Requires Continuous Regulation. J Cell Biol 112:781–783. doi: 10.1083/JCB.112.5.781

Blau HM, Brazelton TR, Weimann JM (2001) The evolving concept of a stem cell: entity or function? Cell 105:829-41. doi: S0092-8674(01)00409-3 [pii]

Blau HM, Pavlath GK, Hardeman EC, et al (1985) Plasticity of the differentiated state. Science 230:758-66. doi: 10.1126/science.2414846

Boesch M, Sopper S, Zeimet AG, et al (2016) Heterogeneity of Cancer Stem Cells: Rationale for Targeting the Stem Cell Niche. Biochim Biophys Acta 1866:276–289. doi: 10.1016/j.bbcan.2016.10.003

Boyd R (1999) Homeostasis, Species, and Higher Taxa. In: Wilson RA (ed) Species: New Interdisciplinary Essays. MIT Press, Cambridge, pp 141–185

Brawley C, Matunis E (2004) Regeneration of male germline stem cells by spermatogonial dedifferentiation in vivo. Science (80-) 304:1331-4. doi: 10.1126/science.1097676 1097676 [pii]

Clevers H, Watt FM (2018) Defining Adult Stem Cells by Function, not by Phenotype. Annu Rev Biochem 87:1015–1027. doi: 10.1146/annurev-biochem-062917-012341

Deininger MWN, Tyner JW, Solary E (2017) Turning the tide in myelodysplastic/myeloproliferative neoplasms. Nat Rev Cancer 17:425–440. doi: 10.1038/nrc.2017.40

Donati G, Watt FM (2015) Stem Cell Heterogeneity and Plasticity in Epithelia. Cell Stem Cell 16:465–476. doi: 10.1016/j.stem.2015.04.014

Ellis B (2010) Causal Powers and Categorical Properties. In: Marmodoro A (ed) The Metaphysics of Powers: Their Grounding and Their Manifestations. Routledge, London, pp 133–142

Fagan MB (2013a) Stem Cell Pluralism: Responding to the "Stemness" Debate. Exeter

Fagan MB (2019) Stem Cell Biology: a conceptual overview. In: Hauskeller C, Manzeschke A, Pichl A (eds) The Matrix of Stem Cell Research. An Approach to Rethinking Science in Society. Routledge

Fagan MB (2013b) Philosophy of Stem Cell Biology. Knowledge in Flesh and Blood. Palgrave-Macmillan, London

Fagan MB (2013c) Philosophy of Stem Cell Biology - an Introduction. Philos Compass 8:1147–1158. doi: 10.1111/phc3.12088

Fagan MB (2014) The Stem Cell Uncertainty Principle. Philos Sci 80:945–957. doi: 10.1086/674014

Fagan MB (2015) Crucial stem cell experiments? Stem cells, uncertainty, and single-cell experiments. Theor An Int J Theory, Hist Found Sci 30:183–205

Friedenstein A (1989) Stromal-hematopoietic interrelationships: Maximov's ideas and modern models. Mod Trends Hum Leuk VIII Haematol Blood Transfus / Hämatologie und Bluttransfusion, 32:159–167. doi: 10.1007/978-3-642-74621-5-27

Giladi A, Paul F, Herzog Y, et al (2018) Single-cell characterization of haematopoietic progenitors and their trajectories in homeostasis and perturbed haematopoiesis. Nat Cell Biol 20:836–846. doi: 10.1038/s41556-018-0121-4

Gupta PB, Fillmore CM, Jiang G, et al (2011) Stochastic state transitions give rise to phenotypic equilibrium in populations of cancer cells. Cell 146:633-44. doi: S0092-8674(11)00824-5 [pii] 10.1016/j.cell.2011.07.026

Gurdon JB (1962) The developmental capacity of nuclei taken from intestinal epithelium cells of feeding tadpoles. J Embryol Exp Morphol 10:622-40

Haas S, Trumpp A, Milsom MD (2018) Causes and Consequences of Hematopoietic Stem Cell Heterogeneity. Cell Stem Cell 22:627–638. doi: 10.1016/j.stem.2018.04.003

Hauskeller C, Manzeschke A, Pichl A (2019) The matrix of stem cell research: an approach to rethinking science in society. Routledge

Hermann PC, Sainz B (2018) Pancreatic cancer stem cells: A state or an entity? Semin Cancer Biol 53:223–231. doi:10.1016/j.semcancer.2018.08.007

Hull DL (1965) The Effect of Essentialism on Taxonomy–Two Thousand Years of Stasis (II). Br J Philos Sci 16:1–18. doi: 10.2307/686135

Kai T, Spradling A (2004) Differentiating germ cells can revert into functional stem cells in Drosophila melanogaster ovaries. Nature 428:564–9. doi: 10.1038/nature02436nature02436 [pii]

Karamitros D, Stoilova B, Aboukhalil Z, et al (2018) Single-cell analysis reveals the continuum of human lympho-myeloid progenitor cells. Nat Immunol 19:85–97. doi: 10.1038/s41590-017-0001-2

Kraft A, Rubin BP (2016) Changing cells: An analysis of the concept of plasticity in the context of cellular differentiation. Biosocieties 11:497–525. doi: 10.1057/s41292-016-0027-y

Lai AG, Aboobaker AA (2018) EvoRegen in animals: Time to uncover deep conservation or convergence of adult stem cell evolution and regenerative processes. Dev Biol 433:118–131. doi: 10.1016/j.ydbio.2017.10.010

Lancaster C (2017) A history of embryonic stem cell research: Concepts, laboratory work, and contexts. Duram University

Lander AD (2009) The "stem cell" concept: is it holding us back? J Biol 8:70. doi: jbiol177 [pii] 10.1186/jbiol177

Laplane L (2011) Stem Cells and the Temporal Boundaries of Development: Toward a Species-Dependent View. Biol Theory 6:48–58. doi: 10.1007/s13752-011-0009-z

Laplane L (2016) Cancer Stem Cells: Philosophy and Therapies. Harvard Univerity Press, Cambridge, MA

Laplane L, Duluc D, Bikfalvi A, et al (2019a) Beyond the tumour microenvironment. Int J Cancer ijc.32343. doi: 10.1002/ijc.32343

Laplane L, Duluc D, Larmonier N, et al (2018) The Multiple Layers of the Tumor Environment. Trends in Cancer 4:802–809. doi: 10.1016/j.trecan.2018.10.002

Laplane L, Mantovani P, Adolphs R, et al (2019b) Why science needs philosophy. Proc. Natl. Acad. Sci. U. S. A. 116:3948–3952

Lawrence HJ (2004) Stem cells and the Heisenberg uncertainty principle. Blood 104:597–598. doi: 10.1182/blood-2004-05-1862

Leychkis Y, Munzer SR, Richardson JL (2009) What is stemness? Stud Hist Philos Biol Biomed Sci 40:312–320. doi: S1369-8486(09)00051-X [pii]10.1016/j.shpsc.2009.09.002

Loeffler M, Roeder I (2002) Tissue stem cells: definition, plasticity, heterogeneity, self-organization and models–a conceptual approach. Cells Tissues Organs 171:8–26. doi: cto71008 [pii]

Maehle AH (2011) Ambiguous cells: the emergence of the stem cell concept in the nineteenth and twentieth centuries. Notes Rec R Soc L 65:359–378

Maienschein J (2003) Who's View of Life? Embryos, Cloning, and Stem Cells. Harvard Univerity Press, Cambridge, MA

Maienschein J (2014) Embryos Under the Microscope: The Diverging Meanings Of Life. Harvard Univerity Press, Cambridge, MA

Melo F de SE, Kurtova A V, Harnoss JM, et al (2017) A distinct role for Lgr5(+) stem cells in primary and metastatic colon cancer. Nature 543:676–680. doi: 10.1038/nature21713

Méndez-Ferrer S, Bonnet D, Steensma DP, et al (2020) Bone marrow niches in haematological malignancies. Nat Rev Cancer 20:285–298. doi: 10.1038/s41568-020-0245-2

Mikkers H, Frisen J (2005) Deconstructing stemness. J Eur Mol Biol Organ 24:2715–2719. doi: 7600749 [pii] 10.1038/sj.emboj.7600749

Morcos MNF, Zerjatke T, Glauche I, et al (2020) Continuous mitotic activity of primitive hematopoietic stem cells in adult mice. J Exp Med 217:e20191284. doi: 10.1084/jem.20191284

Morral C, Stanisavljevic J, Hernando-Momblona X, et al (2020) Zonation of Ribosomal DNA Transcription Defines a Stem Cell Hierarchy in Colorectal Cancer. Cell Stem Cell 0:. doi: 10.1016/j.stem.2020.04.012

Mumford S (1998) Dispositions. Oxford University Press, Oxford

Pichl A (2019) What Keeps an Outdated Stem Cell Concept Alive? A search for traces in science and society. In: Hauskeller C, Manzeschke A, Pichl A (eds) The Matrix of Stem Cell Research. An Approach to Rethinking Science in Society. Routledge, p 214

Post Y, Clevers H (2019) Defining Adult Stem Cell Function at Its Simplest: The Ability to Replace Lost Cells Through Mitosis. Cell Stem Cell 25:174–183. doi: 10.1016/J.STEM.2019.07.002

Potten CS, Loeffler M (1990) Stem cells: attributes, cycles, spirals, pitfalls and uncertainties. Lessons for and from the crypt. Development 110:1001–1020

Reya T, Morrison SJ, Clarke MF, Weissman IL (2001) Stem cells, cancer, and cancer stem cells. Nature 414:105-11. doi: 10.1038/3510216735102167 [pii]

Robert JS (2004) Model systems in stem cell biology. Bioessays 26:1005–1012. doi:10.1002/bies.20100

Saygin C, Matei D, Majeti R, et al (2019) Targeting Cancer Stemness in the Clinic: From Hype to Hope. Cell Stem Cell 24:25–40. doi: 10.1016/j.stem.2018.11.017

Schofield R (1978) The relationship between the spleen colony-forming cell and the haemopoietic stem cell. Blood Cells 4:7–25

Schofield R (1983) The stem cell system. Biomed Pharmacother 37:375-80

Seaberg RM, van der Kooy D (2003) Stem and progenitor cells: the premature desertion of rigorous definitions. Trends Neurosci 26:125-31. doi: S0166223603000316 [pii]

Shostak S (2006) (Re)defining stem cells. Bioessays 28:301–308. doi: 10.1002/bies.20376

Slack JMW (2018) What is a stem cell? Wiley Interdiscip Rev Dev Biol 7:e323. doi: 10.1002/wdev.323

Staerk J, Constantinescu SN (2012) The JAK-STAT pathway and hematopoietic stem cells from the JAK2 V617F perspective. JAK-STAT 1:184–190. doi: 10.4161/jkst.22071

Tajbakhsh S (2009) Stem cell: what's in a name? Nat Reports Stem Cells. doi: 10.1038/stemcells.2009.90

Takahashi K, Yamanaka S (2006) Induction of pluripotent stem cells from mouse embryonic and adult fibroblast cultures by defined factors. Cell 126:663-76. doi: 10.1016/j.cell.2006.07.024

Velten L, Haas SF, Raffel S, et al (2017) Human haematopoietic stem cell lineage commitment is a continuous process. Nat Cell Biol 19:271–281. doi: 10.1038/ncb3493

Wilson A, Laurenti E, Oser G, et al (2008) Hematopoietic stem cells reversibly switch from dormancy to self-renewal during homeostasis and repair. Cell 135:1118–1129. doi: S0092-8674(08)01386-X [pii] 10.1016/j.cell.2008.10.048

Wilson RA, Barker MJ, Brigandt I (2007) When Traditional Essentialism Fails: Biological Natural Kinds. Philos Top 35:189–215. doi: 10.2307/43154503

Wolf NS, Trentin JJ (1968) Hemopoietic colony studies. V. Effect of hemopoietic organ stroma on differentiation of pluripotent stem cells. J Exp Med 127:205–214. doi: 10.1084/jem.127.1.205

Zeng A, Li H, Guo L, et al (2018) Prospectively Isolated Tetraspanin+ Neoblasts Are Adult Pluripotent Stem Cells Underlying Planaria Regeneration. Cell 173:1593-1608.e20. doi: 10.1016/J.CELL.2018.05.006

Zipori D (2004) The nature of stem cells: state rather than entity. Nat Rev Genet 5:873-8. doi: nrg1475 [pii]10.1038/nrg1475

28

Future Outlook

Olivera Miladinovic, Pierre Charbord* and Charles Durand*

Sorbonne Université, CNRS, Inserm U1156, Institut de Biologie Paris Seine, Laboratoire de Biologie du Développement/UMR7622, 9 Quai St-Bernard, 75005 Paris, France
E-mail: charles.durand@sorbonne-universite.fr;
pierre.charbord@sorbonne-universite.fr
*Corresponding authors

28.1 Introduction

The second edition of the book "Stem Cell Biology and Regenerative Medicine" aims at providing the readers with a cutting-edge knowledge on stem cell biology. The first section was dedicated to introduce important concepts and methods used in the field (chapters 1 to 6). These included the biology of embryonic and adult stem cells, the procedures for enriching cell populations with stem cell activity and testing their functionalities, the modes of stem cell proliferation and differentiation, the existence of stem cell niches, and the molecular mechanisms involved in gene expression regulation and DNA repair. Computational approaches and models to investigate stemness were also discussed in details (chapters 5 and 6). The second section continued the discussion on pluripotency with a particular emphasis on reprogramming and disease modeling using hepatic cells as an example, X chromosome inactivation and regeneration in several model organisms such as plants, *Hydra* and anamniotic vertebrates (chapters 7 to 11). The third section covered the biology of diverse adult stem cells including hematopoietic, neural, dental, mesenchymal, epithelial, mammary, and intestinal stem cells (chapters 12 to 21). The fourth section explored some aspects related to cancer stem cells and cell therapy (chapters 22 to 25). Finally, the last section addressed important issues on stem cell biology and regenerative

705

medicine with an epistemological and regulatory perspective (chapters 26 and 27). Collectively, all these contributions invited the readers in understanding what are stem cells, how are they regulated during development and diseases and how could they be used in therapeutical settings. We also believe they provide opening and challenging questions on stem cell research both on the fundamental and translational points of view. Some of these issues are briefly discussed in this concluding chapter.

28.2 Understanding and Capturing Pluripotency

We have seen that embryonic and adult stem cells represent two distinct cellular entities differing in terms of origin and functional properties. Pluripotency defines a unique differentiation potential (i.e., the ability to give rise to all tissues of an organism including the germ line) restricted to a very short period of time during development, the blastocyst stage prior implantation. From the naive epiblast, embryonic stem (ES) cell lines can be isolated and maintained undifferentiated in culture for long periods of time (Nichols and Smith, 2009). While the development of ES cell lines has enabled considerable progress on the identification of the molecular mechanisms and gene networks involved in pluripotency, whether these lines account for the full potential of the cells of the inner cell mass of pre-implantation embryos and capture their genetic and functional heterogeneity remains uncertain. Furthermore, it is unclear how pluripotent stem cells *in vivo* segregate and contribute to the germ line and the diverse somatic cell lineages representative of the three germ layers during early development. Studying pluripotency constitutes an exceptional opportunity to explore in depth the cascade of molecular and cellular mechanisms involved in cell differentiation during early development. It has also major impacts on increasing the efficiency of reprogramming technologies. This field is under intensive investigations and raises major conceptual and methodological issues. For example, if a major gene regulatory network involving *Nanog, Sox2,* and *Oct4* has been shown to regulate pluripotency, how this network is connected with external signaling pathways, epigenetic modifications, and chromatin remodeling remains poorly understood. Moreover, different degrees of pluripotency (naive, primed, and alternative) have been identified and have to be further characterized at the molecular level and also between species (interestingly, human ES cells share many similarities with rodent primed ES cells) (Li and Izpisua Belmonte, 2018; Nichols and Smith, 2009). Contrasting with pluripotency as a transient state *in vivo,* several chapters in this book explored the fascinating issue related to the reprogramming of somatic cells into pluripotent stem cells. Reprogramming and differentiation of pluripotent stem cells into specific cell lineages are also instrumental for modeling human diseases, testing the toxicity and efficacy of novel medicine, and eventually producing cell derivatives that could be used in clinical applications to restore physiological functions. This is a very challenging issue. It requires to properly recapitulate in a petri dish the coordinated molecular and cellular pathways that are active in the developing embryo and that control in time and space complex cell specification processes. *Ex vivo*, defining the appropriate combination of cytokines, growth factors and signaling molecules at precise time points and concentrations is a perquisite for efficiently support cell survival, proliferation and differentiation of specific cell types. Physical parameters such as the elasticity of the extracellular matrix, oxygen tension, blood flow and mechanical forces also play a crucial role and need to be adapted for any cell differentiation pathways. Moreover, as highlighted by chapters 8, 17, 21, and 23, novel 3D technologies and biomaterials are emerging as important strategies for mimicking the structure, mechanical properties and functions of specific tissues. Among the challenges that face stem cell biologists in this field are the availability of specific markers to select from the cell cultures the cell types of interest, the exploration of their molecular and functional features (as compared to their physiological counterparts) and the refinement, in case of cell transplantation protocols, of their mode of delivery.

28.3 Integrating Signals of Intrinsic, Extrinsic and Systemic Origins

Several chapters of this book addressed the fundamental question of where stem cells are and how stemness is controlled. It is now clear that, in a variety of experimental models and stem cell systems, both cell autonomous (active in stem cells themselves) and non-cell autonomous (relying on the functions of surrounding cells) mechanisms are major regulators of stem cell biology. Stem cells coexist and interact in a very dynamic manner with niche cells that constitute a supportive microenvironment for stem cell survival, proliferation and differentiation. To the variety of adult stem cell systems is associated a variety of niche cells. For example, specific somatic cells constitute the major niche cell components of germ stem cells whereas hematopoietic stem cells (HSCs) are closely associated with mesenchymal stem cells (MSCs) and their derivatives in the adult bone marrow. Cell-to-cell communication, cell adhesion and extracellular matrix remodeling are then major cellular processes active at the interplay between stem cells and their niches. What are the similarities and differences in the molecular pathways utilized by niche cells to regulate stemness in different tissues is still unclear. Moreover, accumulating evidence supports the notion that this crosstalk is bidirectional. Niche cells indeed receive signals from stem cells that influence their supporting capacities, phenotypes and molecular identities. How these dialogs are established during development, have emerged during evolution, and are eventually perturbed during diseases and ageing is a major issue, both in fundamental biology and regenerative medicine. Importantly, on top of these intrinsic and local extrinsic factors released by niche cells, systemic factors active at the level of the organisms or provided by the environment are now appearing as critical regulators of stem cell functions. Hormones, neurotransmitters, the circadian clock, and infections are among those factors. In models as diverse as *Hydra*, the Drosophila ovarioles or the vertebrate intestine, it has been demonstrated that microbes are an essential component of stem cell ecosystems. Changes in the composition of the microbiota and in its metabolic activities may also play a critical role in the process of tumor formation. For example, elegant experiments in *Hydra* have revealed that *Fox0*, a gene encoding for a transcription factor highly expressed in *Hydra* stem cells, is involved in the maintenance of stem cells, in the control of stemness genes and antimicrobial peptides expression as well as in bacteria colonization (Boehm et al., 2012; Mortzfeld et al., 2018). In Drosophila females, Wolbachia (maternally transmitted intracellular bacteria) infect germ stem cells, the germ stem cell niche and brain cells secreting insulin-like peptides. Infected females produce more eggs than uninfected females. Thus, by controlling germ stem cell activity through cell autonomous, extrinsic and/or systemic regulation, Wolbachia promotes its own propagation (Fast et al., 2011), raising particularly interesting questions in evolutionary biology. How microbes are integrated within the immune system and the machinery of stemness is a fascinating biological question. Further investigations on the crosstalk between microbes, immune cells and stem cells are really promising and should bring major novel findings on the mechanisms that drive stem cell regulation and tumor formation. Collectively, one of the major challenges in the field of stem cell biology will be to understand how single stem cell integrates such variety of signals to survive and commit into self-renewal or differentiation pathways.

28.4 Stem Cell Heterogeneity and Plasticity

Recent advances and novel technologies such as single cell transcriptomics, lineage tracing and refined functional assays revealed that stem cells are highly heterogeneous. This heterogeneity occurs at multiple levels including cell cycle status, metabolism, gene expression, surface phenotype, niche cell association, responses to radiation and functional characteristics. It has been documented in several stem cell systems such as HSCs, intestinal and skin stem cells (Goodell et al., 2015). In the intestine, crypt basal columnar cells and +4 cells constitute two distinct stem cell compartments that can eventually interconvert, the former highly proliferative and providing robust cell turnover whereas the latter more quiescent and recruited in case of injuries. The classical view of hematopoiesis supports the existence of long-term HSCs that give rise to short-term HSCs and then multipotent progenitor cells, at the origin of the common myeloid and lymphoid progenitors. Single cell transplantation assays, lineage barcoding and transcriptomic approaches revealed that HSCs are lineage-biased, exhibit transcriptional lineage priming and thus are highly heterogeneous (Haas et al., 2018). Several mechanisms may cooperate for establishing cell heterogeneity. These may involve the preferential association of stem cells with specific niche cells, stochastic cellular events, and genetic and epigenetic heterogeneity within single stem cells (Haas et al., 2018). Interestingly, such cell heterogeneity and plasticity have also been reported for both liquid and solid tumors (Meacham and Morrison, 2013; Nassar and Blanpain, 2016). This is well exemplified by the seminal study of Gupta and colleagues based on the isolation by flow cytometry of three distinct epithelial state (luminal, basal and stem-like) from human breast cancer lines (Gupta et al., 2011). After culture, the three cell populations could separately reconstitute tumor heterogeneity at similar proportions to the parental population, demonstrating interconversion and equilibrium between cell states (Gupta et al., 2011). Together, these findings have profound impacts on how tissue homeostasis is maintained *in vivo,* how tissues are regenerated after injuries and how tumors are initiated and propagated. They also provide stem cell biologists with new concepts and methods for better understanding tumor resistance and for elaborating more efficient treatment protocols. Finally, they provide a unique framework for elucidating what stem cells are. As underscored in this book (chapters 1 and 27), stem cells may be referred as either 'entities' or 'states' (Zipori, 2004). According to the 'entity' ontology, a stem cell would be a predetermined cell type with intrinsic stemness ability to self-renew and differentiate in relation to predefined supportive niche cells. By contrast, the 'state' ontology favors the hypothesis that stemness is a property of a cell population transiently acquired following stochastic or deterministic events. According to this hypothesis, stemness may be induced in fully differentiated cells exposed to specific niche factors. Non-hierarchical stem cell models are being developed and should be further refined to account for the 'state' ontology. Lucie Laplane proposed four distinct stem cell properties (categorical, dispositional, relational, and systemic) depending on niche-dependent and dedifferentiation characteristics (Laplane and Solary, 2019).

28.5 Interface Between Stem Cell and Computational Biology

The advent of high throughput technologies and computational approaches are providing researchers with a unique methodological framework for interrogating the biology of stem cells. For example, bulk and single cell transcriptomics have considerably increased our knowledge on the molecular identities of pluripotent and adult stem cells, stem cell heterogeneity, and cell differentiation trajectories (Cahan et al., 2021; Kester and van Oudenaarden, 2018; Li and Izpisua Belmonte, 2018). They are also instrumental for reconstructing molecular networks representative and eventually predictive of a particular cell state or cell behavior (Cahan et al., 2021). This knowledge has also major implications in our understanding of human diseases such as cancer and for recapitulating developmental programs *ex vivo* for producing specific cell types from precursor cells. We have to admit that these fast technological advances are also very challenging since they are accompanied by the development of multiple algorithms and statistical analyses. For example, in the field of single cell transcriptomics, several computational approaches have recently been developed for reducing inherent noise related to low level of gene expression and for improving the identification of cell clusters, extraction of sets of differentially expressed genes and data visualization (Kester and van Oudenaarden, 2018). On our side, we took advantage of some of these concepts and methods for exploring the molecular identity of stromal cells that provide a supportive microenvironment for HSCs. By analyzing the transcriptomes of stromal lines with differing capacity to support HSCs and by integrating numerous gene datasets produced by other laboratories on HSC niche cells, we uncovered a molecular core representative of the HSC support and identified gene networks active in discrete bone marrow niche cell populations (Charbord et al., 2014; Desterke et al., 2020). By combining a subtractive strategy with dimensionality reduction methods and network concepts, our analyses also revealed unexpected novel HSC niche factors. We also believe that computational biology provides stem cell biologists with unique resources such as gene expression datasets that can be utilized to improve the robustness of the *in silico* analyses, to compare and refine molecular signatures and to validate experimental designs. Major perspectives in the field will consist in integrating multiscale datasets (e.g., genomic, epigenomic, transcriptomic, and proteomic) to incorporate transcriptional informations with epigenetic regulation and chromatin accessibility. Since stem cells are strictly controlled by supportive niche cells, integrating the molecular identities of stem cells and surrounding cells also constitutes an immense challenge. We are confident that systems biology approaches using mathematical, statistical and physical tools and concepts to analyze the experimental data will be instrumental to provide the dynamics of the crosstalk between stem cells and their niches in development and diseases, to establish predictive models of stem cell behavior, and to point out molecular hubs that may be promising targets for drug development.

28.6 Therapeutic Applications of Stem Cells

Clinical applications of stem cells have been, and still are, a major boost for stem cell research. Many aspects have been treated throughout this book. We will in this last chapter try and outline the main challenges that therapy using stem cells have to face today.

Transplantation of HSCs has been a remarkable success story. Over time the sources of HSCs have evolved: firstly bone marrow, then mobilized circulating blood and more recently cord (placental) blood. The indications have increased, from treatment of hematopoietic malignancies to treatment of solid tumors. In any case the procedure has been to induce an aplastic condition through chemo and radiotherapy (conditioning), then to provide the HSCs, via the intra-venous route, to reconstitute the blood system of the recipient. Grafts are either allogeneic, HSCs being recovered from an HLA-compatible donor (intra-familial or unrelated), or autologous, when HSCs are collected from the patient himself before conditioning. The major challenges today are to amplify *ex vivo* HSCs, a desirable procedure for cord blood transplantation in the adult, and to generate HSCs from pluripotent stem cells, which would be highly beneficial in patients without HLA-compatible familial donor since the clinical outcome of patients having received unrelated grafts remains of relatively poor prognosis. There is no satisfactory, and therefore standardized, procedure to amplify by culture a cell population containing substantial amount of repopulating HSCs with maintained self-renewal capacity. Although major achievements have been made through cell reprogramming (Lis et al., 2017; Sugimura et al., 2017), generating functional HSCs from ES cells or induced pluripotent stem (iPS) cells still constitutes a major issue.

The second success story of stem cell-based therapy has been the transplantation of autologous skin grafts in heavily burnt patients. One problem is to recover enough intact skin for culture in patients with extensively burnt skin surface. On way to circumvent this problem is to generate HLA-compatible skin equivalents from ES or iPS cells. This is presently feasible but still represents a major biological and logistical challenge.

Whether therapy using MSCs will be comparably successful, and in which indications, has to be definitively demonstrated. As already emphasized in chapters 1 and 23, there are many indications for such cell therapy, in orthopedics and rheumatology (e.g., bone fractures with wide bone defect, osteoarthritic cartilage lesions in the elderly), in cardio-vascular diseases (e.g., heart infarct, ischemia of the lower limbs), in hematology (e.g., double HSC and MSC transplantation in aplastic patients with poor engraftment prognosis), in immunological diseases (e.g., Crohn's bowel disease, graft-versus-host disease [GvHD] post-allogeneic transplantation, systemic sclerosis). Some of the indications (such large fractures, osteoarthritis, and GvHD), are beyond proof of concept and demonstration of feasibility, and await phase 2 and 3 trials for proof of efficacy in well-defined patient populations. However, besides efficacy, which may be due to MSC properties unrelated to stemness, there are many logistical problems to solve. The cost of the culture procedure is also a major issue due to the absolute requirement of Good Manufacturing Practices with all required controls of bacterial and viral non-contamination and of lack of potentially malignant clone selection. One solution would be to use a universal allogeneic donor, which would alleviate the procedure and its cost, MSCs being ready off-the-shelf when necessary. This is possible due to the relative immune tolerance of, if not immune suppression induced by MSCs. Another possibility would be to recover bioactive material from the culture supernatant since MSCs are known to release extracellular vesicles enriched in regulatory molecules such as microRNAs. The development of the different strategies will depend on bio-technological improvements and, of course, proof of concept in pilot studies.

As underlined in chapter 1, the use of other adult tissue stem cells encounters many difficulties. Thus, two major strategies have been envisioned. The first is the generation of ES cell lines that could be differentiated to appropriate lineages before administration to the patient. This strategy is difficult to implement. The procurement of ES cells raises ethical and regulatory issues (see chapter 26). Allogeneic grafting implies the development of banks of cell lines with HLA group diversity representative of the general population. Cells from lines have to be completely differentiated lest residual pluripotent stem cells generate teratomas in the recipient. The second strategy is to use a population of autologous somatic cells reprogrammed into stem cells secondarily differentiated into appropriate lineage before infusion to the patient. The development of iPS cells renders such strategy an attractive possibility. However, as emphasized in chapters 8, 21, 23, and 24, indications have to be carefully selected and the number of hurdles that should be overcome remains considerable. Two situations appear more readily to lead to successful applications. As indicated in chapter 21, one situation is the generation of retinal cells from iPS cells. These cells can be locally implanted without fear of dissemination in the organism, and, most importantly, the clinical outcome of the graft can be closely monitored. As indicated in chapter 24, the other situation is the generation from iPS cells of red blood cells. The procedure has several advantages. The process of differentiation is well known and can be standardized. The graft is safe since consisting in anucleated terminally differentiated red blood cells. A large panel of red blood cell types can be generated, representing the diversity of red blood cell groups. However, whether it is possible to generate, at reasonable cost, amounts of cells large enough to satisfy the requirement of red blood cell transfusions in many patients remains to be demonstrated. A last point to mention is the direct transdifferentiation of a somatic cell type into another, which constitutes an alternative to the iPS cell strategy noticeably attractive due to the generation of appropriate cells in short spans of time without the risk of intervening stem cells. However, the procedure is still far from being standardized and adapted to clinically-safe conditions. Finally, the reprogramming technologies and cellular differentiation strategies will also benefit from novel advances in the field of bioengineering, 3D cultures and biomaterials. These achievements should help stem cell biologists to better mimic the architecture and physical properties of a given tissue *ex vivo* and improve the production of specific functional cell types.

In conclusion, the reader should be aware of the remarkable clinical potential of stem cell-based therapy, but also of the no less remarkable difficulties that such therapy entails, since societal, ethical, and economical issues compound the biological hurdles. However, it may be hoped that a rigorous and perseverant approach, as exemplified by that of Donnall Thomas and collaborators, may lead to novel clinical applications as effective as the transplantation of HSCs.

References

Boehm, A.M., Khalturin, K., Anton-Erxleben, F., Hemmrich, G., Klostermeier, U.C., Lopez-Quintero, J.A., Oberg, H.H., Puchert, M., Rosenstiel, P., Wittlieb, J., et al. (2012). FoxO is a critical regulator of stem cell maintenance in immortal Hydra. Proceedings of the National Academy of Sciences of the United States of America 109, 19697-19702.

Cahan, P., Cacchiarelli, D., Dunn, S.J., Hemberg, M., de Sousa Lopes, S.M.C., Morris, S.A., Rackham, O.J.L., Del Sol, A., and Wells, C.A. (2021). Computational Stem Cell Biology: Open Questions and Guiding Principles. Cell stem cell 28, 20-32.

Charbord, P., Pouget, C., Binder, H., Dumont, F., Stik, G., Levy, P., Allain, F., Marchal, C., Richter, J., Uzan, B., et al. (2014). A systems biology approach for defining the molecular framework of the hematopoietic stem cell niche. Cell stem cell 15, 376-391.

Desterke, C., Petit, L., Sella, N., Chevallier, N., Cabeli, V., Coquelin, L., Durand, C., Oostendorp, R.A.J., Isambert, H., Jaffredo, T., et al. (2020). Inferring Gene Networks in Bone Marrow Hematopoietic Stem Cell-Supporting Stromal Niche Populations. iScience 23, 101222.

Fast, E.M., Toomey, M.E., Panaram, K., Desjardins, D., Kolaczyk, E.D., and Frydman, H.M. (2011). Wolbachia enhance Drosophila stem cell proliferation and target the germline stem cell niche. Science (New York, NY) 334, 990-992.

Goodell, M.A., Nguyen, H., and Shroyer, N. (2015). Somatic stem cell heterogeneity: diversity in the blood, skin and intestinal stem cell compartments. Nature reviews Molecular cell biology 16, 299-309.

Gupta, P.B., Fillmore, C.M., Jiang, G., Shapira, S.D., Tao, K., Kuperwasser, C., and Lander, E.S. (2011). Stochastic state transitions give rise to phenotypic equilibrium in populations of cancer cells. Cell 146, 633-644.

Haas, S., Trumpp, A., and Milsom, M.D. (2018). Causes and Consequences of Hematopoietic Stem Cell Heterogeneity. Cell stem cell 22, 627-638.

Kester, L., and van Oudenaarden, A. (2018). Single-Cell Transcriptomics Meets Lineage Tracing. Cell stem cell 23, 166-179.

Laplane, L., and Solary, E. (2019). Towards a classification of stem cells. eLife 8.

Li, M., and Izpisua Belmonte, J.C. (2018). Deconstructing the pluripotency gene regulatory network. Nature cell biology 20, 382-392.

Lis, R., Karrasch, C.C., Poulos, M.G., Kunar, B., Redmond, D., Duran, J.G.B., Badwe, C.R., Schachterle, W., Ginsberg, M., Xiang, J., et al. (2017). Conversion of adult endothelium to immunocompetent haematopoietic stem cells. Nature 545, 439-445.

Meacham, C.E., and Morrison, S.J. (2013). Tumour heterogeneity and cancer cell plasticity. Nature 501, 328-337.

Mortzfeld, B.M., Taubenheim, J., Fraune, S., Klimovich, A.V., and Bosch, T.C.G. (2018). Stem Cell Transcription Factor FoxO Controls Microbiome Resilience in Hydra. Frontiers in microbiology 9, 629.

Nassar, D., and Blanpain, C. (2016). Cancer Stem Cells: Basic Concepts and Therapeutic Implications. Annual review of pathology 11, 47-76.

Nichols, J., and Smith, A. (2009). Naive and primed pluripotent states. Cell stem cell 4, 487-492.

Sugimura, R., Jha, D.K., Han, A., Soria-Valles, C., da Rocha, E.L., Lu, Y.F., Goettel, J.A., Serrao, E., Rowe, R.G., Malleshaiah, M., et al. (2017). Haematopoietic stem and progenitor cells from human pluripotent stem cells. Nature 545, 432-438.

Zipori, D. (2004). The nature of stem cells: state rather than entity. Nature reviews Genetics 5, 873-878.

Index

A

Acute myeloid leukemia 53, 118, 154, 384
Adipocyte 109, 111, 369, 496
Adult stem cells 2, 8, 71, 82
Ageing 29, 72, 80, 86, 123, 232
Aging 8, 30, 52, 59, 90, 125
AGM 299, 301, 309, 325, 326, 330
Algorithm 109, 139, 140, 150, 710
Alignment 136, 146, 160,
Allogenic 373, 567
AML 53, 109, 154
Amphibian 4, 243, 245, 259, 303
Anamniotic 243, 256, 705
Aorta 9, 298, 302, 304, 308
Apoptosis 12, 72, 79, 80, 155, 217
Appendage regeneration 245, 248
Astrocytes 6, 89, 257, 461
Asymmetrical 2, 126, 280, 396
Autologous 362, 398, 533, 632

B

Barcode 147, 584, 590
Batch 144, 157, 635
Bioengineering 189, 201, 568, 603, 713
Biomaterials 199, 534, 568, 614
Blastema 244, 248, 249
Blastocyst 4, 164, 183, 184, 204, 562
Blood 2, 5–7, 85, 107, 137
BM logette 495, 498

Bone 2, 3, 49, 112, 155, 254
Bone marrow 2, 3, 106, 108, 155, 181
Bone tissue engineering 534
Boolean 153, 154
Brain regeneration 257, 543
Bulge 81, 395, 400–406,

C

Cancer stem cells 15, 79, 95, 389, 411
Cartilage 246, 253, 496, 502, 505
Categorical property 693, 696, 699
Causal 134, 152, 162, 258
Causative 43, 152, 231
Cell adhesion 11, 186, 217, 507, 659
Cell autonomous 9, 21, 276, 323
Cell cycle 8, 24, 73, 126, 209
Cell identity 37, 135, 149, 216, 286
Cell plasticity 19, 39, 418, 421
Cell state 123, 158, 281, 685, 709
Cell therapy 17, 191, 372, 450, 528, 553, 603
Cell transplantation 97, 195, 368, 563, 689, 709
Cell-to-cell communication 13
CFU-f 498, 500
Chemokine 10, 108, 341, 482, 588
Chemotherapy 109, 154, 503, 589, 631
Chicken 298, 301, 310, 336, 500, 560
Chimeric 7, 300, 495, 607,

Cholangiocyte differentiation 189
Chondrocyte 17, 496, 500, 536
Chondrogenic 504, 512
Chromatin 9, 35, 37, 44
Chromosome 2, 53, 163, 173
Classifier 140, 154
Cloning 209, 673, 702
Clonogenic 7, 438, 499
Clustering 143, 162
Cnidaria 208, 230
Collagen hydrogels 534
Colony 7, 84, 297, 312, 363, 629
Computational 1, 123, 155, 710
Computational approaches 9, 705, 710
Computational model 8, 123
Connective tissue 11, 81, 84, 251
Connectivity 149, 151
Constitutive property 693, 696
Correlation 148, 151
Craniofacial bone 524, 533
Culture 4, 86, 133, 286, 312
Cytokines 10, 114, 187, 338

D
Dedifferentiation 15, 252, 260, 699
Definitive hematopoiesis 302, 330, 332
Dental 5, 523, 532
Dental pulp stem cells 525, 540
Deterministic 17, 127, 383
Development 6, 21, 35, 129
Differentiation 1, 187, 277
Disease modeling 181, 189, 191, 705
Diseases 55, 182, 226, 395, 448
Dispositional property 694, 696, 699
DNA damage 71, 72, 280
DNA damage response 77, 80
DNA repair 84, 131, 591

DNA-PK 75, 91
Dorsal aorta 9, 302, 310
Dorsal lateral Plate 304, 321
Dosage compensation 163
Double-Strand Breaks 71
Droplet 147
Dropout 148
Drug 21, 143, 193, 450

E
ECM 5, 341, 604
Ectoderm 4, 184, 209, 210, 260
Embryo 2, 3, 44, 254, 325
Embryonic development 36, 182, 245, 422
Embryonic stem cell 3, 166, 181, 215, 632
Embryonic stem cell line 3, 204, 625
Endoderm 4, 183, 184, 195
Endosteal 110, 499, 502
Endothelial 17, 107, 189, 195, 300
Endothelial cells 107, 186, 304, 329
Endothelial to hematopoietic transition 306, 327, 329
Endothelial-to-haematopoietic transition 107
Endothelium 115, 305, 328
Entity 20, 37, 129, 149
Entropy 150, 152
Environment 10, 46, 125, 148, 166
Environmental cue 210, 219, 279, 611
Epiblast 3, 173, 686, 706
Epidermal 18, 81, 277
Epigenetic 5, 10, 38, 129–132, 217
Epigenetics 1, 126, 182, 451
Epimorphosis 244
Epistemological 697, 706
Epithelial 3, 81, 188, 210, 393

Erythropoiesis 37, 628, 637
Evolution 37, 63, 75, 141
Extracellular matrix 10, 189, 210, 323
Extracellular vesicle 10, 13, 22, 30
Extrinsic 10, 124–128, 196, 369

F

Fetal liver 9, 298, 301, 325, 326, 333
FGF pathway 426, 558
Fibroblast 15, 45, 90, 128, 248
Fibrosis 12, 191, 498, 503
Fish 46, 243, 244, 257
FoxO 211, 220, 221
Framework 9, 46, 141, 665

G

Gamete 6, 210, 220, 677, 681
Gata2 17, 36, 53, 312, 332
Gata3 336, 337, 350
Gaussian 142, 146, 148
Gene expression regulation 705
Gene networks 9, 155, 706, 714
Gene Regulatory Network 40, 55, 153, 706
Gene Set Enrichment Analysis 145, 160, 161
Gene-environment interactions 212, 221, 226
Genetic model 506
Genome instability 105
Genomics 63, 139, 141
Germ layers 183, 184, 562, 611
Germ stem cells 6, 11, 707, 708
Germline 4, 216, 224, 694
Glioblastomas 79, 579, 589
Glioma stem cells 95, 580, 592
Gliomas 579, 594, 596

Good manufacturing practices 565, 671, 712
Graft 7, 196, 300
Grafts 196, 305, 306, 711
Graph 143, 149
GRN 153
Growth factors 5, 10, 305, 338
Growth plate 502, 504, 511
GSEA 145, 147

H

Hair follicle 3, 81, 393, 399, 405
Heart regeneration 258, 259, 265
Hedgehog 11, 250, 314, 334, 427
Hematopoietic 2, 35, 186, 297, 299
Hematopoietic Niche 336, 334, 374
Hematopoietic Progenitors 36, 298, 302, 630
Hematopoietic stem cell 2, 35, 135, 150, 215
Hematopoietic stem cell emergence 350, 356
Hematopoietic stem cells 2, 35, 93, 150, 323, 326
Hemogenic endothelial cell 315, 327, 329, 332
Hepatic cells 181, 192, 197, 705
Hepatocyte differentiation 185, 187
Heterogeneity 13, 18, 110, 146, 428, 529
Hierarchical 13, 151, 360, 510
Hierarchy 298, 325, 363, 377, 421
HLA 650, 711, 713
Homeostasis 3, 20, 86, 183, 221, 225
Homologous recombination (HR) 72
HSC 6, 8, 12, 113, 309, 335, 366
Hub 11, 151, 221
Hydra 207, 210, 212, 215, 221, 224
Hypoxia 45, 342, 344, 374

I

Immortalized erythroblasts 636
Immunosuppression 500, 647
Induced pluripotent stem cells 81, 90, 173, 181, 523
Injury 17, 248, 257, 557
Inner cell mass 4, 164, 301, 562, 686
Interaction 9, 42, 189, 221
Interfollicular epidermis 82, 394, 395
Intestinal 15, 81, 124, 127, 439
Intestine 3, 124, 437, 438
Intra-aortic hematopoietic Clusters 302, 329
Intrinsic 10, 17, 87, 125, 128
Irradiation 5, 81, 111, 297, 365

L

Leukemia 12, 53, 154, 360, 471
Lineage 1, 39, 262, 405
Lineage barcoding 709
Lineage priming 15, 512, 709
Lineage tracing 1, 397, 407, 428
Lineage-biased 16, 709
Linear function 154
Liver diseases 194, 205
Liver organoids 191, 193, 195
Lymphocyte 35, 449, 500, 528, 585
Lymphoid 39, 75, 303, 363, 374

M

Machine learning 140, 154
Macrophages 39, 87, 111, 114, 302
Mammary 81, 406, 415, 417
Mammary gland 82, 406, 416, 417, 420, 421
Mammary stem cells 82, 415, 422
Markers 14, 190, 313, 361, 425
Mechanical 124, 196, 279, 346
Membrane antigen 497, 504

Mesenchymal 11, 195, 334, 500
Mesenchymal stem cells 11, 48, 84, 124, 398
Mesenchymoangioblast 510
Mesensphere 7, 500, 507
Mesoderm 4, 93, 183, 184, 214
Metabolism 72, 94, 141, 261
Microarray 142, 153, 156, 332
Microbes 10, 214, 224, 227, 708
Microbiome 214, 224, 225
Microenvironment 11, 16, 107, 148, 213
Microglia 461, 585, 586, 589
MicroRNAs 9, 276, 289, 712
Mitochondrial 71, 111, 441
Modeling 17, 54, 181
Module 151, 584
Molecular signature 20, 145, 207, 442
Mouse 8, 254, 299, 355
mRNA 112, 147, 150
MSC 15, 109, 114, 372
Multipotency 285, 377, 419, 560
Multipotent 6, 81, 209, 252
Muscle 3, 87, 252, 306
Muscle stem cell 86, 647, 651, 686
Muscular Dystrophy 645, 651
Mutations 53, 395, 443, 553
Mutual information 152, 153
Myeloid 12, 84, 111, 298, 361
Myoblast 86, 608, 645, 647
Myogenic cell 19, 653

N

Naïve 4, 174, 254, 313
Natural kind 685, 689, 698, 704
Nearest neighbor 149, 154
Neoblast 687, 704

Network 9, 10, 40, 108, 128
Neural 3, 48, 89, 135, 183
Neural progenitor cells 492, 581
Neural stem cells 5, 89, 461, 466
Neuron 6, 49, 257, 586
Neurosphere 466, 467
Next Generation Sequencing 139, 145, 214
NGS 140, 145, 152
Niche 9, 107, 275, 323
Non-cell autonomous 9, 323, 707
Non-homologous end-joining (NHEJ) 72
Normalization 142, 148
Notch 11, 127, 128, 133
Notch pathway 128, 189, 330, 374
Notch signaling 127, 330, 340, 388

O

Oncology 1, 685, 698
Operating system 139, 140
Organoids 189, 191, 440, 441
Osteoblast 12, 17, 94, 110
Osteogenic 94, 503, 511, 526
Outlier 143

P

p53 78, 82, 584, 598
PCA 143, 149, 156
Pericyte 320, 481, 510, 531
Periosteal 504, 542
Periosteum 502, 503, 536
Perivascular 93, 324, 369, 479
Phenotype 5, 10, 36, 145, 170
Philosophy 20, 685, 695
Photoreceptors 553, 555, 557, 564
Plant 274, 290, 295, 296

Plasticity 421, 458, 512, 588, 700, 709
Pluripotency 4, 169, 171, 706
Pluripotency factor 171, 175
Pluripotent 90, 181, 186, 190, 655, 704
Pluripotent stem cells 21, 568
Prediction 126, 154
Primed 4, 308, 512, 706, 715
Principal component analysis 142, 156
Progenitor 6, 7, 18, 39, 84
Progenitor cells 7, 12, 84, 315
Proliferation 9, 20, 45, 85, 113
Pseudotime 149, 150
p-value 142, 143, 149

Q

QC 143, 147, 280, 282
Quality control 157
Quiescence 8, 84, 108, 281

R

Radiation 5, 19, 72, 82, 110
Radioresistance 81, 84, 588, 591
Radiotherapy 81, 588, 589, 711
Reads 146, 306
Red blood cells 303, 359, 627, 633
Regeneration 3, 78, 86, 110, 191
Regenerative medicine 5, 181, 195, 407
Regulatory 9, 33, 35, 123, 128
Relational property 694, 696, 699
Repair 17, 71, 76, 77, 80
Reprogramming 4, 312, 562
Resistance 79, 91, 94, 224, 589
Retinal pigmented epithelium 553, 575, 577

Reversibility 14, 558
Reversible 14, 512, 589
RNASeq 146
R-package 142, 146, 147, 153
Runx1 306, 312, 315, 328

S

Satellite cell 6, 93, 607, 647
Scale-free 151
Sebaceous glands 82, 393, 394
Self-renewal 6, 85, 107, 207
Senescence 79, 80, 92, 212
Sex chromosome 163, 300
Signaling 13, 45, 85, 189
Single cell 7, 147, 213, 324
Single-cell RNA sequencing 324, 330, 402, 565
Sinusoid 108, 498, 504
Skeletal Stem Cell 504, 514, 521
Skin 3, 90, 182, 243
Skin stem cells 15, 709
Smooth muscle 306, 308, 320, 406
Somatopleura 304, 305, 306
Somite 299, 304, 306, 308
Splanchnopleura 302, 304, 306, 308
SSC 89, 332, 504, 511
Steady state 2, 112, 130, 359
Stem cell 1–10, 18, 20, 78
Stem cell activation 6, 95, 252, 262
Stem cell lineage 208, 211, 543
Stem cell niches 1, 107, 280, 480
Stem cell organization 123, 125, 127, 133
Stemness 9, 15, 126, 211, 223
Stochastic 16, 149, 157, 314
Stress 5, 81, 112, 248, 342
Stro-1 497, 507
Stromal 7, 327, 365, 374

Stromal cell 109, 312, 327, 364
Subaortic mesenchyme 335, 336
Sub-Aortic Mesenchyme 299, 306, 308, 315
Subventricular zone 89, 464, 467, 469, 480
Surface phenotype 531, 709
Sweat glands 394, 406, 423
Sympathetic nervous system 113, 333, 337
Systemic 1, 10, 323, 649, 696
Systemic property 695, 696, 697, 699

T

Taxonomically restricted genes 218, 219, 235
Telomere 8, 86, 87, 580
Tendon 183, 496
Teratoma 197, 345, 653, 713
TGFβ pathway 185, 222, 223, 426
Therapeutical 539, 565, 615, 706
Tissue homeostasis 5, 71, 94, 111, 211
Tissue-specific stem cells 212, 275, 686
Totipotent 3, 686
Trajectories 16, 149, 157, 213, 512
Trajectory 149, 150, 331, 345
Transcription 9, 16, 33, 35, 41
Transcription factor 16, 33, 35, 41, 47, 154
Transcriptional 9, 45, 288, 342
Transcriptomic 118, 139, 311, 611
Transdifferentiation 19, 250, 558
Transgenesis 212, 212, 215, 217, 620
Transplantation 4, 195, 299, 373
Tsix 164, 167, 170
Tumor 15, 138, 225, 428, 579

U

Unipotency 420, 421

V

Validation 136, 164, 197, 336
Velocity 150, 158, 617, 689
Ventral Blood Island 303, 317
Vision 553, 569, 575

W

Weighted Gene Correlation
  Analysis 151, 157
WGCNA 151, 157, 160
Wnt 11, 30, 68, 121, 127
Wnt pathway 68, 128, 440,
  611, 626

X

X chromosome inactivation
  163, 174
XCI 163–168, 171, 172
XCI activator 171, 172, 173
XCI inhibitor 172
Xenograft models 584
Xist 164, 174, 178

Y

Yolk sac 299, 301, 316, 318

Z

Zebrafish 243, 253, 258, 309

About the Editors

Charles Durand received his PhD in Developmental Biology from the University Pierre and Marie Curie (UPMC). He did his postdoctoral training on the regulation of hematopoietic stem cells (HSCs) at the Institut Cochin (Paris) and the Erasmus Medical Center (Rotterdam). He became associate Professor at UPMC in 2005 and Professor at Sorbonne Université in 2019. He is exploring the molecular dialog between HSCs and niche cells with a developmental and systems biology perspective. He is the founder and chairperson of the master 2 programme 'The Biology of Stem Cells' at Sorbonne Université.

Pierre Charbord is an emeritus Inserm scientist, presently working at the Sorbonne Université in "Institut de Biology Paris-Seine". After his training as MD, he has joined the French National Health Institute (Inserm) in 1979 and since that date has conducted research on both hematopoietic and mesenchymal stem cells, investigating their differentiation potentials, the niche formation and their potential implication in human diseases such as cancer. He has contributed to more that 140 original publications and to several book chapters.

For Product Safety Concerns and Information please contact our EU
representative GPSR@taylorandfrancis.com
Taylor & Francis Verlag GmbH, Kaufingerstraße 24, 80331 München, Germany